THE BIG **R**-BOOK

THE BIG **R**-BOOK

FROM DATA SCIENCE TO LEARNING MACHINES AND BIG DATA

Philippe J.S. De Brouwer

WILEY

This edition first published 2021

Registered Office
John Wiley & Sons, Inc., 111 River Street, Hoboken, NJ 07030, USA

Editorial Office
111 River Street, Hoboken, NJ 07030, USA

For details of our global editorial offices, customer services, and more information about Wiley products visit us at www.wiley.com.

Wiley also publishes its books in a variety of electronic formats and by print-on-demand. Some content that appears in standard print versions of this book may not be available in other formats.

Library of Congress Cataloging-in-Publication Data

Names: De Brouwer, Philippe J. S., author.
Title: The big R-book : from data science to learning machines and big data / Philippe J.S. De Brouwer.
Description: Hoboken, NJ, USA : Wiley, 2020. | Includes bibliographical references and index.
Identifiers: LCCN 2019057557 (print) | LCCN 2019057558 (ebook) | ISBN 9781119632726 (hardback) | ISBN 9781119632764 (adobe pdf) | ISBN 9781119632771 (epub)
Subjects: LCSH: R (Computer program language)
Classification: LCC QA76.73.R3 .D43 2020 (print) | LCC QA76.73.R3 (ebook) | DDC 005.13/3–dc23
LC record available at https://lccn.loc.gov/2019057557
LC ebook record available at https://lccn.loc.gov/2019057558

Cover Design: Wiley
Cover Images: Information Tide series and Particle Geometry series © agsandrew/Shutterstock, Abstract geometric landscape © gremlin/Getty Images, 3D illustration Rendering © MR.Cole_Photographer/Getty Images

Set in 9.5/12.5pt STIXTwoText by SPi Global, Chennai, India

Printed in the United States of America

SKY10023032_120420

To Joanna, Amelia and Maximilian

Short Overview

Contents

Foreword

This book brings together skills and knowledge that can help to boost your career. It is an excellent tool for people working as database manager, data scientist, quant, modeller, statistician, analyst and more, who are knowledgeable about certain topics, but want to widen their horizon and understand what the others in this list do. A wider understanding means that we can do our job better and eventually open doors to new or enhanced careers.

The student who graduated from a science, technology, engineering or mathematics or similar program will find that this book helps to make a successful step from the academic world into a any private or governmental company.

This book uses the popular (and free) software R as leitmotif to build up essential programming proficiency, understand databases, collect data, wrangle data, build models and select models from a suit of possibilities such linear regression, logistic regression, neural networks, decision trees, multi criteria decision models, etc. and ultimately evaluate a model and report on it.

We will go the extra mile by explaining some essentials of accounting in order to build up to pricing of assets such as bonds, equities and options. This helps to deepen the understanding how a company functions, is useful to be more result oriented in a private company, helps for one's own investments, and provides a good example of the theories mentioned before. We also spend time on the presentation of results and we use R to generate slides, text documents and even interactive websites! Finally we explore big data and provide handy tips on speeding up code.

I hope that this book helps you to learn faster than me, and build a great and interesting career.

Enjoy reading!

Philippe De Brouwer
2020

About the Author

Dr. Philippe J.S. De Brouwer leads expert teams in the service centre of HSBC in Krakow, is Honorary Consul for Belgium in Krakow, and is also guest professor at the University of Warsaw, Jagiellonian University, and AGH University of Science and Technology. He teaches both at executive MBA programs and mathematics faculties.

He studied theoretical physics, and later acquired his second Master degree while working. Finishing this Master, he solved the "fallacy of large numbers puzzle" that was formulated by P.A. Samuelson 38 years before and remained unsolved since then. In his Ph.D., he successfully challenged the assumptions of the noble price winning "Modern portfolio Theory" of H. Markovitz, by creating "Maslowian Portfolio Theory."

His career brought him into insurance, banking, investment management, and back to banking, while his specialization shifted from IT, data science to people management.

For Fortis (now BNP), he created one of the first capital guaranteed funds and got promoted to director in 2000. In 2002, he joined KBC, where he merged four companies into one and subsequently became CEO of the merged entity in 2005. Under his direction, the company climbed from number 11 to number 5 on the market, while the number of competitors increased by 50%. In the aftermath of the 2008 crisis, he helped creating a new asset manager for KBC in Ireland that soon accommodated the management of ca. 1000 investment funds and had about €120 billion under management. In 2012, he widened his scope by joining the risk management of the bank and specialized in statistics and numerical methods. Later, Philippe worked for the Royal Bank of Scotland (RBS) in London and specialized in Big Data, analytics and people management. In 2016, he joined HSBC and is passionate about building up a Centre of Excellence in risk management in the service centre in Krakow. One of his teams, the independent model review team, validates the most important models used in the banking group worldwide.

Married and father of two, he invests his private time in the future of the education by volunteering as board member of the International School of Krakow. This way, he contributes modestly to the cosmopolitan ambitions of Krakow. He gives back to society by assuming the responsibility of Honorary Consul for Belgium in Krakow, and mainly helps travellers in need.

In his free time, he teaches at the mathematics departments of AGH University of Science and Technology and Jagiellonian University in Krakow and at the executive MBA programs the Krakow Business School of the University of Economics in Krakow and the Warsaw University. He teaches subjects like finance, behavioural economics, decision making, Big Data, bank management, structured finance, corporate banking, financial markets, financial instruments, team-building, and leadership. What stands out is his data and analytics course: with this course he manages to provide similar content with passion for undergraduate mathematics students and experienced professionals of an MBA program. This variety of experience and teaching experience in both business and mathematics is what lays the foundations of this book: the passion to bridge the gap between theory and practice.

Acknowledgements

Writing a book that is so eclectic and holds so many information would not have been possible without tremendous support from so many people: mentors, family, colleagues, and ex-colleagues at work or at universities. This book is in the first place a condensation of a few decades of interesting work in asset management and banking and mixes things that I have learned in C-level jobs and more technical assignments.

I thank the colleagues of the faculties of applied mathematics at the AGH University of Science and Technology, the faculty of mathematics of the Jagiellonian University of Krakow, and the colleagues of HSBC for the many stimulating discussions and shared insights in mathematical modelling and machine learning.

To the MBA program of the Cracovian Business School, the University of Warsaw, and to the many leaders that marked my journey, I am indebted for the business insight, stakeholder management and commercial wit that make this book complete.

A special thanks goes to Piotr Kowalczyk, FRM and Dr. Grzegorz Goryl, PRM, for reading large chunks of this book and providing detailed suggestions. I am also grateful for the general remarks and suggestions from Dr. Jerzy Dzieża, faculty of applied mathematics at the AGH University of Science and Technology of Krakow and the fruitful discussions with Dr. Tadeusz Czernik, from the University of Economics of Katowice and also Senior Manager at HSBC, Independent Model Review, Krakow.

This book would not be what it is now without the many years of experience, the stimulating discussions with so many friends, and in particular my wife, Joanna De Brouwer who encouraged me to move from London in order to work for HSBC in Krakow, Poland. Somehow, I feel that I should thank the city council and all the people for the wonderful and dynamic environment that attracts so many new service centres and that makes the ones that already had selected for Krakow grow their successful investments. This dynamic environment has certainly been an important stimulating factor in writing this book.

However, nothing would have been possible without the devotion and support of my family: my wife Joanna, both children, Amelia and Maximilian, were wonderful and are a constant source of inspiration and support.

Finally, I would like to thank the thousands of people who contribute to free and open source software, people that spend thousands of hours to create and improve software that others can use for free. I profoundly believe that these selfless acts make this world a better and more inclusive place, because they make computers, software, and studying more accessible for the less fortunate.

A special honorary mentioning should go to the people that have built Linux, LaTeX, R, and the ecosystems around each of them as well as the companies that contribute to those projects, such as Microsoft that has embraced R and RStudio that enhances R and never fails to share the fruits of their efforts with the larger community.

Preface

The author has written this book based on his experience that spans roughly three decades in insurance, banking, and asset management. During his career, the author worked in IT, structured and managed highly technical investment portfolios (at some point oversaw €24 billion in thousand investment funds), fulfilled many C-level roles (e.g. was CEO of KBCTFI SA [an asset manager in Poland], was CIO and COO for Eperon SA [a fund manager in Ireland] and sat on boards of investment funds, and was involved in big-data projects in London), and did quantitative analysis in risk departments of banks. This gave the author a unique and in-depth view of many areas ranging form analytics, big-data, databases, business requirements, financial modelling, etc.

In this book, the author presents a structured overview of his knowledge and experience for anyone who works with data and invites the reader to understand the bigger picture, and discover new aspects. This book also demystifies hype around machine learning and AI, by helping the reader to understand the models and program them in R without spending too much time on the theory.

This book aims to be a starting point for quants, data scientists, modellers, etc. It aims to be the book that bridges different disciplines so that a specialist in one domain can grab this book, understand how his/her discipline fits in the bigger picture, and get enough material to understand the person who is specialized in a related discipline. Therefore, it could be the ideal book that helps you to make career move to another discipline so that in a few years you are that person who understands the whole data-chain. In short, the author wants to give you a short-cut to the knowledge that he spent 30 years to accumulate.

Another important point is that this book is written by and for practitioners: people that work with data, programming and mathematics for a living in a corporate environment. So, this book would be most interesting for anyone interested in data-science, machine learning, statistical learning and mathematical modelling and whomever wants to convey technical matters in a clear and concise way to non-specialists.

This also means that this book is not necessarily the best book in any of the disciplines that it spans. In every specialisation there are already good contenders.

- More formal introductions to statistics are for example in: Cyganowski, Kloeden, and Ombach (2001) and Andersen et al. (1987). There are also many books about specific stochastic processes and their applications in financial markets: see e.g. Wolfgang and Baschnagel (1999), Malliaris and Brock (1982), and Mikosch (1998). While knowledge of stochastic processes and their importance in asset pricing are important, this covers only a very narrow spot of applications and theory. This book is more general, more gently on theoretical foundations and focusses more on the use of data to answer real-life problems in everyday business environment.

- A comprehensive introduction to statistics or econometrics can be found in Peracchi (2001) or Greene (1997). A general and comprehensive introduction in statistics is also in Neter, Wasserman, and Whitmore (1988).
- This is not simply a book about programming and/or any related techniques. If you just want to learn programming in R, then Grolemund (2014) will be get you started faster. Our Part II will also get you started in programming, though it assumes a certain familiarity with programming and mainly zooms in on aspects that will be important in the rest of the book.
- This book is not a comprehensive books about financial modelling. Other books do a better job in listing all types of possible models. No book does a better job here than Bernard Marr's publication: Marr (2016): "Key Business Analytics, the 60+ business analysis tool every manager needs to know." This book will list you all words that some managers might use and what it means, without any of the mathematics nor any or the programming behind. I warmly recommend keeping this book next to ours. Whenever someone comes up with a term like "customer churn analytics" for example, you can use Bernard's book to find out what it actually means and then turn to ours to "get your hands dirty" and actually do it.
- If you are only interested in statistical learning and modelling, you will find the following books more focused: Hastie, Tibshirani, and Friedman (2009) or also James, Witten, Hastie, and Tibshirani (2013) who also uses R.
- A more in-depth introduction to AI can be found in Russell and Norvig (2016).
- Data science is more elaborately treated in Baesens (2014) and the recent book by Wickham and Grolemund (2016) that provides an excellent introduction to R and data science in general. This last book is a great add-on to this book as it focusses more on the data-aspects (but less on the statistical learning part). We also focus more on the practical aspects and real data problems in corporate environment.

A book that comes close to ours in purpose is the book that my friend professor Bart Baetens has compiled "Analytics in a Big Data World, the Essential guide to data science and its applications": Baesens (2014). If the mathematics, programming, and R itself scare you in this book, then Bart's book is for you. Bart's book covers different methods, but above all, for the reader, it is sufficient to be able to use a spreadsheet to do some basic calculations. Therefore, it will not help you to tackle big data nor programming a neural network yourself, but you will understand very well what it means and how things work.

Another book that might work well if the maths in this one are prohibitive to you is Provost and Fawcett (2013), it will give you some insight in what the statistical learning is and how it works, but will not prepare you to use it on real data.

Summarizing, I suggest you buy next to this book also Marr (2016) and Baesens (2014). This will provide you a complete chain from business and buzzwords (Bernard's book) over understanding what modelling is and what practical issues one will encounter (Bart's book) to implementing this in a corporate setting and solve the practical problems of a data scientist and modeller on sizeable data (this book).

In a nutshell, this book does it all, is gentle on theoretical foundations and aims to be a one-stop shop to show the big picture, learn all those things and actually apply it. It aims to serve as a basis when later picking up more advanced books in certain narrow areas. This book will take you on a journey of working with data in a real company, and hence, it will discuss also practical problems such as people filling in forms or extracting data from a SQL database.

It should be readable for any person that finished (or is finishing) university level education in a quantitative field such as physics, civil engineering, mathematics, econometrics, etc. It should also be readable by the senior manager with a technical background, who tries to understand what his army of quants, data scientists, and developers are up to, while having fun learning R. After reading this book you will be able to talk to all, challenge their work, and make most analysis yourself or be part of a bigger entity and specialize in one of the steps of modelling or data-manipulation.

In some way, this book can also be seen as a celebration of FOSS (Free and Open Source Software). We proudly mention that for this book no commercial software was used at all. The operating system is Linux, the windows manager Fluxbox (sometimes LXDE or KDE), Kile and vi helped the editing process, Okular displayed the PDF-file, even the database servers and Hadoop/Spark are FOSS …and of course R and LaTeX provided the icing on the cake. FOSS makes this world a more inclusive place as it makes technology more attainable in poorer places on this world. **FOSS**

Hence, we extend a warm thanks to all people that spend so much time to contributing to free software.

About the Companion Site

This book is accompanied by a companion website:

www.wiley.com/go/De Brouwer/The Big R-Book

The website includes materials for students and instructors:

The Student companion site will contain the R-code, and the Instructor companion site will contain PDF slides based on the book's content.

PART I

Introduction

♣ 1 ♣

The Big Picture with Kondratiev and Kardashev

You have certainly heard the words: "data is the new oil," and you probably wondered "are we indeed on the verge of a new era of innovation and wealth creation or ... is this just hype and will it blow over soon enough?"

Since our ancestors left the trees about 6 million years ago, we roamed the African steppes and we evolved a more upright position and limbs better suited for walking than climbing. However, for about 4 million years physiological changes did not include a larger brain. It is only in the last million years that we gradually evolved a more potent frontal lobe capable of abstract and logical thinking.

The first good evidence of abstract thinking is the Makapansgat pebble, a jasperite cobble – roughly 260 g and 5 by 8 cm – that by geological tear and wear shows a few holes and lines that vaguely resemble (to us) a human face. About 2.5 million years ago one of our australopithecine ancestors not only realized this resemblance but also deemed it interesting enough to pick up the pebble, keep it, and finally leave it in a cave miles from the river where it was found.

This development of abstract thinking that goes beyond vague resemblance was a major milestone. As history unfolded, it became clear that this was only the first of many steps that would lead us to the era of data and knowledge that we live in today. Many more steps towards more complex and abstract thinking, gene mutations and innovation would be needed. **abstract thinking**

Soon we developed language. With language we were able to transform learning from an individual level to a collective level. Now, experiences could be passed on to the next generation or peers much more efficiently, it became possible to prepare someone for something that he or she did not yet encounter and to accumulate more knowledge with every generation.

More than ever before this abstract thinking and accumulation of collective experiences lead to a "knowledge advantage" and smartness became an attractive trait in a mate. This allowed our brain to develop further and great innovations such as the wheel, scripture, bronze, agriculture, iron, specialisation of labour soon started to transform not only our societal coherence but also the world around us.

Without those innovations, we would not be where we are now. While it is discussable to classify these inventions as the fruit of scientific work, it is equally hard to deny that some kind of scientific approach was necessary. For example, realizing the patterns in the movements of the sun, we could predict seasons and weather changes to come and this allowed us to put the grains on the right moment in the ground. This was based on observations and experience.

The Big R-Book: From Data Science to Learning Machines and Big Data, First Edition. Philippe J.S. De Brouwer.

Companion Website: www.wiley.com/go/De Brouwer/The Big R-Book

Science and progress flourished, but the fall of the Western European empire made Europe sink in the dark medieval period where thinking was dominated by religious fear and superstition and hence scientific progress came to grinding halt, and it is wake improvements in medical care, food production and technology.

The Arab world continued the legacy of Aristotle (384–322 BCE, Greece) and Alhazen (Ibn al-Haytham, 965–1039 Iraq), who by many is considered as the father of the modern scientific method.[1] It was this modern scientific method that became a catalyst for scientific and technological development.

scientific method

A class of people that accumulated wealth through smart choices emerged. This was made possible by private enterprise and an efficient way of sharing risks and investments. In 1602, the East Indies Company became the first common stock company and in 1601 the Amsterdam Stock Exchange created a platform where innovative, exploratory and trade ideas could find the necessary capital to flourish.

In 1775, James Watt's improvement of the steam engine allowed to leverage on the progress made around the joint stock company and the stock exchange. This combination powered the raise of a new societal organization, capitalism and fueled the first industrial wave based on automation (mainly in the textile industry).

capitalism

While this first industrial wave brought much misery and social injustice, as a species we were preparing for the next stage. It created wealth as never before on a scale never seen before. From England, the industrialization, spread fast over Europe and the young state in North America. It all ended in "the Panic of 1873," that brought the "Long Depression" to Europe and the United States of America. This depression was so deep that it would indirectly give rise to a the invention of an new economic order: communism.

The same steam engine, however, had another trick up its sleeve: after industrialisation it was able to provide mass transport by the railway. This fuelled a new wave of wealth creation and lasted till the 1900s, where that wave of wealth creation ended in the "Panic of 1901" and the "Panic of 1907" – the first stock market crashes to start in the United States of America. The internal combustion engine, electricity and magnetism became the cornerstones of an new wave of exponential growth based on innovation. The "Wall Street Crash of 1929" ended this wave and started the "Great Depression."

It was about 1935 when Kondratiev noticed these long term waves of exponential growth and devastating market crashes and published his findings in Kondratieff and Stolper (1935) – republished in Kondratieff (1979). The work became prophetic as the automobile industry and chemistry fuelled a new wave of development that gave us individual mobility and lasted till 1973–1974 stock market crash.

Kondratiev

The scenario repeated itself as clockwork when it was the turn of the electronic computer and information technology (IT) to fuel exponential growth till the crashes of 2002–2008.

IT
information technology

Now, momentum is gathering pace with a few strong contenders to pull a new wave of economic development and wealth creation, a new phase of exponential growth. These contenders include in our opinion:

- quantum computing (if we will manage to get them work, that is),
- nanotechnology and development in medical treatments,
- machine learning, artificial intelligence and big data.

[1] With "scientific method" we refer to the empirical method of acquiring knowledge based on observations, scepticism and scrutiny from peers, reproducibility of results. The idea is to formulate a hypothesis, based on logical induction based on observations, then allowing peers to review and publish the results, so that other can falsify of conform the results.

This book is about the last group: machine learning (statistical learning) and data, and if you are reading it then you are certainly playing a significant role in designing the newest wave of exponential growth. If regulators will listen to scientists[2] and stop using incoherent risk measures in legislation (such as the Basel agreement and UCITS IV), then it might even be the first phase of exponential growth that does not end in a dramatic crash of financial markets.

Working with data is helping a new Kondratiev wave of wealth creation to take off. The Kondratiev waves of development seem to bring us step by step closer to a Type I civilisation, as described by Nikolai Kardashev – (see Kardashev, 1964). Nikolai Kardashev describes a compelling model of how intelligent societies could develop. He recognizes three stages of development. The stages are:

Kardashev

1. **Type 0 society** is the tribal organization form that we know today and has lasted since the dawn of the Homo genus.
2. A **Type I society** unites all people of a planet in one community and wields technology to influence weather and harvest energy of that planet.
3. A **Type II society** has colonized more than one planet of one solar system, is able to capture the energy of a star, and in some sense rules its solar system.
4. A **Type III** society has a galaxy in its control, is able to harvest energy from multiple stars, maybe even feeds stars into black holes to generate energy, etc.
5. There is no type **Type IV society** described in Kondratiev, though logically one can expect that the next step would be to spread over the local cluster of galaxies.[3] However, we would argue that more probably we will by then have found that our own species, survival is a legacy problem and that in order to satisfy the deepest self-actualization there is no need for enormous amounts of energy – where Kardashev seems to focus on. Therefore, we argue that the fourth stage would be very different.

While both Kondratiev's and Kardashev's theories are stylised idioms of a much more complex underlying reality, they paint a picture that is recognizable and allows us to realize how important scientific progress is. Indeed, if we do not make sure that we have alternatives to this planet, then our species is doomed to extinction rather soon. A Yellowstone explosion, an asteroid the size of an small country, a rogue planet, a travelling black hole, a nearby supernova explosion and so many more things can make our earth disappear or at least render it unsuitable to sustain life as we have known it for many thousands of years or – in the very best case – trigger a severe mass extinction.[4] Only science will be able to help us to transcend the limitation of the thin crust of this earth.

Only science has brought welfare and wealth, only science and innovation will help us to leave a lasting legacy in this universe.

But what is that "science"?

---

[2]See elaboration on coherent risk measures for example in Artzner et al. (1997), Artzner et al. (1999), and De Brouwer (2012).

[3]Dominating the universe has multiple problems as there is a large part of the universe that will be forever invisible, even with the speed of light, due to the Dark Energy and the expansion of the universe.

[4]With this statement we do not want to to take a stance in the debate whether the emergence of the Homo sapiens coincided with the start of an extinction event – that is till ongoing – nor do we want to ignore that for the last 300 000 years species are disappearing at an alarming rate, nor do we want to take a stance in the debate that the Homo sapiens is designing its own extinction. We only want to point out that dramatic events might occur on our lovely planet and make it unsuitable to sustain intelligent life.

♣ 2 ♣

The Scientific Method and Data

The world around us is constantly changing, making the wrong decisions can be disastrous for any company or person. At the same time it is more than ever important to innovate. Innovating is providing a diversity of ideas that just as in biological evolution might hold the best suited mutation for the changing environment.

There are many ways to come to a view on what to do next. Some of the more popular methods include instinct and prejudice, juiced up with psychological biases both in perception and decision making. Other popular methods include decision by authority ("let the boss decide"), deciding by decibels ("the loudest employee is heard") and dogmatism ("we did this in the past" or "we have a procedure that says this"). While these methods of creating an opinion and deciding might coincidently work out, in general they are sub-optimal by design. Indeed, the best solution might not even be considered or be pre-emptively be ruled out based on flawed arguments.

Looking at scientific development throughout times as well as human history, one is compelled to conclude that the only workable construct so far is also known as the scientific method. No other methods haves brought the world so many innovations and progress, no other methods have stood up in the face of scrutiny.

scientific method

The Scientific Method

Aristotle (384–322 BCE, Greece) can be seen as the father of the scientific method, because of his rigorous logical method which was much more than natural logic. But it is fair to credit Ibn al-Haytham (aka Alhazen — 965–1039, Iraq) for preparing the scientific method for collaborative use. His emphasis on collecting empirical data and reproducibility of results laid the foundation for a scientific method that is much more successful. This method allows people to check each other and confirm or reject previous results.

However, both the scientific method and the word "scientist" only came into common use in the nineteenth century and the scientific method only became the standard method in the twentieth century. Therefore, it should not come as a surprise that this became also a period of inventions and development as never seen before.

scientist

Indeed, while previous inventions such as fire, agriculture, the wheel, bronze and steel might not have followed explicitly the scientific method they created a society ready to embrace the scientific method and fuel an era of accelerated innovation and expansion. The internal combustion engine, electricity and magnetism fuelled the economic growth as never seen before.

The Big R-Book: From Data Science to Learning Machines and Big Data, First Edition. Philippe J.S. De Brouwer.

Companion Website: www.wiley.com/go/De Brouwer/The Big R-Book

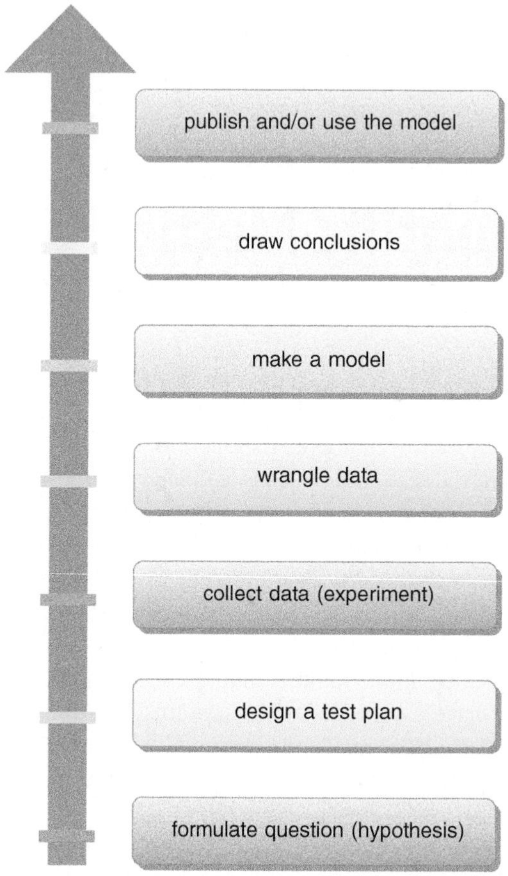

Figure 2.1: *A view on the steps in the scientific method for the data scientist and mathematical modeller, aka "quant." In a commercial company the communication and convincing or putting the model in production bear a lot of importance.*

The electronic computer brought us to the twenty first century and now a new era of growth is being prepared by big data, machine learning, nanotechnology and – maybe – quantum computing.

Indeed, with huge powers come huge responsibility. Once an invention is made, it is impossible to "un-invent" it. Once the atomic bomb exist, it cannot be forgotten, it is forever part of our knowledge. What we can do is promote peaceful applications of quantum technology, such as sensors to open doors, diodes, computers and quantum computers.

singularity

For example, as information and data technology advances, the singularity[1]. It is our responsibility to foresee potential dangers and do all what is in our power to avoid that these dangers

[1] The term "singularity" refers to the point in time where an intelligent system would be able to produce an even more intelligent system that also can create another system that is a certain percentage smarter in a time that is a certain percentage faster. This inevitably leads to exponentially increasing creating of better systems. This time series converges to one point in time, where "intelligence" of the machine would hit its absolute limits. First, record of the subject is by Stanislaw Ulam in a discussion with John Von Neuman in the 1950s and an early and convincing publication is Good (1966). It is also elaborately explored in Kurzweil (2010).

become an extinction event. Many inventions had a dark side and have led to more efficient ways of killing people, degenerating the ozone layer or polluting our ecosystem. Humanity has had many difficult times and very dark days, however, never before humanity became extinct. That would be the greatest disaster for there would be no recovery possible.

So the scientific method is important. This method has brought us into the information age and we are only scratching the surface of possibilities. It is only logical that all corporates try to stay abreast of changes and put a strong emphasis on innovation. This leads to an ever-increasing focus on data, algorithms, mathematical models such as machine learning.

Data, statistics and the scientific method are powerful tools. The company that has the best data and uses its data best is the company that will be the most adaptable to the changes and hence the one to survive. This is not biological evolution, but guided evolution. We do not have to rely on a huge number of companies with random variations, but we can use data to see trends and react to them.

The role of the data-analyst in any company cannot be overestimated. It is the reader of the book on whose shoulders rest not only to read those patterns from the data but also to convince decision makers to act in this fact-based insight.

Because the role of data and analytics is so important, it is essential to follow scientific rigour. This means in the first place following the scientific method for data analysis. An interpretation of the scientific method for data-science is in Figure 2.2 on page 10.

Till now we discussed the role of the data scientists and actions that they would take. But how does it look from the point of view of data itself?

Using that scientific method for data-science, the most important thing is probably to make sure that the one understands the data very well. Data in itself is meaningless. For example, 930 is just a number. It could be anything: from the age of Adamath in Genesis, to the price of chair or the code to unlock your bike-chain. It could be a time and 930 could mean "9:30" (assume "am" if your time-zone habits require so). Knowing that interpretation, the numbers become information, but we cannot understand this information till we know what it means (it could be the time I woke up – after a long party, the time of a plane to catch, a meeting at work, etc.). We can only understand the data if we know that it is a bus schedule of the bus "843-my-route-to-work" for example. This understanding, together with the insight that this bus always runs 15 minutes late and my will to catch the bus can lead to action: to go out and wait for that bus and get on it.

data

information

insight

action

This simple example shows us how the data cycle in any company or within any discipline should work. We first have a question, such as for example "to which customers can we lend money without pushing them into a debt-spiral." Then one will collect data (from own systems or credit bureau). This data can then be used to create a model that allows us to reduce the complexity of all observations to the explaining variables only: a projection in a space of lower dimensions. That model helps us to get the insight from the data and once put in production allows us to decide on the right action for each credit application.

This institution will end up with a better credit approval process, where less loss events occur. That is the role of data-science: to drive companies to the creation of more sustainable wealth in a future where all have a place and plentifulness.

This cycle – visualized in Figure 2.2 on page 10 – brings into evidence the importance of data-science. Data science is a way to bring the scientific method into a private company, so that decisions do not have to be based on gut-feeling alone. It is the role of the data scientist to take data, transform that data into information, create understanding from that data that can lead to actionable insight. It is then up to the management of the business to decide on the actions and follow them through. The importance of being connected to the reality via contact with the business cannot be overstated. In each and every step, mathematics will serve as tools, such as screwdrivers

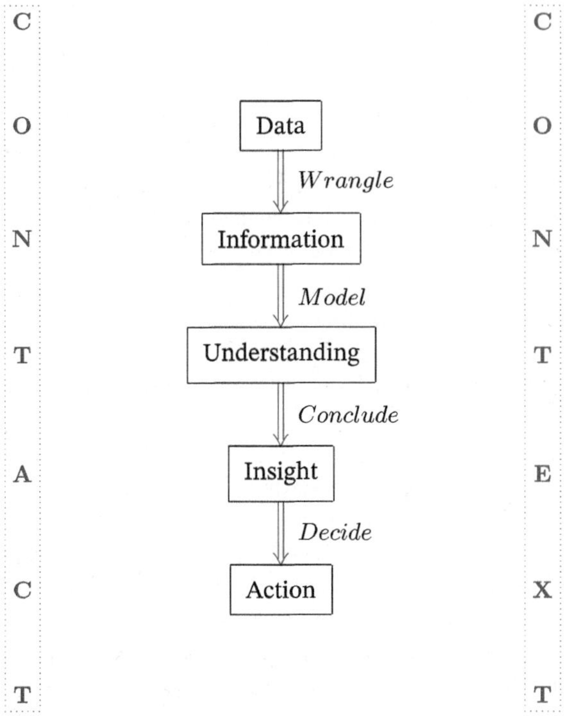

Figure 2.2: *The role of data-science in a company is to take data and turn it into actionable insight. At every step – apart from technical issues that will be discussed in this book – it is of utmost importance to understand the* context *and limitations of data, business, regulations and customers. For effectiveness in every step, one needs to pay attention to communication and permanent* contact *with all stakeholders and environment is key.*

and hammers. However, the choice about which one to use depends on a good understanding what we are working with and what we are trying to achieve.

♣ 3 ♣

Conventions

This book is formatted with LaTeX. The people who know this markup language will have high expectations for the consistency and format of this book. As you can expect there is

1. a table of contents at the start;
2. an index at the end, from page 1103;
3. a bibliography on page 1088;
4. as well as a list of all short-hands and symbols used on page 1117.

This is a book with a programming language as leitmotif and hence you might expect to find a lot of chunks of code. R is an interpreted language and it is usually accessed by opening the software R (simply type R on the command prompt and press enter).[1]

```
# This is code
1+pi
## [1] 4.141593

Sys.getenv(c("EDITOR","USER","SHELL", "LC_NUMERIC"))
##       EDITOR         USER        SHELL    LC_NUMERIC
##         "vi"       "root"  "/bin/bash" "pl_PL.UTF-8"
```

As you can see, the code is highlighted, that means that not all things have the same colour and it is easier to read and understand what is going on. The first line is a "comment" that means that R will not do anything with it, it is for human use only. The next line is a simple sum. In your R terminal, this what you will type or copy after the > prompt. It will rather look like this:

```
> # This is code
> 1+pi
[1] 4.141593
> Sys.getenv(c("EDITOR","USER","SHELL","XDG_SESSION_TYPE")
  EDITOR          USER          SHELL          LC_NUMERIC
    "vi"    "philippe"    "/bin/bash"      "pl_PL.UTF-8"
>
```

[1]You, will of course, first have to install the base software R. More about this in Chapter 4 *"The Basics of R"* on page 21.

The Big R-Book: From Data Science to Learning Machines and Big Data, First Edition. Philippe J.S. De Brouwer.

Companion Website: www.wiley.com/go/De Brouwer/The Big R-Book

In this, book there is nothing in front of a command and the reply of R is preceded by two pound signs: "##."[2] The pound sign (#) is also the symbol used by R to precede a comment, hence R will ignore this line if fed into the command prompt. This allows you to copy and paste lines or whole chunks if you are working from an electronic version of the book. If the > sign would precede the command, then R would not understand if, and if you accidentally copy the output that from the book, nothing will happen because the #-sign indicates to R to ignore the rest of the line (this is a comment for humans, not for the machine).

`Sys.getenv()`

The function `Sys.getenv()` returns us all environment variables if no parameter is given. If it is supplied with a list of parameters, then it will only return those.

In the example above the function got three variables supplied, hence only report on these three. You will also notice that the variables are wrapped in a special function `c(...)`. This is because the function `Sys.getenv()` expects one vector as argument and the function `c()` will create the vector out of a list of supplied arguments.

Note that in this paragraph above name of the function `Sys.getenv()` is mono-spaced. That is our convention to use code within text. Even in the index, at the end of this book, we will follow that convention.

You will also have noticed that in text – such as this line – we refer to code fragments and functions, using fixed width font such as for example "the function `mean()` calculates the average." When this part of the code needs emphasizing or is used as a word in the sentence, we might want to highlight it additionally as follows: `mean(1 + pi)`.

Some other conventions also follow from this small piece of code. We will assume that you are using Linux (unless mentioned otherwise). But do not worry: that is not something that will stand in your way. In Chapter 4 *"The Basics of R"* on page 21, we will get you started in Windows and all other things will be pretty much the same. Also, while most books are United States centric, we want to be as inclusive as possible and not assume that you live in the United States working from United States data.

United States of America

ISO standard

As a rule, we take a country-agnostic stance and follow the ISO standards[3], for dates and dimensions of other variables. For example, we will use meters instead of feet.

Learning works best when you can do something with the knowledge that you are acquiring. Therefore, we will usually even show the code of a plot that is mainly there for illustrative purposes, so you can immediately try everything yourself.

When the code produces a plot (chart or graph), then the plot will appear generally at that point between the code lines. For example, consider we want to show the generator function for the normal distribution.

---

[2]The number sign, #, is also known as the "hash sign" or "pound sign." It probably evolved from the "libra ponda" (a pound weight). It is currently used in any different fields as part of phone numbers, in programming languages (e.g. in an URL it indicates a sub-address, in R it precedes a comment, etc), the command prompt for the root user in Unix and Linux, in set theory ($\#S$ is the cardinality of the set S), in topology ($A\#B$ is the connected sum of manifolds A and B), number theory ($\#n$ is the primorial of n), as keyword in some social media, etc. The pronunciation hence varies widely: "hash" when used to tag keywords (#book would be the hash sign and the tag book). Hence, reading the "#"-sign as "hashtag" is at least superfluous). Most often, it is pronounced as "pound." Note that the musical notation is another symbol, ♯, that is pronounced as "sharp" as in the music (e.g. C♯).

[3]ISO standards refer to the standards published by the International Organization for Standardization (ISO). This is an international standard-defining body, founded on 23 February 1947, it promotes worldwide proprietary, industrial and commercial standards. At this point, 164 countries are member and decisions are made by representatives of those countries. ISO is also recognised by the United Nations.

```
# generate 1000 random numbers between 0 and 100
x <- rnorm(1000, mean = 100, sd = 2)

# to illustrate previous, we show the histogram.
hist(x, col = "khaki3")
```

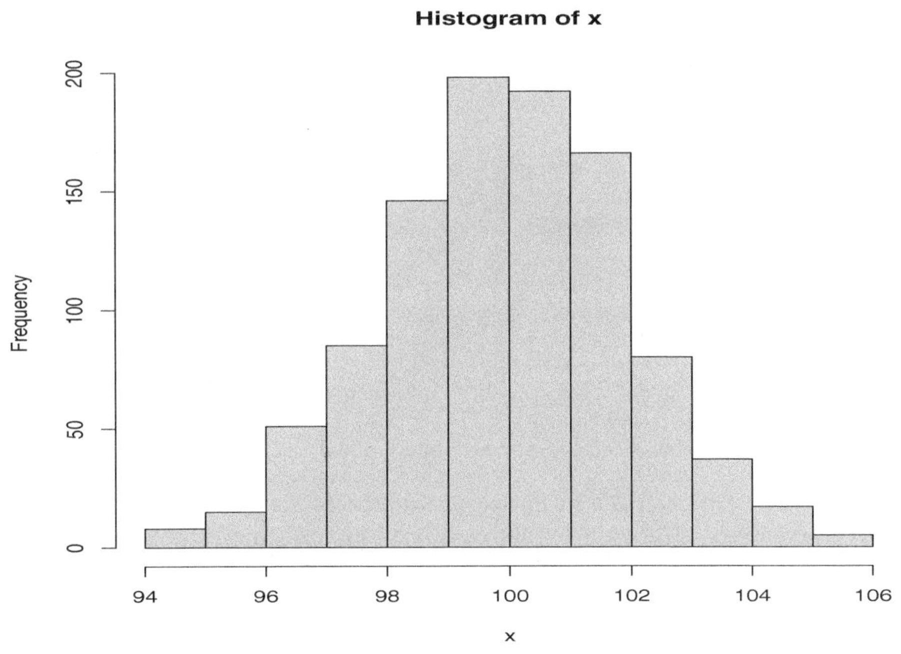

Figure 3.1: *An example showing the histogram of data generated from the normal distribution.*

```
# This code follows the 'hist' command.
# In rare cases the plot will be on the this page
# alone and this comment is the previous page.
```

In most cases, the plot will be just after the code that it generates – even if the code continues after the `plot(...)` command. Therefore, the plot will usually sit exactly where the code creates it. However, in some rare cases, this will not be possible (it would create page layout that would not be aesthetically appealing). The plot will then appear near the code chunk (maybe on the next page). To help you to find and identify the plot in such case, we will usually add a numbered caption to the plot.

The R code is so ubiquitous and integrated in the text that it will appear just where it should be (though charts might move). They are integral part of the text and the comments that appear there might not be repeated in the normal text later.

There is also some other code from the command prompt and/or from SQL environments. That code appears much less, so they are numbered and appear as in Listings 3.1 and 3.2.

```
$ R

R version 3.4.4 (2018-03-15) -- "Someone to Lean On"
Copyright (C) 2018 The R Foundation for Statistical Computing
Platform: x86_64-pc-linux-gnu (64-bit)

R is free software and comes with ABSOLUTELY NO WARRANTY.
You are welcome to redistribute it under certain conditions.
Type 'license()' or 'licence()' for distribution details.

  Natural language support but running in an English locale

R is a collaborative project with many contributors.
Type 'contributors()' for more information and
'citation()' on how to cite R or R packages in publications.

Type 'demo()' for some demos, 'help()' for on-line help, or
'help.start()' for an HTML browser interface to help.
Type 'q()' to quit R.

[Previously saved workspace restored]

>
```

Listing 3.1: *This is what you would see if you start R in the command line terminal. Note that the last sign is the R-prompt, inviting you to type commands. This code fragment is typical for how code that is not in the R-language has been typeset in this book.*

```
$ factor 1492
1492: 2 2 373
$ calc 2*2*373
        1492
$ pi 60
3.14159265358979323846264338327950288419716939937510582097494
```

Listing 3.2: *Another example of a command line instructions: factor, calc, and pi. This example only has CLI code and does not start R.*

Note that in these environments, we do not "comment out" the output. We promise to avoid mixing input and output, but in some cases, the output will just be there. So in general it is only possible to copy line per line the commands to see the output on the screen. Copying the whole block and pasting it in the command prompt leads to error messages, rather than the code being executed. This is unlike the R code, which can be copied as a whole, pasted in the R-command prompt and it should all work fine.

Questions or tasks look as follows:

? Question #1 Histogram

Consider Figure 3.1 on page 13. Now, imagine that you did not generate the data, but someone gave it to you, so that you do not know how it was generated. Then what could this data represent? Or, rephrased, what could `x` be? Does it look familiar?

Questions or tasks can be answered by the tools and methods explained previously. Note that it might require to do some research by your own, such as looking into the help files or other documentation (we will of course explain how to access these). If you are using this book to prepare for an exam or test, it is probably a good preparation, but if you are in a hurry it is possible to skip these (in this book they do not add to the material explained). However, in general thinking about the issues presented will help you to solve your data-problems more efficiently.

Note that the answers to most questions can be found in E "Answers to Selected Questions" on page 1061.The answer might not always be detailed but it should be enough to solve the issue raised.

Definition: This is a definition

This is not an book about exact mathematics. This is a pragmatic book with a focus on practical applications. Therefore, we use the word "definition" also in a practical sense.

Definitions are not always rigorous definitions as a mathematician would be used to. We rather use practical definitions (e.g. how a function is implemented).

The use of a function is – mainly at the beginning of the book – highlighted as follows. For example:

Function use for mean()

`mean (x, na.rm = FALSE, trim = 0, ...)`

Where

- `x` is an R-object,
- `na.rm` is a boolean (setting this to `TRUE` will remove missing values),
- `trim` is the fraction of observations to be removed on both sides of the distribution before the mean is computed – the default is 0 and it cannot be higher than 0.5

From this example, it should be clear how the function `mean()` is used. Note the following:

- The name of the function is in the title.
- On the first line we repeat the function with its most important parameters.
- The parameter `x` is an obligatory parameter.
- The parameter `na.rm` can be omitted. When it is not provided, the default `FALSE` is used. A parameter with a default can be recognised in the first line via the equal sign.
- The three dots indicate that other parameters can be passed that will not be used by the function `mean()`, but they are passed on to further methods.
- Some functions can take a lot of parameters. In some cases, we only show the most important ones.

Later on in the book, we assume that the reader is able to use the examples and find more about the function in its documentation. For example, `?mean` will display information about the function `mean`.

When a new concept or idea is built up with examples they generally appear just like that in the text. Additional examples after the point is made are usually highlighed as follows:

Example: Mean

An example of a function is `mean()`, as the name suggests it calculates the arithmetic mean (average) of data fed into the function.

```
# First, generate some data:
x <- c(1,2,3)

# Then calculate the mean:
mean(x)
## [1] 2
```

Some example environments are split in two parts: the question and the solution as follows:

Example: Mean

What is the mean of all integer numbers from one to 100? Use the function `mean()`.

```
mean(1:100)
## [1] 50.5
```

There are a few more special features in the layout that might be of interest.

A hint is something that adds additional practical information that is not part of the normal flow of the text.

Hint – Using the hint boxes

When first studying a section, skip the hints, and when reading it a second time pay more attention to the hints.

When we want to draw attention to something that might or might not be clear from the normal flow of the text, we put it in a "notice environment." This looks as follows:

Note – Layout details

Note that hints, notes and warnings look all similar, but for your convenience, we have differentiating colours and layout details.

There are more such environments and we let them speak for themselves.

Digression – This is good to know

A digression does what you would expect from it. It is not necessary to read in order to understand the rest of the chapter. However, it provides further insight that is useful to gain a deeper insight of the subject discussed.

Skip the digressions when you read the text first, and come back to them later.

Warning – Read comments in code

When reading the book, always read the comments in the code.

```
# Code and especially the comments in it are part of
# the normal flow of the text!
```

In general, a warning is important to read once you will start working on your own.

Note – Shadow

Note that the boxes with a shadow are "lifted off the page" and are a little independent from the flow of the main text. Those that are no shadow are part of the main flow of the text (definitions, examples, etc.)

PART II

Starting with R and Elements of Statistics

♡

The Basics of R

In this book we will approach data and analytics from a practitioners point of view and our tool of choice is R. R is in some sense a re-implementation of S – a programming language written in 1976 by John Chambers at Bell Labs – with added lexical scoping semantics. Usually, code written in S will also run in R. **S**

R is a modern language with a rather short history. In 1992, the R-project was started by Ross Ihaka and Robert Gentleman at the University of Auckland, New Zealand. The first version was available in 1995 and the first stable version was available in 2000.

Now, the R Development Core Team (of which Chambers is a member) develops R further and maintains the code. Since a few years Microsoft has embraced the project and provides MRAN (Microsoft R Application Network). This package is also free and open source software (FOSS) **FOSS** and has some advantages over standard R such as enhanced performance (e.g. multi-thread support, the checkpoint package that makes results more reproducible).

Essentially, R is ...

- a programming language built for statistical analysis, graphics representation and reporting;
- an interpreted computer language which allows branching, looping, modular programming as well as object and functional oriented programming features.

R offers its users ...

- integration with the procedures written in the C, C++, .Net, Python, or FORTRAN languages for efficiency; **C** **C++** **.Net** **Fortran**
- zero purchase cost (available under the GNU General Public License), and pre-compiled binary versions are provided for various operating systems like Linux, Windows, and Mac; **Linux** **Windows** **Mac**
- simplicity and effectiveness;
- a free and open environment;
- an effective data handling and storage facility;
- a suite of operators for calculations on arrays, lists, vectors, and matrices;
- a large, coherent, and integrated collection of tools for data analysis;

The Big R-Book: From Data Science to Learning Machines and Big Data, First Edition. Philippe J.S. De Brouwer.

Companion Website: www.wiley.com/go/De Brouwer/The Big R-Book

- graphical facilities for data analysis and display either directly at the computer or printing;
- a supportive on-line community;
- the ability for you to stand on the shoulders of giants (e.g. by using libraries).

R is arguably the most widely used statistics programming language and is used from universities to business applications, while it still gains rapidly in popularity.

Hint – Getting more help

If at any point you are trying to solve a particular issues and you are stuck, the online community will be very helpful. To get unstuck, do the following:

- First, look up your problems by adding the keyword "R" in the search string. Most probably, someone else encountered the very same problem before you, and the answer is already posted. Avoid to post a question that has been answered before.
- If you need to ask your question in a forum such as for example `www.stackexchange.com` then you will need to add a minimal reproducible example. The package `reprex` can help you to do just that.

4.1 Getting Started with R

installing R

Before we can start, we need a working installation of R on our computer. On Linux, this can be done via the command line. On Debian and its many derivatives such as Ubuntu or Mint, this looks as follows:[1]

```
sudo apt-get install r-base
```

On Windows or Mac, you want to refer to `https://cran.r-project.org` and download the right package for your system.

To start R, open the command line and type `R` (followed by enter). This is the R interpreter (or R console). You can do all your data crunching here. To leave the environment type `q()` followed by [enter].

> **Hint – Using R Online**
>
> It is also possible to use R online:
>
> - `https://www.tutorialspoint.com/execute_r_online.php`
> - `http://www.r-fiddle.org`

RStudio

For the user, who is not familiar with the command line, it is highly recommendable to use an IDE, such as RStudio (see `https://www.rstudio.com`). Later on – for example in Chapter 32 "*R Markdown*" on page 879 – we will see that RStudio has some unique advantages over the R-console in store, that will convince even the most traditional command-line-users.

IDE
RStudio

Whether you use standard R or MRAN, using RStudio will enhance your performance and help you to be more productive.

MRAN

Rstudio is an integrated development environment (IDE) for R and provides a console, editor with syntax-highlighting, a window to show plots and some workspace management.

IDE
integrated development environment

> **Hint – RStudio is free**
>
> Rstudio can be downloaded from `https://www.rstudio.com` for free.

RStudio is also in the repositories of most Linux distributions. That means that there is no need to go to the website or RStudio, but it is possible to download R, and install it with just one line in the CLI of your OS. For example, in Debian and its derivatives, this would be:

```
# Note that the first 2 lines are just to make sure that
# you are using the latest version of the repository.
sudo apt-get update
sudo apt-get upgrade

# Install RStudio:
sudo apt-get install rstudio
```

[1] There are different package management systems for different flavours of Linux and discussing them all is not only beyond the scope of this book, but not really necessary. We assume that if you use Genttoo, that you will know what to do or where to choose.

This process is very simple: provide your admin password and the script will take care of everything. Then use your preferred way to start the new software. R can be started by typing `R` in the command prompt and RStudio will launch with the command `rstudio`.

An important note is that RStudio will work with the latest version of R that is installed. Also when you install MRAN, RStudio will automatically load this new version.

Basic arithmetic

addition
product
power

The basic operators work as one would expect. Simply type in the R terminal `2+3` followed by `ENTER` and R will immediately display the result.

```
#addition
2 + 3
#product
2 * 3
#power
2**3
2^3
#logic
2 < 3
x <- c(1,3,4,3)
x.mean <- mean(x)
x.mean
y <- c(2,3,5,1)
x+y
```

Hint – White space

Usually, white space is not important for R.
For example, `mean ( c (c + pi , 2 ) )` will do exactly the same as `mean(c(1+pi,2))`. However, it is a good habit to add a white space before and after operators and after the comma. This will improve readability, which makes the debugging process easier and is helpful for other people who want to read your code.

Editing variables interactively

`scan()`

To create a variable `x`, type:

```
x <- scan()
```

will start an interface that invites you to type all values of the vector one by one.

In order to get back to the command prompt: type enter without typing a number (ie. leave one empty to end).

Further information – Other ways to import data

This is only one of the many ways to get data into R. Most probably you will use a mix of defining variables in the code and reading in data from files. See for example Chapter 15 "*Connecting R to an SQL Database*" on page 327.

To modify an existing variable, one can use the `edit()` function

edit()

```
edit(x)
```

Warning – Using CLI tools

The edit function will open the editor that is defined in the options. While in RStudio, this is a specially designed pop-up window with buttons to save and cancel, in the command line interface (CLI) this might be `vi`. The heydays of this fantastic editor are over and you might never have seen it before. It is not really possible to use vi without reading the manual (e.g. via the `man vi` command on the OS CLI or an online tutorial). To get out of `vi`, type: `[ESC]:q![ENTER]`. Note that we show the name of a key within square brackets and that all the other strings are just one keystroke each.

Batch mode

R is an interpreted language and while the usual interaction is typing commands and getting the reply appear on the screen, it is also possible to use R in batch mode.

batch mode

functions

1. create a file `test.R`
2. add the content `print("Hello World")`
3. run the command line `Rscript test.R`
4. now, open R and run the command `source("test.R")`

source()

5. add in the file

```
my_function <- function(a,b)
{
  a + b
}
```

6. now repeat step 4 and run `my_function(4,5)`

In this section we will present you with a practical introduction to R, it is not a formal introduction. If you would like to learn more about the foundations, then we recommend the documentation provided by the R-core team here: `https://cran.r-project.org/doc/manuals/r-release/R-lang.pdf`.

4.2 Variables

variables

As any computer language, R allows to use variables to store information. The variable is referred to by its name. Valid names in R consist of numbers and characters. Even most special characters can be used.

In R, variables

- can contain letters as well as "_" (underscore) and "." (dot), and
- variables must start with a letter (that can be preceded with a dot).

For example, `my_var.1` and `my.Cvar` are valid variables, but `_myVar`, `my%var` and `1.var` are not acceptable.

Assignment

assignment

Assignment can be made left or right:

```
x.1 <- 5
x.1 + 3 -> .x
print(.x)
## [1] 8
```

R-programmers will use the arrow sign `<-` most often, however, R allows left assignment with the `=` sign.

```
x.3 = 3.14
x.3
## [1] 3.14
```

There are also occasions that we must use the `=` operator. For example, when assigning values to named inputs for functions.

```
v1 <- c(1,2,3,NA)
mean(v1, na.rm = TRUE)
## [1] 2
```

There are more nuances and therefore, we will come back to the assignment in Chapter 4.4.4 "*Assignment Operators*" on page 78. These nuances are better understood with some more background, and for now it is enough to be able to assign values to variable.

Variable Management

With what we have seen so far, it is possible to make already simple calculations, define and modify variables. There is still a lot to follow and it is important to have some basic tools to keep things tidy. One of such tools is the possibility to see defined variables and eventually remove unused ones.

```
# List all variables
ls()                    # hidden variable starts with dot
ls(all.names = TRUE)    # shows all

# Remove a variable
rm(x.1)                 # removes the variable x.1
ls()                    # x.1 is not there any more
rm(list = ls())         # removes all variables
ls()
```

ls()

rm()

Note – What are invisible variables

A variable whose name starts with a dot (e.g. `.x`) is in all aspects the same as a variable that starts with a letter. The only difference is that the first will be hidden with the standard arguments of the function `ls()`.

4.3 Data Types

As most computer languages, R has some built-in data-types. While it is possible to do certain things in R without worrying about data types, understanding and consciously using these base-types will help you to write bug-free code that is more robust and it will certainly speed up the debugging process. In this section we will highlight the most important ones.

4.3.1 The Elementary Types

There is no need to declare variables explicitly and tell R what type the variable will be before using it. R will assign them a class whenever this is needed and even change the type when our code implies a change.

class()

long

complex numbers

string

```
# Booleans can be TRUE or FALSE:
x <- TRUE
class(x)
## [1] "logical"

# Integers use the letter L (for Long integer):
x <- 5L
class(x)
## [1] "integer"

# Decimal numbers, are referred to as 'numeric':
x <- 5.135
class(x)
## [1] "numeric"

# Complex numbers use the letter i (without multiplication sign):
x <- 2.2 + 3.2i
class(x)
## [1] "complex"

# Strings are called 'character':
x <- "test"
class(x)
## [1] "character"
```

Warning – Changing data types

While R allows to change the type of a variable, doing so is not a good practice. It makes code difficult to read and understand.

```
# Avoid this:
x <- 3L       # x defined as integer
x
## [1] 3

x <- "test"  # R changes data type
x
## [1] "test"
```

So, keep your code tidy and to not change data types.

Dates

date

Working with dates is a complex subject. We explain the essence of the issues in Section 17.6 "*Dates with lubridate*" on page 407. For now, it is sufficient to know that dates are one of the base types of R.

```
# The function as.Data coerces its argument to a date
d <- as.Date(c("1852-05-12", "1914-11-5", "2015-05-01"))

# Dates will work as expected
d_recent <- subset(d, d > as.Date("2005-01-01"))
print(d_recent)
## [1] "2015-05-01"
```

as.Date()
subset()

Further information – More about dates

Make sure to read Section 17.6 "*Dates with lubridate*" on page 407 for more information about working with dates as well as the inevitable problems related to dates.

4.3.2 Vectors

4.3.2.1 Creating Vectors

vector

Simply put, vectors are lists of objects that are all of the same type. They can be the result of a calculation or be declared with the function c().

```
x <- c(2, 2.5, 4, 6)
y <- c("apple", "pear")
class(x)
## [1] "numeric"

class(y)
## [1] "character"
```

More about lists can be found in Section 4.3.6 "*Lists*" on page 53.

4.3.3 Accessing Data from a Vector

In many situations you will need to address a vector as a whole, but there will also be many occasions where it is necessary to do something with just one element. Since this is such a common situation, R provides more than one way to get this done.

```
# Create v as a vector of the numbers one to 5:
v <- c(1:5)

# Access elements via indexing:
v[2]
## [1] 2

v[c(1,5)]
## [1] 1 5

# Access via TRUE/FALSE:
v[c(TRUE,TRUE,FALSE,FALSE,TRUE)]
## [1] 1 2 5
```

```
# Access elements via names:
v <- c("pear" = "green", "banana" = "yellow", "coconut" = "brown")
v
##     pear   banana  coconut
##  "green" "yellow"  "brown"

v["banana"]
##   banana
## "yellow"

# Leave out certain elements:
v[c(-2,-3)]
##    pear
## "green"
```

4.3.3.1 Vector Arithmetic

The standard behaviour for vector arithmetic in R is element per element. With "standard" we mean operators that do not appear between percentage signs (as in `%.%` for example).

```
v1 <- c(1,2,3)
v2 <- c(4,5,6)
# Standard arithmetic
v1 + v2
## [1] 5 7 9

v1 - v2
## [1] -3 -3 -3

v1 * v2
## [1]  4 10 18
```

> **Warning – Not all operations are element per element**
>
> The dot-product and other non-element-per-element-operators are available via specialized operators such as `%.%`: see Section 4.4.1 "*Arithmetic Operators*" on page 75

4.3.3.2 Vector Recycling

Vector recycling refers to the fact that in case an operation is requested with one too short vector, that this vector will be concatenated with itself till it has the required length.

```
# Define a short and long vector:
v1 <- c(1, 2, 3, 4, 5)
v2 <- c(1, 2)

# Note that R 'recycles' v2 to match the length of v1:
v1 + v2

## Warning in v1 + v2: longer object length is not a multiple of shorter object length
## [1] 2 4 4 6 6
```

Warning – Vector recycling

This behaviour is most probably different from what the experienced programmer will expect. Not only we can add or multiply vectors of different nature (e.g. long and real), but also we can do arithmetic on vectors of different size. This is usually not what you have in mind, and does lead to programming mistakes. Do take an effort to avoid vector recycling by explicitly building vectors of the right size.

4.3.3.3 Reordering and Sorting

sorting

sort()

To sort a vector, we can use the function `sort()`.

```
# Example 1:
v1 <- c(1, -4, 2, 0, pi)
sort(v1)
## [1] -4.000000  0.000000  1.000000  2.000000  3.141593

# Example 2: To make sorting meaningful, all variables are coerced to
# the most complex type:
v1 <- c(1:3, 2 + 2i)
sort(v1)
## [1] 1+0i 2+0i 2+2i 3+0i

# Sorting is per increasing numerical or alphabetical order:
v3 <- c("January", "February", "March", "April")
sort(v3)
## [1] "April"    "February" "January"  "March"

# Sort order can be reversed:
sort(v3, decreasing = TRUE)
## [1] "March"    "January"  "February" "April"
```

Question #2 Temperature conversion

The time series `nottem` (from the package "datasets" that is usually loaded when R starts) contains the temperatures in Notthingham from 1920 to 1939 in Fahrenheit. Create a new object that contains a list of all temperatures in Celsius.

Hint – Addressing the object nottem

temperature

length()

Note that `nottem` is a time series object (see Chapter 10 "*Time Series Analysis*" on page 255) and not a matrix. Its elements are addressed with `nottam[n]` where n is between 1 and `length(nottam)`. However, when printed it will look like a matrix with months in the columns and years in the rows. This is because the print-function will use functionality specific to the time series object.[a]
Remember that $T(C) = \frac{5}{9}(T(F) - 32)$.

[a]This behaviour is caused by the dispatcher-function implementation of an object-oriented programming model. To understand how this works and what it means, we refer to Section 6.2 "*S3 Objects*" on page 122.

4.3.4 Matrices

matrix Matrices are a very important class of objects. They appear in all sorts of practical problems: investment portfolios, landscape rendering in games, image processing in the medical sector, fitting of neural networks, etc.

4.3.4.1 Creating Matrices

matrix() A matrix is in two-dimensional data set where all elements are of the same type. The `matrix()` function offers a convenient way to define it:

```
# Create a matrix.
M = matrix( c(1:6), nrow = 2, ncol = 3, byrow = TRUE)
print(M)
##      [,1] [,2] [,3]
## [1,]    1    2    3
## [2,]    4    5    6

M = matrix( c(1:6), nrow = 2, ncol = 3, byrow = FALSE)
print(M)
##      [,1] [,2] [,3]
## [1,]    1    3    5
## [2,]    2    4    6
```

It is also possible to create a unit or zero vector with the same function. If we supply one matrix() scalar instead a vector to the first argument of the function `matrix()`, it will be recycled as much as necessary.

```
# Unit vector:
matrix (1, 2, 1)
##      [,1]
## [1,]    1
## [2,]    1

# Zero matrix or vector:
matrix (0, 2, 2)
##      [,1] [,2]
## [1,]    0    0
## [2,]    0    0

# Recycling also works for shorter vectors:
matrix (1:2, 4, 4)
##      [,1] [,2] [,3] [,4]
## [1,]    1    1    1    1
## [2,]    2    2    2    2
## [3,]    1    1    1    1
## [4,]    2    2    2    2

# Fortunately, R expects that the vector fits exactly n times in the matrix:
matrix (1:3, 4, 4)

## Warning in matrix(1:3, 4, 4): data length [3] is not a sub-multiple or multiple of
the number of rows [4]
##      [,1] [,2] [,3] [,4]
## [1,]    1    2    3    1
## [2,]    2    3    1    2
## [3,]    3    1    2    3
## [4,]    1    2    3    1

# So, the previous was bound to fail.
```

4.3.4.2 Naming Rows and Columns

While in general naming rows and/or columns is more relevant for datasets than matrices it is possible to work with matrices to store data if it only contains one type of variable.

```
row_names = c("row1", "row2", "row3", "row4")
col_names = c("col1", "col2", "col3")
M <- matrix(c(10:21), nrow = 4, byrow = TRUE,
            dimnames = list(row_names, col_names))
print(M)
##      col1 col2 col3
## row1   10   11   12
## row2   13   14   15
## row3   16   17   18
## row4   19   20   21
```

dimnames

Once the matrix exists, the columns and rows can be renamed with the functions `colnames()` and `rownames()`

`colnames()`

`rownames()`

```
colnames(M) <- c('C1', 'C2', 'C3')
rownames(M) <- c('R1', 'R2', 'R3', 'R4')
M
##    C1 C2 C3
## R1 10 11 12
## R2 13 14 15
## R3 16 17 18
## R4 19 20 21
```

4.3.4.3 Access Subsets of a Matrix

It might be obvious, that we can access one element of a matrix by using the row and column number. That is not all, R has a very flexible – but logical – model implemented. Let us consider a few examples that speak for themselves.

```
M <- matrix(c(10:21), nrow = 4, byrow = TRUE)
M
##      [,1] [,2] [,3]
## [1,]   10   11   12
## [2,]   13   14   15
## [3,]   16   17   18
## [4,]   19   20   21

# Access one element:
M[1,2]
## [1] 11

# The second row:
M[2,]
## [1] 13 14 15

# The second column:
M[,2]
## [1] 11 14 17 20

# Row 1 and 3 only:
M[c(1, 3),]
##      [,1] [,2] [,3]
## [1,]   10   11   12
## [2,]   16   17   18
```

```
# Row 2 to 3 with column 3 to 1
M[2:3, 3:1]
##      [,1] [,2] [,3]
## [1,]   15   14   13
## [2,]   18   17   16
```

4.3.4.4 Matrix Arithmetic

Matrix arithmetic allows both base operators and specific matrix operations. The base operators operate always element per element.

Basic arithmetic on matrices works element by element:

```
M1 <- matrix(c(10:21), nrow = 4, byrow = TRUE)
M2 <- matrix(c(0:11),  nrow = 4, byrow = TRUE)
M1 + M2
##      [,1] [,2] [,3]
## [1,]   10   12   14
## [2,]   16   18   20
## [3,]   22   24   26
## [4,]   28   30   32

M1 * M2
##      [,1] [,2] [,3]
## [1,]    0   11   24
## [2,]   39   56   75
## [3,]   96  119  144
## [4,]  171  200  231

M1 / M2
##          [,1]      [,2]     [,3]
## [1,]      Inf 11.000000 6.000000
## [2,] 4.333333  3.500000 3.000000
## [3,] 2.666667  2.428571 2.250000
## [4,] 2.111111  2.000000 1.909091
```

dot-product

? **Question #3 Dot product**

Write a function for the dot-product for matrices. Add also some security checks. Finally, compare your results with the "%*%-operator."

The dot-product is pre-defined via the `%*%` opeartor. Note that the function `t()` creates the transposed vector or matrix.

```
# Example of the dot-product:
a <- c(1:3)
a %*% a
##      [,1]
## [1,]   14

a %*% t(a)
##      [,1] [,2] [,3]
## [1,]    1    2    3
## [2,]    2    4    6
## [3,]    3    6    9
```

```
t(a) %*% a
##      [,1]
## [1,]   14

# Define A:
A <- matrix(0:8, nrow = 3, byrow = TRUE)

# Test products:
A %*% a
##      [,1]
## [1,]    8
## [2,]   26
## [3,]   44

A %*% t(a) # this is bound to fail!

## Error in A %*% t(a): non-conformable arguments

A %*% A
##      [,1] [,2] [,3]
## [1,]   15   18   21
## [2,]   42   54   66
## [3,]   69   90  111
```

There are also other operations possible on matrices. For example the quotient works as follows:

```
A %/% A
##      [,1] [,2] [,3]
## [1,]   NA    1    1
## [2,]    1    1    1
## [3,]    1    1    1
```

Note – Percentage signs point towards matrix operations

Note that matrices will accept both normal operators and specific matrix operators.

```
# Note the difference between the normal product:
A * A
##      [,1] [,2] [,3]
## [1,]    0    1    4
## [2,]    9   16   25
## [3,]   36   49   64

# and the matrix product %*%:
A %*% A
##      [,1] [,2] [,3]
## [1,]   15   18   21
## [2,]   42   54   66
## [3,]   69   90  111



# However, there is -of course- only one sum:
A + A
##      [,1] [,2] [,3]
## [1,]    0    2    4
## [2,]    6    8   10
## [3,]   12   14   16
```

```
# Note that the quotients yield almost the same:
A %/% A
##      [,1] [,2] [,3]
## [1,]   NA    1    1
## [2,]    1    1    1
## [3,]    1    1    1

A / A
##      [,1] [,2] [,3]
## [1,]  NaN    1    1
## [2,]    1    1    1
## [3,]    1    1    1
```

The same hold for quotient and other operations.

Warning – R consistently works element by element

Note that while $exp(A)$, for example, is well defined for a matrix as the sum of the series:

$$exp(A) = \sum_{n=0}^{+\infty} A^n/n!$$

R will resort to calculating the $exp()$ element by element!
Using the same matrix A as in the aforementioned code:

```
# This is the matrix A:
A
##      [,1] [,2] [,3]
## [1,]    0    1    2
## [2,]    3    4    5
## [3,]    6    7    8

# The exponential of A:
exp(A)
##           [,1]        [,2]        [,3]
## [1,]   1.00000    2.718282    7.389056
## [2,]  20.08554   54.598150  148.413159
## [3,] 403.42879 1096.633158 2980.957987
```

The same holds for all other functions of base R:

```
# The natural logarithm
log(A)
##          [,1]     [,2]      [,3]
## [1,]     -Inf 0.000000 0.6931472
## [2,] 1.098612 1.386294 1.6094379
## [3,] 1.791759 1.945910 2.0794415

sin(A)
##            [,1]       [,2]       [,3]
## [1,]  0.0000000  0.8414710  0.9092974
## [2,]  0.1411200 -0.7568025 -0.9589243
## [3,] -0.2794155  0.6569866  0.9893582
```

Note also that some operations will collapse the matrix to another (simpler) data type.

```
# Collapse to a vectore:
colSums(A)
## [1]  9 12 15

rowSums(A)
## [1]  3 12 21

# Some functions aggregate the whole matrix to one scalar:
mean(A)
## [1] 4

min(A)
## [1] 0
```

We already saw the function `t()` to transpose a matrix. There are a few others available in base R. For example, the function `diag()` diagonal matrix that is a subset of the matrix, `det()` caluclates the determinant, etc. The function `solve()` will solve the equation `A %.% x = b`, but when A is missing, it will assume the idenity vector and return the inverse of A.

diag()

solve()

```
M <- matrix(c(1,1,4,1,2,3,3,2,1), 3, 3)
M
##      [,1] [,2] [,3]
## [1,]    1    1    3
## [2,]    1    2    2
## [3,]    4    3    1

# The diagonal of M:
diag(M)
## [1] 1 2 1

# Inverse:
solve(M)
##            [,1]        [,2]        [,3]
## [1,]  0.3333333 -0.66666667  0.33333333
## [2,] -0.5833333  0.91666667 -0.08333333
## [3,]  0.4166667 -0.08333333 -0.08333333

# Determinant:
det(M)
## [1] -12

# The QR composition:
QR_M <- qr(M)
QR_M$rank
## [1] 3

# Number of rows and columns:
nrow(M)
## [1] 3

ncol(M)
## [1] 3

# Sums of rows and columns:
colSums(M)
## [1] 6 6 6
```

```
rowSums(M)
## [1] 5 5 8

# Means of rows, columns, and matrix:
colMeans(M)
## [1] 2 2 2

rowMeans(M)
## [1] 1.666667 1.666667 2.666667

mean(M)
## [1] 2

# Horizontal and vertical concatenation:
rbind(M, M)
##      [,1] [,2] [,3]
## [1,]    1    1    3
## [2,]    1    2    2
## [3,]    4    3    1
## [4,]    1    1    3
## [5,]    1    2    2
## [6,]    4    3    1

cbind(M, M)
##      [,1] [,2] [,3] [,4] [,5] [,6]
## [1,]    1    1    3    1    1    3
## [2,]    1    2    2    1    2    2
## [3,]    4    3    1    4    3    1
```

4.3.5 Arrays

Matrices are very useful, however there will be times that data has more dimensions than just two. R has a solutions with the base-type "array." Unlike matrices that have always two dimensions, arrays can be of any number of dimensions. However, the requirement that all elements are of the same data-type is also valid for arrays.

array

Note that words like "array" are used as keywords in many computer languages, and that it is important to understand exactly how it is implemented in the language that you want to use. In this section we will introduce you to the practical aspects of working with arrays.

4.3.5.1 Creating and Accessing Arrays

Arrays can be created with the `array()` function; this function takes a "dim" attribute which defines the number of dimension. While arrays are similar to lists, they have to be of one class type (lists can consist of different class types).

`array()`

In the example we create an array with two elements, which are both three by three matrices.

```
# Create an array:
a <- array(c('A','B'),dim = c(3,3,2))
print(a)
## , , 1
##
##      [,1] [,2] [,3]
## [1,] "A"  "B"  "A"
## [2,] "B"  "A"  "B"
## [3,] "A"  "B"  "A"
```

```
##
## , , 2
##
##      [,1] [,2] [,3]
## [1,] "B"  "A"  "B"
## [2,] "A"  "B"  "A"
## [3,] "B"  "A"  "B"

# Access one element:
a[2,2,2]
## [1] "B"

# Access one layer:
a[,,2]
##      [,1] [,2] [,3]
## [1,] "B"  "A"  "B"
## [2,] "A"  "B"  "A"
## [3,] "B"  "A"  "B"
```

4.3.5.2 Naming Elements of Arrays

In most applications it will be enough, to refer to an element in an array by its number. However naming elements makes code easier to read and can make it more robust. For example, if we change the array definition, the numbers of its elements can change, but the name will still be a valid reference.

```
# Create two vectors:
v1 <- c(1,1)
v2 <- c(10:13)
col.names <- c("col1","col2", "col3")
row.names <- c("R1","R2")
matrix.names <- c("Matrix1","Matrix2")

# Take these vectors as input to the array.
a <- array(c(v1,v2),dim = c(2,3,2),
     dimnames = list(row.names,col.names,
     matrix.names))
print(a)

# This allows to address the first row in Matrix 1 as follows:
a['R1',,'Matrix1']
```

4.3.5.3 Manipulating Arrays

Using arrays and accessing its elements is in many aspects similar to working with matrices.

```
M1 <- a[,,1]
M2 <- a[,,2]
M2
##    col1 col2 col3
## R1    1   10   12
## R2    1   11   13
```

4.3.5.4 Applying Functions over Arrays

While it is possible to use a for-loop and cycle through the elements of an array, it is usually faster to use the built-in functions that R provides.

An efficient way to apply the same function over each element of an array is via the function `apply()`: that functions is designed to do exactly that. `apply()`

Function use for apply()

`apply(X, MARGIN, FUN, ...)` with:

1. X: an array, including a matrix.
2. MARGIN: a vector giving the subscripts which the function will be applied over. E.g., for a matrix '1' indicates rows, '2' indicates columns, 'c(1, 2)' indicates rows and columns. Where 'X' has named dimnames, it can be a character vector selecting dimension names.
3. FUN: the function to be applied: see 'Details'. In the case of functions like '+', 'back-quoted or quoted

It is sufficient to provide the data, the dimension of application and the function that has to be applied. To show how this works, we construct a simple example to calculate sums of rows and sums of columns.

`cbind()`

```
x <- cbind(x1 = 3, x2 = c(4:1, 2:5))
dimnames(x)[[1]] <- letters[1:8]
apply(x, 2, mean, trim = .2)
## x1 x2
##  3  3

col.sums <- apply(x, 2, sum)
row.sums <- apply(x, 1, sum)
rbind(cbind(x, Rtot = row.sums),
      Ctot = c(col.sums, sum(col.sums)))
##       x1 x2 Rtot
## a      3  4    7
## b      3  3    6
## c      3  2    5
## d      3  1    4
## e      3  2    5
## f      3  3    6
## g      3  4    7
## h      3  5    8
## Ctot 24 24   48
```

`apply()`

The reader will notice that in the example above the variable `x` is actually not an array but rather a data frame. The function `apply()` works however the same: instead of 2 dimensions, there can be more.

Consider the previous example with the array a, and remember that a has three dimensions: 2 rows, 3 columns, and 2 matrices, then the following should be clear.

```
# Re-create the array a (shorter code):
col.names <- c("col1","col2", "col3")
row.names <- c("R1","R2")
matrix.names <- c("Matrix1","Matrix2")
a <- array(c(1,1,10:13),dim = c(2,3,2),
     dimnames = list(row.names,col.names,
     matrix.names))

# Demonstrate apply:
apply(a, 1, sum)
## R1 R2
## 46 50
```

```
apply(a, 2, sum)
## col1 col2 col3
##    4   42   50

apply(a, 3, sum)
## Matrix1 Matrix2
##      48      48
```

4.3.6 Lists

list

Where vectors, arrays, and matrices only can contain variables of the same sort (numeric, character, integer, etc.), the list object allows to mix different types into one object. The concept of a list is similar to the concept "object" in many programming languages such as C++. Notice, however, that there is no abstraction, only instances.

4.3.6.1 Creating Lists

`list()`

Definition: List

In R, lists are objects which are sets of elements that are not necessarily all of the same type. Lists can mix numbers, strings, vectors, matrices, functions, boolean variables, and even lists.

```
  # List is created using list() function.
  myList <- list("Approximation", pi, 3.14, c)
  print(myList)
## [[1]]
## [1] "Approximation"
##
## [[2]]
## [1] 3.141593
##
## [[3]]
## [1] 3.14
##
## [[4]]
## function (...)  .Primitive("c")
```

Further information – Object-oriented programming in R

List might be reminiscent to how objects work in other languages (e.g. it looks similar to the `struct` in C). Indeed, everything is an object in R. However, to understand how R implements different styles of objects and object-oriented programming, we recommend to read Chapter 6 on page 117.

4.3.6.2 Naming Elements of Lists

While it is perfectly possible to address elements of lists by their number, it is sometimes more meaningful to use a specific name.

```
# Create the list:
L <- list("Approximation", pi, 3.14, c)

# Assign names to elements:
names(L) <- c("description", "exact", "approx","function")
print(L)
## $description
## [1] "Approximation"
##
## $exact
## [1] 3.141593
##
## $approx
## [1] 3.14
##
## $`function`
## function (...)  .Primitive("c")

# Addressing elements of the named list:
print(paste("The difference is", L$exact - L$approx))
## [1] "The difference is 0.00159265358979299"

print(L[3])
## $approx
## [1] 3.14

print(L$approx)
## [1] 3.14

# However, "function" was a reserved word, so we need to use
# back-ticks in order to address the element:
a <- L$`function`(2,3,pi,5)  # to access the function c(...)
print(a)
## [1] 2.000000 3.000000 3.141593 5.000000
```

4.3.6.3 List Manipulations

Since lists can contain objects of different types, it would have been confusing to overload the base operators such as the addition. There are a few other things that make sense.

Lists of Lists Are Also Lists

```
V1 <- c(1,2,3)
L2 <- list(V1, c(2:7))
L3 <- list(L2,V1)
print(L3)
## [[1]]
## [[1]][[1]]
## [1] 1 2 3
##
## [[1]][[2]]
## [1] 2 3 4 5 6 7
##
##
## [[2]]
## [1] 1 2 3

print(L3[[1]][[2]][3])
## [1] 4
```

Note how the list L3 is a list of lists, rather than the concatenation of two lists. Instead of adding the elements of L2 after those of V1 and having nine slots for data, it has two slots. Each of those slots contains the object V1 and L2 respectively.

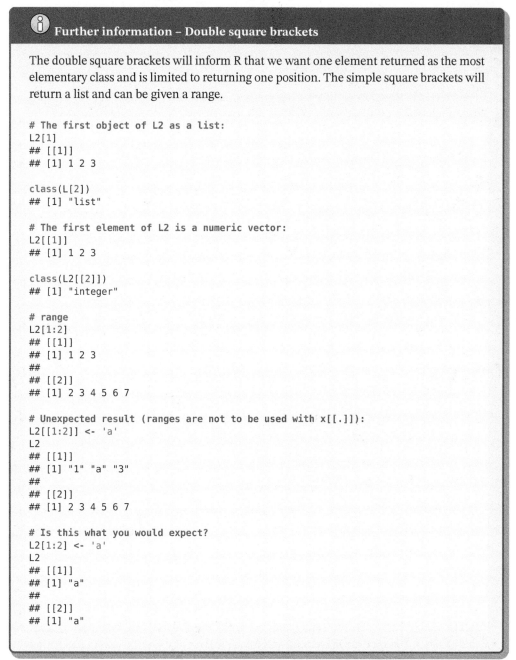

Further information – Double square brackets

The double square brackets will inform R that we want one element returned as the most elementary class and is limited to returning one position. The simple square brackets will return a list and can be given a range.

```
# The first object of L2 as a list:
L2[1]
## [[1]]
## [1] 1 2 3

class(L[2])
## [1] "list"

# The first element of L2 is a numeric vector:
L2[[1]]
## [1] 1 2 3

class(L2[[2]])
## [1] "integer"

# range
L2[1:2]
## [[1]]
## [1] 1 2 3
##
## [[2]]
## [1] 2 3 4 5 6 7

# Unexpected result (ranges are not to be used with x[[.]]):
L2[[1:2]] <- 'a'
L2
## [[1]]
## [1] "1" "a" "3"
##
## [[2]]
## [1] 2 3 4 5 6 7

# Is this what you would expect?
L2[1:2] <- 'a'
L2
## [[1]]
## [1] "a"
##
## [[2]]
## [1] "a"
```

Add and Delete Elements of a List

A numbered element can be added while skipping positions. In the following example the position 3 is left undefined (NULL).

```
L <- list(1,2)
L[4] <- 4  # position 3 is NULL
L
## [[1]]
## [1] 1
##
## [[2]]
## [1] 2
##
## [[3]]
## NULL
##
## [[4]]
## [1] 4
```

Named elements are always added at the end of the list:

```
L$pi_value <- pi
L
## [[1]]
## [1] 1
##
## [[2]]
## [1] 2
##
## [[3]]
## NULL
##
## [[4]]
## [1] 4
##
## $pi_value
## [1] 3.141593
```

Delete an element by assigning NULL to it:

```
L[1] <- NULL
L
## [[1]]
## [1] 2
##
## [[2]]
## NULL
##
## [[3]]
## [1] 4
##
## $pi_value
## [1] 3.141593
```

It is also possible to delete an element via the squared brackets. Note that if we address the elements of a list by their number, we need to recalculate the numbers. If we were addressing the elements of the list by name, nothing needs to be changed.

```
L <- L[-2]
L
## [[1]]
## [1] 2
##
```

```
## [[2]]
## [1] 4
##
## $pi_value
## [1] 3.141593
```

Warning – Deleting elements in lists

When deleting an element in a list, the numbering will change so that it appears that the deleted element was never there. This implies that when accessing elements of the list by number, it is unsafe to delete elements and can lead to unwanted side effects of the code.

Convert list to vectors

Vectors can only contain one type of variable, while a list can be of mixed types. It might make sense to convert lists to vectors, for example, because some operations on vectors will be significantly faster.

unlist()

To do this R, provides the function `unlist()`.

```
L  <- list(c(1:5), c(6:10))
v1 <- unlist(L[1])
v2 <- unlist(L[2])
v2-v1
## [1] 5 5 5 5 5
```

Warning – Silent failing of unlist()

Lists are more complex than vectors, instead of failing with a warning and requiring additional options to be set, the `unlist()` function will silently make some decisions for you.

```
# A list of vectors of integers:
L <- list(1L,c(-10L:-8L))
unlist(L)
## [1]   1 -10  -9  -8

# Note the named real-valued extra element:
L <- list(c(1:2),c(-10:-8), "pi_value" = pi)
unlist(L)
##
##    1.000000   2.000000 -10.000000  -9.000000  -8.000000
##    pi_value
##    3.141593
```

Apart from performance considerations, it might also be necessary to convert parts of a list to a vector, because some functions will expect vectors and will not work on lists.

4.3.7 Factors

Factors are the objects which hold a series of labels. They store the vector along with the distinct values of the elements in the vector as label. Factors are in many ways similar to the `enum` data

factors

type in C, C++ or Java, here they are mainly used to store named constants. The labels are always of the character-type[2] irrespective of data type of the elements in the input vector.

4.3.7.1 Creating Factors

factor() Factors are created using the factor() function.

```
# Create a vector containing all your observations:
feedback <- c('Good','Good','Bad','Average','Bad','Good')

# Create a factor object:
factor_feedback <- factor(feedback)

# Print the factor object:
print(factor_feedback)
## [1] Good    Good    Bad     Average Bad     Good
## Levels: Average Bad Good
```

From the aforementioned example it is clear that the factor-object "is aware" of all the labels for all observations as well as the different levels (or different labels) that exist. The next code fragment makes clear that some functions – such as plot() – will recognize the factor-object and produce results that make sense for this type of object. The following line of code is enough to produce the output that is shown in Figure 4.1.

```
# Plot the histogram -- note the default order is alphabetic
plot(factor_feedback)
```

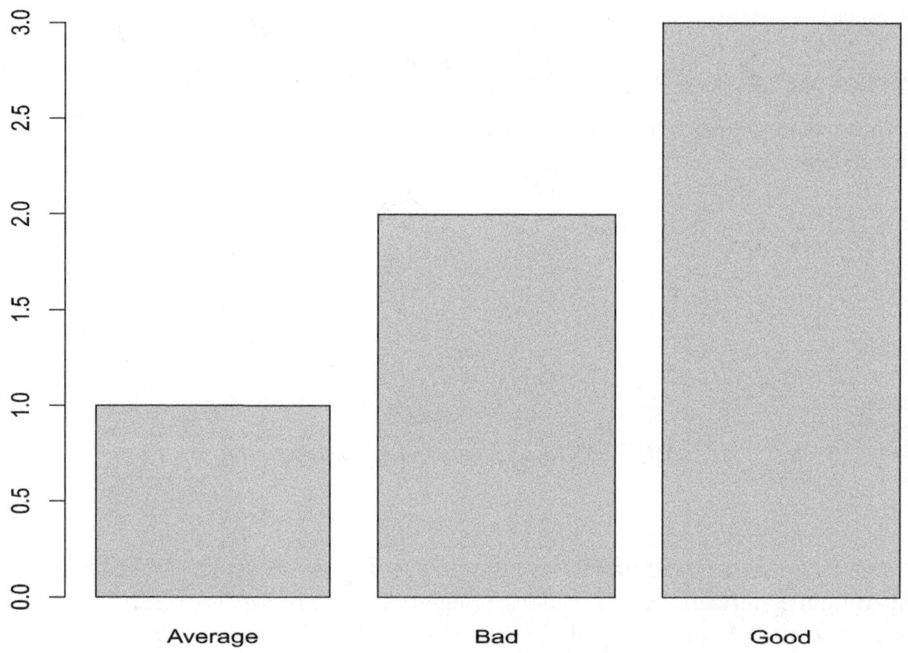

Figure 4.1: *The plot-function will result in a bar-chart for a factor-object.*

[2]The character type in R is what in most other languages would be called a "string." In other words, that is text that – generally – will not allow much arithmetic.

There are a few specific functions for the factor-object. For example, the function `nlevels()` returns the number of levels in the factor object.

nlevels()

```
# The nlevels function returns the number of levels:
print(nlevels(factor_feedback))
## [1] 3
```

Digression – The reduced importance of factors

When R was in its infancy, both computing power and memory were not at the level as today and in most cases it made sense to coerce strings to factors. For example, the base-R functions to load data in a data-frame (i.e. two dimensional data) will silently convert strings to factors. Today, that is most probably not what you need. Therefore, we recommend to make it a habit to use the functions from the `tidyverse` (see Chapter 7 "*Tidy R with the Tidyverse*" on page 161).

4.3.7.2 Ordering Factors

In the example about creating a factor-object for feedback one will have noticed that the plot-function does show the labels in alphabetical order and not in an order that for us – humans – would be logical. It is possible to coerce a certain order in the labels by providing the levels – in the correct order – while creating the factor-object.

```
feedback <- c('Good','Good','Bad','Average','Bad','Good')
factor_feedback <- factor(feedback,
                          levels=c("Bad","Average","Good"))
plot(factor_feedback)
```

In Figure 4.2 on page 63 we notice that the order is now as desired (it is the order that we have provided via the attribute `labels` in the function `factor()`.

Generate Factors with the Function gl()

gl()

Function use for gl()

`gl(n, k, length = n*k, labels = seq_len(n), ordered = FALSE)`
with

- n: The number of levels
- k: The number of replications (for each level)
- length (optional): An integer giving the length of the result
- labels (optional): A vector with the labels
- ordered: A boolean variable indicating whether the results should be ordered.

```
gl(3,2,,c("bad","average","good"),TRUE)
## [1] bad     bad     average average good    good
## Levels: bad < average < good
```

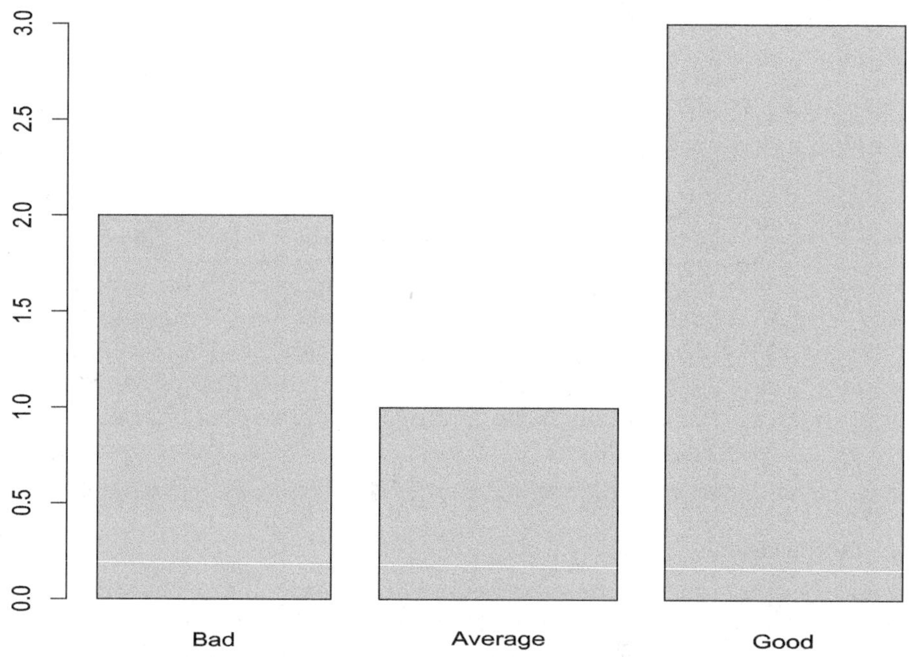

Figure 4.2: *The factor objects appear now in a logical order.*

Question #4

Use the dataset mtcars (from the library MASS) and explore the distribution of number of gears. Then explore the correlation between gears and transmission.

Question #5

Then focus on the transmission and create a factor-object with the words "automatic" and "manual" instead of the numbers 0 and 1.

mtcars

Use the `?mtcars` to find out the exact definition of the data.

Question #6

Use the dataset mtcars (from the library MASS) and explore the distribution of the horsepower (hp). How would you proceed to make a factoring (e.g. Low, Medium, High) for this attribute? Hint: Use the function `cut()`.

`cut()`

4.3.8 Data Frames

4.3.8.1 Introduction to Data Frames

Data frames are the prototype of all two-dimensional data (also known as “rectangular data”). For statistical analysis this is obviously an important data-type.

data frame
rectangular data

Data frames are very useful for statistical modelling; they are objects that contain data in a tabular way. Unlike a matrix in data frame each column can contain different types of data. For example, the first column can be factorial, the second logical, and the third numerical. It is a composite data type consisting of a list of vectors of equal length.

Data frames are created using the `data.frame()` function.

`data.frame()`

```
# Create the data frame.
data_test <- data.frame(
   Name   = c("Piotr", "Pawel","Paula","Lisa","Laura"),
   Gender = c("Male", "Male","Female", "Female","Female"),
   Score  = c(78,88,92,89,84),
   Age    = c(42,38,26,30,35)
   )
print(data_test)
##    Name Gender Score Age
## 1 Piotr   Male    78  42
## 2 Pawel   Male    88  38
## 3 Paula Female    92  26
## 4  Lisa Female    89  30
## 5 Laura Female    84  35

# The standard plot function on a data-frame (Figure 4.3)
# with the pairs() function:
plot(data_test)
```

`pairs()`

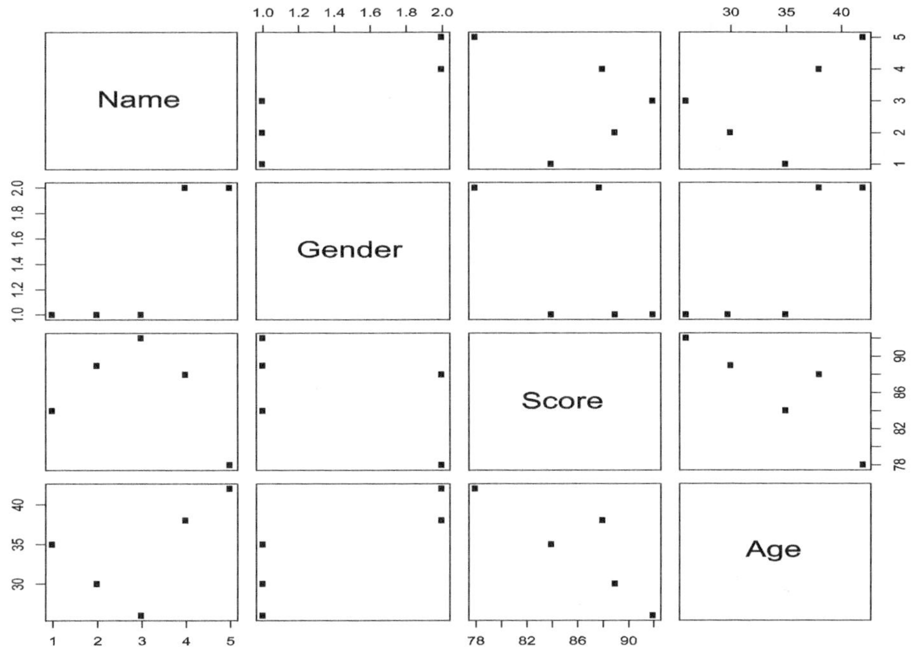

Figure 4.3: *The standard plot for a data frame in R shows each column printed in function of each other. This is useful to see correlations or how generally the data is structured.*

4.3.8.2 Accessing Information from a Data Frame

Most data is rectangular, and in almost any analysis we will encounter data that is structured in a data frame. The following functions can be helpful to extract information from the data frame, investigate its structure and study the content.

summary()
head()
tail()

```
# Get the structure of the data frame:
str(data_test)
## 'data.frame':        5 obs. of  4 variables:
##  $ Name  : Factor w/ 5 levels "Laura","Lisa",..: 5 4 3 2 1
##  $ Gender: Factor w/ 2 levels "Female","Male": 2 2 1 1 1
##  $ Score : num  78 88 92 89 84
##  $ Age   : num  42 38 26 30 35

# Note that the names became factors (see warning below)

# Get the summary of the data frame:
summary(data_test)
##     Name       Gender       Score            Age
##  Laura:1   Female:3   Min.   :78.0   Min.   :26.0
##  Lisa :1   Male  :2   1st Qu.:84.0   1st Qu.:30.0
##  Paula:1              Median :88.0   Median :35.0
##  Pawel:1              Mean   :86.2   Mean   :34.2
##  Piotr:1              3rd Qu.:89.0   3rd Qu.:38.0
##                       Max.   :92.0   Max.   :42.0

# Get the first rows:
head(data_test)
##    Name Gender Score Age
## 1 Piotr   Male    78  42
## 2 Pawel   Male    88  38
## 3 Paula Female    92  26
## 4  Lisa Female    89  30
## 5 Laura Female    84  35

# Get the last rows:
tail(data_test)
##    Name Gender Score Age
## 1 Piotr   Male    78  42
## 2 Pawel   Male    88  38
## 3 Paula Female    92  26
## 4  Lisa Female    89  30
## 5 Laura Female    84  35

# Extract the column 2 and 4 and keep all rows
data_test.1 <- data_test[,c(2,4)]
print(data_test.1)
##   Gender Age
## 1   Male  42
## 2   Male  38
## 3 Female  26
## 4 Female  30
## 5 Female  35

# Extract columns by name and keep only selected rows
data_test[c(2:4),c(2,4)]
##   Gender Age
## 2   Male  38
## 3 Female  26
## 4 Female  30
```

Warning – Avoiding conversion to factors

The default behaviour of R is to convert strings to factors when a data.frame is created. Decades ago this was useful for performance reasons. Now, this is usually unwanted behaviour.[a] To avoid this put `stringsAsFactors = FALSE` in the `data.frame()` function.

```
d <- data.frame(
   Name   = c("Piotr", "Pawel","Paula","Lisa","Laura"),
   Gender = c("Male", "Male","Female", "Female","Female"),
   Score  = c(78,88,92,89,84),
   Age    = c(42,38,26,30,35),
   stringsAsFactors = FALSE
   )
d$Gender <- factor(d$Gender)  # manually factorize gender
str(d)
## 'data.frame':        5 obs. of  4 variables:
##  $ Name  : chr  "Piotr" "Pawel" "Paula" "Lisa" ...
##  $ Gender: Factor w/ 2 levels "Female","Male": 2 2 1 1 1
##  $ Score : num  78 88 92 89 84
##  $ Age   : num  42 38 26 30 35
```

[a] See Chapter 7 "*Tidy R with the Tidyverse*" on page 161 and note that tibbles (the data-frame alternative in the tidyverse) do not coerce text to factors.

4.3.8.3 Editing Data in a Data Frame

While one usually reads in large amounts of data and uses an IDE such as RStudio that facilitates the visualization and manual modification of data frames, it is useful to know how this is done when no graphical interface is available. Even when working on a server, all these functions will always be available.

`de()`
`data.entry()`
`edit()`

```
de(x)                  # fails if x is not defined
de(x <- c(NA))         # works
x <- de(x <- c(NA))    # will also save the changes
data.entry(x)          # de is short for data.entry
x <- edit(x)           # use the standard editor (vi in *nix)
```

Of course, there are also multiple ways to address data directly in R.

```
# The following lines do the same.
data_test$Score[1] <- 80
data_test[3,1]     <- 80
```

4.3.8.4 Modifying Data Frames

Add Columns to a Data-frame

Typically, the variables are in the columns and adding a column corresponds to adding a new, observed variable. This is done via the function `cbind()`.

`cbind()`

```
# Expand the data frame, simply define the additional column:
data_test$End_date <-  as.Date(c("2014-03-01", "2017-02-13",
                    "2014-10-10", "2015-05-10","2010-08-25"))
print(data_test)
##     Name Gender Score Age   End_date
## 1 Piotr   Male    80  42 2014-03-01
## 2 Pawel   Male    88  38 2017-02-13
## 3  <NA> Female    92  26 2014-10-10
## 4  Lisa Female    89  30 2015-05-10
## 5 Laura Female    84  35 2010-08-25

# Or use the function cbind() to combine data frames along columns:
Start_date <- as.Date(c("2012-03-01", "2013-02-13",
                    "2012-10-10", "2011-05-10","2001-08-25"))

# Use this vector directly:
df <- cbind(data_test, Start_date)
print(df)
##     Name Gender Score Age   End_date Start_date
## 1 Piotr   Male    80  42 2014-03-01 2012-03-01
## 2 Pawel   Male    88  38 2017-02-13 2013-02-13
## 3  <NA> Female    92  26 2014-10-10 2012-10-10
## 4  Lisa Female    89  30 2015-05-10 2011-05-10
## 5 Laura Female    84  35 2010-08-25 2001-08-25

# or first convert to a data frame:
df <- data.frame("Start_date" = t(Start_date))
df <- cbind(data_test, Start_date)
print(df)
##     Name Gender Score Age   End_date Start_date
## 1 Piotr   Male    80  42 2014-03-01 2012-03-01
## 2 Pawel   Male    88  38 2017-02-13 2013-02-13
## 3  <NA> Female    92  26 2014-10-10 2012-10-10
## 4  Lisa Female    89  30 2015-05-10 2011-05-10
## 5 Laura Female    84  35 2010-08-25 2001-08-25
```

Adding Rows to a Data-frame

rbind()

Adding rows corresponds to adding observations. This is done via the function rbind().

```
# To add a row, we need the rbind() function:
data_test.to.add <- data.frame(
   Name = c("Ricardo", "Anna"),
   Gender = c("Male", "Female"),
   Score = c(66,80),
   Age = c(70,36),
   End_date = as.Date(c("2016-05-05","2016-07-07"))
   )
data_test.new <- rbind(data_test,data_test.to.add)
print(data_test.new)
##       Name Gender Score Age   End_date
## 1   Piotr   Male    80  42 2014-03-01
## 2   Pawel   Male    88  38 2017-02-13
## 3    <NA> Female    92  26 2014-10-10
## 4    Lisa Female    89  30 2015-05-10
## 5   Laura Female    84  35 2010-08-25
## 6 Ricardo   Male    66  70 2016-05-05
## 7    Anna Female    80  36 2016-07-07
```

Merging data frames

Merging allows to extract the subset of two data-frames where a given set of columns match.

```
data_test.1 <- data.frame(
  Name = c("Piotr", "Pawel","Paula","Lisa","Laura"),
  Gender = c("Male", "Male","Female", "Female","Female"),
  Score = c(78,88,92,89,84),
  Age = c(42,38,26,30,35)
  )
data_test.2 <- data.frame(
  Name = c("Piotr", "Pawel","notPaula","notLisa","Laura"),
  Gender = c("Male", "Male","Female", "Female","Female"),
  Score = c(78,88,92,89,84),
  Age = c(42,38,26,30,135)
  )
data_test.merged <- merge(x=data_test.1,y=data_test.2,
                          by.x=c("Name","Age"),by.y=c("Name","Age"))

# Only records that match in name and age are in the merged table:
print(data_test.merged)
##    Name Age Gender.x Score.x Gender.y Score.y
## 1 Pawel  38     Male      88     Male      88
## 2 Piotr  42     Male      78     Male      78
```

merge()

Short-cuts

short-cut

R will allow the use of short-cuts, provided that they are unique. For example, in the data-frame `data_test` there is a column `Name`. There are no other columns whose name start with the letter "N"; hence. this one letter is enough to address this column.

```
data_test$N
## [1] Piotr Pawel Paula Lisa  Laura
## Levels: Laura Lisa Paula Pawel Piotr
```

> **Warning – Short-cuts can be dangerous**
>
> Use "short-cuts" sparingly and only when working interactively (not in functions or code that will be saved and re-run later). When later another column is added the short-cut will no longer be unique and behaviour is hard to predict and it is even harder to spot the programming error in a part of your code that previously worked fine.

Naming Rows and Columns

In the preceding code, we have named columns when we created the data-frame. It is also possible to do that later or to change column names ...and it is even possible to name each row individually.

```
# Get the rownames.
colnames(data_test)
## [1] "Name"     "Gender"   "Score"    "Age"      "End_date"

rownames(data_test)
## [1] "1" "2" "3" "4" "5"
```

```
colnames(data_test)[2]
## [1] "Gender"

rownames(data_test)[3]
## [1] "3"

# assign new names
colnames(data_test)[1] <- "first_name"
rownames(data_test) <- LETTERS[1:nrow(data_test)]
print(data_test)
##   first_name Gender Score Age   End_date
## A      Piotr   Male    80  42 2014-03-01
## B      Pawel   Male    88  38 2017-02-13
## C       <NA> Female    92  26 2014-10-10
## D       Lisa Female    89  30 2015-05-10
## E      Laura Female    84  35 2010-08-25
```

Question #7

1. Create 3 by 3 matrix with the numbers 1 to 9,
2. Convert it to a data-frame,
3. Add names for the columns and rows,
4. Add a column with the column-totals,
5. Drop the second column.

4.3.9 Strings or the Character-type

string Strings are called the "character-type" in R. They follow some simple rules:

- strings must start and end with single or double quotes,
- a string ends when the same quotes are encountered the next time,
- until then it can contain the other type of quotes.

Example: Using strings

paste()

```
a <- "Hello"
b <- "world"
paste(a, b, sep = ", ")
## [1] "Hello, world"

c <- "A 'valid' string"
```

Note – Paste

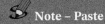

In many cases we do not need anything between strings that are concatenated. We can of course supply an empty string as separator (`sep = ''`), but it is also possible to use the custom function `pate0()`: `past0()`

```
paste0(12, '%')
## [1] "12%"
```

Formatting with `format()`

In many cases, it will be useful to format a date or number consistently and neatly in plot and tables. The function `format()` is a great tool to start formatting. `format()`

Function use for format()

```
format(x, trim = FALSE, digits = NULL, nsmall = 0L,
       justify = c("left", "right", "centre", "none"),
       width = NULL, na.encode = TRUE, scientific = NA,
       big.mark   = "",    big.interval = 3L,
       small.mark = "", small.interval = 5L,
       decimal.mark = getOption("OutDec"),
       zero.print = NULL, drop0trailing = FALSE, ...)
```

- `x` is the vector input.
- `digits` is the total number of digits displayed.
- `nsmall` is the minimum number of digits to the right of the decimal point.
- `scientific` is set to TRUE to display scientific notation.
- `width` is the minimum width to be displayed by padding blanks in the beginning.
- `justify` is the display of the string to left, right or center.

Formatting examples

```
a<-format(100000000,big.mark=" ",
                 nsmall=3,
                 width=20,
                 scientific=FALSE,
                 justify="r")
print(a)
## [1] "       100 000 000.000"
```

Further information – format()

More information about the format-function can be obtained via `?format` or `help(format)`.

Other string functions

- nchar() `nchar()`: returns the number of characters in a string
- toupper() `toupper()`: puts the string in uppercase
- tolower() `tolower()`: puts the string in lowercase
- substring() `substring(x,first,last)`: returns
  a substring from x starting with the "first" and ending with the "last"
- strsplit() `strsplit(x,split)`: split
  the elements of a vector into substrings according to matches of a substring "split."
  grep() there is also a family of search functions: `grep()`,
  grepl() `grepl()`,
  regexpr() `regexpr()`,
  gregexpr() `gregexpr()`,
  regexec() and `regexec()`
  that supply powerful search and replace capabilities.
  sub() `sub()`
  gsub() will replace the first of all matches and `gsub()`
  will replace all matches.

4.4 Operators

While we already encountered operators in previous sections when we introduced the data types, here we give a systematic overview of operators on base types.

operators

4.4.1 Arithmetic Operators

arithmetic – operators

Arithmetic operators act on each element of an object individually.

operator – arithmetic

```
v1 <- c(2,4,6,8)
v2 <- c(1,2,3,5)
v1 + v2     # addition
## [1]  3  6  9 13

v1 - v2     # subtraction
## [1] 1 2 3 3

v1 * v2     # multiplication
## [1]  2  8 18 40

v1 / v2     # division
## [1] 2.0 2.0 2.0 1.6

v1 %% v2    # remainder of division
## [1] 0 0 0 3

v1 %/% v2   # round(v1/v2 -0.5)
## [1] 2 2 2 1

v1 ^ v2     # v1 to the power of v2
## [1]     2    16   216 32768
```

addition

substraction

multiplication

division

power

Warning – Element-wise operations in R

While the result of the sum will not surprise anyone, the result of the multiplication might come as a surprise for users of matrix oriented software such as Mathlab or Octave for example. In R an operations is always element per element – unless explicitly requested. For example, the dot-product can be obtained as follows.

```
v1 %*% v2
##      [,1]
## [1,]   68
```

4.4.2 Relational Operators

Relational Operators compare vectors element by element

relational operators
operator – relational

```
v1 <- c(8,6,3,2)
v2 <- c(1,2,3,5)
```

bigger than

```
v1 > v2      # bigger than
## [1]  TRUE  TRUE FALSE FALSE
```

smaller than

```
v1 < v2      # smaller than
## [1] FALSE FALSE FALSE  TRUE
```

bigger or equal

```
v1 <= v2     # smaller or equal
## [1] FALSE FALSE  TRUE  TRUE

v1 >= v2     # bigger or equal
## [1]  TRUE  TRUE  TRUE FALSE
```

equal

```
v1 == v2     # equal
## [1] FALSE FALSE  TRUE FALSE
```

not equal

```
v1 != v2     # not equal
## [1]  TRUE  TRUE FALSE  TRUE
```

4.4.3 Logical Operators

operator – logical

Logical Operators combine vectors element by element. While logical operators can be applied directly on composite types, they must be able to act on numeric, logical or complex types in order to produce understandable results.

```
v1 <- c(TRUE, TRUE, FALSE, FALSE)
v2 <- c(TRUE, FALSE, FALSE, TRUE)

v1 & v2       # and
## [1]  TRUE FALSE FALSE FALSE

v1 | v2       # or
## [1]  TRUE  TRUE FALSE  TRUE

!v1           # not
## [1] FALSE FALSE  TRUE  TRUE

v1 && v2     # and applied to the first element
## [1] TRUE

v1 || v2     # or applied to the first element
## [1] TRUE

v1 <- c(TRUE,FALSE,TRUE,FALSE,8,6+3i,-2,0,   NA)
class(v1)  # v1 is a vector or complex numbers
## [1] "complex"

v2 <- c(TRUE)
as.logical(v1)  # coerce to logical (only 0 is FALSE)
## [1]  TRUE FALSE  TRUE FALSE  TRUE  TRUE  TRUE FALSE    NA

v1 & v2
## [1]  TRUE FALSE  TRUE FALSE  TRUE  TRUE  TRUE FALSE    NA

v1 | v2
## [1] TRUE TRUE TRUE TRUE TRUE TRUE TRUE TRUE TRUE
```

Note – Numeric equivalent and logical evalutation

Note that numbers different from zero are considered as TRUE, but only zero is considered as FALSE. Further, NA is implemented in a smart way. For example, in order to assess `TRUE & NA` we need to know what the second element is, hence it will yield NA. However, `TRUE | NA` will be true regardless what the second element is, hence R will show the result.

```
FALSE | NA
## [1] NA

TRUE  | NA
## [1] TRUE

FALSE & NA
## [1] FALSE

TRUE  & NA
## [1] NA

FALSE | NA | TRUE | TRUE
## [1] TRUE

TRUE  & NA & FALSE
## [1] FALSE
```

4.4.4 Assignment Operators

operator – assignment
and
or
not

R has multiple ways to express an assignment. While it is possible to mix and match, we prefer to choose just one and stick with it. We will use `<-`.

```
# left assignment
x <- 3
x = 3
x<<- 3

# right assignment
3 -> x
3 ->> x

#chained assignment
x <- y <- 4
```

assignment – left
assignment – right
assignment – chained

The `<<-` or `->>` operators change a variable in the actual environment and the environment above the actual one. Environments will be discussed in Section 5.1 "*Environments in R*" on page 110. Till now we have always been working in the command prompt. This is the root-environment. A function will create a new (subordinated) environment and might for example use a new variable. When the function stops running that environment stops to exist and the variable exists no longer.[3]

[3] If your programming experience is limited, this might seem a little confusing. It is best to accept this and read further. Then, after reading Section 5.1 "*Environments in R*" on page 110 have another look at this section.

Hint – Assignment

Generally it is best to keep it simple, most people will expect to see a left assignment, and while it might make code a little shorter, the right assignment will be a little confusing for most readers. Best is to stick to `=` or `<-` and of course `<<-` whenever assignment in a higher environment is also intended.

In some special cases, such as the definition of parameters of a function, it is not possible to use the "arrow" and one must revert to the `=` sign. This makes sense, because that is not the same as a traditional assignment.

```
mean(v1, na.rm = TRUE)  # works (v1 is defined in previous section)
## [1] 1.75+0.375i

mean(v1, na.rm <- TRUE) # fails

## Error in mean.default(v1, na.rm <- TRUE): 'trim' must be numeric of length one
```

While the `<<-` seems to do exactly the same, it changes also the value of the variable in the environment above the actual one. The following example makes clear how `<-` only changes the value of `x` while the function is active, but `<<-` also changes the value of the variable `x` in the environment where the function was called from.

```
# f
# Assigns in the current and superior environment 10 to x,
# then prints it, then makes it 0 only in the function environment
# and prints it again.
# arguments:
#   x    -- numeric
f <- function(x) {x <<- 10; print(x); x <- 0; print(x)}

x <- 3
x
## [1] 3

# Run the function f():
f(x)
## [1] 10
## [1] 0

# Only the value assigned with <<- is available now:
x
## [1] 10
```

Digression – For C++ programmers

If you are moving from C++, you will miss the speed and functionality of passing variable by their pointer. The `<<-` operator will provide you the ability to change a variable in the environment above the function.

Warning – Sparingly change variables in other environments

While it is certainly cool and most probably efficient to change a variable in the environment above the function, it will make your code harder to read and if the function is hidden in a package it might lead to unexpected side effects. This superpower is best used sparingly.

4.4.5 Other Operators

There are of course more operators, that allow to execute common commands more efficiently or apply to certain specific objects such as matrices. For example, we already have seen the operator `:` that creates a sequence. In R it is possible to define your own operators.

```
# +-+
# This function is a new operator
# arguments:
#   x -- numeric
#   y -- numeric
# returns:
#   x - y
`+-+` <- function(x, y) x - y
5 +-+ 5
## [1] 0

5 +-+ 1
## [1] 4

# Remove the new operator:
rm(`+-+`)
```

Warning – Redefine existing operators

It is even possible to redefine elementary operators such as + with the aforementioned code. This is of course not a wise thing to do, but we understand how it can be a fun practical joke or a tricky job interview question.

The following are some common operators that help working with data.

operator – other

```
# create a list
x <- c(10:20)
x
##  [1] 10 11 12 13 14 15 16 17 18 19 20

# %in% can find an element in a vector
2  %in% x   # FALSE since 2 is not an element of x
## [1] FALSE

11 %in% x   # TRUE since 11 is in x
## [1] TRUE

x[x %in% c(12,13)] # selects elements from x
## [1] 12 13
```

```
x[2:4]   # selects the elements with index
## [1] 11 12 13

          # between 2 and 4
```

4.5 Flow Control Statements

flow control

R is Turing complete and hence offers a range of tools to make choices and repeat certain parts of code. Knowing the different ways to change the flow of a code by if-statements and loops is essential knowledge for each R-programmer.

4.5.1 Choices

4.5.1.1 The if-Statement

`if()`

The workhorse to control the flow of actions is the `if()` function.

The construct is both simple and efficient.

Function use for if()

```
if (logical statement) {
  executed if logical statement is true
} else {
  executed if the logical statement if false
}
```

Note that the else-statement is optional.

This basic construct can also be enriched with `else if` statements. For example, we draw a random number from the normal distribution and check if it is bigger than zero.

```
set.seed(1890)
x <- rnorm(1)
if (x < 0) {
  print('x is negative')
} else if (x > 0) {
  print('x is positive')
} else {
  print('x is zero')
}
## [1] "x is positive"
```

Hint – Extending the if-statement

It is possible to have more than one else-if statement and/or use nested statements.

```
  x <- 122
  if (x < 10) {
    print('less than ten')
    } else if (x < 100) {
    print('between 10 and 100')
    } else if (x < 1000) {
    print('between 100 and 1000')
    } else {
    print('bigger than 1000 (or equal to 1000)')
    }
## [1] "between 10 and 1000"
```

Note that the statements do not necessarily have to be encapsulated by curly brackets if the statement only takes one line.

```
x <- TRUE
y <- pi
y <- if (x) 1 else 2
y  # y is now 1
## [1] 1
```

Note that hybrid forms are possible, but it gets confusing very fast. In the following piece of code the variable y will not get the value one, but rather six.

```
z <- 0
y <- if (x) {1; z <- 6} else 2
y  # y is now 6
## [1] 6

z  # z is also 6
## [1] 6
```

4.5.1.2 The Vectorised If-statement

The function `ifelse()` is the vectorised version of the if-function. It allows to use vectors as input and output. While the if-statement is useful for controlling flow in code, the ifelse-function handy for data manipulation.

`ifelse()`

```
x <- 1:6
ifelse(x %% 2 == 0, 'even', 'odd')
## [1] "odd"  "even" "odd"  "even" "odd"  "even"
```

The `ifelse` function can also use vectors as parameters in the output.

```
x <- 1:6
y <- LETTERS[1:3]
ifelse(x %% 2 == 0, 'even', y)
## [1] "A"    "even" "C"    "even" "B"    "even"

# Note that y gets recycled!
```

4.5.1.3 The Switch-statement

An if-else construct that assigns one value to a variable based on one other variable can be condensed via the `switch()` function.

`switch()`

```
x <- 'b'
x_info <- switch(x,
    'a' = "One",
    'b' = "Two",
    'c' = "Three",
    stop("Error: invalid `x` value")
  )
# x_info should be 'two' now:
x_info
## [1] "Two"
```

The switch statement can always be written a an else-if statement. The following code does the same as the aforementioned code.

```
x <- 'b'
x_info <- if (x == 'a' ) {
    "One"
  } else if (x == 'b') {
    "Two"
  } else if (x == 'c') {
    "Three"
  } else {
    stop("Error: invalid `x` value")
  }
# x_info should be 'two' now:
x_info
## [1] "Two"
```

The `switch()` statement can always be written as with the if-else-if construction, which in its turn can always be written based on with if-else statements. This is same logic also applies for loops (that repeat parts of code). All loop constructs can always be written with a for-loop and an if-statement, but more advanced structures help to keep code clean and readable.

4.5.2 Loops

One of the most common constructs in programming is repeating a certain block of code a few times. This repeating of code is a "loop." Loops can repeat code a fixed number of times or do this only when certain conditions are fulfilled.

4.5.2.1 The For Loop

for

loop – for

As in most programming languages, there is a "for-loop" that repeats a certain block of code a given number of times. Interestingly, the counter does not have to follow a pre-defined increment, the counter will rather follow the values supplied in a vector. R's for-loop is an important tool to add to your toolbox.

The for-loop is useful to repeat a block of code a certain number of times. R will iterate a given variable through elements of a vector.

Function use for for()

```
for (value in vector) {
   statements
}
```

The for-loop will execute the statements for each value in the given vector.

Example: For loop

```
x <- LETTERS[1:5]
for ( j in x) {
   print(j)
}
## [1] "A"
## [1] "B"
## [1] "C"
## [1] "D"
## [1] "E"
```

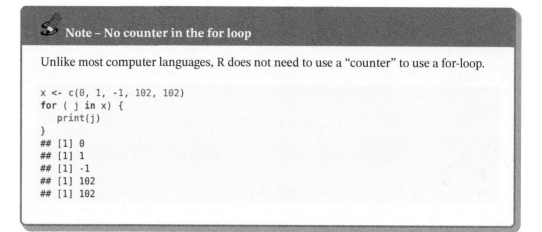

Note – No counter in the for loop

Unlike most computer languages, R does not need to use a "counter" to use a for-loop.

```
x <- c(0, 1, -1, 102, 102)
for ( j in x) {
   print(j)
}
## [1] 0
## [1] 1
## [1] -1
## [1] 102
## [1] 102
```

repeat()
loop – repeat

4.5.2.2 Repeat

The repeat-loop will repeat the block of commands till it executes the `break` command.

Function use for repeat()

```
repeat {
   commands
   if(condition) {break}
}
```

Example: Repeat loop

```
x <- c(1,2)
c <- 2
repeat {
   print(x+c)
   c <- c+1
   if(c > 4) {break}
}
## [1] 3 4
## [1] 4 5
## [1] 5 6
```

Warning – Break out of he repeat loop

Do not forget the `{break}` statement, it is integral part of the repeat loop.

4.5.2.3 While

while
loop – while

The while-loop is similar to the repeat-loop. However, the while-loop will first check the condition and then run the code to be repeated. So, this code might not be executed at all.

Function use for while()

```
while (test_expression) {
   statement
}
```

The statements are executed as long the test_expression is TRUE.

Example: While loop

```
x <- c(1,2); c <- 2
while (c < 4) {
   print(x+c)
   c <- c + 1
}
## [1] 3 4
## [1] 4 5
```

4.5.2.4 Loop Control Statements

When the `break` statement is encountered inside a loop, that loop is immediately terminated and program control resumes at the next statement following the loop. **break**

```
v <- c(1:5)
for (j in v) {
   if (j == 3) {
      print("--break--")
      break
   }
   print(j)
}
## [1] 1
## [1] 2
## [1] "--break--"
```

The `next` statement will skip the remainder of the current iteration of a loop and starts next iteration of the loop.

```
v <- c(1:5)
for (j in v) {
   if (j == 3) {
      print("--skip--")
      next
   }
   print(j)
}
## [1] 1
## [1] 2
## [1] "--skip--"
## [1] 4
## [1] 5
```

Digression – The speed of loops

The code base of R has been improved a lot and loops are faster than they used to be, however, there can still be a significant time difference between using a loop or using a pre-implemented function.

```
n <- 10^7
v1 <- 1:n

# -- using vector arithmetic
t0 <- Sys.time()
v2 <- v1 * 2
t1 <- Sys.time() - t0
print(t1)
## Time difference of 0.06459093 secs

# -- using a for loop
rm(v2)
v2 <- c()
t0 <- Sys.time()
for (k in v1) v2[k] <- v1[k] * 2
t2 <- Sys.time() - t0
print(t2)
## Time difference of 3.053826 secs

# t1 and t2 are difftime objects and only differences
# are defined.
# To get the quotient, we need to coerce them to numeric.
T_rel <- as.numeric(t2, units = "secs") /
         as.numeric(t1, units = "secs")
T_rel
## [1] 47.27949
```

Note that in our simple experiment the for-loop is 47.3 times slower than the vector operation! So, use operators at the highest level (vectors, matrices, etc) and especially do an effort to understand the `apply`-family functions (see Chapter 4.3.5 "*Arrays*" on page 50) or their tidyverse equivalents: the `map`-family of functions of the package `purrr`. An example for the apply-family can be found in Chapter 22.2.6 on page 513 and one for the map-family is in Chapter 19.3 on page 459.

Further information – Speed

Further information about optimising code for speed and more elegant and robust timing of code can be found in Chapter 40 "*The Need for Speed*" on page 997.

4.6 Functions

functions

More than in any other programming language, functions play a prominent role in R. Part of the reason is the implementation of the dispatcher function based object model with the S3 objects — see Chapter 6 "*The Implementation of OO*" on page 117.

The user is both able to use the built-in functions and/or define his own bespoke functions.

4.6.1 Built-in Functions

Right after starting, R some functions are available. We call these the "built-in functions." Some examples are:

- `demo()`: shows some of the capabilities of R
- `q()`: quits R
- `data()`: shows the datasets available
- `help()`: shows help
- `ls()`: shows variables
- `c()`: creates a vector
- `seq()`: creates a sequence
- `mean()`: calculates the mean
- `max()`: returns the maximum
- `sum()`: returns the sum
- `paste()`: concatenates vector elements

demo() q() data() help() ls() c() seq() mean() max() sum() paste()

4.6.2 Help with Functions

If you do not remember exactly what parameters a certain function needs or what type of variable the output will be, then there are multiple ways to get support in R.

apropos()

Help with functions

```
help(c)   # shows help help with the function c
?c        # same result

apropos("cov") # fuzzy search for functions
```

> **Further information on packages**
>
> R has many "packages" that act as a library of functions that can be loaded with one command. For more information refer to Chapter 4.7 "*Packages*" on page 96.

function()

function – create

4.6.3 User-defined Functions

The true flexibility comes from being able to define our own functions. To create a function in R, we will use the function-generator called "function."

Function use for function()

In R a user defined function (UDF) is created via the function `function()`.

```
function_name <- function(arg_1, arg_2, ...) {
   function_body
   return_value
}
```

Example: A bespoke function

```
# c_surface
# Calculates the surface of a circle
# Arguments:
#   radius -- numeric, the radius of the circle
# Returns
#   the surface of the cicle
c_surface <- function(radius) {
  x <- radius ^ 2 * pi
  return (x)
  }
c_surface(2) + 2
## [1] 14.56637
```

Note that it is not necessary to explicitly "return" something. A function will automatically return the last value that is send to the standard output. So, the following fragment would do exactly the same:

```
# c_surface
# Calculates the surface of a circle
# Arguments:
#   radius -- numeric, the radius of the circle
# Returns
#   the surface of the cicle
c_surface <- function(radius) {
  radius ^ 2 * pi
  }

# Test the function:
c_surface(2) + 2
## [1] 14.56637
```

4.6.4 Changing Functions

Usually, we will keep functions in a separate file that is then loaded in our code with the command `source()`. Editing a function is then done by changing this file and reloading it – and hence overwriting the existing function content.

Most probably you will work in a modern environment such as the IDE RStudio, which makes editing a text-file with code and running that code a breeze. However, there might be cases where one has only terminal access to R. In that case, the following functions might come in handy.

edit()

fix()

```
# Edit the function with vi:
fix(c_surface)

# Or us edit:
c_surface <- edit()
```

Hint

The `edit()` function uses the `vi` editor when using the CLI on Linux. This editor is not so popular any more and you might not immediately know how to close it. To get out of it: press `[esc]`, then type `:q` and press `[enter]`.

vi

4.6.5 Creating Function with Default Arguments

Assigning a default value to the argument of a function means that this argument will get the default value, unless another value is supplied – in other words: if nothing is supplied then the default is used.

It is quite handy to have the possibility to assign a default value to a function. It allows to save a lot of typing work and makes code more readable, but it allows also to add a variable to an existing function and make it compatible with all previous code where that argument was not defined.

paste()

Example

The function `paste()` collates the arguments provided and returns one string that is a concatenation of all strings supplied, separated by a separator. This separator is supplied in the function via the argument `sep`. What is the default separator used in `paste()`?

Creating functions with a default value

Example: default value for function

```
c_surface <- function(radius = 2) {
  radius ^ 2 * pi
  }
c_surface(1)
## [1] 3.141593

c_surface()
## [1] 12.56637
```

4.7 Packages

One of the most important advantages of R is that it allows you to stand on the shoulders of giants. It allows you to load a library of additional functionality, so that you do not waste time writing and debugging something that has been solved before. This allows you to focus on your research and analysis.

Unlike environments like spreadsheets, R is more like a programming language that is extremely flexible, modular, and customizable.

4.7.1 Discovering Packages in R

install.packages()

library()
require()

Additional functions come in "packages." To use them one needs to install the package first with the function `install.packages()`; this will connect to a server, download the functions and prepare them for use. Once installed on our computer, they can be loaded in the active environment with the function `library()` or `require()`

Example: loading the package DiagrammeR

```
# Download the package (only once):
install.packages('DiagrammeR')

# Load it before we can use it (once per session):
library(DiagrammeR)
```

The number of packages availabe increases fast. At the time of writing there are about 15 thousand packages available (see the next "Further information" section). We can of course not explain each package in just one book. Below we provide a small selection as illustration and in the rest of the book we will use a selection of 60 packages (which contain a few hundred upstream packages). The choice of packages is rather opinionated and personal. R is free software and there are always many ways to achieve the same result.

Further information – Packages

More information about the packages as well as the packages themselves can be found on the CRAN server `https://cran.r-project.org/web/packages`.

Useful functions for packages

Below we show some useful functions - note that the output is suppressed.

```
# See the path where libraries are stored:
.libPaths()

# See the list of installed packages:
library()

# See the list of currently loaded packages:
search()
```

Further information – All available packages

R provides also functionality to get a list of all packages – there is no need to use a web-crawling or scraper interface.

```
# available.packages() gets a list:
pkgs <- available.packages(filters = "duplicates")
colnames(pkgs)
##  [1] "Package"               "Version"
##  [3] "Priority"              "Depends"
##  [5] "Imports"               "LinkingTo"
##  [7] "Suggests"              "Enhances"
##  [9] "License"               "License_is_FOSS"
## [11] "License_restricts_use" "OS_type"
## [13] "Archs"                 "MD5sum"
## [15] "NeedsCompilation"      "File"
## [17] "Repository"

# We don't need all, just keep the name:
pkgs <- pkgs[,'Package']

# Show the results:
print(paste('Today, there are', length(pkgs), 'packages for R.'))
## [1] "Today, there are 15477 packages for R."
```

available.packages()

Further information – All installed packages

We can use the function `library()` to get a list of all packages that are installed on our machine.

library()

```
# Get the list (only names):
my_pkgs <- library()$results[,1]

## Warning in library(): library '/usr/local/lib/R/site-library' contains no
packages

# Show the results:
print(paste('I have', length(my_pkgs), 'packages for R.'))
## [1] "I have 282 packages for R."
```

Alternatively, you can use the function `installed.packages()`

installed.packages()

4.7.2 Managing Packages in R

In the previous section, we learned how to install a package, and got a flavour of the available packages. It is also a good idea to keep the repository of packages stable during a big project, but from time to time update packages as well as R. Not only there are bug fixes, but also new features.

```
# See all installed packages:
installed.packages()
```

Note – Cold code in this section

While in the rest of the book, most code is "active" in this sense that the output that appears under a line or the plot that appears close to it are generated while the book was compiled, the code in this book is "cold": the code is not executed. The reason is that the commands from this section would produce long and irrelevant output. The lists would be long, because the author's computer has many packages installed, but also little relevant to you, because you have certainly a different configuration. Other commands would even change packages as a side effect of compiling this book.

A first step in managing packages is knowing which packages can be updated.

```
# List all out-dated packages:
old.packages()
```

Once we know which packages can be updated, we can execute this update:

```
# Update all available packages:
update.packages()
```

If you are very certain that you want to update all packages at once, use the `ask` argument:

```
# Update all packages in batch mode:
update.packages(ask = FALSE)
```

During an important project, you will want to update just one package to solve a bug and keep the rest what as they are in order to reduce the risk that code needs to rewritten and debugged while you are struggling to keep your deadline. Updating a package is done by the same function that is used to install packages.

```
# Update one package (example with the TTR package):
install.packages("TTR")
```

4.8 Selected Data Interfaces

Most analysis will start with reading in data. This can be done from many types of electronic formats such as databases, spreadsheet, CSV files, fixed width text-files, etc.

Reading text from a file in a variable can be done by asking R to request the user to provide the file name as follows:

```
t <- readLines(file.choose())
```

file.choose()

or by providing the file name directly:

```
t <- readLines("R.book.txt")
```

readLines()

This will load the text of the file in one character string `t`. However, typically that is not exactly what we need. In order to manipulate data and numbers, it will be necessary to load data in a vector or data-frame for example.

In further sections – such as Chapter 15 "*Connecting R to an SQL Database*" on page 327 – we will provide more details about data-input. Below, we provide a short overview that certainly will come in handy.

4.8.1 CSV Files

CSV
import – csv

For the example we have first downloaded the CSV file with currency exchange rates from `http://www.ecb.europa.eu/stats/policy_and_exchange_rates/euro_reference_exchange_rates/html/index.en.html`.[4] This file is now on a local hard-drive and will be read in from there.[5]

```
# To read a CSV-file it needs to be in the current directory
# or we need to supply the full path.
getwd()                        # show actual working directory
setwd("./data") # change working directory
data <- read.csv("eurofxref-hist.csv")
is.data.frame(data)
ncol(data)
nrow(data)
head(data)
hist(data$CAD, col = 'khaki3')
```

```
plot(data$USD, data$CAD, col = 'red')
```

Hint – Reading files directly from the Internet

In the aforementioned example, we have first copied the file to our local computer, but that is not necessary. The function `read.csv()` is able to read a file directly from the Internet.

[4] A direct link to this file (in zip-format) is here: `http://www.ecb.europa.eu/stats/eurofxref/eurofxref-hist.zip?c6f8f9a0a5f970e31538be5271051b3c`.

[5] With R it is also possible to read files directly from the Internet by supplying the URL in to the function `read.csv()`.

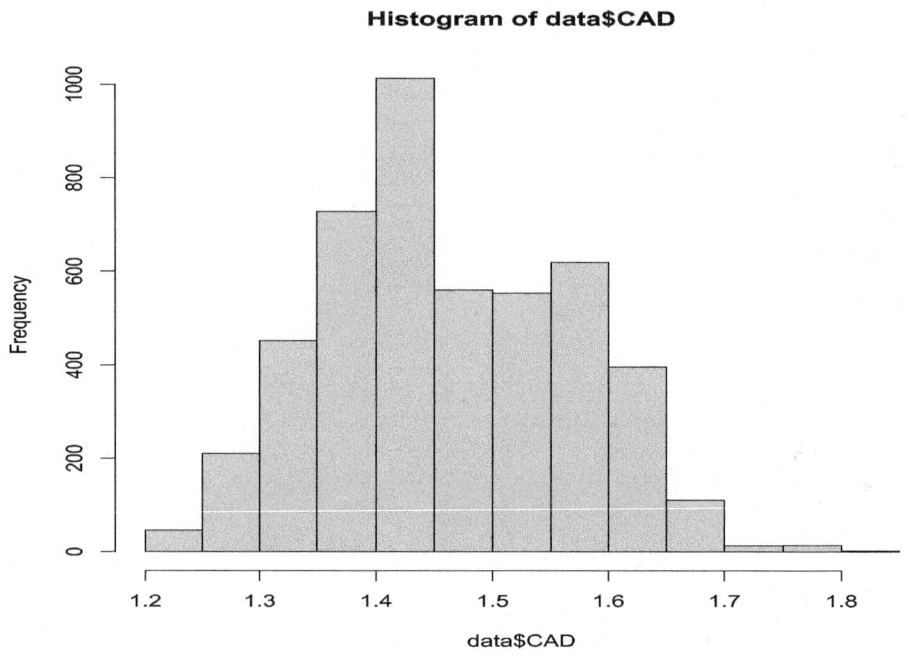

Figure 4.4: *The histogram of the CAD.*

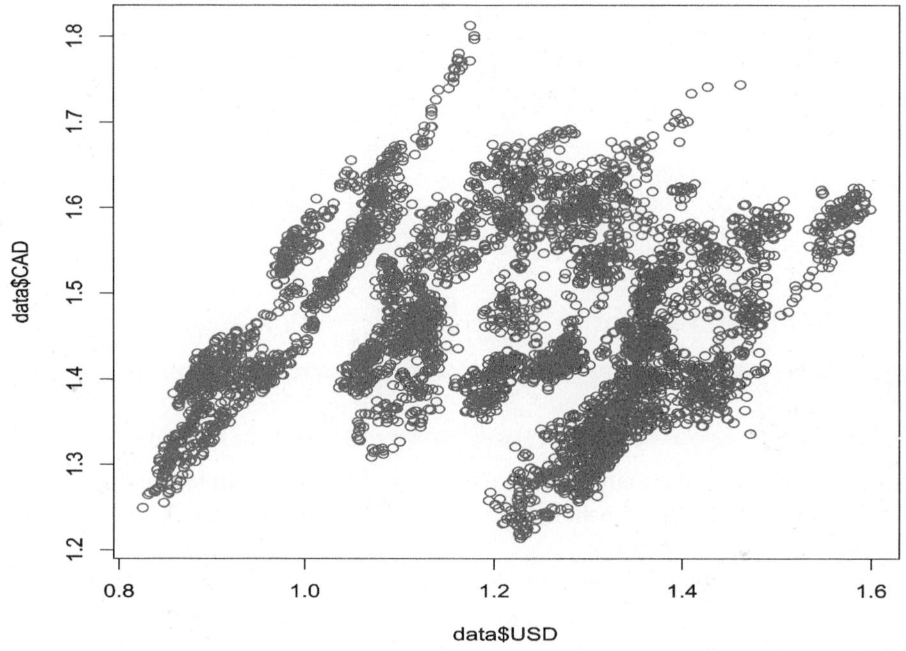

Figure 4.5: *A scatter-plot of one variable with another.*

Finding data

Once the data is loaded in R it is important to be able to make selections and further prepare the data. We will come back to this in much more detail in Part IV "*Data Wrangling*" on page 335, but present here already some essentials.

```
# get the maximum exchange rate
maxCAD <- max(data$CAD)
# use SQL-like selection
d0 <- subset(data, CAD == maxCAD)
d1 <- subset(data, CAD > maxCAD - 0.1)
d1[,1]
##  [1] 2008-12-30 2008-12-29 2008-12-18 1999-02-03
##  [5] 1999-01-29 1999-01-28 1999-01-27 1999-01-26
##  [9] 1999-01-25 1999-01-22 1999-01-21 1999-01-20
## [13] 1999-01-19 1999-01-18 1999-01-15 1999-01-14
## [17] 1999-01-13 1999-01-12 1999-01-11 1999-01-08
## [21] 1999-01-07 1999-01-06 1999-01-05 1999-01-04
## 4718 Levels: 1999-01-04 1999-01-05 ... 2017-06-05

d2<- data.frame(d1$Date,d1$CAD)
d2
##       d1.Date d1.CAD
## 1  2008-12-30 1.7331
## 2  2008-12-29 1.7408
## 3  2008-12-18 1.7433
## 4  1999-02-03 1.7151
## 5  1999-01-29 1.7260
## 6  1999-01-28 1.7374
## 7  1999-01-27 1.7526
## 8  1999-01-26 1.7609
## 9  1999-01-25 1.7620
## 10 1999-01-22 1.7515
## 11 1999-01-21 1.7529
## 12 1999-01-20 1.7626
## 13 1999-01-19 1.7739
## 14 1999-01-18 1.7717
## 15 1999-01-15 1.7797
## 16 1999-01-14 1.7707
## 17 1999-01-13 1.8123
## 18 1999-01-12 1.7392
## 19 1999-01-11 1.7463
## 20 1999-01-08 1.7643
## 21 1999-01-07 1.7602
## 22 1999-01-06 1.7711
## 23 1999-01-05 1.7965
## 24 1999-01-04 1.8004

hist(d2$d1.CAD, col = 'khaki3')
```

Writing to a CSV file

It is also possible to write data back into a file. Best is to use a structured format such as a CSV-file. subset()

```
write.csv(d2,"output.csv", row.names = FALSE)
new.d2 <- read.csv("output.csv")
print(new.d2)
##       d1.Date d1.CAD
## 1  2008-12-30 1.7331
## 2  2008-12-29 1.7408
```

```
## 3  2008-12-18 1.7433
## 4  1999-02-03 1.7151
## 5  1999-01-29 1.7260
## 6  1999-01-28 1.7374
## 7  1999-01-27 1.7526
## 8  1999-01-26 1.7609
## 9  1999-01-25 1.7620
## 10 1999-01-22 1.7515
## 11 1999-01-21 1.7529
## 12 1999-01-20 1.7626
## 13 1999-01-19 1.7739
## 14 1999-01-18 1.7717
## 15 1999-01-15 1.7797
## 16 1999-01-14 1.7707
## 17 1999-01-13 1.8123
## 18 1999-01-12 1.7392
## 19 1999-01-11 1.7463
## 20 1999-01-08 1.7643
## 21 1999-01-07 1.7602
## 22 1999-01-06 1.7711
## 23 1999-01-05 1.7965
## 24 1999-01-04 1.8004
```

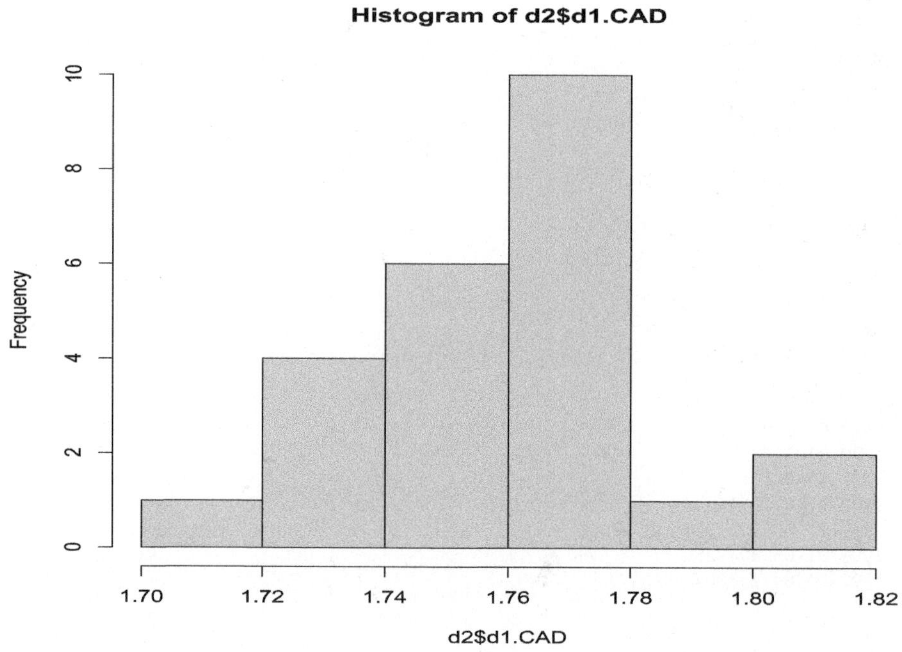

Figure 4.6: *The histogram of the most recent values of the CAD only.*

> **Warning – Silently added rows**
>
> Without the `row.names = FALSE` statement, the function `write.csv()` would add a row that will get the name "X."

4.8.2 Excel Files

Microsoft's Excel is omnipresent in the corporate environment and many people will have some data in that format. There is no need to to first save the data as a CSV file and then upload in R. The package `xlsx` will allow us to directly import a file in xlsx format.

Excel

.xlx

Importing and xlsx-file is very similar to importing a CSV-file.

```
# install the package xlsx if not yet done
if (!any(grepl("xlsx",installed.packages()))){
  install.packages("xlsx")}
library(xlsx)
data <- read.xlsx("input.xlsx", sheetIndex = 1)
```

4.8.3 Databases

Spreadsheets and CSV-file are good carriers for reasonably small datasets. Any company that holds a lot of data will use a database system – see for example Chapter 13 "*RDBMS*" on page 285. Importing data from a database system is somewhat different. The data is usually structured in "tables" (logical units of rectangular data); however, they will seldom contain all information that we need and usually have too much rows. We need also some protocol to communicate with the database: that is the role of the structured query language (SQL) – see Chapter 14 "*SQL* on page 291.

database – MySQL

R can connect to many popular database systems. For example, MySQL: as usual there is a package that will provide this functionality.

MySQL

```
if(!any(grepl("xls", installed.packages()))){
  install.packages("RMySQL")}
library(RMySQL)
```

`RMySQL`

Connecting to the Database

This is explained in more detail in Chapter 15 "*Connecting R to an SQL Database*" on page 327, but the code segment below will already get you the essential ingredients.

```
# The connection is stored in an R object myConnection and
# it needs the database name (db_name), username and password
myConnection = dbConnect(MySQL(),
  user = 'root',
  password = 'xxx',
  dbname = 'db_name',
  host = 'localhost')

# e.g. list the tables available in this database.
dbListTables(myConnection)
```

`dbConnect()`
`dbSendQuery()`

Fetching Data Drom a Database

Once connected to the database, we can select and download the data that we will use for our analysis in R.

```
# Prepare the query for the database
result <- dbSendQuery(myConnection,
  "SELECT * from tbl_students WHERE age > 33")
```

```
# fetch() will get us the results, it takes a parameter n, which
# is the number of desired records.
# Fetch all the records(with n = -1) and store it in a data frame.
data <- fetch(result, n = -1)
```

Update Queries

fetch()

It is also possible to manipulate data in the database directly from R. This allows use to prepare data in the database first and then download it and do our analysis, while we keep all the code in one file.

The dbSendQuery() function can be used to send any query, including UPDATE, INSERT, CREATE TABLE and DROP TABLE queries so we can push results back to the database.

dbSendQuery()

```
sSQL = ""
sSQL[1] <- "UPDATE tbl_students
            SET score = 'A' WHERE raw_score > 90;"
sSQL[2] <- "INSERT INTO tbl_students
            (name, class, score, raw_score)
            VALUES ('Robert', 'Grade 0', 88,NULL);"
sSQL[3] <- "DROP TABLE IF EXISTS tbl_students;"
for (k in c(1:3)){
  dbSendQuery(myConnection, sSQL[k])
  }
```

Create Tables from R Data-frames

Finally, it is possible to write back data from a dataset in R to a table in a database.

```
dbWriteTable(myConnection, "tbl_name",
             data_frame_name[, ], overwrite = TRUE)
```

dbWriteTable()

Warning – Closing the database connection

Even while connections will be automatically closed when the scope of the database object is lost, it is a good idea to close a connection explicitly. Closing a connection explicitly, makes sure that it is closed and does not remain open if something went wrong, and hence the resources of our RDBMS are freed. Closing a database connection can be done with dbDisconnect(myConnection, ...).

dbDisconnect()

♣ 5 ♣

Lexical Scoping and Environments

5.1 Environments in R

Environments can be thought of as a set of existing or declared objects (functions, variables, etc.). When we start R, it will create an environment before the command prompt is available to the user.

The top-level environment is the R command prompt. This is the "global environment" and known as R_GlobalEnv and can be accessed as .GlobalEnv.

global environment
R_GlobalEnv

As mentioned earlier, the function ls() shows which variables and functions are defined in the current environment. The function environment() will tell us which environment is the current one.

environment()

```
environment()  # get the environment
## <environment: R_GlobalEnv>

rm(list=ls())  # clear the environment
ls()           # list all objects
## character(0)

a <- "a"
f <- function (x) print(x)
ls()           # note that x is not part of.GlobalEnv
## [1] "a" "f"
```

When a function starts to be executed this will create a new environment that is subordonated to the environment that calls the function.

```
# f
# Multiple actions and side effects to illustrate environments
# Arguments:
#    x -- single type
f <- function(x){
     # define a local function g() within f()
     g <- function(y){
             b <- "local variable in g()"
             print(" -- function g() -- ")
             print(environment())
             print(ls())
             print(paste("b is", b))
```

The Big R-Book: From Data Science to Learning Machines and Big Data, First Edition. Philippe J.S. De Brouwer.

Companion Website: www.wiley.com/go/De Brouwer/The Big R-Book

```
            print(paste("c is", c))
            }

    # actions in function f:
    a <<- 'new value for a, also in Global_env'
    x <- 'new value for x'
    b <- d    # d is taken from the environment higher
    c <- "only defined in f(), not in g()"
    g("parameter to g")
    print(" -- function f() -- ")
    print(environment())
    print(ls())
    print(paste("a is", a))
    print(paste("b is", b))
    print(paste("c is", c))
    print(paste("x is", x))
    }

# Test the function f():
b <- 0
a <- c <- d <- pi
rm(x)

## Warning in rm(x): object 'x' not found

f(a)
## [1] " -- function g() -- "
## <environment: 0x557538bf9e28>
## [1] "b" "y"
## [1] "b is local variable in g()"
## [1] "c is only defined in f(), not in g()"
## [1] " -- function f() -- "
## <environment: 0x557536bcc808>
## [1] "b" "c" "g" "x"
## [1] "a is new value for a, also in Global_env"
## [1] "b is 3.14159265358979"
## [1] "c is only defined in f(), not in g()"
## [1] "x is new value for x"

# Check the existence and values:
a
## [1] "new value for a, also in Global_env"

b
## [1] 0

c
## [1] 3.141593

x

## Error in eval(expr, envir, enclos): object 'x' not found
```

Each function within a function or environment will create a new environment that has its own variables. Variable names can be the same but the local environment will always take precedence. A few things stand out in the example above.

- The variable a does not appear in the scope of the functions.
- However, a function can access a variable defined the level higher if it is not re-defined in the function itself (see what happens with the variable c: it is not defined in g(), so automatically R will search the environment above.
- The function g() can use the variable b without defining it or without it being passed as an argument. When it changes that variable, it can use that new value, but once we are back in the function f(), the old value is still there.

5.2 Lexical Scoping in R

Just as any programming language, R as rules for lexical scoping. R is extremely flexible and this can be quite intimidating when starting, but it is possible to amass this flexibility.

First, a variable does not necessarily need to be declared, R will silently create it or even change it is type.

```
x <- 'Philippe'
rm(x)          # make sure the definition is removed
x              # x is indeed not there (generates an error message)

## Error in eval(expr, envir, enclos): object 'x' not found

x <- 2L        # now x is created as a long integer
x <- pi        # coerced to double (real)
x <- c(LETTERS[1:5]) # now it is a vector of strings
```

This can, of course, lead to mistakes in our code: we do not have to declare variables, so we cannot group those declarations so that these error become obvious. This means that if there is a mistake, one might expect to see strange results that are hard to explain. In such case, debugging is not easy. However, this is quite unlikely to come into your way. Follow the rule that one function is never longer than half an A4 page and most likely this feature of R will save time instead of increasing debugging time.

Next, one will expect that each variable has a scope.

```
# f
# Demonstrates the scope of variables
f <- function() {
  a <- pi      # define local variable
  print(a)     # print the local variable
  print(b)     # b is not in the scope of the function
}

# Define two variables a and b
a <- 1
b <- 2

# Run the function and note that it knows both a and b.
# For b it cannot find a local definition and hence
# uses the definition of the higher level of scope.
f()
## [1] 3.141593
## [1] 2

# f() did not change the value of a in the environment that called f():
print(a)
## [1] 1
```

This illustrates that the scoping model in R is *dynamical scoping*. This means that when a variable is used, that R in the first place will try to find it in the local scope, if that fails, it goes a step up and continues to do so until R ends up at the root level or finds a definition.

dynamical scoping

Note – Dynamic scoping

Dynamic scoping is possible, because R is an interpreted language. If R would be compiled, then the compiler would have a hard time to create all possible memory allocations at the time of compilation.

To take this a step further we will study how the scoping within S3 objects works.[1] The example below is provided by the R Core Team.

```
# Citation from the R documentation:
# Copyright (C) 1997-8 The R Core Team
open.account <- function(total) {
     list(
        deposit = function(amount) {
            if(amount <= 0)
                stop("Deposits must be positive!\n")
            total <<- total + amount
            cat(amount,"deposited. Your balance is", total, "\n\n")
        },
        withdraw = function(amount) {
            if(amount > total)
                stop("You do not have that much money!\n")
            total <<- total - amount
            cat(amount,"withdrawn.  Your balance is", total, "\n\n")
        },
        balance = function() {
            cat("Your balance is", total, "\n\n")
        }
        )
 }

ross <- open.account(100)

robert <- open.account(200)

ross$withdraw(30)
## 30 withdrawn.  Your balance is 70

ross$balance()
## Your balance is 70

robert$balance()
## Your balance is 200

ross$deposit(50)
## 50 deposited. Your balance is 120

ross$balance()
## Your balance is 120

try(ross$withdraw(500)) # no way..
## Error in ross$withdraw(500) : You do not have that much money!
```

This is a prime example of how flexible R is. At first this is quite bizarre, until we notice the `<<-` operator. This operator indicates that the definition of a level higher is to be used. Also the

[1]To fully appreciate what is going on here, it is best to read the section on the object models (Chapter 6 *"The Implementation of OO"* on page 87) in R first and more especially Section 6.2 *"S3 Objects"* on page 91.

variable passed to the function automatically becomes an attribute of the object. Or maybe it was there because the object is actually defined as a function itself and that function got the parameter "amount" as a variable.

This example is best understood by realizing that this can also be written with a declared value "total" at the top level of our object.

> **Warning – Dynamical scoping**
>
> While dynamical scoping has its advantages when working interactively, it makes code more confusing, and harder to read and maintain. It is best to be very clear with what is intended (if an object has an attribute, then declare it). Also never use variables of a higher level, rather pass them to your function as a parameter, then it is clear what is intended.)

In fact, a lot is going on in this short example and especially in the line `total <<- total + amount`. First, of all, `open.account` is defined as a function. That function has only one line of code in some sense. This is a fortiori the last line of code, and hence, this is what the function will return. So it will try to return a list of three functions: "deposit," "withdraw" and "balance."

What might be confusing in the way this example is constructed is probably is that when the function open.account is invoked, it will immediately execute the first function of that list. Well, that is not what happens. What happens is that the `open.account` object gets a variable `total`, because it is passed as an argument. So, it will exist within the scope of that function.

The odd thing is that rather than behaving as a function this construct behaves like a normal object that has an attribute `amount` and a creator function `open.account()`. This function in sets this attribute to be equal to the value passed to that function.

> **Hint – Write readable code**
>
> R will store variable definitions in its memory, and it may happen that you manage to read in a file for example, but then keep a mistake in your file. If you do not notice the mistake, it will go unnoticed (since your values are still there and you can work with them). The next person who tries to reproduce your code will fail. Start each file that contains your code with:
>
> ```
> rm(list=ls())  # clear the environment
> ```

At this point, it is important to get a better understanding of the OO model implemented in R – or rather the multitude of models that are implemented in R.

♣ 6 ♣

The Implementation of OO

R is an object oriented (OO) language but if you know how objects work in for example C++, it might take some mental flexibility to get your mind around how it works in R. R is not a compiled language, and it is build with versatility in mind. In some sense most objects are a to be considered as an object and can readily be extended without much formalism and without recompiling.[1]

OO
object oriented

Personally, I find it useful to think of R as a functional language with odd object possibilities. This means that if you want to make some simple analysis, then you might skip this section. We did our best to keep the knowledge of OO to a minimum for the whole book.

Programming languages that provide support for object oriented programming, allow code to be data-centric as opposed to functional languages that are in essence logic-oriented. They do that by introducing the concept of a "class." A manifestation of that class then becomes the object that we can work with. Objects represent real life things. For example, if we want to create a software that manages bank accounts, it might be possible to have one object that is the account, another that is the customer, etc.

The main idea is that in other parts of the code we can work with the object "account" and ask that object for its balance. This has multiple advantages. First, of all, the logic of the way a balance is found is only in one place and in every call the same. Second, it becomes easier to pass on information and keep code lean: if you will need the balance all you have to do is import the object account and you inherit all its functionality.

There are other ways to keep code clean, for example creating an object, that is a savings account, will automatically inherit the functionality and data that all accounts share. So it becomes easy to create other types of accounts that are based on one primordial object account. For example, current accounts, savings accounts and investment accounts can all inherit from the object "account." One of the basic things that all accounts will share is for example the way ownership works and how transactions are allowed. This can be programmed once and used in all types of accounts that inherit from this one. If necessary, this can even be overwritten, if there one type of account that uses another logic.

[1] Object oriented programming refers to the programming style that provides a methodology that enables a logical system (real life concept, such as for example "student")) to be modelled an "objects." In further code it will then be possible to address this object via its methods and attributes. For example, the object student can have a method "age" that returns the age of the student based in its attribute birth-date.

The Big R-Book: From Data Science to Learning Machines and Big Data, First Edition. Philippe J.S. De Brouwer.

Companion Website: www.wiley.com/go/De Brouwer/The Big R-Book

Another example could be how we can keep meta-data together with data.

The following code creates, for example, an attribute `data_source` within the object `df`.

```
L <- list(matrix(1:16, nrow=4))
L$data_source <- "mainframe 1"
L$get_data_src <- function(){L$data_source}
print(L$get_data_src())
## [1] "mainframe 1"
```

In many ways, the list object (and many other objects) act as manifestations of objects that can freely be expanded. So already in the Section 4.3.6 *"Lists"* on page 41, we have used the object oriented capabilities of R explicitely. This might be a little bewildering and leaves the reader probably wondering what is the true object model behind R. Well, the answer is that there is not one but rather four types of classes.

Multiple OO systems ready to use in R. R's OO systems differ in how classes and methods are defined:

C

1. **Base types**. This is not a true OO system as it misses critical features like inheritance and flexibility. However, this underlies all other OO systems, and hence, it is worth to understand them a little. They are typically thought of as a `struct` in C.

2. **S3**. S3 is a popular class type in R. It is very different from the OO implementation that is in most programming languages such as C++, C#, PHP, Java, etc. In those languages one would expect to pass a message to an object (for example ask the object `my_account` its balance as via the method `my_curr_acc.balance()`. The object `my_account` is of the type account and hence it is the logic that sits there that will determine what function `balance()` is used. These implementations are called message-passing systems., S3 uses a different logic: it is a generic-function implementation of the OO logic. The S3 object can still have its own methods, but there are generic functions that will decide which method to use. For example, the function `print()` will do different things for a linear model, a dataset, or a scalar.

3. Then there is also **S4**, which works very similarly to S3, but there is a formal declaration of the object (its definition) and it has special helper functions for defining generics and methods. S4 also allows for "multiple dispatch," which means that generic functions can pick methods based on the class of any number of arguments, not just one.

C++
C#

RC – reference class

4. Reference classes (RC) are probably closest to what you would expect based on your C++ or C# experience. RC implements a message-passing OO system in R. This means that a method (function) will belong to a class, and it is not a generic method that will decide how it behaves on different classes. The RC implementation uses the $. This means that a call to the function balance of an object of class account will look like `my_account$dbalance()`. RC objects in R are "mutable." This means that they don't follow R's general logic "copy-on-modify" semantics, but are modified in place. This allows for difficult to read code but is invaluable to solve problems that cannot be solved in S3 or S4.

6.1 Base Types

The base types are build upon the "structures" from the language C that underlies R.[2] Knowing this inheritance, the possibilities and limitations of the base types should not be a mystery. A `struct` is basically a collection of variables that are gathered under one name. In our example (the bank account) it could hold the name of the account holder, the balance as a number, but not the balance as a function. The following works:

struct

```
# Define a string:
acc <- "Philippe"

# Force an attribute, balance, on it:
acc$balance <- 100

## Warning in acc$balance <- 100: Coercing LHS to a list

# Inspect the result:
acc
## [[1]]
## [1] "Philippe"
##
## $balance
## [1] 100
```

This means that the base type holds information on how the object is stored in memory (and hence how much bytes it occupies), what variables it has, etc. The base types are part of R's code and compiled, so it is only possible to create new ones by modifying R's source code and recompiling. When thinking about the base types, one readily recalls all the types that we studied in the previous sections such as integers, vectors, matrices are base types. However, there are more exotic ones such as environments, functions, calls.

Some conventions are not straightforward but deeply embedded in R and many people's code, some things might be somewhat surprising. Consider the following code:

```
# a function build in core R
typeof(mean)
## [1] "closure"

is.primitive(mean)
## [1] FALSE

# user defined function are "closures:
add1 <- function(x) {x+1}
typeof(add1)
## [1] "closure"

is.function(add1)
## [1] TRUE

is.object(add1)
## [1] FALSE
```

`is.primitive()`

`typeof()`

`is.function()`

`is.object()`

[2]The reader that has knowledge of C, might want to know that this is the object-like functionality that is provided by the `struct` keyword in C. R is written in C and to program base-R one used indeed those structures.

As mentioned before, the mainstream OO implementation in R is a generic-function implementation. That means that functions that display different behaviour for different objects will dispatch the action to a more specialized function.[3]

The importance of these `struct`-based base type is that all other object types are built upon these: S3 objects are directly build on top of the base types, S4 objects use a special-purpose base type, and RC objects are a combination of S4 and environments (which is also a base type).

[3]Experienced C-users might want to think of this as something like the statement `switch(TYPEOF(x))` in C.

6.2 S3 Objects

S3 is probably the most simple implementation of an OO system that is still useful. In its simplicity, it is extremely versatile and user friendly (once you get your old C and C++ reflexes under control).

The function `is.object()` returns true both for S3 and S4 objects. There is no base function that allows directly to test if an object is S3, but there is a to test to check if an object is S4. So we can test if something is S3 as follows.

```
# is.S3
# Determines if an object is S3
# Arguments:
#    x -- an object
# Returns:
#   boolean -- TRUE if x is S3, FALSE otherwise
is.S3 <- function(x){is.object(x) & !isS4(x)}

# Create two test objects:
M  <- matrix(1:16, nrow=4)
df <- data.frame(M)

# Test our new function:
is.S3(M)
## [1] FALSE

is.S3(df)
## [1] TRUE
```

However, it is not really necessary to create such function by ourselves. We can leverage the library `pryr`, which provides a function `otype()` that returns the type of object.

pryr
otype()

```
library(pryr)
otype(M)
## [1] "base"

otype(df)
## [1] "S3"

otype(df$X1)            # a vector is not S3
## [1] "base"

df$fac <-factor(df$X4)
otype(df$fac)           # a factor is S3
## [1] "S3"
```

The methods are provided by the generic function.[4] Those functions will do different things for different S3 objects.

If you would like to determine if a function is S3 generic, then you can check the source code for the use of the function `useMethod()`. This function will take care of the dispatching and hence decide which method to call for the given object.

useMethod()

However, this method is not foolproof because some primitive functions have this switch statement embedded in their C-code. For example, `[`, `sum()`, `rbind()`, and `cbind()` are generic functions, but this is not visible in their code in R.

[4] A good way to see a generic function is as an overloaded function with a twist.

Alternatively, it is possible to use the function `ftype` from the package `pryr`:

```
mean
## function (x, ...)
## UseMethod("mean")
## <bytecode: 0x563423e48908>
## <environment: namespace:base>

ftype(mean)
## [1] "s3"      "generic"

sum
## function (..., na.rm = FALSE)  .Primitive("sum")

ftype(sum)
## [1] "primitive" "generic"
```

R calls the functions that have this switch in their C-code `"internal" "generic"`.

The S3 generic function basically decides to what other function to dispatch its task. For example, the function print can be called with any base or S3 object and print will decide what to do based on its class. Try the function `apropos()` to find out what different methods exist (or type `print.` in RStudio.

```
apropos("print.")
##  [1] "print.AsIs"
##  [2] "print.by"
##  [3] "print.condition"
##  [4] "print.connection"
##  [5] "print.data.frame"
##  [6] "print.Date"
##  [7] "print.default"
##  [8] "print.difftime"
##  [9] "print.Dlist"
## [10] "print.DLLInfo"
## [11] "print.DLLInfoList"
## [12] "print.DLLRegisteredRoutines"
## [13] "print.eigen"
## [14] "print.factor"
## [15] "print.function"
## [16] "print.hexmode"
## [17] "print.libraryIQR"
## [18] "print.listof"
## [19] "print.NativeRoutineList"
## [20] "print.noquote"
## [21] "print.numeric_version"
## [22] "print.octmode"
## [23] "print.packageInfo"
## [24] "print.POSIXct"
## [25] "print.POSIXlt"
## [26] "print.proc_time"
## [27] "print.restart"
## [28] "print.rle"
## [29] "print.simple.list"
## [30] "print.srcfile"
## [31] "print.srcref"
## [32] "print.summary.table"
## [33] "print.summaryDefault"
## [34] "print.table"
## [35] "print.warnings"
## [36] "printCoefmat"
## [37] "sprintf"
```

```
apropos("mean.")
##  [1] ".colMeans"     ".rowMeans"     "colMeans"
##  [4] "kmeans"        "mean.Date"     "mean.default"
##  [7] "mean.difftime" "mean.POSIXct"  "mean.POSIXlt"
## [10] "rowMeans"
```

This approach shows all functions that include "print." in their name and are not necessarily the methods of the function print. Hence, a more elegant ways is to use the purpose build function `methods()`.

`methods()`

```
methods(methods)
## no methods found

methods(mean)
## [1] mean.Date     mean.default  mean.difftime
## [4] mean.POSIXct  mean.POSIXlt
## see '?methods' for accessing help and source code
```

Hint – Naming conventions

Do not use the dot "." in function names because it makes them look like S3 functional methods. This might lead to confusion with the convention that the methods are named as `<<generic function>>.<<class name>>`. Especially, if there is more than one dot in the name. For example, `print.data.frame()` is not univocal: is it a data-frame method for the generic function print or is it the frame method for the generic function print.data? Another example is the existence of the function `t.test()` to run t-tests as well as `t.dataframe()`, that is the S3 method for the generic function `t()` to transpose a data frame.

`t.test()`
`t.data.frame()`
`t()`

To access the source code of the class-specific methods, one can use the function `getS3method()`.

`getS3method()`

```
getS3method("print","table")
## function (x, digits = getOption("digits"), quote = FALSE, na.print = "",
##     zero.print = "0", justify = "none", ...)
## {
##     d <- dim(x)
##     if (any(d == 0)) {
##         cat("< table of extent", paste(d, collapse = " x "),
##             ">\n")
##         return(invisible(x))
##     }
##     xx <- format(unclass(x), digits = digits, justify = justify)
##     if (any(ina <- is.na(x)))
##         xx[ina] <- na.print
##     if (zero.print != "0" && any(i0 <- !ina & x == 0))
##         xx[i0] <- zero.print
##     if (is.numeric(x) || is.complex(x))
##         print(xx, quote = quote, right = TRUE, ...)
##     else print(xx, quote = quote, ...)
##     invisible(x)
## }
## <bytecode: 0x5634250f12e8>
## <environment: namespace:base>
```

The other way around it is possible to list all generic functions that have a specific method for a given class.

You can also list all generics that have a method for a given class:

```
methods(class = "data.frame")
##  [1] [             [[            [[<-
##  [4] [<-           $             $<-
##  [7] aggregate     anyDuplicated as.data.frame
## [10] as.list       as.matrix     by
## [13] cbind         coerce        dim
## [16] dimnames      dimnames<-    droplevels
## [19] duplicated    edit          format
## [22] formula       head          initialize
## [25] is.na         Math          merge
## [28] na.exclude    na.omit       Ops
## [31] plot          print         prompt
## [34] rbind         row.names     row.names<-
## [37] rowsum        show          slotsFromS3
## [40] split         split<-       stack
## [43] str           subset        summary
## [46] Summary       t             tail
## [49] transform     unique        unstack
## [52] within
## see '?methods' for accessing help and source code
```

6.2.1 Creating S3 Objects

S3

S3 is a minimalistic and informal OO system; there is not even a formal definition of a class. In S3, you never create a class definition and start from the instance itself. Actually, to create an S3 object, it is sufficient to set its class attribute.

```
my_curr_acc <- list("name" = "Philippe", "balance" <- 100)
class(my_curr_acc) <- "account"  # set the class attribute
otype(my_curr_acc)               # it is an S3 object
## [1] "S3"

class(my_curr_acc)               # the class type is defined above
## [1] "account"

# Note that the class attribute is not visible in the structure:
str(my_curr_acc)
## List of 2
##  $ name: chr "Philippe"
##  $     : num 100
##  - attr(*, "class")= chr "account"
```

`structure()`

It is also possible to create a class and set the attribute simultaneously with the function `structure`.

```
my_object <- structure(list(), class = "boringClass")
```

generic function

To create methods for S3 generic function, all we have to do is follow the syntax: `<<generic function>>.<<class name>>`. R will then make sure that if an object of "class name" calls the generic function that then the generic function will dispatch the action to this specific function.

```
# print.account
# Print an object of type 'account'
# Arguments:
#   x -- an object of type account
```

```
print.account <- function(x){
   print(paste("account holder",x[[1]],sep=": "))
   print(paste("balance        ",x[[2]],sep=": "))
   }
print(my_curr_acc)
## [1] "account holder: Philippe"
## [1] "balance        : 100"
```

S3 objects are always build on other more elementary types. The function `inherits (x, "classname")` allows the user to determine if the class x inherits from the class "classname."

You probably remember that R returned `"internal" "generic"` as the class for some functions; so the class can be a vector. That means that the behaviour of that object can depend on different class-specific methods. The classes have to be listed from from most to least specific, so that the behaviour can follow this cascade and will always execute the most specific behaviour if it is present.

For example, the class of the `glm()` object is `c("glm", "lm")`. This means that the most specific class is the generalised linear model, but that some behaviour they might inherit from linear models. When a generic function will be called, it will first try to find a glm-specific method. If that fails, it will look for the lm-method.

It is possible to provide a constructor function for an S3 class. This constructor function can, for example be used to check if we use the right data-type for its attributes. **constructor**

```
# account
# Constructor function for an object of type account
# Arguments:
#    x -- character (the name of the account holder)
#    y -- numeric (the initial balance of the account
# Returns:
#    Error message in console in case of failure.
account <- function(x,y) {
  if (!is.numeric(y))    stop("Balance must be numeric!")
  if (!is.atomic(x))     stop("Name must be atomic!!")
  if (!is.character(x))  stop("Name must be a string!")
  structure(list("name" = x, "balance" = y), class = "account")
}

# create a new instance for Paul:
paul_account <- account("Paul", 200)

# print the object with print.account():
paul_account
## [1] "account holder: Paul"
## [1] "balance        : 200"
```

The advantage of using the creator function for an instance of an object is obvious: it will perform some checks and will avoid problems later on. Unlike in message-passing OO implementations, the S3 implementation allows to bypass the creator function or worse it allows you to change the class all too easy. Consider the following example. **creator function**

```
class(paul_account) <- "data.frame"
print(paul_account)   # R thinks now it is a data.frame
## [1] name    balance
## <0 rows> (or 0-length row.names)
```

```
paul_account[[2]]       # the data is still correct
## [1] 200

class(paul_account) <- "account"
print(paul_account)   # back to normal: the class is just an attribute
## [1] "account holder: Paul"
## [1] "balance        : 200"
```

6.2.2 Creating Generic Methods

UseMethod()

We already saw how to create a method for a generic function by using simply the right naming convention. To add a new generic function, it is sufficient to call `UseMethod()`.

UseMethod() takes two arguments: the name of the generic function, and the argument to use for the method dispatch. If you omit the second argument it will dispatch on the first argument to the function. There is no need to pass any of the arguments of the generic to `UseMethod()`, R will take care of that for you.

```
# add_balance
# Dispatcher function to handle the action of adding a given amount
# to the balance of an account object.
# Arguments:
#    x      -- account -- the account object
#    amount -- numeric -- the amount to add to the balance
add_balance <- function(x, amount) UseMethod("add_balance")
```

This construct will do nothing else than trying to dispatch the real action to other functions. However, since we did not program them yet, there is nothing to dispatch to. To add those methods, it is sufficient to create a function that has the right naming convention.

```
# add_balance.account
# Object specific function for an account for the dispatcher
# function add_balance()
# Arguments:
#    x      -- account -- the account object
#    amount -- numeric -- the amount to add to the balance
add_balance.account <- function(x, amount) {
   x[[2]] <- x[[2]] + amount;
   # Note that much more testing and logic can go here
   # It is not so easy to pass a pointer to a function so we
   # return the new balance:
   x[[2]]}
my_curr_acc <- add_balance(my_curr_acc, 225)
print(my_curr_acc)
## [1] 325
```

Leaving the code up to this level is not really safe. It is wise to foresee a default action in case the function `add_balance()` is called with an object of another class.

```
# add_balance.default
# The default action for the dispatcher function add_balance
# Arguments:
#    x      -- account -- the account object
#    amount -- numeric -- the amount to add to the balance
add_balance.default <- function(x, amount) {
  stop("Object provided not of type account.")
}
```

6.2.3 Method Dispatch

So in S3, it is not the object that has to know how a call to it has to be handled, but it is the generic function[5] that gets the call and has to dispatch it.

UseMethod()

The way this works in R is by calling `UseMethod()` in the dispatching function. This creates a vector of function names, like
`paste0("generic", ".", c(class(x), "default"))`
and dispatches to the most specific handling function available. The default class is the last in the list and is the last resort: if R does not find a class specific method it will call the default action.

Beneath you can see this in action:

```
# probe
# Dispatcher function
# Arguments:
#    x -- account object
# Returns
#    confirmation of object type
probe <- function(x) UseMethod("probe")

# probe.account
# action for account object for dispatcher function probe()
# Arguments:
#    x -- account object
# Returns
#    confirmation of object "account"
probe.account <- function(x) "This is a bank account"

# probe.default
# action if an incorrect object type is provided to probe()
# Arguments:
#    x -- account object
# Returns
#    error message
probe.default <- function(x) "Sorry. Unknown class"

probe (structure(list(), class = "account"))
## [1] "This is a bank account"

# No method for class 'customer', fallback to 'account'
probe(structure(list(), class = c("customer", "account")))
## [1] "This is a bank account"

# No method for class 'customer', so falls back to default
probe(structure(list(), class = "customer"))
## [1] "Sorry. Unknown class"

probe(df)          # fallback to default for data.frame
## [1] "Sorry. Unknown class"

probe.account(df) # force R to use the account method
## [1] "This is a bank account"

my_curr_acc <- account("Philippe", 150) # real account
probe(my_curr_acc)
## [1] "This is a bank account"
```

[5]Sometimes generic functions are also referred to as "generics."

generic

Note – Avoid direct calls

As you can see from the above, methods are normal R functions with a specific name. So, you might be tempted to call them directly (e.g. call directly `print.data.frame()` when working with a data-frame) Actually, that is not such a good idea. This means that if, for example later you improve the dispatch method that this call will never see those improvements.

Hint – Speed gain

However, you might find that in some cases, there is a significant performance gain when skipping the dispatch method ...well, in that case you might consider to bypass the dispatching and add a remark in the code to watch this instance.[a]

[a] How to measure and improve speed is described in Chapter 40 *"The Need for Speed"* on page 793.

6.2.4 Group Generic Functions

It is possible to implement methods for multiple generic functions with one function via the mechanism of group generics. Group generic methods can be defined for four pre-specified groups of functions in R: "Math," "Ops," "Summary" and "Complex." [6]

The four "group generics" and the functions they include are:

1. Group `Math`: Members of this group dispatch on `x`. Most members accept only one argument, except `log`, `round` and `signif` that accept one or two arguments, while `trunc` accepts one or more. Members of this group are:
   - `abs`, `sign`, `sqrt`, `floor`, `ceiling`, `trunc`, `round`, `signif`
   - `exp`, `log`, `expm1`, `log1p`, `cos`, `sin`, `tan`, `cospi`, `sinpi`, `tanpi`, `acos`, `asin`, `atan` `cosh`, `sinh`, `tanh`, `acosh`, `asinh`, `atanh`
   - `lgamma`, `gamma`, `digamma`, `trigamma`
   - `cumsum`, `cumprod`, `cummax`, `cummin`
2. Group `Ops`: This group contains both binary and unary operators (`+`, `-` and `!`): when a unary operator is encountered, the `Ops` method is called with one argument and `e2` is missing. The classes of both arguments are considered in dispatching any member of this group. For each argument, its vector of classes is examined to see if there is a matching specific (preferred) or `Ops` method. If a method is found for just one argument or the same method is found for both, it is used. If different methods are found, there is a warning about `incompatible methods`: in that case or if no method is found for either argument, the internal method is used. If the members of this group are called as functions, any argument names are removed to ensure that positional matching is always used.
   - `+`, `-`, `*`, `/`, `^`, `\%\%`, `\%/\%`

`methods`

[6] Notice that there are no objects of these names in base R, but for example you will find some in the `methods` package. This package provides formally defined methods and objects for R.

- `&`, `|`, `!`
- `==`, `!=`, `<`, `<=`, `>=`, `>`

3. Group `Summary`: Members of this group dispatch on the first argument supplied.
   - `all`, `any`
   - `sum`, `prod`
   - `min`, `max`
   - `range`
4. Group `complex`: Members of this group dispatch on `z`.
   - `Arg`, `Conj`, `Im`, `Mod`, `Re`

Of course, a method defined for an individual member of the group takes precedence over a method defined for the group as a whole, because it is more specific.

Note – Distinguish groups and functions

`Math`, `Ops`, `Summary`, and `Complex` aren't functions themselves, but instead represent groups of functions. Also note that inside a group generic function a special variable `.Generic` provides the actual generic function that is called.

If you have complex class hierarchies, it is sometimes useful to call the *parent method*. This parent method is the method that would have been called if the object-specific one does not exist. For example, if the object is `savings_account`, which is a child of `account` then calling the function with `savings_account` will return the method associated to `account` if there is no specific method and it will call the specific method if it exists.

Hint – Find what is the next method

More information can be found using `?NextMethod`.

6.3 S4 Objects

The S4 system is very similar to the S3 system, but it adds a certain obligatory formalism. For example, it is necessary to define the class before using it. This adds some lines of code but the payoff is increased clarity.

In S4

1. classes have formal definitions that describe their data fields and inheritance structures (parent classes);
2. method dispatch is more flexible and can be based on multiple arguments to a generic function, not just one; and
3. there is a special operator, `@`, for extracting fields from an S4 object.

All the S4 related code is stored in the methods package.

Hint – Loading the library methods

While the `methods` package is always available when running R interactively (like in RStudio or in the R terminal), it is not necessarily loaded when running R in batch mode. So, you might want to include an explicit `library(methods)` statement in your code when using S4.

6.3.1 Creating S4 Objects

While an S3 object can be used without defining it first, to create a valid S4 object, we need at least:

- *Name:* An alpha-numeric string that identifies the class
- *Representation:* A list of slots (or attributes), giving their names and classes. For example, a person class might be represented by a character name and a numeric age, as follows: representation(name = "character", age = "numeric")
- *Inheritance:* A character vector of classes that it inherits from, or in S4 terminology, contains. Note that S4 supports multiple inheritance, but this should be used with extreme caution as it makes method lookup extremely complicated.

setClass()

S4 objects are created with the function `setClass()`.

```
# Create the object type Acc to hold bank-accounts:
setClass("Acc",
  representation(holder       = "character",
                 branch       = "character",
                 opening_date = "Date"))

# Create the object type Bnk (bank):
setClass("Bnk",
  representation(name = "character", phone = "numeric"))
```

```
# Define current account as a child of Acc:
setClass("CurrAcc",
  representation(interest_rate = "numeric",
                 balance  = "numeric"),
  contains = "Acc")

# Define investment account as a child of Acc
setClass("InvAcc",
  representation(custodian = "Bnk"), contains = "Acc")
```

This will only create the definition of the objects. So, to create a variable in your code that can be used to put your data or models inside, instances have to be created with the function `new()`. `new()`

Note – Difference between inheritance and methods

Note the difference in syntax – for the function `setClass` – between how the argument `representation` and the argument `contains` take values. The `representation` argument takes a function and hence, more arguments can be passed by adding them comma separated. In order to pass more than one parent class to `contains`, one needs to provide a character vector (for example `c("InvAcc","Acc")`).

Both the arguments `slots` and `contains` will readily use S4 classes and the implicit class of a base type. In order to use S3 classes, one needs first to register them with `setOldClass()`. If we do not want type control when an instance of a class is generated, we can provide to the `slots` argument a special class "ANY" (this tell R not to restrict the input).

You might not have noticed right away, but we started off with a complex problem where some objects depend on others (in OO we speak about "parents" and "children") and even where some objects take others as attributes. Those two things are very different and a little tricky to understand.

At this point, the *classes* `Bnk` and `Acc` exist and we can create a first instance for both.

```
# Create an instance of Bnk:
my_cust_bank <- new("Bnk",
                    name = "HSBC",
                    phone = 123456789)

# Create an instance of Acc:
my_acc <- new("Acc",
              holder       = "Philippe",
              branch       = "BXL12",
              opening_date = as.Date("2018-10-02"))
```

6.3.2 Using S4 Objects

Now, we have two S4 objects and we can use them in our code as necessary. For example, we can change the phone number.

```
# Check if it is really an S4 object:
isS4(my_cust_bank)
## [1] TRUE

# Change the phone number and check:
my_cust_bank@phone = 987654321  # change the phone number
print(my_cust_bank@phone)       # check if it changed
## [1] 987654321
```

Note – Compare addressing slots in S4 and S3

In order to access slots of an S4 object, we use `@`, not `$`:

attr() There is also a specific function to get attributes from an object: `attr()`. This function allows to create attributes, change them or even remove them (by setting them to NULL)

```
# This will do the same as my_cust_bank@phone:
attr(my_cust_bank, 'phone')
## [1] 987654321

# The function also allows partial matching:
attr(my_cust_bank, which='ph', exact = FALSE)
## [1] 987654321

# attr can also change the value of an attribute.
attr(my_cust_bank, which='phone') <- '123123123'
# Let us verify:
my_cust_bank@phone
## [1] "123123123"

# It is even possible to create a new attribute or remove one.
attr(my_cust_bank, 'something') <- 'Philippe'
attr(my_cust_bank, 'something')
## [1] "Philippe"

attr(my_cust_bank, 'something') <- NULL
attr(my_cust_bank, 'something')
## NULL

str(my_cust_bank) # the something attribute is totally gone
## Formal class 'Bnk' [package ".GlobalEnv"] with 2 slots
##   ..@ name : chr "HSBC"
##   ..@ phone: chr "123123123"
```

Warning – Partial matching

While the function `attr()` allows partial matching. It is never a good idea to use partial matching in a batch environment. This can lead to hard to detect programming errors.

Some slots – like `class`, `comment`, `dim`, `dimnames`, `names`, `row.names` and `tsp` (for time series objects) – are special: they can only take some values. This knowledge can even be used to change those attributes.

```
x <- 1:9
x         # x is a vector
## [1] 1 2 3 4 5 6 7 8 9

class(x)
## [1] "integer"

attr(x, "dim") <- c(3,3)
x         # is is now a matrix!
```

```
##      [,1] [,2] [,3]
## [1,]    1    4    7
## [2,]    2    5    8
## [3,]    3    6    9

class(x) # but R is not fooled.
## [1] "matrix"
```

Hint – Alternative to address slots

Alternatives to access slots (attributes) include the function `slot()`, that works like `[[` for regular objects.

slot()

```
slot(my_acc, "holder")
## [1] "Philippe"
```

The object `my_acc` is actually not very useful. It is a structure that would be in common for all types of accounts (e.g. investment accounts, savings accounts and current accounts). However, no bank would just sell and empty structure account. So, let us open a current account first.

```
my_curr_acc <- new("CurrAcc",
                   holder = "Philippe",
                   interest_rate = 0.01,
                   balance=0,
                   branch = "LDN12",
                   opening_date= as.Date("2018-12-01"))

# Note that the following does not work and is bound to fail:
also_an_account <- new("CurrAcc",
                       holder = "Philippe",
                       interest_rate = 0.01,
                       balance=0, Acc=my_acc)

## Error in initialize(value, ...): invalid name for slot of class "CurrAcc": Acc
```

Question #8

Why does the second approach fail? Would you expect it to work?

It appears that while the object `my_acc` exist, it is not possible to insert it in the definition of a new object – even while this inherits from the first. This makes sense, because the object "account" is not an attribute of the object "current account," but its attributes become directly attributes of current account.

The object `my_curr_acc` is now ready to be used. For example, we can change the balance.

```
my_curr_acc@balance <- 500
```

Now, we will create an investment account. At this point, it becomes crucial to see that the object "custodian bank" is not a parent class, but rather an attribute. This means that before we can create an investment account, we need to define at least one custodian.

```
my_inv_acc <- new("InvAcc",
                  custodian = my_cust_bank,
                  holder="Philippe",
                  branch="DUB01",
                  opening_date = as.Date("2019-02-21"))

# note that the first slot is another S4 object:
my_inv_acc
## An object of class "InvAcc"
## Slot "custodian":
## An object of class "Bnk"
## Slot "name":
## [1] "HSBC"
##
## Slot "phone":
## [1] "123123123"
##
##
## Slot "holder":
## [1] "Philippe"
##
## Slot "branch":
## [1] "DUB01"
##
## Slot "opening_date":
## [1] "2019-02-21"
```

Question #9

If you look careful at the code fragment above this question, you will notice that it is possible to provide an object `my_cust_bank` as attribute to the object `my_inv_acc`. This situation is similar to the code just above previous question, but unlike in the creation of `also_an_account`, now it works. Why is this?

To understand what happened here, we need to dig a little deeper.

```
my_inv_acc@custodian          # our custodian bank is HSBC
## An object of class "Bnk"
## Slot "name":
## [1] "HSBC"
##
## Slot "phone":
## [1] "123123123"

my_inv_acc@custodian@name    # note the cascade of @ signs
## [1] "HSBC"

my_inv_acc@custodian@name <- "DB"  # change it to DB
my_inv_acc@custodian@name    # yes, it is changed
## [1] "DB"

my_cust_bank@name            # but our original bank isn't
## [1] "HSBC"

my_cust_bank@name <- "HSBC Custody" # try something different
my_inv_acc@custodian@name    # did not affect the account
## [1] "DB"

my_inv_acc@custodian@name <- my_cust_bank@name # change back
```

Hint – List all slots

The function getSlots() will return a description of all the slots of a class:

getSlots()

```
getSlots("Acc")
##       holder       branch opening_date
##  "character"  "character"       "Date"
```

6.3.3 Validation of Input

While S3 provides no mechanism to check if all attributes are of the right type, creating an S4 object with the function new() will check if the slots are of the correct type. For example, if we try to create an account while providing a string for the balance (while a number is expected), then R will not create the new object and inform us of the mistake.

```
# Note the mistake in the following code:
my_curr_acc <- new("CurrAcc",
                   holder = "Philippe",
                   interest_rate = 0.01,
                   balance="0",  # Here is the mistake!
                   branch = "LDN12",
                   opening_date= as.Date("2018-12-01"))

## Error in validObject(.Object): invalid class "CurrAcc" object: invalid object for
slot "balance" in class "CurrAcc": got class "character", should be or extend class
"numeric"
```

If you omit a slot, R coerces that slot to the default value.

```
x_account <- new("CurrAcc",
                   holder = "Philippe",
                   interest_rate = 0.01,
                   #no balance provided
                   branch = "LDN12",
                   opening_date= as.Date("2018-12-01"))
x_account@balance  # show what R did with it
## numeric(0)
```

Warning – Silent setting to default

Did you notice that R is silent about the missing balance? This is something to be careful with. If you forget that a default value has been assigned then this might lead to confusing mistakes.

An empty value for balance is not very useful and it can even lead to errors. Therefore, it is possible to assign default values with the function prototype when creating the class definition. prototype()

```
setClass("CurrAcc",
  representation(interest_rate = "numeric",
                 balance       = "numeric"),
  contains = "Acc",
  prototype(holder        = NA_character_,
            interst_rate = NA_real_,
            balance      = 0))

x_account <- new("CurrAcc",
                 # no holder
                 # no interest rate
                 # no balance
                 branch = "LDN12",
                 opening_date= as.Date("2018-12-01"))
x_account        # show what R did:
## An object of class "CurrAcc"
## Slot "interest_rate":
## numeric(0)
##
## Slot "balance":
## [1] 0
##
## Slot "holder":
## [1] NA
##
## Slot "branch":
## [1] "LDN12"
##
## Slot "opening_date":
## [1] "2018-12-01"
```

Warning – Changing class definitions at runtime

Most programming languages implement an OO system where class definitions are created when the code is compiled and instances of classes are created at runtime. During runtime, it is not possible to change the class definitions.
However, R is an interpreted language that is interactive and functional. The consequence is that it is possible to change class definitions at runtime ("while working in the R-terminal"). So it is possible to call `setClass()` again with the same object name, and R will assume that you want to change the previously defined class definition and silently override it. This can lead, for example, to the situation where different objects pretend to be of the same class, while they are not.

Hint – Locking a class definition

To make sure that a previous class definition cannot be changed add `sealed = TRUE` to the call to `setClass()`

Hint – Typesetting conventions

It is common practice to use "UpperCamelCase" for S4 class names.[a] Using a convention is always a good idea, using one that many other people use is even better. This convention avoids any confusion with the dispatcher functions that use the dot as separator.

[a] UppercamelCase is easy to understand when comparing to lowerCamelCase, the dot.separator and snake_case. It refers to the way long names (of objects, variables and functions) are kept readable in code. They are all good alternatives, and each programmer has his/her preference, though in many communities, there are some unwritten rules. These rules are best followed because that makes your code much easier to read.

UpperCamelCase
lowerCamelCase
snake_case
dot.separator

6.3.4 Constructor functions

Constructor functions should be given the same name as the class and it allows much more testing and action than the standard new() function.

```
This is the constructor function to create a new instance of CurrAcc:
.CurrAcc <- function (holder,
                      interest_rate
                      # branch we know from the user
                      # balance should be 0
                      # opening_date is today
                      ) {

  error_msg = "Invalid input while creating an account\n"
  if (is.atomic(holder) & !is.character(holder)) {
    stop(error_msg, "Invalid holder name.")
    }
  if (!(is.atomic(interest_rate) & is.numeric(interest_rate)
      & (interest_rate >= 0) & (interest_rate < 0.1))) {
    stop(error_msg, "Interest rate invalid.")
    }
  br <- "PAR01"  # pretending to find balance by looking up user
  dt <- as.Date(Sys.Date())
  new("CurrAcc",
                    holder = holder,
                    interest_rate = interest_rate,
                    balance=0,
                    branch = br,
                    opening_date= dt)
  }

# Create a new account:
lisa_curr_acc <- .CurrAcc("Lisa", 0.01)
lisa_curr_acc
## An object of class "CurrAcc"
## Slot "interest_rate":
## [1] 0.01
##
## Slot "balance":
## [1] 0
##
## Slot "holder":
## [1] "Lisa"
##
```

Sys.Date()

```
## Slot "branch":
## [1] "PAR01"
##
## Slot "opening_date":
## [1] "2020-01-30"
```

Hint – Calling the constructor function

C++

Unlike C++, for example a call to `new()` will not automatically invoke the constructor function (its existence is not enough to invoke it automatically). Make it a good habit to always use explicitly the constructor function for an S4 objects (provided it exists of course).

6.3.5 The .Data slot

If an S4 object inherits from an S3 class or a base type, R will give it a special `.Data` slot that contains the data of this underlying object (S3 or base type):

runif()
setClass()
new()

```
# Here is the prototype of a dataset that holds some extra
# information in a structured way.
 setClass("myDataFrame",
          contains = "data.frame",
          slots = list(MySQL_DB    = "character",
                       MySQL_tbl   = "character",
                       data_owner = "character"
                       )
          )
xdf <- new("myDataFrame",
    data.frame(matrix(1:9, nrow=3)),
    MySQL_DB = "myCorporateDB@102.12.12.001",
    MySQL_tbl = "tbl_current_accounts",
    data_owner = "customer relationship team")

xdf@.Data
## [[1]]
## [1] 1 2 3
##
## [[2]]
## [1] 4 5 6
##
## [[3]]
## [1] 7 8 9

xdf@data_owner
## [1] "customer relationship team"
```

6.3.6 Recognising Objects, Generic Functions, and Methods

While we casually used already `isS4()` to check if an object is S4, there are multiple ways to find out if an object is S4:

str()
isS4()
otype()

- `str()` will report it as an S4 class,
- `isS4()` returns TRUE, note that this is not the same as `is.S3()`, this is the class-specific method of the function `is()`,
- `pryr::otype()` returns `S4`.

S4 generics and methods are also easy to identify because they are S4 objects with well-defined classes.

There aren't any S4 classes in the commonly used base packages (stats, graphics, utils, datasets, and base), so we will continue to use our previous example of the bank accounts.

```
str(my_inv_acc)
## Formal class 'InvAcc' [package ".GlobalEnv"] with 4 slots
##   ..@ custodian   :Formal class 'Bnk' [package ".GlobalEnv"] with 2 slots
##   .. .. ..@ name : chr "HSBC Custody"
##   .. .. ..@ phone: chr "123123123"
##   ..@ holder      : chr "Philippe"
##   ..@ branch      : chr "DUB01"
##   ..@ opening_date: Date[1:1], format: "2019-02-21"

isS4(my_inv_acc)
## [1] TRUE

pryr::otype(my_inv_acc)
## [1] "S4"
```

The package `methods` provides the function `is()`. This function takes one object as argument, and lists all classes that the object provided as argument inherits from. Using `is()` with two arguments will test if an object inherits from the class specified in the second argument. `is()`

```
is(my_inv_acc)
## [1] "InvAcc" "Acc"

is(my_inv_acc, "Acc")
## [1] TRUE
```

Note – Nuances in the OO system

The downside of the function centric OO system is that some things become a little subtle. Earlier we explained how to use `isS4()`. There is no function `isS3()`, but one will notice that `is.S3()` exists. Now, you will understand that `is.S3()` is the S3 specific method of the function `is()`.
Looking up the source code can be helpful:

```
is.S3
## function(x){is.object(x) & !isS4(x)}
## <bytecode: 0x5634256e9f60>
```

There are many functions related to S4 objects, and it is not the aim to provide a full list however, the following might be useful for your code.

- `getGenerics()` lists all S4 generics; `getGenerics()`
- `getClasses()` lists all S4 classes (it does however, include shim classes for S3 classes and base types); `getClasses()`
- `showMethods()` shows the methods for one or more generic functions, possibly restricted to those involving specified classes. Note that the argument `where` can be used to restrict the search to the current environment by using `where = search()`; `showMethods()`

6.3.7 Creating S4 Generics

R provides specific tools functions to create new generics and methods:

setGeneric()

- setGeneric() creates a new generic or converts an existing function into a generic.

setMethod()

- setMethod() creates a method for a generic function aligned to a certain class. It takes as argument the function, the signature of the class and the function definition.

We will build further on the example of the bank accounts as used in the previous sections of this chapter. As a first step, we can create methods to credit and debit a current account S4 object.

```
# setGeneric needs a function, so we need to create it first.

# credit
# Dispatcher function to credit the ledger of an object of
# type 'account'.
# Arguments:
#    x -- account object
#    y -- numeric -- the amount to be credited
credit <- function(x,y){useMethod()}

# transform our function credit() to a generic one:
setGeneric("credit")
## [1] "credit"

# Add the credit function to the object CurrAcc
setMethod("credit",
   c("CurrAcc"),
   function (x, y) {
     new_bal <- x@balance + y
     new_bal
     }
   )
## [1] "credit"

# Test the function:
my_curr_acc@balance
## [1] 500

my_curr_acc@balance <- credit(my_curr_acc, 100)
my_curr_acc@balance
## [1] 600
```

While the functionality for credit might seem trivial, in reality crediting an account will require a lot of checks (e.g. sanctioned countries and terrorist financing). So, let us create now a little more useful example with a function debet(), because before debiting an account, one will need to check if there is enough balance.

```
# debet
# Generic function to debet an account
# Arguments:
#    x -- account object
#    y -- numeric -- the amount to be taken from the account
# Returns
#    confirmation of action or lack thereof
debet <- function(x,y){useMethod()}

# Make it a generic function that verifies the balance
# before the account a debet is booked.
setGeneric("debet")
## [1] "debet"
```

```
# Add the debet() function as a method for objects of type CurrAcc
setMethod("debet",
   c("CurrAcc"),
   function (x, y) {
     if(x@balance >= y) {
       new_bal <- x@balance + y} else {
       stop("Not enough balance.")
       }
     new_bal
     }
   )
## [1] "debet"

# Test the construction:
my_curr_acc@balance  # for reference
## [1] 600

my_curr_acc@balance <- debet(my_curr_acc, 50)
my_curr_acc@balance  # the balance is changed
## [1] 650

# We do not have enough balance to debet 5000, so the
# following should fail:
my_curr_acc@balance <- debet(my_curr_acc, 5000)

## Error in debet(my_curr_acc, 5000): Not enough balance.

my_curr_acc@balance  # the balance is indeed unchanged:
## [1] 650
```

Warning – Overloading functions

If you want to overload an existing function such as `union()` and `exp()`, then you should of course *not* define the function first – as in the first line in the aforementioned code. Doing so will make its original definition unavailable.

6.3.8 Method Dispatch

In its most simple form, S4 dispatching is similar to S3 dispatching. However, the implementation for S4 is more powerful. When an S4 generic function dispatches on a single class with a single parent, then S4 method dispatch is no different from the S3 method dispatch.

For practical use, the main difference is how default values are defined. In S4, there is a special class "ANY" that functions as the default.

Similarly to S3, S4 knows about group generics. Further, details can be found in the function documentation (visualize the documentation – as usual – with the command `?S4groupGeneric`. There is also a function to call the "parent method": `callNextMethod()`.

`S4groupGeneric()`
`callNextMethod()`

The main advantage of S4 is that it is possible to dispatch on multiple arguments (though it all becomes a little more complicated and abstract). That is the case when you have multiple parent classes. The rules can be found in the documentation of the function `Methods`.

`Methods()`

These aspects require utmost care. Personally, the author believes that this is where the S4 OO implementation stops to be practically relevant. When one needs further dependencies and complexity the S4 method becomes too easy too complex and it might be hard to predict which method will be called.

For this reason, it might be better to avoid multiple inheritance and multiple dispatch unless absolutely necessary.

Finally, there are methods that allow us to identify which method gets called, given the specification of a generic call:

```
selectMethod("credit", list("CurrAcc"))
## Method Definition:
##
## function (x, y)
## {
##     new_bal <- x@balance + y
##     new_bal
## }
##
## Signatures:
##         x
## target  "CurrAcc"
## defined "CurrAcc"
```

There is a lot more to say about S4 inheritance and method dispatch, though it is not really necessary in the rest of the book. Therefore, we refer to other literature. For example, "Advanced R" by Wickham (2014) is a great source.

6.4 The Reference Class, refclass, RC or R5 Model

RC
R5

refclass

The reference class OO system is also known as "RC," "refclass" or as "R5." It is the most recent[7] OO implementation and it introduces a message-passing OO system for R.

> **Note – Recent developments**
>
> Reference classes are reasonably new to R and therefore, will develop further after the publication of this book. So for the most up-to-date information, we refer to R itself: type `?ReferenceClasses` in the command prompt to see the latests updates in your version of R.

OOP
object oriented programming

People with OOP background will naturally feel more comfortable with RC, it is something what people with C++, C#, PHP, Java, Python, etc., background will be familiar with: a message-passing OO implementation. However, that sense of comfort has to be mended, in many ways, the refclass system in R is a combination of S4 and environments.

That said, the RC implementation brings R programming to a next level. This system is particularly suited for larger projects, and it will seamlessly collaborate with S4, S3 and base types. However, note that the vast majority of packages in R does not use RC, actually none of the most-often used packages do. This is not only because the pre-date the refclass system but also because they do not need it (even while some are rather complex).

Using RC in R will add some complexity to your code and many people advice to use the refclass system only where the mutable state is required. This means that even while using R5, it is still possible to keep most of the code functional.

6.4.1 Creating RC Objects

`setRefClass()`

R provides the function `setRefClass` to create a new R5 class definition that can immediately be used to create an instance of the object.

The RC object has three main components, given by three arguments to its creator function `setRefClass`:

1. *contains:* The classes from which it inherits – note that only other refclass objects are allowed;

2. *fields:* These are the attributes of the class – they are the equivalent of "slots" in S4; one can supply them via a vector of field names or a named list of field types;

3. *methods:* These functions are the equivalent for for dispatched methods in S4, and they operate within the context of the object and can modify its fields. While it is possible to add these later, this is not good programming practice, makes the code less readable and will lead to misunderstandings.

We will illustrate this with a remake of the example about bank accounts.

[7] R5 has recently been added to R (2015). It responds to the need for mutable objects and as such makes packages such as `R.oo`, `proto` and `mutatr` obsolete.

```
# Note that we capture the returned value of the setRefClass
# Give this always the same name as the class.
account <- setRefClass("account",
            fields = list(ref_number   = "numeric",
                          holder       = "character",
                          branch       = "character",
                          opening_date = "Date",
                          account_type = "character"
                          ),
            # no method yet.
            )

x_acc <- account$new(ref_number   = 321654987,
                     holder       = "Philippe",
                     branch       = "LDN05",
                     opening_date = as.Date(Sys.Date()),
                     account_type = "current"
                     )
x_acc
## Reference class object of class "account"
## Field "ref_number":
## [1] 321654987
## Field "holder":
## [1] "Philippe"
## Field "branch":
## [1] "LDN05"
## Field "opening_date":
## [1] "2020-01-30"
## Field "account_type":
## [1] "current"
```

Note

Fields can also be supplied as a vector (we use a list in the example). Using a vector does not allow to specify a type. It looks as follows:

```
setRefClass("account", fields = c("ref_number",
                                  "holder",
                                  "branch",
                                  "opening_date",
                                  "account_type"
                                  )
           )
```

Hint

It is possible to leave the input type undecided by not providing a type in the list environment. This could look like this:

```
setRefClass("account",
    fields = list(holder,       # accepts all types
             branch,            # accepts all types
             opening_date = "Date" # dates only
            )
    )
```

Let us now explore this object that was returned by the function `setRefClass()`.

```
isS4(account)
## [1] TRUE

# account is S4 and it has a lot more than we have defined:
account
## Generator for class "account":
##
## Class fields:
##
## Name:    ref_number        holder        branch
## Class:      numeric     character     character
##
## Name:  opening_date account_type
## Class:         Date    character
##
## Class Methods:
##      "field", "trace", "getRefClass", "initFields",
##      "copy", "callSuper", ".objectPackage", "export",
##      "untrace", "getClass", "show", "usingMethods",
##      ".objectParent", "import"
##
## Reference Superclasses:
##      "envRefClass"
```

generator object

The object returned by `setRefClass` (or retrieved later by `getRefClass`) is called a *generator object*. This object automatically gets the following methods.

- *new* to create instances of the class. Its arguments are the named arguments specifying initial values for the fields;
- *methods* to add methods or modify existing;
- *help* to get help about the methods;
- *fields* to get a list of attributes (fields) of that class;
- *lock* to lock the named fields so that their value can only be set once;
- *accessors* sets up accessor-methods in form of getxxx and setxxx (where "xxx" is replaced by the field names).

Refclass objects are mutable, or in other words, they have reference semantics. To understand this, we shall create the current account class and the investment account class.

```
custBank    <- setRefClass("custBank",
                  fields = list(name =  "character",
                                phone = "character"
                                )
                  )
invAccount  <- setRefClass("invAccount",
                  fields = list(custodian = "custBank"),
                  contains = c("account")
                  # methods go here
                  )
```

Let us now come back to the current account and define it while adding some methods (though as we will see later, it is possible to add them later too).

```
# Definition of RC object currentAccount
currAccount <- setRefClass("currentAccount",
                 fields = list(interest_rate = "numeric",
                               balance       = "numeric"),
                 contains = c("account"),
                 methods = list(
                   credit = function(amnt) {
                         balance <<- balance + amnt
                         },
                   debet =  function(amnt) {
                         if (amnt <= balance) {
                           balance <<- balance - amnt
                           } else {
                           stop("Not rich enough!")
                           }
                     }
                   )
                 )
# note how the class reports on itself:
currAccount
## Generator for class "currentAccount":
##
## Class fields:
##
## Name:     ref_number        holder        branch
## Class:       numeric     character     character
##
## Name:   opening_date  account_type interest_rate
## Class:          Date     character       numeric
##
## Name:        balance
## Class:       numeric
##
## Class Methods:
##      "debet", "credit", "import", ".objectParent",
##      "usingMethods", "show", "getClass", "untrace",
##      "export", ".objectPackage", "callSuper", "copy",
##      "initFields", "getRefClass", "trace", "field"
##
## Reference Superclasses:
##      "account", "envRefClass"
```

Note – Assigning in the encapsulating environment

Notice that, the assignment operator within an object is the local assignment operator: `<<-`. The operator `<<-` will call the accessor functions if defined (via the `object$accessors()` function).

We can now create accounts and use the methods supplied.

```
ph_acc <- currAccount$new(ref_number    = 321654987,
                          holder        = "Philippe",
                          branch        = "LDN05",
                          opening_date  = as.Date(Sys.Date()),
                          account_type  = "current",
                          interest_rate = 0.01,
                          balance       = 0
                          )
```

Now, we can start using the money and withdrawing money.

```
ph_acc$balance      # after creating balance is 0:
## [1] 0

ph_acc$debet(200)   # impossible (not enough balance)

## Error in ph_acc$debet(200): Not rich enough!

ph_acc$credit(200) # add money to the account
ph_acc$balance      # the money arrived in our account
## [1] 200

ph_acc$debet(100)   # this is possible
ph_acc$balance      # the money is indeed gone
## [1] 100
```

Note – Addressing attributes and methods

The Reference Class OO implementation R5 uses the `$` to access attributes and methods.

It is also possible – though not recommendable – to create methods after creation of the class.

```
alsoCurrAccount <- setRefClass("currentAccount",
                   fields = list(
                            interest_rate = "numeric",
                            balance       = "numeric"),
                   contains = c("account")
                   )
alsoCurrAccount$methods(list(
                      credit = function(amnt) {
                          balance <<- balance + amnt
                          },
                      debet = function(amnt) {
                          if (amnt <= balance) {
                             balance <<- balance - amnt
                             } else {
                             stop("Not rich enough!")
                             }
                           }
                      ))
```

Note – No dynamic editing of field definitions

In general, R is very flexible in how any of the OO systems is implemented. However, the refclasses do not allow to add fields after creating. This makes sense because it would make all existing objects invalid, since they would not have those new fields.

6.4.2 Important Methods and Attributes

All refclass object inherit from the same superclass `envRefClass`, and so they get at creation some hidden fields and methods. For example, there is `.self`. This variable refers to the object itself.

Other common methods for R5 objects that are always available (because they are inherited from `envRefClass`) are the following, illustrated for an object of RC class named `RC_obj`:

- `RC_obj$callSuper()`: This method initializes the values defined in the super-class.
- `RC_obj$initFields()`: This method initializes the values fields if the super-class has no initialization methods.
- `RC_obj$copy()`: This method creates a copy of the current object. This is necessary because Reference Classes classes don't behave like most R objects, which are copied on assignment or modification.
- `RC_obj$field()`: This method provides access to named fields, it is the equivalent to slots for S4. `RC_obj$field("xxx")` the same as `RC_obj$xxx`. `RC_obj$field("xxx", 5)`, the same as assigning the value via `RC_obj$xxx <- 5`.
- `RC_obj$import(x)`: This method coerces `x` into this object, and `RC_obj$export(Class)` coerces a copy of `RC_obj` into that class.

6.5 Conclusions about the OO Implementation

The OO system that R provides is unlike what other OO languages provide. In the first place, it offers not only a method-dispatching system but also has a message-passing system. Secondly, it is of great importance that it is possible to use R without even knowing that the OO system exists. In fact, for most of the following chapters in this book, it is enough to know that the generic-function implementation of the OO logic exists.

Digression – R6

Note that we did not discuss the R6 OO system. R6 is largely the same as R5 but adds some important OO features to the mix, such as private and public functions and properties. However, it is not widely used and most probably will become obsolete in further versions of R5.[a] R6

[a] The development of R6 can be followed up here: `https://www.r-project.org/nosvn/pandoc/R6.html`

Three or even four OO systems in one language is a lot, but that complexity does not stand in the way of practical applications. It seems that a little common sense allows the user to take the best of what is available, even when mixing systems.

So what system to use?

1. For a casual calculation of a few hundred lines of code, it is probably not necessary to define your own classes. You probably will use S3 implicitly in the packages that you will load, but you will not have to worry about it at all: just use `plot()` and `print()` and expect it to work.

2. If your inheritance structure is not too complex (and so you do not need multi-argument method signatures) and if your objects will not change themselves then S3 is the way to go. Actually, most packages on CRAN use S3.

3. If still your objects are static (not self-modifying), but you need multi-argument method signatures, then S4 is the way to go. You might want to opt for S4 also to use the formalism, as it can make code easier to understand for other people. However, be careful with the inheritance structure, it might get quite complex.

4. If your objects are self-modifying, then RC is the best choice. In S3 and S4, you would need to use replacement methods. For example:

```
my_replacement_method(object) <- new_value
```

5. For larger projects, where many people work on the same code it makes sense to have a look at R6 (it allows private methods, for example).

In fact, the OO system implemented in R is so that it does not come in the way of what you want to do, and you can do all your analysis of great complexity and practical applicability without even knowing about the OO systems. For casual use and the novice user alike, the OO system takes much of the complexity away. For example, one can type `summary(foo)`, and it will work regardless of what this "foo" is, and you will be provided with a summary that is relevant for the type of object that foo represents.

♣ 7 ♣

Tidy R with the Tidyverse

7.1 The Philosophy of the Tidyverse

R is Free and Open Source Software (FOSS), that implies that it is free to use, but also that you have access to the code – if desired. As most FOSS projects, R is also easy to expand. Fortunately, it is also a popular language and some of these millions of R users[1] might have created a packages and enhance R's functionality to do just what you need. This allows any R users to stand on the shoulders of giants: you do not have to re-invent the wheel, but you can just pick a package and expand your knowledge and that of humanity. That is great, and that is one of the most important reasons to use R. However, this has also a dark side: the popularity and the ease to expand the language means that there are literally thousands of packages available. It is easy to be overwhelmed by the variety and vast amount of packages available and this is also one of the key weaknesses of R.

Most of those packages will require one or more other packages to be loaded first. These packages will in their turn also have dependencies on yet other (or the same) packages. These dependencies might require a certain version of the upstream package. This package maintenance problem used to be known as the "dependency hell." The package manager of R does, however, a good job and it usually will work as expected.

Using the same code again after a few years, is usually more challenging. In the meanwhile you might have updated R to a newer version and most packages will be updated too. It might happen that some packages have become obsolete and are not maintained any more and therefore, the new version is not available. This can cause some other packages to fail.

Maintaining code is not a big challenge if you just write a project for a course at the university and will never use it again. Code maintenance becomes an issue when you want to use the code later ... but it becomes a serious problem if other colleagues need to review your work, expand it and change it later (while you might not be available).

Another issue is that because of this flexibility, core R is not very consistent (though people will argue that while Linux does even a worse job here and still is the best OS).

OS
operating system

Consistency does matter and it follows from a the choice of a programming philosophy. For example, R is a software to do things with data, so each function should have a first argument that refers to the data. Many functions will follow this rule, but not all. Similar issues exist for arguments to functions, names of objects and classes (e.g. there is `vector` and `Date`, etc.)

[1] According to the Tiobe-index (see `https://www.tiobe.com/tiobe-index`), R is the 14th most popular programming language and still on the rise.

The Big R-Book: From Data Science to Learning Machines and Big Data, First Edition. Philippe J.S. De Brouwer.

Companion Website: www.wiley.com/go/De Brouwer/The Big R-Book

Then there is the `tidyverse`. It is a recent addition to R that is both a collection of often used functionalities and a philosophy.

The developers of `tidyverse` promote[2]:

- Use existing and common data structures. So all the packages in the tidyverse will share a common S3 class types; this means that in general functions will accept data frames (or `tibbles`). More low-level functions will work with the base R vector types.

- Reuse data structures in your code. The idea here is that there is a better option than always over-writing a variable or create a new one in every line: pass on the output of one line to the next with a "pipe": `%>%`. To be accepted in the tidyverse, the functions in a package need to be able to use this pipe.[3]

pipe

- Keep functions concise and clear. For example, do not mix side-effects and transformations, function names should be verbs where ever possible (unless they become too generic or meaningless of course), and keep functions short (they do only one thing, but do it well).

- Embrace R as a functional programming language. This means that reflexes that you might have from say C++, C#, python, PHP, etc., will have to be mended. This means for example that it is best to use immutable objects and copy-on-modify semantics and avoid using the refclass model (see Section 6.4 *"The Reference Class, refclass, RC or R5 Model"* on page 113). Use where possible the generic functions provided by S3 and S4. Avoid writing loops (such as `repeat` and `for` but use the apply family of functions (or refer to the package `purrr`).

- Keep code clean and readable for humans. For example, prefer meaningful but long variable names over short but meaningless ones, be considerate towards people using auto-complete in RStudio (so add an id in the first and not last letters of a function name), etc.

Tidyverse is in permanent development as core R itself and many other packages. For further and most up-to-date information we refer to the website of the Tidyverse: `http://tidyverse.tidyverse.org`.

Tidy Data

Tidy data is in essence data that is easy to understand by people and is formatted and structured with the following rules in mind.

1. a tibble/data-frame for each dataset,

2. a column for each variable,

3. a row for each observation,

4. a value (or NA) in each cell (a "cell" is the intersection between row and column).

The concept of tidy data is so important that we will devote a whole section to tidy data (Section 17.2 *"Tidy Data"* on page 275) and how to make data tidy (Chapter 17 *"Data Wrangling in the tidyverse"* on page 265). For now, it is sufficient to have the previous rules in mind. This will allow us to introduce the tools of the tidyverse first and then later come back to making data tidy by using these tools.

[2]More information can be found in this article of Hadley Wickham: `https://tidyverse.tidyverse.org/articles/manifesto.html`.

[3]A notable exception here is `ggplot2` This package uses operator overloading instead of piping (overloading of the + operator).

Tidy Conventions

The tidyverse also enforces some rules to keep code tidy. The aims are to make code easier to read, reduce the potential misunderstandings, etc.

For example, we remember the convention that R uses to implement it is S3 object oriented programming framework from Chapter 6.2 *"S3 Objects"* on page 91. In that section we have explained how R finds for example the right method (function) to use when printing an object via the generic dispatcher function `print()`. When an object of class "glm" is passed to `print()`, then the function will dispatch the handling to the function `print.glm()`.

However, this is also true for data-frames: the handling is dispatched to `print.data.frame()`. This example illustrate how at this point it becomes unclear if the function `print.data.frame()` is the specific case for a data.frame for the `print()` function or if it is the special case to print a "frame" in the framework of a call to "print.data()." Therefore, the tidyverse recommends naming conventions to avoid the dot (`.`). And use the `snake_style` or `UpperCase` style instead.

Further information – Tidyverse philosophy

More about programming style in the tidyverse can be found in the online manifesto of the tidyverse website: `https://tidyverse.tidyverse.org/articles/manifesto.html`.

7.2 Packages in the Tidyverse

tidyverse

Loading the tidyverse will report on which packages are included:

```
# we assume that you installed the package before:
# install.packages("tidyverse")
# so load it:
library(tidyverse)

## - Attaching packages ----------- tidyverse 1.3.0 -
## v ggplot2 3.2.1     v purrr   0.3.3
## v tibble  2.1.3     v dplyr   0.8.3
## v tidyr   1.0.0     v stringr 1.4.0
## v readr   1.3.1     v forcats 0.4.0
## - Conflicts ------------- tidyverse_conflicts() -
## x purrr::compose() masks pryr::compose()
## x dplyr::filter()  masks stats::filter()
## x dplyr::lag()     masks stats::lag()
## x purrr::partial() masks pryr::partial()
```

So, loading the library `tidyverse`, loads actually a series of other packages. The collection of these packages are called "core-tidyverse."

Further, loading tidyverse also informs you about which potential conflicts may occur. For example, we see that calling the function `filter()` will dispatch to `dplyr::filter()` (ie. "the function `filter` in the package `dplyr`," while before loading tidyverse, the function `stats::filter()` would have been called).[4]

filter()

Digression – Calling methods of not loaded packages

When a package is not loaded, it is still possible to call its member functions. To call a function from a certain package, we can use the `::` operator.
In other words, when we use the `::` operator, we specify in which package this function should be found. Therefore it is possible to use a function from a package that is not loaded or is superseded by a function with the same name from a package that got loaded later.

R allows you to stand on the shoulders of giants: when making your analysis, you can rely on existing packages. It is best to use packages that are part of the tidyverse, whenever there is choice. Doing so, your code can be more consistent, readable, and it will become overall a more satisfying experience to work with R.

7.2.1 The Core Tidyverse

The core tidyverse includes some packages that are commonly used in data wrangling and modelling. Here is a word of explanation already. Later we will explore some of those packages more in detail.

tidyr

- `tidyr` provides a set of functions that help you get to tidy up data and make adhering to the rules of tidy data easier.

[4] Here we use the notation `package1::function1()` to make clear that the `function1` is the one as defined in `package1`.

The idea of tidy data is really simple: it is data where every variable has its own column, and every column is a variable. For more information, see Chapter 17.3 *"Tidying Up Data with tidyr"* on page 277.

- `dplyr` provides a grammar of data manipulation, providing a consistent set of verbs that solve the most common data manipulation challenges. For more information, see Chapter 17 *"Data Wrangling in the tidyverse"* on page 265.
- `ggplot2` is a system to create graphics with a philosophy: it adheres to a "Grammar of Graphics" and is able to create really stunning results at a reasonable price (it is a notch more abstract to use than the core-R functionality). For more information, see Chapter 31 *"A Grammar of Graphics with ggplot2"* on page 687. `ggplot2`

  For both reasons, we will talk more about it in the sections about reporting: see Chapter 31 on page 687.
- `readr` expands R's standard[5] functionality to read in rectangular[6] data. `readr`

  It is more robust, knows more data types and is faster than the core-R functionality. For more information, see Chapter 17.1.2 *"Importing Flat Files in the Tidyverse"* on page 267 and its subsections.
- `purrr` is casually mentioned in the section about the OO model in R (see Chapter 6 on page 87), and extensively used in Chapter 25.1 *"Model Quality Measures"* on page 476. `purrr`

  It is a rather complete and consistent set of tools for working with functions and vectors. Using `purrr` it should be possible to replace most loops with call to `purr` functions that will work faster.
- `tibble` is a new take on the data frame of core-R. It provides a new base type: `tibbles`. `tibble`

  Tibbles are in essence data frames, that do a little less (so there is less clutter on the screen and less unexpected things happen), but rather give more feedback (show what went wrong instead of assuming that you have read all manuals and remember everything). Tibbles are introduced in the next section.
- `stringr` expands the standard functions to work with strings and provides a nice coherent set of functions that all start with `str_`. `stringi`

  The package is built on top of `stringi`, which uses the ICU library that is written in C, so it is fast too. For more information, see Chapter 17.5 *"String Manipulation in the tidyverse"* on page 299. `stringr`
- `forcats` provides tools to address common problems when working with categorical variables[7]. `forcats`

7.2.2 The Non-core Tidyverse

Besides the core tidyverse packages – that are loaded with the command `library(tidyverse)`, there are many other packages that are part of the tidyverse. In this section we will describe briefly the most important ones.

---

[5]The standard functions to read in data are covered in Section 4.8 *"Selected Data Interfaces"* on page 75.

[6]Rectangular data is data that – when printed – looks like a rectangle, for example movies and pictures are not rectangular data, while a CSV file or a database table are rectangular data.

[7]Categorical variables are variables that have a fixed and known set of possible values. These values might or might not have a (strict) order relation. For example, "sex" (M or F) would not have an order, but salary brackets might have.

readxl
xlsx
xls

- **Importing data:** `readxl` for .xls and .xlsx files) and `haven` for SPSS, Stata, and SAS data.[8]

lubridate
hms
blob

- **Wrangling data:** `lubridate` for dates and date-times, `hms` for time-of-day values, `blob` for storing binary data. `lubridate` –for example – is discussed in Chapter 17.6 *"Dates with lubridate"* on page 314.

purrr
magrittr
paste()
glue

- **Programming:** `purrr` for iterating within R objects, `magrittr` provides the famous pipe, `%>%` command plus some more specialised piping operators (like `%$%` and `%<>%`), and `glue` provides an enhancement to the `paste()` function.

recipes
rsample
modelr
broom

- **Modelling:** this is not really ready, though `recipes` and `rsample` are already operational and show the direction this is taking. The aim is to replace `modelr`[9]. Note that there is also the package `broom` that turns models into tidy data.

Warning – Work in progress

While the core-tidyverse is stable, the packages that are not core tend still to change and improve. Check their online documentation when using them.

[8] Of course, if you need something else you will want to use the package that does exactly what you want. Here are some good ones that adhere largely to the tidyverse philosophy: jsonlite for JSON, xml2 for XML, httr for web APIs, rvest for web scraping, DBI for relational databases — a good resources is http://db.rstudio.com.

[9] The lack of coherent support for the modelling and reporting area makes clear that the tidyverse is not yet a candidate to service the whole development cycle of the company yet. Modelling departments might want to have a look at the `tidymodels` package.

tidymodels

7.3 Working with the Tidyverse

7.3.1 Tibbles

Tibbles are in many aspects a special type of data frames. The do the same as data frames (i.e. store rectangular data), but they have some advantages.

Let us dive in and create a tibble. Imagine for example that we want to show the sum of the sine and cosine functions. The output of the code below is in Figure 7.1 on this page.

```
x <- seq(from = 0, to = 2 * pi, length.out = 100)
s <- sin(x)
c <- cos(x)
z <- s + c
plot(x, z, type = "l",col="red", lwd=7)
lines(x, c, col = "blue",  lwd = 1.5)
lines(x, s, col = "darkolivegreen", lwd = 1.5)
```

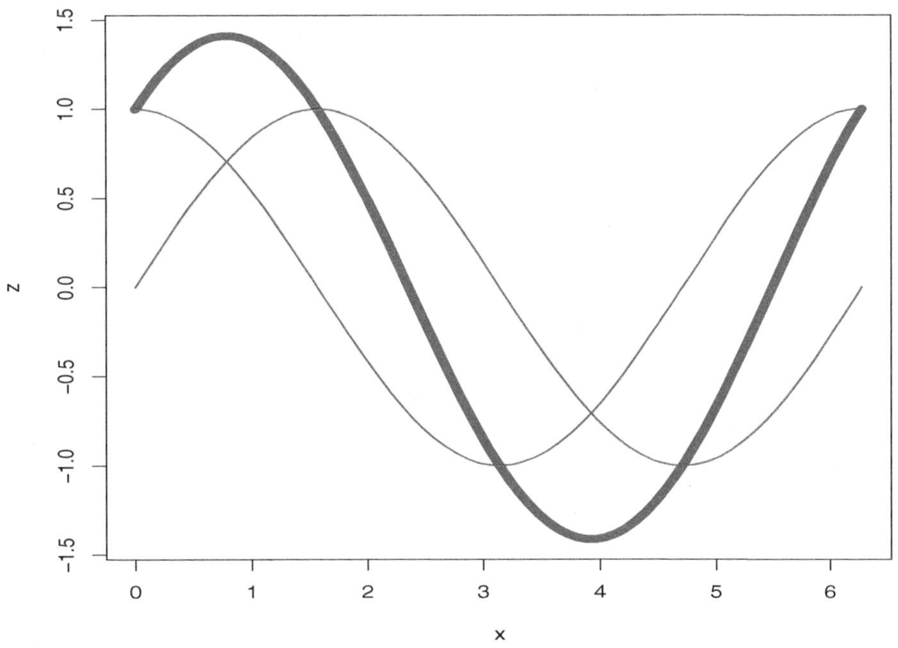

Figure 7.1: *The sum of sine and cosine illustrated.*

Imagine further that our purpose is not only to plot these functions, but to use them in other applications. Then it would make sense to put them in a data, frame. The following code does exactly the same using a data frame.

```
x <- seq(from = 0, to = 2 * pi, length.out = 100)
#df <- as.data.frame((x))
df <- rbind(as.data.frame((x)),cos(x),sin(x), cos(x) + sin(x))
# plot etc.
```

This is already more concise. With the tidyverse, it would look as follows (still without using the piping):

```
library(tidyverse)
x <- seq(from = 0, to = 2 * pi, length.out = 100)
tb <- tibble(x, sin(x), cos(x), cos(x) + sin(x))
```

The code below first prints the tibble in the console and then plots the results in Figure 7.2 on this page.

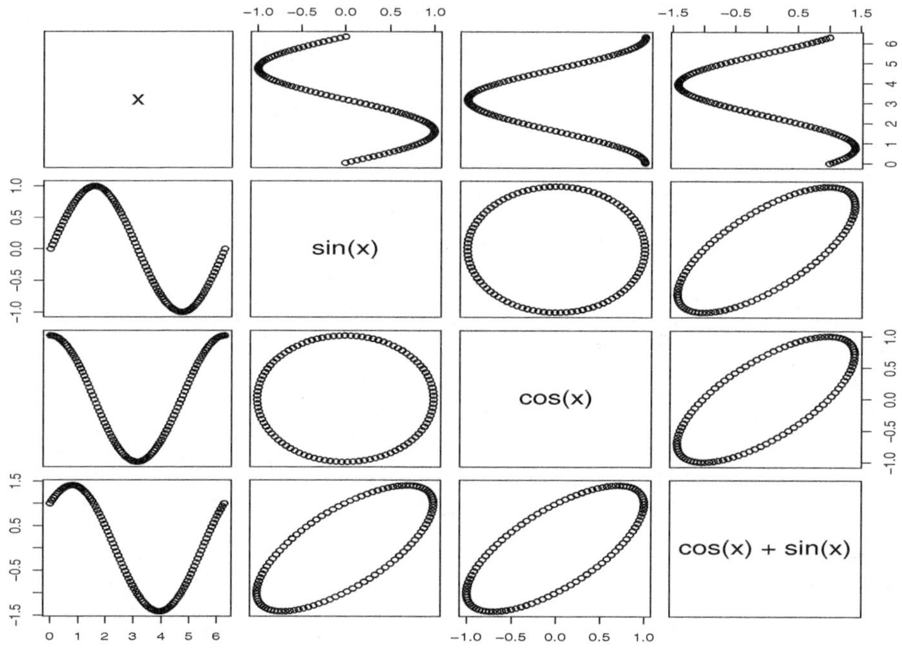

Figure 7.2: *A tibble plots itself like a data-frame.*

The code with a tibble is just a notch shorter, but that is not the point here. The main advantage in using a tibble is that it will usually do things that make more sense for the modern R-user. For example, consider how a tibble prints itself (compared to what a data frame does).

```
# Note how concise and relevant the output is:
print(tb)
## # A tibble: 100 x 4
##          x `sin(x)` `cos(x)` `cos(x) + sin(x)`
##      <dbl>     <dbl>     <dbl>             <dbl>
##  1 0         0         1                 1
##  2 0.0635    0.0634    0.998             1.06
##  3 0.127     0.127     0.992             1.12
##  4 0.190     0.189     0.982             1.17
##  5 0.254     0.251     0.968             1.22
##  6 0.317     0.312     0.950             1.26
##  7 0.381     0.372     0.928             1.30
##  8 0.444     0.430     0.903             1.33
##  9 0.508     0.486     0.874             1.36
## 10 0.571     0.541     0.841             1.38
## # ... with 90 more rows

# This does the same as for a data-frame:
plot(tb)
```

```
# Actually a tibble will still behave as a data frame:
is.data.frame(tb)
## [1] TRUE
```

Digression – Special characters in column names

Note the back-ticks in `` `sin(x)` `` when the tibble reports on itself. That is of course because in R variables are not allowed to use brackets in their names. The tibble does allow in the names of columns non-R-compliant variable names. To address this column by name, we need to refer to the column by its number or use back-ticks.

```
tb$`sin(x)`[1]
## [1] 0
```

This convention is not specific to tibbles, it is used throughout R (e.g. the same back-ticks are needed in ggplot2, tidyr, dyplr, etc.).

Hint

Be aware of the saying “They have to recognize that great responsibility is an inevitable consequence of great power.”[a] It is not because you can do something that you must. Indeed, you can use a numeric column names in a tibble and the following is valid code.

```
tb <- tibble(`1` = 1:3, `2` = sin(`1`), `1`*pi, 1*pi)
tb
## # A tibble: 3 x 4
##     `1`   `2` `\`1\` * pi` `1 * pi`
##   <int> <dbl>        <dbl>    <dbl>
## 1     1 0.841         3.14     3.14
## 2     2 0.909         6.28     3.14
## 3     3 0.141         9.42     3.14
```

However, is this good practice?

[a]This quote is generally attributed to the Voltaire (pen-name of Jean-Marie Arouet; 1694–1778) and is published in the French National Convention of 8 May, 1793 (see con (1793) – page 72). After that many leaders and writers of comic books have used many variants of this phrase.

So, why use a tibble instead of a data frame?

1. It will do less things (such as changing strings into factors, creating row names, change names of variables, no partial matching, but a warning message when you try to access a column that does not exist, etc.).

2. A tibble will report more errors instead of doing something silently (data type conversions, import, etc.), so they are safer to use.

3. The specific print function for the tibble, `print.tibble()`, will not overrun your screen with thousands of lines, it reports only on the ten first. If you need to see all columns, then the traditional `head(tibble)` will still work, or you can tweak the behaviour of the print function via the function `options()`.

`print()` `head()`

4. The name of the class itself is not confusing. Where the function `print.data.frame()` potentially can be the specific method for the `print` function for a `data.frame`, it can also be the specific method for the `print.data` function for a `frame` object. The name of the class `tibble` does not use the dot and hence cannot be confusing.

To illustrate some of these differences, consider the following code:

```
# -- data frame --
df <- data.frame("value" = pi, "name" = "pi")
df$na          # partial matching of column names
## [1] pi
## Levels: pi

# automatic conversion to factor, plus data frame
# accepts strings:
df[,"name"]
## [1] pi
## Levels: pi

df[,c("name", "value")]
##   name    value
## 1   pi 3.141593

# -- tibble --
df <- tibble("value" = pi, "name" = "pi")
df$name        # column name
## [1] "pi"

df$nam         # no partial matching but error msg.

## Warning: Unknown or uninitialised column: 'nam'.
## NULL

df[,"name"]   # this returns a tibble (no simplification)
## # A tibble: 1 x 1
##   name
##   <chr>
## 1 pi

df[,c("name", "value")] # no conversion to factor
## # A tibble: 1 x 2
##   name  value
##   <chr> <dbl>
## 1 pi     3.14
```

This partial matching is one of the nicer functions of R, and certainly was an advantage for interactive use. However when using R in batch mode, this might be dangerous. Partial matching is especially dangerous in a corporate environment: datasets can have hundreds of columns and many names look alike, e.g. BAL180801, BAL180802, and BAL180803. Till a certain point it is safe to use partial matching since it will only work when R is sure that it can identify the variable uniquely. But it is bound to happen that you create new rows and suddenly someone else's code will stop working (because now R got confused).

Digression – Changing how a tibble is printed

To adjust the default behaviour of print on a tibble, run the function `options` as follows:

options()

```
options(
  tibble.print_max=n, # If there are more than n
  tibble.print_min=m, # rows, only print the m first
                      # (set n to Inf to show  all)
  tibble.width = l    # max nbr of columns to print
                      # (set to Inf to show  all)
  )
```

Tibbles are also data frames, and most older functions – that are unaware of tibbles – will work just fine. However, it may happen that some function would not work. If that happens, it is possible to coerce the tibble back into data frame with the function `as.data.frame()`.

```
tb <- tibble(c("a", "b", "c"), c(1,2,3), 9L,9)
is.data.frame(tb)
## [1] TRUE

# Note also that tibble did no conversion to factors, and
# note that the tibble also recycles the scalars:
tb
## # A tibble: 3 x 4
##   `c("a", "b", "c")` `c(1, 2, 3)`  `9L`   `9`
##   <chr>                     <dbl> <int> <dbl>
## 1 a                             1     9     9
## 2 b                             2     9     9
## 3 c                             3     9     9

# Coerce the tibble to data-frame:
as.data.frame(tb)
##   c("a", "b", "c") c(1, 2, 3) 9L 9
## 1                a          1  9 9
## 2                b          2  9 9
## 3                c          3  9 9

# A tibble does not recycle shorter vectors, so this fails:
fail <- tibble(c("a", "b", "c"), c(1,2))

## Error: Tibble columns must have consistent lengths, only values of length one are
recycled:
## * Length 2: Column 'c(1, 2)'
## * Length 3: Column 'c("a", "b", "c")'

# That is a major advantage and will save many programming errors.
```

Hint – Viewing the content of a tibble

The function `view(tibble)` works as expected and is most useful when working with RStudio where it will open the tibble in a special tab.

While on the surface a tibble does the same as a data.frame, they have some crucial advantages and we warmly recommend to use them.

7.3.2 Piping with R

pipe
magrittr
%>%

This section is not about creating beautiful music, it explains an argument passing system in R. Similar to the pipe in Linux, the pipe operator, `|`, the operator `%>%` from the package `magrittr` allows to pass the output of one line to the first argument of the function on the next line.[10]

When writing code, it is common to work on one object for a while. For example, when we need to import data, then work with that data to clean it, add columns, delete some, summarize data, etc.

To start, consider a simple example:

```
t <- tibble("x" = runif(10))
t <- within(t, y <- 2 * x + 4 + rnorm(10, mean=0,sd=0.5))
```

This can also be written with the piping operator from `magrittr`

```
t <- tibble("x" = runif(10))  %>%
     within(y <- 2 * x + 4 + rnorm(10, mean=0,sd=0.5))
```

What R does behind the scenes, is feeding the output left of the pipe operator as main input right of the pipe operator. This means that the following are equivalent:

```
# 1. pipe:
a %>% f()
# 2. pipe with shortened function:
a %>% f
# 3. is equivalent with:
f(a)
```

Example: – Pipe operator

```
a <- c(1:10)
a %>% mean()
## [1] 5.5

a %>% mean
## [1] 5.5

mean(a)
## [1] 5.5
```

Hint – Pronouncing the pipe

It might be useful to pronounce the pipe operator, `%>%` as "then" to understand what it does.

[10] R's piping operator is very similar to the piping command that you might know from the most of the CLI shells of popular *nix systems where messages like the following can go a long way: `dmesg | grep "Bluetooth"`, though differences will appear in more complicated commands.

Note – Equivalence of piping and nesting

```
# The following line
c <- a    %>%
     f()
# is equivalent with:
c <- f(a)

# Also, it is easy to see that
x <- a %>% f(y) %>% g(z)
# is the same as:
x <- g(f(a, y), z)
```

7.3.3 Attention Points When Using the Pipe

This construct will get into problems for functions that use lazy evaluation. Lazy evaluation is a feature of R that is introduced in R to make it faster in interactive mode. This means that those functions will only calculate their arguments when they are really needed. There is of course a good reason why those functions have lazy evaluation and the reader will not be surprised that they cannot be used in a pipe. So there are many functions that use lazy evaluation, but most notably are the error handlers. These are functions that try to do something, but when an error is thrown or a warning message is generated, they will hand it over to the relevant handler. Examples are `try`, `tryCatch`, etc. We do not really discuss error handling in any other parts of this book, so here is a quick primer.

`try()`
`tryCatch()`

handler

```
# f1
# Dummy function that from which only the error throwing part
# is shown.
f1 <- function() {
    # Here goes the long code that might be doing something risky
    # (e.g. connecting to a database, uploading file, etc.)
    # and finally, if it goes wrong:
    stop("Early exit from f1!")  # throw error
    }

tryCatch(f1(),     # the function to try
         error   = function(e) {paste("_ERROR_:",e)},
         warning = function(w) {paste("_WARNING_:",w)},
         message = function(m) {paste("_MESSSAGE_:",m)},
         finally="Last command"     # do at the end
         )
## [1] "_ERROR_: Error in f1(): Early exit from f1!\n"
```

As can be understood from the example above, the error handler should not be evaluated if f1 does not throw an error. That is why they use error handling. So the following will not work:

```
# f1
# Dummy function that from which only the error throwing part
# is shown.
f1 <- function() {
    # Here goes the long code that might be doing something risky
    # (e.g. connecting to a database, uploading file, etc.)
```

```
    # and finally, if it goes wrong:
    stop("Early exit from f1!")  # something went wrong
    }   %>%
tryCatch(
        error   = function(e) {paste("_ERROR_:",e)},
        warning = function(w) {paste("_WARNING_:",w)},
        message = function(m) {paste("_MESSSAGE_:",m)},
        finally="Last command"     # do at the end
        )
# Note that it fails in silence.
```

> **Further information – Error catching**
>
> There is a lot more to error catching than meets the eye here. We recommend to read the documentation of the relevant functions carefully. Another good place to start is "Advanced R," page 163, Wickham (2014).

Another issue when using the pipe operator `%>%` occurs when functions use explicitely the current environment. In those functions, one will have to be explicit which environment to use. More about environments and scoping can be found in Chapter 5 on page 81.

7.3.4 Advanced Piping

7.3.4.1 The Dollar Pipe

Below we create random data that has a linear dependency and try to fit a linear model on that data.[11]

```
# This will not work, because lm() is not designed for the pipe.
lm1 <- tibble("x" = runif(10))                              %>%
       within(y <- 2 * x + 4 + rnorm(10, mean=0, sd=0.5)) %>%
       lm(y ~ x)

## Error in as.data.frame.default(data): cannot coerce class ""formula"" to a data.frame
```

The aforementioned code fails. This is because R will not automatically add something like `data = t` and use the "t" as far as defined till the line before. The function `lm()` expects as first argument the formula, where the pipe command would put the data in the first argument. Therefore, `magrittr` provides a special pipe operator that basically passes on the variables of the data frame of the line before, so that they can be addressed directly: the `%$%`.

```
# The Tidyverse only makes the %>% pipe available. So, to use the
# special pipes, we need to load magrittr
library(magrittr)

##
## Attaching package: 'magrittr'
## The following object is masked from 'package:purrr':
##
##     set_names
## The following object is masked from 'package:tidyr':
##
##     extract
```

[11] The function `lm()` generates a linear model in R of the form $y = a_0 + \sum_i^n a_i x_i$. More information can be found in Section 21.1 *"Linear Regression"* on page 375. The functions `summary()` and `coefficients()` that are used on the following pages are also explained there.

```
lm2 <- tibble("x" = runif(10))                              %>%
       within(y <- 2 * x + 4 + rnorm(10, mean=0,sd=0.5)) %$%
       lm(y ~ x)
summary(lm2)
##
## Call:
## lm(formula = y ~ x)
##
## Residuals:
##     Min      1Q  Median      3Q     Max
## -0.6101 -0.3534 -0.1390  0.2685  0.8798
##
## Coefficients:
##             Estimate Std. Error t value Pr(>|t|)
## (Intercept)   4.0770     0.3109  13.115 1.09e-06 ***
## x             2.2068     0.5308   4.158  0.00317 **
## ---
## Signif. codes:
## 0 '***' 0.001 '**' 0.01 '*' 0.05 '.' 0.1 ' ' 1
##
## Residual standard error: 0.5171 on 8 degrees of freedom
## Multiple R-squared:  0.6836,Adjusted R-squared:  0.6441
## F-statistic: 17.29 on 1 and 8 DF,  p-value: 0.003174
```

This can be elaborated further:

```
coeff <- tibble("x" = runif(10))                              %>%
         within(y <- 2 * x + 4 + rnorm(10, mean=0,sd=0.5)) %$%
         lm(y ~ x)                                            %>%
         summary                                              %>%
         coefficients
coeff
##             Estimate Std. Error   t value     Pr(>|t|)
## (Intercept) 4.131934  0.2077024 19.893534 4.248422e-08
## x           1.743997  0.3390430  5.143882 8.809194e-04
```

Note – Using functions without brackets

Note how we can omit the brackets for functions that do not take any argument.

7.3.4.2 The T-Pipe

This works nice, but now imagine that we want to keep "t" as the tibble, but still add some operations on it – for example plot it. In that case, there is the special `%T>%` "T-pipe" that will rather pass on the left side of the expression than the right side. The output of the code below is the plot in Figure 7.3 on page 136.

```
library(magrittr)
t <- tibble("x" = runif(100))                              %>%
     within(y <- 2 * x + 4 + rnorm(10, mean=0, sd=0.5)) %T>%
     plot(col="red")    # The function plot does not return anything
                        # so we used the %T>% pipe.

lm3 <-   t                    %$%
         lm(y ~ x)            %T>% # pass on the linear model
         summary              %T>% # further pass on the linear model
```

```
        coefficients

tcoef <- lm3 %>% coefficients  # we anyhow need the coefficients

# Add the model (the solid line) to the previous plot:
abline(a = tcoef[1], b=tcoef[2], col="blue", lwd=3)
```

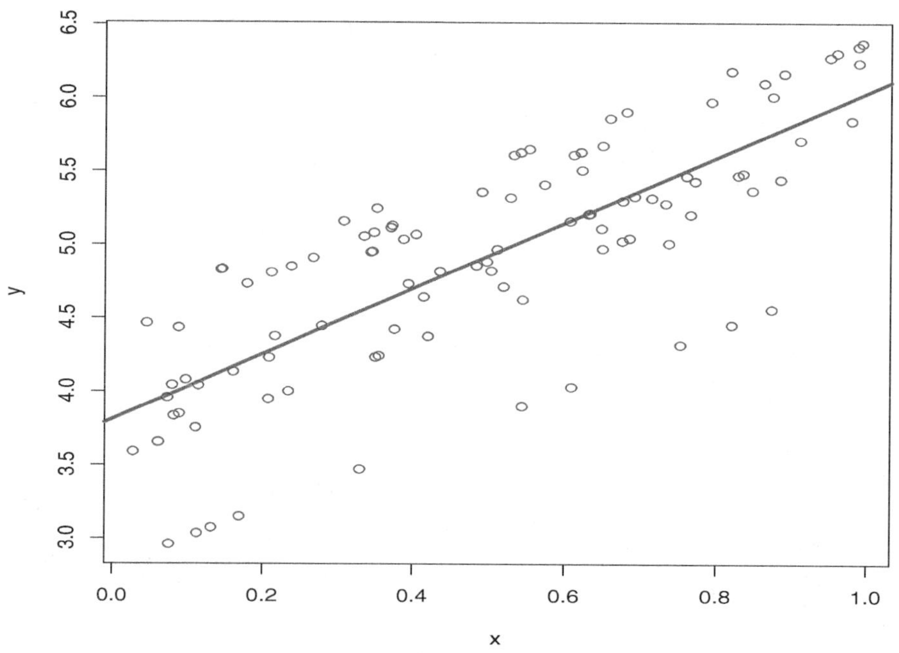

Figure 7.3: *A linear model fit on generated data to illustrate the piping command.*

7.3.4.3 The Assignment Pipe

This last variation of the pipe operator allows us to simplify the first line, by providing an assignment with a special piping operator.

```
x <- c(1,2,3)

# The following line:
x <- x %>% mean

# is equivalent with the following:
x %<>% mean

# Show x:
x
## [1] 2
```

Note that the original meaning of "x" is gone.

Warning – Assignment pipe

We recommend to use this pipe operator only when no confusion is possible. We also argue that this pipe operator makes code less readable, while not really making the code shorter.

7.3.5 Conclusion

When you come from a background of compiled languages that provides fine graded control over memory management (such as C or C++), you might not directly see the need for pipes that much. However, it does reduce the amount of text that needs to be typed and makes the code more readable.

Indeed, the piping operator will not provide a speed increase nor memory advantage even if we would create a new variable at every line. R has a pretty good memory management and it does only copy columns when they are really modified. For example, have a careful look at the following:

```
library(pryr)
x <- runif(100)
object_size(x)
## 840 B

y <- x

# x and y together do not take more memory than only x.
object_size(x,y)
## 840 B

y <- y * 2

# Now, they are different and are stored separately in memory.
object_size(x,y)
## 1.68 kB
```

The piping operator can be confusing at first and is not really necessary (unless to read code that is using it). However, it has the advantage to make code more readable – once used to it – and it also makes code shorter. Finally, it allows the reader of the code to focus more on what is going on (the actions instead of the data, since that is passed over invisibly).

Hint – Use pipes sparingly

Pipes are as spices in the kitchen. Use them, but do so with moderation. A good rule of thumb is that five lines is enough, and simple one-line commands do not need to be broken down in more lines in order to use a pipe.

♣ 8 ♣

Elements of Descriptive Statistics

statistics

8.1 Measures of Central Tendency

central tendency
measure – central tendency

A measure of central tendency is a single value that attempts to describe a set of data by identifying the central position within that set of data. As such, measures of central tendency are sometimes called measures of central location. They are also classed as summary statistics. The mean (often called the average) is most likely the measure of central tendency that you are most familiar with, but there are others, such as the median and the mode.

The mean, median, and mode are all valid measures of central tendency, but under different conditions, some measures of central tendency become more appropriate to use than others. In the following sections, we will look at the mean, mode, and median, and learn how to calculate them and under what conditions they are most appropriate to be used.

8.1.1 Mean

mean

central tendency – mean

Probably the most used measure of central tendency is the "mean." In this section we will start from the arithmetic mean, but illustrate some other concepts that might be more suited in some situations too.

8.1.1.1 The Arithmetic Mean

mean – arithmetic

mean – arithmetic

The most popular type of mean is the "arithmetic mean." It is the average of a set of numerical values; and it is calculated by adding those values first together and then dividing by the number of values in the aforementioned set.

Definition: Arithmetic mean

$$\begin{aligned}\bar{x} &= \sum_{n=1}^{N} P(x).x && \text{(for discrete distributions)}\\ &= \int_{-\infty}^{+\infty} x.f(x)\,\mathrm{d}x && \text{(for continuous distributions)}\end{aligned}$$

The Big R-Book: From Data Science to Learning Machines and Big Data, First Edition. Philippe J.S. De Brouwer.

Companion Website: www.wiley.com/go/De Brouwer/The Big R-Book

$\bar{x}$
mean
P()
probability
probability
f()
probability density function
`mean()`

The unbiased estimator of the mean for K observations x_k is:

$$E[\bar{x}] = \frac{1}{K}\sum_{k=1}^{K} x_k$$

Not surprisingly, the arithmetic mean in R is calculated by the function `mean()`.

This is a dispatcher function[1] and it will work in a meaningful way for a variety of objects, such as vectors, matrices, etc.

```
# The mean of a vector:
x <- c(1,2,3,4,5,60)
mean(x)
## [1] 12.5

# Missing values will block the override the result:
x <- c(1,2,3,4,5,60,NA)
mean(x)
## [1] NA

# Missing values can be ignored with na.rm = TRUE:
mean(x, na.rm = TRUE)
## [1] 12.5

# This works also for a matrix:
M <- matrix(c(1,2,3,4,5,60), nrow=3)
mean(M)
## [1] 12.5
```

Hint – Outliers

The mean is highly influenced by the outliers. To mitigate this to some extend the parameter `trim` allows to remove the tails. It will sort all values and then remove the x% smallest and x% largest observations.

```
v <- c(1,2,3,4,5,6000)
mean(v)
## [1] 1002.5

mean(v, trim = 0.2)
## [1] 3.5
```

mean – generalized

8.1.1.2 Generalised Means

More generally, a mean can be defined as follows:

Definition: f-mean

$$\bar{x} = f^{-1}\left(\frac{1}{n} . \sum_{k=1}^{K} f(x_k)\right)$$

[1]More information about the concept “dispatcher function” is in Chapter 6 *“The Implementation of OO”* on page 87.

Popular choices for `f()` are:

- $f(x) = x$: arithmetic mean,
- $f(x) = \frac{1}{x}$: harmonic mean,
- $f(x) = x^m$: power mean,
- $f(x) = \ln x$: geometric mean, so $\bar{x} = \left(\prod_{k=1}^{K} x_k\right)^{\frac{1}{K}}$

arithmetic mean
mean – harmonic
harmonic mean

mean – power
power mean

mean – geometric
geometric mean

The Power Mean

One particular generalized mean is the power mean or Hölder mean. It is defined for a set of K positive numbers x_k by

holder mean
mean – holder

$$\bar{x}(m) = \left(\frac{1}{n} \cdot \sum_{k=1}^{K} x_k^m\right)^{\frac{1}{m}}$$

by choosing particular values for m one can get the quadratic, arithmetic, geometric and harmonic means.

mean – quadratic

- $m \to \infty$: maximum of x_k
- $m = 2$: quadratic mean
- $m = 1$: arithmetic mean
- $m \to 0$: geometric mean
- $m = 1$: harmonic mean
- $m \to -\infty$: minimum of x_k

Example: Which mean makes most sense?

What is the average return when you know that the share price had the following returns: $-50\%, +50\%, -50\%, +50\%$. Try the arithmetic mean and the mean of the log-returns.

```
returns <- c(0.5,-0.5,0.5,-0.5)

# Arithmetic mean:
aritmean <- mean(returns)

# The ln-mean:
log_returns <- returns
for(k in 1:length(returns)) {
  log_returns[k] <- log( returns[k] + 1)
  }
logmean <- mean(log_returns)
exp(logmean) - 1
## [1] -0.1339746

# What is the value of the investment after these returns:
V_0 <- 1
V_T <- V_0
for(k in 1:length(returns)) {
  V_T <- V_T * (returns[k] + 1)
  }
V_T
## [1] 0.5625

# Compare this to our predictions:
## mean of log-returns
V_0 * (exp(logmean) - 1)
## [1] -0.1339746

## mean of returns
V_0 * (aritmean + 1)
## [1] 1
```

median

8.1.2 The Median

central tendency – median

While the mean (and the average in particular) is widely used, it is actually quite vulnerable to outliers. It would therefore, make sense to have a measure that is less influenced by the outliers and rather answers the question: what would a typical observation look like. The median is such measure.

The median is the middle-value so that 50% of the observations are lower and 50% are higher.

```
x <- c(1:5,5e10,NA)
x
## [1] 1e+00 2e+00 3e+00 4e+00 5e+00 5e+10    NA

median(x)              # no meaningful result with NAs
## [1] NA

median(x,na.rm = TRUE) # ignore the NA
## [1] 3.5
```

```
# Note how the median is not impacted by the outlier,
# but the outlier dominates the mean:
mean(x, na.rm = TRUE)
## [1] 8333333336
```

8.1.3 The Mode

mode
central tendency – mode

The mode is the value that has highest probability to occur. For a series of observations, this should be the one that occurs most often. Note that the mode is also defined for variables that have no order-relation (even labels such as "green," "yellow," etc. have a mode, but not a mean or median — without further abstraction or a numerical representation).

In R, the function `mode()` or `storage.mode()` returns a character string describing how a variable is stored. In fact, R does not have a standard function to calculate mode, so let us create our own:

`mode()`
`storage.mode()`

`unique()`
Linux
FreeBSD
`tabulate()`
`uniqv()`

```
# my_mode
# Finds the first mode (only one)
# Arguments:
#   v -- numeric vector or factor
# Returns:
#   the first mode
my_mode <- function(v) {
   uniqv <- unique(v)
   tabv  <- tabulate(match(v, uniqv))
   uniqv[which.max(tabv)]
   }

# now test this function
x <- c(1,2,3,3,4,5,60,NA)
my_mode(x)
## [1] 3

x1 <- c("relevant", "N/A", "undesired", "great", "N/A",
        "undesired", "great", "great")
my_mode(x1)
## [1] "great"

# text from https://www.r-project.org/about.html
t <- "R is available as Free Software under the terms of the
Free Software Foundation's GNU General Public License in
source code form. It compiles and runs on a wide variety of
UNIX platforms and similar systems (including FreeBSD and
Linux), Windows and MacOS."
v <- unlist(strsplit(t,split=" "))
my_mode(v)
## [1] "and"
```

While this function works fine on the examples provided, it only returns the first mode encountered. In general, however, the mode is not necessarily unique and it might make sense to return them all. This can be done by modifying the code as follows:

```
# my_mode
# Finds the mode(s) of a vector v
# Arguments:
#   v -- numeric vector or factor
#   return.all -- boolean -- set to true to return all modes
```

```
# Returns:
#   the modal elements
my_mode <- function(v, return.all = FALSE) {
  uniqv  <- unique(v)
  tabv   <- tabulate(match(v, uniqv))
  if (return.all) {
    uniqv[tabv == max(tabv)]
  } else {
    uniqv[which.max(tabv)]
  }
}

# example:
x <- c(1,2,2,3,3,4,5)
my_mode(x)
## [1] 2

my_mode(x, return.all = TRUE)
## [1] 2 3
```

Hint – Use default values to keep code backwards compatible

We were confident that it was fine to over-ride the definition of the function my_mode. Indeed, if the function was already used in some older code, then one would expect to see only one mode appear. That behaviour is still the same, because we chose the default value for the optional parameter `return.all` to be `FALSE`. If the default choice would be `TRUE`, then older code would produce wrong results and if we would not use a default value, then older code would fail to run.

8.2 Measures of Variation or Spread

measures of spread

Variation or spread measures how different observations are compared to the mean or other central measure. If variation is small, one can expect observations to be closer to each other.

Definition: Variance

variance

$$\mathrm{VAR}(X) = E\left[(X - \bar{X})^2\right]$$

Standard Deviation

Definition: Standard deviation

spread – standard deviation
standard deviation

$$\mathrm{SD}(X) = \sqrt{\mathrm{VAR}(X)}$$

The estimator for standard deviation is:

$$\widehat{\mathrm{SD}(X)} = \sqrt{\frac{1}{N-1}\sum_{n=1}^{N}(X_n - \bar{X})^2}$$

```
t <- rnorm(100, mean=0, sd=20)
var(t)
## [1] 248.2647
```

sd()

```
sd(t)
## [1] 15.75642

sqrt(var(t))
## [1] 15.75642

sqrt(sum((t - mean(t))^2)/(length(t) - 1))
## [1] 15.75642
```

Median absolute deviation

Definition: mad

mad
median absolute deviation

$$\mathrm{mad}(X) = \frac{1}{1.4826}\mathrm{median}\left(|X - \mathrm{median}(X)|\right)$$

mad()

```
mad(t)
## [1] 14.54922

mad(t,constant=1)
## [1] 9.813314
```

The default “constant = 1.4826” (approximately $\frac{1}{\Phi^{-1}(\frac{3}{4})} = \frac{1}{\text{qnorm}(3/4)}$ ensures consistency, i.e.,

$$E[\text{mad}(X_1, ..., X_n)] = \sigma$$

for X_i distributed as N(μ, σ^2) and large n.

8.3 Measures of Covariation

covariation

When there is more than one variable, it is useful to understand what the interdependencies of variables are. For example when measuring the size of peoples hands and their length, one can expect that people with larger hands on average are taller than people with smaller hands. The hand size and length are positively correlated.

The basic measure for linear interdependence is covariance, defined as

$$\begin{aligned}\text{covar}(X) &= E\left[(X - E[X])\,(Y - E[Y])\right] \\ &= E[XY] - E[X]E[Y]\end{aligned}$$

8.3.1 The Pearson Correlation

correlation – Pearson

An important metric for linear relationship is the Pearson correlation coefficient ρ.

Definition: Pearson Correlation Coefficient

$$\begin{aligned}\rho_{XY} &= \frac{\text{covar}(X,Y)}{\sigma_X \sigma_Y} \\ &= \frac{(X - E[X])(Y - E[Y])}{\sqrt{(X - E[X])(Y - E[Y])}} \\ &=: \text{covar}(x,y)\end{aligned}$$

cor()

```
cor(mtcars$hp,mtcars$wt)
## [1] 0.6587479
```

Of course, we also have functions that provide the covariance matrix and functions that convert the one in the other.

var()

cov()

cor()

```
d <- data.frame(mpg = mtcars$mpg, wt = mtcars$wt, hp = mtcars$hp)
# Note that we can feed a whole data-frame in the functions.
var(d)
##             mpg        wt          hp
## mpg   36.324103 -5.116685 -320.73206
## wt    -5.116685  0.957379   44.19266
## hp  -320.732056 44.192661 4700.86694

cov(d)
##             mpg        wt          hp
## mpg   36.324103 -5.116685 -320.73206
## wt    -5.116685  0.957379   44.19266
## hp  -320.732056 44.192661 4700.86694

cor(d)
##            mpg         wt         hp
## mpg  1.0000000 -0.8676594 -0.7761684
## wt  -0.8676594  1.0000000  0.6587479
## hp  -0.7761684  0.6587479  1.0000000
```

cov2cor()

```
cov2cor(cov(d))
##            mpg         wt         hp
## mpg  1.0000000 -0.8676594 -0.7761684
## wt  -0.8676594  1.0000000  0.6587479
## hp  -0.7761684  0.6587479  1.0000000
```

8.3.2 The Spearman Correlation

The measure for correlation, as defined in previous section, actually tests for a linear relation. This means that even the presence of a strong non-linear relationship can go undetected.

```
x  <- c(-10:10)
df <- data.frame(x=x, x_sq=x^2, x_abs=abs(x), x_exp=exp(x))
cor(df)
##              x      x_sq     x_abs     x_exp
## x     1.000000 0.0000000 0.0000000 0.5271730
## x_sq  0.000000 1.0000000 0.9671773 0.5491490
## x_abs 0.000000 0.9671773 1.0000000 0.4663645
## x_exp 0.527173 0.5491490 0.4663645 1.0000000
```

The correlation between x and x^2 is zero, and the correlation between x and $exp(x)$ is a meagre 0.527173.

correlation – Spearman

The Spearman correlation is the correlation applied to the ranks of the data. It is one if an increase in the variable X is always accompanied with an increase in variable Y.

```
cor(rank(df$x), rank(df$x_exp))
## [1] 1
```

The Spearman correlation checks for a relationship that can be more general than only linear. It will be one if X increases when Y increases.

? Question #10

Consider the vectors

1. $x = c(1, 2, 33, 44)$ and $y = c(22, 23, 100, 200)$,
2. $x = c(1 : 10)$ and $y = 2 * x$,
3. $x = c(1 : 10)$ and $y = exp(x)$,

Plot y in function of x. What is their Pearson correlation? What is their Spearman correlation? How do you understand that?

Warning – Correlation is more specific than relation

Not even the Spearman correlation will discover all types of dependencies. Consider the example above with x^2.

```
x  <- c(-10:10)
cor(rank(x), rank(x^2))
## [1] 0
```

8.3.3 Chi-square Tests

Chi-square test is a statistical method to determine if two categorical variables have a significant correlation between them. Both those variables should be from same population, and they should be categorical like "Yes/No," "Male/Female," "Red/Amber/Green," etc.

test – chi square

For example, we can build a dataset with observations on people's ice-cream buying pattern and try to correlate the gender of a person with the flavour of the ice-cream they prefer. If a correlation is found, we can plan for appropriate stock of flavours by knowing the number of gender of people visiting.

Chi-Square test in R

Function use for chisq.test()

`chisq.test(data)`
where `data` is the data in form of a table containing the count value of the variables

For example, we can use the `mtcars` dataset that is most probably loaded when R was initialised.

```
# we use the dataset mtcars from MASS
df <- data.frame(mtcars$cyl,mtcars$am)
chisq.test(df)
```

`chisq.test()`

```
## Warning in chisq.test(df): Chi-squared approximation may be incorrect
##
##      Pearson's Chi-squared test
##
## data:  df
## X-squared = 25.077, df = 31, p-value = 0.7643
```

The chi-square test reports a p-value. This p-value is the probability that the correlations is actually insignificant. It appears that in practice a correlation lower than 5% can be considered as insignificant. In this example, the p-value is higher than 0.05, so there is no significant correlation.

8.4 Distributions

R is a statistical language and most of the work in R will include statistics. Therefore we introduce the reader to how statistical distributions are implemented in R and how they can be used.

The names of the functions related to statistical distributions in R are composed of two sections: the first letter refers to the function (in the following) and the remainder is the distribution name.

pdf
probability density function
cdf
cumulative density function
quantile function
random

- `d`: The pdf (probability density function)
- `p`: The cdf (cumulative probability density function)
- `q`: The quantile function
- `r`: The random number generator.

distribution – normal
distribution – exponential
distribution – log-normal
distribution – logistic
distribution – geometric
distribution – Poisson
distribution – t
distribution – f
distribution – beta
distribution – weibull
distribution – binomial
distribution – negative binomial
distribution – chi-squared
distribution – uniform
distribution – gamma
distribution – cauchy
distribution – hypergeometric

Distribution	R-name	Distribution	R-name
Normal	norm	Weibull	weibull
Exponential	exp	Binomial	binom
Log-normal	lnorm	Negative binomial	nbinom
Logistic	logis	χ^2	chisq
Geometric	geom	Uniform	unif
Poisson	pois	Gamma	gamma
t	t	Cauchy	cauchy
f	f	Hypergeometric	hyper
Beta	beta		

Table 8.1: *Common distributions and their names in R.*

As all distributions work in a very similar way, we use the normal distribution to show how the logic works.

distribution – normal

8.4.1 Normal Distribution

One of the most quintessential distributions is the Gaussian distribution or Normal distribution. Its probability density function resembles a bell. The centre of the curve is the mean of the data set. In the graph, 50% of values lie to the left of the mean and the other 50% lie to the right of the graph.

The Normal Distribution in R

R has four built-in functions to work with the normal distribution. They are described below.

`dnorm()`

- `dnorm(x, mean, sd)`: The height of the probability distribution

`pnorm()`

- `pnorm(x, mean, sd)`: The cumulative distribution function (the probability of the observation to be lower than x)

- `qnorm(p, mean, sd)`: Gives a number whose cumulative value matches the given probability value p qnorm()
- `rnorm(n, mean, sd)`: Generates normally distributed variables, rnorm()

with

- x: A vector of numbers
- p: A vector of probabilities
- n: The number of observations(sample size)
- $mean$: The mean value of the sample data (default is zero)
- sd: The standard deviation (default is 1).

Illustrating the Normal Distribution

In the following example we generate data with the random generator function `rnorm()` and then compare the histogram of that data with the ideal probability density function of the Normal distribution. The output of the following code is Figure 8.1 on this page.

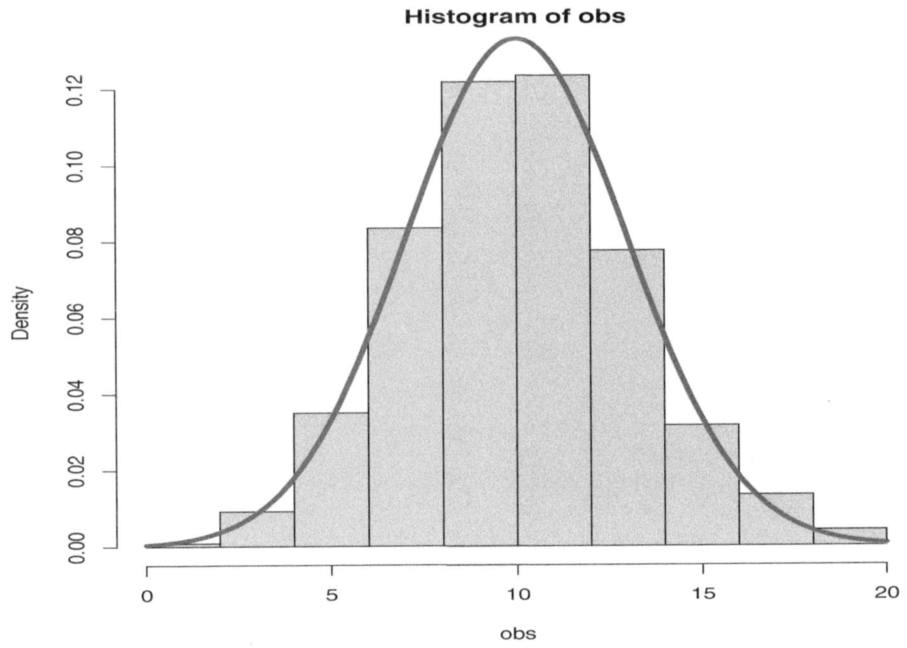

Figure 8.1: *A comparison between a set of random numbers drawn from the normal distribution (khaki) and the theoretical shape of the normal distribution in blue.*

```
obs <- rnorm(600,10,3)
hist(obs,col="khaki3",freq=FALSE)
x <- seq(from=0,to=20,by=0.001)
lines(x, dnorm(x,10,3),col="blue",lwd=4)
```

Case Study: Returns on the Stock Exchange

In this simple illustration, we will compare the returns of the index S&P500 to the Normal distribution. The output of the following code is Figure 8.2 on this page.

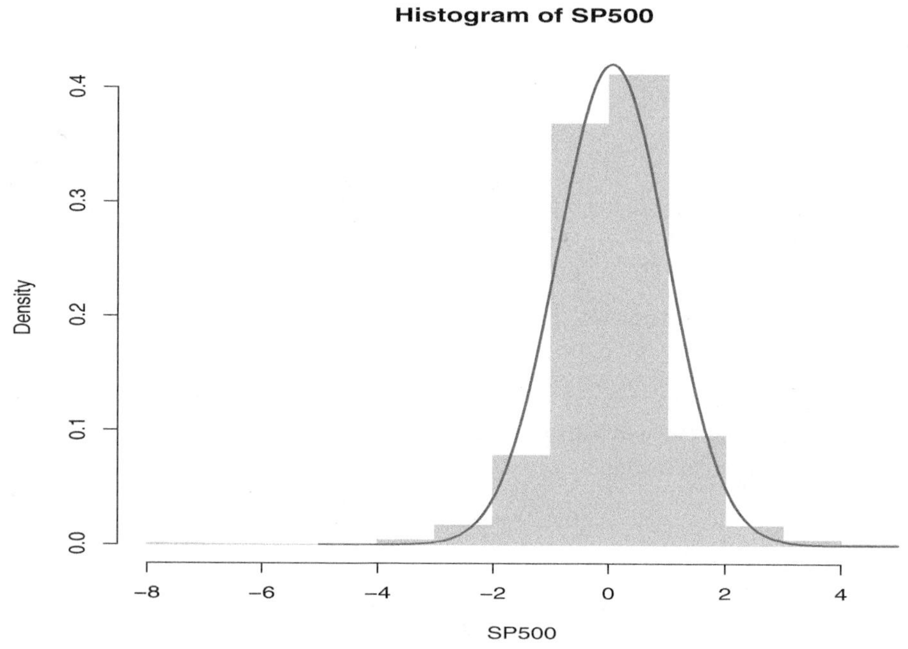

Figure 8.2: *The same plot for the returns of the SP500 index seems acceptable, though there are outliers (where the normal distribution converges fast to zero).*

```
library(MASS)

## ## Attaching package: 'MASS'
## The following object is masked from 'package:dplyr':## ##    select

hist(SP500,col="khaki3",freq=FALSE,border="khaki3")
x <- seq(from=-5,to=5,by=0.001)
lines(x, dnorm(x,mean(SP500),sd(SP500)),col="blue",lwd=2)
```

Q-Q plot

A better way to check for normality is to study the Q-Q plot. A Q-Q plot compares the sample quantiles with the quantiles of the distribution and it makes very clear where deviations appear.

```
library(MASS)
qqnorm(SP500,col="red"); qqline(SP500,col="blue")
```

From the Q-Q plot in Figure 8.3 on page 153 (that is generated by the aforementioned code block), it is clear that the returns of the S&P-500 index are not Normally distributed. Outliers far from the mean appear much more often than the Normal distribution would predict. In other words: returns on stock exchanges have "fat tales."

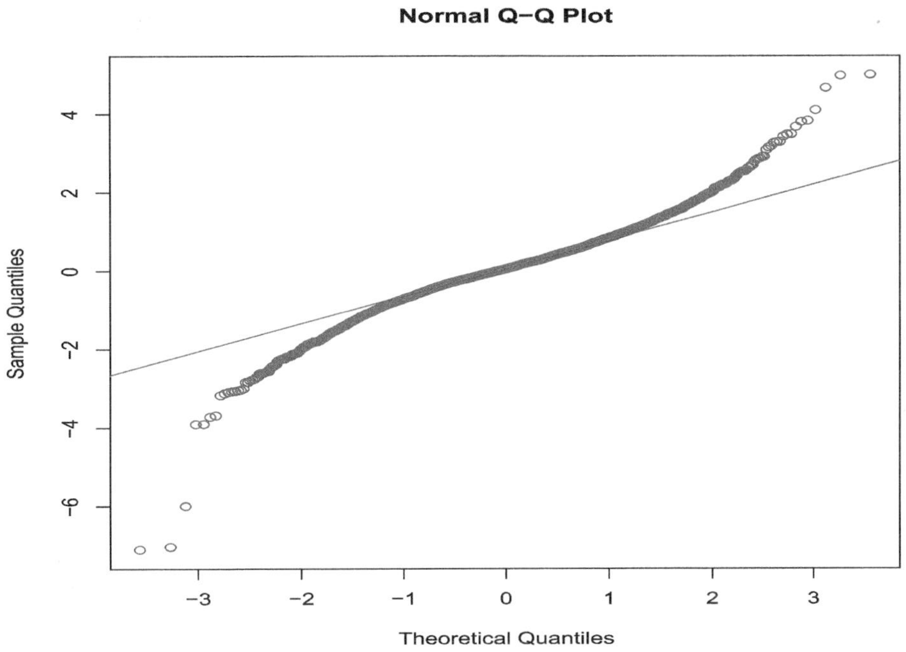

Figure 8.3: *A Q-Q plot is a good way to judge if a set of observations is normally distributed or not.*

8.4.2 Binomial Distribution

The Binomial distribution models the probability of an event which has only two possible outcomes. For example, the probability of finding exactly 6 heads in tossing a coin repeatedly for 10 times is estimated during the binomial distribution.

distribution – binomial

The Binomial Distribution in R

As for all distributions, R has four in-built functions to generate binomial distribution:

dbinom()
pbinom()
dbinom()

- `dbinom(x, size, prob)`: The density function
- `pbinom(x, size, prob)`: The cumulative probability of an event
- `qbinom(p, size, prob)`: Gives a number whose cumulative value matches a given probability value
- `rbinom(n, size, prob)`: Generates random variables following the binomial distribution.

pbinom()
qbinom()
rbinom()

Following parameters are used:

- x: A vector of numbers
- p: A vector of probabilities
- n: The number of observations
- $size$: The number of trials
- $prob$: The probability of success of each trial

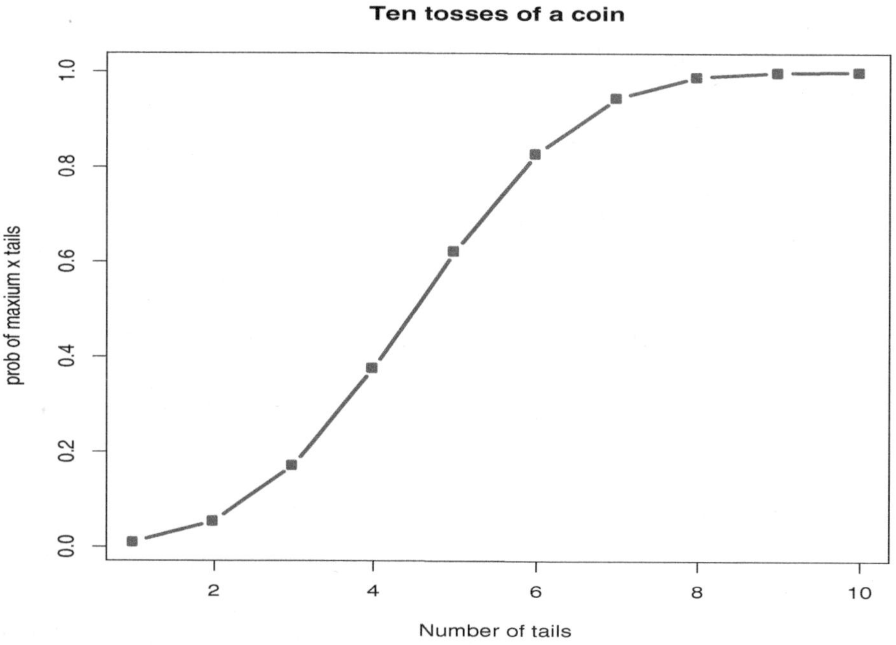

Figure 8.4: *The probability to get maximum x tails when flipping a fair coin, illustrated with the binomial distribution.*

An Example of the Binomial Distribution

The example below illustrates the biniomial distribution and generates the plot in Figure 8.4.

```
# Probability of getting 5 or less heads from 10 tosses of
# a coin.
pbinom(5,10,0.5)
## [1] 0.6230469

# visualize this for one to 10 numbers of tosses
x <- 1:10
y <- pbinom(x,10,0.5)
plot(x,y,type="b",col="blue", lwd=3,
    xlab="Number of tails",
    ylab="prob of maxium x tails",
    main="Ten tosses of a coin")
```

```
# How many heads should we at least expect (with a probability
# of 0.25) when a coin is tossed 10 times.
qbinom(0.25,10,1/2)
## [1] 4
```

Similar to the Normal distribution, random draws of the Binomial distribution can be obtained via a function that starts with the letter 'r': `rbinom()`.

`rbinom()`

```
# Find 20 random numbers of tails from and event of 10 tosses
# of a coin
rbinom(20,10,.5)
##  [1] 5 7 2 6 7 4 6 7 3 2 5 9 5 9 5 5 5 5 5 6
```

8.5 Creating an Overview of Data Characteristics

In the Chapter 4 *"The Basics of R"* on page 21, we presented some of the basic functions of R that – of course – include the some of the most important functions to describe data (such as mean and standard deviation).

Mileage may vary, but in many research people want to document what they have done and will need to include some summary statistics in their paper or model documentation. The standard `summary` of the relevant object might be sufficient.

```
N <- 100
t <- data.frame(id = 1:N, result = rnorm(N))
summary(t)
##        id              result
##  Min.   :  1.00   Min.   :-1.8278
##  1st Qu.: 25.75   1st Qu.:-0.5888
##  Median : 50.50   Median :-0.0487
##  Mean   : 50.50   Mean   :-0.0252
##  3rd Qu.: 75.25   3rd Qu.: 0.4902
##  Max.   :100.00   Max.   : 2.3215
```

This already produces a neat summary that can directly be used in most reports.[2]

Note – A tibble is a special form of data-frame

A tibble and data frame will produce the same summaries.

We might want to produce some specific information that somehow follows the format of the table. To illustrate this, we start from the dataset `mtcars` and assume that we want to make a summary per brand for the top-brands (defined as the most frequent appearing in our database).

```
library(tidyverse) # not only for %>% but also for group_by, etc.
# In mtcars the type of the car is only in the column names,
# so we need to extract it to add it to the data
n <- rownames(mtcars)

# Now, add a column brand (use the first letters of the type)
t <- mtcars %>%
     mutate(brand = str_sub(n, 1, 4))    # add column
```

To achieve this, the function `group_by()` from `dplyr` will be very handy. Note that this function does not change the dataset as such, it rather adds a layer of information about the grouping. `group_by()`

```
# First, we need to find out which are the most abundant brands
# in our dataset (set cutoff at 2: at least 2 cars in database)
top_brands <- count(t, brand) %>% filter(n >= 2)

# top_brands is not simplified to a vector in the tidyverse
print(top_brands)
## # A tibble: 5 x 2
##   brand     n
```

[2]In the sections Chapter 32 *"R Markdown"* on page 699 and Chapter 33 *"knitr and LaTeX"* on page 703 it will be explained how these results from R can directly be used in reports without the need to copy and paste things.

brand	avgDSP	avgCYL	minMPG	medMPG	avgMPG	maxMPG
Fiat	78.9	4.0	27.3	29.85	29.85	32.4
Horn	309.0	7.0	18.7	20.05	20.05	21.4
Mazd	160.0	6.0	21.0	21.00	21.00	21.0
Merc	207.2	6.3	15.2	17.80	19.01	24.4
Toyo	95.6	4.0	21.5	27.70	27.70	33.9

Table 8.2: *Summary information based on the dataset* `mtcars`.

```
##    <chr> <int>
## 1 Fiat       2
## 2 Horn       2
## 3 Mazd       2
## 4 Merc       7
## 5 Toyo       2
```

summarise()

```
grouped_cars <- t                                %>% # start with cars
   filter(brand %in% top_brands$brand) %>% # only top-brands
   group_by(brand)                               %>%
   summarise(
       avgDSP = round(mean(disp), 1),
       avgCYL = round(mean(cyl),  1),
       minMPG = min(mpg),
       medMPG = median(mpg),
       avgMPG = round(mean(mpg),2),
       maxMPG = max(mpg),
     )
print(grouped_cars)
## # A tibble: 5 x 7
##    brand avgDSP avgCYL minMPG medMPG avgMPG maxMPG
##    <chr>  <dbl>  <dbl>  <dbl>  <dbl>  <dbl>  <dbl>
## 1 Fiat     78.8     4     27.3   29.8   29.8   32.4
## 2 Horn    309       7     18.7   20.0   20.0   21.4
## 3 Mazd    160       6     21     21     21     21
## 4 Merc    207.      6.3   15.2   17.8   19.0   24.4
## 5 Toyo     95.6     4     21.5   27.7   27.7   33.9
```

The sections on `knitr` and `rmarkdown` (respectively Chapter 33 on page 703 and Chapter 32 on page 699) will explain how to convert this output via the function `kable()` into Table 8.2.

There are a few things about `group_by()` and `summarise()` that should be noted in order to make working with them easier. For example, `summarize` works opposite to `group_by` and hence will peel back any existing grouping, it is possible to use expression in group by, new groups will preplace by default existing ones, etc. These aspects are illustrated in the following code.

```
# Each call to summarise() removes a layer of grouping:
by_vs_am <- mtcars %>% group_by(vs, am)
by_vs <- by_vs_am %>% summarise(n = n())
by_vs
## # A tibble: 4 x 3
## # Groups:   vs [2]
##       vs    am     n
##    <dbl> <dbl> <int>
## 1      0     0    12
## 2      0     1     6
## 3      1     0     7
## 4      1     1     7
```

```
by_vs %>% summarise(n = sum(n))
## # A tibble: 2 x 2
##      vs     n
##   <dbl> <int>
## 1     0    18
## 2     1    14

# To removing grouping, use ungroup:
by_vs %>%
  ungroup() %>%
  summarise(n = sum(n))
## # A tibble: 1 x 1
##       n
##   <int>
## 1    32

# You can group by expressions: this is just short-hand for
# a mutate/rename followed by a simple group_by:
mtcars %>% group_by(vsam = vs + am)
## # A tibble: 32 x 12
## # Groups:   vsam [3]
##      mpg   cyl  disp    hp  drat    wt  qsec    vs    am
##    <dbl> <dbl> <dbl> <dbl> <dbl> <dbl> <dbl> <dbl> <dbl>
##  1  21       6  160    110  3.9   2.62  16.5     0     1
##  2  21       6  160    110  3.9   2.88  17.0     0     1
##  3  22.8     4  108     93  3.85  2.32  18.6     1     1
##  4  21.4     6  258    110  3.08  3.22  19.4     1     0
##  5  18.7     8  360    175  3.15  3.44  17.0     0     0
##  6  18.1     6  225    105  2.76  3.46  20.2     1     0
##  7  14.3     8  360    245  3.21  3.57  15.8     0     0
##  8  24.4     4  147.    62  3.69  3.19  20       1     0
##  9  22.8     4  141.    95  3.92  3.15  22.9     1     0
## 10  19.2     6  168.   123  3.92  3.44  18.3     1     0
## # ... with 22 more rows, and 3 more variables:
## #   gear <dbl>, carb <dbl>, vsam <dbl>

# By default, group_by overrides existing grouping:
mtcars            %>%
  group_by(cyl)    %>%
  group_by(vs, am) %>%
  group_vars()
## [1] "vs" "am"

# Use add = TRUE to append grouping levels:
mtcars                          %>%
  group_by(cyl)                 %>%
  group_by(vs, am, add = TRUE)  %>%
  group_vars()
## [1] "cyl" "vs"  "am"
```

♣ 9 ♣

Visualisation Methods

This section demonstrates – in no particular order – some of the most useful plotting facilities in R.[1]

The most important function to plot anything is `plot()`. The OO implementation of R is function centric and the plot-function will recognize what object is fed to it and then dispatch to the relevant function. The effect is that you can provide a wide variety of objects to the function `plot()` and that R will "magically" present something that makes sense for that particular object. `plot()`

We illustrate this with the two very basic and simple objects, a vector and a data frame (the plots appear respectively in Figure 9.1 on page 160 and Figure 9.2 on page 160):

```
x <- c(1:20)^2
plot(x)
```

```
df <- data.frame('a' = x, 'b' = 1/x, 'c' = log(x), 'd' = sqrt(x))
plot(df)
```

If the standard plot for your object is not what you need, you can choose one of the many specific plots that we illustrate in the remainder of this chapter.

[1]The modular structure of R allows to build extensions that enhance the functionality of R. The plotting facility has many exentions and arguably the most popular is `ggplot2`, which is described in Chapter 31 *"A Grammar of Graphics with ggplot2"* on page 687.

The Big R-Book: From Data Science to Learning Machines and Big Data, First Edition. Philippe J.S. De Brouwer.

Companion Website: www.wiley.com/go/De Brouwer/The Big R-Book

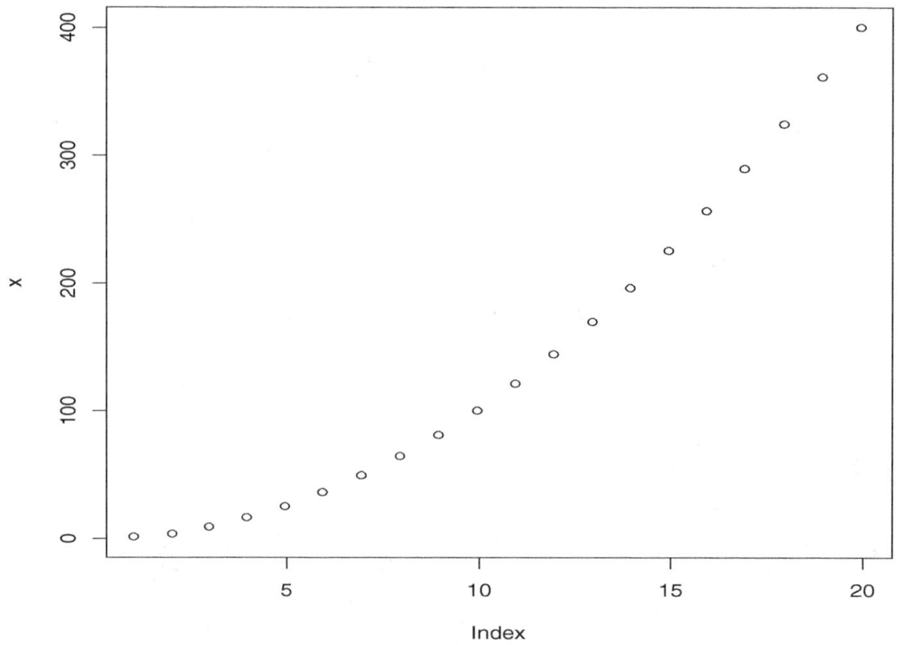

Figure 9.1: *The plot-function will generate a scatter-plot for a vector. Note also that the legend is automatically adapted. The xis is the index of the number in the vector and the y-axis is the value of the corresponding number in the vector.*

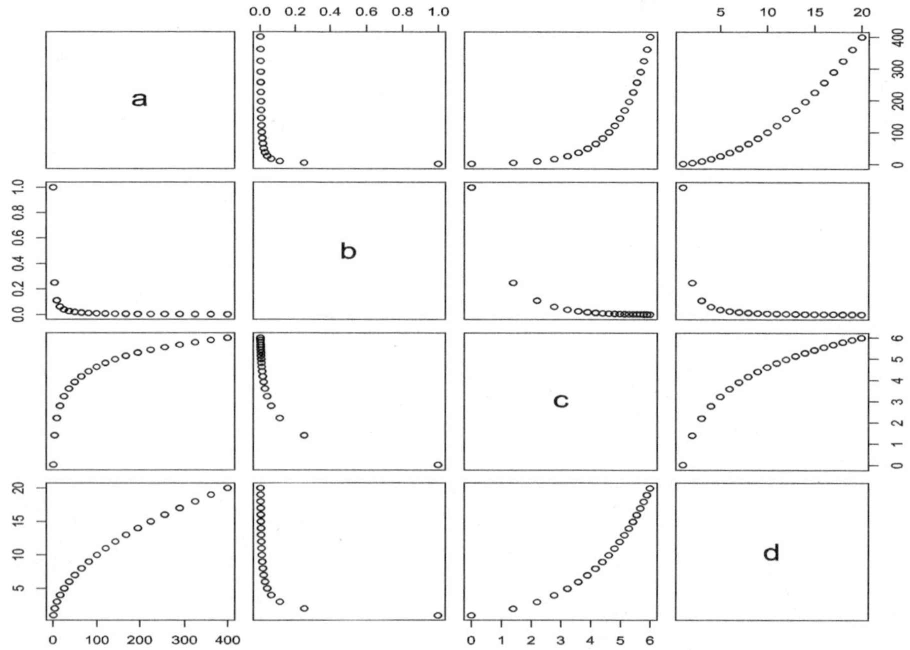

Figure 9.2: *The plot-function will generate a scatter-plot of each column in function of each other column for a data frame. The main diagonal of the matrix has the names of the columns. Note also that the x and y axis are labelled only on one side and that each row shares the same y-axis, while each column shares one x-axis.*

9.1 Scatterplots

Scatterplots are probably the first type of plot one can think of, they show points plotted on the Cartesian plane. Each point represents the combination of two variables. One variable is chosen in the horizontal axis and another in the vertical axis.

scatterplot
plot – scatter
chart – scatterplot

Function use for plot() – for a scatterplot

`plot(x, y, main, xlab, ylab, xlim, ylim, axes, ...)` with

- `x`: the data set for the horizontal axis
- `y`: the data set for the vertical axis
- `main`: the tile of the graph
- `xlab`: the title of the x-axis
- `ylab`: the title of the x-axis
- `xlim`: the range of values on the x-axis
- `ylim`: the range of values on the y-axis
- `pch`: the display symbol
- `axes`: indicates whether both axes should be drawn on the plot.

`plot()`

With the argument `pch` (that is short for "plot character"), it is possible to change the symbol that is displayed on the scatterplot. Integer values between 0 to 25 specify a symbol as shown in Figure 9.3. It is possible to change the colour via the argument `col`. pch values from 21 to 25 are filled symbols that allow you to specify a second color bg for the background fill. Most other characters supplied to pch other than this will plot themselves.

Further information – See the code

Would you like to see the code that generated the plot Figure 9.3 on page 162? Please refer to Chapter D *"Code Not Shown in the Body of the Book"* on page 840.

To illustrate the scatterplot, we use the dataset `mtcars`, and try to gain some insight in how the fuel consumption depends on the horse power of a car – the output is in Figure 9.4 on page 162.

```
# Import the data
library(MASS)

# To make this example more interesting, we convert mpg to l/100km

# mpg2l
# Converts miles per gallon into litres per 100 km
# Arguments:
#    mpg -- numeric -- fuel consumption in MPG
# Returns:
#    Numeric -- fuel consumption in litres per 100 km
mpg2l <- function(mpg = 0) {
100 * 3.785411784 / 1.609344 / mpg}
```

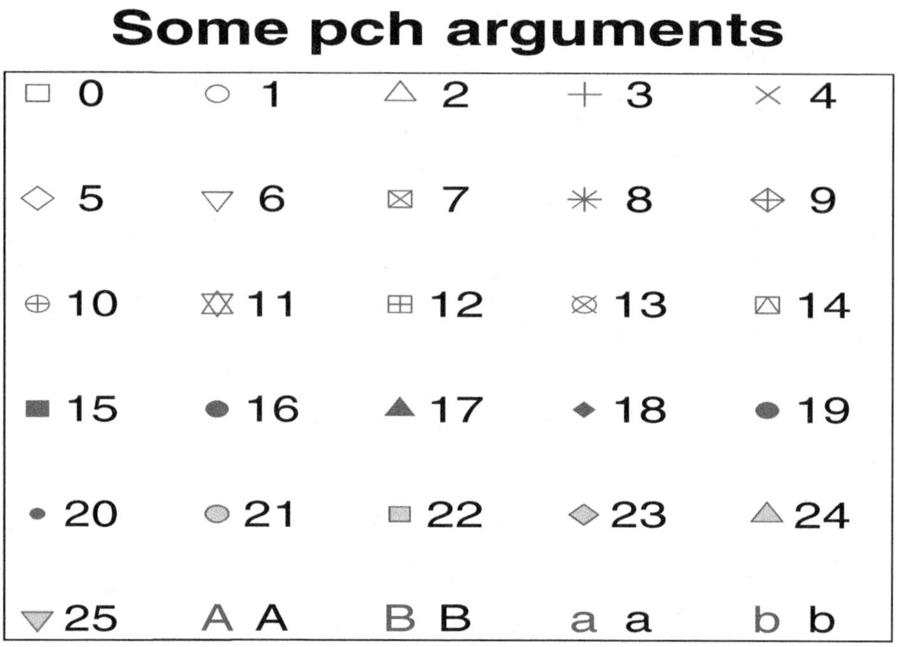

Figure 9.3: *Some plot characters. Most other characters will just plot themselves.*

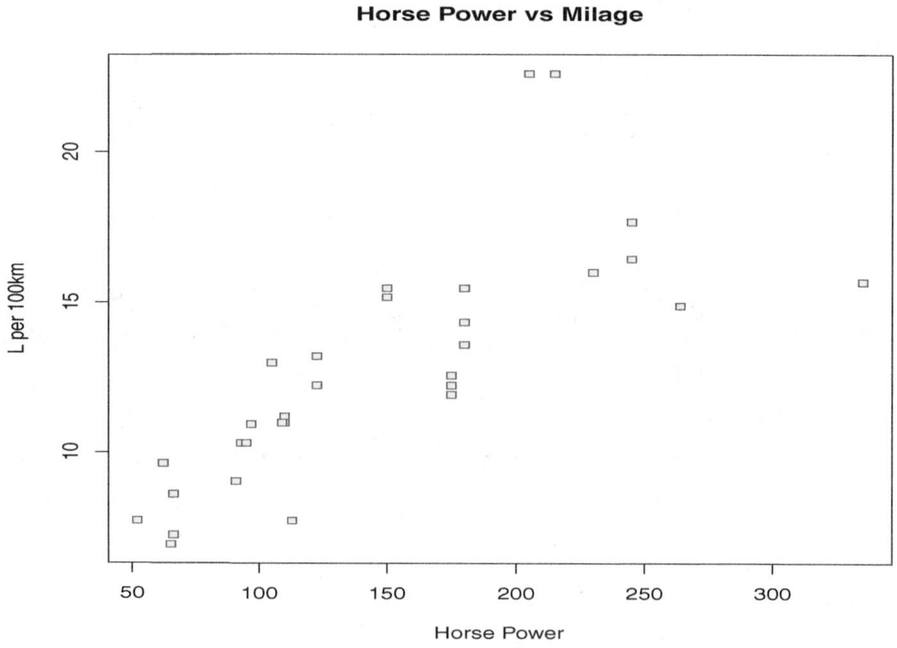

Figure 9.4: *A scatter-plot needs an x and a y variable.*

```
mtcars$l <- mpg2l(mtcars$mpg)
plot(x = mtcars$hp,y = mtcars$l, xlab = "Horse Power",
 ylab = "L per 100km", main = "Horse Power vs Milage",
 pch = 22, col="red", bg="yellow")
```

9.2 Line Graphs

plot – line
chart – line
line plot

It might be sufficient to add lines to a scatterplot (with the `lines()` function), but other methods to create line plots are available and will be discussed below. One of the most useful is the function `plot()`.

Function use for plot() – for line plots

`plot(x, type , main, xlab, ylab, xlim, ylim, axes, sub, asp ...)` with

- `x`: the data set for the horizontal axis
- `y`: the data set for the vertical axis (optional)
- `type`: indicates the type of plot to be made:
  - "p" for *p*oints,
  - "l" for *l*ines,
  - "b" for *b*oth,
  - "c" for the lines part alone of "b",
  - "o" for both '*o*verplotted',
  - "h" for '*h*istogram' like (or 'high-density') vertical lines,
  - "s" for stair *s*teps,
  - "S" for other *s*teps, see 'Details' in the documentation,
  - "n" for no plotting.
- `main`: the tile of the graph
- `xlab`: the title of the x-axis
- `ylab`: the title of the x-axis
- `xlim`: the range of values on the x-axis
- `ylim`: the range of values on the y-axis
- `axes`: indicates whether both axes should be drawn on the plot.
- `sub`: the sub-title
- `asp`: the y/x aspect ratio

A line-plot example

The following code first genearates data and then plots it in a line chart. The output of the code is in Figure 9.5 on the next page.

```
# Prepare the data:
years <- c(2000,2001,2002,2003,2004,2005)
sales <- c(2000,2101,3002,2803,3500,3450)
plot(x = years,y = sales, type = 'b',
     xlab = "Years", ylab = "Sales in USD",
     main = "The evolution of our sales")
points(2004,3500,col="red",pch=16) # highlight one point
text(2004,3400,"top sales")         # annotate the highlight
```

points()
text()

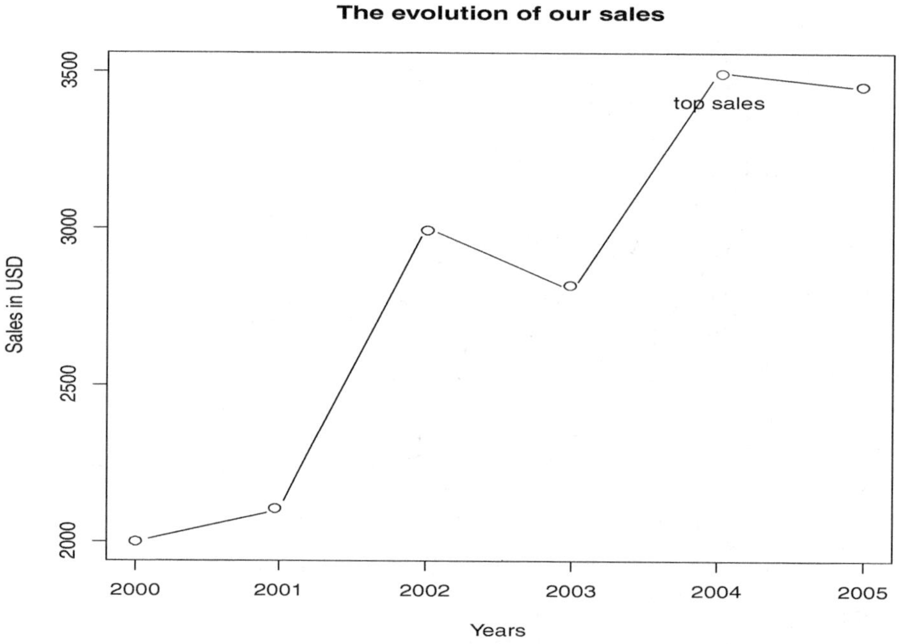

Figure 9.5: *A line plot of the type b.*

9.3 Pie Charts

A pie chart is a simple, but effective way to represent data when the total is 100%. The total circumference of the circle is that 100% and the angle of each section corresponds to the percentage of one of the categories.

chart – pie
pie chart
plot – pie

To use this plot, we need a variable that is categorical (a "factor" in R) and a corresponding numerical value. For example, the sales (numerical) per country (the categorical variable).

Avoid this plot when it is not clear what the 100% actually is or when there are too much categories.

Function use for pie()

`pie()`

```
pie(x, labels = names(x), edges = 200, radius = 0.8, clockwise =
FALSE, init.angle = if(clockwise) 90 else 0, density = NULL, angle =
45, col = NULL, border = NULL, lty = NULL, main = NULL, ...)
```

where the most important parameters are

- `x`: a vector of non-negative numerical quantities. The values in 'x' are displayed as the areas of pie slices
- `labels`: strings with names for the slices
- `radius`: the radius of the circle of the chart (value between âffh1 and +1)
- `main`: indicates the title of the chart
- `col`: the colour palette
- `clockwise`: a logical value indicating if the slices are drawn clockwise or anti clockwise

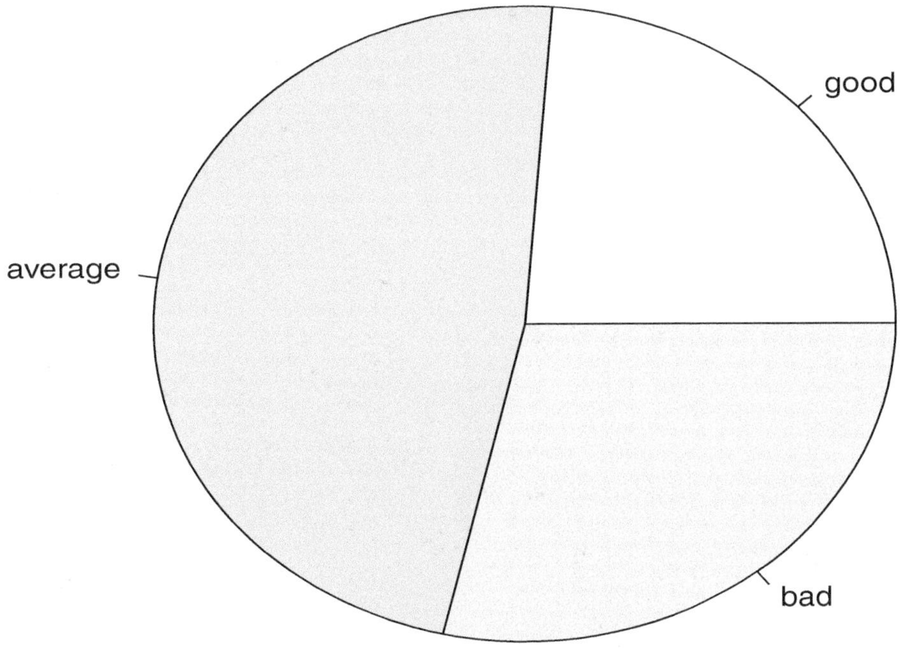

Figure 9.6: *A pie-chart in R.*

Pie chart example

The following code segment produces the pie chart that is in Figure 9.6.

dev.off()

pie()

```
x <- c(10, 20, 12)                # Create data for the graph
labels <- c("good", "average", "bad")
#for saving to a file:
  png(file = "feedback.jpg")  # Give the chart file a name
  pie(x,labels)               # Plot the chart
  dev.off()                   # Save the file
## pdf
##   2

pie(x,labels)                 # Show in the R Graphics screen
```

9.4 Bar Charts

Bar charts visualise data via the length of rectangles. It is one of the most suited plots if the data is numerical for different categories (that are not numerical).

The function barplot()

Function use for barplot()

`barplot(height, width=1, xlab=NULL, ylab=NULL, main=NULL, names.arg=NuLL, col=NULL, ...)`

Some parameters:

- `height`: is the vector or matrix containing numeric values used in chart
- `xlab`: the label for the x-axis
- `ylab`: is the label for y-axis
- `main`: is the title of the chart
- `names.arg`: is a vector of names of each bar
- `col`: is used to give colors to the bars in the graph.

barplot()

The following example creates Figure 9.7 on page 168:

```
sales <- c(100,200,150,50,125)
regions <- c("France", "Poland", "UK", "Spain", "Belgium")
barplot(sales, width=1,
      xlab="Regions", ylab="Sales in EUR",
        main="Sales 2016", names.arg=regions,
        border="blue", col="brown")
```

barplot()

The function `barplot()` is the key-function for bar plots, but it can do more than the simple plot of Figure 9.8 on page 168. For example, it is also possible to stack the bars as shown in Figure 9.8 on page 168. This figure is produced by the following code:

Stacked bar charts

chart – stacked barcharts
plot – stacked barplots

```
# Create the input vectors.
colours <- c("orange","green","brown")
regions <- c("Mar","Apr","May","Jun","Jul")
product <- c("License","Maintenance","Consulting")

# Create the matrix of the values.
values <- matrix(c(20,80,0,50,140,10,50,80,20,10,30,
     10,25,60,50), nrow = 3, ncol = 5, byrow = FALSE)

# Create the bar chart.
barplot(values, main = "Sales 2016",
 names.arg = regions, xlab = "Region",
 ylab = "Sales in EUR", col = colours)

# Add the legend to the chart.
legend("topright", product, cex = 1.3, fill = colours)
```

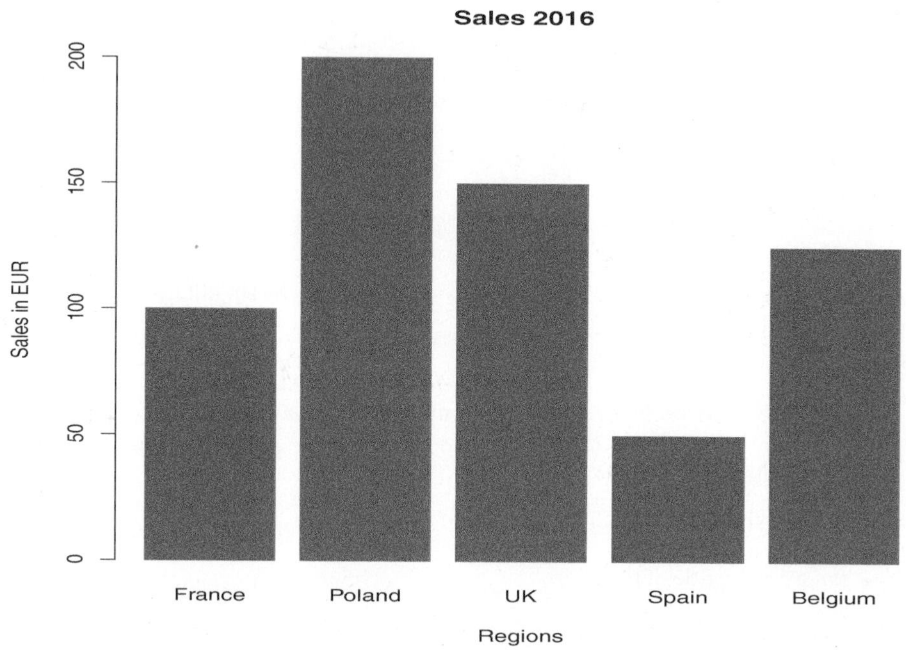

Figure 9.7: *A standard bar-chart based on a vector.*

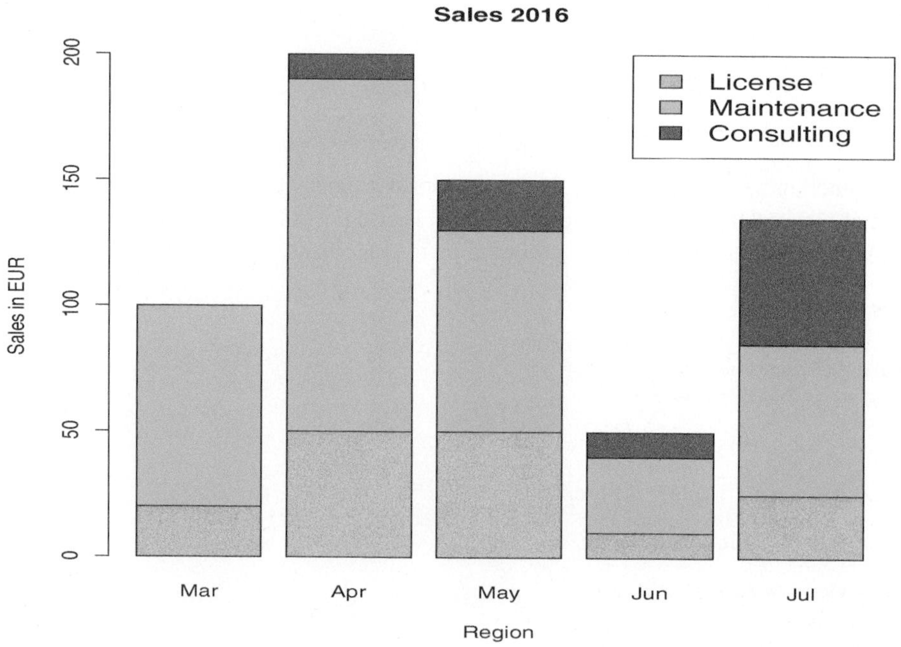

Figure 9.8: *A bar-chart based on a matrix will produce stacked bars. Note how nicely this plot conveys the seasonal trend in the data.*

Barplots With Total of 100 Procent

The function `barplot` has no option to simply force all bars to 100% (equal length), we can do this with the function `barplot()`[2]. The function takes one argument `margin`, which is the index, or vector of indices for which the margins need to be calculated. prop.table()

The following piece of code illustrates how this can work (the output is in Figure 9.9):

```
# We reuse the matrix 'values' from previous example.

# Add extra space to right of plot area; change clipping to figure
par(mar = c(5, 4, 4, 8) + 0.1, # default margin was c(5, 4, 4, 2) + 0.1
    xpd = TRUE)       # TRUE to restrict all plotting to the plot region

# Create the plot with all totals coerced to 1.0:
barplot(prop.table(values, 2), main = "Sales 2016",
   names.arg = regions, xlab = "Region",
   ylab = "Sales in EUR", col = colours)

# Add the legend, but move it to the right with inset:
legend("topright", product, cex = 1.0, inset=c(-0.3,0), fill = colours,
       title="Business line")
```

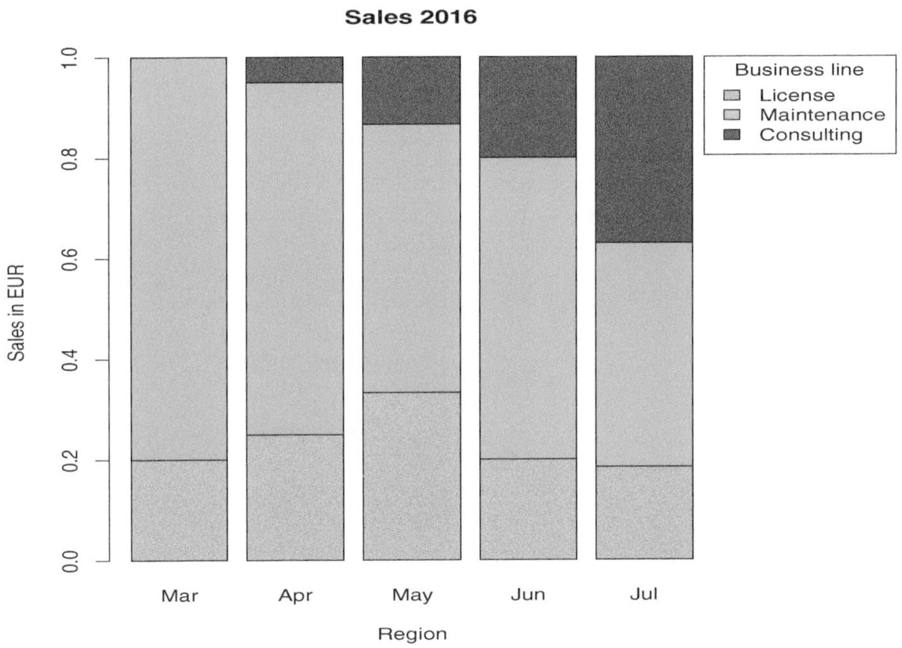

Figure 9.9: *A boxplot where the total of each bar equals 100%. Note how the seasonal trend is obscured in this plot, but how it now tells the story of how the consulting business is growing and the maintenance business, relatively, cannot keep up wih that growth.*

---

[2]This function is a wrapper that does the same as `sweep(x, margin, margin.table(x, margin), "/")`. The only difference is that if the argument `margin` has length zero, then one gets `for newbies, except that if 'margin' has length zero, then one x/sum(x).` sweep()

Warning – Scaled boxplots

This type of plot should only be used if the total indeed is 100%, or if that concept really makes sense. In any case, make sure that the legend of your plot describes what is going on.

9.5 Boxplots

boxplot
chart – boxplot
plot – boxplot

Boxplotsare a particular form of plot that do an excellent job in summarising data. They will show quartiles, outliers, and median all in one plot.

An example can be seen in Figure 9.10 on page 172. Each category (4, 6, and 8) has one boxplot. the centre is a box that spans from the first quartile to the third. The median is the horizontal line in the box. The two bars that stick out on the top are called the "whiskers." The whiskers stretch from the lowest value at the bottom to the highest value on the top. However, the size of the whiskers is limited to 1.5 times the interquartile range[3]. This behaviour can be changed via the parameter `range`.

boxplot()
`boxplo()`

Function use for boxplot()

`boxplot(formula, data = NULL,notch = FALSE, varwidth = FALSE, names, main = NULL, ...` with: Following is the description of the parameters used âffh

- `formula`: a vector or a formula.
- `data`: the data frame.
- `notch`: a logical value (set to TRUE to draw a notch)
- `varwidth`: a logical value (set to true to draw width of the box proportionate to the sample size)
- `names`: the group labels which will be printed under each boxplot.
- `main`: the title to the graph.
- `range`: this number determines how far the plot whiskers can reach. If it is positive, then the whiskers extend to the most extreme data point which is no more than "range" times the interquartile range from the box. If range is set to 0, then the whiskers extend to the data extremes.

To illustrate boxplots, we consider the dataset `ships` (from the library "MASS"), and use the following code to generate Figure 9.10 on page 172:

```
library(MASS)
boxplot(mpg ~ cyl,data=mtcars,col="khaki3",
        main="MPG by number of cylinders")
```

[3]The interquartile range is defined as the difference of quartile 3 minus quartile 1.

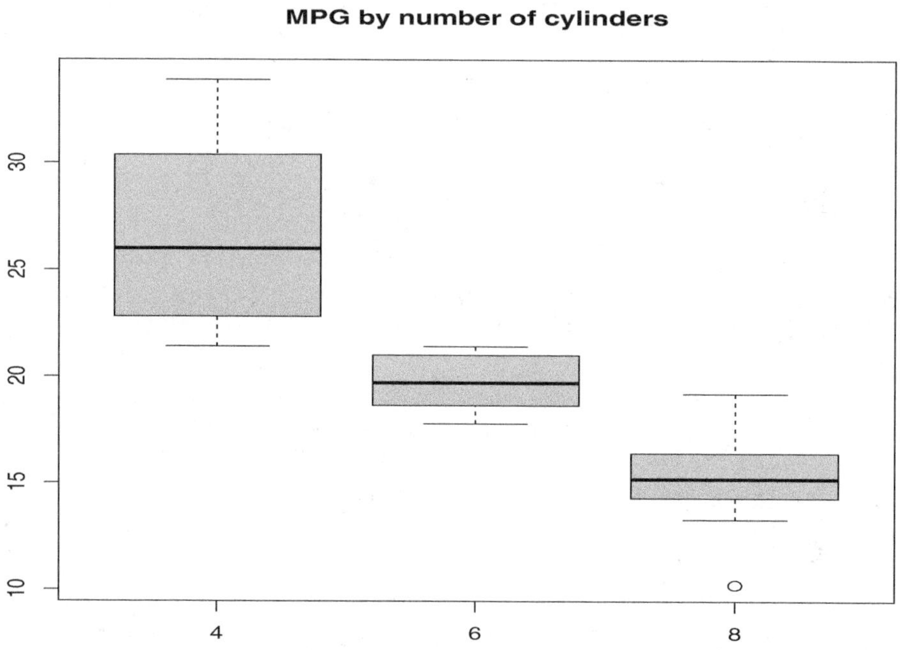

Figure 9.10: *Boxplots show information about the central tendency (median) as well as the spread of the data.*

9.6 Violin Plots

Boxplots are great, but have some limitations. For example, the shape of the distribution is only shown by five points. This implies that it might hide that data is bimodal[4] for example.

bimodal

As described by Hintze and Nelson (1998), boxplots marry the advantages of boxplots with the added information of a an approximation of the distribution. A violin plot is a boxplot with a rotated kernel density plot on each side.

In R, making boxplots is made easy by the package `vioplot`.

The function `vioplot()` will only plot violin plot.

So unlike the boxplots, we have to use the function `with()` to get more of them.

vioplot
vioplot()
with()

This is illustrated in the following code and the output is in Figure 9.11 on page 174:

```
# install.packages('vioplot') # only do once
library(vioplot)              # load the vioplot library

## Loading required package: sm
## Package 'sm', version 2.2-5.6: type help(sm) for summary information
##
## Attaching package: 'sm'
## The following object is masked from 'package:MASS':
##
##     muscle
## Loading required package: zoo
##
## Attaching package: 'zoo'
## The following objects are masked from 'package:base':
##
##     as.Date, as.Date.numeric

with(mtcars , vioplot(mpg[cyl==4] , mpg[cyl==6], mpg[cyl==8],
                      col=rgb(0.1,0.4,0.7,0.7) ,
                      names=c("4","6","8")
                      )
    )
```

Tracing violin plots is also possible with the package `ggplot2`[5]. Below we show one possibility and in Chapter 22.2.3 *"The AUC"* on page 396 we use another variation of violin plots. The output of the following code is in Figure 9.12 on page 174 and Figure 9.13 on page 175.

```
# Library
library(ggplot2)

# First, type of color
ggplot(mtcars, aes(factor(cyl), mpg)) +
  geom_violin(aes(fill = cyl))


# Second type
ggplot(mtcars, aes(factor(cyl), mpg)) +
  geom_violin(aes(fill = factor(cyl)))
```

[4]Bimodal means that the probability density function shows two maxima. This can mean that there are two distinct populations in the data. For example, in `mtcars` the group of eight cylinder cars seems to be trimodal (composed of three difference groups of cars: SUVs, sports cars, luxury limousines.)

[5]The package `ggplot2` will be described in more detail in Chapter 31 *"A Grammar of Graphics with ggplot2"* on page 687.

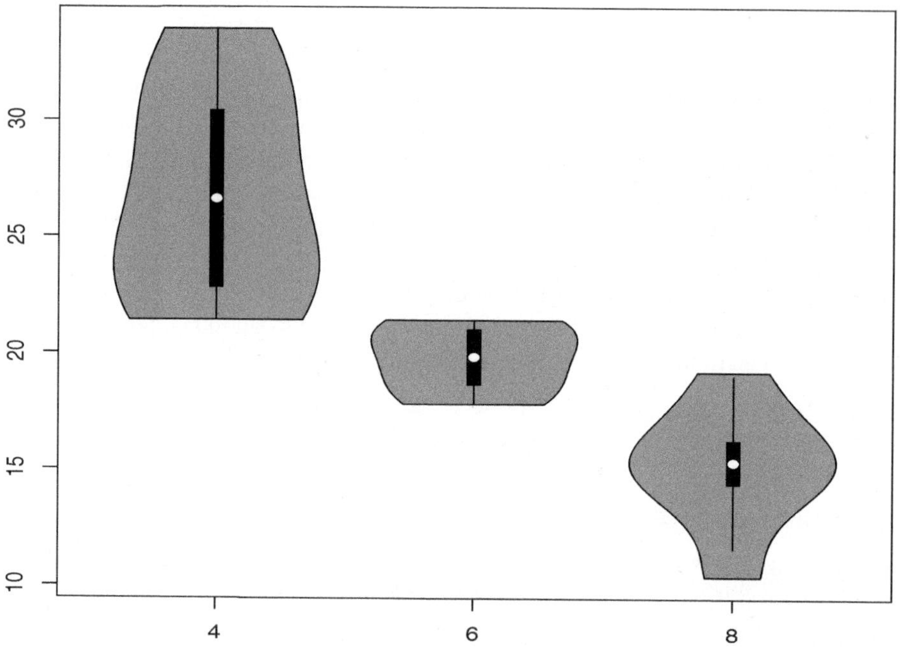

Figure 9.11: *Violin plot as provided by the function* `vioplot` *from the package of the same name.*

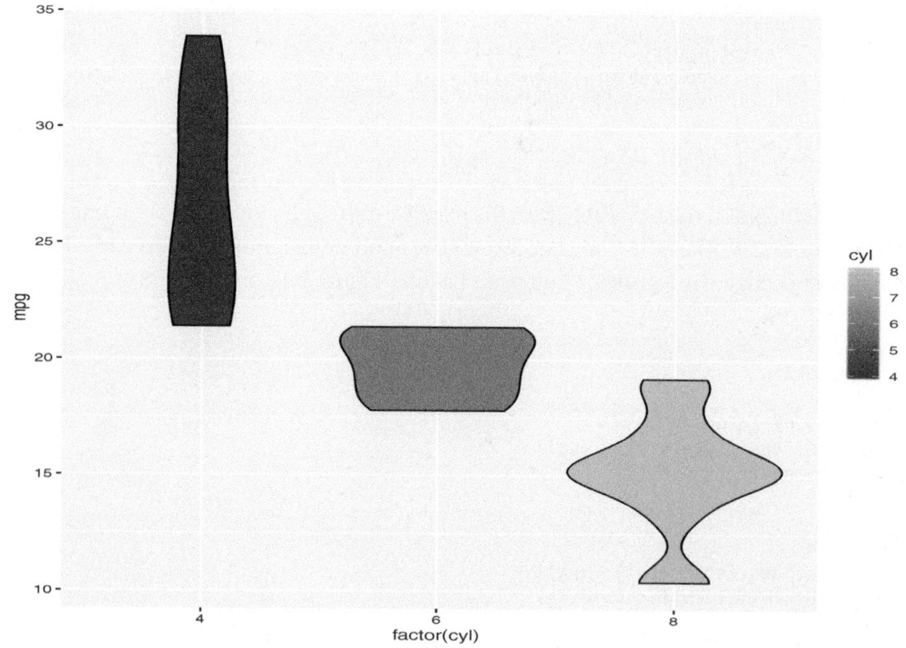

Figure 9.12: *Violin plot as traced by* `geom_violin` *provided by the library* `ggplot2`*; with the colouring done according to the number of cylinders.*

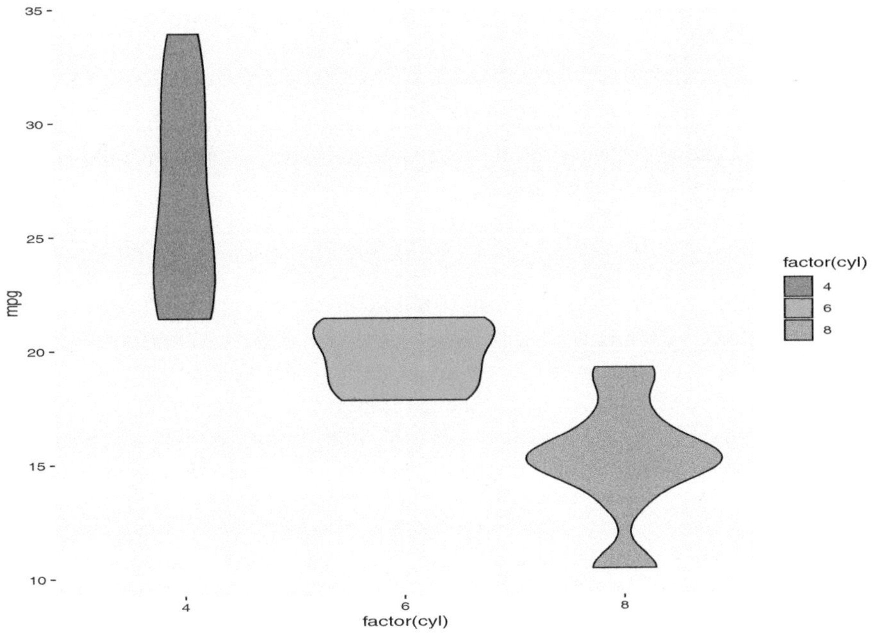

Figure 9.13: *Violin plot as traced by* `geom_violin` *provided by the library* `ggplot2`*; with the colouring done according to the factoring of the number of cylinders.*

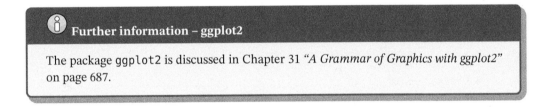

Further information – ggplot2

The package `ggplot2` is discussed in Chapter 31 *"A Grammar of Graphics with ggplot2"* on page 687.

histogram

9.7 Histograms

chart – histogram
plot – histogram

Every person, who works with data, will know a histogram. Histograms to a good job in visualising the probability distribution of observations. In base-R, the function `hist()` can produce a histogram with just one command.

hist()

Function use for hist()

`hist(x, breaks = "Sturges", freq = NULL, probability = !freq, include.lowest = TRUE, right = TRUE, density = NULL, angle = 45, col = NULL, border = NULL, main = paste("Histogram of" , deparse(substitute(x))), xlim = range(breaks), ylim = NULL, xlab = deparse(substitute(x)), ylab, axes = TRUE, plot = TRUE, labels = FALSE, nclass = NULL, warn.unused = TRUE, ...)` with the most important parameters:

- `x`: the vector containing numeric values to be used in the histogram
- `main`: the title of the chart
- `col`: the color of the bars
- `border`: the border color of each bar
- `xlab`: the title of the x-axis
- `xlim`: the range of values on the x-axis
- `ylim`: the range of values on the y-axis
- `breaks`: one of
  - a vector giving the breakpoints between histogram cells,
  - a function to compute the vector of breakpoints,
  - a single number giving the number of cells for the histogram,
  - a character string naming an algorithm to compute the number of cells,
  - a function to compute the number of cells
- `freq`: TRUE for frequencies, FALSE for probability density

hist()

To illustrate how this function works, we will use the database `ships`, that is provided in the package `MASS`. The code below shows two variants in Figure 9.14 on page 177 and Figure 9.15 on page 177.

```
library(MASS)
incidents <- ships$incidents
# figure 1: with a rug and fixed breaks
hist(incidents,
   col=c("red","orange","yellow","green","blue","purple"))
rug(jitter(incidents))  # add the tick-marks
```

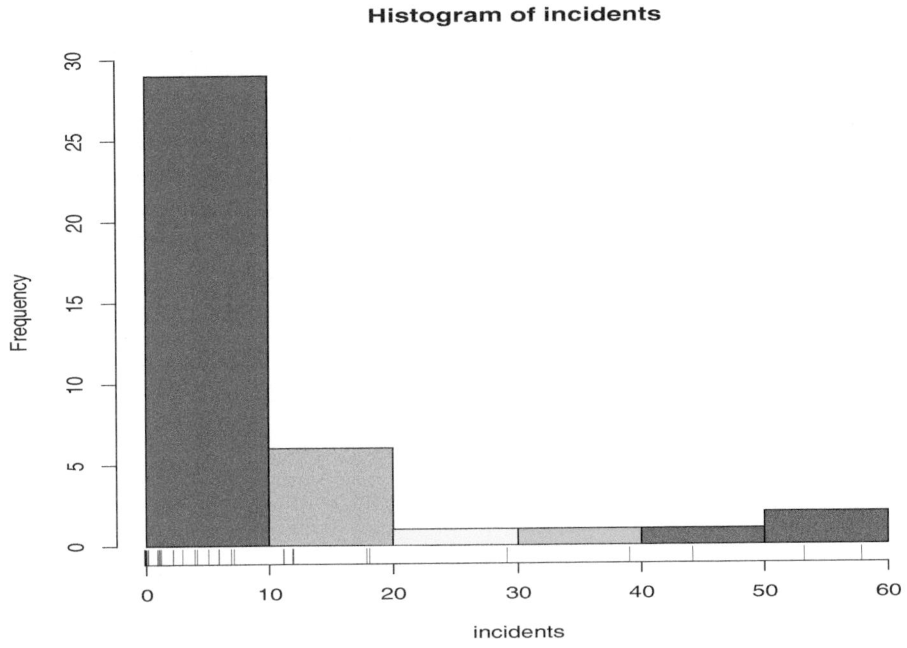

Figure 9.14: *A histogram in R is produced by the hist() function.*

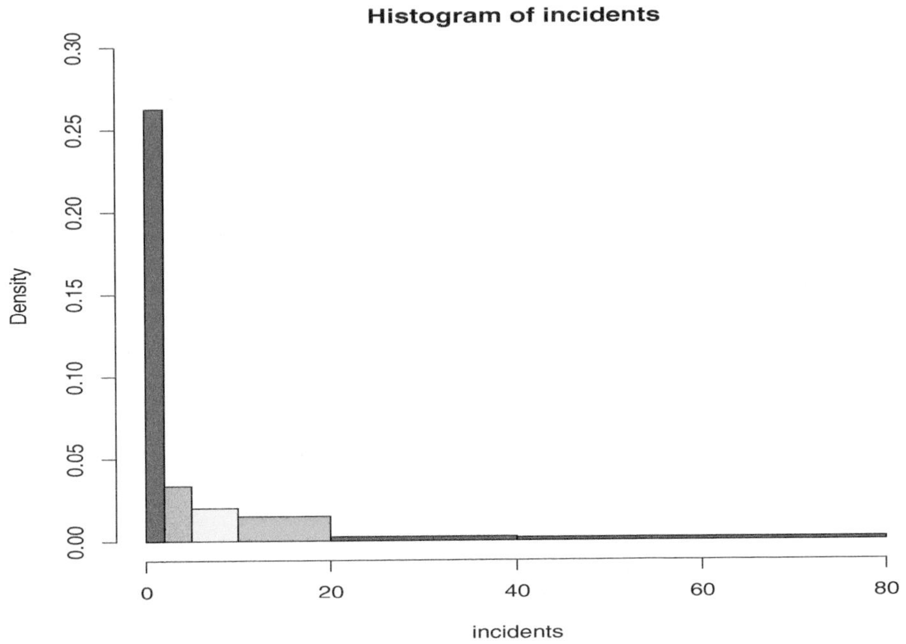

Figure 9.15: *In this histogram, the breaks are changed, and the y-axes is now calibrated as a probability. Note that leaving freq=TRUE would give the wrong impression that there are more observations in the wider brackets.*

```
# figure 2: user-defined breaks for the buckets
hist(incidents,
  col=c("red","orange","yellow","green","blue","purple"),
  ylim=c(0,0.3), breaks=c(0,2,5,10,20,40,80),freq=FALSE)
```

In the aforementioned code, we also used the function `rug()` and `jitter()`.

`rug()`

The function `rug()` produces a small vertical line per observation along the x-axis. However, when some observations have the same value, the lines would overlap and only one line would

`jitter()`

be visible. To provide some insight in the density, the function `jitter()` adds random noise to the position.

9.8 Plotting Functions

While the function plot() allows to draw functions, there is a specific function `curve()`, that will plot functions. The following code illustrate this function by creating Figure 9.16 and makes also clear how to add mathematical expressions to the plot:

plot – functions(of)

```
fn1 <- function(x) sqrt(1-(abs(x)-1)^2)
fn2 <- function(x) -3*sqrt(1-sqrt(abs(x))/sqrt(2))
curve(fn1,-2,2,ylim=c(-3,1),col="red",lwd=4,
      ylab = expression(sqrt(1-(abs(x)-1)^2) +++ fn_2))
curve(fn2,-2,2,add=TRUE,lw=4,col="red")
text(0,-1,expression(sqrt(1-(abs(x)-1)^2)))
text(0,-1.25,"++++")
text(0,-1.5,expression(-3*sqrt(1-sqrt(abs(x))/sqrt(2))))
```

`expression()`

`curve()`

`text()`

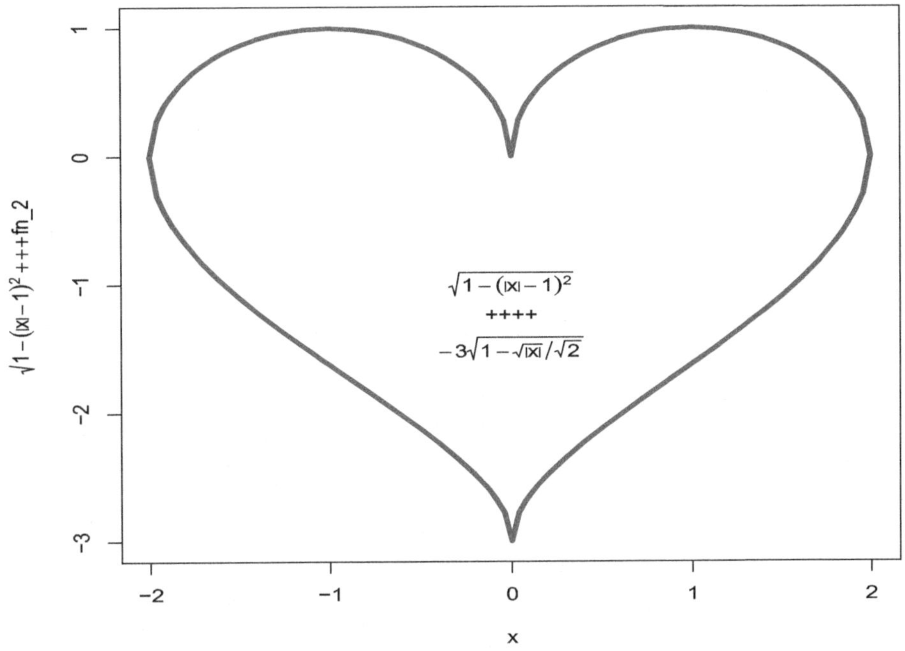

Figure 9.16: *Two line plots plotted by the function curve().*

This also shows the standard capacity of R to include mathematical formulae into its plots and even format annotations in a LaTeX-like markup language.

LaTeX

9.9 Maps and Contour Plots

In this section we show how to make a plot to visualise real values in a 2D projection. This could be a heat-map, a transition matrix for a Markov Chain, an elevation-map, etc. For street maps, we refer to Section 36.2.1 *"HTML-widgets"* on page 719.

The function `image()` from the package `graphics` is a generic function to plot coloured rectangles. The following uses the dataset `volcano` from the package `datasets` and the examples inspired by the documentation (see R Core Team (2018)). The output is generated by the last section in Figure 9.17

```
# A basic plot is possible with the function image()

# We present here improved plot as per documentation of image().
x <- 1:nrow(volcano)
y <- 1:ncol(volcano)

# the mapping of colours.
image(x, y, volcano, col = terrain.colors(100),
      axes = FALSE, xlab='', ylab='')

# add the contour plot
contour(x, y, volcano, levels = seq(90, 200, by = 5),
        add = TRUE, col = "brown")

# add axis, a box and a title:
axis(1, at = seq(10, 80, by = 10))
axis(2, at = seq(10, 60, by = 10))
box()
title(main = "Auckland's Maunga Whau Volcano", font.main = 4)
```

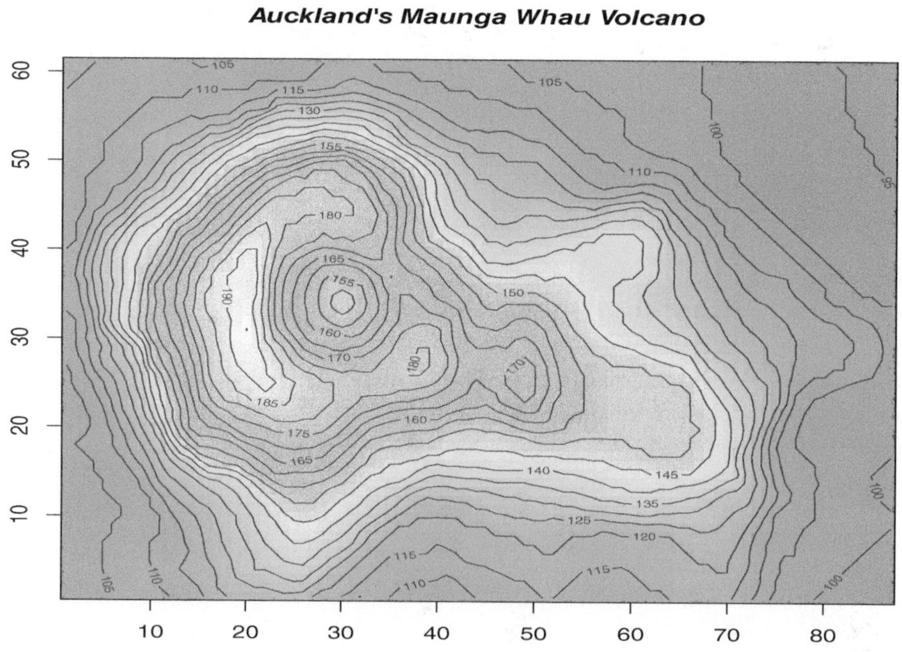

Figure 9.17: *A colour mapping combined with a contour plot provides a nice image of the heights of Auckland's Maunga Whau Volcano.*

9.10 Heat-maps

heat-map

Long gone are Galileo's times where data was scarce and had to be produced carefully. We live in a time where there is a lot of data around. Typically, commercial and scientific institutions have too much data to handle easily, or to see what is going on. Imagine for example to have the data of a book of one million loans.

heatmap()

One way to get started is a heat-map. A heat-map is in essence a visualization of a matrix and is produced by the function `heatmap()`, as illustrated below. The output is in Figure 9.18.

```
d=as.matrix(mtcars,scale="none")
heatmap(d)
```

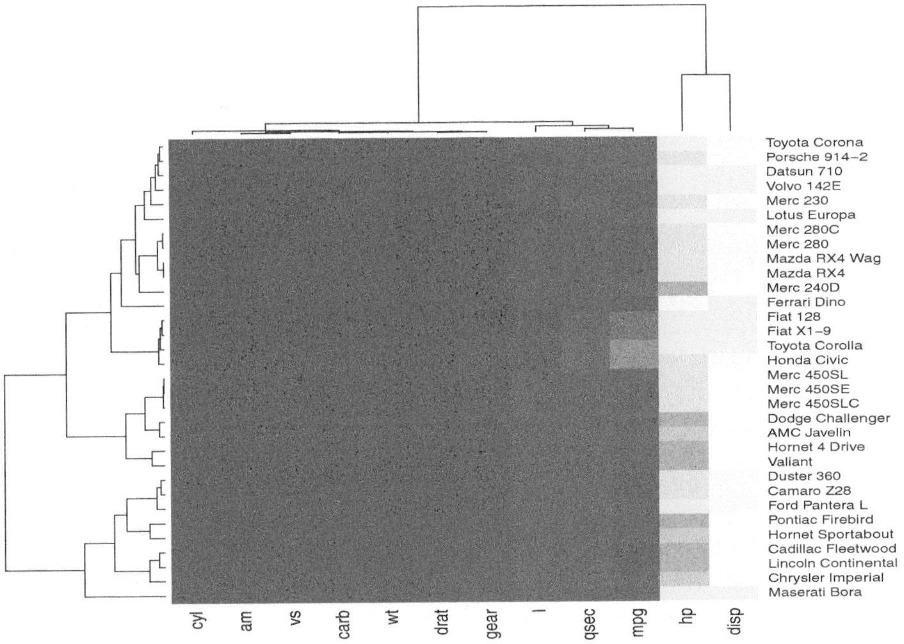

Figure 9.18: *Heatmap for the "mtcars" data.*

dendrogram

Note that this function by default will change the order of rows and columns in order to be able to create a pattern, and the heuristics are visualized by the dendrograms (top and left in Figure 9.18).

While this is interesting, the results for the low numbers are all red. This is because the numbers are not of the same nature (have different scales), so we might want to rescale and reveal the hidden patterns per variable. This is done with the variable `scale` and we tell how we want the scaling to be done (rows or columns), as illustrated in the following code. The result is in Figure 9.19 on page 182.

```
heatmap(d,scale="column")
```

The function `heatmap()` has many more useful parameters. Below we describe the most relevant.

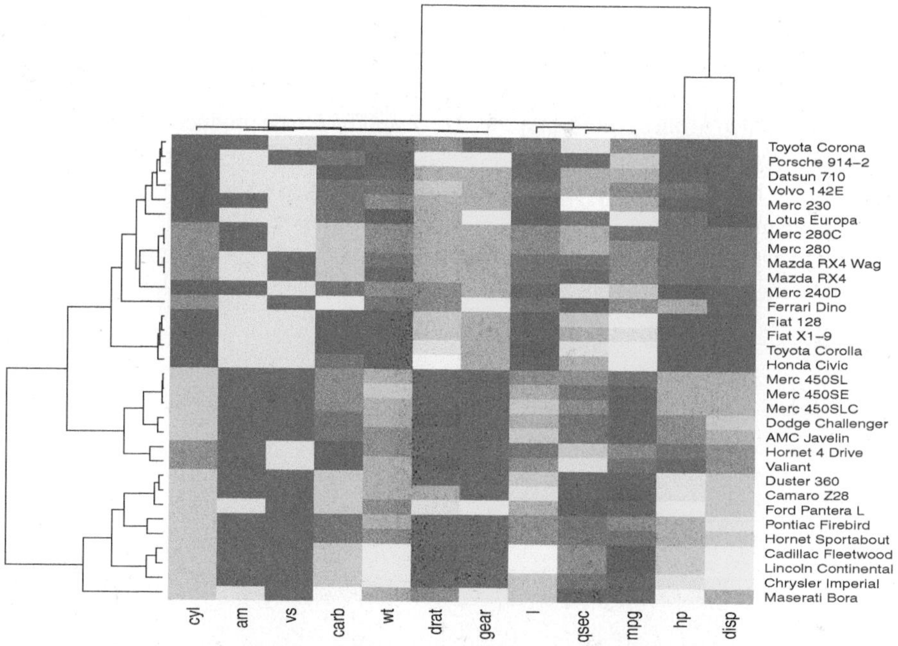

Figure 9.19: *Heatmap for the "mtcars" data with all columns rescaled*

Function use for heatmap()

```
heatmap(x, Rowv = NULL, Colv = if(symm) "Rowv" else NULL,
        distfun = dist, hclustfun = hclust,
        reorderfun = function(d, w) reorder(d, w),
        add.expr, symm = FALSE, revC = identical(Colv, "Rowv"),
        scale = c("row", "column", "none"), na.rm = TRUE,
        margins = c(5, 5), ColSideColors, RowSideColors,
        cexRow = 0.2 + 1/log10(nr), cexCol = 0.2 + 1/log10(nc),
        labRow = NULL, labCol = NULL, main = NULL,
        xlab = NULL, ylab = NULL,
        keep.dendro = FALSE, verbose = getOption("verbose"),
        ...)
```

with the most important.

- *x* the data;
- *scale* the default is "row," it can be turned off by using "none" or switched to "column";
- *Rowv* and *Colv:* determines if and how the row or column dendrogram should be computed and re-ordered. Use "NA" to suppress re-ordering;
- *distfun:* the function to calculate distances;
- *na.rm:* a logical value that indicates how missing values have to be treated;
- *labCol* and *labRow:* these can be character vectors with row and column labels to use in the plot. Their default is `rownames(x)` or `colnames(x)` respectively (with x the matrix supplied to the function);

- *main, xlab, ylab:* have their usual meaning;
- *keep.dendro:* logical indicating if the dendrogram(s) should be kept (when 'Rowv' and/or 'Colv' are not NA).
- other parameters can be explored by the command `help(heatmap)`

text mining

9.11 Text Mining

9.11.1 Word Clouds

word cloud

Word clouds are one of the best ways to convey context. The idea is simple: use a text-file and count all occurrences of all words, then put the most popular ones in a plot, so that the larger text corresponds to the most ubuquitous ones.

To achieve this, we will need a few packages:

tm
SnowballC
RColorBrewer
wodcloud

```
# If neccesary first download the packages:
install.packages("tm")           # text mining
install.packages("SnowballC")    # text stemming
install.packages("RColorBrewer") # colour palettes
install.packages("wordcloud")    # word-cloud generator
```

Then, we load the packages:

```
# Then load the packages:
library("tm")

## Loading required package: NLP
## ## Attaching package: 'NLP'
## The following object is masked from 'package:ggplot2':## ##    annotate
## ## Attaching package: 'tm'
## The following object is masked from 'package:pryr':## ##    inspect

library("SnowballC")
library("RColorBrewer")
library("wordcloud")
```

Example: – The text of this book

This section will be developed around one example: the text version of this book.

This book is made in LaTeX, and hence it is easy to convert to a text-file. The program `detex` comes with your LaTeX (on Linux), and it is sufficient to use the following line on the Linux CLI:

```
detex r-book.tex > data/r-book.txt
```

readLines()

An now, we can walk through to the work-flow. The first step is of course, reading in the text to be analysed. We will first import a text-file with the function `readLines()`:

Step 1: Importing the Text

```
# In this example we use a text version of this very book:
t <- readLines("data/r-book.txt")

# Then create a corpus of text
doc <- Corpus(VectorSource(t))
```

tm
Corpus()

The function `Corpus()` from the package `tm` creates corpora. Corpora are collections of documents that contain human language. The variable `doc` is now an S3 object of class `Corpus`.

Step 2: Cleaning the Text

In this step, we remove all text that should not appear in our word-cloud. For example, if we would convert the PDF file to text and not the LaTeX-file, then also the headers and footers of every page would end up in the text. This means that in our example, the name of the author would appear at least thousand times and hence this would become the dominant words in the word-could. We know who wrote this book, and hence in this step we would have to remove this name.

We did start from the LaTeX file, and hence

```
# The file has still a lot of special characters
# e.g. the following replaces '\', '#', and '|' with space:
toSpace <- content_transformer(function (x , pattern )
                                gsub(pattern, " ", x))
doc <- tm_map(doc, toSpace, "\\\\")
doc <- tm_map(doc, toSpace, "#")
doc <- tm_map(doc, toSpace, "\\|")
# Note that the backslash needs to be escaped in R
```

> **Hint – Visualize the text-file**
>
> The command `inspect(doc)` can display information about the corpus or any text document.

The `tm_map()` function is used to remove unnecessary white space, to convert the text to lower case, and to remove common stop-words like "the" and "we."

`tm_map()`

The information value of "stop-words" is near zero due to the fact that they are so common in a language. Removing these kinds of words is useful before further analyses. For "stop-words," supported languages are Danish, Dutch, English, Finnish, French, German, Hungarian, Italian, Norwegian, Portuguese, Russian, Spanish, and Swedish. Language names are case sensitive and are provided without capital.

We will also show you how to make your own list of stop-words, and how to remove them from the text. You can also remove numbers and punctuation with "removeNumbers" and "removePunctuation" arguments.

Another important preprocessing step is to make a "text stemming." This is reducing words to their root form. So this process will remove suffixes from words to make it simple and to get the common origin. For example, the words "functional" and "functions," would become "function."

The package `SnowballC`, that we loaded before is able to do this stemming. It works as follows:

```
# Convert the text to lower case
doc <- tm_map(doc, content_transformer(tolower))

# Remove numbers
doc <- tm_map(doc, removeNumbers)

# Remove english common stopwords
doc <- tm_map(doc, removeWords, stopwords("english"))

# Remove your own stop words
# specify your stopwords as a character vector:
doc <- tm_map(doc, removeWords, c("can","value","also","price",
             "cost","option","call","need","possible","might",
             "will","first","etc","one","portfolio", "however",
             "hence", "want", "simple", "therefore"))
```

```
# Remove punctuations
doc <- tm_map(doc, removePunctuation)

# Eliminate extra white spaces
doc <- tm_map(doc, stripWhitespace)

# Text stemming
#doc <- tm_map(doc, stemDocument)
```

Note however, that the word "stemming" is not perfect. For example, it replaces "probability density function" with 'probabl densiti function," "example" with "exampl," "variable" with "variabl," etc. That is why we choose not to use it and comment it out.

The downside of leaving it out in our example is that the both the terms "function" and "functions" will be in the top-list. To avoid this, we need to apply some corrections. This is similar to replacing words according to a dictionary.

```
library(stringi)

wordToReplace <- c('functions', 'packages')
ReplaceWith   <- c('function',  'package')

doc <- tm_map(doc,  function(x) stri_replace_all_fixed(x,
           wordToReplace, ReplaceWith, vectorize_all = FALSE))

## Warning in tm_map.SimpleCorpus(doc, function(x) stri_replace_all_fixed(x, :
transformation drops documents
```

Step 3: Build a Term-document Matrix

A document matrix is a table containing the frequency of the words. Column names are words and row names are documents. The function `TermDocumentMatrix()` from text mining package can be used as follows.

```
dtm <- TermDocumentMatrix(doc)
m <- as.matrix(dtm)
v <- sort(rowSums(m),decreasing=TRUE)
d <- data.frame(word = names(v),freq=v)
head(d, 10)
##              word freq
## function function  378
## data         data  240
## use           use  190
## model       model  180
## example   example  153
## code         code  142
## package   package  125
## company   company  116
## method     method  103
## market     market   96
```

Now, we can visualize how many times certain words appear. The frequency of the first 10 frequent words are plotted in Figure 9.20 on page 187, that is produced via the following code fragment.

```
barplot(d[1:10,]$freq, las = 2, names.arg = d[1:10,]$word,
        col ="khaki3", main ="Most frequent words",
        ylab = "Word frequencies")
```

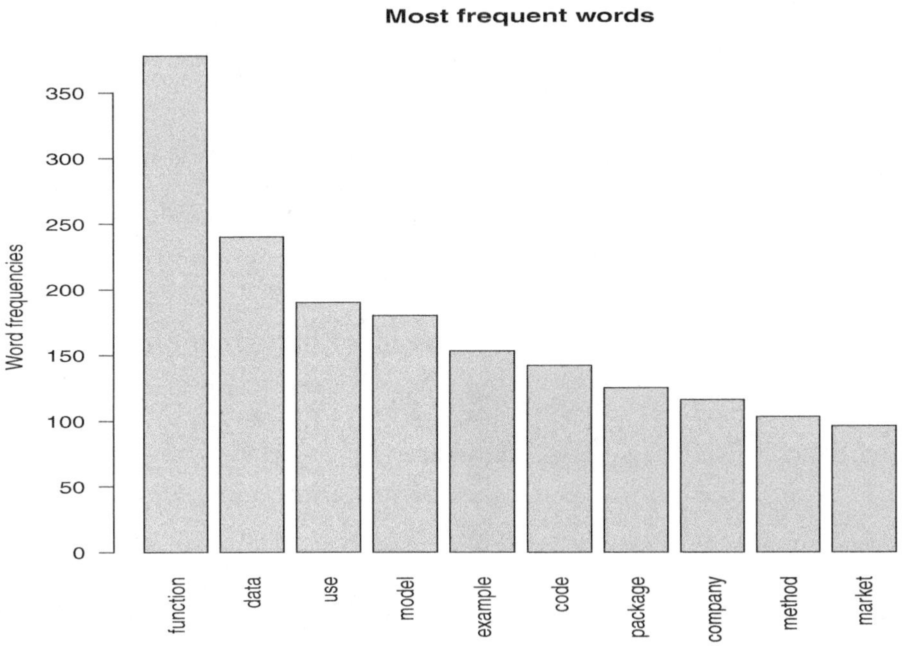

Figure 9.20: *The frequency of the ten most occurring words in this text. Note tha the name of the software, R, is only one letter, and hence not retained as a word. So R is not not part of this list.*

Step 4: Generate the Word-cloud

Finally, we can generate the word-could and produce Figure 9.21 on page 188.

```
set.seed(1879)
wordcloud(words = d$word, freq = d$freq, min.freq = 10,
          max.words=200, random.order=FALSE, rot.per=0.35,
          colors=brewer.pal(8, "Dark2"))
```

Function use for wordcloud()

wordcloud()

```
wordcloud(words, freq, scale = c(4,.5), min.freq = 3,
          max.words=Inf, random.order = TRUE,
          random.color=FALSE, rot.per=.1,
          colors = "black", ordered.colors = FALSE,
          use.r.layout=FALSE, fixed.asp=TRUE, ...)
```

The important parameters of the word-cloud function are:

- words : the words to be plotted
- freq : their frequencies
- min.freq : words with frequency below min.freq will not be plotted
- max.words : maximum number of words to be plotted

Figure 9.21: *A word-could for the text of this book. This single image gives a good idea of what this book is about.*

- random.order : plot words in random order. If false, they will be plotted in decreasing frequency
- rot.per : proportion words with 90 degree rotation (vertical text)
- colors : color words from least to most frequent. Use, for example, `colors = 'black'` for single color.

9.11.2 Word Associations

Word-clouds are not the only way to visualize large amounts of natural language. You can have a look at the frequent terms in the term-document matrix as follows.

In a first step, we might want to identify the words that appear frequently. The function `findFreqTerms()` of the package `tm` finds frequent terms in a document-term or term-document matrix. In the example below we extract words that occur at least 150 times:

Word Associations in R

```
findFreqTerms(dtm, lowfreq = 150)
## [1] "example"  "model"    "function" "data"     "use"
```

One can analyze the association between frequent terms (i.e., terms which correlate) using findAssocs() function. function `findAssocs()` of the same package can identifey the functions

```
# e.g. for the word "function"
findAssocs(dtm, terms = "function", corlimit = 0.15)
## $`function`
##               recycled                   list
##                   0.27                   0.25
##              arguments               argument
##                   0.24                   0.22
##               sequence                strings
##                   0.22                   0.22
##                 second             dispatcher
##                   0.21                   0.21
##                calling          functionality
##                   0.21                   0.20
##                    via                   code
##                   0.19                   0.19
##                 unique                  calls
##                   0.19                   0.19
##                element           clusterapply
##                   0.18                   0.18
##                   work                 allows
##                   0.17                   0.17
##                    run                cluster
##                   0.17                   0.17
##                  apply                wrapped
##                   0.17                   0.16
##                shorter               dispatch
##                   0.16                   0.16
##                 except                   note
##                   0.16                   0.15
##                density                 cosine
##                   0.15                   0.15
##              currently              indicator
##                   0.15                   0.15
##                 summed            dynamically
##                   0.15                   0.15
##         sparkdataframe                   cast
##                   0.15                   0.15
##               foresees                   join
##                   0.15                   0.15
## windowpartitionbypclass                binned
##                   0.15                   0.15
##            elementwise              exception
##                   0.15                   0.15
##         sqltransformer           transformers
##                   0.15                   0.15
##              certainty                confirm
##                   0.15                   0.15
##                    nas                  slows
```

```
##                   0.15                 0.15
##      collapsedeparsef             converts
##                   0.15                 0.15
##                deparse               bquote
##                   0.15                 0.15
##                 messed                nicer
##                   0.15                 0.15
##               plotmath               titles
##                   0.15                 0.15
```

These word associations show us in what context the word "function" is used.

9.12 Colours in R

As a data scientist, you might not worry so much about colours; however, a good selection of colours makes a plot easy to read and helps the message to stand out, and a limited consistent choice of colours will make your document look more professional. Certainly we do not want to worry too much about colours, and R does a good job of making colour management easy, but still allow full customization.

To start with, R has 657 named colours built in. A list of those names can be obtained with the function `colours()`.

Hint – Using American English

While many programming languages will force the user to use one of the flavours of the English language, R allows everywhere (and even most package contributors respect this) both "color" and "colour."

This list of colours can be used to search for a colour whose name contains a certain string.

```
# find colour numbers that contain the word 'khaki'
grep("khaki",colours())
## [1]  83 382 383 384 385 386

# find the names of those colours
colors()[grep("khaki",colours())]
## [1] "darkkhaki" "khaki"     "khaki1"    "khaki2"
## [5] "khaki3"    "khaki4"
```

R allows also to define colours in different ways: named colours, RGB colours, hexadecimal colours, and it also allows to convert the one to the other.

```
# extract the rgb value of a named colour
col2rgb("khaki3")
##       [,1]
## red    205
## green  198
## blue   115
```

To illustrate the way colours can be addressed, we show a few plots. They do use a specific plotting library `ggplot2` that is introduced in Chapter 31 *"A Grammar of Graphics with ggplot2"* on page 687. For now, we suggest to focus on the colour definitions. The following code creates different plots and pareses them together into one set of three by two plots in one figure. That figure is shown in Figure 9.22 on page 192.

```
library(ggplot2)
library(gridExtra)

##
## Attaching package: 'gridExtra'
## The following object is masked from 'package:dplyr':
##
##     combine
```

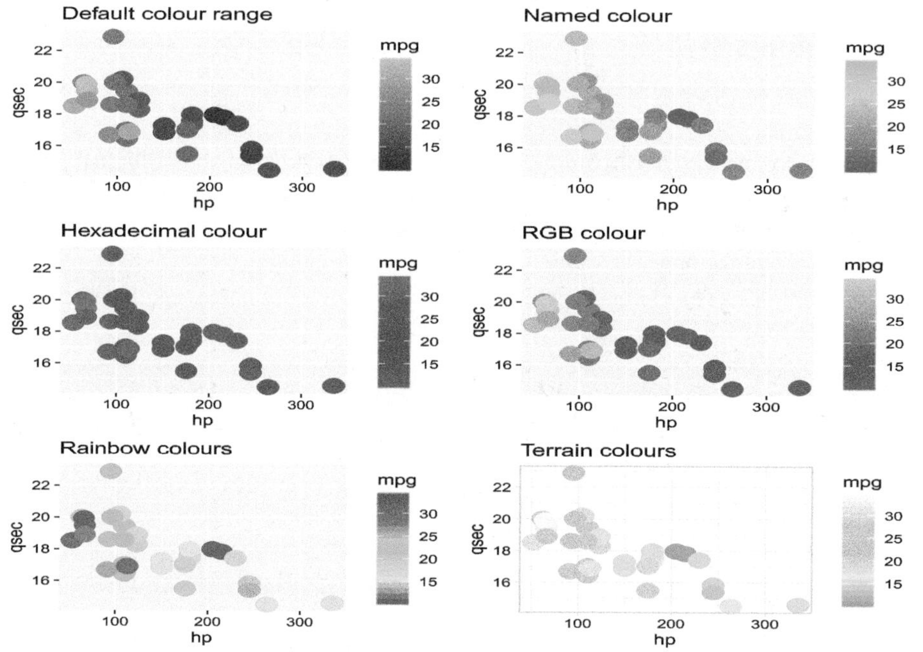

Figure 9.22: *An illustration of six predefined colour schemes in R. This figure uses the dataset* mtcars *and shows in each sub-plot the number of seconds the car needs to speed up to one fourth of a mile in function of its power (*hp*). The colouring of each observation (car type) is done in function of the fuel economy* mpg. *This presentation allows to visualize more than one relationship in one plot.*

```
p <- ggplot(mtcars, aes(x = hp, y = qsec, color = mpg)) +
      geom_point(size=5)

# no manual colour specification:
p0 <- p + ggtitle('Default colour range')

# using named colours:
p1 <- p + scale_color_gradient(low="red", high="green") +
      ggtitle('Named colour')

# using hexadecimal representation
p2 <- p + scale_color_gradient(low="#0011ee", high="#ff55a0") +
      ggtitle('Hexadecimal colour')

# RGB definition of colour
p3 <- p + scale_color_gradient(low=rgb(0,0,1), high=rgb(0.8,0.8,0)) +
      ggtitle('RGB colour')

# rainbow colours
p4 <- p + scale_color_gradientn(colours = rainbow(5)) +
      ggtitle('Rainbow colours')

# using another colour set
p5 <- p + scale_color_gradientn(colours = terrain.colors(5)) +
      ggtitle('Terrain colours') +
      theme_light()  # turn off the grey background

grid.arrange(p0, p1, p2, p3, p4, p5, ncol = 2)
```

We used ggplot2 as an example here – more about ggplot2 is in Chapter 31 *"A Grammar of Graphics with ggplot2"* on page 687. For ggplot2, the colour effects are provided by the functions:

scale_colour_grey()

- scale_colour_xxx() for points and lines;
- scale_fill_xxx() for surfaces in box plots, bar plots, violin plots, histograms, etc.

scale_fill_grey()

Hint – Online list of colours for R

A list of colours can for example be found here: http://www.stat.columbia.edu/~tzheng/files/Rcolor.pdf.

```
N   <- length(colours())  # this is 657
df  <- data.frame(matrix(1:N, nrow = 73, byrow = TRUE))
image(1:(ncol(df)), 1:(nrow(df)), as.matrix(t(df)),
      col = colours(),
      xlab = "X", ylab = "Y")
```

The plot is in Figure 9.23 on page 194. In this plot, the number of the colour can be found by the following formula:

$$nbr = (y - 1)9 + x$$

With y, the value on the y-axis, and x, the number on the x-axis. Once the formula is applied, the name of the colour is the colour with that number in R's list. Here are a few examples:

```
colours()[(3 - 1)  * 9 + 8]
## [1] "blue"

colours()[(50 - 1) * 9 + 1]
## [1] "lightsteelblue4"
```

Colour sets

The philosophy of R is that it is modular and easy to extend. There are many extensions that make working with colours easier and produce really nice results. For example, we can use colour sets as in the following example; the output of this code is in Figure 9.24 on page 195.

```
p <- ggplot(mtcars) +
     geom_histogram(aes(cyl, fill=factor(cyl)), bins=3)

# no manual colour specification:
p0 <- p + ggtitle('default colour range')


# using a built-in colour scale
p1 <- p + scale_fill_grey() +
      ggtitle('Shades of grey')


library(RColorBrewer)
p2 <- p + scale_fill_brewer() +
      ggtitle('RColorBrewer')

p3 <- p + scale_fill_brewer(palette='Set2') +
```

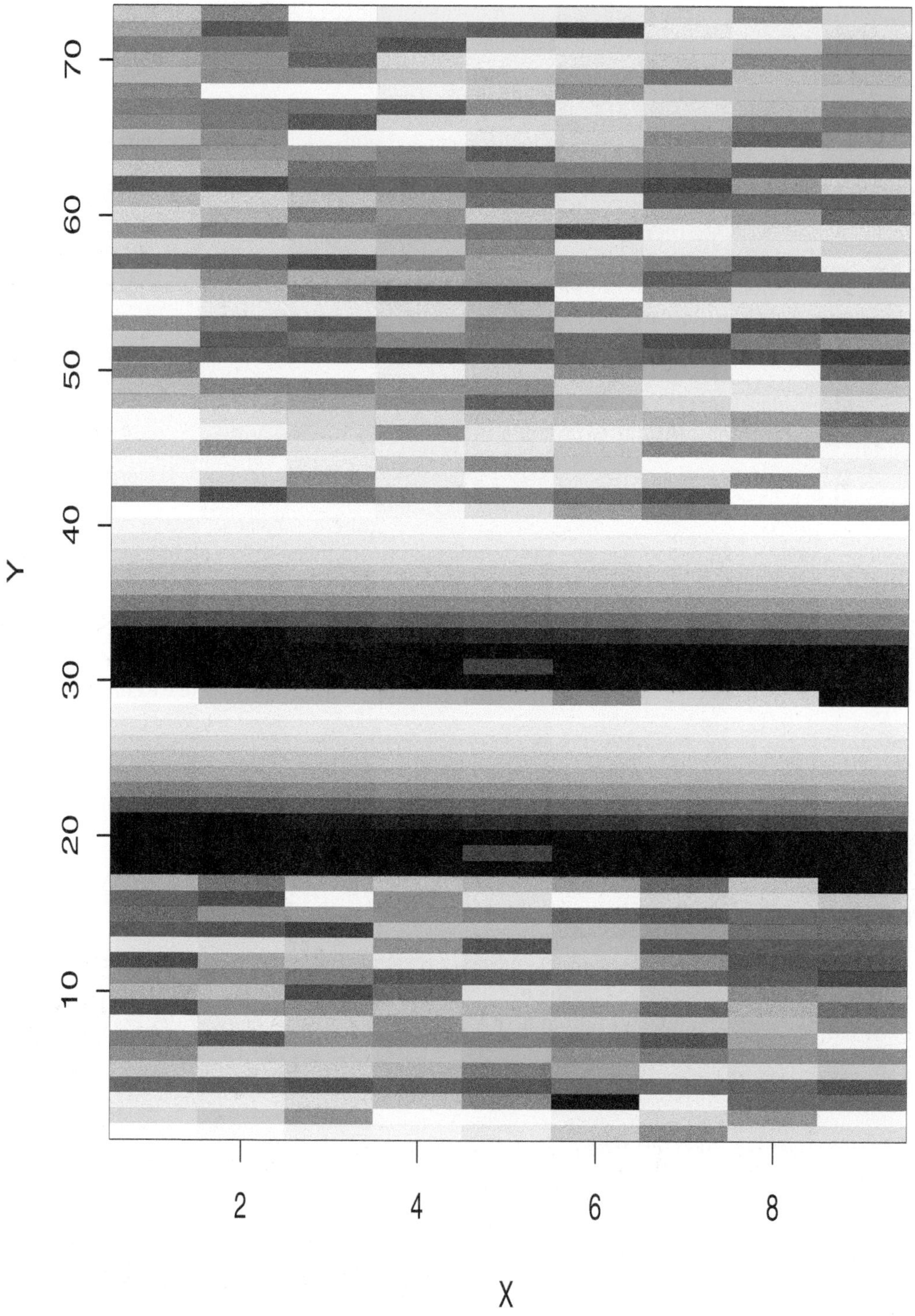

Figure 9.23: *A visualisation of all built in colours in R. Note that the number of the colour can be determined as by taking the y-value minus one times nine plus the x-value.*

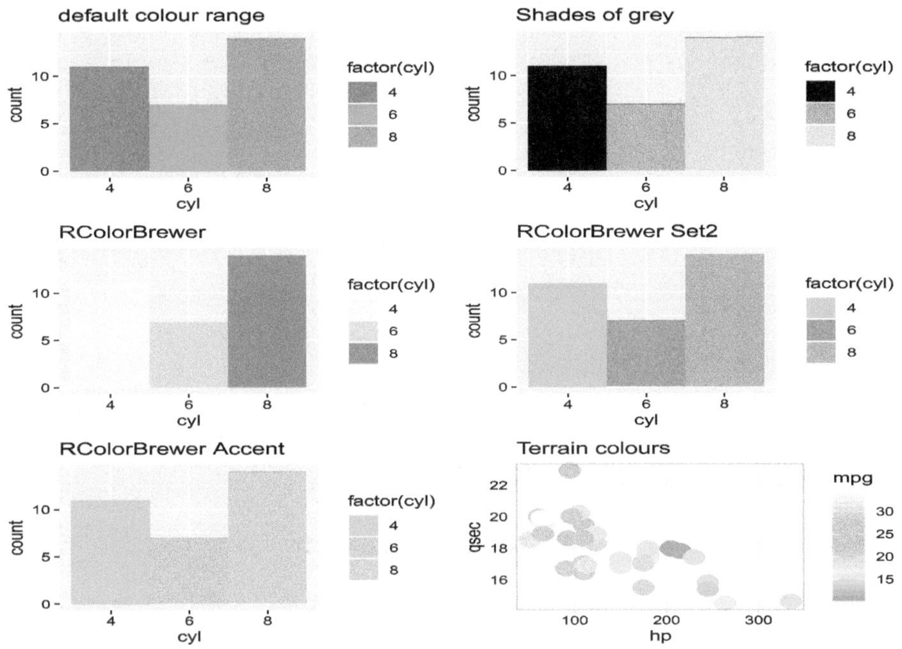

Figure 9.24: *Examples of discrete colour sets. The name of the colour-set is the title of each plot.*

```
    ggtitle('RColorBrewer Set2')

p4 <- p + scale_fill_brewer(palette='Accent') +
    ggtitle('RColorBrewer Accent')

grid.arrange(p0, p1, p2, p3, p4, p5, ncol = 2)
```

The scales provided by the library `RColorBrewer` provide colour scales that cater for different needs: sequential, diverging, and qualitative colour schemes are available to underline the pattern in the data or lack thereof. More information is on `http://colorbrewer2.org`.

Further information – More plots

We hope that with this section you have a solid toolbox to make plots and charts to illustrate your work. However, we did not cover everything and maybe you want some more eye-candy? We suggest to have a look at `https://www.r-graph-gallery.com`.

♣ 10 ♣

Time Series Analysis

10.1 Time Series in R

Time series are lists of data points in which each data point is associated with a time-stamp. A simple example is the price of a stock in the stock market at different points of time on a given day. Another example is the amount of rainfall in a region at different months of the year. R language uses many functions to create, manipulate and plot the time series data. The data for the time series is stored in an R object called time series object. It is also a R data object like a vector or data frame.

time series

time-stamp

10.1.1 The Basics of Time Series in R

10.1.1.1 The Function ts()

The time series object is created by using the `ts()` function.

`ts()`

Function use for ts()

```
ts(data = NA, start = 1, end = numeric(), frequency = 1,
        deltat = 1, ts.eps = getOption("ts.eps"), class = ,
        names = )
```

with

- `data`: A vector or matrix containing the values used in the time series.
- `start`: The start time for the first observation in time series.
- `end`: The end time for the last observation in time series.
- `frequency` The number of observations per unit time.
    - frequency = 12: pegs the data points for every month of a year.
    - frequency = 4: pegs the data points for every quarter of a year.
    - frequency = 6: pegs the data points for every 10 minutes of an hour.
    - frequency = 24×6: pegs the data points for every 10 minutes of a day.

The Big R-Book: From Data Science to Learning Machines and Big Data, First Edition. Philippe J.S. De Brouwer.

Companion Website: www.wiley.com/go/De Brouwer/The Big R-Book

Except the parameter "data" all other parameters are optional. To check if an object is a time series, we can use the function `is.ts()`, and `as.ts(x)` will coerce the variable x into a time series object.

To illustrate the concept of time series, we use an example of the stock exchange: the "S&P500." The S&P500 is short for the Standard and Poors stock exchange index of 500 most important companies in the USA. It is the evolution of a virtual basket, consisting of 500 shares. It is, of course, managed so that it is at any point a fair reflection of the 500 most significant companies. Since it creation in 1926, it has shown an year-over-year increase in 70% of the years. To a reasonably extend it is a barometer for the economy of the USA.

S&P500 It is possible to import this data in R, but the library MASS, that comes at no cost with the your free installation of R, has a dataset that contains the returns of the S&P500 of the 1990s: the dataset SP500:

```
library(MASS)
# The SP500 is available as a numeric vector:
str(SP500)
##  num [1:2780] -0.259 -0.865 -0.98 0.45 -1.186 ...
```

Now, we convert it to a time series object with the function `ts()`:

```
# Convert it to a time series object.
SP500_ts <- ts(SP500,start = c(1990,1),frequency = 260)
```

The object SP500 is not a time series class object:

```
# Compare the original:
class(SP500)
## [1] "numeric"

# with:
class(SP500_ts)
## [1] "ts"
```

The time series object (class ts) has its own S3 plotting functionality. All we need to do is use the standard `plot()` function to produce the plot in Figure 10.1 on page 199:

```
plot(SP500_ts)
```

10.1.1.2 Multiple Time Series in one Object

In many examples, time series are related. For example we can keep track of a patient's blood pressure, temperature and pulse at the same intervals. Also on the stock exchange, we have thousands of stocks that have a quotation at the same time. Therefore, it is useful to have the capability of treating those observations in one structure.

The time series object type ts, allows, just as a data-frame, multiple "columns." The following code illustrates how this works:

```
val = c(339.97)
for (k in 2:length(SP500)){
  val[k] = val[k-1] * (SP500[k-1] / 100 + 1)
}
# Convert both series to a matrix:
M <- matrix(c(SP500,val),nrow=length(SP500))

# Convert the matrix to a time series object:
SP <- ts(M, start=c(1990,1),frequency=260)
colnames(SP) <- c("Daily Return in Pct","Value")
```

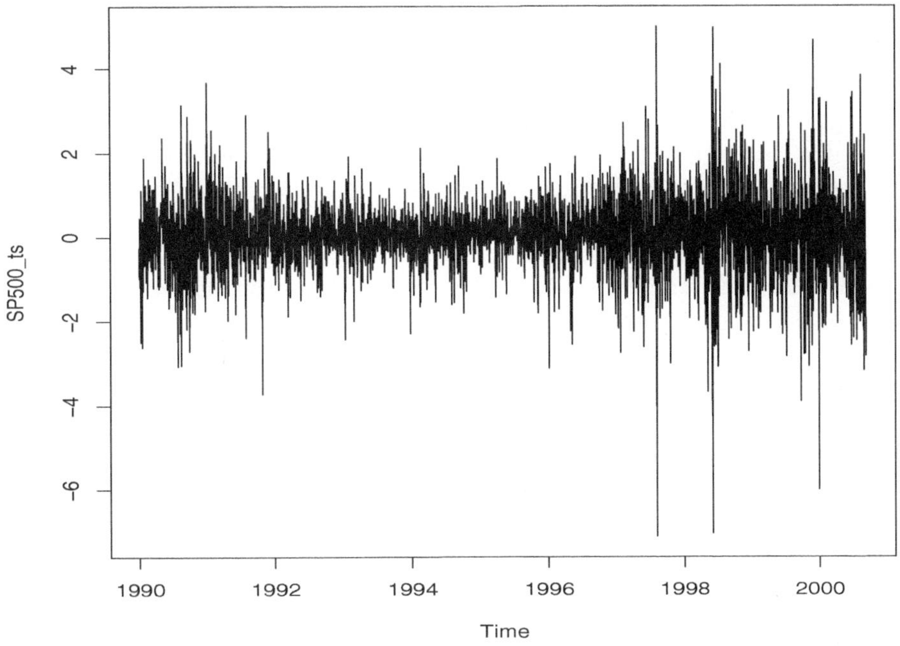

Figure 10.1: *The standard plot for a time series object for the returns of the SP500 index in the 1990s.*

For example, when plotting the ts object, we will see all its observations in stacked plots that share a common time-axis (see Figure 10.2)

```
plot(SP, type = "l", main = "SP500 in the 1990s")
```

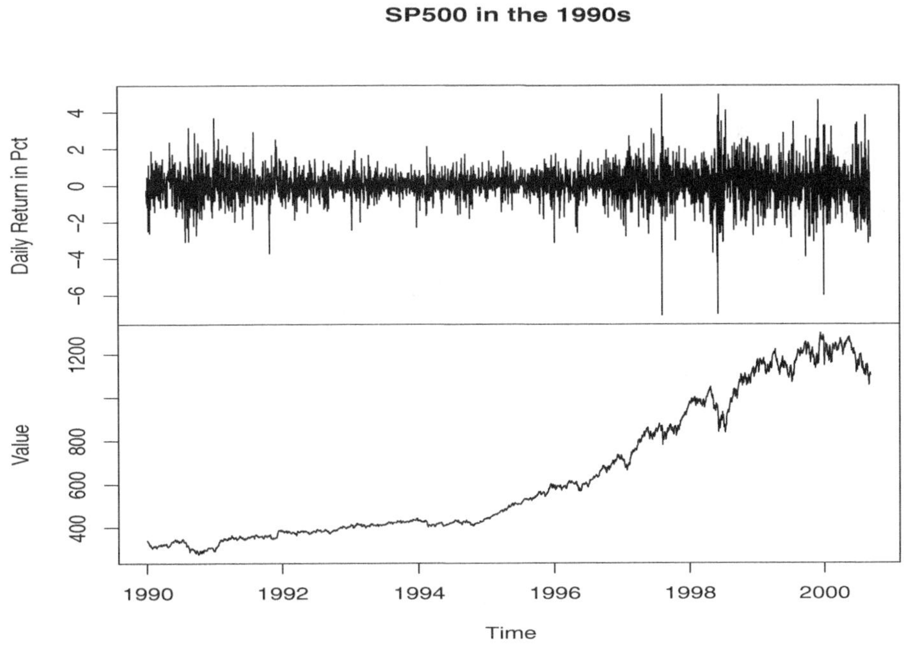

Figure 10.2: *The standard plot functionality of time series will keep the z-axis for both variables the same (even use one common axis).*

10.2 Forecasting

forecasting Forecasting is the process of making predictions about the future and it is one of the main reasons to do statistics. The idea is that the past holds valuable clues about the future and that by studying the past one can make reasonable suggestions about the future.

Every company will need to make forecasts in order to plan, every investor needs to forecast the market and company performance in order to make decisions, etc. For example, if the profit grew the last four years every year between 15% and 20%, it is reasonable to forecast a hefty growth rate based on this endogenous data. Of course, this makes abstraction of exogenous data, such as an economic depression in the making. Indeed global economic crisis or pandemics do not warn us in advance. They are outside most models ("exogeneous").

No forecast is absolutely accurate, and the only thing that we know for sure is that the future will hold surprises. It is of course possible to attach some degree of certainty to our forecast. This is referred to as "confidence intervals," and this is more valuable than a precise prediction.

There are a number of quantitative techniques that can be utilized to generate forecasts. While some are straightforward, others might for example incorporate exogenous factors and become necessarily more complex.

Our brain is in essence an efficient pattern-recognition machine that is so efficient that it will even see patterns where there are none.[1]

Typically, one relies on the following key concepts:

- *Trend:* A trend is a long-term increase or decrease (despite short-term fluctuations). This is a specific form of auto-correlation (see below).
- *Seasonality:* A seasonal pattern is a pattern that repeats itself (e.g. sales of ice-cream in the summer, temperatures in winter compared to summer).
- *Autocorrelation:* This refers to the phenomena whereby values at time t are influenced by previous values. R provides an autocorrelation plot that helps to find the proper lag structure and the nature of auto correlated values.
- *Stationarity:* A time series is said to be stationary if there is no systematic trend, no systematic change in variance, and if strictly periodic variations or seasonality do not exist.
- *Randomness:* A time series that shows no specific pattern (e.g. a random walk or Brownian motion). Note that there can still be a systematic shift (a long term trend).

10.2.1 Moving Average

10.2.1.1 The Moving Average in R

moving average In absence of a clear and simple trend (such as a linear or exponential trend) the moving average is a versatile tool. it is a non-parametric model that simply "forecasts" the near future based on the average observations of the near past.

[1]This refers to the bias "hot hand fallacy" – see e.g. De Brouwer (2012)

Example: – GDP data

When it comes to macro economical data, the World Bank is a class apart. It's website `https://data.worldbank.org` has thousands of indicators that can be downloaded and analysed. Their data catalogue is here: `https://datacatalog.worldbank.org`. We have downloaded the GDP data of Poland and stored it in a csv-file on our hard-disk. In the example that we use to explain the concepts in the following sections, we will use that data.

To start, we load in the data stored in a csv file on our local hard-drive, and plot the data in Figure 10.3

```
g <- read.csv('data/gdp/gdp_pol_sel.csv') # import the data
attach(g) # the names of the data are now always available
plot(year, GDP.per.capitia.in.current.USD, type='b',
     lwd=3, xlab = 'Year', ylab = 'Polish GDP per Capita in USD')
```

The next step is creating the moving average forecast. To do this we use the package `forecast`: `forecast`

```
require(forecast)

## Loading required package: forecast

# make the forecast with the moving average (ma)
g.data <- ts(g$GDP.per.capitia.in.current.USD,start=c(1990))
g.movav = forecast(ma(g.data, order=3), h=5)
```

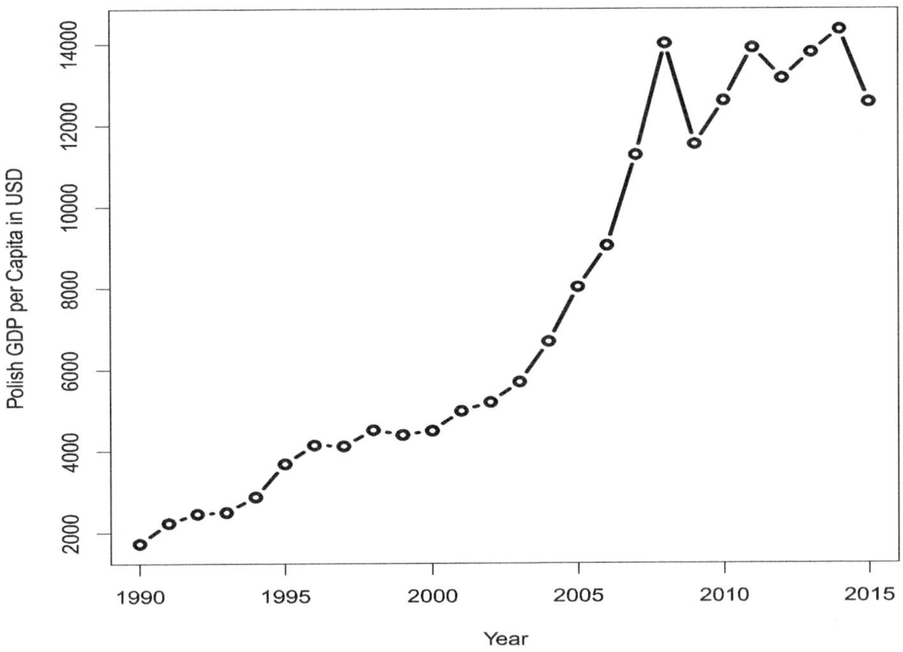

Figure 10.3: *A first plot to show the data before we start. This will allow us to select a suitable method for forecasting.*

Now, we can plot the forecasted data together with the source data as follows with the following code. The plot is in Figure 10.4 on page 202.

```
# show the result:
plot(g.movav,col="blue",lw=4,
     main="Forecast of GDP per capita of Poland",
     ylab="Income in current USD")
lines(year,GDP.per.capitia.in.current.USD,col="red",type='b')
```

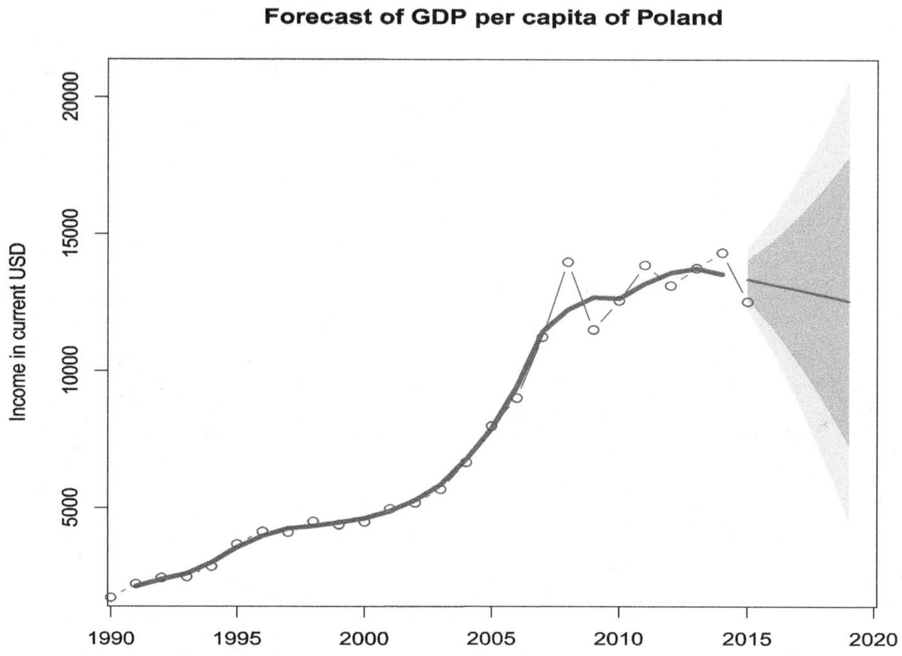

Figure 10.4: *A forecast based on moving average.*

10.2.1.2 Testing the Accuracy of the Forecasts

accuracy() The package `forecast` also provides us the tools to check the accuracy of those forecasts: the function `forecast()`, will do this in an elegant way:

Testing the Accuracy of Forecasts – Backtesting

backtesting

sampling

```
# Testing accuracy of the model by sampling:
g.ts.tst <- ts(g.data[1:20],start=c(1990))
g.movav.tst <- forecast(ma(g.ts.tst,order=3),h=5)
accuracy(g.movav.tst, g.data[22:26])
##                      ME      RMSE       MAE        MPE
## Training set    32.0006  342.1641  217.8447  0.7619824
## Test set      -1206.5948 1925.5738 1527.0227 -9.2929075
##                    MAPE      MASE        ACF1
## Training set  3.229795 0.3659014 -0.06250102
## Test set     11.599237 2.5648536          NA
```

It is also easy to plot the forecast together with the observations and confidence intervals. This is shown in the following code and the result is plotted in Figure 10.5 on page 203.

```
plot(g.movav.tst,col="blue",lw=4,
     main="Forecast of GDP per capita of Poland",
     ylab="Income in current USD")
lines(year, GDP.per.capitia.in.current.USD, col="red",type='b')
```

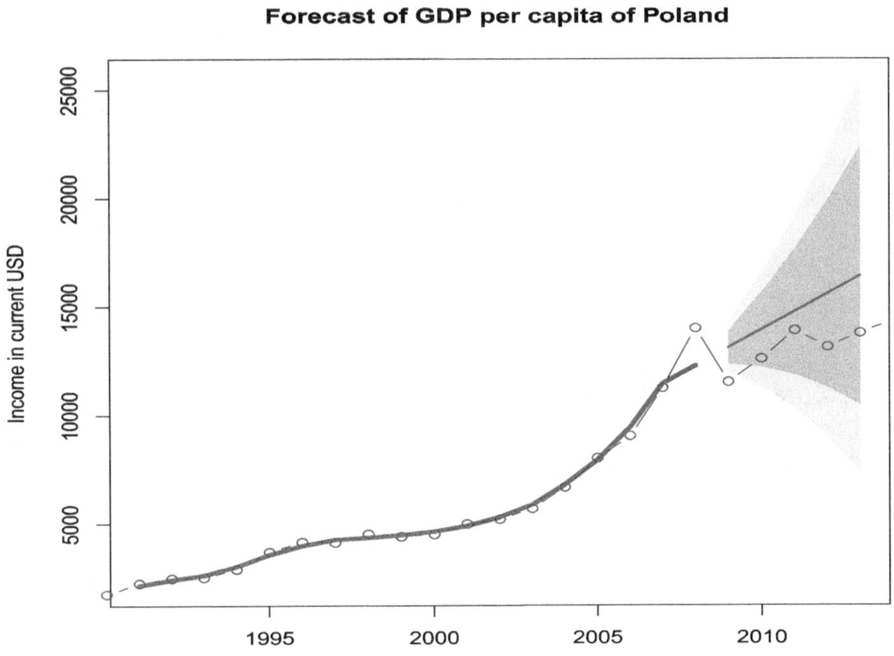

Figure 10.5: *A backtest for our forecast.*

In the forecast package, there is an automated forecasting function that will run through possible models and select the most appropriate model give the data. This could be an auto regressive model of the first order (AR(1)), an ARIMA model (autoregressive integrated moving average model) with the right values for p, d, and q, or even something else that is more appropriate. The following code uses those functions to plot a forecast in Figure 10.6 on page 204. **ARIMA**

```
train = ts(g.data[1:20],start=c(1990))
test  = ts(g.data[21:26],start=c(2010))
arma_fit <- auto.arima(train)
arma_forecast <- forecast(arma_fit, h = 6)
arma_fit_accuracy <- accuracy(arma_forecast, test)
arma_fit; arma_forecast; arma_fit_accuracy
## Series: train
## ARIMA(0,1,0) with drift
##
## Coefficients:
##          drift
##       515.5991
## s.e.  231.4786
##
## sigma^2 estimated as 1074618:  log likelihood=-158.38
## AIC=320.75   AICc=321.5   BIC=322.64
##      Point Forecast    Lo 80    Hi 80     Lo 95    Hi 95
## 2010       12043.19 10714.69 13371.70 10011.419 14074.97
## 2011       12558.79 10680.00 14437.58  9685.431 15432.15
## 2012       13074.39 10773.35 15375.43  9555.257 16593.52
```

```
## 2013        13589.99 10932.98 16247.00  9526.444 17653.54
## 2014        14105.59 11134.96 17076.22  9562.406 18648.77
## 2015        14621.19 11367.03 17875.35  9644.381 19598.00
##                       ME      RMSE      MAE        MPE
## Training set  0.06078049  983.4411 602.2902 -2.3903997
## Test set     54.36338215 1036.0989 741.8024  0.1947719
##                  MAPE      MASE        ACF1 Theil's U
## Training set 8.585820 0.7612222 -0.03066223        NA
## Test set     5.668504 0.9375488  0.04756832 0.9928242
```

auto.arima()
forecast()

```
plot(arma_forecast, col="blue",lw=4,
     main="Forecast of GDP per capita of Poland",
     ylab="income in current USD")
lines(year,GDP.per.capitia.in.current.USD,col="red",type='b')
```

ARIMA
autoregressive integrated moving average model
ARMA
autoregressive moving average model

Note that ARIMA stands for autoregressive integrated moving average model. It is a generalization of an autoregressive moving average (ARMA) model. Both of these models are fitted to time series data either to better understand the data or to predict future points in the series (forecasting). ARIMA models are applied in some cases where data shows evidence of non-statrionarity, where an initial differencing step (corresponding to the "integrated" part of the model) can be applied one or more times to eliminate the non-stationarity.

10.2.1.3 Basic Exponential Smoothing

exponential smoothing

Exponential smoothing assigns higher weights to the most recent observations (the weight will decrease exponentially for older observations). The effect will be that a new dramatic event has a much faster impact, and that the “memory of it” will decrease exponentially.

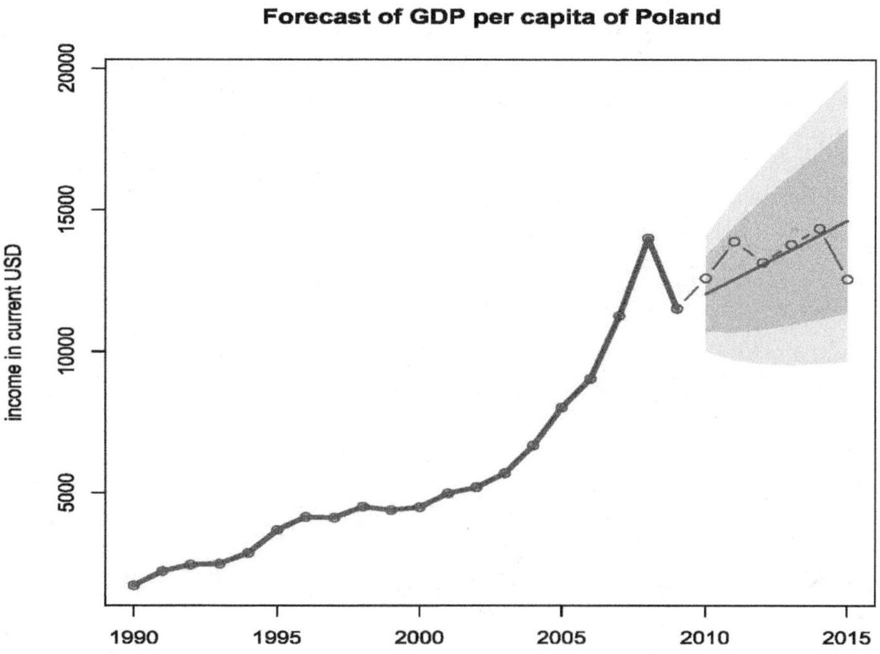

Figure 10.6: *Optimal moving average forecast.*

The package `forecast` provides the function `ses` to execute this as follows:

```
g.exp <- ses(g.data,5,initial="simple")
g.exp  # simple exponential smoothing uses the last value as
##      Point Forecast       Lo 80    Hi 80     Lo 95    Hi 95
## 2016       12558.87 11144.352 13973.39 10395.550 14722.19
## 2017       12558.87 10558.438 14559.30  9499.473 15618.27
## 2018       12558.87 10108.851 15008.89  8811.889 16305.85
## 2019       12558.87  9729.832 15387.91  8232.229 16885.51
## 2020       12558.87  9395.909 15721.83  7721.538 17396.20

       # the forecast and finds confidence intervals around it
```

Plotting the results is made easy by R's function based OO model, and we can simply use the standard plot functionality – the results are in Figure 10.7: `ses()`

```
plot(g.exp,col="blue",lw=4,
     main="Forecast of GDP per capita of Poland",
     ylab="income in current USD")
lines(year,GDP.per.capitia.in.current.USD,col="red",type='b')
```

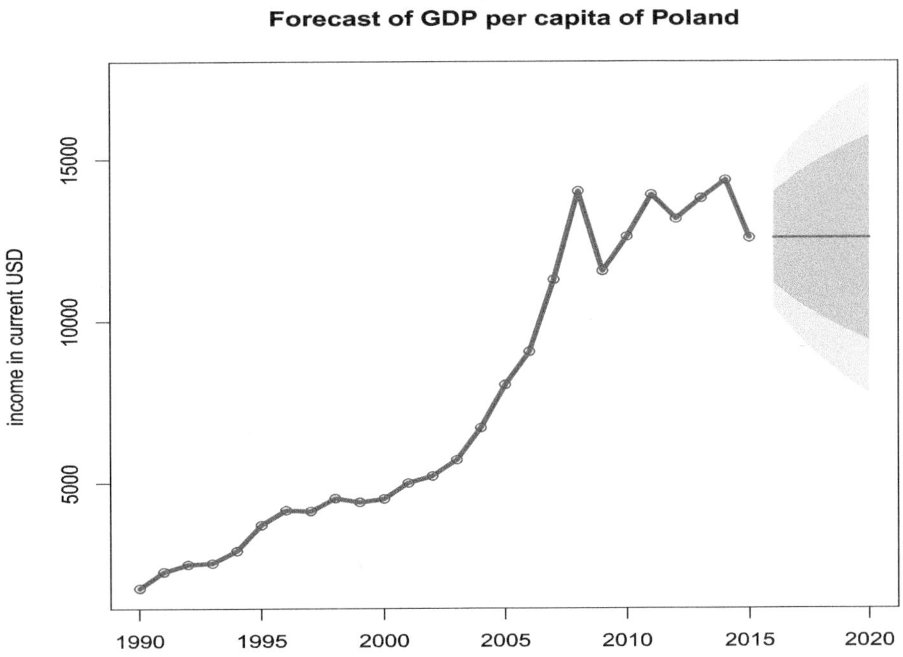

Figure 10.7: *Forecasting with an exponentially smoothed moving average.*

10.2.1.4 Holt-Winters Exponential Smoothing

Triple Exponential Smoothing, or Holt-Winters Exponential Smoothing, is of particular interest for time-series, because it is able to take into account a seasonal trend. To do this in R, the package `forecast` brings us the function `holt()`: `holt()`

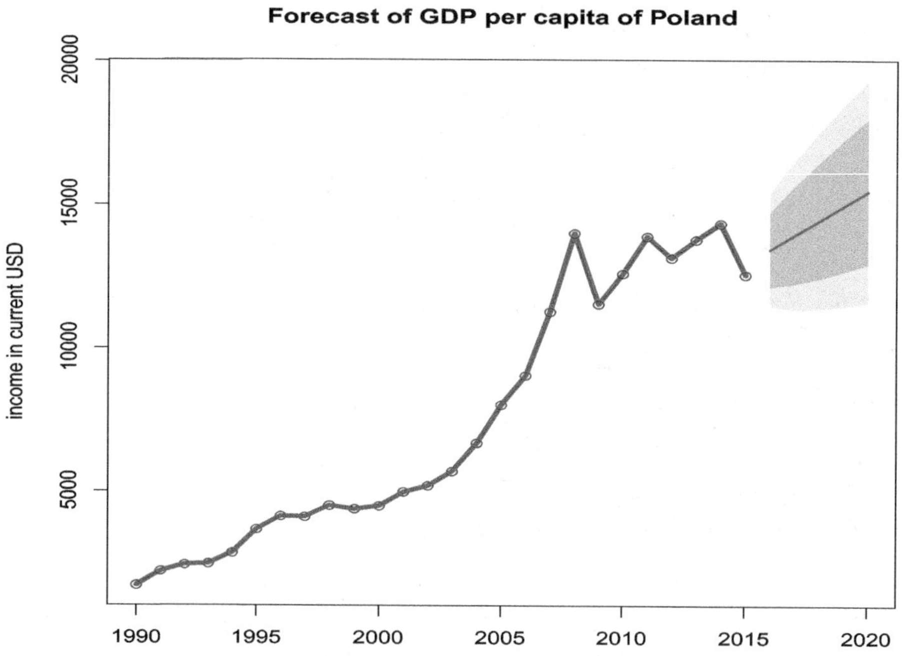

Figure 10.8: *Holt exponentially smoothed moving average.*

Holt Exponential smoothing

Holt Exponential Smoothing

```
g.exp <- holt(g.data,5,initial="simple")
g.exp  # Holt exponential smoothing
##         Point Forecast     Lo 80    Hi 80    Lo 95    Hi 95
## 2016          13445.71 12144.40 14747.02 11455.53 15435.89
## 2017          13950.04 12257.07 15643.01 11360.86 16539.22
## 2018          14454.37 12444.67 16464.07 11380.80 17527.94
## 2019          14958.70 12675.80 17241.60 11467.31 18450.10
## 2020          15463.04 12936.30 17989.77 11598.73 19327.34
```

Again, plotting is made simple. The result is show in Figure 10.8

```
plot(g.exp,col="blue",lw=4,
     main="Forecast of GDP per capita of Poland",
     ylab="income in current USD")
lines(year,GDP.per.capitia.in.current.USD,col="red",type='b')
```

STL

10.2.2 Seasonal Decomposition

Loess

STL
seasonal trend

`stl()`

The seasonal trend decomposition using Loess (STL) is an algorithm that was developed to help to divide up a time series into three components namely: the trend, seasonality and remainder. The methodology was presented by Robert Cleveland, William Cleveland, Jean McRae, and Irma Terpenning in the *Journal of Official Statistics* in 1990. The STL is available within R via the `stl()` function.

The use of the function `stl` is straightforward and shown in the following code. In the last line, we plot the data – the plot is in Figure 10.9.

```
# we use the data nottem
# Average Monthly Temperatures at Nottingham, 1920-1939
nottem.stl = stl(nottem, s.window="periodic")
plot(nottem.stl)
```

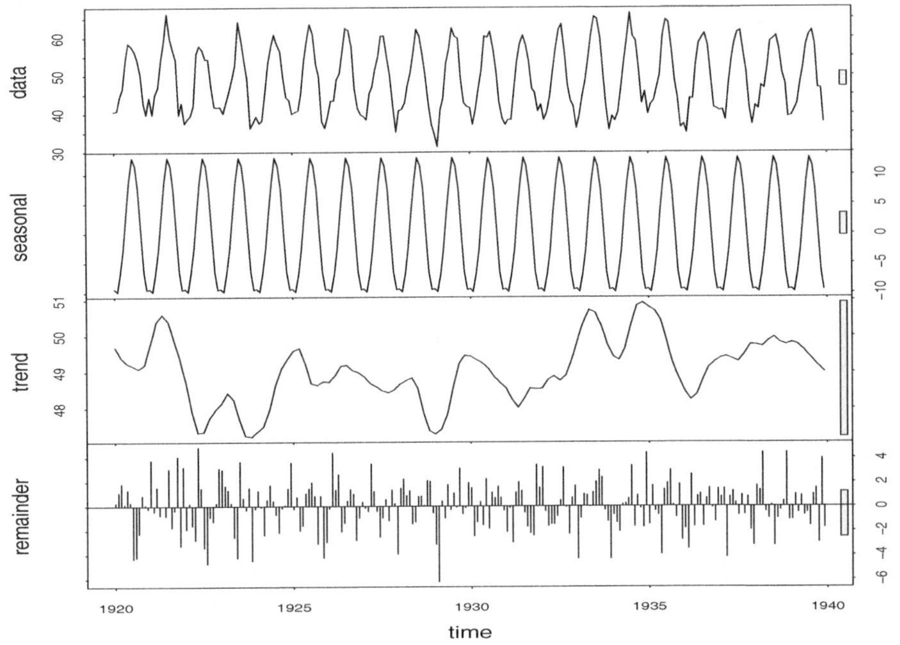

Figure 10.9: *Using the stl-function to decompose data in a seasonal part and a trend.*

The four graphs are the original data, seasonal component, trend component and the remainder and this shows the periodic seasonal pattern extracted out from the original data and the trend that moves around between 47 and 51 degrees Fahrenheit. There is a bar at the right hand side of each graph to allow a relative comparison of the magnitudes of each component. For this data the change in trend is less than the variation doing to the monthly variation.

Note – Exponential trends

Note that a series with multiplicative effects can often by transformed into series with additive effects through a `log` transformation. For example, assume that we have a time series object `exp_ts`, then we can transform it into one with additional interdependence (`add_ts`) via:

```
add_ts <- log(exp_ts)
```

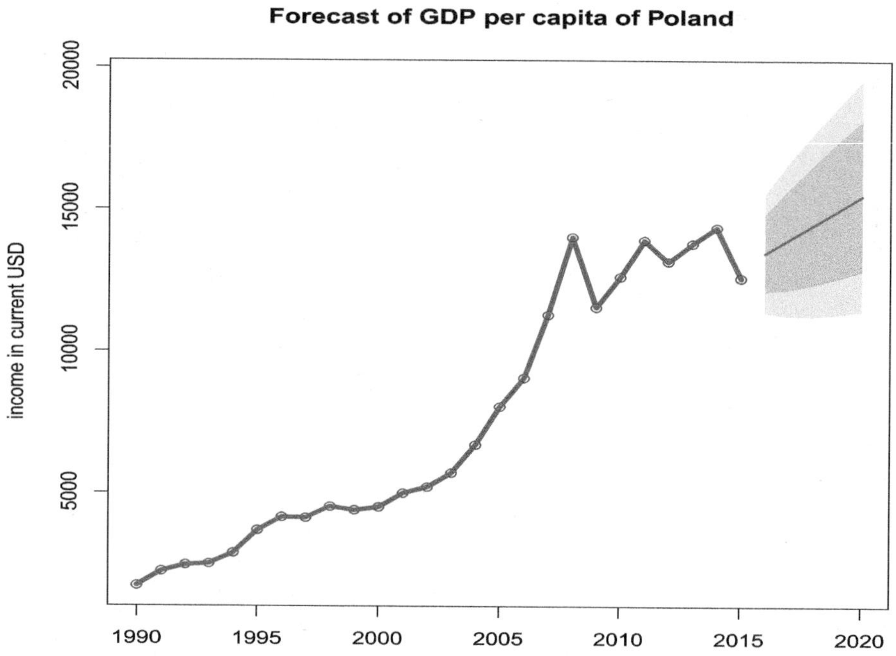

Figure 10.10: *The Holt-Winters model fits an exponential trend. Here we plot the double exponential model.*

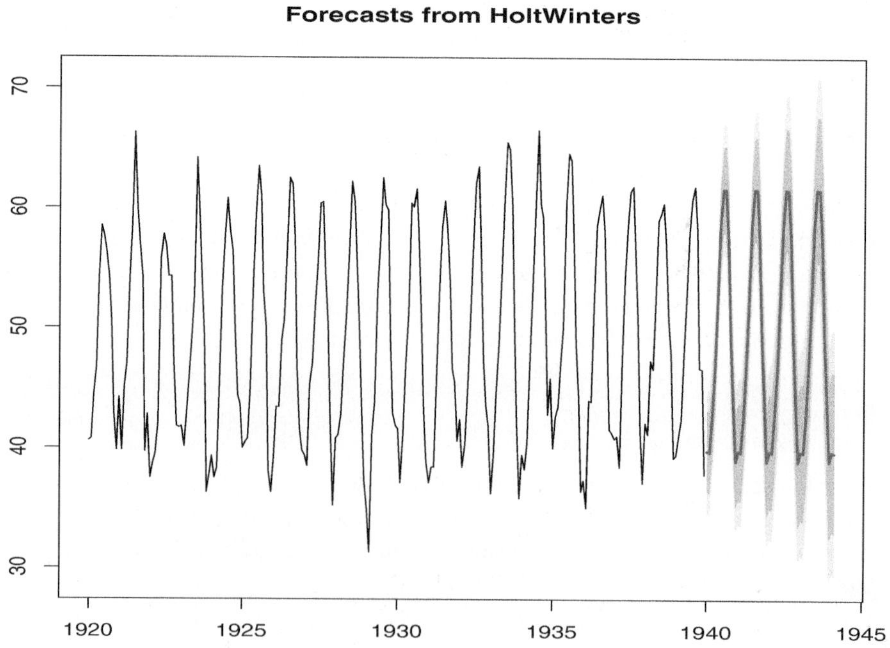

Figure 10.11: *The Holt-Winters model applied to the temperatures in Nottingham.*

Exponential Models

Both the `HoltWinters()` function from the package `stats`, and the `ets()` function in the forecast package, can be used to fit exponential models.

model – exponential

```
# Simple exponential: models level
fit <- HoltWinters(g.data, beta=FALSE, gamma=FALSE)

# Double exponential: models level and trend
fit <- HoltWinters(g.data, gamma=FALSE)

# Triple exponential: models level, trend, and seasonal
# components. This fails on the example, as there is no
# seasonal trend:
#fit <- HoltWinters(g.data)

# Predictive accuracy
library(forecast)
accuracy(forecast(fit,5))
##                    ME     RMSE      MAE       MPE
## Training set -69.84485 1051.488 711.7743 -2.775476
##                  MAPE      MASE        ACF1
## Training set 9.016881 0.8422587 0.008888197
```

Again, we can use the function `forecast()` to generate forecasts, based on the model:

```
# predict next 5 future values
forecast(fit, 5)
##      Point Forecast    Lo 80    Hi 80    Lo 95    Hi 95
## 2016       13457.63 12084.15 14831.11 11357.07 15558.18
## 2017       13961.96 12179.69 15744.23 11236.21 16687.71
## 2018       14466.29 12352.87 16579.71 11234.10 17698.49
## 2019       14970.62 12571.33 17369.91 11301.23 18640.02
## 2020       15474.95 12820.41 18129.50 11415.17 19534.74
```

We will plot the forecast together with the existing data (that we add via the function `lines()`). The result is in Figure 10.10 on page 208

```
plot(forecast(fit, 5),col="blue",lw=4,
     main="Forecast of GDP per capita of Poland",
     ylab="income in current USD")
lines(year,GDP.per.capitia.in.current.USD,col="red",type='b')
```

Since, the GDP data did not show a seasonal trend, and hence it was not possible to use the triple exponential mode, we will consider another example. Temperatures should certainly display a seasonal trend. The dataset `nottem` – from the package `datasets`, that is loaded with base R – contains the daily temperatures in Nottingham between 1920 and 1939. In the example, we use the `HolWinters` function and plot the results in Figure 10.11 on page 208.

```
# Use the Holt-Winters method for the temperatures
n.hw <- HoltWinters(nottem)
n.hw.fc <- forecast(n.hw,50)
plot(n.hw.fc)
```

Question #11

Use the moving average method on the temperatures in Nottingham (nottam). Does it work? Which model would work better?

♣ 11 ♣

Further Reading

Hopefully you got a taste of R and you liked it so far. The rest of the book will use R to dive into data (see Part III *"Data Import"* on page 213), data wrangling (see Part IV *"Data Wrangling"* on page 259), modelling and machine learning (see Part V *"Modelling"* on page 373), companies and corporate environment (see Part VI *"Introduction to Companies"* on page 565) and reporting (see Part VII *"Reporting"* on page 685). If you want to learn more about the foundations of R then there are many excellent books and online documentation available.

Hint – What if you are stuck?

- `https://stackexchange.com` and `http://www.google.com` are your friends for questions when you are stuck with something in particular. With a user base of millions of people it is more than likely that someone else experienced the same problem as you and that the answer is already there.
- `https://cran.r-project.org/doc` is also a great place to do further thematic research and grab free documentation.

Further information – CRAN

The website of the R is `https://cran.r-project.org` and it carries also plenty of documentation. Here is a selection:

- `/doc/manuals/R-intro.html:` another take on what we have covered in this part.
- `/doc/manuals/R-FAQ.html` A faq with very specific information that never will be included in a book like these but that might be relevant if you have happen to have certain backgrounds (such as S for example)
- `/doc/manuals/R-lang.html` describes in much more detail the internal workings of the language.

The Big R-Book: From Data Science to Learning Machines and Big Data, First Edition. Philippe J.S. De Brouwer.

Companion Website: www.wiley.com/go/De Brouwer/The Big R-Book

- `/doc/manuals/R-data.html`: covers the data-import and data-export functions in great detail. Best to read after next part if you happen to be on the data-side of the development cycle.
- `/doc/manuals/R-exts.html` will help you to create your very own packages. This makes a lot of sense, even if your company want to keep them propriety to the company. Capturing all the logic of your specifics in a package will save enormous amounts of time and money. Just imagine that you write once how to import data from your corporate mainframe or central data-warehouse. Hundreds of programmers and thousands of modellers can use the same package. Model validators only need to review this module once, etc.

The information about R and its ecosystem is growing fast and probably a search engine like DuckDuckGo or Qwant is an ideal starting point, rather than a book that is more static. I will maintain a web-page for this book at `http://www.de-brouwer.com/r-book`, where you can find all the code for this book but also further references and examples.

PART III

Data Import

♣ 12 ♣

A Short History of Modern Database Systems

Before we dive into extracting data and modelling, it is worth spending a few minutes to understand how the data is stored. There are many data storage systems available and most are in use. Many companies started investing in computers in the late 1950s or in the 1960s. IBM was an important player in the market of electronic computers and software and developed a particular useful computer language in the 1950s: FORTRAN. FORTRAN became soon the software of choice for many companies. Later developments allowed for structured programming (in a nutshell this is less using the `goto lineNbr` command and more use constructs such as `if`, `for`, etc.)[1]

FORTRAN

Many large companies have since been building on their code and adding functionality. This means that some code is about retirement age (60 years old).[2]

DBMS

DBMS – navigational

punched cards

Fortran outlived the punched cards, the tape readers and also encountered the navigational database management systems (DBMS) in the 1960s. In the age of tapes, any data was simply a list of things. It is at the advent of the disk and drum readers that "random access"[3] to data became possible and that the term "database" came in use. The technologies from the 1960s were a leap ahead and allowed more complex data and more relations between data to be stored and accessed.

CODASYL

COBOL

As the number of computers, computer users and applications grew faster and faster, it can of course be expected that the number of database systems also grew faster. Soon the need for a standard imposed itself and the "Database Task Group" was formed within CODASYL (the same group that created the COBOL programming language).

IBM

That standard was designed for a simple sequential approach of data search. In that early standard, there was already the concept of a primary key (called "the CALC key") as well as the concept of relations. However, all records should be searched in a sequential order. Soon there were many variations and implementations of the CODASYL standard. IBM had a slightly different system: IMS (Information Management System) – designed for the Apollo program on their System/360.

BASIC
Fortran II

[1] Maybe the reader will not know FORTRAN, but most probably you are familiar with "BASIC." BASIC is derived from Fortran II. BASIC improved on Fortran by having improved logical structures.

[2] One of the author's previous employers and once the bank with the largest balance sheet in the world has still such gems in their payment systems. Indeed, why would one replace code that works and is literally interlinked with thousands of other systems and software.

[3] Random access means that each piece of data can be accessed when it is needed, as opposed to a tape that has to be read sequentially. So if for example addresses are stored on a tape and you need the address of a person, then there is no other way than reading name after name till we find the right match.

The Big R-Book: From Data Science to Learning Machines and Big Data, First Edition. Philippe J.S. De Brouwer.

Companion Website: www.wiley.com/go/De Brouwer/The Big R-Book

The 1970s saw the dawn of the relational database management system (RDBMS) or the relations database system that till today dominates data storage. The standard of RDBMS improved on the CODASYL standard by adding an efficient search facility that made the best use of a random access storage. The focus on records that were stored in a hierarchical system that was the core of the CODASYL system also was replaced by the concept "table," and it allowed the content of the database to evolve without too much overhead to rewrite links and pointers.

RDBMS

table

Also in theory and application, the RDBMS was superior as it allows both hierarchical and navigational models to be implemented. The design can be tuned for performance or storage optimization and ease to update data.[4]

Edgard Codd not only invented the RDBMS but also provided a set-oriented language based on tuple calculus (a field to which he contributed greatly). This language would later be transformed into SQL.

INGRES
System R
QUEL

Soon RDBMS was implemented in different systems. It is rare to encounter systems from the 1970s such as INGRES or System R (which has its own query language: QUEL).

Later IBM created SQL/DS based on their System R and later on introduced DB/2 in the 1980s. In those days, the author worked in the IT department of an insurance company, who was one of the early adaptors of the DB2 system. The new database system of IBM was extremely fast and powerful compared to the other mainframe solutions that were used till then. Many systems from those days are still in use today, and it is not uncommon to encounter a DB2 database server in the server rooms.

Oracle

Oracle was founded in 1977, and its database solutions started to conquer the world. For a long time the strategy of that company that is now (2018) the world's fourth largest software company was based on one product: a RDBMS that had the same name as the company. Oracle started also to supply additions to SQL with their SQL-PL[5] that is to be Turing complete.

PC
personal computer

The 1980s saw also the advent of a personal computer (henceforth, PC) that was powerful and affordable enough to conquer the workfloor of any company. At the beginning of the 1980s, most employees would not have a computer on their desk, but when the 1980s came to their end, everyone (but the security guard) who had a desk would have had a computer on it. Once the 1990s ended, also the security guard got his/her PC.

This also meant that many databases would be held on a PC, and soon solutions like Lotus 1-2-3 and dBase conquered the market. However, today they are all replaced by their heirs MS Access and MS Excel (or eventually free alternatives such as Base and LibreOffice).

OO

After the release of C++ in 1985 by Bjarne Stoustrup, we had to wait till the 1990s to see object oriented programming (henceforth, OO) really take off. Data-storage had become more affordable and available there was more data than ever, and software became more and more complex. Since in many applications data plays the principle role and the rest of the code only transforms data, it made sense to wrap code around data instead of around procedural logic. The OO concept had also many other advantages, for example it made code easier to maintain.

ORM
object-relational mappings

Programmers preferred not to follow the strict "normal form" logic of building a relational database in its purest form. Rather it made sense to build databases that followed the objects. While it is possible to implement this directly in a RDBMS, a cleaner mapping is possible. In fact the relations between data become relations between objects and their attributes – as opposed to relations between individual fields. This gap with the relational data model was called "the object-relational impedance mismatch." In fact it is the inconvenience of translating between

---

[4] We will show how the RDBMS can work in next section.

[5] The "PL" in SQL-PL translates as "procedural language." It is an additional set of instructions that allows to create functions, use variables, and creates a programming language around SQL. Users of MySQL probably will find the SQL-PL environment similar in functionality (though will need to adapt to the other syntax).

programmed objects and database tables made explicit. This gave rise to "object databases" and "object-relational databases." Other approaches involved software libraries that programmers could use to map objects to tables. These were called object-relational mappings (ORMs) attempt to solve the same problem.

object databases

The amount of data available continued to grow exponentially: the Internet became more popular and as soon as every household had its personal computer we saw mobile phones even faster conquering the world and producing data at ever increasing rates. Recently the Internet of Things (henceforth, IoT) came to further enhance this trend.

IoT
Internet of things

This made it impractical to bring data to the processing unit, and we needed to turn around the concept and bring the processors closer to the data: Big Data was born. Distributed computing such as Hadoop allows to use cheap and redundant hardware and dispatch calculations to those CPUs that are close to the data that is needed for a certain calculation and then bring the results together for the user. This revived the NoSQL concept. The concept actually dates from the 1950s, but was largely neglected till it became impossible to recalculate indexes because databases were so large and data was coming in faster than ever before.

Big Data

NoSQL

NoSQL allows for horizontal scaling as well as for relational rules to be broken, so they do not require table schemes to be well defined or waste too much time on recalculating indices.[6]

A recent development is NewSQL. It is a new approach to a RDBMS that aims to provide the same scalable performance as NoSQL systems for transaction processing (read-write) workloads; still using SQL and maintaining the atomicity, consistency, isolation, and durability (ACID) guarantees of a traditional database system.[7]

NewSQL

ACID

When working with data in a corporate setting, it is common to encounter older systems (or "legacy systems") or RDBMS that understand (a dialect of) SQL, desktop databases (in the form of MS Access or MS Excel), and big data systems mixed together. However, the most popular are those systems that are an RDMS and understand SQL.

Because of the importance of relational database systems, we will study them in more detail in next chapter: Chapter 13 "RDBMS" on page 1.

[6]As distributed databases grew in popularity the search for high partition tolerance was on. However, we found that (due to the "CAP theorem") it is impossible for a distributed database system to simultaneously provide (1) consistency, (2) availability, and (3) partition tolerance guarantee. It seems that a distributed database system can satisfy two of these guarantees but not all three at the same time.

[7]A database system that is ACID compliant will respect rules of atomicity, consistency, isolation and durability. This means that the database system would guarantee intended to guarantee validity even when errors occur or power failures happen.

RDBMS

In a nutshell, a relational database management system (RDBMS) is a set of tables (rectangular data), where each piece of information is only stored once and where relations are used to find the information needed. Usually, they are accompanied by an intelligent indexing system that allows searching in logarithmic times.

RDBMS

relational database management system

Consider a simple example that will demonstrate the basics of a relational database system. Imagine that we want to create a system that governs a library of books. There are multiple ways to do this, but the following tables are a good choice to get started:

- Authors, with their name, first name, eventually year of birth and death (if applicable) – A table of authors: Table 13.1 on page 220;
- Books, with title, author, editor, ISIN, year, number of pages, subject code, etc. – A table of books: Table 13.2 on page 220;
- Subject codes, with a description – A table of genres: Table 13.3 on page 220.

The tables have been constructed so that they already apply with some principles that underbuild relational database systems. For example, we see that certain field from the Author table are used in the Books table. This allows us to store information related to the author (such as birth date) only once. The birth-date is necessarily the same for all books authored by this person, because the birth-date depends only on the person.

Indeed, this simple example, with only three tables, shows right away some interesting aspects:

- The unit of interest is the "table": rectangular data – "data frame" in R's terminology. **table**
- Each table has fields (in R that would be referred to as "columns"). **field**
- Each table has a primary key (identified by the words PK in the table). The primary key is unique for each record (the record is the row), or otherwise stated each record can be uniquely identified with the primary key. **PK** **primary key**

The Big R-Book: From Data Science to Learning Machines and Big Data, First Edition. Philippe J.S. De Brouwer.

Companion Website: www.wiley.com/go/De Brouwer/The Big R-Book

tbl_authors				
id PK	**pen_name**	**full_name**	**birth**	**death**
1	Marcel Proust	Valentin Louis G. E. Marcel Proust	1871-07-10	1922-11-18
2	Miguel de Cervantes	Miguel de Cervantes Saavedra	1547-09-29	1616-04-22
3	James Joyce	James Augustine Aloysius Joyce	1882-02-02	1941-01-13
4	E.L. James	Erika Leonard	1963-03-07	
5	Isaac Newton	Isaac Newton	1642-12-25	1726-03-20
7	Euclid	Euclid of Alexandria	Mid-4th C BC	Mid-3rd C BC
11	Bernard Marr	Bernard Marr		
13	Bart Baesens	Bart Baesens	1975-02-27	
17	Philippe De Brouwer	Philippe J.S. De Brouwer	1969-02-21	

Table 13.1: *The table of authors for our simple database system.*

tbl_books				
id PK	**author FK**	**year**	**title**	**genre FK**
1	1	1896	Les plaisirs et les jour	LITmod
2	1	1927	Albertine disparue	LITmod
4	1	1954	Contre Sainte-Beuve	LITmod
5	1	1871–1922	À la recherche du temps perdu	LITmod
7	2	1605 and 1615	El Ingenioso Hidalgo Don Quijote de la Mancha	LITmod
9	2	1613	Novelas ejemplares	LITmod
10	4	2011	Fifty Shades of Grey	LITero
15	5	1687	PhilosophiNaturalis Principia Mathematica	SCIphy
16	7	300 BCE	Elements	SCImat
18	13	2014	Big Data World	SCIdat
19	11	2016	Key Business Analytics	SCIdat
20	17	2011	Malowian Portfolio Theory	SCIfin

Table 13.2: *The table that contains information related to books.*

tbl_genres			
id (PK) PK	**type**	**sub_type**	**location FK**
LITmod	literature	modernism	001.45
LITero	literature	erotica	001.67
SCIphy	science	physics	200.43
SCImat	science	mathematics	100.53
SCIbio	science	biology	300.10
SCIdat	science	data science	205.13
FINinv	financial	investments	405.08

Table 13.3: *A simple example of a relational database system or RDBMS for a simple system for a library. It shows that each piece of information is only stored once and that tables are rectangular data.*

- The primary key can be something meaningful (e.g. "LITmod" can be understood as the subsection "modern" within the larger group of "literature") or just a number.[1] The nice thing about numbers is that most RDBMS will manage those for you (for example when you create a new record, you do not have to know which numbers are taken and the system will automatically choose for your the smallest available number).

- There might be other fields that are unique besides the primary key. For example, our locations (in the genres) are unique as well as our dates and years in the table of both books and authors. However, in this particular case, they should not define them as unique, because we might have more than one author that is born on the same date.

- Some tables have a foreign key (FK). A foreign key is a reference to the primary key of another table. That is called a relation.

FK
foreign key

- In many cases, the relationship PK–FK will be a "one-to-many" (one author can write many books). However, thinking about it, we could have the situation where one book is written by more than one person . . . so actually we should foresee a "many-to-many" relationship. But that would not work exactly the way we intend it to work. To achieve that, we should build an auxiliary table that sits between the authors and the books—but let us revisit this idea later.

- Not all primary keys have to be used as foreign keys.

- This small example contains a few issues that you are bound to encounter in real life with real data:

  - Some values are missing. While this can happen randomly, it also might have a particular meaning (such as no death-date for an author can mean that he or she is still alive) or it can be simply a mistake.

  - Some columns contain non-consistent data types (e.g. years and dates).

  - Some values are self-explanatory, but some are cryptic (e.g. location in tbl_genre seems to consist of two numbers). We need a *data dictionary* to really understand the data.[2] to find out what it means,

data dictionary

  - We need intimate and close understanding of what happened in order to understand why things are the way they are.

  - That intimate knowledge will help us understand what is wrong. For example, Alain Proust's famous work "'AÌŁ la recherche du temps perdu" is not just one book. It is a heptalogy and the first volume was printed in 1871, the last in 1922.

  - To solve this we have to go to the library and see what really happened there. The data does not have all information. The data needs correction, the system was designed to

[1]When we refer to an author in the table of books for example we do not use the name (the reason is that (a) this might not be unique, but also (b) because this would lead to errors and confusion and (c) it would be slower to index and more difficult to manage for the RDMBS). In our case we could be reasonably sure that there are no two authors that share the same pen-name, real name and birth-date (though somewhere in distant future this might happen), so we could define the combination of pen-name, real name and birth-date as unique and eventually use this as primary key.

[2]A data dictionary is the documentation of the data, in reality you will find that many columns have cryptic names and in most cases there is some logic behind.

hold one book in tbl_books and not a series of seven in one entry. If we have the series of seven books, we need to find the details of all seven; if we have one big book containing all, then it will have one year where it was printed.[3]

There seem to be many challenges with the data. Let us address them when we build our database in the next section.

[3]A similar issue occurs with the work of Miquel de Servantes. His work is a diptych, one book published in two parts. Most probably we have a modern edition and there is just one book with one publishing date. It seems that the programmers intended to have publishing dates here and our librarian has put in the original first publishing dates.

♣ 14 ♣

SQL

14.1 Designing the Database

Before starting to create tables, it is a good idea to reflect on how the tables should look like and what is optimal. Good database design starts with an *entity-relationship* diagram (ER diagram). The ER diagram is a structured way of representing data and relations and the example of the library from Chapter 13 *"RDBMS"* on page 219 is represented in Figure 14.1 on page 224.

ER
entity relationship

The ER diagram is designed to be intuitive and understandable with just a few words of explanation. In our ER diagram, we notice that:

- the entities are in the rectangles;
- their attributes are in ellipsoids, where the optional fields have a dotted line and the unique identifiers (the primary keys (PKs)) are underlined;
- the relations between entities are the green diamonds.

PK
primary key

This ER diagram is the ideal instrument for the analyst to talk to the data owner or subject matter expert (SME)[1]. So the ER diagram is the tool to make sure the analyst understands all dependencies and relations.

For example, it is critical to understand that one author has only one pen-name in our design. Is this a good choice now and in the near future?

Once the ER diagram is agreed, we can start focussing on the database design. The first step of the database design is a layout of the tables and relations. In our – very simple – case this follows directly from the tables Table 13.1 on page 220, Table 13.2 on page 220, and Table 13.3 on page 220. This database design is visualised in Figure 14.2 on page 224.

The next step is to agree which fields can be empty and understand what type of searches will be common. This might result in altering this design and adding tables with indexes that allow to find the result of those common queries really quick.

[1] Corporates generally like acronyms and the subject matter expert is also referred to as SME, which is not a great acronym as it is also used for "small and medium enterprises." A general rule of advice here is to limit the use of acronyms as much as possible in daily operations. It leads to confusion, less efficient communication and exclusion since other teams might have other acronyms for the same concept or even use the same acronym for something else.

The Big R-Book: From Data Science to Learning Machines and Big Data, First Edition. Philippe J.S. De Brouwer.

Companion Website: www.wiley.com/go/De Brouwer/The Big R-Book

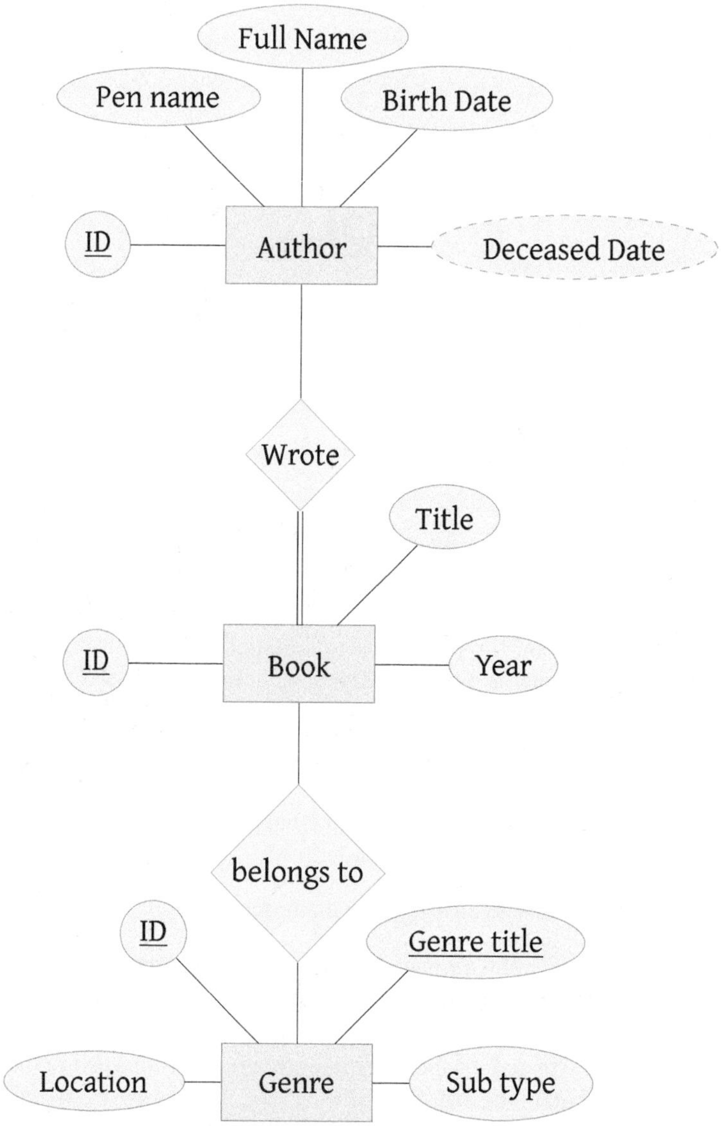

Figure 14.1: *The entity relationship (ER) diagram for our example, the library of books.*

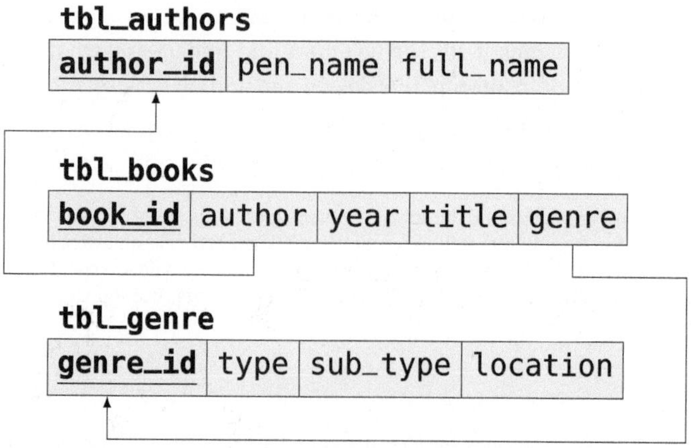

Figure 14.2: *The database scheme.*

Alternatively, the database could be tuned for fast inputs. In a payment system for a bank[2], for example it is important that payments go through fast and that there is no delay. In such systems, there will be no time for building indexes while the payments go through.

Now, we have sufficient information to build our database.

[2]A typical bank will handle anything between ten and hundred thousand payments per second.

14.2 Building the Database Structure

14.2.1 Installing a RDBMS

It is not within the scope of this book to recommend a RDBMS. There are many good choices, but we need to have one to get started. We will choose for a very powerful, yet free database system: MySQL[3], which had its own development company, SQL AB, that got acquired by SUN Microsystems, that in its turn got acquired by Oracle Corporation. MySQL is available under the GNU General Public Licence, as well as under a wide choice of propriety licences. This means that you can use it for free, even for commercial purposes, you can even change it ... but if you really want, you can pay for it too. This has the obvious advantage that the contract can include responsibilities for Oracle Corporation.

MySQL

MySQL is arguably the most popular database system to power websites on the Internet. It is the standard choice for the most popular content management systems (CMS), such as Drupal, Joomla, phpBB, and WordPress; and it is used by some of the largest service providers on the Internet, such as YouTube, Facebook, MediaWiki, Flickr, etc.

How to install MySQL can be found on `https://www.mysql.com`. However, on most of the Linux distributions, it is simple. We show it here for a Debian based distribution:

```
sudo apt-get install mysql-server -y
sudo apt-get install mysql-client -y

# Phpadmin provides an GUI to manage the database
# we will not use it, but you can install it as follows:
sudo apt-get install phpmyadmin -y
```

Listing 14.1: *Installing MySQL on a Linux computer is easy and straightforward. Here shown for a Debian based system.*

The installation will ask you for a master-password for the database ("root password"). Make sure to remember that password.[4] We will refer to it as `MySQLrootPassword`. Whenever you encounter this in our code, you will need to replace this word with your password of choice.

Further information – MariaDB

MariaDB

Alternatively, one can use MariaDB. This is a fork of MySQL that was created when Oracle took over MySQL.

```
sudo apt-get install mariadb-server
sudo apt-get install mariadb-client

# and eventually
sudo apt-get libmariadbclient-dev libmariadbd-dev
sudo apt-get phpmyadmin
```

The popular opinion that MariaDB works as a plug-in replacement for MySQL, is not really true. For example, (in 2019) making a full SQL-dump in MySQL and restoring it

[3]The name "MySQL" is a contraction of "My," which was the name of the daughter of one of the co-founders, and "SQL," which is the acronym for "Structured Query Language."

[4]If you have root access to the operating system, it is not impossible to reset the password. However, it is obviously less hassle and safer to remember the password.

in MariaDB leads to errors on technical tables that do not have the expected number of fields. However, some things are still the same: you also use the command `mysql` to start MariaDB, the same connectors from C++ or R that were created for MySQL still work, etc. MariaDB and MySQL are now separate projects, and the people making decisions are not the same, so expect them to diverge more in the future.

Now, the database server exists, and we can connect to it via the client (that was installed in the line `mysql-client` or `mariadb-client`).

```
mysql -u root -p
Enter password:
Welcome to the MySQL monitor.  Commands end with ; or \g.
Your MySQL connection id is 13
Server version: 5.7.23-0ubuntu0.16.04.1 (Ubuntu)

Copyright (c) 2000, 2018, Oracle and/or its affiliates. All rights reserved.

Oracle is a registered trademark of Oracle Corporation and/or its
affiliates. Other names may be trademarks of their respective
owners.

Type 'help;' or '\h' for help. Type '\c' to clear the current input statement.

mysql>
```

Listing 14.2: *Starting MySQL as root user. The first line is the command in the CLI, the last line is the MySQL prompt, indicating that we are now in the MySQL shell.*

Hint – Hardening the database server

In modern installations of MySQL and MariaDB, only the root user of the server has access to the root account of the database ...and this without extra password. This can be changed with the following approach. The following will get you in:

```
sudo mysql -u root
```

This will bring you in MariaDB, and the prompt is

```
MariaDB [(none)]>
```

If it is appropriate that each user with root access to the computer can have root access to the database, then there is nothing further to do. If, you want a separate password for the database server, then run these commands:

```
use mysql;
update user set plugin='' where User='root';
flush privileges;
exit
```

Now, back in the Linux command prompt run the following, set the root password and answer the questions.

```
sudo systemctl restart mariadb.service
sudo mysql_secure_installation
```

Note that you only type the first line `MySQL -u root -p`, then the line `Enter password:` invites you to type the password. Type it followed by enter and only then the rest of the screen appears. At this point, you have left the (bash) shell and you are in MySQL environment.

Carefully read the text that appears. It explains some commands and also informs that Oracle does not take any responsibility if something goes wrong.[5]

For example, it is a good idea to type `\h` and see the commands that are available at this level. This will also inform you that each line has to be ended by a semicolon `;`. This is noteworthy as in R this is not necessary.

Note – Similarities between the R CLI and the MySQL CLI

The MySQL monitor provides an environment that is very similar to the R terminal. It is an interactive environment where the commands that you type will be executed immediately. However, R will try to execute the command when you type enter. MySQL will do so when you type a semicolon `;`. Then again ... R also understands the semicolon and the following is understood by R as two separate commands.

```
x <- 4; y <- x + 1
```

14.2.2 Creating the Database

To create the database, we need some understanding of how MySQL organizes all data. MySQL is a powerful system and it can do much more than just hold a few tables. It knows the concept "database" as a collection of tables that belong logically together." For example, our library could have a database about books (including authors, locations, etc.), an accounting database, etc. Those databases can access each other, thought you need to make that explicit and user rights need to be granted. Actually, it is even possible to access a database that sits on another server in another part of the world: simply replace "localhost" with the IP-address of that server.

Unlike R, a database is designed to be used by many people at the same time and managing user access rights is crucial. We will not spend much time on it because SQL implementations can be different. In any case, we will define a new user "libroot" and give to that user all rights within our database "library."

```
-- first create a database
CREATE DATABASE library;

-- create a superuser for that database:
GRANT ALL PRIVILEGES ON library.* To 'libroot'@'localhost' IDENTIFIED BY '
    librootPWD';

-- create also a user who can only update data:
```

[5] Rest assured that the MySQL code is still available as under the terms of the GNU General Public License. There are also other licence options possible. However, MySQL was owned and sponsored by the Swedish for-profit company "MySQL AB" and this company has been bought by Oracle Corporation. This basically means that it is free to use, however, if you want to pay (and potentially hold a private company liable for something) then you can do that and you will get the same software, but with another start-up message.

```
CREATE USER librarian@localhost IDENTIFIED BY 'librarianPWD';
GRANT SELECT, INSERT, UPDATE, DELETE ON library.* TO librarian@localhost;

-- display a list of tables:
show databases;
+--------------------+
| Database           |
+--------------------+
| information_schema |
| library            |
| mysql              |
| performance_schema |
| sys                |
+--------------------+
-- Note that we did not create the other databases, they are used by MySQL
-- to manage everything.

-- leave the MySQL terminal
\q
```

Listing 14.3: *Create the database in which all tables will be created.*

From this listing, we can see that there are more databases than only the one that we created. Some might be there from earlier use, but some belong to MySQL itself and are managed by the database manager. For example, "information_schema" will hold information about the structure of all other databases, their tables, their relations and their indexes. Unless, you know really well what you do and why you do it: never ever touch them.

> **Hint – Comments in SQL**
>
> Note that a comment-line in SQL is preceded by two dashes: `--`. Longer comments can be put between `/* ... */`

The last command is `\q`. This command will leave the MySQL terminal and return to the Linux shell. Alternatively, one can use `exit`.

14.2.3 Creating the Tables and Relations

At this point, the database administrator for our library, "libroot," exits. Using this user is safer, because you will not be able to accidentally delete any other database, table or record in a table. We will now log in once more, but with that user-id:

```
mysql -u libroot -p
```

Listing 14.4: *Starting MySQL, as user "libroot." Note that this is done from the Linux CLI.*

Now, that we are logged in as `libroot`, we will start to create the tables in which later all data will reside:

```
-- First, we need to tell MySQL which database to use.
USE library;

-- Check if the table exists and if so delete it:
DROP TABLE IF EXISTS `tbl_authors`;

-- Then we can start to create tables
CREATE TABLE `tbl_authors`
  (
  author_id          INT UNSIGNED  auto_increment PRIMARY KEY not null,
  pen_name           VARCHAR(100) NOT NULL,
  full_name          VARCHAR(100),
  birth_date         DATE,
  death_date         DATE
  )
  ENGINE INNODB COLLATE 'utf8_unicode_ci';
```

Listing 14.5: *Create the table tbl_authors.*

First, we check if the table exist, and then delete it. It takes, indeed, just one line to delete a table and there is *no* warning message nor confirmation method. So, one small mistake might delete very important information. We add this line so that the next line will always work (creating an existing table will fail).[6]

The following line creates the table, and there is a lot going on here. Note that in SQL there is no such thing line one command that operates like a function such as in R. Rather there is a sentence. A sentence that starts with `CREATE TABLE` will attempt to create a table. It is followed by the name of the table. The back-ticks around the table name are not always necessary, but it is needed when one wants to use special characters in the name (such as spaces).

Then – between round brackets – we create all the fields (or "columns" in R-speak). The first words are the type of data that will be stored there. Since database systems are designed to hold millions of records and find rows in fractions of a second, they are very picky about the format of those variables that will be stored. Note the following.

signed

unsigned

- `INT` stands for integer and will hold integers. INT will be encoded in 4 bytes. Since each byte is 4 bits, this will hold 32 bits. We need one bit for the sign, so the range is from $-2^{(32-1)}$ to $+2^{(32-1)}$. INT can be signed or unsigned (ie. allowing negative numbers or not). The default is signed, so that does not have to be mentioned. If you will only have positive numbers, one can choose for `UNSIGNED`. MysQL supports the following:
  1. TINYINT = 1 byte (8 bit): from -128 to 127 signed and from 0 to 255 unsigned
  2. SMALLINT = 2 bytes (16 bit): from $-32\ 768$ to $32\ 767$ signed and from 0 to $65\ 535$ unsigned
  3. MEDIUMINT = 3 bytes (24 bit): from $-8\ 388\ 608$ to $8\ 388\ 607$ signed and from 0 to 16 777 215 unsigned
  4. INT = 4 bytes (32 bit): from $-2\ 147\ 483\ 648$ to $2\ 147\ 483\ 647$ signed and from 0 to 4 294 967 295 unsigned
  5. BIGINT = 8 bytes (64 bit): from -2^{63} to $-2^{63} - 1$ signed and from 0 to $-2^{64} - 1$ unsigned;
- `PRIMARY KEY` indicates – not surprisingly – that this field is the primary key (PK), hence, it will have to be unique and that is the field that will uniquely define the author;

[6]This idea is useful when creating a database and changes occur often, it is absolutely not safe when there is data in the database. Be very careful with the `DELETE TABLE` command.

- `auto_increment` tells MySQL to manage the value of this field by itself: if the user does not provide a unique number, then MySQL will automatically allocate a number that is free when a record is created;
- `VARCHAR(100)` will hold a string up to 100 characters;
- `DATE` will hold dates between 1000-01-01 and 9999-12-31;
- We also provide a "collation." A collation in MySQL is a set of rules that defines how to compare and sort character strings, and is somehow comparable to the typical regional settings. For example, the `utf8_unicode_ci"` implements the standard Unicode Collation Algorithm, it supports expansions and ligatures, for example: German letter ß (U+00DF LETTER SHARP S) is sorted near "ss" Letter œ (U+0152 LATIN CAPITAL LIGATURE OE) is sorted near "OE," etc. We opt for this collation, because we expect to see many international names in this table.

Digression – SQL is not case sensitive

As opposed to R, SQL is essentially case insensitive. For example, the aforementioned code mixes the statement `NOT NULL` in lower and in upper cases. However, MySQL on a case sensitive operating system (OS) will be case sensitive for names of tables and fields because they will be stored on disk. Note that this behaviour can be changed when the *server* is initialized (at installation). Once we start using our database this behaviour cannot be changed.

Now, we understand enough to create the second table:

```
-- If the table already exists, delete it first:
DROP TABLE IF EXISTS `tbl_books`;

-- Create the table tbl_books:
CREATE TABLE `tbl_books`
  (
  book_id           INT unsigned auto_increment
                    PRIMARY KEY not null,
  author            INT unsigned NOT NULL REFERENCES
                    tbl_authors(author_id)
        ON DELETE RESTRICT,
  year              SMALLINT,      -- provides 30~000 years
  title             VARCHAR(50),   -- maximum 50 characters
  genre             CHAR(6) NOT NULL REFERENCES -- always 6
                    tbl_genres(genre_id) ON DELETE RESTRICT
  )
  ENGINE INNODB COLLATE 'utf8_unicode_ci';

-- Create an index for speedy lookup on those fields:
CREATE INDEX idx_book_author ON tbl_books (author);
CREATE INDEX idx_book_genre ON tbl_books (genre);
```

Listing 14.6: *This SQL code block creates the table* `tbl_books` *and then define an index on two of its fields.*

The field `author` is a "foreign key" (FK). This field will refer to the author in the table `tbl_authors`. This is the way that an RDBMS will find the information in the author. From the book, we find the author-id and then we look this id up that in the table of authors to access the rest of the information.

FK
foreign key

A foreign key (FK) is indicated with the keywords `REFERENCES`, followed by table name and field name between round brackets. Then there is the statement `ON DELETE RESTRICT`. This means that as long as there is a book of an author in our library that we cannot delete the author. In order to delete the author, we need firs to delete all the books of that author. The alternative is `ON DELETE CASCADE`, this means that if we delete an author that automatically all his/her books will be deleted.

Also here we made some hard choices. The field `year` is implemented as `SMALLINT`, this means that we will not be able to add a range of dates. However, since the keyword `UNSIGNED` is not provided, it is possible to use negative numbers to code the book written before common era.[7]

The difference between `VARCHAR(n)` and `CHAR(n)` is that the first will hold up to n characters, while the second will encode always n characters. For example, the word "it" in a `CHAR(4)` field will be encoded as "space-space-it," while in a `VARCHAR(4)` field the spaces would not be part of the data in the table.[8]

Then there are two additional lines that start with `CREATE INDEX`. These commands create an additional small table that contains just this keyword (ordered) and a reference to the position in the table. This allows the database system to look up the index logarithmic time[9], and then via the reference find the record in the table. The database server expects many lookups on the primary key, so it will always index this field (therefore, we do not need to specify that).

When the data comes in very fast, when records hold a lot of information, or when the table becomes really long then it is not practically possible to rewrite the whole table so that all data sits ordered via the primary key. So, the relational database management system (RDBMS) will create a thin table that holds only the index and the place where the record is in the physical table. Whenever a new record comes in, the index is re-written.

An index can hence be seen as a thin table that helps MySQL to find data fast. Not only primary keys will have their index, but it is also possible to ask MySQL to index any other field so that lookup operations on that field will become faster. The syntax is:

```
--  Create an index where the values must be unique
-- (except for the NULL values, which may appear multiple times):
ALTER TABLE tbl_name ADD UNIQUE index_name (column_list);

-- Create an index in which any value may appear more than once:
ALTER TABLE tbl_name ADD INDEX index_name (column_list);

-- A special index that helps searching in full text:
ALTER TABLE tbl_name ADD FULLTEXT index_name (column_list);

-- We can also add a primary key:
ALTER TABLE tbl_name ADD PRIMARY KEY (column_list);
```

[7] Note that this year-field is not a date. It is just a number.

[8] If the length of the string can vary, then there must be some way to keep track of the length of the string. This will lead to a small overhead, that will be accumulated during reach read or write operation. These options allow us to choose between speed and flexibility.

[9] Searching in structured data takes $O(\log(N))$ (for N entries in the database), while searching in unordered data will take $O(N)$. For the longer tables, this becomes a massive difference in time. Note also that the searching will be done in a table with less columns (fields), also this contributes to the speed.

```
-- Drop an index:
ALTER TABLE table_name DROP INDEX index_name;

-- Drop a primary key:
ALTER TABLE tbl_name DROP PRIMARY KEY;
```

Listing 14.7: *Manage indexes in MySQL*

Note – Impossible definitions are possible

Note that MySQL allows us to create `tbl_books` and define the field `genre` as a foreign key that refers to `tbl_genres` before that table exists. This is useful when creating the tables, however, we will get problems at this point we would put in data.

Finally, we can create the table of the literary genres:

```
-- In case it would already exist, delete it first:
DROP TABLE IF EXISTS `tbl_genres`;

-- Create the table:
CREATE TABLE `tbl_genres`
  (
  genre_id          CHAR(6) PRIMARY KEY not null,
  type              VARCHAR(20),
  sub_type          VARCHAR(20),
  location          CHAR(7)
  )
  ENGINE INNODB COLLATE 'utf8_unicode_ci';

-- Show the tables in our database:
show tables;
+-------------------+
| Tables_in_library |
+-------------------+
| tbl_authors       |
| tbl_books         |
| tbl_genres        |
+-------------------+
3 rows in set (0,01 sec)
```

Listing 14.8: *This SQL code creates the table tbl_genres and then checks if it is really there.*

Digression – UTF8 collation

We used the collation `utf8_general_ci` because we do not expect any special symbols, and this collation is a little faster.

It is possible to check our work with `DESCRIBE TABLE`. That command will show relevant details about the table and its field:

```
mysql> DESCRIBE tbl_genres;
+----------+-------------+------+-----+---------+-------+
| Field    | Type        | Null | Key | Default | Extra |
+----------+-------------+------+-----+---------+-------+
| genre_id | char(6)     | NO   | PRI | NULL    |       |
| type     | varchar(20) | YES  |     | NULL    |       |
| sub_type | varchar(20) | YES  |     | NULL    |       |
| location | char(7)     | YES  |     | NULL    |       |
+----------+-------------+------+-----+---------+-------+
4 rows in set (0,00 sec)
```

Listing 14.9: *Checking the structure of the table* `tbl_books`.

We have now a minimal viable relational database, and we could start adding data into this structure: we can start adding our books. This will also make clear what the impact is of choices that we made earlier.

14.3 Adding Data to the Database

Since we did the effort to create a username for updating the data, let us use it. In MySQL, type `exit` or `\q` followed by enter and then login again from the command prompt.

```
mysql -u librarian -p
```

Listing 14.10: *Logging in as user "librarian."*

To insert data, we use the query (or "sentence in SQL") that starts with `INSERT INTO`. This statement has generally two forms: one for adding one record and one for adding many records.

Let us start by adding one record:

```
INSERT INTO tbl_authors
  VALUES (1, "Philppe J.S. De Brouwer", "Philppe J.S. De Brouwer", "1969-02-21"
    , NULL);
```

Listing 14.11: *Adding the first author to the database.*

Note – Providing values for automatically incremented fields

While MySQL was supposed to manage the `author_id`, it did not complain when it was coerced in using a certain value (given that the value is in range, of good type and of course still free). Without this lenience, uploading data or restoring and SQL-dump would not always be possible.

This form can only be used if we provide a value for exactly each column (field). Usually, that is not the case and then we have to tell MySQL what fields we have:

```
-- First, remove the record that we just created:
DELETE FROM tbl_authors WHERE author_id = 1;

-- Then add a new one:
INSERT INTO tbl_authors (pen_name, full_name, birth_date)
  VALUES ("Philppe J.S. De Brouwer", "Philppe J.S. De Brouwer", "1969-02-21");
```

Listing 14.12: *An alternative way to add the author to the book by specifying the fields provided.*

Note – Missing values

Note that we did not provide a deceased date since this author still lives and that MySQL decided that this is anyhow NULL. NULL means "nothing," not even an empty string. In R this is concept is called "NA". To see how MySQL represents this, run the following command.

```
mysql> SELECT * FROM tbl_authors;
+-----------+--------------...--+-----------+
| author_id | pen_name      ...  | death_date |
+-----------+--------------...--+-----------+
|         2 | Philppe J.S. ...1 | NULL       |
+-----------+--------------...--+-----------+
1 row in set (0,00 sec)
```

Of course, it is also possible to add multiple records in one statement:

```
-- First, remove all our testing (1 is equivalent with TRUE):
DELETE FROM tbl_authors WHERE 1;

-- Since we provide a value for each row, we can omit the
-- fields, though it is better to make it explicit when
-- inserting the data:
INSERT INTO tbl_authors (author_id, pen_name, full_name, birth_date, death_date
    )
  VALUES
  (1 , "Marcel Proust",
       "Valentin Louis G. E. Marcel Proust",
       "1871-07-10", "1922-11-18"),
  (2 , "Miguel de Cervantes",
       "Miguel de Cervantes Saavedra",
       "1547-09-29", "1616-04-22"),
  (3 , "James Joyce",
       "James Augustine Aloysius Joyce",
       "1882-02-02", "1941-01-13"),
  (4 , "E. L. James",            "Erika Leonard",
       "1963-03-07",  NULL),
  (5 , "Isaac Newton",           "Isaac Newton",
       "1642-12-25", "1726-03-20"),
  (7 , "Euclid",                 "Euclid of Alexandria",
        NULL,         NULL),
  (11, "Bernard Marr",           "Bernard Marr",
        NULL,         NULL),
  (13, "Bart Baesens",           "Bart Baesens",
       "1975-02-27",  NULL),
  (14, "Philippe J.S. De Brouwer",
       "Philippe J.S. De Brouwer",
       "1969-02-21",  NULL)
  ;
```

Listing 14.13: *This SQL code adds all books in one statement.*

Hint – Input errors

If you enter a wrong query (for example forget closing quotes of a string), then – just as R – MySQL will not understand that your command is finished but expect more input (it starts a newline with `+`). In that case, you can press `CTRL+c` followed by `ENTER`.

Here we decided to leave dates earlier than 1000-01-01 blank (NULL). In fact, MysQL will accept some earlier dates, but since the documentation does not specify this, it is safer not to use them. Also note that negative years will not be accepted in a field of type `DATE`.

The result is that we might have missing dates for multiple reasons: the person did not die yet or the date was before 1000-01-01 ... or we do not know (the case of Mr. Marr).

Hint – View the data in a table

While MySQL will give feedback about the success of the query, you might want to check if the data is really there. We will discuss these type of queries in the next section, but in the meanwhile you might want to use this simple code:

```
SELECT * FROM tbl_authors;
```

it is possible to see how it worked out.

Since we defined the field `genre` in `tbl_books` as `NOT NULL`, we will have to add the data to the table with genres first, before we can add any books. This is because MySQL will guard the *referential integrity* of our data. This means that MySQL will make sure that the rules that we have defined when the tables were created are respected at all time (e.g. the PK is unique, a FK refers to an existing record in another table, etc.). The following code adds the necessary records to the table `tbl_genres`.

```
INSERT INTO tbl_genres (genre_id, type, sub_type, location)
  VALUES
  ("LITmod", "literature", "modernism",    "001.45"),
  ("LITero", "literature", "erotica",      "001.67"),
  ("SCIphy", "science",    "physics",      "200.43"),
  ("SCImat", "science",    "mathematics",  "100.53"),
  ("SCIbio", "science",    "biology",      "300.10"),
  ("SCIdat", "science",    "data science", "205.13"),
  ("FINinv", "financial",  "investments",  "405.08")
  ;
```

Listing 14.14: *Add the data to the table tbl_genres.*

For the books, we can leave the `book_id` up to MySQL or specify it ourselves.

```
INSERT INTO tbl_books (author, year, title, genre)
  VALUES
  (1,  1896, "Les plaisirs et les jour",       "LITmod"),
  (1,  1927, "Albertine disparue",             "LITmod"),
  (1,  1954, "Contre Sainte-Beuve",            "LITmod"),
  (1,  1922, "AÌÀ la recherche du temps perdu", "LITmod"),
  (2,  1615, "El Ingenioso Hidalgo Don Quijote de la Mancha", "LITmod"),
  (2,  1613, "Novelas ejemplares",             "LITmod"),
  (4,  2011, "Fifty Shades of Grey",           "LITero"),
  (5,  1687, "PhilosophiÃę Naturalis Principia Mathematica", "SCIphy"),
  (7,  -300, "Elements (translated )",         "SCImat"),
  (13, 2014, "Big Data World",                 "SCIdat"),
  (11, 2016, "Key Business Analytics",         "SCIdat"),
  (14, 2011, "Maslowian Portfolio Theory",     "FINinv")
  ;
```

Listing 14.15: *Add the data to the table tbl_books.*

Note how it was possible to encode a year before common era – for the book – as a negative number. Any software using this data will have to understand this particular implementation, otherwise undesired results will occur. Again, we see how important it is to understand data. The careful reader will notice that actually "300 BCE" might better be implemented as "-299" because there was never a year "0." The year before "1 CE" is the year "-1" (not zero). In our case, we did not do this, because we go for what seems most natural to the librarian. The point is that data in itself is meaningless and that guessing what it means is dangerous, we need to get the data together with the explanation (the data dictionary).

14.4 Querying the Database

Now, the hard work is done: MySQL has our data and it knows our database design and will enforce referential integrity, this means that if a field is defined as a foreign key it will not allow any entries that are not in the table that hold the primary key (PK).

referential integrity

Digression – Can a FK be NULL?

In our simple model, we have no cases where a foreign key (FK) can be `NULL`. This is very well possible, in that sense that the RDBMS will allow this (unless specified that it cannot be null when the table was created). So, unless we have incomplete data, it is a good idea to specify that a FK cannot be NULL.

To get data back from the RDBMS that uses SQL, it is sufficient to run a `SELECT` query. The SELECT-query is constructed as an intuitive sentence.

14.4.1 The Basic Select Query

Below we demonstrate how the select query works. Note that the output would be too long and is hence suppressed.

```
-- Show all info about all authors:
SELECT * from tbl_authors;

-- Show all pen_names and birth_dates from tbl_authors:
SELECT pen_name, birth_date FROM tbl_authors;

-- Show all authors from the last two centuries
SELECT pen_name FROM tbl_authors WHERE birth_date > DATE("1900-01-01");

-- Include also the ones that have no birth data in the system"
SELECT pen_name FROM tbl_authors
    WHERE (
           (birth_date > DATE("1900-01-01"))
           OR
           (ISNULL(birth_date))
           ) ;
```

Listing 14.16: *Some example of SELECT-queries. Note that the output is not shown here, simply because it would be too long.*

Note – Working with dates in SQL

Note that MySQL will understand both

- `birth_date > DATE("1900-01-01")` and
- `birth_date > "1900-01-01"`.

However, typing the number without quotes will be interpreted as a numerical expression and does not do exactly what we want to obtain. Nor is the second expression guaranteed to work in all implementations of SQL.

So far the queries that we have presented only pull information from one table, while the whole idea of a relational database is just to store information in the relations. Well, just as we can ask information from multiple fields by listing them separated by commas and we can do the same for tables. Adding tables after the FROM keyword will actually prompt SQL to compute the Cartesian product (ie. all possible combinations).

Cartesian product

Test this by running:

```
SELECT pen_name, title FROM tbl_authors, tbl_books;
```

This is *not* what you need.

One has to specify which records of all the possible combinations are desired. This is done by adding a "WHERE-clause." For example:

```
SELECT pen_name, title FROM tbl_authors, tbl_books
  WHERE author = author_id;
```

This query will only show those combinations where the author number is equal in both tables. This is the information that we were looking for.

This type of filtering appears so often that it has its own name: "inner join." An inner join will only show those where both fields are equal. This – just as in R – excludes all rows that would have a NULL in that field, because the system has no way to tell if the values match.

14.4.2 More Complex Queries

In the previous section, we explored the basic SELECT-statement in SQL, but just like in human language, one statement can be used in many different sentences. It is possible to change what we select, from what tables, combine tables, and –of course– exploit the fact that the relations are well defined. In this section, we will show some of those possibilities.

While the inner join will not include fields where one table has a NULL value, it might make sense to list for example all the books, including those from which we do not have the information of the author. This result can be obtained by an asymmetric join. They are called left join or right join and include all records from one table at least once and eventually repeat for multiple matches from the other table.

We illustrate briefly the most important join types in Figure 14.3 on page 241.

The SQL-language is quite flexible and Figure 14.3 on page 241 shows only some possibilities (the ones that are likely to occur a lot). It is of course possible for example to link more tables at once, add more conditions, or even select the set difference $A - B$ with:

```
SELECT * FROM A LEFT JOIN B on a = b AND WHERE ISNULL(b)
```

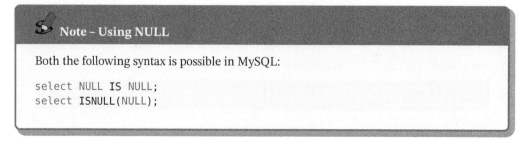

Note – Using NULL

Both the following syntax is possible in MySQL:

```
select NULL IS NULL;
select ISNULL(NULL);
```

$A \cap B$

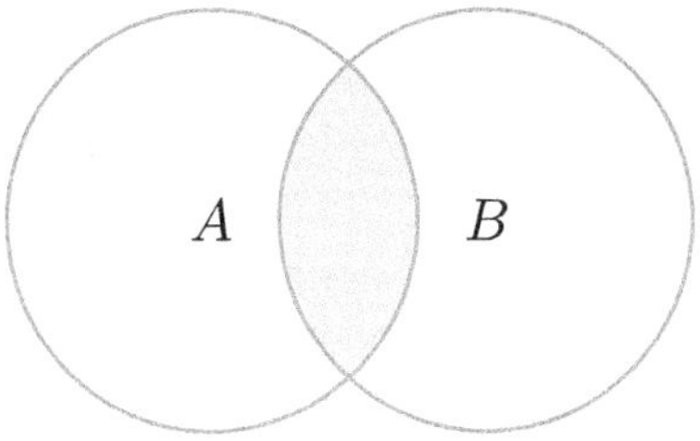

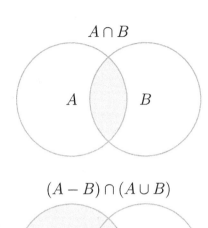

```
SELECT * FROM A INNER JOIN B ON A.a
    = B.b;
```

Inner join selects only the records that match in both tables A and B.

$(A - B) \cap (A \cup B)$

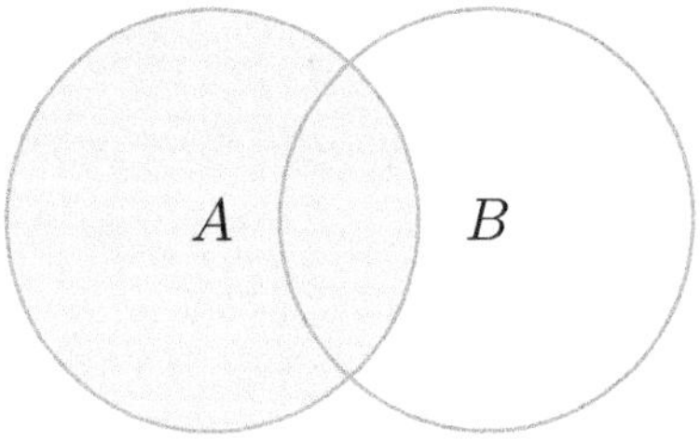


```
SELECT * FROM A LEFT JOIN B ON A.a
   = B.b;
```

A *left outer join* from A to B will select each row of A at least once; if there are multiple matchings in B, the A will be repeated once per matching B.

$A \cup B$

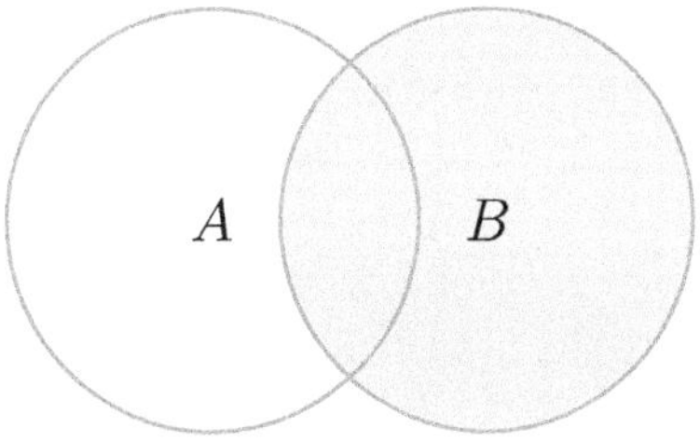


```
SELECT * FROM A RIGHT JOIN B ON A.a
    = B.b;
```

A *right outer join* from A to B will select each row of B at least once; if there are multiple records in A, that match a particular record in B, then B will be repeated once per matching A.

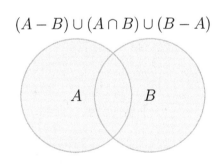

$(A - B) \cup (A \cap B) \cup (B - A)$

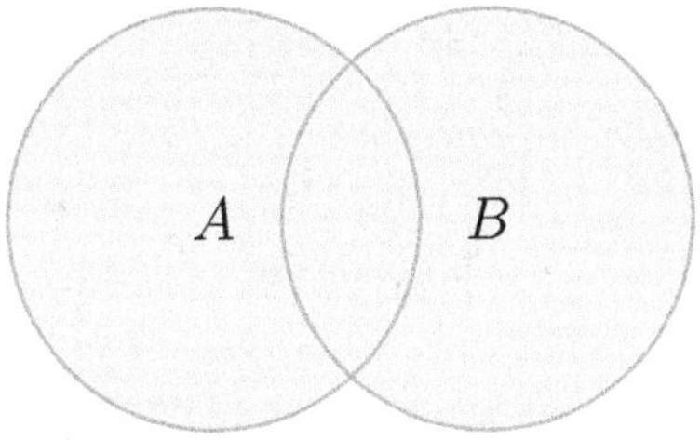


```
SELECT * FROM A OUTER JOIN B ON A.a
    = B.b;
```

A *full outer join* will select each row of A and each row of B at least once. If a row in A matches multiple ones in B, it will be repeated once per match; and the same for B.

```
SELECT * FROM A CROSS JOIN B;
```

A *cross join* will select the Cartesian product from A and B.

$A \times B$

Figure 14.3: *Different join types illustrated. Note that the Venn-diagrams are only for illustration purposes, they do not truly show the complexity. Also note that we used a table "A" with field "a" and a table "B" with field "b."*

Digression – MySQL specific note

In the SQL implementation of MySQL, the words OUTER, INNER, and FULL can be omitted. They will not make a difference. Be careful though, this might depend on the particular SQL implementation that is being used in your RDBMS.

Even with our very simple database design, the difference in syntax can be demonstrated:

```
SELECT pen_name , location FROM tbl_authors A, tbl_books B, tbl_genres B
  WHERE
    (A.author_id = B.author) AND
    (B.genre     = G.genre_id)
    ;

-- Yields the same records as:
SELECT pen_name, location FROM tbl_authors A
  LEFT JOIN tbl_books B  ON A.author_id = B.author
  LEFT JOIN tbl_genres G ON B.genre     = G.genre_id
```

Listing 14.17: *This code can be used to show that a manual linking of fields leads to the same records as a left join. Note that the output is not provided.*

So, why would we use these joins instead of the first simple form with table names separated by commas. Indeed, in our simple example, there is no noticeable difference in speed or complexity. Having a special keyword for joins has two other major advantages. First, it allows the database vendor to speed up joins and it will become significantly faster than the explicit manual linking of tables. Second, the join syntax remains clean when joining multiple tables. Doing this with a WHERE clause becomes quickly very unyielding.

Hint – Automated linking

If you want to join on fields that carry the same name in both tables, then you can use the keyword USING:

```
USING (author)

is equivalent to
ON A.author = B.author
```

While the keyword USING is a handy shorthand, in many cases, it makes the code less transparent, and in some cases, this might not be entirely equivalent.

Note – The difference between right an left joins

Is there a difference between a right and a left joins when the order of the tables are inverted?

```
-- Left join:
SELECT cols_list FROM A LEFT  JOIN ON B USING X;

-- Right join:
SELECT cols_list FROM B RIGHT JOIN ON A USING X;
```

In MySQL, this should return the same, hence there is no difference in both aforementioned statements.

There is a lot more about SQL and queries and it is beyond the scope of this book to dig even deeper. So, let us conclude with a simple example of finding values in table A that have no match in table B (assuming that both tables A and B share a column with the name x).

```
-- Option 1 (slower)
SELECT  A.* FROM A LEFT JOIN B USING x
  WHERE B.x IS NULL;

-- Option 2:
SELECT  A.* FROM A
  WHERE A.x NOT IN (SELECT x FROM  B);
```

Note that the `CROSS JOIN` does the same as the Cartesian product[10], so here is no performance improvement or any other gain to choose one form over the other. Note that it is possible to omit the `ON A.a = B.b` statement.

Hint – Tidy queries

As you will gradually make more and more complex queries, you will end up with many tables and fields in your query. Here are some hints to keep it tidy:

1. output fields (column headings) can be kept tidy by naming them `long_funtion_of_many_fields AS MyField` (now the column heading will be "MyField";
2. tables can be named to be used in the query in a very similar way with the keyword `AS`: `long_table_name AS t` allows us to address the table using the alias "t";
3. many fields from many tables can have the same name, in case there can be doubt, one should make clear from what table the field is with the dot-separator. For example, to make clear that field "fld" is from table A and not B, one types `A.fld`

For example:

```
SELECT pen_name, title AS BookTitle
  FROM tbl_authors AS A, tbl_books as B
  WHERE B.author = A.author_id;
```

Further information – MySQL

Please note that the SQL implementation in MySQL is also Turing complete. It knows for example functions, can read from files, etc. Further, information can be found at `https://dev.mysql.com/doc`.

[10] The Cartesian product is the same as "all possible combinations." So if A has 100 rows and B has 200 rows the Cartesian product has $100 \times 200 = 20\,000$ rows.

14.5 Modifying the Database Structure

This is a great day today. We received finally our copy of Hadley Wickham and Garret Gerolemund's book "R for Data Science" and we want to add it to our library. However, we can enter only one reference to one author in our library. After a brainstorm meeting, we come up with the following solutions:

1. Pretend that this book did not arrive, send it back or make it disappear: adapt reality to the limitations of our computer system.
2. Just put one of the two authors in the system and hope this specific issue does not occur to often.[11]
3. Add a second `author`-field to the table `tbl_books`. That would solve the case for two authors, but not for three or more.
4. Add 10 additional fields as described above. This would indeed solve most cases, but we still would have to re-write all queries in a non-obvious way. Worse, most queries will just run and we will only find out later that something was not as expected. Also, we feel that this solution is not elegant at all.
5. Add a table that links the authors and the books. This will solution would allow us to record between zero and a huge amount of authors. This would be a fundamentally different database design and if the library software would already be written[12] this solution might not pass the Pareto rule.

We will demonstrate how the last solution (solution 5) could work. We choose this one because it will allow us to show how complicated such a seemingly simple thing would be for the database admin. However, rewriting all applications that use the data would be a lot more work. This underlines the importance of a good database design up-front and this in its turn demonstrates how important subject matter knowledge is.[13]

First, we need to update the database design. The database design is a framework for that will guide us on what to do. The solution is illustrated below in Figure 14.4 on page 245:

Digression – More than one solution

There are more designs possible that obtain the same effect. For example, the primary key (PK) `ab_id` is not really necessary. We can also define the pair author/book as a composite primary key in the table `tbl_author_book`. For our needs this will work just fine. However, it makes it more tricky to use the pair as a foreign key (FK). This would occur if – for example – we would like to add information about that instance of author/book. Simply adding who is co-author would go directly in the table `tbl_autor_book`, but for example adding dates with signing sessions would require a separate table referring to the instance author/book.

[11] Indeed, that is not necessarily a bad and lazy approach. In business it is worth to keep the "Pareto rule" (or the "80/20 rule") in mind. It seems that software development does not escape this rule and that by solving the top-20 cases, 80% of the real world problems are solved.

[12] In Section 14.4 *"Querying the Database"* on page 239 we have shown how to query the database. Any software that uses the tables will hence depend on the structure of the tables. Therefore, changing database design is not something that can be done lightly and will have a massive impact on all software using the database.

[13] People that are subject matter experts are often referred to as "SME" – this confusing as the same company will typically also use SME as "small and medium enterprises."

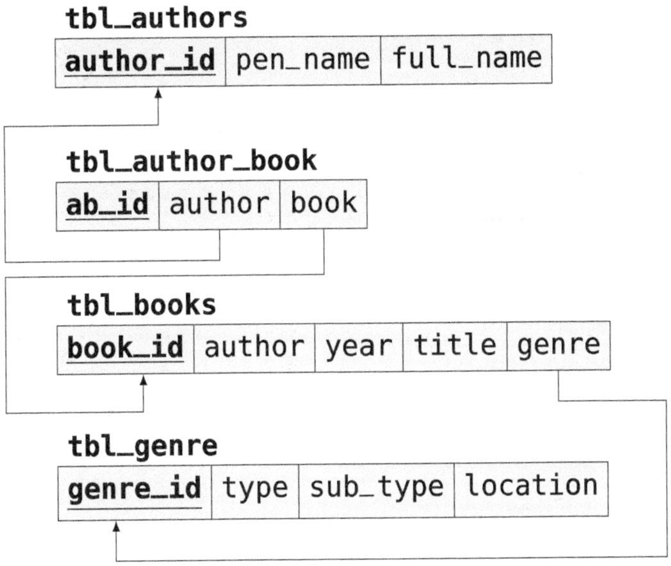

Figure 14.4: *The improved database scheme that allows multiple authors to co-author one book by adding a table* `tbl_author_book` *between the table with books and the table with authors. Now, only the pair author/book has to be unique.*

Note that while many RDBMS are ACID compliant and hence do a good job in making sure that no conflicts occur, in our specific case, we first have to delete the relationship PK/FK and then rebuild it in another table. The type of operations that we are performing here (deleting tables, redefining links, etc.) are not covered by ACID compliance. Redesign of the database will leave **ACID** the database vulnerable in the time in between. So, it would be wise to suspend all transactions in the meanwhile in one way or another. (e.g. allow only one user from a specific IP to access the database, etc.).

However, we think that – for our purpose here – there is another elegant workaround: first build the new tables, insert the relevant data then enforce referential integrity and only then disconnect the old fields. We can do this during lunch break and hope that all will be fine.

Further information

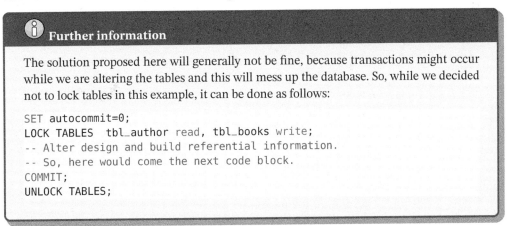

The solution proposed here will generally not be fine, because transactions might occur while we are altering the tables and this will mess up the database. So, while we decided not to lock tables in this example, it can be done as follows:

```
SET autocommit=0;
LOCK TABLES  tbl_author read, tbl_books write;
-- Alter design and build referential information.
-- So, here would come the next code block.
COMMIT;
UNLOCK TABLES;
```

```
-- Note that this has to be done as libroot or root,
-- the user librarian cannot do this!
```

```
-- Note also the different cascading rules. Why did we do so?
DROP TABLE IF EXISTS `tbl_author_book`;
CREATE TABLE `tbl_author_book`
  (
  ab_id  INT unsigned auto_increment PRIMARY KEY not null,
  author INT unsigned NOT NULL
         REFERENCES tbl_authors(author_id) ON DELETE RESTRICT,
  book   INT unsigned NOT NULL
         REFERENCES tbl_books(book_id) ON DELETE CASCADE
  );

-- Ensure the combination of author/book appears only once
ALTER TABLE `tbl_author_book`
                  ADD UNIQUE `unique_index`(`author`, `book`);


-- Insert all pairs of authors and books that we already know.
INSERT INTO tbl_author_book (author, book)
  (SELECT author_id, book_id FROM tbl_authors, tbl_books
  WHERE tbl_books.author = author_id
  );

-- It appears we can just drop the field author from the
-- table tbl_books and the link automatically disappears.
ALTER TABLE tbl_books DROP author;
```

Listing 14.18: *This code first creates the table tbl_author_books and then inserts the necessary information that was already into the database also in that table. Finally, it discards the old information.*

It is, of course, also necessary to update all software that uses these tables. For example, the field `author` can not longer be a simple text-field. The user interface must now allow to select more than one author and the software must push this information no longer to the table `tbl_books`, but rather to the table `tbl_author_book`.

Also retrieving information will be different. For example, finding all the books of a given author can work as follows:

```
SELECT pen_name, title FROM tbl_authors, tbl_author_book, tbl_books
  WHERE
    (author_id = tbl_author_book.author) AND
    (book_id   = tbl_author_book.book)   AND
    (pen_name LIKE '%oust%')
    ;

-- MySQL will then reply this:
+---------------+--------------------------------+
| pen_name      | title                          |
+---------------+--------------------------------+
| Marcel Proust | Les plaisirs et les jour       |
| Marcel Proust | Albertine disparue             |
| Marcel Proust | Contre Sainte-Beuve            |
| Marcel Proust | A la recherche du temps perdu  |
+---------------+--------------------------------+
```

Digression – Matching unknown strings

The LIKE operator is a string matching operator. Note that unlike in most operating systems the % will expand as "0 or more characters."[a] Of course, it allows character escaping with \, fitting of exactly one character, adding the collation, etc. There are much more string functions. Please check the documentation of your SQL server.

[a] Note that this special character that can be any number of other characters is different from the asterisk that you might know from the CLI in Linux, Windows NT, or DOS. If you would be working with MS Access instead, you will be able to use the asterisk, because the particualar dialect implemented in MS Access does not follow the SQL standard.

Note – Composite PKs

tbl_author_book holds not two PKs, but the two fields combined form one PK. It is entirely possible to use this combined key as a FK in other tables. However, this makes querying the database more difficult. That is the reason why we introduced the ab_id field.

Now, it is finally possible to add the book of Hadley and Garrett. To do this, we will need to add the book and the authors and then in a second step add two entries in tbl_author_book:

```
-- Add the book:
INSERT INTO tbl_books (book_id, year, title, genre)
  VALUES (NULL, 2016, "R for Data Science", "SCIdat");
SET @bookID = LAST_INSERT_ID();

  -- Add the authors:
INSERT INTO tbl_authors (author_id, pen_name, full_name)
  VALUES (NULL, "Hadley Wickham", "Hadley Wickham");
SET @auth1 = LAST_INSERT_ID();

INSERT INTO tbl_authors (author_id, pen_name, full_name)
  VALUES (NULL, "Garrett Grolemund", "Garrett Grolemund");
SET @auth2 = LAST_INSERT_ID();

-- Finally, update tbl_author_book
INSERT INTO tbl_author_book (author, book)
  VALUES (@auth1, @bookID), (@auth2, @bookID);
```

Listing 14.19: *Finally, we can add our book that has more than one author to our database.*

Note that the aforementioned code might not be entirely foolproof in a production environment where e.g. 50 000 payments per second are processed. The function LAST_INSERT_ID() will keep the value of the last insert into an auto_increment variable that has been fed with NULL. So, even if some other process added a book or but did not feed NULL into the field book_id, this will keep its value. Since there is only one register, we need to store the result in variables.

Digression – Removing a variable in SQL

Unlike R that allows use to remove a variable with `rm(x)`, MySQL does not allow to remove a variable. It is possible to set it to NULL of course in order to save space – if that would be relevant. This works with `SET @id = NULL;`

14.6 Selected Features of SQL

Mileage may vary on the particular dialect of your SQL server, however, MySQL, MariaDB, Oracle, PostgreSQL, etc. have all many more features. In fact the SQL standard is Turing complete, which means that each process can be simulated in SQL. In this section we will have a quick walk-through and illustrate some features that come in handy.

14.6.1 Changing Data

Apart from querying data that has been inserted, it is of course essential to be able to update the data in our database. Just as SELECT is the workhorse for querying data, SET is the workhorse for changing data.

The command SET also allows to update all selected variables in a table. For example, we can capitalize all author names as follows:

```
UPDATE tbl_authors
  SET full_name = CONCAT(UPPER(SUBSTRING(full_name,1,1)),
                         SUBSTRING(full_name,2,LENGTH(full_name)))
  WHERE 1;
```

Listing 14.20: *Capitalize all first letter of all full names of authors.*

14.6.2 Functions in SQL

While it is not possible to use R without using a function, in SQL a lot can be done without functions. However, you will surely hit that moment where it makes more sense to create a function in order to make code lighter.

Creating a function is just one command, that usually is terminated by `;`, but inside the function one usually has many more commands. Therefore, we first change the delimiter to something else, declare the function, and then change it back to `;`.

A simple example could be a function that calculates the average number of books per author, stores this in a variable @myAVG and then uses this variable:

```
delimiter //
CREATE PROCEDURE CalcAvgBooks (OUT avgBooks INT)
  BEGIN
  SELECT AVG(nbrBooks) INTO avgBooks FROM (SELECT COUNT(*) AS nbrBooks  FROM
    tbl_author_book GROUP BY author);
  END//
delimiter ;

-- Now use the function:
CALL CalcAvgBooks(@myAVG);

-- Use now the parameter myAVG
SELECT CONCAT('The average number of books per author in our libaray is: ',
    @myAvg);
```

Listing 14.21: *Creating a function and using it in SQL.*

Note that functions can also be used within queries to transform data or update data:

```
-- Functions can also be used in queries:
DROP FUNCTION tooSimple;
CREATE FUNCTION tooSimple (str1 CHAR(10), str2 CHAR(50))
 RETURNS CHAR(60) DETERMINISTIC
 RETURN CONCAT(str1,str2,'!');

 -- A UNION query combines two separate queries (column names must match)
 SELECT tooSimple('Hello, ', full_name) AS myMSG FROM tbl_authors WHERE ISNULL(
    death_date)
  UNION
  SELECT tooSimple('RIP, ', full_name) AS myMSG FROM tbl_authors WHERE NOT
    ISNULL(death_date);

-- Output:
+----------------------------------------+
| myMSG                                  |
+----------------------------------------+
| Hello,Erika Leonard!                   |
| Hello,Euclid of Alexandria!            |
| Hello,Bernard Marr!                    |
| Hello,Bart Baesens!                    |
| Hello,Philippe J.S. De Brouwer!        |
| Hello,Hadley Wickham!                  |
| Hello,Garrett Grolemund!               |
| RIP,Valentin Louis G. E. Marcel Proust! |
| RIP,Miguel de Cervantes Saavedra!      |
| RIP,James Augustine Aloysius Joyce!    |
| RIP,Isaac Newton!                      |
+----------------------------------------+
```

Listing 14.22: *Just a little taste of some additional features in SQL. We encourage you to learn more about SQL. This piece of code introduces functions, variable, and the UNION-query.*

Hint – Delimiter in SQL

In SQL, it makes sense to change the delimiter before creating a function. That allows you to use the delimiter `;` inside the function without triggering the server to execute the command that is followed by `;`.

With this section, we hope that we encouraged you to understand databases a little better; that we were able to illustrate the complexity of setting up, maintaining, and using a professional database system, but above all that it is now clear how to get data out of those systems with SQL.

However, it would be inefficient to have to go to the SQL server, select data, export it in a CSV file, and then import this file in R. In Chapter 15 *"Connecting R to an SQL Database"* on page 253 we will see how to import data from MySQL directly to R, and use further our knowledge of data manipulation on the database server and/or within R.

Hint – Make a backup of the database

It is very time-consuming to rebuild a database, and making a backup is a little more tricky than copying a file to another hard disk. Apart from solutions that involve backup servers for redundancy, for the private user an "sql-dump" is the way to go. An sql-dump will copy all tables – complete with their definitions and data – to a text-file on your computer. A complete SQL-dump will even include users and access rights. This backup comes in handy when something goes wrong: restoring this is a work of minutes instead of weeks. The following code shows how this can be done for MySQL from the Linux command prompt:

```
# Make the dump for your databases only:
mysqldump -u root -p -- databases library another_DB > dump_file.sql
# Or for all databases:
mysqldump -u root -p -- all-databases > dump_file.sql

# Then install your new computer, update your system,
# or recover from serious mistakes ... and then do:
mysql -u <user> -p < dump_file.sql
# Note that if the backup was for one database only
# you need to add in the first line of the dump file
# USE my_database_name or add it to the command line:
mysql -u root -p database_name_here < dump_file.sql
```

The sql-dump is a text-file, we encourage you to open it and learn from it.

Further information – More about SQL

This section barely scratched the surface of databases and SQL. It is a good idea to start your own simple project and learn while you go or find a book, online resource or course about SQL and discover many more functions such as ordering data with `ORDER BY`, understand the grouping commands `GROUP BY` with the additional `HAVING` (the `WHERE` clause for grouped data), the ways to upload files with data and commands (the command `SOURCE`), etc.
As mentioned earlier, the documentation of MySQL is a good place to start.

♣ 15 ♣

Connecting R to an SQL Database

In this section we will study one example of one RDBMS: MySQL. There are good reasons for this choice, it is widely used as a back-end of many websites, it is fast, reliable, and free. However, there are many databases available. In general, a good starting point is the package RODBC.

MySQL

RODBC

Hint – RODBC

Some databases have their own drivers for R, others not. It might be a good idea to have a look at the package RODBC. It is a package that allows to connect to many databases via the ODBC protocol.[a]

[a] ODBC stands for Open Database Connectivity and is the standard API for many DBMS.

In many cases, one will find specific drivers (or rather APIs) for the DBMS that is being used. These interfaces come in the familiar form of packages. For MySQL or MariaDB, this is RMySQL.

RMySQL

Digression – MariaDB

As mentioned in previous section, MariaDB is – almost – a drop-in replacement for MySQL. There are some differences in the configuration tables, how data is displayed, the command prompt (in MariaDB you will see the name of the database), CTRL+C is to be avoided in the MariaDB client), etc. However, till now, most other things that only rely on APIs such as phpmyadmin and RMySQL will work for both database engines.

With the package RMySQL, it is possible to both connect to MariaDB and MySQL in a convenient way and copy the data to R for further analysis. The basics of the package are to create a connection variable first and then use that connection to retrieve data.

```
# install.packages('RMySQL')
library(RMySQL)
# connect to the library
con <- dbConnect(MySQL(),
                 user     = "librarian",
                 password = "librarianPWD",
```

The Big R-Book: From Data Science to Learning Machines and Big Data, First Edition. Philippe J.S. De Brouwer.

Companion Website: www.wiley.com/go/De Brouwer/The Big R-Book

```
                dbname   = "library",
                host     = "localhost"
                )

# in case we would forget to disconnect:
on.exit(dbDisconnect(con))
```

Now, we have the connection stored in the object con and can use this to display data about the connection, run queries, and retrieve data.

```
# show some information

show(con)
summary(con, verbose = TRUE)
# dbGetInfo(con)  # similar as above but in list format
dbListResults(con)
dbListTables(con) # check: this might generate too much output

# get data
df_books <- dbGetQuery(con, "SELECT COUNT(*) AS nbrBooks
                  FROM tbl_author_book GROUP BY author;")

# Now, df_books is a data frame that can be used as usual.

# close the connection
dbDisconnect(con)
```

Warning – Batch environment

The aforementioned code will work in the interactive environment of R. However, it fails in a batch environment – for example while compiling this book. It generates the following error: `## Error in .local(dbObj, ...): internal error in RS DBI getConnection: corrupt connection handle`. If this happens, the solution is to wrap the connection and the query in one function.

There are a few other good reasons to wrap the database connection in functions:

1. This will make sure that the connection will always get closed.
2. We can design our own error messages with the functions `try()` and `tryCatch()`.
3. We can assure that we do not keep a connection open too long. This is important because RDBMS are multi-user environments that can get under heavy loads and hence it is always bad to keep connections to SQL servers open too long.

The code below does the same as the aforementioned code, but with our own custom functions that are a wrapper for opening the connection, running the query, returning the data and closing the connection. We strongly recommend to use this version.

```
# Load the package:
library(RMySQL)

## Loading required package: DBI
```

```
# db_get_data
# Get data from a MySQL database
# Arguments:
#    con_info -- MySQLConnection object -- the connection info to
#                                          the MySQL database
#    sSQL     -- character string       -- the SQL statement that
#                                          selects the records
# Returns
#    data.frame, containing the selected records
db_get_data <- function(con_info, sSQL){
  con <- dbConnect(MySQL(),
                   user     = con_info$user,
                   password = con_info$password,
                   dbname   = con_info$dbname,
                   host     = con_info$host
                   )
  df <- dbGetQuery(con, sSQL)
  dbDisconnect(con)
  df
}

# db_run_sql
# Run a query that returns no data in an MySQL database
# Arguments:
#    con_info -- MySQLConnection object -- open connection
#    sSQL     -- character string       -- the SQL statement to run
db_run_sql <-function(con_info, sSQL)
{
  con <- dbConnect(MySQL(),
                   user     = con_info$user,
                   password = con_info$password,
                   dbname   = con_info$dbname,
                   host     = con_info$host
                   )
  rs <- dbSendQuery(con,sSQL)
  dbDisconnect(con)
}
```

Assume that we want to generate a histogram of how many books our authors have in our library. The functions that we defined in the code segment above will help us to connect to the database, get the data to R and finally we can use the functionality of R to produce the histogram as in Figure 15.1 on page 256 with the following code:

```
# use the wrapper functions to get data.

# step 1: define the connection info
my_con_info <- list()
my_con_info$user     <- "librarian"
my_con_info$password <- "librarianPWD"
my_con_info$dbname   <- "library"
my_con_info$host     <- "localhost"


# step 2: get the data
my_query <- "SELECT COUNT(*) AS nbrBooks
                     FROM tbl_author_book GROUP BY author;"
df <- db_get_data(my_con_info, my_query)

# step 3: use this data to produce the histogram:
hist(df$nbrBooks, col='khaki3')
```

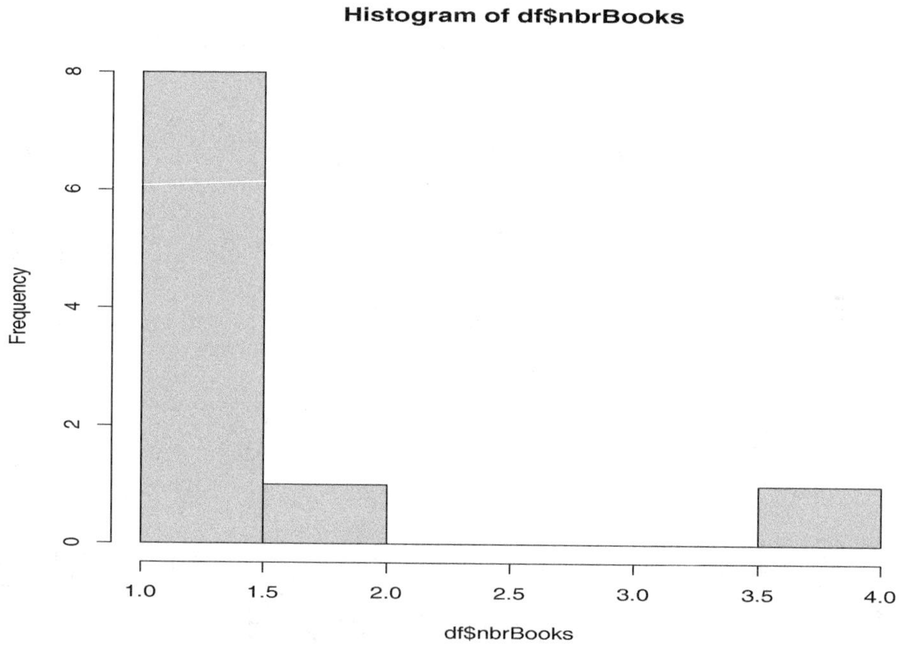

Figure 15.1: *Histogram generated with data from the MySQL database.*

Hint – Clearing the query cache

In case MySQL starts responding erratically and slows down it might be useful to clear the query cache. The following function does this for the whole server and for MySQL.

```
# -- reset query cache
sql_reset_query_cache <- function (con_info)
{
  con <- dbConnect(MySQL(),
                   user     = con_info$user,
                   password = con_info$password,
                   dbname   = con_info$dbname,
                   host     = con_info$host
                   )
# remove all cache:
system("sync && echo 3 | sudo tee /proc/sys/vm/drop_caches")
# clear MySQL cache cache and disconnect:
rc <- dbSendQuery(con, "RESET QUERY CACHE;")
dbDisconnect(con)
# once more remove all cache:
system("sync && echo 3 | sudo tee /proc/sys/vm/drop_caches")
}
```

PART IV

Data Wrangling

data wrangling

Data wrangling refers to the process of transforming a dataset into a form that is easier to understand and easier to work with. Data Wrangling skills will help you to work faster and more efficient. Good preparation of data will also lead to better models.

We will use data wrangling in its narrow sense: transforming data. Some businesses will use the word in a larger sense that also includes data visualization and modelling. This might be useful if the "Data and Analytics Team" will work on data end-to-end (from collecting the data from the database systems up to the final presentation). We will treat this in a separate section of this book: Part VII *"Reporting"* on page 685.

In many companies, "data wrangling" is used as somehow equivalent to "building a data-mart." A data-mart is the something like a supermarket for data, where the modeller can pick up the data in a format that is ready to use. The data-mart can also be seen as the product of data-wrangling.

Data wrangling, just as modelling and writing code, is as much a form of art as it is a science. Wrangling in particular cannot be done without knowing the steps before and the steps after: we need to understand the whole story before we can do a good job. The main goal is transforming the data that is obtained from the transaction system or data-warehouse in such form that it becomes directly useful for making a model.

For example, consider that we are making a credit scorecard in a bank. Then our data might contain the balances of current accounts of the last 24 months. These will be just numbers with the amount that people have on their account. While it is not impossible to feed this directly in a neural network or even linear regression, the results will probably not be robust, and the model is probably over-fit.[1] It makes much more sense to create new parameters, that contain a certain meaningful type of information. For example, we could create a variable `was_overdrawn` that is 1 when at least one of the balances was negative, 0 otherwise:

```
IF any_balance < 0
   THEN was_overdrawn = 0
   ELSE was_overdrawn = 1
   END IF
```

Or maybe, you want to create a parameter that indicates if a person was overdrawn last year as opposed to this year (maybe the financial problems were something from the past); or maybe, you prefer to have a parameter that shows the longest period that someone was overdrawn (which will not impact people that just forget to pa and do pay when the reminder arrives).

The choices that we make here, do not only depend on what data we have (upstream reality), but also on what the exact business model is (downstream reality). For example, if the bank earns a nice fee from someone that forgets to pay for a week, then it is worth not to sanction this behaviour.

In any case, you will have to "wrangle" your data before it is ready for modelling.[2]

However, more often you will be faced with the issue that real data is not homogeneous. For example, your bank might have very little credits of people older than 80 years old. Imagine that those were all produced by one agent that was active in retirement homes. First, of all that agent

---

[1] In wide data (data that contains many rows) it is quite possible to find spurious correlations – correlations that appear to be strong but are rather incidental and limited to that particular cut of data.

data-mart
data-warehouse

[2] At this moment, the practice of fitting a logistic model for creditworthiness resulting in a "credit score" is so long the standard way of working that there is even some kind of understanding of what standard variables would be needed in such case. This has lead to the concept "data-mart" as a standard solution. It is a copy of the data-warehouse of the company with data transformed in such way that it is directly usable for credit-modellers. In these logistic regressions it is also important to "bin" data in a meaningful way.

now is gone, and second this overly pushing of loans has lead to worse credits. So, how should we treat these customers?

"Data-binning" (also known as "discrete binning" or "bucketing") can alleviate these issues. It is a data pre-processing technique used to reduce the effects of minor observation errors. The original values, which fall in a given small interval, a "bin," are replaced by a value representative of that interval, often the central value. It is a form of quantization of continuous data.

Statistical data binning is a way to group a number of more or less continuous values into a smaller number of "bins." For example, if you have data about a group of people, you might want to arrange their ages into a smaller number of age intervals (for example, grouping every five years together). It can also be used in multivariate statistics, binning in several dimensions at once.

For many statistical methods, you might want to normalize your data (have all variables ranging between 0 and 1). This makes sure that the model is easy to understand and can even help in some cases make algorithms converge faster.

All this involves changing all records or adding new and all these examples are a form of "data wrangling" or data manipulation.

Hint – RDBMS and R

Usually, you or someone else will have to get data out of an RDBMS. These systems sit on very efficient servers and are optimized for speed. Just as R they have their strenghts and weaknesses. Doing simple operations on a large amount of data might be faster than your R application. So, use your knowledge from Chapter 14 *"SQL"* on page 223 to prepare as much as possible data on these servers. This might make a significant difference on how fast your will get the model ready.

In this part, we will briefly discuss all of those aspects, but we will start with a particular issue that is most important for privacy: anonymous data.

♣ 16 ♣

Anonymous Data

Before we can start the real work of manipulating data in order to gain more information from it, there might be the need to reduce the information content first and anonymise it. This step should be done before anything else, otherwise copies of sensitive data can be lingering around and will be found by people who should not have access to it. The golden rule to remember is: only take data that you might need and get rid of confidential or personal data as soon as possible in the process.

It is best to anonymise data even before importing it in R. Typically, it will sit in an SQL database and then it is easy to scramble before the data even leaves this database:

```
-- Using AES 256 for example:
MariaDB [(none)]> SELECT AES_ENCRYPT("Hello World", "secret_key_string");
+-----------------------------------------------+
| AES_ENCRYPT("Hello World", "secret_key_string") |
+-----------------------------------------------+
| ï£iEwï£i*0ï£iï£iWï£iï£iï£i5%ï£i                  |
+-----------------------------------------------+
1 row in set (0.00 sec)

-- Example:
SELECT AES_ENCRYPT(name, "secret_key_string"), AES_ENCRYPT(phone number, "
    secret_key_string"),        number_purchases, satifaction_rating,
    customer_since, etc.
       FROM tbl_customers;

-- To decrypt, use AES_DECRYPT(crypt_str, key_string)
```

Listing 16.1: *SQL code for MySQL (or MariaDB) to encrypt using AES256. Note that those relational database systems (RDBMSs) provide much more methods for encryption. It is worth to go through the documentation of your particular system for more support.*

Imagine that despite your request for anonymised data you still got data with data that would allow you to identify a person. In that case, R will also allow us to hash confidential data. This can be the moment to get creative[1] (the least conventional your encryption is, the more challenging it might be to crack it), or as usual in R "stand on the shoulders of giants" and use for example the package `anonymizer`. This package has all the tools to encrypt and decrypt to a very high standard, and might be just what you need.

anonymizer

[1] A nice example are the ideas of Jan Górecki; published here: https://jangorecki.github.io/blog/2014-11-07/Data-Anonymization-in-R.html

The Big R-Book: From Data Science to Learning Machines and Big Data, First Edition. Philippe J.S. De Brouwer.

Companion Website: www.wiley.com/go/De Brouwer/The Big R-Book

OS
operating system

There is also the package `sodium`, created by Jeroen Ooms. It is a wrapper around `libsodium`, which is a standard library. So you will need to install this first on your operating system (OS).

- deb: libsodium-dev (Debian, Ubuntu, etc.)
- rpm: libsodium-devel (Fedora, EPEL)
- csw: libsodium_dev (Solaris)
- brew: libsodium (OSX)

This means – most probably – your will first need to open a terminal and run the following commands in the CLI (command line interface of your OS):

```
sudo apt-get install libsodium-dev
```

sodium

Then we can open R and install the `sodium` library for R.

```
# install.packages('sodium') # do only once
                             # fails if you do not have libsodium-dev
library(sodium)

# Create the SHA256 key based on a secret password:
key <- sha256(charToRaw("My sweet secret"))

# Serialize the data to be encrypted:
msg <- serialize("Philippe J.S. De Brouwer", NULL)

# Encrypt:
msg_encr <- data_encrypt(msg, key)


orig <- data_decrypt(msg_encr, key)
stopifnot(identical(msg, orig))

# Tag the message with your key (HMAC):
tag <- data_tag(msg, key)
```

If you are a programmer or work on a large R-project and you need a secret space in a public code library, then a good place to start looking is the package `secure` from Hadley Wickham.

rcrypt

When there is the need to encrypt entire files, then you might use `rcrypt`, which provides high standard symmetric encryption using GNU Privacy Guard (GPG).[2]. The default encryption mechanism is AES256, but at the user's request it can also do "Camellia256," "TWOFISH," or "AES128." However, last time we checked it did not work with the last version of R and since it is only a wrapper around gpg it makes sense to use the OS and use gpg directly.

AES128
Camellia256
TWOFISH
GPG

This is only a tiny sliver of what encryption is and how to use it. Encryption can be a challenging subject and is a specialisation in itself. For most data scientists and modellers, it is far better to make sure that we get anonymised data.

AES256
GNUPG

Also note that encryption is a vast territory that is in rapid expansion. This is a place to watch on the Internet and make sure that you are up-to-date with the latest developments. Especially now that we are at the verge of a revolution spurred by the dawn of quantum computers, one might expect rapid developments in the field of encryption.

PGP
pretty good privacy

[2]GPG or GnuPG (which stands for "GNU Privacy Guard") is a complete and free implementation of the OpenPGP standard as defined by RFC4880 (also known as PGP – which stands for "pretty good privacy"). GnuPG allows to encrypt and sign your data and communications; it features key management and access modules for all kinds of public key directories.

Further information – Cryptology

It is beyond the scope of this book to provide a solid introduction to the subject of cryptology. There are many good books about cryptography, and some provide great overviews such as Van Tilborg and Jajodia (2014). Don't forget to read Phil Zimmermann's paper "Why I wrote PGP" (see `https://www.philzimmermann.com/EN/essays/WhyIWrotePGP.html`). But above all, I'm sure that you will enjoy reading "The joy of cryptography" by Mike Rosulek – it is a free book and available online and part of the "open textbook initiative" of the Oregon State University – `http://web.engr.oregonstate.edu/~rosulekm/crypto`.

♣ 17 ♣

Data Wrangling in the tidyverse

An iconic effort to obtain a certain form of standardization is the work around the `tidyverse`. The tidyverse defines "tidy data" and "tidy code" in a logical and compelling matter. Adhering to those rules makes code more readable, easier to maintain and more fun to build.

It is possible to program in R without the tidyverse – as we did in Chapter 4 *"The Basics of R"* on page 21 – because it is important to know and be able to read and understand base-R code. Knowing both, we invite you to make your own, informed choice.

In Chapter 7 *"Tidy R with the Tidyverse"* on page 121 we already introduced the tidyverse and largely explained operators and ideas. In this chapter, we focus on obtaining "tidy data."

The Big R-Book: From Data Science to Learning Machines and Big Data, First Edition. Philippe J.S. De Brouwer.

Companion Website: www.wiley.com/go/De Brouwer/The Big R-Book

17.1 Importing the Data

17.1.1 Importing from an SQL RDBMS

In this section we will assume that the reader is familiar with concepts such as RDBMS, relational data, primary key, relation, foreign key, etc. from Chapter 14 *"SQL"* on page 223. We also assume that the reader already has data in R (tibble or data frame) as explained in Chapter 15 *"Connecting R to an SQL Database"* on page 253. Further, we will also use the database design of Figure 14.4 on page 245.

First, we remind the reader of the functions to connect to a database and retrieve data from it. These functions were already presented on page 254 in Chapter 15.

```
# --
library(RMySQL)

# -- The functions as mentioned earlier:
# db_get_data
# Get data from a MySQL database
# Arguments:
#    con_info -- MySQLConnection object -- containing the connection
#                                          info to the MySQL database
#    sSQL     -- character string       -- the SQL statement that selects
#                                          the records
# Returns
#    data.frame, containing the selected records
db_get_data <- function(con_info, sSQL){
  con <- dbConnect(MySQL(),
                   user     = con_info$user,
                   password = con_info$password,
                   dbname   = con_info$dbname,
                   host     = con_info$host
                   )
  df <- dbGetQuery(con, sSQL)
  dbDisconnect(con)
  df
}

# db_run_sql
# Run a query that returns no data in an MySQL database
# Arguments:
#    con_info -- MySQLConnection object -- containing the connection
#                                          info to the MySQL database
#    sSQL     -- character string       -- the SQL statement to be run
db_run_sql <-function(con_info, sSQL)
{
  con <- dbConnect(MySQL(),
                   user     = con_info$user,
                   password = con_info$password,
                   dbname   = con_info$dbname,
                   host     = con_info$host
                   )
  rs <- dbSendQuery(con,sSQL)
  dbDisconnect(con)
}
```

We can now use those functions to connect to the database mentioned above. Since we want to use the tidyverse and its functionalities, we load this first. At the end of this code, the user will notice the function `as_tibble()`. That function coerces a two dimensional dataset into a `tibble`. A tibble is the equivalent of data.frame in the tidyverse.

`as_tibble()`

dplyr

```
## Load dplyr via tidyverse:
library(tidyverse)

# Define the wrapper functions:

# Step 1: define the connection info.
my_con_info <- list()
my_con_info$user     <- "librarian"
my_con_info$password <- "librarianPWD"
my_con_info$dbname   <- "library"
my_con_info$host     <- "localhost"


# -- The data import was similar to what we had done previously.
# -- However, now we import all tables separately
# Step 2: get the data
my_tables <- c("tbl_authors", "tbl_author_book",
               "tbl_books", "tbl_genres")
my_db_names <- c("authors", "author_book",
               "books", "genres")

# Loop over the four tables and download their data:
for (n in 1:4) {
  my_sql <- paste("SELECT * FROM `",my_tables[n],"`;", sep = "")
  df <- db_get_data(my_con_info, my_sql)
  # the next line uses tibbles are from the tidyverse
  as_tibble(assign(my_db_names[n],df))
 }

# Step 3: do something with the data
# -- This will follow in the remainder of the section
```

Now, we have four new tibbles containing our data.

To ilustrate this, we investigate what exactly we have imported (remember that unlike data frames, tibbles will not change data-structure and assume that something is a factor for example).

```
str(authors)
## 'data.frame' :11 obs. of  5 variables:
##  $ author_id :num 1 2 3 4 5 7 11 13 14 15...
##  $ pen_name  :chr "Marcel Proust" "Miguel de Cervantes" "James Joyce" "E. L. James"...
##  $ full_name :chr "Valentin Louis G. E. Marcel Proust" "Miguel de Cervantes Saavedra"
                  "James Augustine Aloysius Joyce" "Erika Leonard"...
##  $ birth_date:chr "1871-07-10" "1547-09-29" "1882-02-02" "1963-03-07"...
##  $ death_date:chr "1922-11-18" "1616-04-22" "1941-01-13" NA...
```

17.1.2 Importing Flat Files in the Tidyverse

R provides via the package `utils` functionality to read flat text files into the working memory or R with the functions of the read-family (e.g. `read.delim()`, `read.csv()`, `read.table()`, `read.fwf()`, etc.). Indeed, reading in own data is the first step of any serious analysis. R provides this functionality right out of the box, and those functions are among the oldest functions of R.

readr

Via the package `readr` we get in the tidyverse a complete new set of functions. The reason to provide new functions is that in the meanwhile R evolved from strictly an interactive experience to a solution that could be trusted for large projects. R was now to a large extend used in batch mode in complex environments. This implied that some of the choices made earlier would not necessarily be the best any more.

utils

Compared to the base R (or as provided by the package `utils`), the functions from `readr` have the following distinct advantages.

- They are faster (which is important when we are reading in a larger dataset – for example high-frequency financial markets data, with multiple millions of lines, or the four billion of payments that the larger banks has pushed through today).
- When things take a while the `readr` functions will have a progress bar (that is useful because we see that something is progressing and should not be doubting that the computer is stuck and interrupt the process for example).
- They are more "strong and independent": they leave less choices to the operating system and allow to fix encodings for example. That helps to make code reproducible. This is especially important in regions such as Belgium where many languages are spoken or in large organisations where one model reviewer or modeller might get one day a model from China – prepared on Windows 10 – and the next day from France, prepared on Linux.
- They adhere to the tidyverse philosophy and hence produce `tibbles`. Doing so, they also adhere to the philosophy of not naming rows[1], not shorten column names, not coercing strings to factors, etc.

In an industrial environment, most of those issues are important, and it did indeed make sense to implement new functions. If you have the choice, and there is more new code than legacy code, then it is worth to move towards the tidyverse and embrace its philosophy, use all its packages and rules. It is even worth to rewrite older code, so that in ten years from now, it is still possible to maintain it.

The functions of `readr` adhere to the tidyverse philosophy and hence work is passed on from more general functions to more specified functions. The functions that the typical reader will use (from the `read`-family that read in a specific file type) will all rely on the basic functions to convert text to data types: the parser functions.

There is a series of parser functions with functions to generate columns that are then combined in a series of read-functions. These read-functions are specialized for a certain type of file.

The parser functions will convert a string (from a flat file or any other input given to them) into a more specialized type. This can be a double, integer, logical, string, etc.

```
# example to illustrate the parser functions

v       <- c("1.0", "2.3", "2.7", ".")
nbrs    <- c("$100,00.00", "12.4%")
s_dte   <- "2018-05-03"

# The parser functions can generate from these strings
# more specialized types.

# The following will generate an error:
parse_double(v)         # reason: "." is not a number

## Warning: 1 parsing failure.
## row col expected actual
##  4   - a double     .
```

[1] The issue with naming rows is illustrated in Chapter 8.5 *"Creating an Overview of Data Characteristics"* on page 155. In that section, we need the brand name of the cars in the `mtcars` database. This is data and hence should be in one of the columns, but it is stored as the names of the columns.

```
## [1] 1.0 2.3 2.7  NA
## attr(,"problems")
## # A tibble: 1 x 4
##     row   col expected actual
##   <int> <int> <chr>    <chr>
## 1     4    NA a double .

parse_double(v, na = ".") # Tell R what the encoding of NA is
## [1] 1.0 2.3 2.7  NA

parse_number(nbrs)
## [1] 10000.0    12.4

parse_date(s_dte)
## [1] "2018-05-03"
```

parse_double()

parse_number()

parse_date()

There is even a function to guess what type suits best: guess_parser(), that is used in the parse_guess() parser-function.

guess_parser()

parse_guess()

```
parse_guess(v)
## [1] "1.0" "2.3" "2.7" "."

parse_guess(v, na = ".")
## [1] "1.0" "2.3" "2.7" NA

parse_guess(s_dte)
## [1] "2018-05-03"

guess_parser(v)
## [1] "character"

guess_parser(v[1:3])
## [1] "double"

guess_parser(s_dte)
## [1] "date"

guess_parser(nbrs)
## [1] "character"
```

There are also parsers for logical types, factors, dates, times and date-times. Each parse_*() function will collaborate with a col_*() function to puzzle tibbles together from the pieces of the flat file read in.

Further information – readr

This is only the top of the iceberg. There are a lot more functions catering for a lot more complicated situations (such as different encodings, local habits such as using a point or comma as decimal. A good place to start is the documentation of the tidyverse: https://readr.tidyverse.org/articles/readr.html

These parser functions are used in the functions that can read in entire files. These functions are all of the read-family. For example, the function to read in an entire csv-file is read_csv().

17.1.2.1 CSV Files

read.csv() utils read_csv() readr

CSV files are probably the most common file types used, they are to some extend readable by humans (at least they are not binary encoded), they are lighter than fixed-with files and more flexible (they allow for really long strings). The function `read.csv()` is replaced by the `read_*` family and hence `readr` has a function `read_csv()` to read csv-files.

```
s_csv = "'a','b','c'\n001,2.34,.\n2,3.14,55\n3,.,43"
read_csv(s_csv)
## # A tibble: 3 x 3
##   `'a'` `'b'` `'c'`
##   <chr> <chr> <chr>
## 1 001   2.34  .
## 2 2     3.14  55
## 3 3     .     43

read_csv(s_csv, na = '.')  # Tell R how to understand the '.'
## # A tibble: 3 x 3
##   `'a'` `'b'` `'c'`
##   <chr> <dbl> <dbl>
## 1 001    2.34    NA
## 2 2      3.14    55
## 3 3     NA       43
```

read_csv2()

Note – Separator specific functions

There is a function `read_csv2()` that will use the semi-colon as separator. This is useful for countries that use the comma as thousand separator.

If the guesses of `readr` are not what you would want to have, then it is possible to over-ride these guesses. A good workflow is to start is to ask `readr` to report on the guesses and modify only where necessary. Below we illustrate how this can be achieved.

spec_csv()

```
# Method 1: before the actual import
spec_csv(s_csv, na = '.')
## cols(
##   `'a'` = col_character(),
##   `'b'` = col_double(),
##   `'c'` = col_double()
## )
```

spec()

```
# Method 2: check after facts:
t <- read_csv(s_csv, na = '.')
spec(t)
## cols(
##   `'a'` = col_character(),
##   `'b'` = col_double(),
##   `'c'` = col_double()
## )
```

Hint – Check data-type before importing

Checking the data-type before data-import makes a lot of sense for large databases when the import can take very long time.

The output obtained via the function spec_csv can serve as a template that can be fed into the function read_csv() to over-ride the defaults guessed by readr. For example we can coerce column "c" to double.

```
read_csv(s_csv, na = '.',
  col_names = TRUE,
  cols(
    `'a'` = col_character(),
    `'b'` = col_double(),
    `'c'` = col_double()       # coerce to double
    )
  )
## # A tibble: 3 x 3
##    `'a'` `'b'` `'c'`
##    <chr> <dbl> <dbl>
## 1 001     2.34    NA
## 2 2       3.14    55
## 3 3      NA       43
```

Question #12 Importing difficult files

The package readr comes with a particularly challenging csv-file as example. Import this file to R without losing data and with no errors in the variable types.
Here is a hint on how to get started:

```
# Start with:
t <- read_csv(readr_example("challenge.csv"))

# Then, to see the issues, do:
problems(t)


# Notice that the problems start in row 1001, so
# the first 1000 rows are special cases. The first improvement
# can be obtained by increase the guesses
## compare
spec_csv(readr_example("challenge.csv"))
## with
spec_csv(readr_example("challenge.csv"), guess_max = 1001)
```

17.1.2.2 Making Sense of Fixed-width Files

Fixed width tables were used because that is how people would present data naturally, so are easy to read by humans and offered a good compromise between limited computing power and functionality. Fixed width tables are in essence text files where each column has a fixed width. They look a little like the output that MySQL gave us in Chapter 15 *"Connecting R to an SQL Database"* on page 253.

utils
read.fwf()

The package `utils` provides `read.fwf()` and in the tidyverse, it is the package `readr` that comes to the rescue with the function `read_fwf()`. This function is coherent with the philosophy of the tidyverse, e.g. using an underscore in the name rather than a dot, allowing for piping and the output is a tibble.[2]

readr

Importing data is a common task to any statistical modelling and hence `readr` is part of the core-tidyverse and will be loaded with the command `library(tidyverse)`. However, it is also possible to load just `readr`

```
# load readr
library(readr)
# Or load the tidyverse with library(tidyverse), it includes readr.
```

We also need a text file. `read_fwf()` can of course read in a file from any disk and you can use any text editor to make the text-file with the following content:

book_id	year	title	genre
1	1896	Les plaisirs et les jour	LITmod
2	1927	Albertine disparue	LITmod
3	1954	Contre Sainte–Beuve	LITmod
4	1922	A la recherche du temps perdu	LITmod
5	1615	El Ingenioso Hidalgo Don Quijote de la Mancha	LITmod
6	1613	Novelas ejemplares	LITmod
7	2011	Fifty Shades of Grey	LITero
8	1687	Philosophiæ Naturalis Principia Mathematica	SCIphy
9	−300	Elements (translated)	SCImat
10	2014	Big Data World	SCIdat
11	2016	Key Business Analytics	SCIdat
12	2011	Maslowian Portfolio Theory	FINinv
13	2016	R for Data Science	SCIdat

It is also possible to define a string variable in R with the same content. this can be done as follows.

```
# Make a string that looks like a fixed-width table (shortened):
txt <- " book_id  year  title                                          genre
       1  1896  Les plaisirs et les jour                        LITmod
       2  1927  Albertine disparue                              LITmod
       3  1954  Contre Sainte-Beuve                             LITmod
       8  1687  Philosophiæ Naturalis Principia Mathematica     SCIphy
       9  -300  Elements (translated )                          SCImat
      10  2014  Big Data World                                  SCIdat
      11  2016  Key Business Analytics                          SCIdat
      12  2011  Maslowian Portfolio Theory                      FINinv
      13  2016  R for Data Science                              SCIdat"
```

Starting from this string variable, we will create a text file that has data in the fixed-width format.

```
fileConn <- file("books.txt")
writeLines(txt, fileConn)
close(fileConn)

my_headers <- c("book_id","year","title","genre")
```

read_fwf()

The previous code chunk has created the text file `book.txt` in the working path of R. Now, we can read it back in to illustrate how the `read_fwf()` function works.

2 More about this issue is in Chapter 17.1.2 *"Importing Flat Files in the Tidyverse"* on page 267.

```
# Reading the fixed-width file
# -- > by indicating the widths of the columns
t <- read_fwf(
  file = "./books.txt",
  skip = 1,                 # skip one line with headers
  fwf_widths(c(8, 6, 48, 8), my_headers)
  )

## Parsed with column specification:
## cols(
##   book_id = col_double(),
##   year = col_double(),
##   title = col_character(),
##   genre = col_character()
## )

# Inspect the input:
print(t)
## # A tibble: 9 x 4
##     book_id  year title                                genre
##       <dbl> <dbl> <chr>                                <chr>
## 1         1  1896 Les plaisirs et les jour             LITm~
## 2         2  1927 Albertine disparue                   LITm~
## 3         3  1954 Contre Sainte-Beuve                  LITm~
## 4         8  1687 Philosophiæ Naturalis Principia Mat~ SCIp
## 5         9  -300 Elements (translated )               SCIm~
## 6        10  2014 Big Data World                       SCId~
## 7        11  2016 Key Business Analytics               SCId~
## 8        12  2011 Maslowian Portfolio Theory           FINi~
## 9        13  2016 R for Data Science                   SCId~

# -- > same but naming directly
t <- read_fwf(
  file = "./books.txt",
  skip = 1,                 # skip one line with headers
  fwf_cols(book_id = 8, year = 6,
           title = 48, genre = 8)
  )

## Parsed with column specification:
## cols(
##   book_id = col_double(),
##   year = col_double(),
##   title = col_character(),
##   genre = col_character()
## )

# -- > by selecting columns (by indicating begin and end):
t2 <- read_fwf(
  file = "books.txt",
  skip = 1,
  fwf_cols(year = c(11, 15),
           title = c(17, 63))
  )

## Parsed with column specification:
## cols(
##   year = col_double(),
##   title = col_character()
## )

# -- > by guessing the columns
# The function fwf_empty can help to guess where the columns start
```

```
# based on white space
t3 <- read_fwf(
  file = "books.txt",
  skip = 1,
  fwf_empty("books.txt")
  )

## Parsed with column specification:
## cols(
##   X1 = col_double(),
##   X2 = col_double(),
##   X3 = col_character(),
##   X4 = col_character(),
##   X5 = col_character()
## )
```

Note that this last method fails: it identifies a separate column for the word "Mathematica", while this is actually part of the column "title":

```
print(t3)
## # A tibble: 9 x 5
##        X1    X2 X3                          X4         X5
##     <dbl> <dbl> <chr>                       <chr>      <chr>
## 1       1  1896 Les plaisirs et les jour    <NA>       LITm~
## 2       2  1927 Albertine disparue          <NA>       LITm~
## 3       3  1954 Contre Sainte-Beuve         <NA>       LITm~
## 4       8  1687 Philosophi\(\ae\) Naturalis Prin~ Mathemati~ SCIp~
## 5       9  -300 Elements (translated )      <NA>       SCIm~
## 6      10  2014 Big Data World              <NA>       SCId~
## 7      11  2016 Key Business Analytics      <NA>       SCId~
## 8      12  2011 Maslowian Portfolio Theory  <NA>       FINi~
## 9      13  2016 R for Data Science          <NA>       SCId~
```

Note – Automated downloading and decompressing

It is even possible to read in compressed files and/or files from the Internet. Actually, files ending in `.gz`, `.bz2`, `.xz`, or `.zip` will be automatically decompressed, where files starting with `http://`, `https://`, `ftp://`, or `ftps://` will be automatically downloaded. Compressed files that are on the Internet will be first downloaded and then decompressed.

Now, that we have seen how to import data from an RDMS directly (Chapter 15 *"Connecting R to an SQL Database"* on page 253), import flat files (text files that contain data in a delimited format or fixed width) in this section, it is time to make sure that the data that we have is tidy.

17.2 Tidy Data

While we already introduced the tidyverse in Chapter 7 *"Tidy R with the Tidyverse"* on page 121, in this section we will focus on the concept "tidy data."

The tidyverse is a collection of packages for R that all adhere to a certain rules. It is quite hard to make functions, scripts, and visualizations work universally and it is even harder to make code readable for everyone. Most people will encode a dataset as columns. The header of the columns is the name of the variable and each row has in that column a value (or NA). But why not storing variables in rows? In R (nor any other programming language) there are no barriers to do so, but it will make your code really hard to read and functionalities such as plotting, and matrix multiplication might not deliver the results that you are seeking.

For example, in the dataset `mtcars`, one will notice that the names of the cars are not stored in a column, but rather are the names of the rows.

```
head(mtcars)
##                    mpg cyl disp  hp drat    wt  qsec vs
## Mazda RX4         21.0   6  160 110 3.90 2.620 16.46  0
## Mazda RX4 Wag     21.0   6  160 110 3.90 2.875 17.02  0
## Datsun 710        22.8   4  108  93 3.85 2.320 18.61  1
## Hornet 4 Drive    21.4   6  258 110 3.08 3.215 19.44  1
## Hornet Sportabout 18.7   8  360 175 3.15 3.440 17.02  0
## Valiant           18.1   6  225 105 2.76 3.460 20.22  1
##                   am gear carb        l
## Mazda RX4          1    4    4 11.20069
## Mazda RX4 Wag      1    4    4 11.20069
## Datsun 710         1    4    1 10.31643
## Hornet 4 Drive     0    3    1 10.99134
## Hornet Sportabout  0    3    2 12.57832
## Valiant            0    3    1 12.99528

mtcars[1,1]          # mpg is in the first column
## [1] 21

rownames(mtcars[1,]) # the name of the car is not a column
## [1] "Mazda RX4"
```

Tidy data looks like the mtcars data frame, but rather will have the car names – which is data – stored into a column. In summary tidy data will always have

1. a tibble/data-frame for each dataset,
2. a column for each variable,
3. a row for each observation,
4. a value (or NA) in each cell – the intersection between row and column.

So, this is what tidy data looks like, but it is harder to describe what messy data looks like. There are many ways to make data messy or "untidy". Each data provider can have his/her own way to produce untidy data.

A common mistake is for example putting data in rownames. Imagine for example a data frame that looks as follows:

```
Unit      Q1    Q2    Q3   Q4
Europe   200   100    88  270
```

```
Asia      320   315   300  321
...
```

This dataset has columns that contain data: the moment of observation (quarter 1, etc.) in the column names. Its tidy equivalent would look like:

```
Unit      Quarter Sales
Europe         Q1 200
Asia           Q1 320
Europe         Q2 100
Asia           Q2 315
...
```

This is the type of data-frame that can be directly used in ggplot2 (that is part of the tidyverse family). In Section 17.3.2 *"Convert Headers to Data"* on page 281 we will see how to make this particular conversion with the function `gather()`.

Another typical mistake is to use variable names in cells

```
Unit   Dollar  Q1
Europe sales  200
Europe profit  55
Asia   sales  320
Asia   profit 120
...
```

This dataset has stored variable names. In Chapter 17.3.3 *"Spreading One Column Over Many"* on page 284 tidyverse provides the function `spread()` to clean up such data.

There are much more ways to produce untidy data. So, we will revisit this subject and devote a whole section – Chapter 17.3 *"Tidying Up Data with tidyr"* on page 277 – to tidying up data.

17.3 Tidying Up Data with tidyr

tidyr

Now, we have some data in R, but long before we can start the cool work and build a neural network or a decision tree, there is still a lot of "heavy lifting" to be done. At this point, the data-wrangling can really start and needs to precede the modelling. In the first place, we want our data to be tidy. Tidy data will lead to neater code, the syntax will be easier to read for your future self or someone else, and it will be more recognisable and that in its turn will lead to more efficient work and more satisfaction.

In defining tidy data we tie in with the ideas proposed by Hadley Wickham – Wickham et al. (2014) – as explained in Chapter 17.2 *"Tidy Data"* on page 275.

Definition: – Tidy data

Tidy Data is rectangular data that is organised so that:

- each variable corresponds to exactly one column,
- each observation corresponds to exactly one row, and
- each type of observational unit corresponds to one table.

Hint – The tidyverse philosophy

At this point, it is important to make sure that the reader knows what the tidyverse is and has a solid understanding of its philosophy and conventions. These ideas can be found in Chapter 7 *"Tidy R with the Tidyverse"* on page 121.

In many cases, our data will not be very tidy. We often observe that:

- there are not columns that can be addressed (this means it is a "flat text file" or a fixed-width table);
- a single table contains more than one observational units (for example we receive the inner join of the database of customers, accounts and balances);
- a single observational unit is spread out over multiple tables (e.g. the customers address details are in the table of current accounts and other details are in the table of credit cards);
- the column headers are also values (and not variable names), for example "sales Q1," sales Q2" are actually different observations (rows: Q1 and Q2) of one variable "sales" seen by the credit card application, customers from core-banking system, customers from the complaints database;
- more than one variable are combined in one column (e.g. postcode and city-name, first name and last name, year-month, etc.);
- etc.

Notice that the above table is not in order of frequency of appearing in real life, but rather in the order that you would want to address it. Also, we did not mention missing data or wrong data for example. Those are not issues that can be solved by following the concept of tidy data.

The good news is that cleaning up data is not an activity that requires a lot of tools. The package `tidyr` provides a small set of functions to get the job done.

17.3.1 Splitting Tables

Assume that we have one big table that actually mixes more than one logical concept. For example, imagine that we downloaded the Cartesian product or our tables from the library[3] example via the following code:

```
## --
## -- Load dplyr via tidyverse
library(tidyverse)
library(RMySQL)

## -- The functions as mentioned earlier:

# db_get_data
# Get data from a MySQL database
# Arguments:
#    con_info -- MySQLConnection object -- the connection info
#                                          to the MySQL database
#    sSQL     -- character string       -- the SQL statement that
#                                          selects the records
# Returns
#    data.frame, containing the selected records
db_get_data <- function(con_info, sSQL){
  con <- dbConnect(MySQL(),
                   user     = con_info$user,
                   password = con_info$password,
                   dbname   = con_info$dbname,
                   host     = con_info$host
                   )
  df <- dbGetQuery(con, sSQL)
  dbDisconnect(con)
  df
}

# db_run_sql
# Run a query that returns no data in an MySQL database
# Arguments:
#    con_info -- MySQLConnection object -- the connection info
#                                          to the MySQL database
#    sSQL     -- character string       -- the SQL statement
#                                          to be run
db_run_sql <-function(con_info, sSQL)
{
  con <- dbConnect(MySQL(),
                   user     = con_info$user,
                   password = con_info$password,
                   dbname   = con_info$dbname,
                   host     = con_info$host
                   )
  rs <- dbSendQuery(con,sSQL)
  dbDisconnect(con)
```

[3]The example of the library is developed in Section 14 *"SQL"* on page 223 and also used in Chapter 15 *"Connecting R to an SQL Database"* on page 253.

```
}

# use the wrapper functions to get data.

# step 1: define the connection info
my_con_info <- list()
my_con_info$user     <- "librarian"
my_con_info$password <- "librarianPWD"
my_con_info$dbname   <- "library"
my_con_info$host     <- "localhost"


## -- Import 2 tables combined
# step 2: get the data
my_sql <- "SELECT * FROM tbl_authors
   JOIN tbl_author_book ON author_id = author
   JOIN tbl_books       ON book      = book_id
   JOIN tbl_genres      ON genre     = genre_id;"
t_mix <- db_get_data(my_con_info, my_sql)
t_mix <- as.tibble(t_mix)  %>%  print
## # A tibble: 14 x 16
##    author_id pen_name full_name birth_date death_date
##        <dbl> <chr>    <chr>     <chr>      <chr>
##  1         1 Marcel ~ Valentin~ 1871-07-10 1922-11-18
##  2         1 Marcel ~ Valentin~ 1871-07-10 1922-11-18
##  3         1 Marcel ~ Valentin~ 1871-07-10 1922-11-18
##  4         1 Marcel ~ Valentin~ 1871-07-10 1922-11-18
##  5         2 Miguel ~ Miguel d~ 1547-09-29 1616-04-22
##  6         2 Miguel ~ Miguel d~ 1547-09-29 1616-04-22
##  7         4 E. L. J~ Erika Le~ 1963-03-07 <NA>
##  8         5 Isaac N~ Isaac Ne~ 1642-12-25 1726-03-20
##  9         7 Euclid   Euclid o~ <NA>       <NA>
## 10        11 Bernard~ Bernard ~ <NA>       <NA>
## 11        13 Bart Ba~ Bart Bae~ 1975-02-27 <NA>
## 12        14 Philipp~ Philippe~ 1969-02-21 <NA>
## 13        15 Hadley ~ Hadley W~ <NA>       <NA>
## 14        16 Garrett~ Garrett ~ <NA>       <NA>
## # ... with 11 more variables: ab_id <dbl>, author <dbl>,
## #   book <dbl>, book_id <dbl>, year <int>, title <chr>,
## #   genre <chr>, genre_id <chr>, type <chr>,
## #   sub_type <chr>, location <chr>
```

It is not uncommon to get a dataset as `t_mix`. In some cases, this might be the most useful form. However, imagine that we get a mix of the customer table and the table of loans. This table would only allow us a to make a prediction based on the rows in it. So, we would actually look at what customers declare, what the loan is. But it is of course more powerful to get a true view of the customer by aggregating all loans, and calculate how much loan instalments the person pays every month as compared to his/her income.

To achieve that we need to split the tables in two separate tables. Assuming that the data is correct and that there are no inconsistencies, then this is not too difficult. In real life, this becomes more difficult. For example, to get a loan a customer declares an income of x. Two years later he/she applies again for a loan and declares $x + \delta$. Is he or she mistaken? Is the new income the correct one? Maybe there is another postcode? Maybe we need to make customer snapshots at the moment of loan application but maybe these are mistakes.

Of course, it is even possible that the current address as reported by the core banking system is different from the address as reported by the loan system. Data is important and it is impossible to create a good model when the data has too much noise.

The problem is often in communication. The agent needs to know the importance of correct data. But also the modeller needs to understand how the data is structured in the transactional system. That is the only way to ask for a suitable set of data. In many cases, it is sufficient to ask for two separate tables to avoid getting a mix of tables. Although the author has met crisis situations where the bank simply had convoluted data and did not have the technical capacity to go back and we had to work from that.

Maybe the best solution is to upload your data back into a RDBMS and build a logical structure there before downloading again to R. However, R can do this also as follows:

1. Understand the data structure, eventually talk to the data owners and understand what is the job at hand. In this case, it is a mix of four tables: authors, a link-table to books, books, and genres.

```
# Make a table of how much each author_id occurs:
nbr_auth <- t_mix  %>% count(author_id)

# Do the same and include all fields that are assumed to
# be part of the table authors.
nbr_auth2 <- t_mix    %>%
  count(author_id, pen_name, full_name, birth_date, death_date, book)

nbr_auth$n - nbr_auth2$n
##  [1] 3 1 0 0 0 0 0 0 0 0 3 1 0 0
```

2. Learn from experiments till we find the right structure. In our case "book" is not unique for an "author," so we try again.

```
# Try without book:
nbr_auth2 <- t_mix    %>%
  count(author_id, pen_name, full_name, birth_date, death_date)

# Now these occurrences are the same:
nbr_auth$n - nbr_auth2$n
##  [1] 0 0 0 0 0 0 0 0 0 0
```

3. This looks better. But note that this exact match is only possible because our data is clean (because we took care and/or because we asked MySQL to help us to guard referential integrity). We still have to determine now which table takes which fields.

4. Now, the heavy lifting is done and we can simply extract all data.

```
my_authors <- tibble(author_id = t_mix$author_id,
                     pen_name   = t_mix$pen_name,
                     full_name  = t_mix$full_name,
                     birth_date = t_mix$birth_date,
                     death_date = t_mix$death_date
                     )       %>%
              unique         %>%
              print
## # A tibble: 10 x 5
##    author_id pen_name    full_name     birth_date death_date
##        <dbl> <chr>       <chr>         <chr>      <chr>
##  1         1 Marcel Pr~ Valentin Lo~ 1871-07-10 1922-11-18
##  2         2 Miguel de~ Miguel de C~ 1547-09-29 1616-04-22
##  3         4 E. L. Jam~ Erika Leona~ 1963-03-07 <NA>
##  4         5 Isaac New~ Isaac Newton 1642-12-25 1726-03-20
```

```
## 5           7 Euclid      Euclid of A~ <NA>       <NA>
## 6          11 Bernard M~ Bernard Marr <NA>       <NA>
## 7          13 Bart Baes~ Bart Baesens 1975-02-27 <NA>
## 8          14 Philippe ~ Philippe J.~ 1969-02-21 <NA>
## 9          15 Hadley Wi~ Hadley Wick~ <NA>       <NA>
## 10         16 Garrett G~ Garrett Gro~ <NA>       <NA>
```

5. Repeat this process for all other tables.
6. Check the data and see once more if it all makes sense. In our case we will want to correct some of the data that has been imported and coerce them to the right type.

```
auth <- tibble(
            author_id = as.integer(my_authors$author_id),
            pen_name   = my_authors$pen_name,
            full_name  = my_authors$full_name,
            birth_date = as.Date(my_authors$birth_date),
            death_date = as.Date(my_authors$death_date)
               )         %>%
        unique           %>%
        print
## # A tibble: 10 x 5
##    author_id pen_name   full_name    birth_date death_date
##        <int> <chr>      <chr>        <date>     <date>
## 1          1 Marcel Pr~ Valentin Lo~ 1871-07-10 1922-11-18
## 2          2 Miguel de~ Miguel de C~ 1547-09-29 1616-04-22
## 3          4 E. L. Jam~ Erika Leona~ 1963-03-07 NA
## 4          5 Isaac New~ Isaac Newton 1642-12-25 1726-03-20
## 5          7 Euclid     Euclid of A~ NA         NA
## 6         11 Bernard M~ Bernard Marr NA         NA
## 7         13 Bart Baes~ Bart Baesens 1975-02-27 NA
## 8         14 Philippe ~ Philippe J.~ 1969-02-21 NA
## 9         15 Hadley Wi~ Hadley Wick~ NA         NA
## 10        16 Garrett G~ Garrett Gro~ NA         NA
```

In our particular case, it seems that we still should clean up the names of books. There seem to be some random quote signs as well as newline (`\n`) characters. However, the quotes are due to how `print()` works on a tibble, so only the newlines should be eliminated. The tools to do this are described in Chapter 17.5 *"String Manipulation in the tidyverse"* on page 299. So if the following is not immediately clear, we refer to that chapter.

```
auth$full_name <- str_replace(auth$full_name, "\n", "")      %>%
   print
##  [1] "Valentin Louis G. E. Marcel Proust"
##  [2] "Miguel de Cervantes Saavedra"
##  [3] "Erika Leonard"
##  [4] "Isaac Newton"
##  [5] "Euclid of Alexandria"
##  [6] "Bernard Marr"
##  [7] "Bart Baesens"
##  [8] "Philippe J.S. De Brouwer"
##  [9] "Hadley Wickham"
## [10] "Garrett Grolemund"
```

17.3.2 Convert Headers to Data

Another very common problem is that users confuse data with fields. Since the early days of MS Excel the prototype of this is maybe sales figures presented as in the following example:

```
# First read in some data (using a flat file to remind
# how this works):
 x <- " January   100        102        108
February  106        105        105
March     104        104        106
April     120        122        118
May       130        100        133
June      141        139        135
July      175        176        180
August    170        188        187
September 142        148        155
October   133        137        145
November  122        128        131
December  102        108        110"

# Read in the flat file via read_fwf from readr:
t <- read_fwf(x,  fwf_empty(x, col_names = my_headers))

# Set the column names:
colnames(t) <-  c("month", "Sales2017", "Sales2018", "Sales2019")

# Finally, we can show the data as it appeared in the spreadsheet
# from the sales department:
print(t)
## # A tibble: 12 x 4
##    month     Sales2017 Sales2018 Sales2019
##    <chr>         <dbl>     <dbl>     <dbl>
##  1 January         100       102       108
##  2 February        106       105       105
##  3 March           104       104       106
##  4 April           120       122       118
##  5 May             130       100       133
##  6 June            141       139       135
##  7 July            175       176       180
##  8 August          170       188       187
##  9 September       142       148       155
## 10 October         133       137       145
## 11 November        122       128       131
## 12 December        102       108       110
```

gather() tidyr In this example the years are part of the headers (column names). The function `gather()` from `tidyr` helps to correct such situation by moving multiple columns into one long column.

```
t2 <- gather(t, "year", "sales", 2:4)
t2$year <- str_sub(t2$year,6,9)  # delete the sales word
t2$year <- as.integer(t2$year)   # convert to integer
t2
## # A tibble: 36 x 3
##    month      year sales
##    <chr>     <int> <dbl>
##  1 January    2017   100
##  2 February   2017   106
##  3 March      2017   104
##  4 April      2017   120
##  5 May        2017   130
##  6 June       2017   141
##  7 July       2017   175
##  8 August     2017   170
##  9 September  2017   142
## 10 October    2017   133
## # ... with 26 more rows
```

The function gather takes the following arguments:

- the tibble to be modified,
- the label of the new column to be created (the headers of the columns mentioned further will go there),
- the label of the row that will contain the data that was in the old tibble,
- the indexes of the columns to be used.

In this case, the problem is not yet really solved. We still have the dates in a format that is not usable. So we still need to convert the dates to a date-format. This will be explained further in Chapter 17.6 *"Dates with lubridate"* on page 314, but the following is probably intuitive. lubridate

The following code uses `lubridate` to convert the dates to the correct format and then plots the results in Figure 17.1:

```
library(lubridate)

##
## Attaching package: 'lubridate'
## The following object is masked from 'package:base':
##
##     date

t2$date <- parse_date_time(paste(t2$year,t$month), orders = "ym")
plot(x = t2$date, y = t2$sales, col = "red")
lines(t2$date, t2$sales, col = "blue")
```

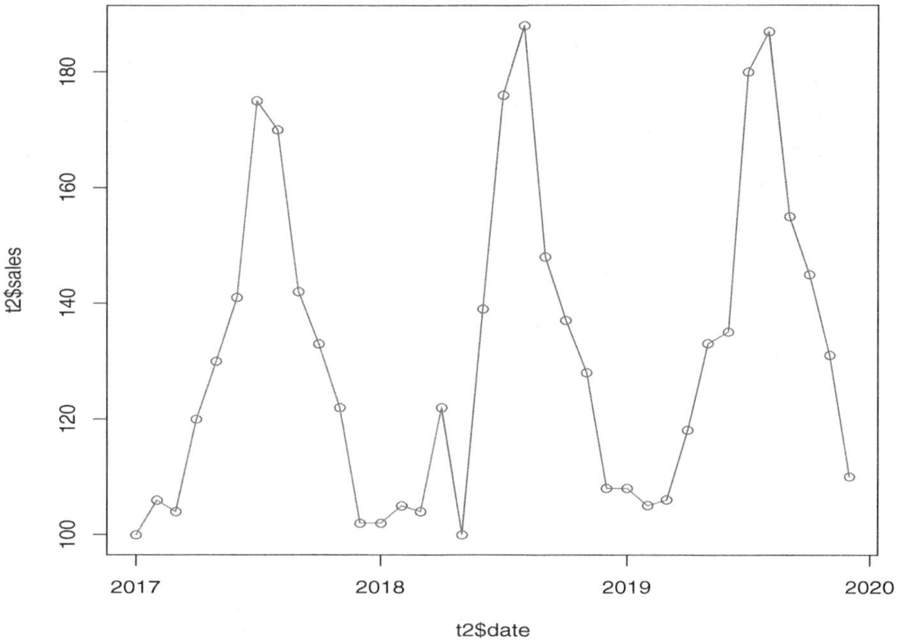

Figure 17.1: *Finally, we are able to make a plot of the tibble in a way that makes sense and allows to see the trends in the data.*

17.3.3 Spreading One Column Over Many

spread()

The function spread() does the opposite. In many cases, users put variables in cells. For example, consider the following data:

We will now create a data frame "sales_info" that has two entries for each date: a sales and a purchase. This is not easy to work with and it would be better to have a column for sales and one for purchases. The function spread() is very helpful to do this.

```
library(dplyr)
sales_info <- data.frame(
       time = as.Date('2016-01-01') + 0:9 + rep(c(0,-1), times = 5),
       type  = rep(c("bought","sold"),5),
       value = round(runif(10, min = 0, max = 10001))
       )
sales_info
##            time    type value
## 1  2016-01-01 bought  5582
## 2  2016-01-01    sold  1998
## 3  2016-01-03 bought  7951
## 4  2016-01-03    sold  6388
## 5  2016-01-05 bought  4382
## 6  2016-01-05    sold  1836
## 7  2016-01-07 bought  3675
## 8  2016-01-07    sold  4028
## 9  2016-01-09 bought   974
## 10 2016-01-09    sold  8081

spread(sales_info, type, value)
##          time bought sold
## 1 2016-01-01    5582 1998
## 2 2016-01-03    7951 6388
## 3 2016-01-05    4382 1836
## 4 2016-01-07    3675 4028
## 5 2016-01-09     974 8081
```

Note – spread() and gather()

The functions spread and gather are each other's opposite and when applied in sequentially on the same data and variables they cancel each other out.

```
sales_info                  %>%
  spread(type, value)       %>%
  gather(type, value, 2:3)
##            time    type value
## 1  2016-01-01 bought  5582
## 2  2016-01-03 bought  7951
## 3  2016-01-05 bought  4382
## 4  2016-01-07 bought  3675
## 5  2016-01-09 bought   974
## 6  2016-01-01    sold  1998
## 7  2016-01-03    sold  6388
## 8  2016-01-05    sold  1836
## 9  2016-01-07    sold  4028
## 10 2016-01-09    sold  8081
```

The tibble that is the output of spread() is indeed what is easier to work with in a sensible way. Things like the average of "bought" and "sold" are now easier to calculate as. We have indeed achieved that each column is a variable and each row and observation so that each cell in our table can be one data point.

17.3.4 Split One Columns into Many

In our initial example we had a column that contained something like "Sales2019." In this case, we did not need to keep the word sales, and the number of letters was always the same. However, in general, the number of characters might vary and the word "sales" might be something that we need to keep.

In those cases, the function separate () from tidyr will come in handy. The following code first produces an example where one column mixes actually multiple values and then solves the issue.

separate()

tidyr

```
library(tidyr)
turnover <- data.frame(
       what = paste(as.Date('2016-01-01') + 0:9 + rep(c(0,-1), times = 5),
                    rep(c("HSBC","JPM"),5), sep = "/"),
       value = round(runif(10, min = 0, max = 50))
       )
turnover
##                 what value
## 1  2016-01-01/HSBC    34
## 2   2016-01-01/JPM    23
## 3  2016-01-03/HSBC    31
## 4   2016-01-03/JPM    46
## 5  2016-01-05/HSBC    44
## 6   2016-01-05/JPM    30
## 7  2016-01-07/HSBC    11
## 8   2016-01-07/JPM    27
## 9  2016-01-09/HSBC     9
## 10  2016-01-09/JPM     1

separate(turnover, what, into = c("date","counterpart"), sep = "/")
##          date counterpart value
## 1  2016-01-01        HSBC    34
## 2  2016-01-01         JPM    23
## 3  2016-01-03        HSBC    31
## 4  2016-01-03         JPM    46
## 5  2016-01-05        HSBC    44
## 6  2016-01-05         JPM    30
## 7  2016-01-07        HSBC    11
## 8  2016-01-07         JPM    27
## 9  2016-01-09        HSBC     9
## 10 2016-01-09         JPM     1
```

Note – Fixed width separation

When the function separate() gets a number as separator (via the argument sep = n), then it will split after n characters. The separator actually takes a regular expression (see more about this in Chapter 17.5.2 *"Pattern Matching with Regular Expressions"* on page 302). Note also that the separator itself will disappear!

```
# Use as separator a digit followed by a forward slash
# and then a capital letter.
separate(turnover, what, into = c("date","counterpart"),
         sep = "[0-9]/[A-Z]")
##          date counterpart value
## 1  2016-01-0         SBC    34
## 2  2016-01-0          PM    23
## 3  2016-01-0         SBC    31
## 4  2016-01-0          PM    46
## 5  2016-01-0         SBC    44
## 6  2016-01-0          PM    30
## 7  2016-01-0         SBC    11
## 8  2016-01-0          PM    27
## 9  2016-01-0         SBC     9
## 10 2016-01-0          PM     1
```

17.3.5 Merge Multiple Columns Into One

The opposite of spreading one column over many is uniting them into one column. This might be handy when combing multiple columns into one

```
library(tidyr)

# Define a data frame:
df <- data.frame(year = 2018, month = 0 + 1:12, day = 5)
print(df)
##    year month day
## 1  2018     1   5
## 2  2018     2   5
## 3  2018     3   5
## 4  2018     4   5
## 5  2018     5   5
## 6  2018     6   5
## 7  2018     7   5
## 8  2018     8   5
## 9  2018     9   5
## 10 2018    10   5
## 11 2018    11   5
## 12 2018    12   5

# Merge the columns to one variable:
unite(df, 'date', 'year', 'month', 'day', sep = '-')
##          date
## 1    2018-1-5
## 2    2018-2-5
## 3    2018-3-5
## 4    2018-4-5
## 5    2018-5-5
## 6    2018-6-5
## 7    2018-7-5
## 8    2018-8-5
## 9    2018-9-5
## 10 2018-10-5
## 11 2018-11-5
## 12 2018-12-5
```

17.3.6 Wrong Data

Many data is gathered manually, and each manual process is prone to mistakes. Software can filter out some mistakes, but not all of them.

For example, there are some non-obvious mistakes, such as the age is 26 for a person that is 62, the postcode is that of another city, or the city-name is missing that are hard to spot or correct. These data mistakes are much harder to spot since they are no outliers (they are reasonable values, but wrong), or missing and we have no information at all.

Solving these issues is not trivial and requires a deep understanding of the data and its meaning. This is not part of the definition of "tidy data" as we have introduced it above, but it has to be addressed as soon as possible. Better still is make sure that the people that gather the data understand what is wrong and how important it is to provide correct data.

We can, of course, try to spot values that are obviously wrong and maybe transform these observations into NA. Indeed, wrong data can be handled with filtering out impossible values or outliers and converting them into missing data.

Missing data is another important issue that should be addressed. We need to understand why data is missing and how the remaining data is biassed. More details as well as some solutions are discussed in Chapter 18 *"Dealing with Missing Data"* on page 333.

17.4 SQL-like Functionality via dplyr

Now, that our data is "tidy," we can start the real work and transform data to insight. This step involves selecting, filtering, and changing variables, as well as combining different tables.

In this section, we will use the example of the simple library that was created in Chapter 14 *"SQL"* on page 223 and downloaded to R in Chapter 15 *"Connecting R to an SQL Database"* on page 253. It was a simple model of a library, with a table for authors, books, genres, and a table between authors and books.

This functionality is provided by the library `dplyr` of the `tidyverse`. So, we will load it here and not repeat this in every sub-section.

```
library(dplyr)
```

Hint – Mix all data wrangling techniques

We described first how to merge tables and columns and how to split them before we show selecting and filtering. Usually, it is necessary to use all techniques simultaneously. For example, merging tables can lead to very large data frames and hence it is best to select immediately only the columns that we will need.

17.4.1 Selecting Columns

Just as in SQL, it is not necessary to print or keep the whole table and it is possible to ask R to keep only certain columns. The function `select()` from `dplyr` does this in a consistent manner.
`select()` The first variable to the function `select()` is always the data frame (or tibble) and then we can feed it with one or more columns that we want to keep.

```
# Using the example of the library:
dplyr::select(genres,       # the first argument is the tibble
       genre_id, location) # then a list of column names
##   genre_id location
## 1   FINinv   405.08
## 2   LITero   001.67
## 3   LITmod   001.45
## 4   SCIbio   300.10
## 5   SCIdat   205.13
## 6   SCImat   100.53
## 7   SCIphy   200.43
```

Note – Using pipes

Note that all the functions of `dplyr` will always return a tibble, so they can be chained with the pipe operator. Piping in the R (with the `tidyverse`) is explained in Section 7.3.2 *"Piping with R"* on page 132.

17.4.2 Filtering Rows

The WHERE clause of SQL is made available in R with the function `filter()`.

filter()

```
a1 <- filter(authors, birth_date > as.Date("1900-01-01"))
paste(a1$pen_name,"--",a1$birth_date)
## [1] "E. L. James -- 1963-03-07"
## [2] "Bart Baesens -- 1975-02-27"
## [3] "Philippe J.S. De Brouwer -- 1969-02-21"
```

Just as in SQL, the result will not include any of the rows that have missing birth dates.

Note – Name-space conflict

When we loaded tidyverse, it informed us that the package `stats` has also a function filter and that it will be "masked." This means that when we use `filter(...)`, then the function from `dplyr` will be used. The function `filter` from `stats` can be accessed as follows: `stats::filter()`.

Another thing that we might want to check is if our assumed primary keys are really unique.[4]

```
authors          %>%
 count(author_id) %>%
 filter(n > 1)    %>%
 nrow()
## [1] 0
```

There are indeed zero rows that would have more than one occurrence for `author_id`.

Which authors do have more than one book in our database? The answer can be found by using `count()` and `filter()`.

count()

```
author_book     %>%
  count(author) %>%
  filter(n > 1)
## # A tibble: 2 x 2
##   author     n
##    <dbl> <int>
## 1      1     4
## 2      2     2
```

Hint – Equivalence between dplyr and SQL

Note that

```
filter(count(author_book, author), n  > 1}
```

is equivalent with the following in SQL

```
SELECT COUNT(author) FROM tbl_author_book
   HAVING COUNT(author) > 1;
```

And, while you are at it, also pay attention that the aforementioned R-code does the same as the one in the text but does not use the piping command.

[4]Note that we use the piping operator that is explained in Section 7.3.2 *"Piping with R"* on page 132.

17.4.3 Joining

join()
left_join()
inner_loin()
right_join()
full_join()

If we want to find out which books can be found where, we would have to join two tables. While this can be done with the WHERE clause or the multiple variants of the JOIN clause in SQL in R there are also functions in dplyr to do this. dplyr provides a complete list of joins that are similar to the familiar clauses in SQL. A list of them is in the documentation and can be accessed by typing ?join at the R-prompt.

join

First, we will distinguish *mutating joins* that output fields of both data frames and *filtering joins* that only output the fields (columns) of the left data frame. Within those two categories there is a series of possibilities that are fortunately very recognizable from the SQL environment.

dplyr

In the tidyverse , dplyr provides a series of join-functions, that all share a similar synthax:

```
*_join(x, y, by = NULL, copy = FALSE, ...)
```

mutating join

1. **mutating joins** output fields of both data frames. Contrary to what the name suggests, they do not mutate the tibbles on which they opearate. These joins output fields of both data frames. dplyr provides the following "mutating joins".

   - *inner_join()* returns all the columns for x and y, but only those rows that have matching values in their respective field/columns mentioned in the by-clause and all columns from "x" and "y." Note that it is possible that some of the join-fields are not unique and hence there can be multiple matches for the same record, then all combinations are all returned.
   - *left_join()* returns all the columns for x and y, so that all rows of x will be returned at least once (with a match of y if it exists, otherwise with a match to NA (or NULL in SQL vocabulary) (matches are defined by the by-clause). Note that it is possible that some of the join-fields are not unique and hence there can be multiple matches for the same record or x, then all combinations are all returned.
   - *right_join()* is similar to the previous but roles of x and y are inverted. Hence, it returns all rows from y, and all columns from x and y. Rows in y with no match in x will still be returned but have NA values in those rows of the y data frame.
   - *full_join()* returns all rows and all columns from both data frames x and y. Where there are not matching values, returns "NA" for the one missing.

filtering join

2. **filtering joins** that only output the fields (columns) of the left data frame.

   - *semi_join()* returns all rows from x but only if there is a matching values in the field of y, while only keeping the columns of x. Note that unlike an inner join, the semi join will never duplicate rows of x
   - *anti_join()* returns all rows from x that do not have a matching value in y, while keeping only the columns of x.

Since we took care that our library database respects referential integrity most all "mutating join" types will provide similar results.

```
a2 <- books                                                   %>%
  inner_join(genres, by = c("genre" = "genre_id"))
paste(a2$title, "-->", a2$location)
```

```
##  [1] "Les plaisirs et les jour --> 001.45"
##  [2] "Albertine disparue --> 001.45"
##  [3] "Contre Sainte-Beuve --> 001.45"
##  [4] "A? la recherche du temps perdu --> 001.45"
##  [5] "El Ingenioso Hidalgo Don Quijote de la Mancha --> 001.45"
##  [6] "Novelas ejemplares --> 001.45"
##  [7] "Fifty Shades of Grey --> 001.67"
##  [8] "Philosophi\xe6 Naturalis Principia Mathematica --> 200.43"
##  [9] "Elements (translated ) --> 100.53"
## [10] "Big Data World --> 205.13"
## [11] "Key Business Analytics --> 205.13"
## [12] "Maslowian Portfolio Theory --> 405.08"
## [13] "R for Data Science --> 205.13"
```

In SQL, we would use nested joins to find out which authors can be found in which location. In dplyr it works similar and the functions are conveniently named along the SQL keywords. On top of that, the piping operator makes the work-flow easy to read.

```
a3 <- authors                                                    %>%
  inner_join(author_book, by = c("author_id" = "author"))        %>%
  inner_join(books,       by = c("book"      = "book_id"))       %>%
  inner_join(genres,      by = c("genre"     = "genre_id"))      %>%
  dplyr::select(pen_name, location)                              %>%
  arrange(location)                                              %>%
  unique()                                                       %>%
  print()
##                  pen_name location
## 1           Marcel Proust   001.45
## 5     Miguel de Cervantes   001.45
## 7             E. L. James   001.67
## 8                  Euclid   100.53
## 9            Isaac Newton   200.43
## 10           Bernard Marr   205.13
## 11           Bart Baesens   205.13
## 12         Hadley Wickham   205.13
## 13      Garrett Grolemund   205.13
## 14 Philippe J.S. De Brouwer   405.08

# Note the difference with the base-R code below!
b <- merge(authors, author_book, by.x = "author_id",
                                 by.y = "author")
b <- merge(b, books,  by.x = "book",  by.y = "book_id")
b <- merge(b, genres, by.x = "genre", by.y = "genre_id")
b <- cbind(b$pen_name, b$location)  # colnames disappear
colnames(b) <- c("pen_name", "location")
b <- as.data.frame(b)
b <- b[order(b$location), ] # sort for data frames is order
b <- unique(b)
print(b)
##                  pen_name location
## 3           Marcel Proust   001.45
## 7     Miguel de Cervantes   001.45
## 2             E. L. James   001.67
## 13                 Euclid   100.53
## 14           Isaac Newton   200.43
## 9            Bart Baesens   205.13
## 10           Bernard Marr   205.13
## 11         Hadley Wickham   205.13
## 12      Garrett Grolemund   205.13
## 1  Philippe J.S. De Brouwer   405.08
```

Hint – Sort order

The default sort order in the function `arrange()` is increasing, to make that decreasing, use the function `desc()` as follows:

```
arrange(desc(location))
```

Note – Remove duplicates

Note that if we want to tell R only to look at unique location values and only keep those rows. Then the function `unique()` is not the best tool. For example, we might have a list of authors and locations (for one location we will find many authors in this case), but we might want to keep just one sample author per location. To achieve that we can do:

```
a3[!duplicated(a3$location), ]
##                    pen_name location
## 1             Marcel Proust   001.45
## 7              E. L. James    001.67
## 8                   Euclid    100.53
## 9             Isaac Newton    200.43
## 10            Bernard Marr    205.13
## 14 Philippe J.S. De Brouwer   405.08
```

duplicated() Note that the function `duplicated()` will keep only one line per location, and is from base-R and hence always available.

group_by() Note that what we are doing here is similar to the `GROUP BY` statement of SQL. This might make the reader think that the function `group_by()` in R will do the same. However, this is misleading, this function only changes the appearance of the data frame (the way it is printed for example). Therefore, the `group_by` function does something different. It groups data so that data manipulations can be done per group.

Note – Short-cuts

Both in SQL and R there are many short-cuts available. Imagine that we have a table A and B, and that *both* have a field `z` that is linked.[a] In that case, it is not necessary to write:

```
inner_join(A, B, by = c("z", "z")  # ambiguous, but works
inner_join(A, B, by = "z")         # shorter
inner_join(A, B)                   # shortest
```

However, the last forms are more vulnerable. If the argument `by` is not supplied, it defaults to `NULL`, and for `dplyr`, this is a sign to try to make an inner join on *all* columns that have the same name. So if later, we add matching columns, this selection will also change.

[a] We have chosen to add `_id` after a primary key, and where the same field is used as a foreign key, we used it without that suffix. So in our tables, foreign and primary keys never have the same names.

17.4.4 Mutating Data

Not all data will be right away in the right format. Maybe we want to combine the views of thousands analysts and millions of transactions to distil a "market sentiment," or maybe we simply want to convert "miles per gallon" into "litres per 100 km." To achieve this, we will have to make our own calculations and add columns with this new data.[5]

This idea is similar to the SET clause in SQL, however, it works very differently. An SQL database has a rigid definition of tables, adding a column is a big deal and in a production environment many people will be involved and a lot of plans will be drafted before this happens. R is an interactive language designed to experiment, explore, and discover data. Therefore, changing or adding columns in R should be straightforward.

Indeed, mutating data in R is one of the first things we discovered in this book. dplyr and the tidyverse make it also convenient and structured. Consider, for example the idea of adding to each author the number of books that we hold in our library.

```
ab <- authors                                          %>%
  inner_join(author_book, by = c("author_id" = "author"))  %>%
  inner_join(books,       by = c("book"      = "book_id")) %>%
  add_count(author_id)
ab$n
## [1] 4 4 4 4 2 2 1 1 1 1 1 1 1 1
```

add_count()

In fact, we were lucky that the particular thing that we wanted to do has its specific function. In general, however, it is wise to rely on the function mutate() from dplyr. This function provides a framework to add columns that contain new information calculated based on existing columns.

In the simplest of all cases, we can just copy an existing column under a new heading (maybe to make joining tibbles easier?)

```
genres                        %>%
   mutate(genre = genre_id)   %>% # add column genre
   inner_join(books)          %>% # leave out the "by="
   dplyr::select(c(title, location))

## Joining, by = "genre"
##                                          title
## 1                    Maslowian Portfolio Theory
## 2                          Fifty Shades of Grey
## 3                       Les plaisirs et les jour
## 4                            Albertine disparue
## 5                           Contre Sainte-Beuve
## 6                  A? la recherche du temps perdu
## 7    El Ingenioso Hidalgo Don Quijote de la Mancha
## 8                            Novelas ejemplares
## 9                                Big Data World
## 10                      Key Business Analytics
## 11                           R for Data Science
## 12                        Elements (translated )
## 13 Philosophi\xe6 Naturalis Principia Mathematica
##    location
## 1    405.08
## 2    001.67
## 3    001.45
## 4    001.45
## 5    001.45
## 6    001.45
```

[5] An example is mentioned earlier: using the weekly balances of current accounts to calculate one parameter that is one if the person was ever overdrawn, zero otherwise.

```
## 7     001.45
## 8     001.45
## 9     205.13
## 10    205.13
## 11    205.13
## 12    100.53
## 13    200.43
```

mutate()

The function mutate() allows, of course, for much more complexity. Below we present an example that makes an additional column for a short author name (one simple and one more complex) and finds out if someone is alive (or if we cannot tell).

```
t <- authors                                                    %>%
     mutate(short_name = str_sub(pen_name,1,7))                 %>%
     mutate(x_name = if_else(str_length(pen_name) > 15,
                             paste(str_sub(pen_name,1,8),
                                   "...",
                                   str_sub(pen_name,
                                           start = -3),
                                   sep = ''),
                             pen_name,
                             "pen_name is NA"
                             )
            )                                                   %>%
     mutate(is_alive =
       if_else(!is.na(birth_date) & is.na(death_date),
            "YES",
            if_else(death_date < Sys.Date(),
                "no",
                "maybe"),
             "NA")
            )                                                   %>%
    dplyr::select(c(x_name, birth_date, death_date, is_alive)) %>%
    print()
##             x_name birth_date death_date is_alive
## 1    Marcel Proust 1871-07-10 1922-11-18       no
## 2   Miguel d...tes 1547-09-29 1616-04-22       no
## 3      James Joyce 1882-02-02 1941-01-13       no
## 4      E. L. James 1963-03-07       <NA>      YES
## 5     Isaac Newton 1642-12-25 1726-03-20       no
## 6           Euclid       <NA>       <NA>     <NA>
## 7     Bernard Marr       <NA>       <NA>     <NA>
## 8     Bart Baesens 1975-02-27       <NA>      YES
## 9   Philippe...wer 1969-02-21       <NA>      YES
## 10 Hadley Wickham       <NA>       <NA>     <NA>
## 11 Garrett ...und       <NA>       <NA>     <NA>
```

It is probably is obvious what the functions if_else() and str_sub() do, however, if you want more information we refer to Section 17.5 *"String Manipulation in the tidyverse"* on page 299.

Of course, the function mutate() also does arithmetic such as new_column = old_column * 2. The function mutate() understands all basic operators, and especially the following functions are useful in mutate statements:

- arithmetic, relational and logical and operators, such as +,-, *, /, >, & (see Section 4.4 *"Operators"* on page 57);
- any base function such as log(), exp(), etc.
- lead(), lag() will find the previous or next value (e.g. to calculate returns between dates) — also see the section about quantmod.

- dense_rank(), min_rank(), percent_rank(), row_number()), ntile(), cume_dist() are handy functions provided by dplyr that all use rank() under the hood.
- cumsum(), cummean(), cummin(), cummax(), cumany() and cumall() from base-R and dplyr together form complet set of functions related to the cumulative distribution.
- na_if() (converts matches to NA — inspired on the NULL_IF statement of SQL), coalesce() (finds the first non-missing value at each position—similar to the SQL clause COALESCE provided by dplyr it helps to complete the functionality that is also available in SQL.
- if_else() (as used before), recode() (a vectorized version of the switch command), case_when() (an elegant way to construct a list of "else if" statements.

If in your particular case, you prefer to drop existing columns, then the function transmute() is what you need.

transmute()

Hint – Advanced mutating

Do you need even more granular functionality over the data transformations, such as when and where the transformation is applied? Then consider looking at the documentation of the scoped variants of mutate(): mutate_all(), mutate_if() and mutate_at() as well as their sister functions of the above mentioned transmute(): transmute_all(), transmute_if(), and transmute_at().

mutate_all()
mutate_if()
mutate_at()
transmute_all()
transmute_if()
transmute_at()

Consider for example the situation where we want a list of authors and a list of dates that are five days before their birthday (so we can put those dates in our calendar and send them a postcard).

```
authors    %>%
  transmute(name = full_name, my_date = as.Date(birth_date) -5)
##                                name    my_date
## 1  Valentin Louis G. E. Marcel Proust 1871-07-05
## 2        Miguel de Cervantes Saavedra 1547-09-24
## 3      James Augustine Aloysius Joyce 1882-01-28
## 4                       Erika Leonard 1963-03-02
## 5                        Isaac Newton 1642-12-20
## 6                 Euclid of Alexandria       <NA>
## 7                        Bernard Marr       <NA>
## 8                        Bart Baesens 1975-02-22
## 9             Philippe J.S. De Brouwer 1969-02-16
## 10                     Hadley Wickham       <NA>
## 11                  Garrett Grolemund       <NA>
```

The list of birthday warnings in previous example is not very useful as it contains also warnings for deceased people as well as missing dates. The function filter() allows us to select only the rows of interest.

filter()

So, let us improve our work.

```
authors    %>%
  filter(!is.na(birth_date) & is.na(death_date)) %>%
  transmute(name = full_name, my_date = as.Date(birth_date) -5)
```

```
##                        name    my_date
## 1             Erika Leonard 1963-03-02
## 2              Bart Baesens 1975-02-22
## 3 Philippe J.S. De Brouwer 1969-02-16
```

Warning – Difference between filter() and joins

It is important not to confuse the function `filter()` with the filtering joins (`semi_join()` and `anti_join()`). These are respectively an equivalent of the `LEFT JOIN` from SQL and the a left-join that would only return rows where there is no match. The function documentation, accessed via `?anti_join`, is kept together so that you have all functions in one place.

17.4.5 Set Operations

While all join types can be expressed as set operations, it is worth to mention that `dplyr` also silently (not mentioned when the package is loaded) overrides R's base functions for set operations on data frames. They still do the same, but are made more generic so that they work on more classes than only data frames.

These functions are:

intersect()
- `intersect(x, y)`: $A \cap B$ but with duplicates removed,

union()
- `union(x, y)`: $A \cup B$ but with duplicates removed,

union_all()
- `union_all(x, y)`: $A \cup B$,

setdiff()
- `setdiff(x, y)`: $A - B$ but with duplicates removed,

setequal()
- `setequal(x, y)`: $A \cap B$.

The similarity with set-operators makes it easy to understand how these operators work. The simple example below helps to make clear what they do.

```
# Define two sets (with one column):
A <- tibble(col1 = c(1L:4L))
B <- tibble(col1 = c(4L,4L,5L))

# Study some of the set-operations:
dplyr::intersect(A,B)
## # A tibble: 1 x 1
##    col1
##   <int>
## 1     4

union(A,B)
## [[1]]
## [1] 1 2 3 4
##
## [[2]]
## [1] 4 4 5
```

```
union_all(A,B)
## # A tibble: 7 x 1
##    col1
##   <int>
## 1     1
## 2     2
## 3     3
## 4     4
## 5     4
## 6     4
## 7     5

setdiff(A,B)
## # A tibble: 4 x 1
##    col1
##   <int>
## 1     1
## 2     2
## 3     3
## 4     4

setequal(A,B)
## [1] FALSE

# The next example uses a data-frame with two columns:
A <- tibble(col1 = c(1L:4L),
            col2 = c('a', 'a', 'b', 'b'))
B <- tibble(col1 = c(4L,4L,5L),
            col2 = c('b', 'b', 'c'))

# Study the same set-operations:
dplyr::intersect(A,B)
## # A tibble: 1 x 2
##    col1 col2
##   <int> <chr>
## 1     4 b

union(A,B)
## [[1]]
## [1] 1 2 3 4
##
## [[2]]
## [1] "a" "a" "b" "b"
##
## [[3]]
## [1] 4 4 5
##
## [[4]]
## [1] "b" "b" "c"

union_all(A,B)
## # A tibble: 7 x 2
##    col1 col2
##   <int> <chr>
## 1     1 a
## 2     2 a
## 3     3 b
## 4     4 b
## 5     4 b
## 6     4 b
## 7     5 c
```

```
setdiff(A,B)
## # A tibble: 4 x 2
##    col1 col2
##   <int> <chr>
## 1     1 a
## 2     2 a
## 3     3 b
## 4     4 b

setequal(A,B)
## [1] FALSE
```

Note – Column headings in data-frames

To use these functions both data frames must have the same column headings, but not necessarily the same length!

17.5 String Manipulation in the tidyverse

Base R provides solid tools to work with strings and for string manipulation – see Section 4.3.9 *"Strings or the Character-type"* on page 54, however, as with many things in free software they have been contributed by different people in different times and hence lack consistency.

The tidyverse includes `stringr` that in its turn is based upon `stringi` and it provides a solid and fast solution that is coherent in naming conventions and on top of that they will follow the philosophy of the tidyverse.

`stringr`
`stringi`

The package `stringr` is installed when installing the `tidyverse`, but it is not loaded when loading the tidyverse. So assuming that we did at some point `install.packages('tidyverse')` we will in the following example omit `install.packages('stringr')`.

`tidyverse`

The following code illustrates some of the functionality of `srtingr`: returning the length of a string with `str_length()` and concatenating strings with `str_c`. Note also that strings can be defined with double quotes or single quotes.

`str_length()`
`str_c()`

```
library(tidyverse)
library(stringr)

# define strings
s1 <- "Hello"  # double quotes are fine
s2 <- 'world.' # single quotes are also fine

# Return the length of a string:
str_length(s1)
## [1] 5

# Concatenate strings:
str_c(s1, ", ", s2)        # str_c accepts many strings
## [1] "Hello, world."

str_c(s1, s2, sep = ", ") # str_c also has a
## [1] "Hello, world."
```

Hint – Naming convention of functions

Did you notice that the string functions of `stringr` in previous example started with `str_`. Well, the good news is that they all start with the same letters. So if you work in an environment such as RStudio, you can see a list of all string functions by typing `str_`.

Warning – str_c() does not return the C-string

Do not confuse the functions `str_c()` with `s.c_str()` in C++. For example, the following R-code

```
s <- 'World'
str_c('Hello, ', s, '.')
## [1] "Hello, World."
```

does the same as the following C++ code:

```
#include <iostream>
#include <string>
using namespace std;

int main ()
 {
  string s ("World");
  std::cout << "Hello, " << s.c_str() << "." << std::endl;
  // We admit that c_str() is not necessary here :-)
 return 0;
 }
```

While the name of the functions `c_str()` and `str_c()` look very similar, they do something very different. Of course, the concept of a C-string makes little sense in a highl level language such as R.

17.5.1 Basic String Manipulation

Combined with previous functions, we get very powerful tools to edit strings and recombine them in the way that makes most sense.

`str_sub()`
`str_to_lower()`
`str_to_upper()`
`str_length()`
`str_flatten()`

Below we illustrate how to extract sub-strings with `str_sub()`, convert to lower or upper case letters with `str_to_lower()` and `str_to_upper()`, get the length of a string with `str_length()`, and flatten a string via `str_flatten()`

```
library(stringr)                     # or library(tidyverse)
sVector <- c("Hello", ", ", "world", "Philippe")

str_sub (sVector,1,3)                # the first 3 characters
## [1] "Hel" ", "  "wor" "Phi"

str_sub (sVector,-3,-1)              # the last 3 characters
## [1] "llo" ", "  "rld" "ppe"

str_to_lower(sVector[4])             # convert to lowercase
## [1] "philippe"

str_to_upper(sVector[4])             # convert to uppercase
## [1] "PHILIPPE"

str_c(sVector, collapse = " ")         # collapse into one string
## [1] "Hello ,  world Philippe"

str_flatten(sVector, collapse = " ") # flatten string
## [1] "Hello ,  world Philippe"

str_length(sVector)                  # length of a string
## [1] 5 2 5 8

# Nest the functions:
str_c(str_to_upper(str_sub(sVector[4],1,4)),
      str_to_lower(str_sub(sVector[4],5,-1))
     )
## [1] "PHILippe"
```

```
# Use pipes:
sVector[4]        %>%
   str_sub(1,4)   %>%
   str_to_upper()
## [1] "PHIL"
```

Hint – Replacing sub-strings

It is also possible to reverse the situation and modify strings with `str_sub()`

```
str <- "abcde"

# Replace from 2nd to 4th character with "+"
str_sub(str, 2, 4) <- "+"
str
## [1] "a+e"
```

There are a few more special purpose functions in `stringr`. Below we list some of the most useful ones.

Duplicate Strings

One of the most simple string manipulations is duplicating them to form a longer string. Here we ask `stringr` to produce a dark shade of grey.

`str_dup()`

```
str <- "F0"
str_dup(str, c(2,3))  # duplicate a string
## [1] "F0F0"    "F0F0F0"
```

Manage White Space

This will help to format nicely tables on the screen, or even in printed documents when it is both unnecessary and not aesthetic to show the complete long strings (in a table).

```
str <- c(" 1 ", "  abc", "Philippe De Brouwer    ")
str_pad(str, 5)  # fills with white-space to x characters
## [1] "   1 "                     "  abc"
## [3] "Philippe De Brouwer    "

# str_pad never makes a string shorter!
# So to make all strings the same length we first truncate:
str         %>%
  str_trunc(10) %>%
  str_pad(10,"right")    %>%
  print
## [1] " 1        " "  abc     " "Philipp..."

# Remove trailing and leading white space:
str_trim(str)
## [1] "1"                     "abc"
## [3] "Philippe De Brouwer"

str_trim(str,"left")
## [1] "1 "                     "abc"
## [3] "Philippe De Brouwer    "
```

```
# Modify an existing string to fit a line length:
"The quick brown fox jumps over the lazy dog. "  %>%
   str_dup(5)     %>%
   str_c          %>% # str_flatten also removes existing \n
   str_wrap(50)   %>% # Make lines of 50 characters long.
   cat                # or writeLines (print shows "\n")
## The quick brown fox jumps over the lazy dog. The
## quick brown fox jumps over the lazy dog. The quick
## brown fox jumps over the lazy dog. The quick brown
## fox jumps over the lazy dog. The quick brown fox
## jumps over the lazy dog.
```

Determining Order and Sorting Strings

```
str <- c("a", "z", "b", "c")

# str_order informs about the order of strings (rank number):
str_order(str)
## [1] 1 3 4 2

# Sorting is done with str_sort:
str_sort(str)
## [1] "a" "b" "c" "z"
```

17.5.2 Pattern Matching with Regular Expressions

regex
regular expression

Pattern matching in `stringr` is done with regular expressions. While this has a steep – but not too high – learning curve regular expressions are extremely powerful. Regular expression allow to match any pattern that one can think of and most importantly it is possible to use them in almost any important programming language.

Regular expressions are to some extend a syntax for pattern matching and look a little like a cryptic programming language (but they are of course not Turing complete). The following example shows the most basic use of matching an exact pattern:

```
library(stringr)    # or library(tidyverse)
sV <- c("philosophy", "physiography", "phis",
        "Hello world", "Philippe", "Philosophy",
        "physics", "philology")

# Extracting substrings that match a regex pattern:
str_extract(sV, regex("Phi"))
## [1] NA    NA    NA    NA    "Phi" "Phi" NA    NA

str_extract(sV, "Phi")          # the same, regex assumed
## [1] NA    NA    NA    NA    "Phi" "Phi" NA    NA
```

So far the regex works as a normal string matching: the algorithm looks for an exact match to the letters "Phi." However, there is a lot more that we can do with regular expressions. For example, we notice that some words are capitalized. Of course, we can coerce all to capital and then match, but that is not the point here. In regex we would use the syntax of "one letter from the set ('p', 'P')" and then exactly "hi." The way to instruct a regex engine to select one or more possibilities is to separate them with `|`.

```
str_extract(sV, "(p|P)hi")
## [1] "phi" NA    "phi" NA    "Phi" "Phi" NA    "phi"

# Or do it this way:
str_extract(sV, "(phi|Phi)")
## [1] "phi" NA    "phi" NA    "Phi" "Phi" NA    "phi"

# Other example:
str_extract(sV, "(p|P)h(i|y)")
## [1] "phi" "phy" "phi" NA    "Phi" "Phi" "phy" "phi"

# Is equivalent to:
str_extract(sV, "(phi|Phi|phy|Phy)")
## [1] "phi" "phy" "phi" NA    "Phi" "Phi" "phy" "phi"
```

Note that it extracts now both "Phi" and "phi," but does not change how it was written.

Now, imagine that we want all matches as before, except those that precede and "l." It is possible to exclude characters by putting them in square brackets and precede them with ^

```
str_extract(sV, "(p|P)h(i|y)[^lL]")
## [1] NA     "phys" "phis" NA     NA     NA     "phys"
## [8] NA
```

Regular expressions are a very rich subject, and the simple rules explained so far can already match very complex patterns. For example, note that there is a shorthand to make the matching case insensitive.

```
str_extract(sV, "(?i)Ph(i|y)[^(?i)L]")
## [1] NA     "phys" "phis" NA     NA     NA     "phys"
## [8] NA
```

17.5.2.1 The Syntax of Regular Expressions

Regular expressions are something like a "syntax for pattern matching." They are not a programming language and are far from being Turing complete, they do just one thing and they do it thorough. In their simplest from a regex like "x" will find a match if a string contains the letter "x." However, things can become quickly very complex, and difficult to read for humans.

Regular expressions allow to check if there is a match for a regex pattern in a string – as in the example above – or we can check if the a string is a valid email for example. To achieve this we would check the string with the pattern

```
^([a-zA-Z0-9_\-\.]+)@([a-zA-Z0-9_\-\.]+)\.([a-zA-Z]{2,5})$
```

Note that – when supplying the pattern to R – we have to escape the escape character `\`, because also R uses the backslash (`\`) as escape character and will remove one before passing to the function. So matching an email would look in R as follows:

```
emails <- c("god@heaven.org", "philippe@de-brouwer.com",
            "falsemaail@nothingmy", "mistaken.email.@com")
regX <-
  "^([a-zA-Z0-9_\\-\\.]+)@([a-zA-Z0-9_\\-\\.]+)\\.([a-zA-Z]{2,5})$"
str_extract(emails, regX)
## [1] "god@heaven.org"            "philippe@de-brouwer.com"
## [3] NA                          NA
```

The regex

```
^([a-zA-Z0-9_\-\.]+)@([a-zA-Z0-9_\-\.]+)\.([a-zA-Z]{2,5})$
```

reads as follows.

- `^([a-zA-Z0-9_\\-\\.]+)` : at the start of the string (indicated by "`^`") find a choice of numbers, letters, underscores, hyphens, and/or points (made clear by "`[a-zA-Z0-9_\\-\\.]`"), but make sure to find at least one of those. This concept "one or more" is expressed by "`(...)+`"
- then find the `@` symbol
- then find at least one more occurence of a letter (lower- or uppercase), underscore, hyphen or point – `([a-zA-Z0-9_\\-\\.]+)`
- followed by a point – `.`
- finally followed by minimum two and maximum five letters (lower- or uppercase) – `([a-zA-Z]{2,5}`
- and that last match, must be the end of the string – `$`.

Digression – Advanced email matching

The careful reader will notice that this is not conclusive to validate an email. For example, what if it starts with a point? There is actually no regex that can exactly capture all emails and reject all non-valid emails. More information is here: `https://www.regular-expressions.info/email.html`. This website shows the following following regex, that should match almost all emails that follow the RFC 5322 standard.

```
\A(?:[a-z0-9!#$%&'*+/=?^_'{|}~-]+(?:\.[a-z0-9!#$%&'*
+/=?^_'{|}~-]+)*|"(?:[\x01-\x08\x0b\x0c\x0e-\x1f
\x21\x23-\x5b\x5d-\x7f]\\[\x01-\x09\x0b\x0c\x0e-\x7f])
*")@ (?:(?:[a-z0-9](?:[a-z0-9-]*[a-z0-9])?\.)+[a-z0-9]
(?:[a-z0-9-]*[a-z0-9])?|\[(?:(?:25[0-5]|2[0-4][0-9]|
[01]?[0-9][0-9]?)\.){3}(?:25[0-5]|2[0-4][0-9]|[01]?
[0-9][0-9]?|[a-z0-9-]*[a-z0-9]:(?:[\x01-\x08\x0b\x0c
\x0e-\x1f\x21-\x5a\x53-\x7f]|\\[\x01-\x09\x0b\x0c
\x0e-\x7f])+)\])\z
```

Can you read this and conceptualise it?

Probably it is already more or less clear by now how a regular expression works and how to read it. Note in particular the following:

- Some characters need to be escaped because they have a special meaning in regular expressions (for example the hyphen, `-`, indicates a range like `[a-z]`. This expression matches all letters whose ASCII values are between those of "a" and "z")
- Some signs are anchors and indicate a position in the string (e.g. "∧" is the start of the word and "$" is the end of the word — so putting the first at the start of the pattern and the latter at the end forces the pattern to match the whole string, this avoids that a valid email is surrounded by other text.
- There are different types of brackets used. In general, the round brackets group characters, the square ones indicate ranges, and the curly brackets indicate a range of matches.

Note – Single or double escape characters

Note that in the remainder of this section, we leave out the double escape characters. If you use this in R, then you will need to replace all \ with \\ wherever it occurs.

Now that we understand how a regex works, we can have a closer look at their syntax and provide an overview of the different symbols and their special meaning. Below we show the main building blocks.

Anchors

`^`	begin of string or line
`$`	end of string (or line)
`\<`	beginning of a word
`\>`	end of a word

Special characters

`\n`	newline
`\r`	carriage return)
`\t`	tab
`\v`	vertical tab
`\f`	form feed

Character groups

`.`	any character, but `\n`
`[abc]`	accepted characters
`[a-z]`	character range
`(...)`	characters group

Quantifiers

`?`	0 or 1 times
`*`	0 or more
`+`	1 or more
`{n}`	n times
`{n,m}`	between n and m times
`{n,}`	n or more times
`{,m}`	m or less times

Logic

`|`	"OR," e.g. `(a|b)` matches a or b
`\1`	content of group one, e.g. `r(\w)g(\1)x` matches "regex"
`\2`	group two, e.g. `r(\w)g(\1)x(\2)xpr` matches "regexepr"
`(?:..)`	non capturing group = ignore that match in the string to return
`[^a-d]`	"not": no character in range a to d

Lookaround – requires PERL = TRUE

`a(?!b)`	a not followed by b
`a(?=b)`	a if followed by b
`(?<=b)a`	a if preceded by b
`(?<!b)a`	a if not preceded by b

Other

`\Qa\E`	treat a verbatim, e.g. `\QC++?\E` matches "C++?"
`\K`	drop match so far, e.g. `x\K\d`returns from `x1` only `1`

Line modifiers

`(?i)`	makes all matches case insensitive
`(?s)`	single line mode: `.` also matches `\n`
`(?m)`	multi line mode: `^` and `$` become begin and end of line

POSIX Character classes	
[[:digit:]] or \d	digit: [0-9]
\\D	not a digit: [^0-9]
[[:xdigit:]] or \x	hexadec. digits: [0-9A-Fa-f]
[[:lower:]]	lower-case: [a-z]
[[:upper:]]	upper-case: [A-Z]
[[:alnum:]]	alphanumeric: [A-z0-9]
\\w	word characters: [A-z0-9_]
\\W	not word characters: [^A-z0-9_]
[[:blank:]]	blank: [\\s\\t]
[[:space:]] or \s	space : \\s
\\S	not space: [^\\s]
[[:punct:]]	punctuation character : !"#$%&'()*+,-./:;<=>?@[]^_`{}~|
[[:graph:]]	graphical character : [[:alnum:]] [[:punct:]]
[[:print:]]	printable character : [[:graph:]] [[:space:]]
[[:cntrl:]] or \c	control characters: e.g. [\\n\\rt]

Lazy and Greedy Quantifiers

The basic rule is that a quantifier applies to whatever is immediately left of it. For example:

- `abcd+` matches "abcdddd" but not "abcdabcd" (the + applies only to the last letter);
- this behaviour can be modified with grouping characters: `x(F1)+` will match "xF1F1F1," but also note that
- `\QC++\E+` matches "C+++++' but not "C+C+C+"

However, there are more nuances that need to be understood.

- The default quantifiers are **greedy:** `\d+` will match `123` (as many digits as possible, not necessarily all the same). In other words, a greedy quantifier gives you the longest possible match (eg, `^\.*` will match always the whole line). However, quantifiers are actually **greedy, but with good manners**. We mean with that the engine will swallow as many matches as possible, but if that would hinder the rest of the pattern to be matched, it will back-track to allow for a match. That is why `^\.*ippe` will still match `Philippe`.
- A quantifier can be made **reluctant or lazy** by adding ? to it. For example, `^P\.*?` will match as little as possible within the possibilities of * (which is "zero or more" and hence defaults to "zero").
- Actually, quantifiers might be **reluctant or lazy but still benevolent**. Meaning if the match was so small that this would hinder the rest of the match to be made, then they will start matching more in order for the further match to be made possible.

```
str_extract("Philippe", "Ph\\w*")  # is greedy
## [1] "Philippe"

str_extract("Philippe", "Ph\\w*?") # is lazy
## [1] "Ph"
```

Other Regex Aspects

There is a lot more in regex than we can explain here. With this section we hope to get you started and hope to activate your curiosity. Reading a regex that is half a page long is not something humans should try to master, but making a regex is very useful and a skill that will come in handy in about every computer language of any value today.

Hint – General methods in R

R uses POSIX extended regular expressions. You can switch to PCRE regular expressions by using `PERL = TRUE` for base-R or by wrapping patterns with `perl()` for `stringr`. perl()

All functions can be used with literal searches using `fixed = TRUE` for base or by wrapping patterns with `fixed ()` for `stringr`. Note also that base functions can be made case insensitive by specifying `ignore.cases = TRUE`.

Further information – Regex

Showing all possibilities and providing a full documentation of the regex implementation in R is a book in itself. Hence, we just showed here the tip of the iceberg and refer for example to the excellent "cheat sheets" published by RStudio: `https://www.rstudio.com/resources/cheatsheets`.
Good sources about regex itself are here in `https://www.regular-expressions.info` and `https://www.rexegg.com`.

Regex for Humans with rex

Regex expressions easily get hard to read. To solve that, there is a library `rex` that provides a function `rex()` to make the process of creating a regular expression a lot easier and a lot more readable. rex rex()

```
library(rex)

##
## Attaching package: 'rex'
## The following object is masked from 'package:stringr':
##
##     regex
## The following object is masked from 'package:dplyr':
##
##     matches
## The following object is masked from 'package:tidyr':
##
##     matches

valid_chars <- rex(one_of(regex('a-z0-9\u00a1-\uffff')))

# In this example we construct the regex to match a valid URL.
expr <- rex(
  start,         # start of the string: ^

  # protocol identifier (optional) + //
  group(list('http', maybe('s')) %or% 'ftp', '://'),

  # user:pass authentication (optional)
  maybe(non_spaces,
    maybe(':', zero_or_more(non_space)),
    '@'),

  #host name
  group(zero_or_more(valid_chars,
        zero_or_more('-')),
```

url

```
        one_or_more(valid_chars)),

    #Domain name:
    zero_or_more('.',
                 zero_or_more(valid_chars,
                 zero_or_more('-')),
                 one_or_more(valid_chars)),

    # Top Level Domain (TLD) identifier:
    group('.', valid_chars %>% at_least(2)),

    # Server port number (optional):
    maybe(':', digit %>% between(2, 5)),

    # Resource path (optional):
    maybe('/', non_space %>% zero_or_more()),

    end
  )

# The rest is only to print it nice (expr can be used):
substring(expr, seq(1,  nchar(expr)-1, 40),
                seq(41, nchar(expr),   40))      %>%
  str_c(sep = "\n")
## [1] "^(?:(?:http(?:s)?|ftp)://)(?:[^[:space:]]"
## [2] "]+(?::(?:[^[:space:]])*)?@)?(?:(?:[a-z0-9"
## [3] "9Âą-\uffff](?:-)*)*(?:[a-z0-9Âą-\uffff])+)(?:\\.(?:[a-"
## [4] "-z0-9Âą-\uffff](?:-)*)*(?:[a-z0-9Âą-\uffff])+)*(?:\\.("
## [5] "(?:[a-z0-9Âą-\uffff]){2,})(?::(?:[[:digit:]]){2"
## [6] ""
```

17.5.2.2 Functions Using Regex

Both `base`-R and `stringr` have a complete suite of string manipulating functions based on regular expressions. Now, that we understand how a regex works, they do not need a lot of explanation. The functions of base-R are conveniently named similar to the functions that one will see in a Unix or Linux shell such as bash and hence might be familiar. The functions of `stringr` follow a more logical naming convention, are more consistent, and allow to use the pipe-command, because the data is the first argument of the function.

In the examples below, we mention both the functions for base-R and `stringr`. The functions of `stringr` can be recognized, because they all start with `str_`. In all examples of this section, we use the same example where we try to match a digit within a string. Let us define it here:

```
string <- c("one:1", "NO digit", "c5c5c5", "d123d", "123", 6)
pattern <- "\\d"
```

The pattern, `\d`, is a shorthand notation that will match any digit. Only the second element of the vector of strings does not have a digit, so only there will be no match.

Detect a Match

These functions will only report if a match is found, no information about starting positions of he match is given.

```
# grep() returns the whole string if a match is found:
grep(pattern, string, value = TRUE)
## [1] "one:1"  "c5c5c5" "d123d"  "123"    "6"

# The default for value is FALSE -> only returns indexes:
grep(pattern, string)
```

```
## [1] 1 3 4 5 6

# L for returning a logical variable:
grepl(pattern, string)
## [1]  TRUE FALSE  TRUE  TRUE  TRUE  TRUE

# --- stringr ---
# similar to grepl (note order of arguments!):
str_detect(string, pattern)
## [1]  TRUE FALSE  TRUE  TRUE  TRUE  TRUE
```

Locate

In many cases, it is not enough to know if there is a match, but also where the match occurs in the string; that is what we call "locating" a match in a string.

```
# Locate the first match (the numbers are the position in the string):
regexpr (pattern, string)
## [1]  5 -1  2  2  1  1
## attr(,"match.length")
## [1]  1 -1  1  1  1  1
## attr(,"useBytes")
## [1] TRUE

# grepexpr() finds all matches and returns a list:
gregexpr(pattern, string)
## [[1]]
## [1] 5
## attr(,"match.length")
## [1] 1
## attr(,"useBytes")
## [1] TRUE
##
## [[2]]
## [1] -1
## attr(,"match.length")
## [1] -1
## attr(,"useBytes")
## [1] TRUE
##
## [[3]]
## [1] 2 4 6
## attr(,"match.length")
## [1] 1 1 1
## attr(,"useBytes")
## [1] TRUE
##
## [[4]]
## [1] 2 3 4
## attr(,"match.length")
## [1] 1 1 1
## attr(,"useBytes")
## [1] TRUE
##
## [[5]]
## [1] 1 2 3
## attr(,"match.length")
## [1] 1 1 1
## attr(,"useBytes")
## [1] TRUE
##
```

```
## [[6]]
## [1] 1
## attr(,"match.length")
## [1] 1
## attr(,"useBytes")
## [1] TRUE

# --- stringr ---
# Finds the first match and returns a matrix:
str_locate(string, pattern)
##      start end
## [1,]     5   5
## [2,]    NA  NA
## [3,]     2   2
## [4,]     2   2
## [5,]     1   1
## [6,]     1   1

# Finds all matches and returns a list (same as grepexpr):
str_locate_all(string, pattern)
## [[1]]
##      start end
## [1,]     5   5
##
## [[2]]
##      start end
##
## [[3]]
##      start end
## [1,]     2   2
## [2,]     4   4
## [3,]     6   6
##
## [[4]]
##      start end
## [1,]     2   2
## [2,]     3   3
## [3,]     4   4
##
## [[5]]
##      start end
## [1,]     1   1
## [2,]     2   2
## [3,]     3   3
##
## [[6]]
##      start end
## [1,]     1   1
```

Note – Using locating functions as boolean

The locating functions make the "detect match functions" almost redundant: if no match is found, the the indexes of the locating function are -1. We can then use this number to check if a match is found.

Replace

Often we want to do more than just finding where a match occurs, but we want to change it with something else. This process is called "replacing" matches with strings.

```
# First, we need additionally a replacement (repl)
repl <- "___"

# sub() replaces the first match:
sub(pattern, repl, string)
## [1] "one:___"  "NO digit" "c___c5c5" "d___23d"  "___23"
## [6] "___"

# gsub() replaces all matches:
gsub(pattern, repl, string)
## [1] "one:___"        "NO digit"       "c___c___c___"
## [4] "d_________d"    "_________"      "___"

# --- stringr ---
# str_replace() replaces the first match:
str_replace(string, pattern, repl)
## [1] "one:___"  "NO digit" "c___c5c5" "d___23d"  "___23"
## [6] "___"

# str_replace_all() replaces all mathches:
str_replace_all(string, pattern, repl)
## [1] "one:___"        "NO digit"       "c___c___c___"
## [4] "d_________d"    "_________"      "___"
```

Extract

If it is not our aim to replace the match, then it might be the case that we want to extract it for further use and manipulation in other sections or functions. The following functions allow to extract matches the regular expressions. from strings. The output of these functions can be quite verbose such as the functions to locate matches.

```
# regmatches() with regexpr() will extract only the first match:
regmatches(string, regexpr(pattern, string))
## [1] "1" "5" "1" "1" "6"

# regmatches() with gregexpr() will extract all matches:
regmatches(string, gregexpr(pattern, string)) # all matches
## [[1]]
## [1] "1"
##
## [[2]]
## character(0)
##
## [[3]]
## [1] "5" "5" "5"
##
## [[4]]
## [1] "1" "2" "3"
##
## [[5]]
## [1] "1" "2" "3"
##
## [[6]]
## [1] "6"

# --- stringr ---
# Extract the first match:
str_extract(string, pattern)
## [1] "1" NA  "5" "1" "1" "6"

# Similar as str_extract, but returns column instead of row:
str_match(string, pattern)
```

```
##      [,1]
## [1,] "1"
## [2,] NA
## [3,] "5"
## [4,] "1"
## [5,] "1"
## [6,] "6"

# Extract all matches (list as return):
str_extract_all(string, pattern)
## [[1]]
## [1] "1"
##
## [[2]]
## character(0)
##
## [[3]]
## [1] "5" "5" "5"
##
## [[4]]
## [1] "1" "2" "3"
##
## [[5]]
## [1] "1" "2" "3"
##
## [[6]]
## [1] "6"

# To get a neat matrix output, add simplify = T:
str_extract_all(string, pattern, simplify = TRUE)
##      [,1] [,2] [,3]
## [1,] "1"  ""   ""
## [2,] ""   ""   ""
## [3,] "5"  "5"  "5"
## [4,] "1"  "2"  "3"
## [5,] "1"  "2"  "3"
## [6,] "6"  ""   ""

# Similar to str_extract_all (but returns column instead of row):
str_match_all(string, pattern)
## [[1]]
##      [,1]
## [1,] "1"
##
## [[2]]
##      [,1]
##
## [[3]]
##      [,1]
## [1,] "5"
## [2,] "5"
## [3,] "5"
##
## [[4]]
##      [,1]
## [1,] "1"
## [2,] "2"
## [3,] "3"
##
## [[5]]
##      [,1]
## [1,] "1"
## [2,] "2"
```

```
## [3,] "3"
##
## [[6]]
##      [,1]
## [1,] "6"
```

Split strings Using the Match as Separator

Finally, it might be useful to split strings based on a separator (for example file-names and file-extensions, dates, etc.). This can be done with the function `strsplit()`

strsplit()

```
# --- base-R ---
strsplit(string, pattern)
## [[1]]
## [1] "one:"
##
## [[2]]
## [1] "NO digit"
##
## [[3]]
## [1] "c" "c" "c"
##
## [[4]]
## [1] "d" ""  ""  "d"
##
## [[5]]
## [1] "" "" ""
##
## [[6]]
## [1] ""

# --- stringr ---
str_split(string, pattern)
## [[1]]
## [1] "one:" ""
##
## [[2]]
## [1] "NO digit"
##
## [[3]]
## [1] "c" "c" "c" ""
##
## [[4]]
## [1] "d" ""  ""  "d"
##
## [[5]]
## [1] "" "" "" ""
##
## [[6]]
## [1] "" ""
```

Further information about regex

Not all regex implementations are created equal. When going through documentation make sure it is relevant for R and `stringr`. For example, Java has a slightly different implementation. Also note that PERL has a rich implementation and has more functionality than the POSIX regex. This functionality is understood by many functions when the additional argument PERL=TRUE is added. It allows for example for `if` statements.

17.6 Dates with lubridate

Working with dates and times looks deceivingly easy. Sure there is that fact that hours are not decimals but duodecimal, but that is something that we got used to as children. We easily forget that not all years are 365 days, not all days are 24 hours, and that not all years exist.

The Egyptians used a year of exactly 365 days. The Julian Calendar was introduced by Gaius Julius Caesar, and it skips one day every four years. However, by 1582 the difference with the tropical year was about 10 days and Pope Gregorius XIII introduced the Gregorian Calendar and doing so skipped 10 days. This calendar was immediately adopted by European Roman Catholic countries since 1582; however, Greece for example only adopted it in 1923. The Gregorian Calendar that uses the 24 hours day and 365 day years needs adjustment every four years (February 29 is added for every year that is divisible by 4, unless they are divisible by 100, in which case they also need to be divisible by 400 to get that leap year. So, 1900, 2100, and 2200 are not leap years, but 1600, 2000, and 2400 are leap years.

So, in a period of four centuries, the Gregorian calendar will be missing 3 of its 100 Julian leap years, leaving 97. This brings the average year to $365 + 97/400 = 365.2425$. This is close but not exactly the same as the astronomical year, which is 365.2422 days. There are propositions to leave out one leap year every 450 years or rather skip one every 4000 years. We might be worrying about that in 2500 years. But even if we get that right than in a few million years the rotation of the earth will be slowed down too significantly. Anyhow, let us hope that by then we would consider the rotation of the earth as a mundane detail not worth bothering. In the meanwhile, all countries that converted to the Gregorian calendar somehow have to adjust count.

Also note that for example there is no year 0, the year before year 1 CE is -1 or 1 BCE. For all those reasons, even the simple task of calculating the difference between two dates can be daunting.

time-zone

Would time be easier? Would 1 p.m. not always be the same? Well, first of all, there are the time-zones. Then there are occasions where countries change time-zone. This can complicate matters if we want to calculate a difference in hours.

DST
daylight savings time

There is also the daylight savings time (DST). Canada used daylight savings time first in 1908, and Germany introduced DST in 1916: the clocks in the German Empire, and its ally Austria, were turned ahead by one hour on 30 April 1916 in the middle of the first World War. The assumed logic was to reduce the use of artificial lighting and save fuel. Within a few weeks, the idea was copied by the United Kingdom, France, and many other countries. Most countries reverted to standard time after World War I, and it was the World War II that made DST dominate in the Europe. Today, the European Union realizes that only a fool would believe that one gets a longer blanket by cutting of a piece on the bottom and sowing it on the top … and hence considers to leave DST.

IERS

At least every day has 24 hours and hence 86 400 seconds, isn't it? Well, even that is not correct. Now, and then, we have to introduce a leap second or seconds as the rotation of the earth slows down. We have now atomic clocks that are so precise that the difference between atomic time and astronomic time (earth rotation) becomes visible as the earth slows down. Since 1972, we have been adding leap seconds whenever the difference reaches 0.9 seconds. The last leap second was on 31 December 2016, the next one has still to be announced by the IERS in Paris.[6]

Since we only consider time on the earth, we discard the complexity of interstellar travel, special and general relativity. But even doing so it will be clear by now that we do not want to deal with all those details if we only want to make some neural network based on time differences of stock prices for example. We need R to help us.

`lubridate`

The package `lubridate` – announced in Grolemund and Wickham (2011) – is part of the `tidyverse` and does an excellent job and provides all functionality that one would need. It does all that while being compliant with the tidyverse philosophy.

---

[6]The IERS is the International Earth Rotation and Reference Systems Service.

We will load the package here and show this part of the code only once. All sub-sections that follow will use this package.

```
# Load the tidyverse for its functionality such as pipes:
library(tidyverse)

# Lubridate is not part of the core-tidyverse, so we need
# to load it separately:
library(lubridate)
```

17.6.1 ISO 8601 Format

ISO 8601

The first key concept is that of a date and a date-time. For most practical purposes, a date is something that can be stored as `yyyy-mmm-dd`.

It can be noted that R follows the ISO 8601 Notation and so will we do. Any person who believes in inclusion and not in imposing historical nation bound standards to the rest of the world will embrace the ISO standards, a fortiori any programmer or modeller with an inclusive world-view will also use the ISO 8601 standards. But there are many people who will not do this, and it is not uncommon to get dates in other formats or have to report dates in those formats. So, the format that we will use is the ISO format: `yyyy-mm-dd`, but we will also show how to convert to other systems.

Digression – R's internal date-format

Internally, R will store date-times as a Unix timestamp or POSIXct format:

unix timestamp
POSIXct

```
as.numeric(Sys.time())  # the number of seconds passed since 1 January 1970
## [1] 1580338862

as.numeric(Sys.time()) / (60 * 60 * 24 * 365.2422)
## [1] 50.07899
```

```
# There is a list of functions that convert to a date
mdy("04052018")
## [1] "2018-04-05"

mdy("4/5/2018")
## [1] "2018-04-05"

mdy("04052018")
## [1] "2018-04-05"

mdy("4052018")  # ambiguous formats are refused!

## Warning: All formats failed to parse. No formats found.
## [1] NA

dmy("04052018") # same string, different date
## [1] "2018-05-04"
```

mdy()

dmy()

Warning – Dates as numbers can be confusing

ymd()

The functions of the family ymd() do not only take strings as input, they can also can take a numerical input. This might lead to confusion as it is not what one would expect: the internal representation of a date.

```
dt <- ymd(20180505)  %>% print
## [1] "2018-05-05"

as.numeric(dt)
## [1] 17656
```

ydm()

```
ymd(17656)

## Warning: All formats failed to parse. No formats found.
## [1] NA
```

R even knows about quarters:

```
yq("201802")
## [1] "2018-04-01"
```

All the functions above can use dates combined with a time. In other words, they can not only be used with dates, but also with a date-time.[7] For example:

```
ymd_hms("2018-11-01T13:59:00")
## [1] "2018-11-01 13:59:00 UTC"

dmy_hms("01-11-2018T13:59:00")
## [1] "2018-11-01 13:59:00 UTC"

ymd_hm("2018-11-01T13:59")
## [1] "2018-11-01 13:59:00 UTC"

ymd_h("2018-11-01T13")
## [1] "2018-11-01 13:00:00 UTC"

hms("13:14:15")
## [1] "13H 14M 15S"

hm("13:14")
## [1] "13H 14M 0S"
```

Hint – Other date formats

The functions above also support other date formats. This can be done by tuning the argument tz (time-zone) and locale. A locale is a set of local customs to represent dates, numbers and lists. In Linux we can get a list of all installed locales via `system('locale -a')`.

Essentially, a date is stored as a number of days since 1970-01-01 and a time is stored as the number of seconds lapsed since 00 : 00 : 00.

[7]There is one exception: the function that relates to quarters.

```
as_date("2018-11-12")
## [1] "2018-11-12"

as_date(0)
## [1] "1970-01-01"

as_date(-365)
## [1] "1969-01-01"

as_date(today()) - as_date("1969-02-21")
## Time difference of 18605 days
```

Note – Today's date

The package `lubridate` also provides an alternative to `Sys.Date(()` with the function `today()` and `now()`.

```
today()
## [1] "2020-01-30"

now()
## [1] "2020-01-30 00:01:02 CET"
```

17.6.2 Time-zones

Date-time can be seen as "a moment in time," and it is a given moment within a given day. Its standard format is `yyyy-mm-dd hh:mm XXX`, where XXX is the time-zone like "UTC." R knows about 600 time-zones. From each time-zone, it keeps a full history, so it will know when Alaska for example changed time-zone and takes that into account when showing time differences.

time-zone

Hint – Available time-zones

To see all time-zones available, use the functions `OlsonNames()`.

`OlsonNames()`

We mentioned before that R knows about 600 time-zones. That is much more than the time-zones that actually exist. However, the simplicity of time-ones can be deceiving. For R to be able to calculate accurately time lapses and differences, it is not even enough to keep a different and unique entry in its database for each time-zone history profile, but also for each DST history.

```
# Note it converts the system time-zone to UTC:
as_datetime("2006-07-22T14:00")
## [1] "2020-06-07 22:14:00 UTC"

# Force time-zone:
as_datetime("2006-07-22T14:00 UTC")
## [1] "2020-06-07 22:14:00 UTC"

as_datetime("2006-07-22 14:00 Europe/Warsaw") #Fails silently!
## [1] "2020-06-07 22:14:00 UTC"

dt <- as_datetime("2006-07-22 14:00",tz = "Europe/Warsaw") %>%
    print
```

```
## [1] "2020-06-07 22:14:00 CEST"

# Get the same date-time numerals in a different time-zone:
force_tz(dt, "Pacific/Tahiti")
## [1] "2020-06-07 22:14:00 -10"

# Get the same cosmic moment in a new time-zone
with_tz(dt,  "Pacific/Tahiti")
## [1] "2020-06-07 10:14:00 -10"
```

force_tz()

with_tz()

Many functions – even when not designed for the purpose of managing time-zones – take a time-zone as input. For example:

today()

```
today(tzone = "Pacific/Tahiti")
## [1] "2020-01-29"
```

date_decimal()

```
date_decimal(2018.521, tz = "UTC")
## [1] "2018-07-10 03:57:35 UTC"
```

Hint – Create date-time from split data

make_datetime()

If you get years, months and days in separate columns then you can put them together with make_datetime(). For example:

```
dt1 <- make_datetime(year = 1890, month = 12L, day = 29L,
                     hour = 8L, tz = 'MST')
dt1
## [1] "1890-12-29 08:00:00 MST"
```

17.6.3 Extract Date and Time Components

The functions that extract a date-time component are named conveniently and the following example does not need much explanation:

year() month() week() day() wday() qday()

```
# We will use the date from previous hint:
dt1
## [1] "1890-12-29 08:00:00 MST"

year(dt)     # extract the year
## [1] 2020

month(dt)    # extract the month
## [1] 6

week(dt)     # extract the week
## [1] 23

day(dt)      # extract the day
## [1] 7

wday(dt)     # extract the day of the week as number
## [1] 1

qday(dt)     # extract the day of the quarter as number
```

```
## [1] 68

yday(dt)       # extract the day of the year as number
## [1] 159

hour(dt)       # extract the hour
## [1] 22

minute(dt)     # extract the minutes
## [1] 14

second(dt)     # extract the seconds
## [1] 0

quarter(dt)    # extract the quarter
## [1] 2

semester(dt)   # extract the  semester
## [1] 1

am(dt)         # TRUE if morning
## [1] FALSE

pm(dt)         # TRUE if afternoon
## [1] TRUE

leap_year(dt)  # TRUE if leap-year
## [1] TRUE
```

yday()

hour()

minute()

second()

quarter()

semester()

am()

pm()

leap_year()

The function update provides a flexible way to change date-times to other known values.

```
# We will use the date from previous example:
dt1
## [1] "1890-12-29 08:00:00 MST"

# Experiment changing it:
update(dt, month = 5)
## [1] "2020-05-07 22:14:00 CEST"

update(dt, year = 2018)
## [1] "2018-06-07 22:14:00 CEST"

update(dt, hour = 18)
## [1] "2020-06-07 18:14:00 CEST"
```

update()

Note – No side effects

All functions of the tidyverse will return a value that contains the requested modifications, but will never as a side effect change the variables that are provided to the function.

17.6.4 Calculating with Date-times

Due to the complexity around dates and times, the function update() together with the internal representation in seconds is not sufficient to calculate differences in time in a way that is both meaningful and correct for humans.

First, consider the following example in a time-zone using DST.

```
moment1 <- as_datetime("2018-10-28 01:59:00", tz = "Europe/Warsaw")
moment2 <- as_datetime("2018-10-28 02:01:00", tz = "Europe/Warsaw")
moment2 - moment1  # Is it 2 minutes or 1 hour and 3 minutes?
## Time difference of 1.033333 hours

moment3 <- as_datetime("2018-10-28 03:01:00", tz = "Europe/Warsaw")

# The clocks were put back in this tz from 3 to 2am.
# So, there is 2 hours difference between 2am and 3am!
moment3 - moment1
## Time difference of 2.033333 hours
```

Sure, the above example makes sense, but it is confusing at least. The difference between the two date-times (simply used as numbers of seconds since 1970-01-01 00:00) gave us the real time that was lapsed between the moments in time, even if humans make things unnecessarily complicated with DST.

So, when calculating time differences, there are a few different concepts that have to be distinguished. We define the following.

1. **Duration:** A duration is the physical amount of time that has been elapsed between two events.
2. **Periods:** Track changes in clock times (so pretend that DST, leap seconds, and leap years do not exist).
3. **Intervals:** Periods of time defined by start and end date-time (duration or period can be extracted)

Those three definitions are connected. The time interval only makes sense if one reduces it to a period or a duration. For example, a time interval that starts before an hour skip in a daylight savings time-zone might be two hours in duration but the "period" measured only one hour. Depending on the application one might need the one or the other.

17.6.4.1 Durations

Durations are the time intervals that model the cosmic time differences regardless the human definitions such as "9 a.m. in CET time-zone." Examples are natural events such as tides, eruptions of geysers, the time it took to fill out a test, etc.

Durations can be created with a range of functions starting with d (for duration). The following examples illustrate what these functions do and how they interact:

```
# Calculate the duration in seconds:
dyears(x = 1/365)
## [1] "86400s (~1 days)"

dweeks(x = 1)
## [1] "604800s (~1 weeks)"

ddays(x = 1)
## [1] "86400s (~1 days)"

dhours(x = 1)
## [1] "3600s (~1 hours)"

dminutes(x = 1)
## [1] "60s (~1 minutes)"

dseconds(x = 1)
## [1] "1s"

dmilliseconds(x = 1)
## [1] "0.001s"
```

```
dmicroseconds(x = 1)
## [1] "1e-06s"

dnanoseconds(x = 1)
## [1] "1e-09s"

dpicoseconds(x = 1)
## [1] "1e-12s"

# Note that a duration object times a number is again a
# Duration object and it allows arithmetic:
dpicoseconds(x = 1) * 10^12
## [1] "1s"

# Investigate the object type:
dur <- dnanoseconds(x = 1)
class(dur)
## [1] "Duration"
## attr(,"package")
## [1] "lubridate"

str(dur)
## Formal class 'Duration' [package "lubridate"] with 1 slot
##   ..@ .Data: num 1e-09

print(dur)
## [1] "1e-09s"
```

If the duration is not given in one number, but for example in with units expressed as a string, we can use the function `duration()`. There is also a series of functions that can coerce to a duration, check if something is a duration:

```
# Useful for automation:
duration(5, unit = "years")
## [1] "157680000s (~5 years)"

# Coerce and logical:
dur <- dyears(x = 10)
as.duration(60 * 60 * 24)
## [1] "86400s (~1 days)"

as.duration(dur)
## [1] "315360000s (~9.99 years)"

is.duration(dur)
## [1] TRUE

is.difftime(dur)
## [1] FALSE

as.duration(dur)
## [1] "315360000s (~9.99 years)"

make_difftime(60, units = "minutes")
## Time difference of 1 mins
```

17.6.4.2 Periods

Periods model time intervals between events that happen at specific clock times. For example, the opening of a stock exchange is always 9 a.m., regardless of DST or a leap second.

```
years(x = 1)
## [1] "1y 0m 0d 0H 0M 0S"

months(x = 1)
## [1] "1m 0d 0H 0M 0S"

weeks(x = 1)
## [1] "7d 0H 0M 0S"

days(x = 1)
## [1] "1d 0H 0M 0S"

hours(x = 1)
## [1] "1H 0M 0S"

minutes(x = 1)
## [1] "1M 0S"

seconds(x = 1)
## [1] "1S"

milliseconds(x = 1)
## [1] "0.001S"

microseconds(x = 1)
## [1] "1e-06S"

nanoseconds(x = 1)
## [1] "1e-09S"

picoseconds(x = 1)
## [1] "1e-12S"

# Investigate the object type:
per <- days(x = 1)
class(per)
## [1] "Period"
## attr(,"package")
## [1] "lubridate"

str(per)
## Formal class 'Period' [package "lubridate"] with 6 slots
##   ..@ .Data : num 0
##   ..@ year  : num 0
##   ..@ month : num 0
##   ..@ day   : num 1
##   ..@ hour  : num 0
##   ..@ minute: num 0

print(per)
## [1] "1d 0H 0M 0S"
```

```
# For automations:
period(5, unit = "years")
## [1] "5y 0m 0d 0H 0M 0S"

# Coerce timespan to period:
as.period(5, unit = "years")
## [1] "5y 0m 0d 0H 0M 0S"

as.period(10)
```

```
## [1] "10S"

p <- seconds_to_period(10) %>%
    print
## [1] "10S"

period_to_seconds(p)
## [1] 10
```

Note – Period functions starting letter

The functions that create periods have no specific starting letter.

These functions return period-objects, that have their own arithmetic defined. This allows us to define periods simply as follows:

```
years(1) + months(3) + days(13)
## [1] "1y 3m 13d 0H 0M 0S"
```

17.6.4.3 Intervals

Time intervals are periods or durations limited by two date-times. This means that in order to define them we need two dates (or two date-times). Intervals can be seen as a start and end-moment in one object, they are nor periods nor durations, but by dividing them by a period or a duration we can find out how long they take.

interval()
%−−%

```
d1 <- ymd_hm("1939-09-01 09:00", tz = "Europe/Warsaw")
d2 <- ymd_hm("1945-08-15 12:00", tz = "Asia/Tokyo")

interval(d1, d2)  # defines the interval
## [1] 1939-09-01 09:00:00 CET--1945-08-15 05:00:00 CEST

# Or use the operator %--%:
ww2 <- d1 %--% d2 # defines the same interval

ww2 / days(1)    # the period expressed in days
## [1] 2174.833

ww2 / ddays(1)   # duration in terms of days
## [1] 2174.792

# The small difference is due to DST and equals one hour:
(ww2 / ddays(1) - ww2 / days(1)) * 24
## [1] -1

# Allow the interval to report on its length:
int_length(ww2) / 60 / 60 / 24
## [1] 2174.792
```

The package `lubridate` provides a set of functions that allow to check if a date is in an interval, move the interval forward, etc.

```
d_date <- ymd("19450430")

# Is a date or interval in another:
d_date %within% ww2
## [1] TRUE
```

%within%

int_aligns()

int_shift()

int_flip()

int_standardize()

int_start()

int_end()

```
ph <- interval(ymd_hm("1941-12-07 07:48", tz = "US/Hawaii"),
                    ymd_hm("1941-12-07 09:50", tz = "US/Hawaii")
                    )
ph %within% ww2    # is ph in ww2?
## [1] TRUE

int_aligns(ph, ww2) # do ww2 and ph share start or end?
## [1] FALSE

# Shift forward or backward:
int_shift(ww2, years(1))
## [1] 1940-09-01 09:00:00 CEST--1946-08-15 05:00:00 CEST

int_shift(ww2, years(-1))
## [1] 1938-09-01 09:00:00 CET--1944-08-15 05:00:00 CEST

# Swap start and end moment:
flww2 <- int_flip(ww2)

# Coerce all to "positive" (start-date before end-date):
int_standardize(flww2)
## [1] 1939-09-01 09:00:00 CET--1945-08-15 05:00:00 CEST

# Modify start or end date:
int_start(ww2) <- d_date; print(ww2)
## [1] 1945-04-30 02:00:00 CEST--1945-08-15 05:00:00 CEST

int_end(ww2)  <-  d_date; print(ww2)
## [1] 1945-04-30 02:00:00 CEST--1945-04-30 02:00:00 CEST
```

17.6.4.4 Rounding

Occasionally, it is useful to round dates. It might make sense to summarize moments to the closest start of the month instead of just dropping the day and coercing all occurrences to the first of that month.

round_date()

floor_date()

ceiling_date()

```
dts <- c(ymd("2000-01-10"), ymd("1999-12-28"),
         ymd("1492-01-01"), ymd("2100-10-15")
         )
round_date(dts, unit = "month")
## [1] "2000-01-01" "2000-01-01" "1492-01-01" "2100-10-01"

floor_date(dts, unit = "month")
## [1] "2000-01-01" "1999-12-01" "1492-01-01" "2100-10-01"

ceiling_date(dts, unit = "month")
## [1] "2000-02-01" "2000-01-01" "1492-02-01" "2100-11-01"

# Change a date to the last day of the previous month or
# to the first day of the month with rollback()
rollback(dts, roll_to_first = FALSE, preserve_hms = TRUE)
## [1] "1999-12-31" "1999-11-30" "1491-12-31" "2100-09-30"
```

Of course, this will work also with years, days, and minutes by using a different value for the `unit` attribute.

17.7 Factors with Forcats

Categorical data in R are called "factors": they help to present discrete labels in a particular order that is not necessarily alphabetical and have been discussed in Section 4.3.7 *"Factors"* on page 45. In the past when computing time was in much more limited supply, they offered a gain in performance and that is why in base R they pop up even where they aren't very helpful. For example, when importing files or coercing data to data frames strings will become factors. This is one of the things that the tidyverse will not allow: silently changing data types.

We already have discussed the base-R functions related to factors in Section 4.3.7 *"Factors"* on page 45. Hence, the reader will be acquainted with the subject of factors itself, and therefore, in this section we can focus on the specificities of the package `forcats`, which is part of the core-tidyverse. "Forcats" is an anagram of "factors."

forcats

In this section, we will focus on an example of a fictional survey (with fabricated data). Imagine that we have the results of a survey that asked about the satisfaction with our service[8] rated as high/medium/low. First, we generate the results for that survey:

```
set.seed(1911)
s <- tibble(reply = runif(n = 1000, min = 0, max = 13))
hml <- function (x = 0) {
  if (x < 0)  return(NA)
  if (x <= 4) return("L")
  if (x <= 8) return("M")
  if (x <= 12) return("H")
  return(NA)
  }
surv <- apply(s, 1, FUN = hml)  # output is a vector
surv <- tibble(reply = surv)  # coerce back to tibble
surv
## # A tibble: 1,000 x 1
##    reply
##    <chr>
##  1 H
##  2 M
##  3 L
##  4 M
##  5 L
##  6 H
##  7 L
##  8 H
##  9 <NA>
## 10 L
## # ... with 990 more rows
```

To put the labels in the right orders, we have to make clear to R that they are factors and that we have a specific order for our factors. This can be done with the argument `levels` in the function `parse_factor()`.

parse_factor()

```
# 1. Define the factor-levels in the right order:
f_levels <- c("L", "M", "H")

# 2. Define our data as factors:
survey <- parse_factor(surv$reply, levels = f_levels)
```

[8] We do not recommend to measure customer satisfaction as high, medium and low. This is not actionable. Good KPIs are actionable. Best is to ask "how likely are you to recommend our service to a friend (on a scale from 1 to 10)." This would allow us to find net-detractors, neutral users and net promoters. This is useful, but not needed for our example to use factors.

Finally, we have our survey data gathered and organised for use. To study the survey, we can ask for its summary via the function `summary()` and plot the results. We can plot with the function `plot()` and R will recognise that the data is a factor-object and hence produce the plot in Figure 17.2.

summary()

```
summary(survey)
##     L    M    H <NA>
##   295  313  310   82

plot(survey, col = "khaki3",
     main = "Customer Satisfaction",
     xlab = "Response to the last survey"
     )
```

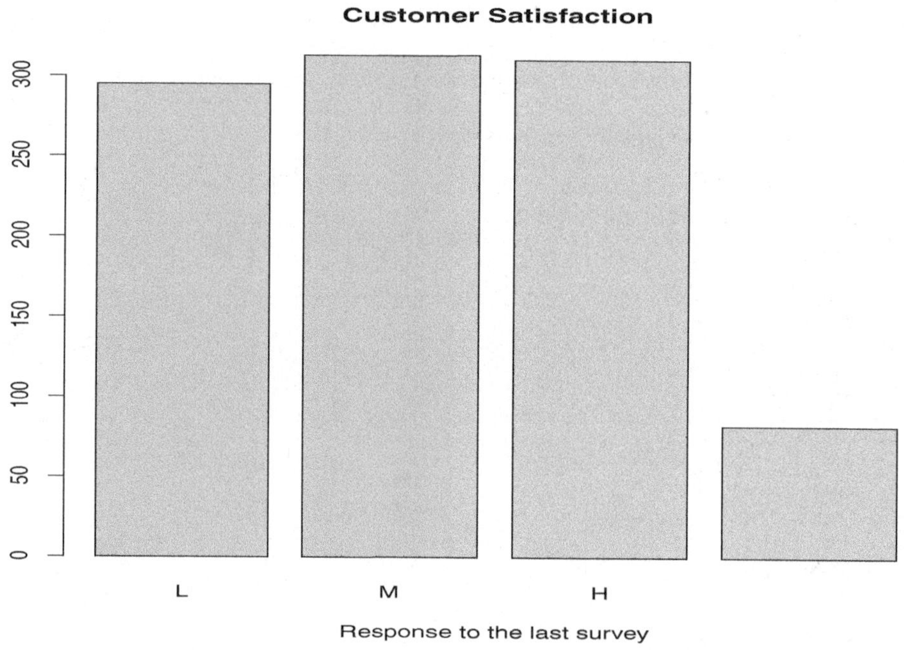

Figure 17.2: *The standard plot function on a factored object with some values NA (last block without label).*

factor()

Warning – Unmatched labels

You will remember that the function `factor()` does exactly the same: converting to factors. However, it will *silently* convert all strings that are not in the set of `levels` to NA. For example, a typo "m" (instead of "M") would become NA.
Therefore, we recommend to use `readr`'s function `parse_factor()`.

Hint – Find out which factor levels exist

If you do not know the factor levels in advance (or need to find out), then you can use the function `unique()` to find out which levels there are and feed that into `parse_factor()`. This works as follows:

```
surv2 <- parse_factor(surv$reply, levels = unique(surv$reply))

# Note that the labels are in order of first occurrence
summary(surv2)
##    H    M    L <NA>
##  310  313  295   82
```

unique()

Other functions from forcats

The package `forcats` provides many functions to work with labels. First, let us count the occurrences of the labels with `fct_count()`.

fct_count()

```
# Count the labels:
fct_count(survey)
## # A tibble: 4 x 2
##   f         n
##   <fct> <int>
## 1 L       295
## 2 M       313
## 3 H       310
## 4 <NA>     82
```

Another useful thing is that labels can be remodelled after creation. The function `fct_relabel()` provides a powerful engine to change labels and remodel the way the data is modelled and/or presented. The following code illustrates how this is done and displays the results in Figure 17.3 on page 328.

fct_relabel()

```
# Relabel factors with fct_relabel:
HML <- function (x = NULL) {
  x[x == "L"] <- "Low"
  x[x == "M"] <- "Medium/High"
  x[x == "H"] <- "Medium/High"
  x[!(x %in% c("High", "Medium/High", "Low"))] <- NA
  return(x)
  }
f <- fct_relabel(survey, HML)
summary(f)
##         Low Medium/High        <NA>
##         295         623          82

plot(f, col = "khaki3",
     main = "Only one third of customers is not happy",
     xlab = "Response to the expensive survey"
     )
```

Hint – Use regex

The open mechanism that allows to pass a function as parameter to another function is a very flexible tool. For example, it allows to use the power of regular expressions – see Section 17.5.2 *"Pattern Matching with Regular Expressions"* on page 302 to achieve the same result.

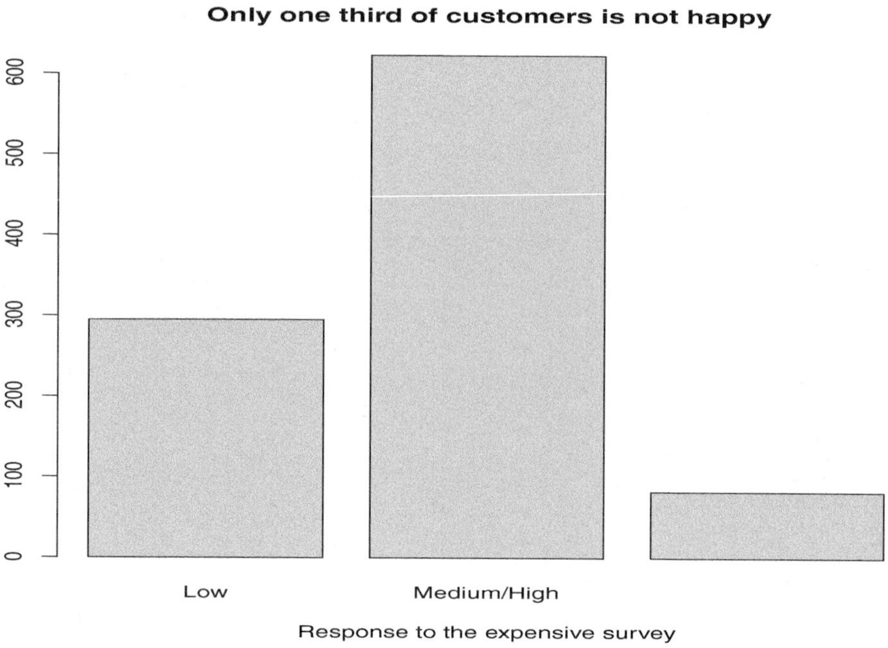

Figure 17.3: *Maybe you would prefer to show this plot to the board meeting? This plot takes the two best categories together and creates the impression that more people are happy. Compare this to previous plot.*

```
HMLregex <- function (x = NULL) {
  x[grepl("^L$", x)] <- "Low"
  x[grepl("^M$", x)] <- "Medium/High"
  x[grepl("^H$", x)] <- "Medium/High"
  x[!(x %in% c("High", "Medium/High", "Low"))] <- NA
  return(x)
  }
# This would do exactly the same, but it is a powerful
# tool with many other possibilities.
```

Besides changing labels, it might occur that the order needs to be changed. This makes most sense when more variables are present and when we want to show the interdependence structure between those two variables. Assume that we have a second variable: age. The following code provides this analysis and plots the results in Figure 17.4 on page 330

```
num_obs <- 1000  # the number of observations in the survey
# Start from a new survey: srv
srv <- tibble(reply = 1:num_obs)
srv$age <- rnorm(num_obs, mean = 50,sd = 20)
srv$age[srv$age < 15] <- NA
srv$age[srv$age > 85] <- NA

hml <- function (x = 0) {
  if (x < 0)  return(NA)
  if (x <= 4) return("L")
  if (x <= 8) return("M")
  if (x <= 12) return("H")
  return(NA)
  }

for (n in 1:num_obs) {
  if (!is.na(srv$age[n])) {
    srv$reply[n] <- hml(rnorm(n = 1, mean = srv$age[n] / 7, sd = 2))
   }
   else {
    srv$reply[n] <- hml(runif(n = 1, min = 1, max = 12))
   }
}
f_levels <- c("L", "M", "H")
srv$fct <- parse_factor(srv$reply, levels = f_levels)

# From most frequent to least frequent:
srv$fct                      %>%
fct_infreq(ordered = TRUE) %>%
  levels()
## [1] "M" "H" "L" NA

# From least frequent to more frequent:
srv$fct      %>%
 fct_infreq  %>%
  fct_rev    %>%
  levels
## [1] NA  "L" "H" "M"

# Reorder the reply variable in function of median age:
fct_reorder(srv$reply, srv$age) %>%
   levels
## [1] "H" "L" "M"

# Add the function min() to order based on the minimum
# age in each group (instead of default median):
fct_reorder(srv$reply, srv$age, min) %>%
   levels
## [1] "H" "L" "M"

# Show the means per class of satisfaction in base-R style:
by(srv$age, srv$fct, mean, na.rm = TRUE)
## srv$fct: L
## [1] 30.65112
## -------------------------------------------
## srv$fct: M
## [1] 44.41898
## -------------------------------------------
## srv$fct: H
## [1] 60.01358
## -------------------------------------------
## srv$fct: NA
## [1] 62.67211
```

fct_infreq()
levels()
fct_rev()
fct_reorder()
by()

group_by()

```
# Much more accessible result with the dplyr:
satisf <- srv                   %>%
          group_by(fct)  %>%
          summarize(
             age = median(age, na.rm = TRUE),
             n = n()
             )              %>%
          print
## # A tibble: 4 x 3
##   fct      age     n
##   <fct> <dbl> <int>
## 1 L      29.9   173
## 2 M      43.6   432
## 3 H      61.0   328
## 4 <NA>   72.4    67

# Show the impact of age on satisfaction visually:
par(mfrow = c(1,2))
barplot(satisf$age,  horiz = TRUE, names.arg = satisf$fct,
        col = c("khaki3","khaki3","khaki3","red"),
        main = "Median age per group")
barplot(satisf$n,  horiz = TRUE, names.arg = satisf$fct,
        col = c("khaki3","khaki3","khaki3","red"),
        main = "Frequency per group")
```

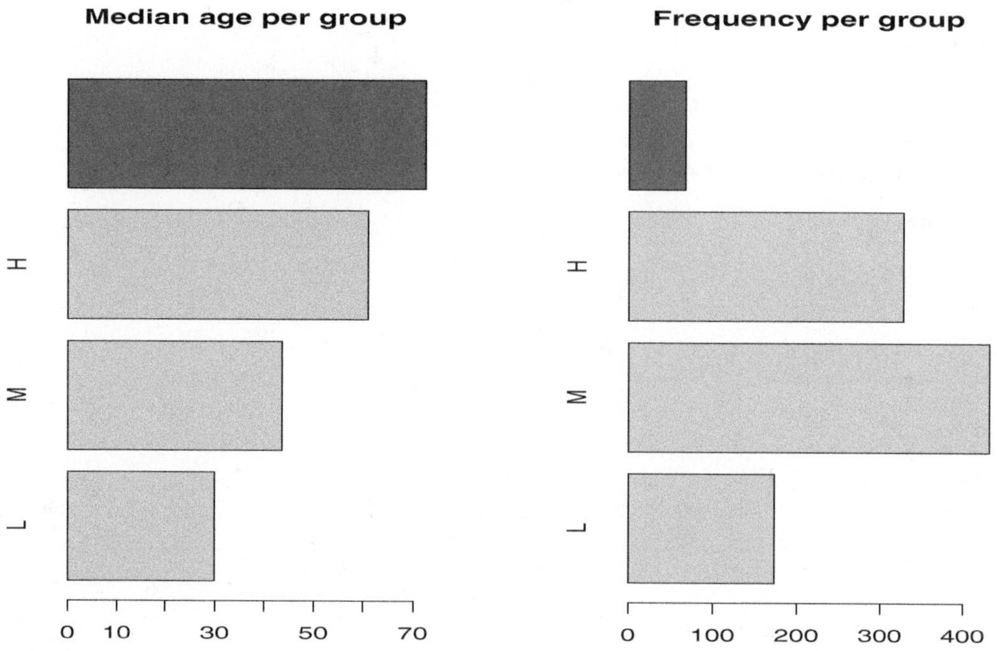

Figure 17.4: *A visualisation of how the age of customers impacted the satisfaction in our made-up example. The NA values have been highlighted in red.*

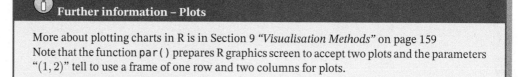

Further information – Plots

More about plotting charts in R is in Section 9 *"Visualisation Methods"* on page 159
Note that the function `par()` prepares R graphics screen to accept two plots and the parameters "(1, 2)" tell to use a frame of one row and two columns for plots.

In some sense, we were "lucky" that the order produced by the medians of the age variable somehow make sense for the satisfaction variable. In general, this will mix up the order of the factorized variable. So, these techniques make most sense for a categorical variable that is a "nominal scale,"[9] such as for example countries, colours, outlets, cities, people, etc.

Finally, we mention that `forcats` foresees a function to anonymise factors. It will change the order and the labels of the levels. For example, consider our survey data (and more in particular the tibble `srv` that has the variable `fct` that is a factor.

```
srv                                  %>%
  mutate("fct_ano" = fct_anon(fct))  %>%
  print
## # A tibble: 1,000 x 4
##    reply    age fct   fct_ano
##    <chr> <dbl> <fct> <fct>
##  1 L      59.5 L     4
##  2 L      30.7 L     4
##  3 M      48.9 M     1
##  4 L      NA   L     4
##  5 H      50.7 H     2
##  6 M      57.7 M     1
##  7 M      45.1 M     1
##  8 H      75.1 H     2
##  9 L      30.0 L     4
## 10 M      41.2 M     1
## # ... with 990 more rows
```

? Question #13

The package `forcats` comes with a dataset `gss_cat` containing census data that allows to link number of hours television watched, political preference, marital status, income, etc. Many of the variables can – and should be – considered as categorical variables and hence can be converted to factors. Remake the previous analysis in a sensible way for that dataset.

[9] A nominal scale is a measurement scale that has no strict order-relation defined – see Chapter B *"Levels of Measurement"* on page 829

♣ 18 ♣

Dealing with Missing Data

Most datasets will have missing values, and missing values will make any modelling a lot more complicated. This section will help you to deal with those missing values.

However, it is – as usual – better to avoid missing data than solve the problem later on. In the first place, we need to be careful when collecting data. The person who collects the data, is usually not the one reading this book. So, the data scientist reading this book, can create awareness to the higher management that can then start measuring data quality and improve awareness or invest in better systems.

We can also change software so that it helps to collect correct data. If, for example, the retail staff systematically leaves the field "birth date" empty, then we can via software refuse that birth dates that are left empty, pop up a warning if the customer is over 100 years old and simple do not accept birth dates that infer an age over 150. The management can also show numbers related to losses due to loans that were accepted on wrong information. Also the procedures can be adapted, and for example require a copy of the customer's ID card, and the audit department can then check compliance with this procedure, etc.

However, there will still be cases where some data is missing. Even if the data quality is generally fine, and we still have thousands of observations left after leaving out the small percentage of missing data, then it is still essential to find out why the data is missing. If data is missing at random, leaving out missing observations will not harm the model and we can expect the model not to loose power when applied to real and new cases.[1]

[1]Even if there is only a small percentage of data missing, then still it is worth to investigate why the data is missing. For example, on a book of a million loans we notice that 0.1% of the fields "ID card number" is missing. When we check the default rate it appears that if that field is missing we have a default rate of 20%, while in the rest of the population this is only 4%. This is a strong indication that not filling in that field is correlated with the intention of fraud. We can, indeed, leave this small percentage out to build our model. However, we need to point out what is happening and prevent further fraud.

The Big R-Book: From Data Science to Learning Machines and Big Data, First Edition. Philippe J.S. De Brouwer.

Companion Website: www.wiley.com/go/De Brouwer/The Big R-Book

18.1 Reasons for Data to be Missing

If data is missing, there is usually an identifiable reason why this is so. Finding out what that reason is, is most important, because it can indicate fraud, or be systematic, and hence influence the model itself.

For example, missing data can be due to:

1. input and pre-processing (e.g. conversion of units: some dates were in American format, others in UK format, some dates got misunderstood and others rejected; a decimal comma is not understood, data is copied to a system that does not recognize large number, etc.);

2. unclear or incomplete formulated questions (e.g. asking "are you male or female?", while having only a "yes" and "no" answer possible.);

3. fraud or other intend (if we know that young males will pay higher for a car insurance we might omit the box where the gender is put);

4. random reason (e.g. randomly skipped a question, interruption of financial markets due to external reason, mistake, typing a number too much, etc.)

Besides missing data it happens also all too often that data is useless. For example, if a school wants to gather feedback about a course, they might ask "How satisfied are you with the printed materials." Students can put a low mark if they did not get the materials, if they thought that the printed materials were too elaborate, too short, low quality, or even because they do not even like to get printed materials and prefer electronic version. Imagine now that the feedback is low for the printed materials. What can we do with this information? Did we even bother to ask if the printed materials are wanted, and if so, how important are they in the overall satisfaction?

The typical mistake of people that formulate a questionnaire is that they are unable to imagine that other people might have another set of beliefs and values, or even have a very different reality.

Example: Unclear questions

Often, data is missing because questionnaires are written carelessly and formulated ambiguously. What to think about questions such as these:

1. rate the quality of the printed materials (1 ... 5);
2. parent name, student name, study level;
3. it is never so that the teacher is too late: yes/no;
4. is the DQP sufficient to monitor the IMRDP and in line with GADQP? yes/no
5. I belong to a minority group (when applying for a job).

These example questions have the following issues.

1. The reader will assume that this question is there because the teacher will be assessed on the quality of printed materials and it also assumes that the student cares. Which is an assumption, that should be asked first. What would you do with this question if you did not want printed materials in the first place?
2. Whose study level are we asking here? That of the student or that of the parent?
3. Assume that the teacher is often late, then you can answer both "yes" or "no," assume the opposite and the same holds. So what to answer?
4. This question has two common problems. First, it uses acronyms that might not be clear to everyone; second, it combines two separate issues – being sufficient and compliance with some rules – into one yes/no question. What would you answer if only one of the two is true?
5. Some people might not want to fill in this question because they feel it should be irrelevant. Would you expect the propensity to leave this question open to depend on the racial group to which one belongs?

It also might be that there is a specific reason why it is missing. This could be because the provider of the data does not like to divulge information. For example, when applying for a loan, the young male might know that being young and being male works in his disadvantage. Maybe he would like to identify as a lady for one day or simply leave the question open? In this example it is possible that most of the missing gender information is that from young males. Leaving out missing values will lead to a model that discriminates less young males, but might lead to more losses for the lender and inadvertently lure many young males into unsustainable debt.

Assume that there is no possible systematic reason why data is missing. Then we have some more options to deal with the missing data points.

18.2 Methods to Handle Missing Data

18.2.1 Alternative Solutions to Missing Data

Since missing data is intertwined with the reasons why the data is missing and the impact that will have on the model that we are building, dealing with missing data is not exact science. Below we present some of the most used methods:

1. Leave out the rows with missing data. If there is no underlying reason why data is missing, then leaving out the missing data is most probably harmless. This will work fine if the dataset is sufficiently large and of sufficient quality (even a dataset with hundred thousand lines but with 500 columns can lead to problems when we leave out all rows that miss one data point).

2. Carefully leave out rows with missing data. Same as above, but first make up our mind which variable will be in the model and then only leave out those rows that miss data on those rows.

3. Somehow fill in the missing data based on some rules or models. For example, we can replace the missing value by:

   (a) the mean value for that column;

   (b) median value for that column;

   (c) conditional mean or median (e.g. fill in missing value for height with gender based mean) where some pre-existing logic or clear and well accepted rules hold;

   (d) an educated guess (e.g. someone who scored all requested dimensions as "4/5" probably intended the missing value for "quality of printed materials" to be a "4/5" as well);

   (e) the mid-value (if that makes sense for example the "3/5" could be a mid-value for "rate on a scale from 1 to 5"), this means choosing the middle of the possible values regardless which values occur more;

   (f) replace the missing value by a more complex model, eventually based on machine learning, such as:

      - regression substitution, which tries to guess the missing value based on a multiple linear regression on other variables,
      - multiple imputation, which uses statistical methods to guess plausible values based on the data that is not missing (linear regression or machine learning) and then reset averages of the variables back by adding random errors in the predictions.

Whatever we do or not do with missing data will influence the outcome of the model and needs to be handled with the utmost care. Therefore, it is important to try to understand the data and make an educated choice in handling missing data. As usual, R comes to the rescue with a variety of libraries.

Example

For the remainder of this section, we will use a concrete example, using the database `iris`, that is provided by the library `datasets` and is usually loaded at startup of R. This database provides information of 150 measurements of petal and sepal measurements as well as the type of Iris (setosa, versicolor, or virginica). This database is most useful for training classification models. For the purpose of this example, we will use this database and introduce some missing values.

```
set.seed(1890)
d1       <- d0  <- iris
i        <- sample(1:nrow(d0), round(0.20 * nrow(d0)))
d1[i,1]  <- NA
i        <- sample(1:nrow(d0), round(0.30 * nrow(d0)))
d1[i,2]  <- NA
head(d1, n=10L)
##     Sepal.Length Sepal.Width Petal.Length Petal.Width
## 1            5.1         3.5          1.4         0.2
## 2            4.9         3.0          1.4         0.2
## 3            4.7         3.2          1.3         0.2
## 4            4.6         3.1          1.5         0.2
## 5            5.0          NA          1.4         0.2
## 6            5.4         3.9          1.7         0.4
## 7            4.6          NA          1.4         0.3
## 8            5.0         3.4          1.5         0.2
## 9            4.4         2.9          1.4         0.2
## 10            NA         3.1          1.5         0.1
##     Species
## 1    setosa
## 2    setosa
## 3    setosa
## 4    setosa
## 5    setosa
## 6    setosa
## 7    setosa
## 8    setosa
## 9    setosa
## 10   setosa
```

For the remainder of this section, we will use the example of the iris flowers as defined above.

Note – The randomness in our example

In order to draw conclusions about the different methods, it would make sense to run a certain model like a neural network, decision tree, or other. Then apply some methods to impute missing data and then compare results. However, since we delete data in a certain way – random – we would expect to see that reflected in the results and in fact not really being able to determine which method is best. In fact there is no such thing as "the best method," since we do not know why the data is missing and if we knew, then it would not be missing.

18.2.2 Predictive Mean Matching (PMM)

PMM
predictive mean matching

One of the oldest methods – see Little (1988) – is predictive mean matching (PMM). For each observation that has a variable with a missing value, the method finds an observation (that has no missing value on this variable) with the closest predictive mean to that variable. The observed value from this observation is used as imputed value. This means that it preserves automatically many important characteristics such as skew, boundness (e.g. only positive data), base type (e.g. integer values only), etc.

The PMM process is as follows:

1. Take all observations that have no missing values and fit a linear regression of variable x – that has the missing values – to one or more variables y, and produce a set of coefficients b.
2. Draw random coefficients b^* from the posterior predictive distribution of b. Typically, this would be a random draw from a multivariate normal distribution with mean b and the estimated covariance matrix of b (with an additional random draw for the residual variance). This step will ensure that there is sufficient variability in the imputed values.
3. Using b^*, generate predicted values for x for all cases (as well for those that have missing values in x as those that do not.
4. For each observation with missing x, identify a set of cases with observed x whose predicted values are close to the predicted value for the observation with missing data.
5. From those observations, randomly choose one and assign its observed value as value to be imputed to the missing x.
6. Repeat steps 2 – 5 till all the missing variables for x have an impute candidate.
7. Repeat steps 1 – 6 for all variables that have missing values.

Interestingly, and unlike many methods of imputation, the purpose of the linear regression is not to find the values to be imputed. Rather, the regression is used to construct a metric for what is a matching observation (whose observation can be borrowed).

Many of the packages that we will describe below have an implementation of PMM.

18.3 R Packages to Deal with Missing Data

18.3.1 mice

If data is missing at random (MAR) – which is defined so that the probability that a value is missing depends only on observed value and can be predicted using only that value, the package `mice`[2] is a great way to start. Multivariate imputation via chained equations (mice) imputes data on a variable by variable basis by specifying an imputation model per variable. This code loads the package and uses its function `md.pattern()` to visualise the missing values. This output is in Figure 18.1.

MAR
missing at random
`mice`

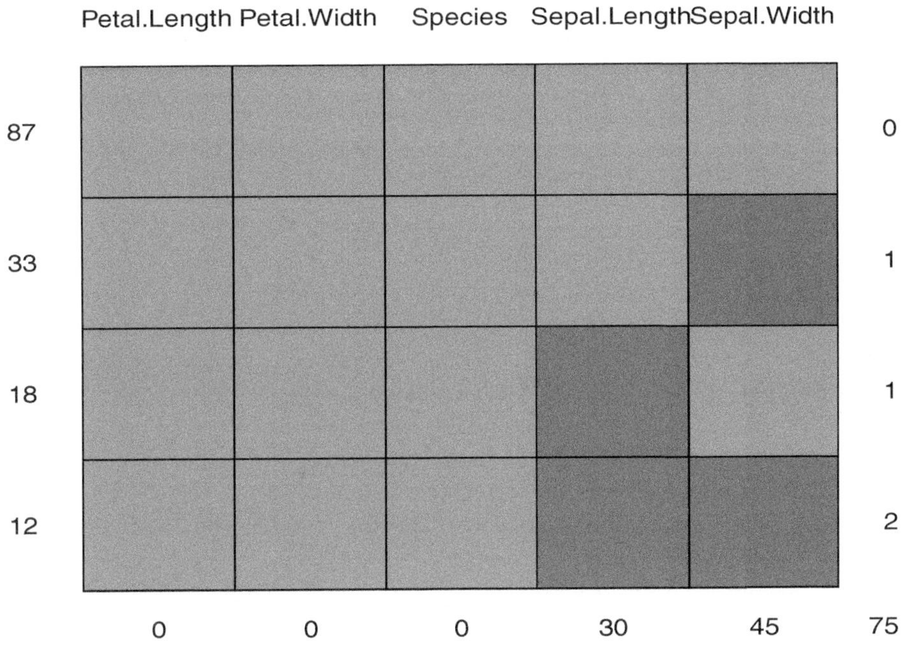

Figure 18.1: *The visualization of missing data with the function* `md.pattern()`.

```
#install.packages('mice')  # uncomment if necessary
library(mice)              # load the package

## Loading required package: lattice
## ## Attaching package: 'mice'
## The following object is masked from 'package:tidyr':## ##    complete
## The following objects are masked from 'package:base':## ##    cbind, rbind

# mice provides the improved visualization function md.pattern()
md.pattern(d1)  # function provided by mice
```

```
##    Petal.Length Petal.Width Species Sepal.Length
## 87            1           1       1            1
## 33            1           1       1            1
## 18            1           1       1            0
## 12            1           1       1            0
##               0           0       0           30
```

[2]Mice is an acronym and stands for "Multivariate Imputation via Chained Equations."

```
##    Sepal.Width
## 87           1  0
## 33           0  1
## 18           1  1
## 12           0  2
##             45 75
```

The table shows that the dataset *d1* has 87 complete cases, 33 missing observations in Sepal.Width, 18 observations, where Sepal.Length is missing, and 12 cases where both are missing.

Now, that we used the capabilities of `mice` to study and visualize the missing data, we can also use it to replace the missing values with a guess. This is done by the function `mice()` as follows:

mice()

```
d2_imp <- mice(d1, m = 5, maxit = 25, method = 'pmm', seed = 1500)
```

This created five possible datasets and we can select one completed set as follows.

```
# e.g. choose set number 3:
d3_complete <- complete(d2_imp, 3)
```

18.3.2 missForest

missForest uses under the hood a random forest algorithm and hence is a non-parametric imputation method. A non-parametric method does not make explicit assumptions about functional form of the distribution. missForest will build a random forest model for each variable and uses this model to predict missing values in the variable with the help of observed values.

It yields an out of bag (OOB) imputation error estimate and provides fine grained control over the imputation process. It even has the possibility to return OOB separately (for each variable) instead of aggregating over the whole data matrix. This allows to assess how accurately the model has chosen values for each variable.

Note also that the process works well on categorical variables.

```
# install.packages('missForest') # only first time
library(missForest)              # load the library
d_mf <- missForest(d1)           # using the same data as before
##   missForest iteration 1 in progress...done!
##   missForest iteration 2 in progress...done!
##   missForest iteration 3 in progress...done!
##   missForest iteration 4 in progress...done!

# access the imputed data in the ximp attribute:
head(d_mf$ximp)
##   Sepal.Length Sepal.Width Petal.Length Petal.Width
## 1          5.1    3.500000          1.4         0.2
## 2          4.9    3.000000          1.4         0.2
## 3          4.7    3.200000          1.3         0.2
## 4          4.6    3.100000          1.5         0.2
## 5          5.0    3.281536          1.4         0.2
## 6          5.4    3.900000          1.7         0.4
##   Species
## 1  setosa
## 2  setosa
## 3  setosa
## 4  setosa
## 5  setosa
## 6  setosa

# normalized MSE of imputation:
d_mf$OOBerror
##     NRMSE       PFC
## 0.1067828 0.0000000
```

Hint – Fine-tuning

Errors can usually be reduced by fine-tuning the parameters of the function `missForest`:

- `mtree` – the number of variables being randomly sampled at each split, and
- `ntree` – number of trees in the forest.

18.3.3 Hmisc

Hmisc does actually a lot more than dealing with missing data. It provides functions for data analysis, plots, complex table making, model fitting, and diagnostics (for linear regression, logistic regression, and cox regression) as well as functionality to deal with missing data.

For missing values `Hmisc` provides the following functions:

Hmisc

- `impute()`: Imputes missing values based on median (default), mean, max, etc.

impute()

- `aregImpute()`: Imputes using additive regression, bootstrapping, and predictive mean matching (PMM).

aregImpute()

In the bootstrapping method, different bootstrap samples are used for each of multiple imputations. Then, an additive model (non parametric regression method) is fitted on samples taken with replacements from original data and missing values elected using non-missing values as independent variables. Finally, it uses PMM by default to impute missing values.

```
# Install the package first via:
# install.packages('Hmisc')
library(Hmisc)

## Loading required package: survival
## Loading required package: Formula
## ## Attaching package: 'Hmisc'
## The following objects are masked from 'package:dplyr':## ##    src, summarize
## The following objects are masked from 'package:base':## ##    format.pval, units

# impute using mean:
SepLImp_mean <- with(d1, impute(Sepal.Length, mean))

# impute a randomly chosen value:
SepLImp_rand <- with(d1, impute(Sepal.Length, 'random'))

# impute the maximum value:
SepLImp_max <- with(d1, impute(Sepal.Length, max))

# impute the minimum value:
SepLImp_min <- with(d1, impute(Sepal.Length, min))

# note the '*' next to the imputed values"
head(SepLImp_min, n = 10L)
##   1   2   3   4   5   6   7   8   9   10
## 5.1 4.9 4.7 4.6 5.0 5.4 4.6 5.0 4.4 4.3*
```

More complex and flexible models can be obtained by `aregImpute()`:

```
aregImp <- aregImpute(~ Sepal.Length + Sepal.Width
                        + Petal.Length + Petal.Width + Species,
                        data = d1, n.impute = 4)
```

```
## Iteration 1
Iteration 2
Iteration 3
Iteration 4
Iteration 5
Iteration 6
Iteration 7

print(aregImp)
##
## Multiple Imputation using Bootstrap and PMM
##
## aregImpute(formula = ~Sepal.Length + Sepal.Width + Petal.Length +
##     Petal.Width + Species, data = d1, n.impute = 4)
##
## n: 150  p: 5  Imputations: 4   nk: 3
##
## Number of NAs:
## Sepal.Length  Sepal.Width Petal.Length  Petal.Width
##           30           45            0            0
##      Species
##            0
##
##              type d.f.
## Sepal.Length    s    2
## Sepal.Width     s    1
## Petal.Length    s    2
## Petal.Width     s    2
## Species         c    2
##
## Transformation of Target Variables Forced to be Linear
##
## R-squares for Predicting Non-Missing Values for Each Variable
## Using Last Imputations of Predictors
## Sepal.Length  Sepal.Width
##        0.901        0.714

# n.impute = 4 produced 4 sets of imputed values
# Access the second imputed data set as follows:
head(aregImp$imputed$Sepal.Length[,2], n = 10L)
##  10  12  21  23  25  29  36  41  48  57
## 4.9 5.1 4.9 5.4 5.5 5.1 5.4 4.4 4.8 5.4
```

Further information – The package mi

The mi package provides multiple imputation with PMM method – as explained earlier. You might want to consider the mi package for one of the following reasons:

- it detects certain issues with the data such as – near – collinearity of variables;
- it provides visual diagnostics of imputation models;
- it adds noise to imputation process.

The subject of missing data is by no means closed with this short overview. It is a complex issue that is hard to tackle. We find that the best and most powerful approach is to start at the source: improve data quality. In any case, imputation methods have to be used with utmost care to avoid influencing the model being fitted.

♣ 19 ♣

Data Binning

19.1 What is Binning and Why Use It

Histograms are well known, and they are an example of "data-binning" (also known as "discrete binning," "bucketing," or simply "binning"). They are used in order to visualize the underlying distributions for data that has a limited number of observations. **binning**

Consider the following simple example where we start with data drawn from a known distribution and plot the histogram (the output of this code is in Figure 19.1 on page 344):

```
set.seed(1890)
d <- rnorm(90)
par(mfrow=c(1,2))
hist(d, breaks=70, col="khaki3")
hist(d, breaks=12, col="khaki3")
```

Other examples of data binning are in image processing. When small shifts in the spectral dimension from mass spectrometry (MS) or nuclear magnetic resonance (NMR) could be falsely interpreted as representing different components, binning will help. Binning allows to reduce the spectrum in resolution to a sufficient degree to ensure that a given peak remains in its bin despite small spectral shifts between analyses. Also, several digital camera systems use a pixel binning function to improve image contrast and reduce noise.

Binning reduces the effects of minor observation errors, especially when the observations are sparse, binning will bring more stability. The original data values which fall in a given small interval, a bin, are replaced by a value representative of that interval, often the central value. It is a form of quantization. Statistical data binning is a way to group a number of more or less continuous values into a smaller number of "bins."

More often one will be faced with the issue that real data relates to a very diverse population. Consider for example a bank that has a million consumer loans[1]. That data will typically have more people in their twenties and thirties than in any group – largely because of the population curves but also because those people have some creditworthiness and need to build up their lives. Even if it has thousands of customers above 80 years, they will still form a minority.

[1] A "consumer loan" is understood as a loan with no collateral. Examples are the overdraft on a current account, a credit card, a purpose loan (e.g. to buy a car – but without the car being a collateral), etc.

The Big R-Book: From Data Science to Learning Machines and Big Data, First Edition. Philippe J.S. De Brouwer.

Companion Website: www.wiley.com/go/De Brouwer/The Big R-Book

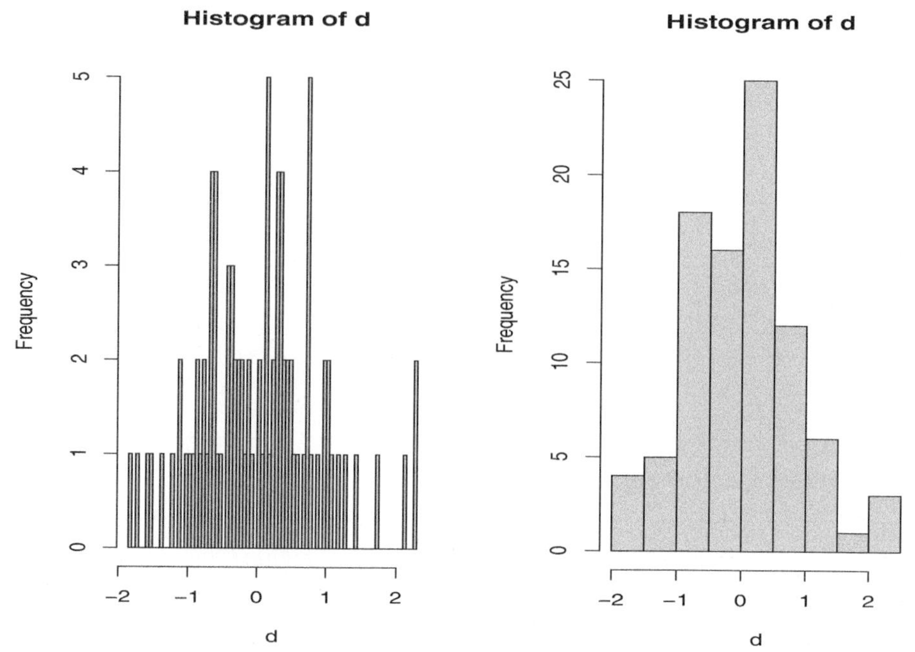

Figure 19.1: *Two histograms of the same dataset. The histogram with less bins (right) is easier to read and reveals more clearly the underlying model (Gaussian distribution). Using more bins (as in the left plot) can overfit the dataset and obscure the true distribution.*

Further information – Bias

Most datasets will have a bias. For example, in databases with people, we might find biases towards, age, postcode, country, nationality, marital status, etc. If the data is about companies, one cane expect a bias against postcode, sector, size, nationality, etc. If the data concerns returns on stock exchanges one will – hopefully – have little data such as 2008 (and hence the data has a bias against dramatic stock market crashes).

bias

The problem is that our observations might be a true trend (for example mining will have more mortal accidents than banking) or might be something that is only in this database (for example creditworthiness and nationality). The problem with this last example is that nationality could indeed be predictive to the creditworthiness in a given place. For example, many poor people emigrated from country X to country Y. These people display a higher default rate in country Y. However, in country X this might not be true. This is yet another important subject: discrimination. With discrimination we refer to judgement based on belonging to certain groups (typically these are factors that are determined by birth and not by choice).

For the bank, it might seem sensible to assume that people from country X will not pay back loans. However, if people from country X will systematically be denied loans, they cannot start a business and they will stay poor. This might lead to further social stigmatisation, social unrest, unproductive labour force, increased criminality, etc. Thinking further, this also implies that the country Y misses out on GDP created by the entrepreneurial people from country X. This will have the most negative impact for the people of country X, but impact the economy of Y, and hence, also our bank.

> More than any other institution, banks, can never be neutral. Whether they do something, or whether they do not, they are partial. More than any other institution, banks carry a serious responsibility.

Imagine that the purpose is to calculate a creditworthiness for further loans to the same customers. In such case, there are generally a few possibilities:

1. the dependency on age is linear and then a linear model will do fine
2. the dependency on age is of non-linear nature, but a simple model – that can be explained – can be fitted
3. the dependency on age is more complex. Young people have on average a less stable situation and hence have worse credit score. Later people get their lives in order and are able to fulfil the engagements they take. However, when they retire the income drops significantly. This new situation can again lead to increased propensity to financial problems.

Sure, we need to make another assumption at this point: is the lender willing to discriminate for age? If age is included in the explaining variables of the models, then the result will be that a people of certain age groups will face more difficulties to get a loan.[2]

In the following example, we assume that we want to build a model that is based on the customer profile, and tries to predict customer churn.[3]

We already encountered in the answer to Question 6 on page 48 how to perform binning in R. Using the example of data from the normal distribution mentioned above, and building further on the histogram in Figure 19.1 on page 344, we are fully armed to make a binning for the data in the data frame d. In the first place, we want to make sure that no bin has too little observations in it.

To achieve this in R, we will use the function `cut()` as our workhorse: `cut()`

```
# Try a possible cut
c <- cut(d, breaks = c(-3, -1, 0, 1, 2, 3))
table(c)
## c
## (-3,-1]  (-1,0]   (0,1]   (1,2]   (2,3]
##       9      34      37       7       3

# This is not good, it will not make solid predictions for the last bin.
# Hence we need to use other bins:
c <- cut(d, breaks = c(-3, -0.5, 0.5, 3))
table(c)
## c
##  (-3,-0.5] (-0.5,0.5]    (0.5,3]
##         27         41         22

# We have now a similar number of observations in each bin.
# Is that the only thing to think about?
```

`table()`

[2]While one might think that "this is the right thing to do because the data tells me to," it in fact a delicate balancing act between risk (and cost to the lender) and social justice. Indeed, if we live in a country where social security for old people is weak, then old people will systematically be considered as worse risk, regardless of the true characteristics of the individual. Replace age with gender, social status, race, postcode, etc. Actually, for most of the observable parameters, the same dilemma exists. More information is in the insert "Bias" on page 344.

[3]"Customer churn" refers to customers that do not come back.

Executing a binning in R is simple enough. However, this is not the whole story: taking care that all bins have a similar size will be conducive for a strong model. Sure, we do not want bins so small that they do not have a statistically significant sample size, but it is not so important that all bins have the same size. Much more important is how different observations in the different bins are.

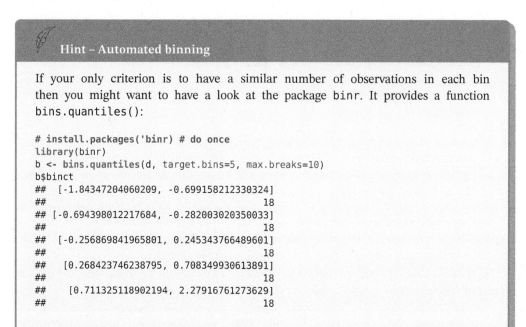

Hint – Automated binning

binr

bins.quantiles()

If your only criterion is to have a similar number of observations in each bin then you might want to have a look at the package `binr`. It provides a function `bins.quantiles()`:

```
# install.packages('binr) # do once
library(binr)
b <- bins.quantiles(d, target.bins=5, max.breaks=10)
b$binct
##  [-1.84347204060209, -0.699158212330324]
##                                       18
## [-0.694398012217684, -0.282003020350033]
##                                       18
##  [-0.256869841965801, 0.245343766489601]
##                                       18
##   [0.268423746238795, 0.708349930613891]
##                                       18
##    [0.711325118902194, 2.27916761273629]
##                                       18
```

In the next section, Section 19.2 *"Tuning the Binning Procedure"* on page 347, we will elaborate on this idea more and will aim to make bins so that they make the model more stable and give it more predictive power.

19.2 Tuning the Binning Procedure

In this section, we will tune our binning so that it will actually make the model both robust and predictive. We will look for patterns in the data and make sure that these patterns are captured.

Let us start from previous dataset generated by the normal distribution and make some further assumptions. The code below generates this data. Look at it carefully and you will see what bias is built in.

```
set.seed(1890)
age <- rlnorm(1000, meanlog = log(40), sdlog = log(1.3))
y <- rep(NA, length(age))
for(n in 1:length(age)) {
  y[n] <- max(0,
              dnorm(age[n], mean= 40, sd=10)
                 + rnorm(1, mean = 0, sd = 10 * dnorm(age[n],
                   mean= 40, sd=15)) * 0.075)
}
y <- y / max(y)
plot(age, y,
     pch = 21, col = "blue", bg = "red",
     xlab = "age",
     ylab = "spending ratio"
     )

# Assume this data is:
#   age            = age of customer
#   spending_ratio = R : = S_n/ (S_{n-1} + S_n)
#                          (zero if both are zero)
#      with S_n the spending in month n
dt <- tibble (age = age, spending_ratio = y)
```

This data could be from an Internet platform and describe how active customers are on our service (game, shop, exchange, etc.). It seems that – just by looking at the data – that our service is particularly attractive and addictive to customers in their thirties and forties. Younger and older people tend to stop being active and get removed from this data after a few months. Of course, we cannot see from this data what the reason is (money, value proposition, presentation, etc.).

Assume now that we want to model this propensity to buy next month, based on the data of this month and the month before. This could lead to a targeted marketing campaign to people that are predicted to spend very little (e.g. offer them a discount maybe).

In this example, we started from fabricating data, so we know very well what patterns are in the data. However, with real data, we need to discover those patterns. Below we show a few tools that can make this work easier and will plot a loess estimation and the histogram.

```
# Leave out NAs (in this example redundant):
d1 <- dt[complete.cases(dt),]

# order() returns sorted indices, so this orders the vector:
d1 <- d1[order(d1$age),]

# Fit a loess:
d1_loess <- loess(spending_ratio ~ age, d1)

# Add predictions:
d1_pred_loess <- predict(d1_loess)

# Plot the results:
par(mfrow=c(1,2))
plot(d1$age, d1$spending_ratio, pch=16,
```

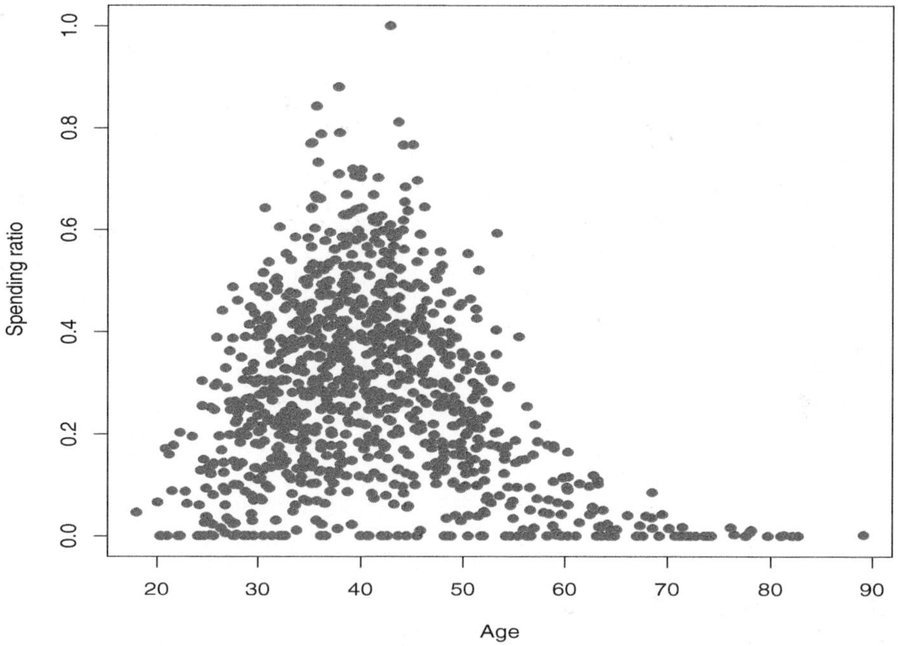

Figure 19.2: *A plot of the fabricated dataset with the spending ratio in function of the age of the customers. The spending ratio is defined as $\frac{S_n}{S_{n-1}+S_n}$, where S_n is the spending in period n. If both spends are 0, then the spending ratio is defined as 0.*

```
    xlab = 'age', ylab = 'spending ratio')
lines(d1$age, d1_pred_loess, lwd = 7, col = 'dodgerblue4')
hist(d1$age, col = 'dodgerblue4', xlab = 'age')
```

```
par(mfrow=c(1,1))
```

> **Further information – ggplot2**
>
> Later, in Chapter 31 *"A Grammar of Graphics with ggplot2"* on page 687, we will introduce the plotting capabilities of the library `ggplot2`. The code that produces a similar visualization with `ggplot2` is here: Section D *"Code Not Shown in the Body of the Book"* on page 839.

From the histogram and loess estimate in Figure 19.3 on page 349, we can see that:

heteroscedastic

- the spending ratio does not simply increase or decrease with age – the relation is non-linear;
- the local volatility is not constant (the dataset is "heteroscedastic");
- we have little young customers and little older ones (it even looks as if some of those have a definite reason to be inactive on our Internet-shop).

So when choosing binning, we need to capture that relationship and make sure that we have bins of the variable age where the spending ratio is high and other bins where the spending ratio is low. Before we do so, we will fit a model without binning so that afterwards we can compare the results.

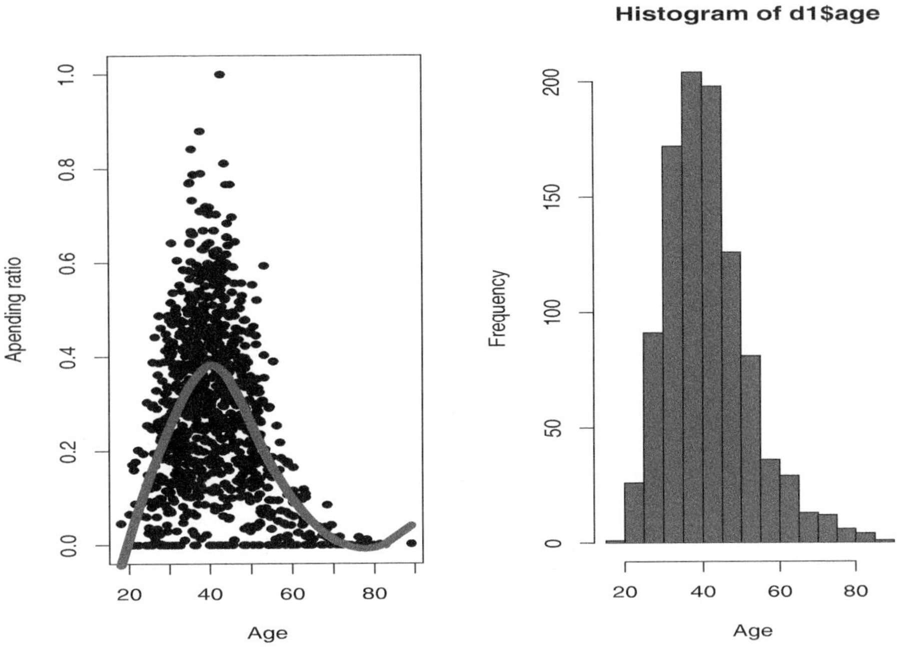

Figure 19.3: *A simple aid to select binning borders is plotting a non-parametric fit (left) and the histogram (right). The information from both plots combined can be used to decide on binning.*

A Model Without Binning

To illustrate the effect of binning, we will use a logistic regression.[4] First, without binning and then with binning.

Fitting the logistic regression worsk as follows:

```
# Fit the logistic regression directly on the data without binning:
lReg1 <- glm(formula = spending_ratio ~ age,
             family = quasibinomial,
             data = dt)

# Investigate the model:
summary(lReg1)
##
## Call:
## glm(formula = spending_ratio ~ age, family = quasibinomial, data = dt)
##
## Deviance Residuals:
##      Min        1Q    Median        3Q       Max
## -0.96344  -0.35725  -0.03202   0.25994   1.62106
##
## Coefficients:
##             Estimate Std. Error t value Pr(>|t|)
## (Intercept) -0.09892    0.11733  -0.843    0.399
## age         -0.02107    0.00282  -7.473 1.71e-13 ***
## ---
## Signif. codes:
## 0 '***' 0.001 '**' 0.01 '*' 0.05 '.' 0.1 ' ' 1
##
```

[4]The logistic regression is explained in Section 22.1 *"Logistic Regression"* on page 388.

```
## (Dispersion parameter for quasibinomial family taken to be 0.1652355)
##
##     Null deviance: 195.63  on 999  degrees of freedom
## Residual deviance: 185.93  on 998  degrees of freedom
## AIC: NA
##
## Number of Fisher Scoring iterations: 4

# Calculate predictions and means square error:
pred1 <- 1 / (1+ exp(-(coef(lReg1)[1] + dt$age * coef(lReg1)[2])))
SE1  <-  (pred1 - dt$spending_ratio)^2
MSE1 <- sum(SE1) / length(SE1)
```

The function `summary(lReg1)` tells us what the estimated parameters are (for example the intercept is −0.099 and the coefficient of the variable age is −0.021). Most importantly it tells us how significant the variables are. For example, the intercept has no stars and the coefficient has 2 stars. The legend for the star is explained below the numbers: two start means that their is only a probability of 1% that the coefficient would be 0 in this model.

A Model with Binning

Inspired by Figure 19.3 on page 349, we can make an educated guess of what bins would make sense. We choose bins that capture the dynamics of our data and make sure to have a bin for values with high spending rations and bins that have low spending ratios.

Now, we will introduce a simple data binning, calculate the logistic mode and show the results:

```
# Bin the variable age:
c <- cut(dt$age, breaks = c(15, 30, 55, 90))

# Check the binning:
table(c)
## c
## (15,30] (30,55] (55,90]
##     118     781     101

# We have one big bucket and two smaller (with the smallest
# more than 10% of our dataset.

lvls <- unique(c)        # find levels
lvls                     # check levels order
## [1] (30,55] (15,30] (55,90]
## Levels: (15,30] (30,55] (55,90]

# Create the tibble (a data-frame also works):
dt <- as_tibble(dt)                                %>%
      mutate(is_L = if_else(age <= 30, 1, 0))  %>%
      mutate(is_H = if_else(age > 55 , 1, 0))

# Fit the logistic regression with is_L and is_H:
# (is_M is not used because it is correlated with the previous)
lReg2 <- glm(formula = spending_ratio ~ is_L + is_H,
             family = quasibinomial, data = dt)

# Investigate the logistic model:
summary(lReg2)
##
## Call:
## glm(formula = spending_ratio ~ is_L + is_H, family = quasibinomial,
##     data = dt)
##
```

```
## Deviance Residuals:
##      Min        1Q    Median       3Q       Max
## -0.88247  -0.31393  -0.03812  0.22173   1.50439
##
## Coefficients:
##             Estimate Std. Error t value Pr(>|t|)
## (Intercept) -0.74222    0.02791 -26.595   <2e-16 ***
## is_L        -0.85871    0.09404  -9.132   <2e-16 ***
## is_H        -2.20235    0.16876 -13.050   <2e-16 ***
## ---
## Signif. codes:
## 0 '***' 0.001 '**' 0.01 '*' 0.05 '.' 0.1 ' ' 1
##
## (Dispersion parameter for quasibinomial family taken to be 0.132909)
##
##     Null deviance: 195.63  on 999  degrees of freedom
## Residual deviance: 144.92  on 997  degrees of freedom
## AIC: NA
##
## Number of Fisher Scoring iterations: 5

# Calculate predictions for our model and calculate MSE:
pred2 <- 1 / (1+ exp(-(coef(lReg2)[1] + dt$is_L * coef(lReg2)[2]
                     + dt$is_H * coef(lReg2)[3])))
SE2 <-  (pred2 - dt$spending_ratio)^2
MSE2 <- sum(SE2) / length(SE2)

# Compare the MSE of the two models:
MSE1
## [1] 0.03294673

MSE2
## [1] 0.02603179
```

We see that indeed the mean square error (MSE) is improved.[5] That is great: our model will make better predictions. However, what is even more important: the significance of our coefficients is up: we have now 3 stars for each *and* the significance of the intercept is up from 0 to 3 stars. That means that the model 2 is much more significant and hence robust to predict the future.

Note – Reasons for Binning

In our example the MSE is improved by binning. However, the main reason why binning is used is that it makes the model less prone to over-fitting and hence should improve the out-of-sample (future) performance. Further, information about this logic and concept can be found in Section 25.4 *"Cross-Validation"* on page 483.

[5]The concept of mean square error is introduced in Section 21.3.1 *"Mean Square Error (MSE)"* on page 384. It is a measure that increases when the differences between the predictions and data are larger – so smaller is better.

19.3 More Complex Cases: Matrix Binning

Now, imagine that we have a dataset where both male and female customers have similar averages on what we want to predict (e.g. spending ratio[6], propensity to have a car accident for insurance claims, etc.) and that there is an underlying reason that causes different relationships between the explained variable for males and for females. Matrix binning can be the answer to this issue.

To illustrate this, we will construct a dataset that has such built-in structure. In the following block of code this data is generated so that males and females have the same average spending ratio, but for males the ratio increases with age and for females it decreases. The code ends with plotting the data in a scatter-plot (see Figure 19.4 on page 353):

```
# Load libraries and define parameters:
library(tidyverse) # provides tibble (only used in next block)
set.seed(1880)     # to make results reproducible
N <- 500           # number of rows

# Ladies first:
# age will function as our x-value:
age_f   <- rlnorm(N, meanlog = log(40), sdlog = log(1.3))
# x is a temporary variable that will become the propensity to buy:
x_f <- abs(age_f + rnorm(N, 0, 20))    # Add noise & keep positive
x_f <- 1 - (x_f - min(x_f)) / max(x_f) # Scale between 0 and 1
x_f <- 0.5 * x_f / mean(x_f)           # Coerce mean to 0.5
# This last step will produce some outliers above 1
x_f[x_f > 1] <- 1   # Coerce those few that are too big to 1

# Then the gentlemen:
age_m   <- rlnorm(N, meanlog = log(40), sdlog = log(1.3))
x_m <- abs(age_m + rnorm(N, 0, 20))    # Add noise & keep positive
x_m <- 1 - (x_m - min(x_m)) / max(x_m) # Scale between 0 and 1
x_m <- 0.5 * x_m / mean(x_m)           # Coerce mean to 0.5
# This last step will produce some outliers above 1
x_m[x_m > 1] <- 1   # Coerce those few that are too big to 1
x_m <- 1 - x_m                         # relation to be increasing

# Rename (p_x is not the gendered propensity to buy)
p_f <- x_f
p_m <- x_m

# We want a double plot, so change plot params & save old values:
oldparams <- par(mfrow=c(1,2))
plot(age_f, p_f,
     pch = 21, col = "blue", bg = "red",
     xlab = "Age",
     ylab = "Spending probability",
     main = "Females"
     )
plot(age_m, p_m,
     pch = 21, col = "blue", bg = "red",
     xlab = "Age",
     ylab = "Spending probability",
     main = "Males"
     )
```

[6]The concept "spending ratio" is introduced in Section 19.2 *"Tuning the Binning Procedure"* on page 347. We will in this section construct a new dataset that is inspired on the one used in the aforementioned section.

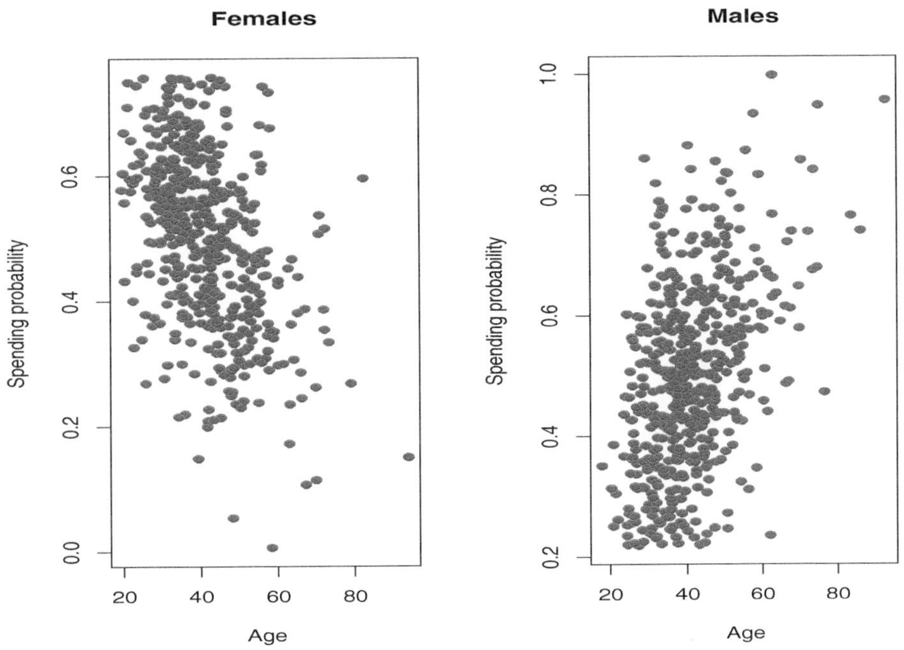

Figure 19.4: *The underlying relation between spending probability for females (left) and males (right) in our fabricated example.*

```
par(oldparams)    # Reset the plot parameters after plotting
```

In Figure 19.4, we can see that the data indeed has the properties that we were looking for. Age is somehow normally distributed, but there are more young than old people, but also the propensity to buy (or spending probability) is decreasing with age for females and for males, it works opposite.

This first step was only to prepare the data and show what is exactly inside. In the next step, we will merge the data, and assume that this merged data set is what we got to work with. The following block of code will do this and then plot the histogram for the all observations (combined males and females) in Figure 19.5 on page 354:

```
# Now, we merge the data and consider this as our input-data:
tf <- tibble("age" = age_f, "sex" = "F", "is_good" = p_f)
tm <- tibble("age" = age_m, "sex" = "M", "is_good" = p_m)
t  <- full_join(tf, tm, by = c("age", "sex", "is_good"))

# Change plot parameters and capture old values:
oldparams <- par(mfrow=c(1,2))
plot(t$age, t$is_good,
     pch  = 21, col = "black", bg = "khaki3",
     xlab = "Age",
     ylab = "Spending probability",
     main = "Dependence on age"
     )
fct_sex <- factor(t$sex, levels=c("F","M"), labels=c(0,1))
t$sexM  <- as.numeric(fct_sex)    # store for later use
plot(fct_sex, t$is_good,
     col="khaki3",
     main="Dependence on sex",
     xlab="Female       Male")
```

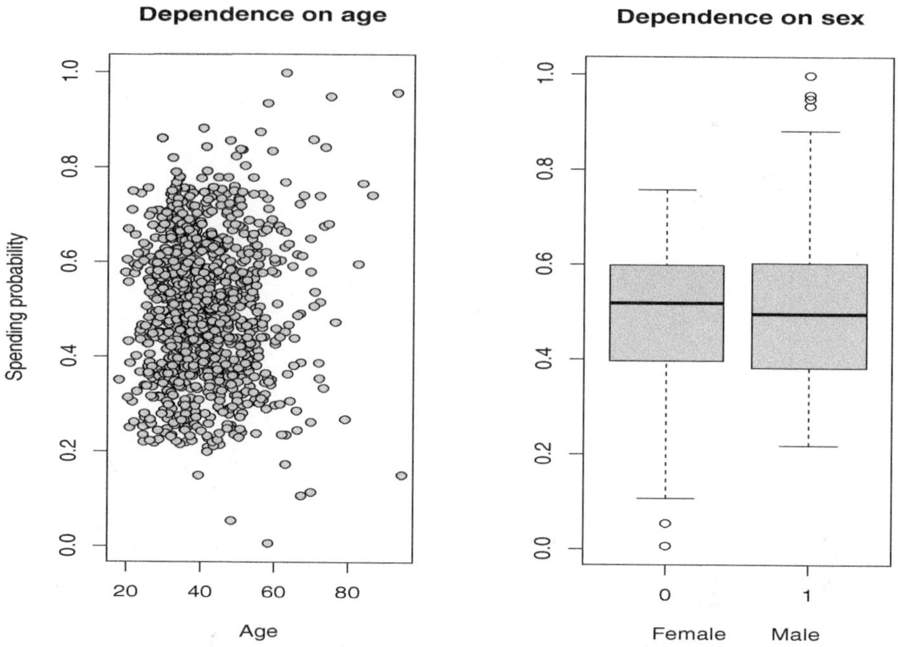

Figure 19.5: *The dataset "as received from the customer service department" does not show any clear relationship between Age or Sex and the variable that we want to explain: the spending ratio.*

```
par(oldparams)    # Reset the plot parameters
```

Now, we are in a situation that is similar as to the starting point of a real situation: we have got data and we need to investigate what is inside. To do this, we will follow the approach that we used earlier in this chapter, and more in particular the simple methods proposed in Section 19.2 *"Tuning the Binning Procedure"* on page 347. The following code does this, and plots the results in Figure 19.6 on page 355:

```
d1 <- t[complete.cases(t),]

d1 <- d1[order(d1$age),]
d1_age_loess <- loess(is_good ~ age, d1)
d1_age_pred_loess <- predict(d1_age_loess)

d1 <- d1[order(d1$sexM),]
d1_sex_loess <- loess(is_good ~ sexM, d1)
d1_sex_pred_loess <- predict(d1_sex_loess)

# Plot the results:
par(mfrow=c(2,2))
d1 <- d1[order(d1$age),]
plot(d1$age, d1$is_good, pch=16,
     xlab = 'Age', ylab = 'Spending probability')
lines(d1$age, d1_age_pred_loess, lwd = 7, col = 'dodgerblue4')
hist(d1$age, col = 'khaki3', xlab = 'age')

d1 <- d1[order(d1$sexM),]
plot(d1$sexM, d1$is_good, pch=16,
     xlab = 'Gender', ylab = 'Spending probability')
lines(d1$sexM, d1_sex_pred_loess, lwd = 7, col = 'dodgerblue4')
hist(d1$sexM, col = 'khaki3', xlab = 'gender')
```

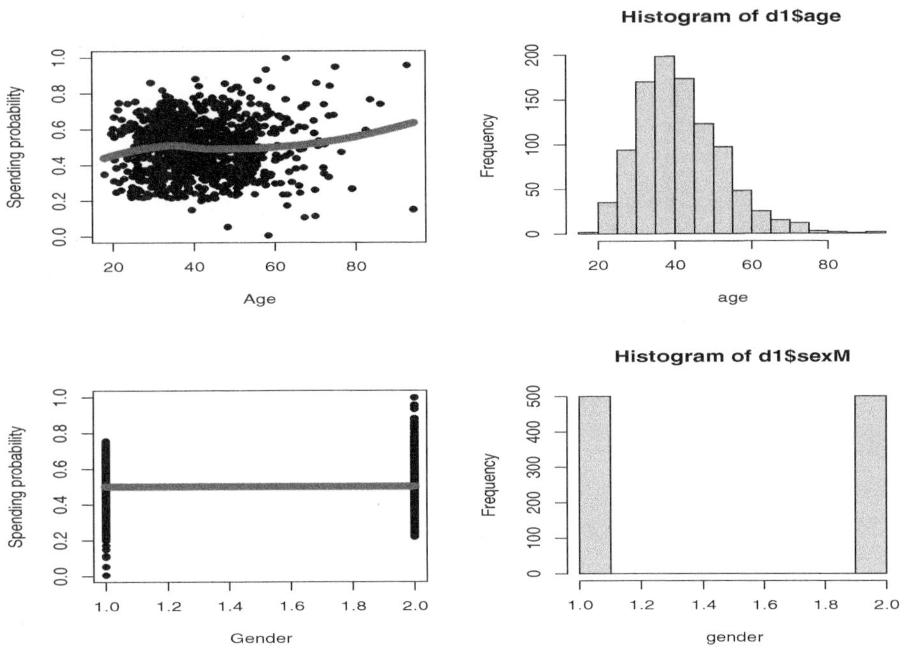

Figure 19.6: *The data does not reveal much patterns for any of the variables (Gender and Age).*

```
par(mfrow=c(1,1))
```

The data – as prepared in the aforementioned code – has a particular relation between the dependent variable and one of the exploratory variables. The spending probability for males and females is on average the same, and the averages for all age groups are comparable. On the surface of it, nothing significant happens. However, for males it will increase with age and for females the opposite happens.

In the case of the logistic regression[7], normalization matters mainly for interpretability of the model. This is not our major concern here so we will skip it. And try directly to fit a first naive model.

```
# Note that we can feed "sex" into the model and it will create
# for us a variable "sexM" (meaning the same as ours)
# To avoid this confusion, we put in our own variable.
regr1 <- glm(formula = is_good ~ age + sexM,
             family = quasibinomial,
             data = t)

# assess the model:
summary(regr1)
##
## Call:
## glm(formula = is_good ~ age + sexM, family = quasibinomial, data = t)
##
## Deviance Residuals:
##      Min        1Q    Median        3Q       Max
## -1.15510  -0.21981   0.00556   0.20597   1.14979
##
```

[7]More information about the model itself is in Section 22.1 *"Logistic Regression"* on page 388.

```
## Coefficients:
##               Estimate Std. Error t value Pr(>|t|)
## (Intercept) -6.990e-02  9.133e-02  -0.765    0.444
## age          1.684e-03  1.680e-03   1.002    0.316
## sexM         6.015e-05  3.730e-02   0.002    0.999
##
## (Dispersion parameter for quasibinomial family taken to be 0.08694359)
##
##     Null deviance: 91.316  on 999  degrees of freedom
## Residual deviance: 91.229  on 997  degrees of freedom
## AIC: NA
##
## Number of Fisher Scoring iterations: 3

pred1 <- 1 / (1+ exp(-(coef(regr1)[1] + t$age * coef(regr1)[2]
                     + t$sexM * coef(regr1)[3])))
SE1 <-  (pred1 - t$is_good)^2
MSE1 <- sum(SE1) / length(SE1)
```

As expected, this did not go well. From the explanatory variables, only the intercept has three stars. Age got one star, but the value is really small. The MSE is 0.022. This is not good for a variable that has a range roughly between 0 and 2.5.

Now, we will try a binning method that combines Age and Sex. This new variables should be able to reveal the interactions between the variables.

```
# 1. Check the potential cut
c <- cut(t$age, breaks = c(min(t$age), 35, 55, max(t$age)))
table(c)
## c
## (17.9,35]   (35,55] (55,94.2]
##       300       591       108

# 2. Create the matrix variables
t <- as_tibble(t)                                            %>%
    mutate(is_LF = if_else((age <= 35) & (sex == "F"), 1L, 0L)) %>%
    mutate(is_HF = if_else((age >  50) & (sex == "F"), 1L, 0L)) %>%
    mutate(is_LM = if_else((age <= 35) & (sex == "M"), 1L, 0L)) %>%
    mutate(is_HM = if_else((age >  50) & (sex == "M"), 1L, 0L)) %>%
    print
## # A tibble: 1,000 x 8
##      age sex   is_good  sexM is_LF is_HF is_LM is_HM
##    <dbl> <chr>   <dbl> <dbl> <int> <int> <int> <int>
##  1  44.3 F       0.564     1     0     0     0     0
##  2  38.3 F       0.636     1     0     0     0     0
##  3  38.9 F       0.552     1     0     0     0     0
##  4  58.5 F       0.351     1     0     1     0     0
##  5  31.5 F       0.623     1     1     0     0     0
##  6  48.4 F       0.487     1     0     0     0     0
##  7  28.9 F       0.552     1     1     0     0     0
##  8  29.9 F       0.493     1     1     0     0     0
##  9  30.1 F       0.549     1     1     0     0     0
## 10  51.1 F       0.241     1     0     1     0     0
## # ... with 990 more rows

# 3. Check if the final bins aren't too small
t[,5:8] %>% map_int(sum)
## is_LF is_HF is_LM is_HM
##   154   104   147   101
```

map_int()

Now, we have created four variables that group a combination of age and sex. There are two groups that we could have calculated too: the middle-aged men and women. However, since these are correlated to the existing categories, it is not wise to add them.

In the next, step we will fit a linear regression on the existing variables:

```
regr2 <- glm(formula = is_good ~ is_LF + is_HF + is_LM + is_HM,
             family = quasibinomial,
             data = t)

# assess the model:
summary(regr2)
##
## Call:
## glm(formula = is_good ~ is_LF + is_HF + is_LM + is_HM, family = quasibinomial,
##     data = t)
##
## Deviance Residuals:
##      Min        1Q    Median        3Q       Max
## -0.98606  -0.18858  -0.00424   0.18159   0.98651
##
## Coefficients:
##             Estimate Std. Error t value Pr(>|t|)
## (Intercept) -0.01768    0.02467  -0.716    0.474
## is_LF        0.27945    0.05094   5.485 5.23e-08 ***
## is_HF       -0.35564    0.06002  -5.925 4.29e-09 ***
## is_LM       -0.22844    0.05183  -4.408 1.16e-05 ***
## is_HM        0.45028    0.06106   7.375 3.46e-13 ***
## ---
## Signif. codes:
## 0 '***' 0.001 '**' 0.01 '*' 0.05 '.' 0.1 ' ' 1
##
## (Dispersion parameter for quasibinomial family taken to be 0.07518569)
##
##     Null deviance: 91.316  on 999  degrees of freedom
## Residual deviance: 78.274  on 995  degrees of freedom
## AIC: NA
##
## Number of Fisher Scoring iterations: 3

pred2 <- 1 / (1+ exp(-(coef(regr2)[1] +
                       + t$is_LF * coef(regr2)[2]
                       + t$is_HF * coef(regr2)[3]
                       + t$is_LM * coef(regr2)[4]
                       + t$is_HM * coef(regr2)[5]
                       )))
SE2 <-  (pred2 - t$is_good)^2
MSE2 <- sum(SE2) / length(SE2)
```

Finally, we also note that the MSE has improved too:

```
MSE1
## [1] 0.02166756

MSE2
## [1] 0.01844601
```

Question #14 Binary dependent variables

In many cases, the dependent variable will be binary (0 or 1, "yes" or "no"). That means that we are trying to model a yes/no decision. For example, 1 can be a customer that defaulted on a loan, a customer to receive a special offer, etc.

```
t <- mutate(t, "is_good" = if_else(is_good >= 0.5, 1L, 0L))
```

Remake model 1 (logistic regression in function of age and sexM) and model 2 (logistic regression for the variables is_LF, is_HF, is_LM, and is_HM). What does this change? Are the conclusions different?

Question #15 Think outside the box

In this particular case, what other approach would you suggest?

Question #16 Think deeper

Why do we use logistic regression instead of linear regression in this chapter?

19.4 Weight of Evidence and Information Value

We tried to make our binning as good as possible: we made sure that no bucket is too small, that we are able to capture the important dynamics in the data, etc. This is essential, but it would be better if somehow we could quantify our success.

Weight of evidence (WOE) and information value (IV) do exactly this in a simple and intuitive way. The idea is to see if proportions of "good" outcomes and "bad" outcomes is really that different in the different bins.[8] This result in a value for WOE or IV and we can compare this value with another one obtained from another possible binning.

19.4.1 Weight of Evidence (WOE)

WOE
weight of evidence

For each bin i of variable j (or for each binary variable j) is defined as

$$WOE_{ij} = \log \left\{ \frac{\frac{\#G_{ij}}{\#G}}{\frac{\#B_{ij}}{\#B}} \right\}$$

Where $\#G_{ij}$ is the number of "good observations" (binary variable is 1) in bin i for variable j. $\#G$ is the number of good observations for the whole dataset, and "B" refers to the "bad observations" (i.e. where the dependent variable is 0).

This makes WOE a measure of predicting power of a binned variable.

WOE will range between 0 and infinity. It will be zero for a bin that has no "good" observations and it will be infinite if a bin has no "bad" variables.

19.4.2 Information Value (IV)

It is easier to work with a number that ranges from zero to one instead of between zero and infinity. Therefore, it is useful to define "information value."

IV
information value

The information value of bin i for variable j is defined as

$$IV_{ij} = \left(\frac{\#G_{ij}}{\#G} - \frac{\#B_{ij}}{\#B} \right) .WOE_{ij}$$

It appears that regardless the functional domain (e.g. banking, oncology, geology, etc.) similar values of information value carry a similar message. The following rule of thumb can be used in most cases:

IV	Predictability
< 0.02	Not predictive
$0.02 - 0.3$	Weak
$0.1 - 0.3$	Medium
$0.3 - 0.5$	Strong
> 0.5	Suspicious

Table 19.1: *Different levels of information value and their commonly accepted interpretation – which works good in the environment of credit data for example.*

19.4.3 WOE and IV in R

We build further on the example developed in Section 19.3 *"More Complex Cases: Matrix Binning"* on page 352. We will use this example further to illustrate how Information Value works. We choose the package `InformationValue` to get the job done elegantly.

`InformationValue`

[8] In many cases, there is indeed something like a good or a bad outcome, but in fact, these are only other words for the dependent variable being 1 (good) or 0 (bad).

kable()

> **Further information – Other packages used**
>
> Just for the sake of illustration and presentation purposes, we use knitr to present any tables via its function kable(). More information is in Section 33 *"knitr and LaTeX"* on page 703.
> We also use the package tidyverse, from which we use the pipe operator. More information is in Chapter 7 *"Tidy R with the Tidyverse"* on page 121.

In this example, we will use the dataset that we have prepared in Section 19.3 *"More Complex Cases: Matrix Binning"* on page 352:

```
# We start from this dataset used in previous section:
print(t)
## # A tibble: 1,000 x 8
##      age sex   is_good  sexM is_LF is_HF is_LM is_HM
##    <dbl> <chr>   <int> <dbl> <int> <int> <int> <int>
##  1  44.3 F           1     1     0     0     0     0
##  2  38.3 F           1     1     0     0     0     0
##  3  38.9 F           1     1     0     0     0     0
##  4  58.5 F           0     1     0     1     0     0
##  5  31.5 F           1     1     1     0     0     0
##  6  48.4 F           0     1     0     0     0     0
##  7  28.9 F           1     1     1     0     0     0
##  8  29.9 F           0     1     1     0     0     0
##  9  30.1 F           1     1     1     0     0     0
## 10  51.1 F           0     1     0     1     0     0
## # ... with 990 more rows
```

This dataset contains a specific property where males and females have a similar propensity to spend as an average population. However, this propensity is decreasing for females and increasing for males. This situation is particularly difficult, since at first glance the variables Age and Sex will not be predictive at all. We need to look at the interactions between the variables in order to find the underlying relations.

Now, that we have data, we can load the package InformationValue, create a weight of evidence table and calculate the information value for a given variable:

```
#install.packages("InformationValue")
library(InformationValue)

WOETable(X = factor(t$sexM), Y = t$is_good, valueOfGood=1) %>%
  knitr::kable(format.args = list(big.mark = " ", digits=2))
```

CAT	GOODS	BADS	TOTAL	PCT_G	PCT_B	WOE	IV
1	267	233	500	0.52	0.48	0.088	0.0039
2	245	255	500	0.48	0.52	-0.088	0.0039

```
## also functions WOE() and IV(), e.g.
# IV of a categorical variable is the sum of IV of its categories
IV(X = factor(t$sexM), Y = t$is_good, valueOfGood=1)
## [1] 0.007757952
## attr(,"howgood")
## [1] "Not Predictive"
```

Digression – The function kable()

The function `kable` does the formatting of the table. It makes a table look a lot nicer. This function is part of the package `knitr`... and it only makes sense if you are compiling the document with LaTeX. More information is in Section 33 *"knitr and LaTeX"* on page 703.

Indeed, just dividing the data based on gender is not sufficient, and it does not work. Using our variables, such as `is_LF` (female from the lower age group), which combine the information of age and gender should work better.

```
WOETable(X = factor(t$is_LF), Y = t$is_good, valueOfGood=1) %>%
  knitr::kable(digits=2)
```

CAT	GOODS	BADS	TOTAL	PCT_G	PCT_B	WOE	IV
0	396	450	846	0.77	0.92	-0.18	0.03
1	116	38	154	0.23	0.08	1.07	0.16

```
IV(X = factor(t$is_LF), Y = t$is_good, valueOfGood=1)
## [1] 0.1849507
## attr(,"howgood")
## [1] "Highly Predictive"
```

Question #17

Consider the dataset `mtcars` and investigate if the gearbox type (the variable `am` is a good predictor for the layout of the motor (the variable `vs`, V-motor or not). Do this by using WOE and IV.

20

Factoring Analysis and Principle Components

Factor analysis and principal component analysis are mathematically related: they both rely on calculating eigenvectors (on a correlation matrix or on a covariance matrix of normalized data), both are data reduction techniques that help to reduced the dimensionality of the data and outputs will look very much the same. Despite all these similarities, they solve a different problem: principal component analysis (PCA) is a linear combination of variables (so that the principal components (PCs) are orthogonal); factor analysis is a measurement model of a latent variable.

PCA
principal component analysis
PCs
principal components
factor analysis

The Big R-Book: From Data Science to Learning Machines and Big Data, First Edition. Philippe J.S. De Brouwer.

Companion Website: www.wiley.com/go/De Brouwer/The Big R-Book

20.1 Principle Components Analysis (PCA)

PCA is a data reduction technique that calculates new variables from the set of the measured variables. These new variables are linear combinations (think of it as a weighted average) of those measured variables (columns). The index-variables that result from this process are called "components."

The process relies on finding eigenvalues and their eigenvectors and – unless the covariance matrix is singular – this results in as many variables as the datasets has measured variables. However, a PCA will also give us the tools to decide how much of those components are really necessary. So we can find an optimal number of components, which implies an optimal choice of measured variables for each component, and their optimal weights in those principal components.

The results is that any regression or other model can be build on this – limited – number of components, so it will be less complex and more stable (since all variables are orthogonal) and one can even hope that it is less over-fit since we left out the components that explain only a small portion of the variance in the dataset. As a rule of thumb – but that depends on your data and purpose – one tries to capture at least 85% of the variance.

The function `princomp` of the package `stats` allows to execute an un-rotated principal component analyses (PCA).

The following code executes the PCA analysis, shows how to access the details of the principal components, and produces the plots Figure 20.1 on page 366 (see the line with `plot(...)`) and Figure 20.2 on page 367 (see the line with `biplot(...)`):

princomp()

```
# PCA: extracting PCs from the correlation matrix
fit <- princomp(mtcars, cor=TRUE)

summary(fit)       # print the variance explained by PC
## Importance of components:
##                           Comp.1    Comp.2     Comp.3
## Standard deviation     2.7291248 1.6282264 0.81487601
## Proportion of Variance 0.6206769 0.2209268 0.05533524
## Cumulative Proportion  0.6206769 0.8416036 0.89693885
##                            Comp.4     Comp.5     Comp.6
## Standard deviation     0.53912166 0.47970508 0.47112904
## Proportion of Variance 0.02422101 0.01917641 0.01849688
## Cumulative Proportion  0.92115987 0.94033628 0.95883316
##                            Comp.7     Comp.8     Comp.9
## Standard deviation     0.41394996 0.35875849 0.27757334
## Proportion of Variance 0.01427955 0.01072564 0.00642058
## Cumulative Proportion  0.97311271 0.98383835 0.99025893
##                            Comp.10    Comp.11     Comp.12
## Standard deviation     0.231111809 0.20918039 0.140441337
## Proportion of Variance 0.004451056 0.00364637 0.001643647
## Cumulative Proportion  0.994709983 0.99835635 1.000000000

loadings(fit)     # show PC loadings
##
## Loadings:
##      Comp.1 Comp.2 Comp.3 Comp.4 Comp.5 Comp.6 Comp.7
## mpg   0.345        -0.208                      -0.527
## cyl  -0.349        -0.219        -0.156  0.158 -0.112
## disp -0.347               0.265  0.279  0.231 -0.399
## hp   -0.310 -0.252       -0.126  0.168  0.520 -0.163
## drat  0.275 -0.272  0.208  0.722 -0.433  0.278
## wt   -0.330  0.141  0.313  0.208  0.121 -0.229 -0.357
## qsec  0.184  0.465  0.408               -0.266 -0.307
## vs    0.284  0.235  0.440 -0.182  0.275  0.568  0.236
```

```
## am    0.221 -0.427 -0.145  0.221  0.579 -0.251  0.117
## gear  0.195 -0.460  0.291 -0.231  0.235        -0.279
## carb -0.202 -0.415  0.474 -0.291 -0.352 -0.202
## l    -0.340          0.249  0.341  0.270 -0.104  0.390
##      Comp.8 Comp.9 Comp.10 Comp.11 Comp.12
## mpg  -0.307  0.233 -0.358   0.447  -0.273
## cyl  -0.139         0.750   0.372  -0.186
## disp  0.112  0.198                  0.672
## hp   -0.179 -0.576 -0.214  -0.165  -0.230
## drat                0.118
## wt           0.360         -0.434  -0.458
## qsec -0.158 -0.529  0.219   0.186   0.158
## vs   -0.150  0.359  0.177
## am   -0.448         0.253  -0.122
## gear  0.645         0.168   0.146
## carb -0.408  0.170 -0.118   0.109   0.302
## l                  -0.241   0.599  -0.218
##
##                Comp.1 Comp.2 Comp.3 Comp.4 Comp.5 Comp.6
## SS loadings     1.000  1.000  1.000  1.000  1.000  1.000
## Proportion Var  0.083  0.083  0.083  0.083  0.083  0.083
## Cumulative Var  0.083  0.167  0.250  0.333  0.417  0.500
##                Comp.7 Comp.8 Comp.9 Comp.10 Comp.11
## SS loadings     1.000  1.000  1.000   1.000   1.000
## Proportion Var  0.083  0.083  0.083   0.083   0.083
## Cumulative Var  0.583  0.667  0.750   0.833   0.917
##                Comp.12
## SS loadings      1.000
## Proportion Var   0.083
## Cumulative Var   1.000
```

```
head(fit$scores)  # the first principal components
##                       Comp.1    Comp.2      Comp.3
## Mazda RX4          0.76571314 -1.736029 -0.59718129
## Mazda RX4 Wag      0.73700298 -1.550604 -0.38460216
## Datsun 710         2.82972458  0.161250 -0.08771846
## Hornet 4 Drive     0.44528713  2.358261 -0.24262379
## Hornet Sportabout -1.83426210  0.735143 -1.31278386
## Valiant            0.02193262  2.787107  0.15132829
##                      Comp.4     Comp.5      Comp.6
## Mazda RX4          0.1392233 -0.4196776 -0.85595790
## Mazda RX4 Wag      0.2102023 -0.3684001 -1.00145261
## Datsun 710         0.1831589  0.8400435  0.02793731
## Hornet 4 Drive    -0.5332976  0.3618086  0.46352938
## Hornet Sportabout -0.1200421 -0.2106460  0.33627472
## Valiant           -0.7561817  0.7201174 -0.08223269
##                        Comp.7      Comp.8      Comp.9
## Mazda RX4          0.33057594 -0.28510214  0.14936331
## Mazda RX4 Wag      0.13843506 -0.35391932  0.07649975
## Datsun 710         0.65001018  0.05268582 -0.13152126
## Hornet 4 Drive    -0.06403088 -0.08107365  0.22414184
## Hornet Sportabout -0.42919573  0.10494889 -0.02173963
## Valiant            0.36191440  0.01034000 -0.03188577
##                      Comp.10     Comp.11     Comp.12
## Mazda RX4         0.10705038 -0.05404947  0.21089835
## Mazda RX4 Wag     0.19163106 -0.10975408  0.13993575
## Datsun 710        0.09844544 -0.31450104 -0.03079258
## Hornet 4 Drive    0.10580202 -0.08940771  0.20026656
## Hornet Sportabout 0.06324138 -0.01014054  0.18581313
## Valiant           0.24864767 -0.04265171  0.04261024
```

```
# plot the loadings (output see figure):
plot(fit,type="b", col='khaki3')
```

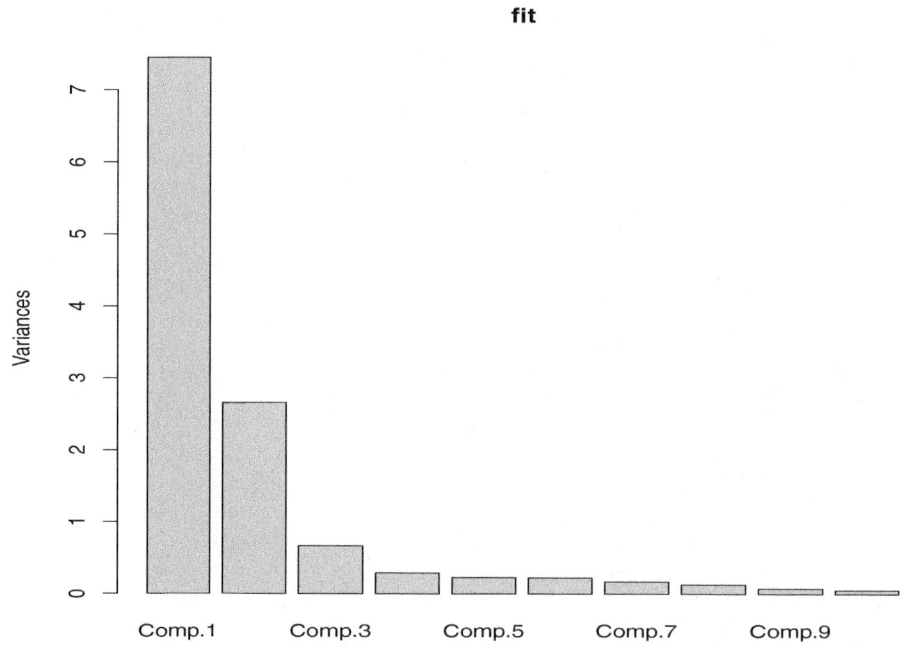

Figure 20.1: *A visualization of the loadings of the principal components of the dataset mtcars.*

```
# show the biplot:
biplot(fit)
```

Hint – Fine tuning the function princomp()

In case we did not normalize (scale) our data, all variables will have a different range, and hence, and the PCA method will be biased towards the variables with the largest range. That is why we have set the option `cor = TRUE`.
If data is normalized (all ranges are similar or the same), then principal components can be based on the covariance matrix and the option can be set to `FALSE`. Also note that in case there are constant variables, the correlation matrix is not defined and cannot be used. Further, the `covmat`-option can be used to enter a correlation or covariance matrix directly. If you do so, be sure to specify the option `n.obs`.

The aforementioned code is only a few short lines, but a lot can be learned from its output. For example, note that:

- `summary(fit)` shows that the first two principal components explain about 85% of the variance. That is really high, adding a third PC brings this to almost 90%. This means that while our dataset has 11 columns, there are in fact only 3 truly independent underlying dimensions. To some extend other dimensions also exist, but they are less relevant.
- `biplot(fit)` shows the structure of the dataset as projected in the plane spanned by the two first principal components. This 2D project explains 85% of the variance – in our case – and hence can be expected to include most of the information of the dataset. Note for example that

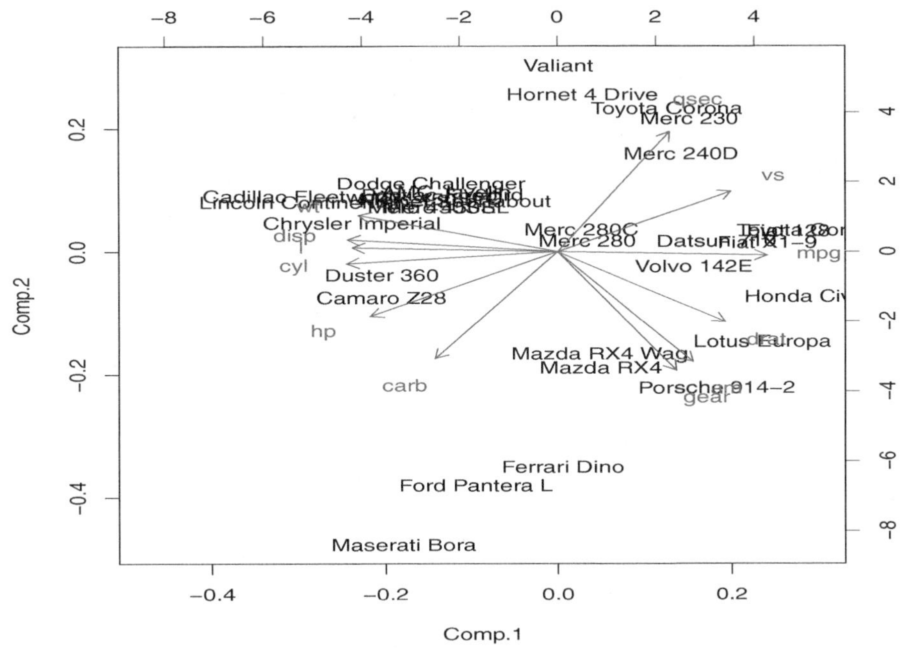

Figure 20.2: *The biplot of the dataset mtcars: all observation and dimensions projected in the plane span by the first two principal components.*

- the cars group in four clusters: this means that there are four underlying car concepts;
- in the Northwest part of the plot, we see the cluster that represents "muscle cars" and it is very well aligned with the dimension of weight, displacement, cylinders, and horse power;
- dimensions such as fuel consumption, line motor, and rear axle ratio seem to be opposite to the previous groups of dimensions;
- indeed, the output of `loadings(fit)` confirms this finding.

• A lot of more detailed observations can be made.

Hint – Executing PCA before fitting a model

Since a lot of variance is explained in the first PCs, it is a good idea to fit any model (such as a logistic regression or any other model) not directly on `mtcars`, but rather on its principal components.
This will make the model more stable and one can expect the model to perform better out of sample, the only cost is the loss of transparency of the model.[a]

[a] If the first principal components can be summarizes as a certain concept, then there is little to no loss of transparency. However, usually, the PCs are composed of too many variables and do not summarize as one concept.

20.2 Factor Analysis

A factor analysis also will lead to data reduction, but it answers a fundamentally different question. Factor analysis is a model that tries to identify a "latent variable." The assumption is that this latent variable cannot be directly measured with a single variable. The latent variable is made visible through the correlations that it causes in the observed variables.

For example, factor analyses can help to determine the number of dimensions of the human character. What is really an independent dimension in our character? How to name it? In a commercial setting, it can try to measure something like customer satisfaction, that is in the questionnaire comprised of many variables rather than one question "how satisfied are you?."

The idea is that one latent factor that summarises some of our initial parameters might exisist. For example, in psychology, the dimensions "agreeability" seems to be a fundamentally independent dimension that build up our character. This underlying latent factor cannot be asked directly. Rather we will ask to rate on a scale of one to five questions such as "When someone tells me to clean my room, I'm inclined to do so right away.", "When someone of authority ask me to do something, this is motivating for me," etc. The underlying, latent factor agreeability will cause answers to those questions to be correlated. Factor analysis aims to identify the underlying factor.

Compared to PCA one could say that the assumed causality is opposite. Factor analysis assumes that the latent factor is the cause, while in PCA, the component is a result of the observed variables.

The function `factanal()` of the package `stats` can be used to execute a maximum likelihood factor analysis on a covariance matrix of on the raw data. This is illustrated in the next code segment, which produces the figure that follows the code.

```
# Maximum Likelihood Factor Analysis

# Extracting 3 factors with varimax rotation:
fit <- factanal(mtcars, 3, rotation = "varimax")
print(fit, digits = 2, cutoff = .3, sort = TRUE)
##
## Call:
## factanal(x = mtcars, factors = 3, rotation = "varimax")
##
## Uniquenesses:
##  mpg  cyl disp   hp drat   wt qsec   vs   am gear carb
## 0.11 0.06 0.09 0.13 0.29 0.06 0.06 0.22 0.21 0.12 0.18
##    l
## 0.11
##
## Loadings:
##      Factor1 Factor2 Factor3
## mpg   0.60   -0.45   -0.58
## disp -0.69    0.51    0.42
## drat  0.80
## wt   -0.72            0.61
## am    0.87
## gear  0.93
## cyl  -0.59    0.69    0.34
## hp            0.70    0.56
## qsec         -0.94
## vs           -0.79
## carb          0.54    0.71
## l    -0.59    0.40    0.62
##
```

```
##                 Factor1 Factor2 Factor3
## SS loadings       4.49    3.50    2.36
## Proportion Var    0.37    0.29    0.20
## Cumulative Var    0.37    0.67    0.86
##
## Test of the hypothesis that 3 factors are sufficient.
## The chi square statistic is 44.41 on 33 degrees of freedom.
## The p-value is 0.0887

# plot factor 1 by factor 2
load <- fit$loadings[,1:2]
plot(load, type = "n")                          # plot the loads
text(load, labels = colnames(mtcars),
     cex = 1.75, col = 'blue')                  # add variable names
```

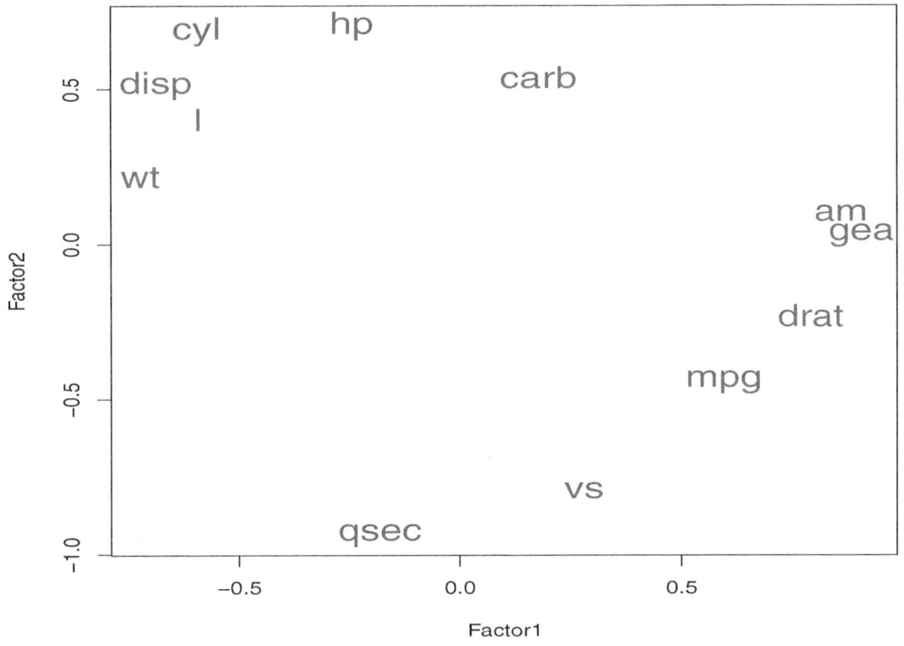

Hint – Customisation of factanal()

The options for the parameter `rotation` include "varimax," "promax," and "none." The option `scores` accepts the values "regression" or "Bartlett" and determines the method to calculate the factor scores. The option `covmat` can be used to specify directly a correlation or covariance matrix directly. When entering a covariance matrix, do not forget to provide the option `n.obs`.

A crucial decision in exploratory factor analysis is how many factors to extract. The nFactors package offers a suite of functions to aid in this decision. The code below will demonstrate the use of this package and finally plot the Scree test in Figure 20.3 on page 370. First, we load the package:

```
# load the library nFactors:
library(nFactors)
```

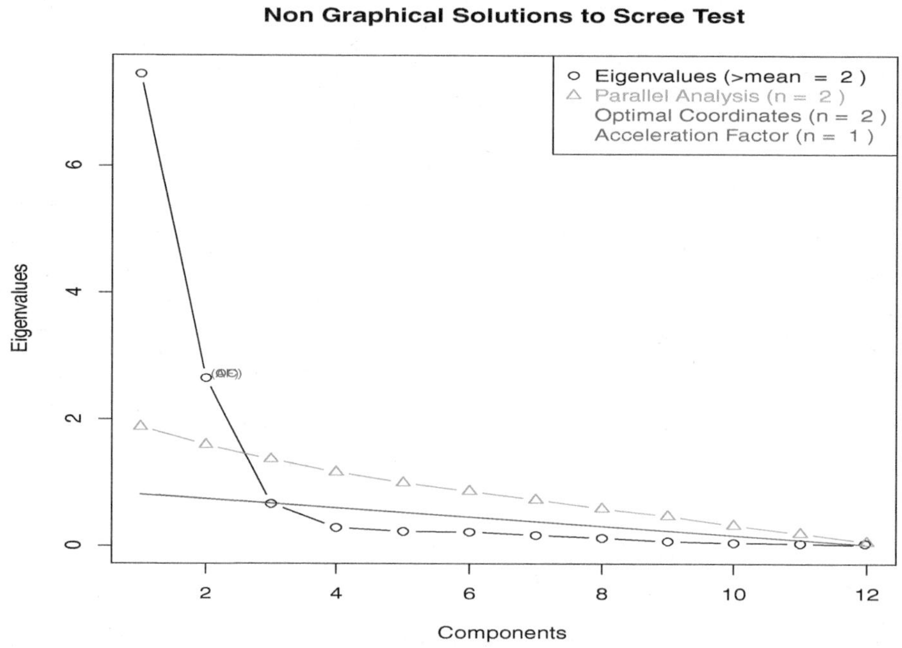

Figure 20.3: *Visual aids to select the optimal number of factors.*

Then we can perform the analysis, get the optimal number of factors, and plot a visualisation:

parallel()

```
# Get the eigenvectors:
eigV <- eigen(cor(mtcars))


# Get mean and selected quantile of the distribution of eigen-
# values of correlation or a covariance matrices of standardized
# normally distributed variables:
aPar <- parallel(subject = nrow(mtcars),var = ncol(mtcars),
                 rep = 100, cent = 0.05)

# Get the optimal number of factors analysis:
nScr <- nScree(x = eigV$values, aparallel = aPar$eigen$qevpea)

# See the result
nScr
##    noc naf nparallel nkaiser
## 1   2   1         2       2

# and plot it.
plotnScree(nScr)
```

We will not discuss those rules here, but – for your reference – the rules used are the Kaiser-Guttman rule, parallel analysis, and the nScree does the nScree test to find an optimal number of factors. The documentation of the nScree() function has all references.

nScree()

Further information – More tools for factor analysis

The FactoMineR package offers a large number of additional functions for exploratory factor analysis. This includes the use of both quantitative and qualitative variables, as well as the inclusion of supplementary variables and observations. Here is an example of the types of graphs that you can create with this package – note that the output is not shown.

```
# PCA Variable Factor Map
library(FactoMineR)

# PCA will generate two plots as side effect (not shown here)
result <- PCA(mtcars)
```

PART V

Modelling

♣ 21 ♣

Regression Models

21.1 Linear Regression

Linear Regression

With a linear regression we try to estimate an unknown variable y based on a known variable x and some constants (a and b). Its form is

regression – linear
linear regression

$$y = ax + b$$

To illustrate the linear regression, we will use data of the dataset `survey` from the package `MASS`. This dataset contains some physical measures, such as the span of the hand (variable `Wr.Hand`), the height of the person (variable `Height`, and the gender (variable `Sex`). Let us first illustrate the data by plotting the hand size in function of the height (results in Figure 21.1 on page 376):

```
library(MASS)

# Explore the data:
plot(survey$Height, survey$Wr.Hnd)
```

The package `stats`, that is loaded at start of R, has a function called `lm()` that can handle a linear regression. Its use is not difficult: we provide a data-frame (argument `data`) and a formula (argument `formula`). The formula has the shape of

lm()

```
< dependent variable> tilde <sum of independent variables>
```

```
# Create the model:
lm1 <- lm (formula = Wr.Hnd ~ Height, data = survey)
summary(lm1)
##
## Call:
## lm(formula = Wr.Hnd ~ Height, data = survey)
##
## Residuals:
```

The Big R-Book: From Data Science to Learning Machines and Big Data, First Edition. Philippe J.S. De Brouwer.

Companion Website: www.wiley.com/go/De Brouwer/The Big R-Book

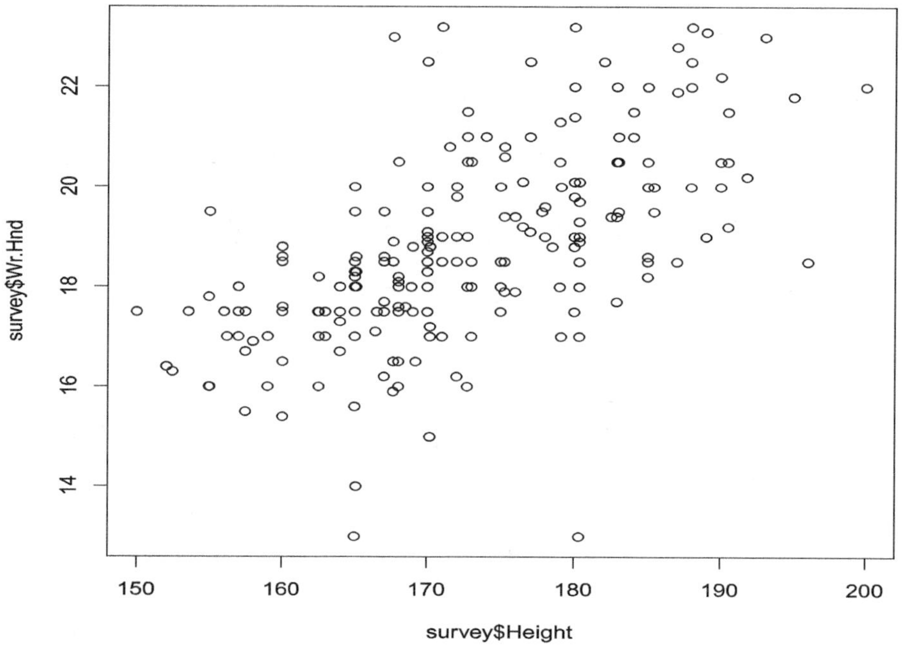

Figure 21.1: *A scatter-plot generated by the line "plot(survey$Height, survey$Wr.Hnd)."*

```
##      Min      1Q  Median      3Q     Max
## -6.6698 -0.7914 -0.0051  0.9147  4.8020
##
## Coefficients:
##              Estimate Std. Error t value Pr(>|t|)
## (Intercept) -1.23013    1.85412  -0.663    0.508
## Height       0.11589    0.01074  10.792   <2e-16 ***
## ---
## Signif. codes:
## 0 '***' 0.001 '**' 0.01 '*' 0.05 '.' 0.1 ' ' 1
##
## Residual standard error: 1.525 on 206 degrees of freedom
##   (29 observations deleted due to missingness)
## Multiple R-squared:  0.3612,Adjusted R-squared:  0.3581
## F-statistic: 116.5 on 1 and 206 DF,  p-value: < 2.2e-16
```

The linear model is of class `lm` and the method based OO system in R takes care of most of the specific issues. As seen above, the `summary()` method prints something meaningful for this class. Plotting the linear model is illustrated below, the output is in Figure 21.2 on page 377:

```
# predictions
h <- data.frame(Height = 150:200)
Wr.lm <- predict(lm1, h)
plot(survey$Height, survey$Wr.Hnd,col="red")
lines(t(h),Wr.lm,col="blue",lwd=3)
```

`abline()`

In previous code, we visualized the model by adding the predictions of the linear model (completed with lines between them, so the result is just a line). The function `abline()` provides another elegant way to draw straight lines in plots. The function takes as arguments the intercept and slope. The following code illustrates its use, and the result is in Figure 21.3 on page 377

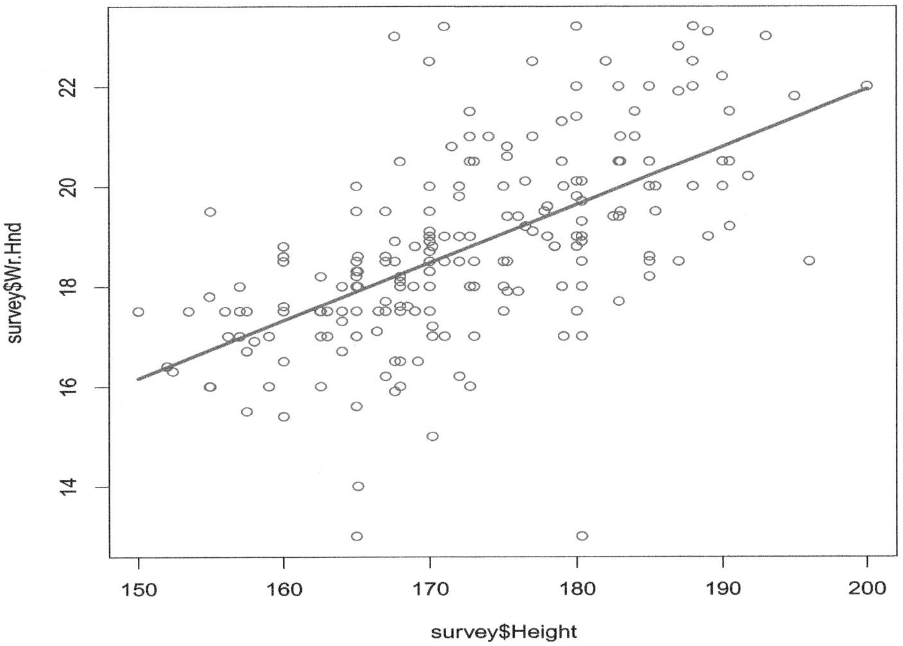

Figure 21.2: *A plot visualizing the linear regression model (the data in red and the regression in blue).*

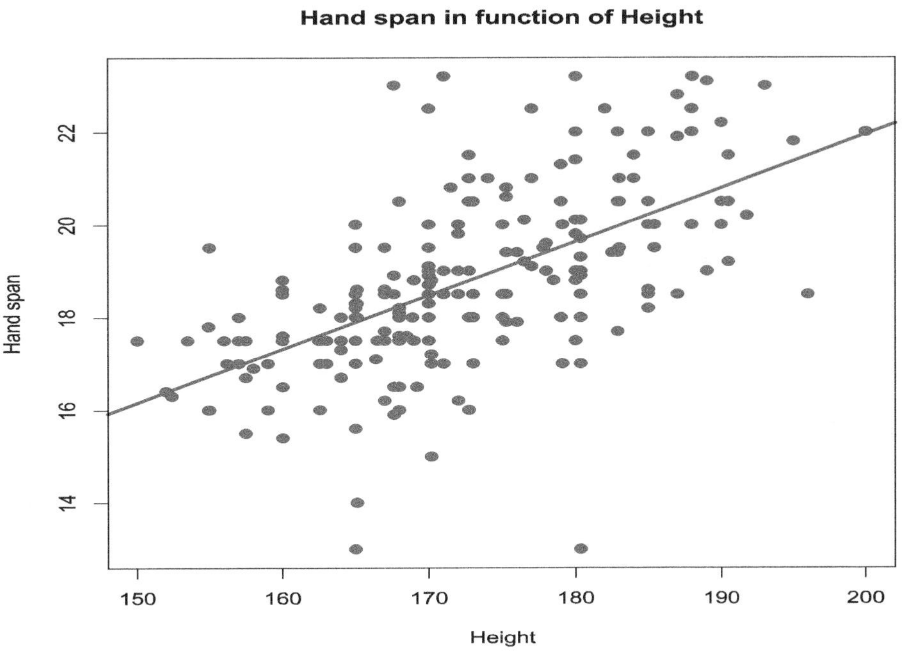

Figure 21.3: *Using the function abline() and cleaning up the titles.*

```
# Or use the function abline()
plot(survey$Height, survey$Wr.Hnd,col = "red",
     main = "Hand span in function of Height",
     abline(lm(survey$Wr.Hnd ~ survey$Height ),
            col='blue',lwd=3),
     cex = 1.3,pch = 16,
     xlab = "Height",ylab ="Hand span")
```

? Question #18 – Build a linear model

Consider the data set `mtcars` from the library MASS. Make a linear regression of the fuel consumption in function of the parameter that according to you has the most explanatory power. Study the residuals. What is your conclusion?

21.2 Multiple Linear Regression

Multiple regression is a relationship between more than two known variables to predict one variable.

regression – multiple linear
multiple linear regression

$$y = b + a_1x_1 + a_2x_2 + \cdots + a_nx_n$$

In R, the `lm()` function will handle this too. All we need to do is update the parameter `formula`:

`lm()`

```
# We use mtcars from the library MASS
model <- lm(mpg ~ disp + hp + wt, data = mtcars)
print(model)
##
## Call:
## lm(formula = mpg ~ disp + hp + wt, data = mtcars)
##
## Coefficients:
## (Intercept)         disp           hp           wt
##   37.105505    -0.000937    -0.031157    -3.800891
```

Note also that all coefficients and intercept can be accessed via the function `coef()`:

`coef()`

```
# Accessing the coefficients
intercept <- coef(model)[1]
a_disp    <- coef(model)[2]
a_hp      <- coef(model)[3]
a_wt      <- coef(model)[4]

paste('MPG =', intercept, '+', a_disp, 'x disp +',
     a_hp,'x hp +', a_wt, 'x wt')
## [1] "MPG = 37.1055052690318 + -0.000937009081489667 x disp + -0.0311565508299456 x hp
## + -3.80089058263761 x wt"
```

The variables that are defined in this way, can be used to implement the model (for example in a production system). It works as follows:

```
# This allows us to manually predict the fuel consumption
# e.g. for the Mazda Rx4
2.23 + a_disp * 160 + a_hp * 110 + a_wt * 2.62
##      disp
## -11.30548
```

? Question #19 – Build a multiple linear regression

Consider the data set `mtcars` from the library `MASS`. Make a linear regression that predicts the fuel consumption of a car. Make sure to include only significant variables and remember that the significance of a variable depends on the other variables in the model.

MASS

21.2.1 Poisson Regression

regression – Poisson
Poisson
distribution – Poisson
model – Poisson
model – log-linear

The Poisson regression assumes that the response variable Y has a Poisson distribution, and that the logarithm of its expected value can be modeled by a linear combination of certain variables. The Poisson regression model is also known as a "log-linear model."

response variable
variable – unknown

In particular the Poisson regression can only be useful where the unknown variable (response variable) can never be negative, for example in the case of predicting counts (e.g. numbers of events).

Definition: Poisson Regression

The general form of the Poisson Regression is

$$\log(y) = b + a_1x_1 + a_2x_2 + b_nx_n$$

with:

- `y`: the predicted variable (aka response variable or unknown variable)
- `a` and `b` are the numeric coefficients.
- `x` is the known variable (or the predictor variable)

The Poisson Regression can be handled by the function `glm()` in R, its general form is as follows.

Function use for glm()

```
glm(formula, data, family)
```

where:

- `formula` is the symbolic representation the relationship between the variables,
- `data` is the dataset giving the values of these variables,
- `family` is R object to specify the details of the model and for the Poisson Regression is value is "Poisson".

`glm()`

Note – The function glm()

`glm()`

Note that the function `glm()` is also used for the logistic regression; see Chapter 22.1 on page 388. The function `glm()` has mutliple types of models build in. For example: linear, gaussian, inverse gaussian, gamma, quasi binomial, quasi poisson. To access its full documentation type in R: `?family`.

Consider a simple example, where we want to check if we can estimate the number of cylinders of a car based on its horse power and weight, using the dataset `mtcars`

```
m <- glm(cyl ~ hp + wt, data = mtcars, family = "poisson")
summary(m)
##
## Call:
## glm(formula = cyl ~ hp + wt, family = "poisson", data = mtcars)
##
```

```
## Deviance Residuals:
##      Min        1Q    Median        3Q       Max
## -0.59240  -0.31647  -0.00394   0.29820   0.68731
##
## Coefficients:
##             Estimate Std. Error z value Pr(>|z|)
## (Intercept) 1.064836   0.257317   4.138  3.5e-05 ***
## hp          0.002220   0.001264   1.756    0.079 .
## wt          0.124722   0.090127   1.384    0.166
## ---
## Signif. codes:
## 0 '***' 0.001 '**' 0.01 '*' 0.05 '.' 0.1 ' ' 1
##
## (Dispersion parameter for poisson family taken to be 1)
##
##     Null deviance: 16.5743  on 31  degrees of freedom
## Residual deviance:  4.1923  on 29  degrees of freedom
## AIC: 126.85
##
## Number of Fisher Scoring iterations: 4
```

Weight does not seem to be relevant, so we drop it and try again (only using horse power):

```
m <- glm(cyl ~ hp, data = mtcars, family = "poisson")
summary(m)
##
## Call:
## glm(formula = cyl ~ hp, family = "poisson", data = mtcars)
##
## Deviance Residuals:
##      Min        1Q    Median        3Q       Max
## -0.97955  -0.30748  -0.03387   0.28155   0.73433
##
## Coefficients:
##              Estimate Std. Error z value Pr(>|z|)
## (Intercept) 1.3225669  0.1739422   7.603 2.88e-14 ***
## hp          0.0032367  0.0009761   3.316 0.000913 ***
## ---
## Signif. codes:
## 0 '***' 0.001 '**' 0.01 '*' 0.05 '.' 0.1 ' ' 1
##
## (Dispersion parameter for poisson family taken to be 1)
##
##     Null deviance: 16.5743  on 31  degrees of freedom
## Residual deviance:  6.0878  on 30  degrees of freedom
## AIC: 126.75
##
## Number of Fisher Scoring iterations: 4
```

This works better, and the Poisson model seems to work fine for this dataset.

21.2.2 Non-linear Regression

regression – non-linear

In many cases, one will observe that the relation between the unknown variable and the known variables is not simply linear. Whenever we plot the data and see that the relation is not a straight line, but rather a curve, the relation is non-linear. It is possible to model this by applying a function such as squaring, a sine, a logarithm, or an exponential to the known variable(s) and then running a linear regression. However, R, has a specific function for this: `nls()`.

In least square regression, we establish a regression model in which the sum of the squares of the vertical distances of different points from the regression curve is minimized. We generally start with a defined model and assume some values for the coefficients. We then apply the nls() function of R to get the more accurate values along with the confidence intervals.

Function use for nls()

`nls(formula, data, start)` with

1. `formula` a non-linear model formula including variables and parameters,
2. `data` the data-frame used to optimize the model,
3. `start` a named list or named numeric vector of starting estimates.

nls()

The use of the function `nls()` is best clarified by a simple example. We will fabricate an example in the following code, and plot the data in Figure 21.4 on page 383

```
# Consider observations for dt = d0 + v0 t + 1/2 a t^2:
t  <- c(1,2,3,4,5,1.5,2.5,3.5,4.5,1)
dt <- c(8.1,24.9,52,89.2,136.1,15.0,37.0,60.0,111.0,8)

# Plot these values:
plot(t,dt,xlab="time",ylab="distance")

# Take the assumed values and fit into the model:
model <- nls(dt ~ d0 + v0 * t + 1/2 * a * t^2,
             start = list(d0 = 1,v0 = 3,a = 10))

# Plot the model curve:
simulation.data <- data.frame(t = seq(min(t),max(t),len = 100))
lines(simulation.data$t,predict(model,
      newdata = simulation.data), col="red", lwd = 3)
```

The model seems to fit quite well the data. As usual, we can extract more information from the model object via the functions `summary()` and/or `print()`.

```
# Learn about the model:
summary(model)               # the summary
##
## Formula: dt ~ d0 + v0 * t + 1/2 * a * t^2
##
## Parameters:
##    Estimate Std. Error t value Pr(>|t|)
## d0    4.981      4.660   1.069    0.321
## v0   -1.925      3.732  -0.516    0.622
## a    11.245      1.269   8.861 4.72e-05 ***
## ---
## Signif. codes:
## 0 '***' 0.001 '**' 0.01 '*' 0.05 '.' 0.1 ' ' 1
##
## Residual standard error: 3.056 on 7 degrees of freedom
##
## Number of iterations to convergence: 1
## Achieved convergence tolerance: 1.822e-07

print(sum(residuals(model)^2))# squared sum of residuals
## [1] 65.39269
```

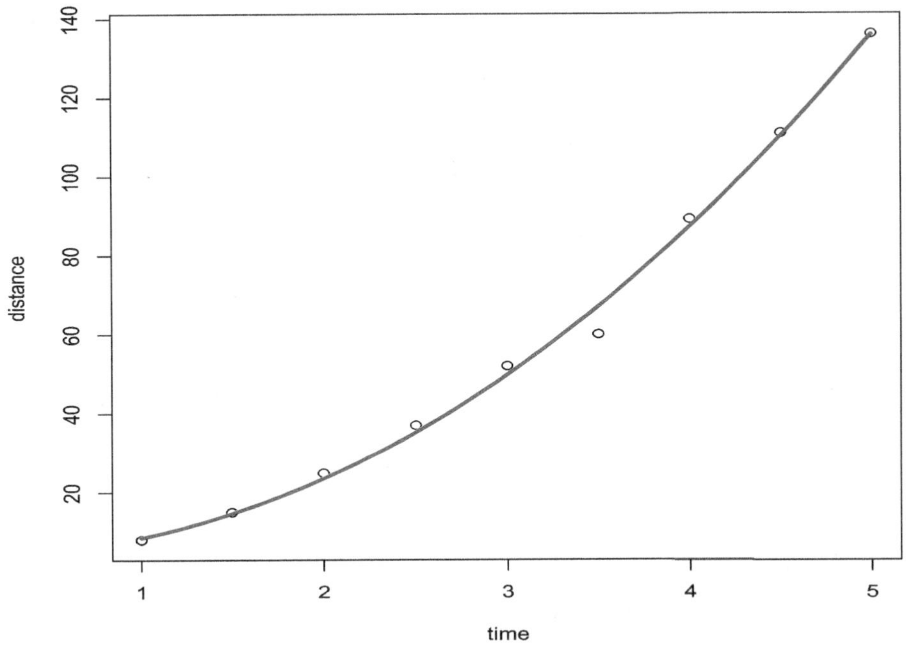

Figure 21.4: *The results of the non-linear regression with nls(). This plot indicates that there is one outlier and you might want to rerun the model without this observation.*

```
print(confint(model))          # confidence intervals

## Waiting for profiling to be done...
##            2.5%     97.5%
## d0   -6.038315 15.999559
## v0  -10.749091  6.899734
## a     8.244167 14.245927
```

Hint – Shorthand notation

Note that R has a shorthand notation built in for the function `residuals()`: `resid()` will do the same.

`resid()`
`residuals()`
`confint()`

21.3 Performance of Regression Models

model performance

The choice of a model is a complicated task. One needs to consider:

- Simple predictive model quality: is the model able to make systhematically better predictions than a random model (i.e. Height of the lift curve / AUC for classification models and mean square error for regression models).
- Generalization ability of the model (difference of model quality when compared on the training data (creation) and testing data. In other words: can we be sure that the model is not over-fit?)
- Explanatory power of the model (does the model "make sense" of the data? Can it explain something?)
- Model Stability (What is the confidence interval of the model on the lift curves?)
- Is the model robust against the erosion of time?
- Clarity (can the model be explained?).

In this section, we will focus on the first issue: the intrinsic predictive quality of the model and present some tools that are frequently used.

21.3.1 Mean Square Error (MSE)

mean square error
MSE

One of the best loved and known measures is the means square error. It is the sum of all differences between observations and model, where we first square all differences. The square has two functions: make all contributions positive (and avoid compensation of an underestimation by an overestimation), and penalise more the large deviations.

Definition: Mean Square Error (MSE)

The means square error is the average residual variance. The following is a predictor:

$$\text{MSE}(y, \hat{y}) = \frac{1}{N} \sum_{k=1}^{N} (y_k - \hat{y})^2$$

21.3.2 R-Squared

R-squared

While MSE provides a reliable measure of variation that is not explained by the model, it is sensitive to units (e.g. the use of millimetres or kilometres will result in a different MSE for the same model).

There is another, equally important measure, R^2 (pronounced "R-sqaured"), that does not have this problem. In some way it is a normalized version of MSE.

Definition: R-squared

R-squared is the proportion of the variance in the dependent variable that is predictable from the independent variable(s). We can calculate R-squared as:

$$R^2 = 1 - \frac{\sum_{k=1}^{N} (y_k - \hat{y})^2}{\sum_{k=1}^{N} (y_k - \bar{y})^2}$$

with $\hat{y_k}$ the estimate for observation y_k based on our model, and $\bar{y_k}$ the mean of all observations y_k.

Let us consider a simple example of a linear model that predicts the fuel consumption of a car based on its weight, and use the dataset `mtcars` to do so. The R-squared is such important measure, that it shows up in the summary and it is immediately available after fitting the linear model:

```
m <- lm(data = mtcars, formula = mpg ~ wt)
summary(m)
##
## Call:
## lm(formula = mpg ~ wt, data = mtcars)
##
## Residuals:
##     Min      1Q  Median      3Q     Max
## -4.5432 -2.3647 -0.1252  1.4096  6.8727
##
## Coefficients:
##             Estimate Std. Error t value Pr(>|t|)
## (Intercept)  37.2851     1.8776  19.858  < 2e-16 ***
## wt           -5.3445     0.5591  -9.559 1.29e-10 ***
## ---
## Signif. codes:
## 0 '***' 0.001 '**' 0.01 '*' 0.05 '.' 0.1 ' ' 1
##
## Residual standard error: 3.046 on 30 degrees of freedom
## Multiple R-squared:  0.7528,Adjusted R-squared:  0.7446
## F-statistic: 91.38 on 1 and 30 DF,  p-value: 1.294e-10

summary(m)$r.squared
## [1] 0.7528328
```

Further information – About the summary

hint: get more information via `help(summary.lm)`

Question #20 – Find a better model

Use the dataset `mtcars` (from the library MASS), and try to find the model that best explains the consumption (mpg).

Another way of understanding R^2, is noting that

$$R^2 = 1 - \frac{\text{SS}_{\text{res}}}{\text{SS}_{\text{tot}}}$$

sum of squares of the residuals SS_{res}
total sum of squares SS_{tot}

with

- $\text{SS}_{\text{res}} := \sum_{k=1}^{N}(y_k - \hat{y})$ the sum of squares of the residuals, and
- $\text{SS}_{\text{tot}} := \sum_{k=1}^{N}(y_k - \bar{y})^2$ the total sum of squares.

For OLS regressions it is so that

$$\text{SS}_{\text{res}} + \text{SS}_{\text{reg}} = \text{SS}_{\text{tot}}$$

sum of squares of the regression SS_{reg}

where $\text{SS}_{\text{reg}} = \sum_{k=1}^{N}(\hat{y_k} - \bar{y})^2$, the variance of the predictions of the model. Hence we have the formula

$$R^2 = \frac{\text{SS}_{\text{reg}}}{\text{SS}_{\text{tot}}}$$

21.3.3 Mean Average Deviation (MAD)

mean average deviation MAD

In the case where outliers matter less, or where outliers distort the results, one can rely on more robust methods that are based on the median. Many variations of these measures can be useful. We present a selection.

Definition: Mean average deviation (MAD)

$$\text{MAD}(y, \hat{y}) := \frac{1}{N}\sum_{k=1}^{N}|y_k - \hat{y}|$$

♣ 22 ♣

Classification Models

Classification models do not try to predict a continuous variable, but rather try to predict if an observation belongs to a certain discrete class. For example, we can predict if a customer would be creditworthy or not, or we can train a neural network to classify animals or plants on a picture. These models with discrete outcome have an equally large domain of use and most essential to any business application. The first one that we will discuss, the logistic regression, does only binary prediction but is very transparent and allows to build strong models.

The Big R-Book: From Data Science to Learning Machines and Big Data, First Edition. Philippe J.S. De Brouwer.

Companion Website: www.wiley.com/go/De Brouwer/The Big R-Book

22.1 Logistic Regression

regression – logistic
logit

binomial

Logistic regression (aka logit regression) is a regression model where the unknown variable is categorical (can have only a limited number of values): it can either be "0" or "1." In reality if you can refer to any mutually exclusive concept such as: repay/default, pass/fail, win/lose, survive/die, or healthy/sick.

Cases where the dependent variable has more than two outcome categories may be analysed in multinomial logistic regression, or, if the multiple categories are ordered, in ordinal logistic regression. In the terminology of economics, logistic regression is an example of a qualitative response/discrete choices.

In its most general form, the logistic regression is defined as follows.

Definition: – Generalised logistic regression

A logistic regression, is a regression of the log-odds:

$$\ln \left\{ \frac{P[Y=1|X]}{P[Y=0|X]} \right\} = \alpha + \sum_{n=1}^{N} f_n(X_n)$$

with $X = (X_1, X_2, \ldots, X_N)$ the set of prognostic factors.

This type of model can be used to predict the probability that $Y = 1$ or to study the $f_n(X_n)$ and hence understand the dynamics of the problem.

The general, additive logistic regression can be solved by estimating the $f_n()$ via a back-fitting algorithm within a Newton-Rapshon procedure. The linear additive logistic regression is defined as follows.

Logistic Regression

Definition: – Additive logistic regression

Assuming a linear model for the f_n such that, the probability that $Y = 1$ is modelled as:

$$y = \frac{1}{1 + e^{-(b + a_1 x_1 + a_2 x_2 + a_3 x_3 + \cdots)}}$$

glm()

This regression can be fitted with the function `glm()`, that we encountered earlier.

```
# Consider the relation between the hours studied and passing
# an exam (1) or failing it (0):
hours <- c(0,0.50, 0.75, 1.00, 1.25, 1.50, 1.75,
           1.75, 2.00, 2.25, 2.50, 2.75, 3.00, 3.25,
           3.50, 4.00, 4.25, 4.50, 4.75, 5.00, 5.50)
pass  <- c(0,0, 0, 0, 0, 0, 0, 1, 0, 1, 0, 1, 0,
           1, 0, 1, 1, 1, 1, 1, 1)
d <- data.frame(cbind(hours,pass))
m <- glm(formula = pass ~ hours, family = binomial,
         data = d)
```

Note – Multiple uses for glm()

The function `glm()` is also used for the Poisson regression; see Chapter 21.2.1 on page 379.

Plotting the observations and the logistic fit, can be done with the following code. The plot is in Figure 22.1.

```
# Visualize the results:
plot(hours, pass, col = "red", pch = 23, bg = "grey",
     xlab = 'Hours studied',
     ylab = 'Passed exam (1) or not (0)')
pred <- 1 / (1+ exp(-(coef(m)[1] + hours * coef(m)[2])))
lines(hours,pred,col="blue",lwd=3)
```

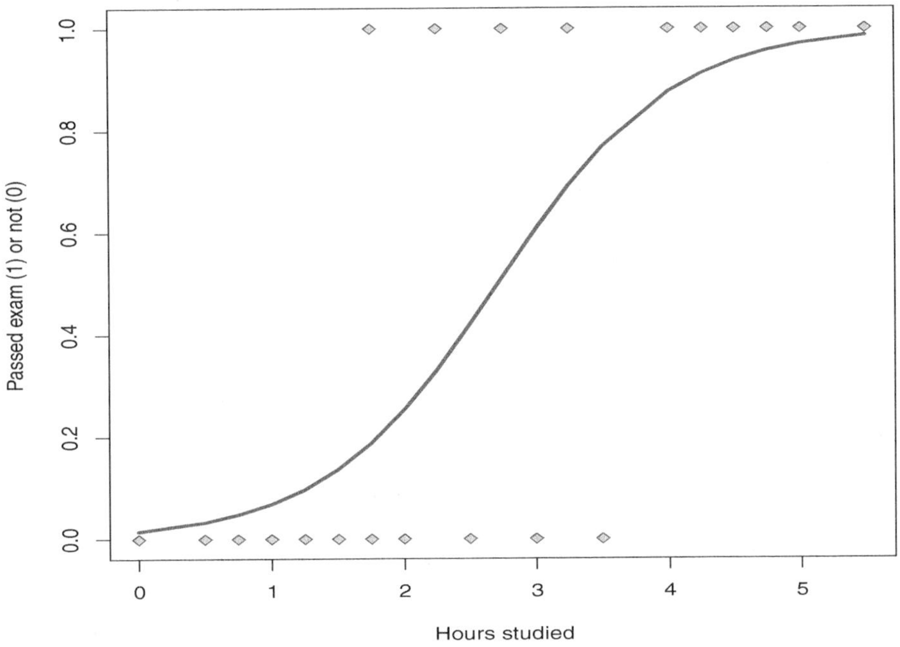

Figure 22.1: *The grey diamonds with red border are the data-points (not passed is 0 and passed is 1) and the blue line represents the logistic regression model (or the probability to succeed the exam in function of the hours studied.*

22.2 Performance of Binary Classification Models

Regression models fit a curve through a dataset and hence the performance of the model can be judged by how close the observations are from the fit. Classification models aim to find a yes/no prediction. For example, in the dataset of the Titanic disaster we can predict if a given passenger would survive the disaster or not.

Pseudo-R^2 values have been developed for binary logistic regression. While there is value in testing for predictive discrimination, the pseudo R^2 for logistic regressions have to be handled with care: there might be more than one reason for them to be high or low. A better approach is to present any of the goodness of fit tests available. For example, the Hosmer–Lemeshow test. This test provides an overall calibration error test. However, it does not properly take over fitting into account, nor the choice of bins, and has other issues. Therefore, Hosmer et al. provide another test: the omnibus test of fit, which is implemented in the `rms` package and its `residuals.lrm()` function.

When selecting the model for the logistic regression analysis, another important consideration is the model fit. Adding independent variables to a logistic regression model will always increase the amount of variance explained in the log odds (for example expressed as R^2). However, adding too many variables to the model can result in over-fitting, which means that while the R^2 improves, the model loses on predictive power on new data.

In the following sections we will use the dataset from the package `titanic`. This is data of the passengers on the RMS Titanic, that sunk in 1929 in the Northern Atlantic Ocean after a collision with an iceberg.

The data can be unlocked as follows:

```
# if necessary: install.packages('titanic')
library(titanic)

# This provides a.o. two datasets titanic_train and titanic_test.
# We will work further with the training-dataset.
t <- titanic_train
colnames(t)
##  [1] "PassengerId" "Survived"    "Pclass"
##  [4] "Name"        "Sex"         "Age"
##  [7] "SibSp"       "Parch"       "Ticket"
## [10] "Fare"        "Cabin"       "Embarked"
```

Fitting the logistic model is done with the function `glm()`, where we provide the "binomial" for the argument `family`, and the rest works the same as for other regressions.

```
# fit provide a simple model
m <- glm(data    = t,
         formula = Survived ~ Pclass + Sex + Pclass * Sex +
                   Age + SibSp,
         family  = binomial)
summary(m)
##
## Call:
## glm(formula = Survived ~ Pclass + Sex + Pclass * Sex + Age +
##     SibSp, family = binomial, data = t)
##
## Deviance Residuals:
##     Min       1Q   Median       3Q      Max
## -3.3507  -0.6574  -0.4438   0.4532   2.3450
```

```
##
## Coefficients:
##                  Estimate Std. Error z value Pr(>|z|)
## (Intercept)      8.487528   0.996601   8.516  < 2e-16 ***
## Pclass          -2.429192   0.330221  -7.356 1.89e-13 ***
## Sexmale         -6.162294   0.929696  -6.628 3.40e-11 ***
## Age             -0.046830   0.008603  -5.443 5.24e-08 ***
## SibSp           -0.354855   0.120373  -2.948   0.0032 **
## Pclass:Sexmale   1.462084   0.349338   4.185 2.85e-05 ***
## ---
## Signif. codes:
## 0 '***' 0.001 '**' 0.01 '*' 0.05 '.' 0.1 ' ' 1
##
## (Dispersion parameter for binomial family taken to be 1)
##
##     Null deviance: 964.52  on 713  degrees of freedom
## Residual deviance: 614.22  on 708  degrees of freedom
##   (177 observations deleted due to missingness)
## AIC: 626.22
##
## Number of Fisher Scoring iterations: 6
```

22.2.1 The Confusion Matrix and Related Measures

A simple and clear way to show what the model does is a "confusion matrix": simply show how much of the observations are classified correctly by the model.

The following are useful measures for how good a classification model fits its data:

- *Accuracy:* The proportion of predictions that were correctly identified. **accuracy**
- *Precision* (or positive predictive value): The proportion of positive cases that correct. **precision**
- *Negative predictive value:* The proportion of negative cases that were correctly identified.
- *Sensitivity* or Recall: The proportion of actual positive cases which are correctly identified. **sensitivity**
- *Specificity:* The proportion of actual negative cases which are correctly identified. **specificity**

Let us use the following definitions:

- Objective concepts (depends only on the data):
  - *P:* The number of positive observations ($y = 1$)
  - *N:* The number of negative observations ($y = 0$)
- Model dependent definitions:
  - True positive (TP) the positive observations ($y = 1$) that are by the model correctly classified as positive; **TP** **true positive**
  - False positive (FP) the negative observations ($y = 0$) that are by the model incorrectly classified as positive – this is a false alarm (Type I error); **FP** **false positive** **Type I error**
  - True negative (TN) the negative observations ($y = 0$) that are by the model correctly classified as negative; **TN** **true negative**
  - False negative (FN) the positive observations ($y = 1$) that are by the model incorrectly classified as negative – miss (Type II error). **FN** **false negative** **Type II error**

Using these definitions, we construct a matrix that summarizes that information: the confusion matrix. This confusion matrix is visualised in Table 22.1.

	Observed pos.	Observed neg.	
Pred. pos.	TP	FP	Pos.pred.val $= \frac{TP}{TP+FP}$
Pred. neg.	FN	TN	Neg.pred.val $= \frac{TN}{FN+TN}$
	Sensitivity	Specificity	Accuracy
	$= \frac{TP}{TP+FN}$	$= \frac{TN}{FP+TN}$	$= \frac{TP+TN}{TP+FN+FP+TN}$
	$= \frac{TP}{TP+FN}$	$= \frac{TN}{FP+TN}$	$= \frac{TP+TN}{TP+FN+FP+TN}$

Table 22.1: *The confusion matrix, where "pred." refers to the predictions made by the model, "pred." stands for "predicted," and the words "positive" and "negative" are shortened to three letters.*

More than any metric, the confusion matrix is a good instrument to make clear to lay persons that any model has inherent risk, and that there is no ideal solution for cut-off value. One can quite easy observer what happens when

table()

A prediction object and the function `table()` is all we need in R to produce a confusion matrix.

```
# We build further on the model m as defined earlier.
# Predict scores between 0 and 1 (odds):
t2 <- t[complete.cases(t),]
predicScore <- predict(object=m,type="response", newdat = t2)

# Introduce a cut-off level above which we assume survival:
predic <- ifelse(predicScore > 0.7, 1, 0)

# The confusion matrix is one line, the headings 2:
confusion_matrix <- table(predic, t2$Survived)
rownames(confusion_matrix) <- c("predicted_death",
                                "predicted_survival")
colnames(confusion_matrix) <- c("observed_death",
                                "observed_survival")
print(confusion_matrix)
##
## predic              observed_death observed_survival
##    predicted_death             414               134
##    predicted_survival           10               156
```

Building further on the idea of classification errors as a measure of suitability of the model, we will explore in the following sections measures that focus on separation of the distribution of the observations that had a 0 or 1 outcome and how well the model is able to discriminate between them.

Unfortunately, there are multiple words that refer to the same concepts. Here we list a few important ones.

- **TPR** = True Positive Rate = sensitivity = recall = hit rate = probability of detection

$$\text{TPR} = \frac{\text{TP}}{P} = \frac{\text{TP}}{\text{TP} + \text{FN}} = 1 - \text{FNR}$$

- **FPR** = False Positive Rate = fallout = 1 - Specificity

$$\text{FPR} = \frac{\text{FP}}{N} = \frac{\text{FP}}{\text{FP} + \text{TN}} = 1 - \text{TNR}$$

- **TNR** = specificity = selectivity = true negative rate

$$\text{TNR} = \frac{\text{TN}}{N} = \frac{\text{TN}}{\text{FP} + \text{TN}} = 1 - \text{FPR}$$

- **FNR** = false negative rate = miss rate

$$\text{FNR} = \frac{\text{FN}}{P} = \frac{\text{FN}}{\text{TP} + \text{FN}} = 1 - \text{TPR}$$

- **Precision** = positive predictive value = PPV

$$\text{PPV} = \frac{\text{TP}}{\text{TP} + \text{FP}}$$

- **Negative predictive value** = NPV

$$NPV = \frac{\text{TN}}{\text{TN} + \text{FN}}$$

- **ACC** = accuracy

$$\text{ACC} = \frac{\text{TP} + \text{TN}}{N + P} = \frac{\text{TP} + \text{TN}}{\text{TP} + \text{TN} + \text{FP} + \text{FN}}$$

- **F_1 score** = harmonic mean of precision and sensitivity

$$\text{F}_1 = \frac{\text{PPV} \times \text{TPR}}{\text{PPV} + \text{TPR}} = \frac{2\,\text{TP}}{2\,\text{TP} + \text{FP} + \text{FN}}$$

There are more measures used from time to time, though, we believe these to be the most important.

22.2.2 ROC

Engineers working on improving the radar during World War II developed the ROC for detecting (or not detecting) enemy objects. They called it the "receiver operator characteristic" (ROC). Since then it became one of the quintessential tools for assessing the discriminative power of binary models.

ROC
receiver operator characteristic

The ROC curve is formed by plotting the true positive rate (TPR) against the false positive rate (FPR) at various cut-off levels.[1] Formally, the ROC curve is the interpolated curve made of points whose coordinates are functions of the threshold: threshold $= \theta \in \mathbb{R}$, here $\theta \in [0, 1]$

$$\text{ROC}_x(\theta) = \text{FPR}(\theta) = \frac{\text{FP}(\theta)}{\text{FP}(\theta) + \text{TN}(\theta)} = \frac{\text{FP}(\theta)}{\#N}$$
$$\text{ROC}_y(\theta) = \text{TPR}(\theta) = \frac{\text{TP}(\theta)}{\text{FN}(\theta) + \text{TP}(\theta)} = \frac{\text{FP}(\theta)}{\#P} = 1 - \frac{\text{FN}(\theta)}{\#P} = 1 - \text{FNR}(\theta)$$

An alternative way of obtaining the ROC curve is using the probability density function of the true positives (TP, or "detection" as in the original radar problem) and false positives (FP, or "false alarm" in the war terminology). The ROC is then obtained by plotting the cumulative distribution

[1] In other words, the ROC curve is the sensitivity plotted as a function of fall-out.

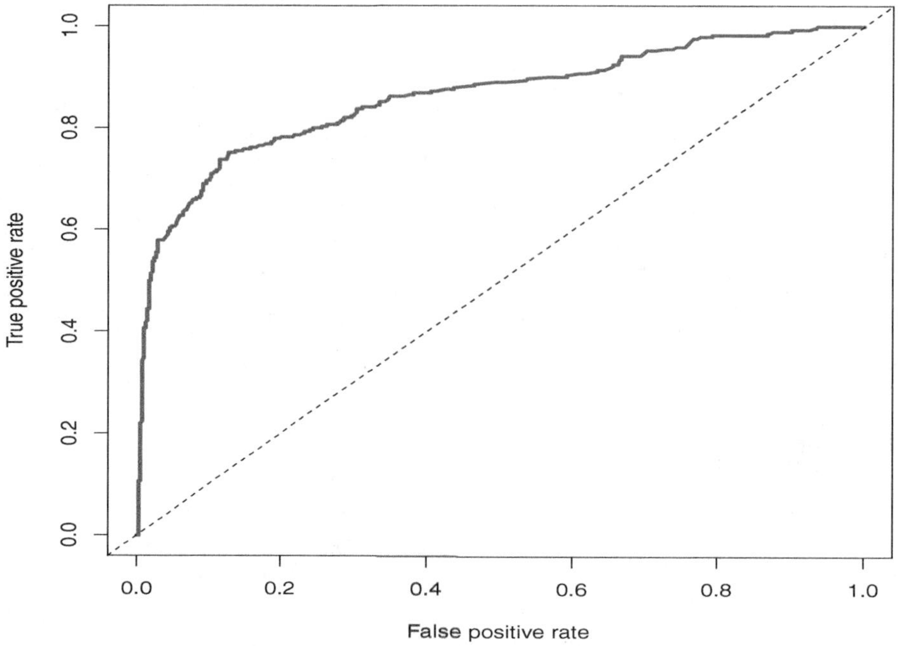

Figure 22.2: *The ROC curve of a logistic regression.*

function of the TP probability in the y-axis versus the cumulative distribution function of the FP probability on the x-axis.

Later we will see how the ROC curve can help to identify which models are (so far) optimal and which ones are inferior. Also, it can be related to the cost functions typically associated with TP and FP.

ROCR

We visualize the ROC curve with the aid of the `ROCR` package. First, we need to add predictions to the model, and then we can use the function `plot()` on a performance object that uses the predictions. The plot is generated in the last lines of the following code, and is in Figure 22.2.

plot()

```
library(ROCR)
# Re-use the model m and the dataset t2:
pred <- prediction(predict(m, type = "response"), t2$Survived)

# Visualize the ROC curve:
plot(performance(pred, "tpr", "fpr"), col="blue", lwd=3)
abline(0, 1, lty = 2)
```

The function `performance()` from the package `ROCR` generates an S4 object with (see Section 6.3 *"S4 Objects"* on page 100). It typically gets slots with `x` and `y` values and their names (as well as in this case slots for `alpha`). This object will know how it can be plotted. If necessary, then it can be converted to an a data frame as follows:

```
S4_perf <- performance(pred, "tpr", "fpr")
df <- data.frame(
   x = S4_perf@x.values,
   y = S4_perf@y.values,
   a = S4_perf@alpha.values
   )
```

```
colnames(df) <- c(S4_perf@x.name, S4_perf@y.name, S4_perf@alpha.name)
head(df)
##   False positive rate True positive rate    Cutoff
## 1         0.000000000        0.000000000       Inf
## 2         0.002358491        0.000000000 0.9963516
## 3         0.002358491        0.003448276 0.9953019
## 4         0.002358491        0.013793103 0.9950778
## 5         0.002358491        0.017241379 0.9945971
## 6         0.002358491        0.024137931 0.9943395
```

In a final report, it might be desirable to use the power of ggplot2 consistently. In the following code we illustrate how this a ROC curve can be obtained in ggplot2.[2] The plot is in Figure 22.3.

ggplot2

```
library(ggplot2)
p <- ggplot(data=df,
          aes(x = `False positive rate`, y = `True positive rate`)) +
          geom_line(lwd=2, col='blue')  +
          # The next lines add the shading:
          aes(x = `False positive rate`, ymin = 0,
              ymax = `True positive rate`) +
          geom_ribbon(, alpha=.5)
p
```

The performance object can also provide the accuracy of the model, and this can be plotted as follows – note that the plot is in Figure 22.4 on page 396.

```
# Plotting the accuracy (in function of the cut-off):
plot(performance(pred, "acc"), col="blue", lwd=3)
```

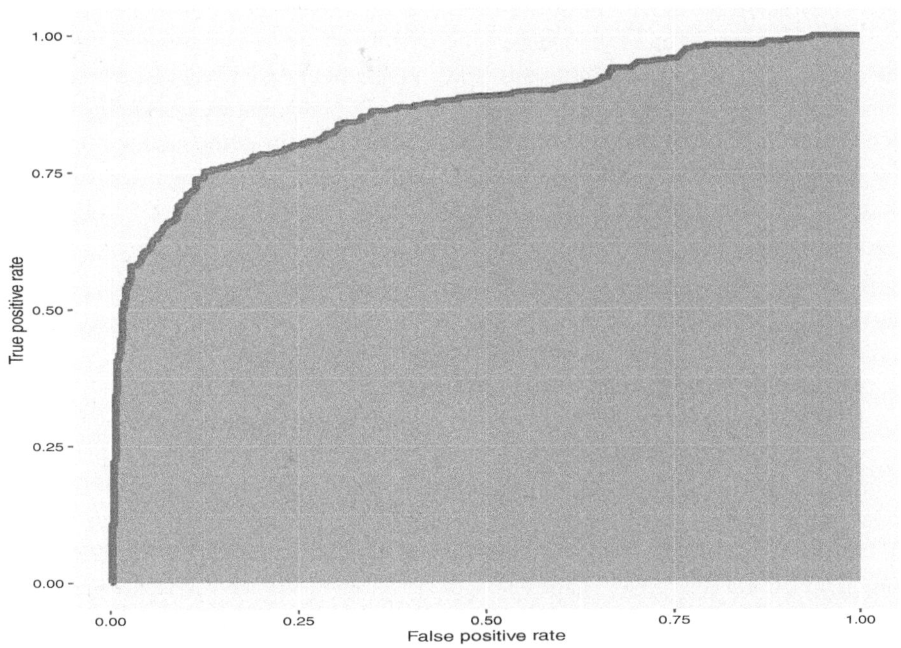

Figure 22.3: *The ROC curve plotted with ggplot2.*

[2]While the example should be clear, it might be useful to read through Chapter 31 *"A Grammar of Graphics with ggplot2"* on page 687 to get a better understanding of how ggplot2 works.

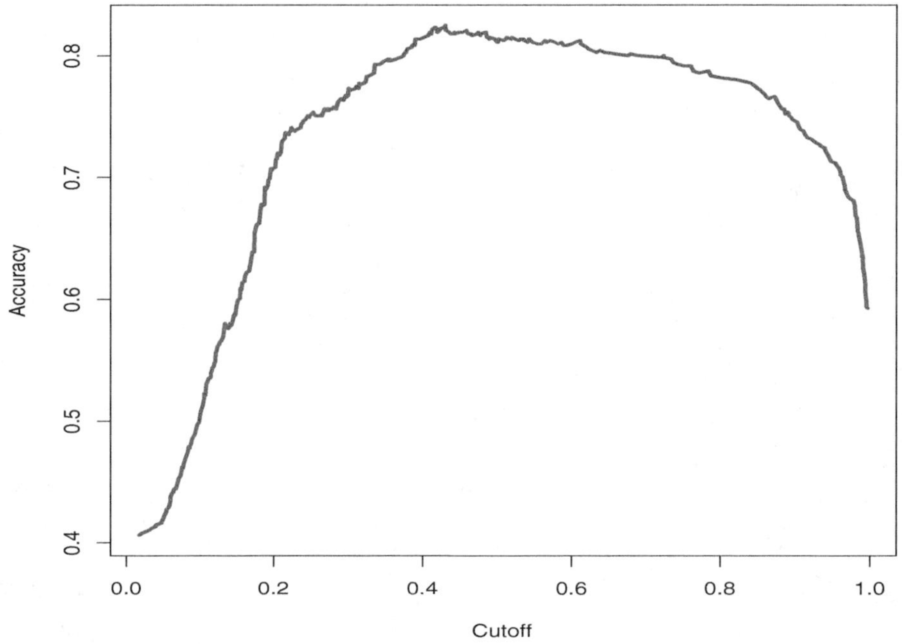

Figure 22.4: *A plot of the accuracy in function of the cut-off (threshold) level.*

AUC

22.2.3 The AUC

area under curve

When increasing the cut-off from 0, a really good model will gain fast true positives (and slower false positives). So the ROC curve must increase faster for a better model and hence this provides a first way to identify good and bad models.

The optimal point on the ROC curve is (FPR, TPR) = (0,1). No false positives and all true positives. So the closer we get there the better, this would imply that we could use the distance between the ROC curve and the point $(0, 1)$ as a measure of quality of the model.

The second essential observation is that the curve is by definition monotonically increasing.

$$\theta \leq \theta' \implies \text{TPR}(\theta) \leq \text{TPR}(\theta')$$

A good way to understand this inequality is by remembering that the TPR is also the fraction of observed survivors that is by the model predicted to survive (in "lending terms": the fraction of "goods accepted"). This fraction can only increase and never decrease.

One also observes that a random prediction would follow the identity line $\text{ROC}_y = \text{ROC}_x$, because if the model has no prediction power, then the TPR will increase equally with TNR.

Therefore, a reasonable model's ROC is located above the identity line as a point below it would imply a prediction performance worse than random. If that would happen, it is possible to do the invert the prediction.

All those features combined make it reasonable to summarize the ROC into a single value by calculating the area of the convex shape below the ROC curve – this is the area under the curve (AUC). The closer the ROC gets to the optimal point of perfect prediction, the closer the AUC gets to 1.

AUC

To illustrate this, we draw the ROC and label the areas in Figure 22.5 on page 397, which can be obtained by executing the following code.

```
# Assuming that we have the predictions in the prediction object:
plot(performance(pred, "tpr", "fpr"), col="blue", lwd=3)
abline(0,1,lty=2)
text(0.3,0.5,"A")
text(0.1,0.9,"B")
text(0.8,0.3,"C")
```

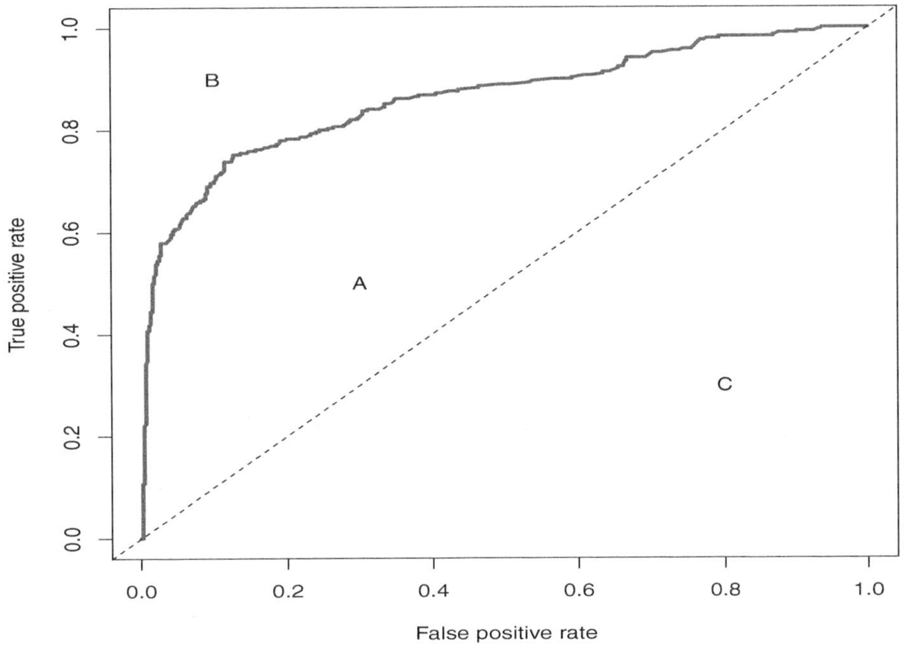

Figure 22.5: *The area under the curve (AUC) is the area A plus the area C. In next section we characterise the Gini coeffient, which equals area A divided by area B.*

The AUC in R is provided by the `performance()` function of `ROCR` and stored in the performance object. It is an S4 object, and hence we can extract the information as follows.

```
AUC <- attr(performance(pred, "auc"), "y.values")[[1]]
AUC
## [1] 0.8615241
```

22.2.4 The Gini Coefficient

Gini

The Gini coefficient is an alternative measure based on the same plot as used for the AUC. The name refers to Corrado Gini (1912) and is a standard measure when talking about income inequality. Since it is so much related to AUC, we can illustrate the concept by referring to the same plot (Figure 22.5).

Gini is the area under the ROC curve compared to the area above ($\frac{A}{A+B} = \frac{A}{0.5} = 2A$ in the figure). Since $A + B = C = 0.5$, the Gini is related to the AUC as follows.

$$\text{Gini} = 2 \times \text{AUC} - 1$$

So where a really bad model has an AUC close to 0.5, the same really bad model will have a Gini that is close to 0. However, a model that is really good will have an AUC close to 1 and a Gini also close to 1.

In R, extracting the Gini coefficient from the performance object is trivial, given the AUC that we calculated before. In fact, we can use the AUC to obtain the Gini:

```
paste("the Gini is:",round(2 * AUC - 1, 2))
## [1] "the Gini is: 0.72"
```

KS

22.2.5 Kolmogorov-Smirnov (KS) for Logistic Regression

Kolmogorov Smirnov

The KS test is another measure that aims to summarize the power of a model in one parameter. In general, the KS is the largest distance between two cumulative distribution functions:

$$\mathrm{KS} = \sup |F_1(x) - F_2(x)|$$

This KS of a logistic regression model is the KS applied to the distribution of the "bads" (negatives) and the "goods" (positives) in function of their score. The higher the KS, the better model we have.[3]

Since these cumulative distribution functions are related to the TPR and the FPR, there are two ways of understanding the KS: as the maximum distance between the ROC curve and the bisector or as the maximum vertical distance between the cumulative distribution functions of positives and negatives in our dataset.

A visualization of the KS measure, based on the definition of the cumulative distribution functions can be found in Figure 22.6.

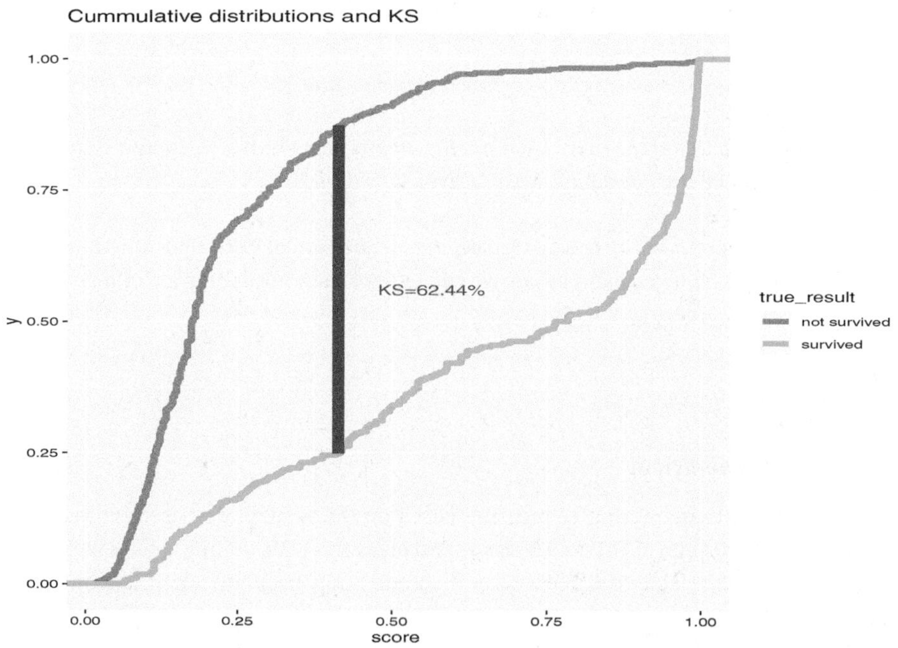

Figure 22.6: *The KS as the maximum distance between the cumulative distributions of the positive and negative observations.*

[3]Note that the wording "goods" and "bads" are maybe not correct English words, they mean respectively "an observation with y-value 1" and "an observation with y-value 0."

Further information – code for plots

The code for the plots in Figure 22.6 on page 398 is not included here, but you can find it in the appendix: Chapter D *"Code Not Shown in the Body of the Book"* on page 841. That code includes an interesting alternative to calculate the KS measure.

The package `stats` from base R provides the functions `ks.test()` to calculate the KS.

ks.test()
stats

```
pred <- prediction(predict(m,type="response"), t2$Survived)
ks.test(attr(pred,"predictions")[[1]],
        t2$Survived,
        alternative = 'greater')

## Warning in ks.test(attr(pred, "predictions")[[1]], t2$Survived, alternative =
"greater"): p-value will be approximate in the presence of ties
##
##  Two-sample Kolmogorov-Smirnov test
##
## data:  attr(pred, "predictions")[[1]] and t2$Survived
## D^+ = 0.40616, p-value < 2.2e-16
## alternative hypothesis: the CDF of x lies above that of y
```

As you can see in the aforementioned code, this does not work in some cases. Fortunately, it is easy to construct an alternative:

```
perf <- performance(pred, "tpr", "fpr")
ks <- max(attr(perf,'y.values')[[1]] - attr(perf,'x.values')[[1]])
ks
## [1] 0.6243656

# Note: the following line yields the same outcome
ks <- max(perf@y.values[[1]] - perf@x.values[[1]])
ks
## [1] 0.6243656
```

Now, that we have the value of the KS, we can visualize it on the ROC curve in Figure 22.7 on page 400. This is done as follows:

```
pred <- prediction(predict(m,type="response"), t2$Survived)
perf <- performance(pred, "tpr", "fpr")
plot(perf, main = paste0(' KS is',round(ks*100,1),'%'),
     lwd = 4, col = 'red')
lines(x = c(0,1),y=c(0,1), col='blue')

# The KS line:
diff <- perf@y.values[[1]] - perf@x.values[[1]]
xVal <- attr(perf,'x.values')[[1]][diff == max(diff)]
yVal <- attr(perf,'y.values')[[1]][diff == max(diff)]
lines(x = c(xVal, xVal), y = c(xVal, yVal),
      col = 'khaki4', lwd=8)
```

22.2.6 Finding an Optimal Cut-off

So far we have assessed how good the model is able to separate the good outcomes from the bad outcomes. We even moved the cut-off level θ from 0 to 1 to obtain the ROC curve and from that the area under the curve (AUC).

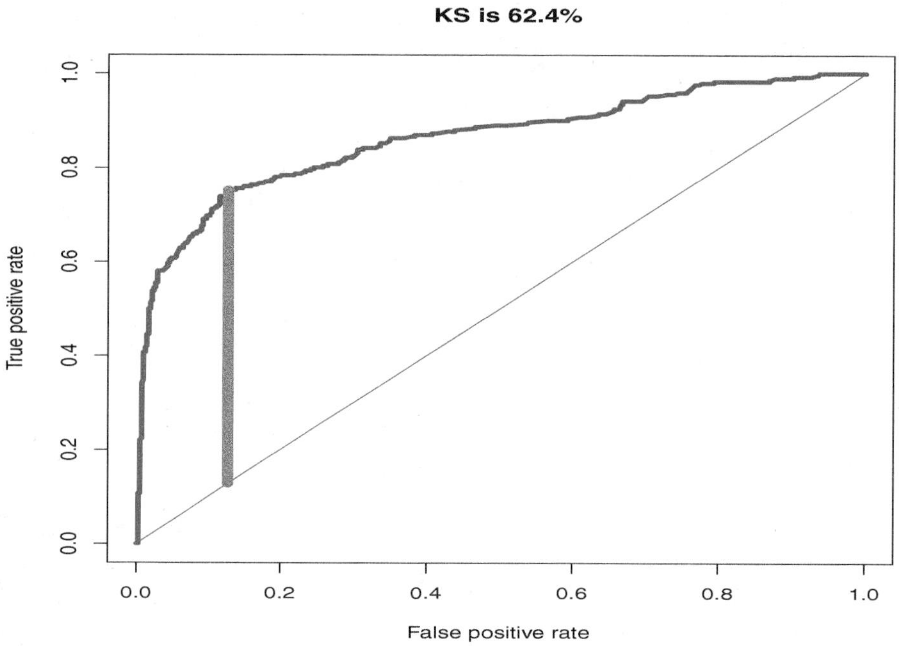

Figure 22.7: *The KS as the maximum distance between the model and a pure random model.*

However, to make predictions we need to choose one cut-off level. In a first naive approach, one might want to select an optimal threshold by optimizing specificity and sensitivity.

```
# First, we need a logistic regression. We use the same as before:
library(titanic)
t <- titanic_train
t2 <- t[complete.cases(t),]
m <- glm(data     = t,
         formula = Survived ~ Pclass + Sex + Pclass*Sex + Age + SibSp,
         family  = binomial)
library(ROCR)
pred <- prediction(predict(m,type="response", newdat = t2),
                   t2$Survived)
perf <- performance(pred, "tpr", "fpr")
```

We are now ready to present a simple function that optimizes sensitivity and specificity.

```
# get_best_cutoff
# Finds a cutof for the score so that sensitivity and specificity
# are optimal.
# Arguments
#   fpr    -- numeric vector -- false positive rate
#   tpr    -- numeric vector -- true positive rate
#   cutoff -- numeric vector -- the associated cutoff values
# Returns:
#   the cutoff value (numeric)
get_best_cutoff <- function(fpr, tpr, cutoff){
        cst <- (fpr - 0)^2 + (tpr - 1)^2
        idx = which(cst == min(cst))
        c(sensitivity = tpr[[idx]],
          specificity = 1 - fpr[[idx]],
          cutoff = cutoff[[idx]])
```

```
    }

# opt_cut_off
# Wrapper for get_best_cutoff. Finds a cutof for the score so that
# sensitivity and specificity are optimal.
# Arguments:
#    perf -- performance object (ROCR package)
#    pred -- prediction object (ROCR package)
# Returns:
#   The optimal cutoff value (numeric)
opt_cut_off = function(perf, pred){
    mapply(FUN=get_best_cutoff,
           perf@x.values,
           perf@y.values,
           pred@cutoffs)
   }

# Test the function:
opt_cut_off(perf, pred)
##                  [,1]
## sensitivity 0.7517241
## specificity 0.8726415
## cutoff      0.4161801
```

However, in general, this is a little naive for it assumes that sensitivity and specificity come at the same cost. Indeed, in general, the cost of Type I error is not the same as the cost of a Type II error. For example, in banking, it is hundreds of times more costly to have one customer that defaults on a loan than rejecting a good customer unfairly. In medical image recognition, it would be costly to start cancer treatment for someone who does not require it, but it would be a lot worse to tell a patient to go home while in fact he or she needs treatment.

To take this asymmetry into account one can introduce a cost-function that is this multiplier between the cost of type I and the cost of type II errors:

```
# We introduce cost.fp to be understood as a the cost of a
# false positive, expressed as a multiple of the cost of a
# false negative.

# get_best_cutoff
# Finds a cutof for the score so that sensitivity and specificity
# are optimal.
# Arguments
#   fpr     -- numeric vector -- false positive rate
#   tpr     -- numeric vector -- true positive rate
#   cutoff  -- numeric vector -- the associated cutoff values
#   cost.fp -- numeric        -- cost of false positive divided
#                                by the cost of a false negative
#   (default = 1)
# Returns:
#   the cutoff value (numeric)
get_best_cutoff <- function(fpr, tpr, cutoff, cost.fp = 1){
        cst <- (cost.fp * fpr - 0)^2 + (tpr - 1)^2
        idx = which(cst == min(cst))
        c(sensitivity = tpr[[idx]],
          specificity = 1 - fpr[[idx]],
          cutoff = cutoff[[idx]])
    }

# opt_cut_off
# Wrapper for get_best_cutoff. Finds a cutof for the score so that
# sensitivity and specificity are optimal.
```

```
# Arguments:
#    perf    -- performance object (ROCR package)
#    pred    -- prediction object (ROCR package)
#    cost.fp -- numeric -- cost of false positive divided by the
#                          cost of a false negative (default = 1)
# Returns:
#   The optimal cutoff value (numeric)
opt_cut_off = function(perf, pred, cost.fp = 1){
    mapply(FUN=get_best_cutoff,
           perf@x.values,
           perf@y.values,
           pred@cutoffs,
           cost.fp)
  }

# Test the function:
opt_cut_off(perf, pred, cost.fp = 5)
##                  [,1]
## sensitivity 0.5793103
## specificity 0.9716981
## cutoff      0.6108004
```

Hint – Backwards compatibility

When over-writing functions, it is always a good idea to make sure that new parameters have a default value that results in a behaviour that is exactly the same is in the previous version. That way we make sure that code that was written with the older version of the function in mind will still work.

While it is insightful to program this function ourselves, it is not really necessary. ROCR provides a more convenient way to obtain this. All we need to do is provide this `cost.fp` argument to the `performance()` function of ROCR on beforehand. Then, all that remains is finding the minimal fpr.

```
# e.g. cost.fp = 1 x cost.fn
perf_cst1 <- performance(pred, "cost", cost.fp = 1)
str(perf_cst1) # the cost is in the y-values
## Formal class 'performance' [package "ROCR"] with 6 slots
##   ..@ x.name      : chr "Cutoff"
##   ..@ y.name      : chr "Explicit cost"
##   ..@ alpha.name  : chr "none"
##   ..@ x.values    :List of 1
##   .. ..$ : Named num [1:410] Inf 0.996 0.995 0.995 0.995 ...
##   .. .. ..- attr(*, "names")= chr [1:410] "" "298" "690" "854" ...
##   ..@ y.values    :List of 1
##   .. ..$ : num [1:410] 0.406 0.408 0.406 0.402 0.401 ...
##   ..@ alpha.values: list()

# the optimal cut-off is then the same as in previous code sample
pred@cutoffs[[1]][which.min(perf_cst1@y.values[[1]])]
##       738
## 0.4302302

# e.g. cost.fp = 5 x cost.fn
perf_cst2 <- performance(pred, "cost", cost.fp = 5)

# the optimal cut-off is now:
pred@cutoffs[[1]][which.min(perf_cst2@y.values[[1]])]
##       306
## 0.7231593
```

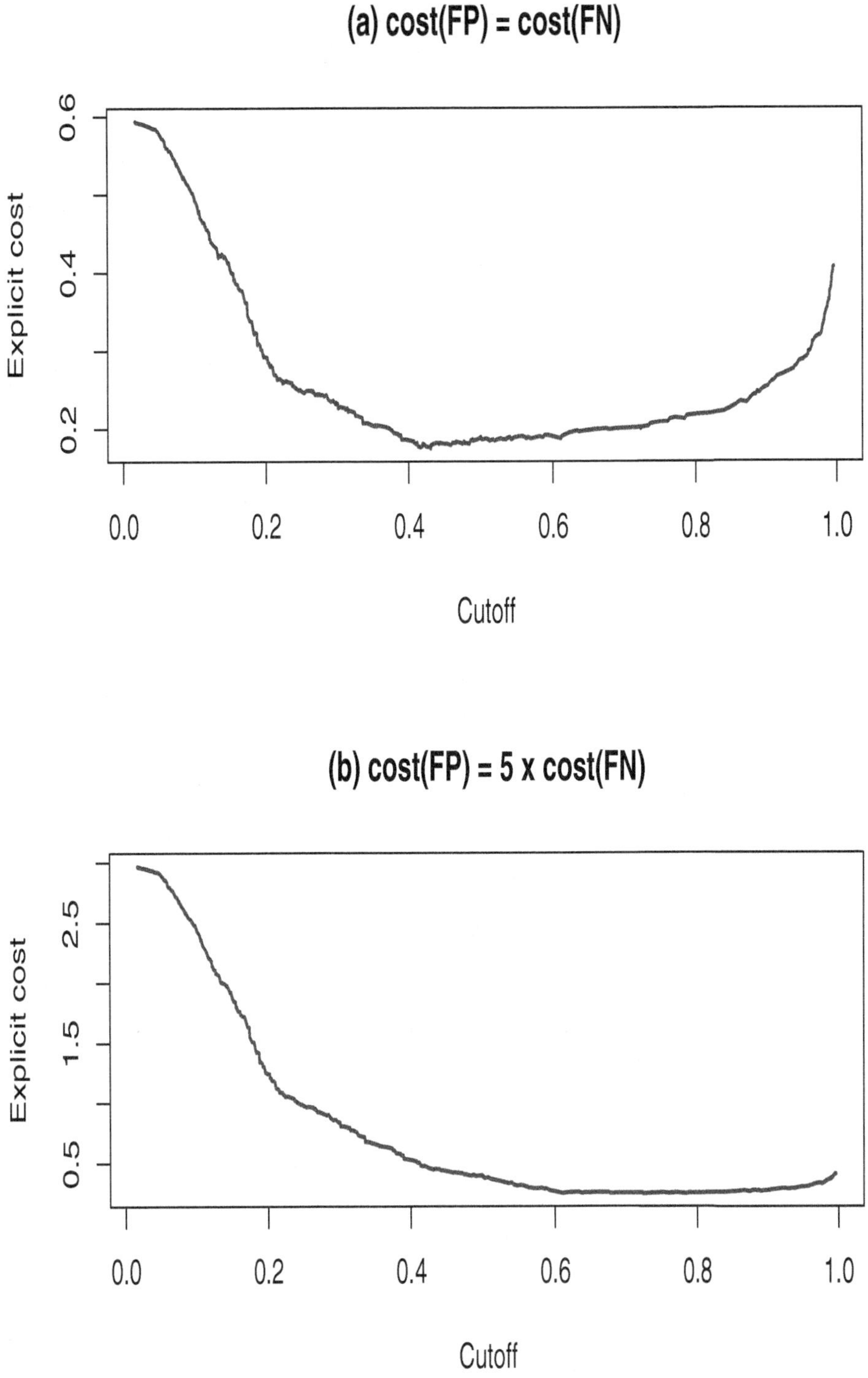

Figure 22.8: *The cost functions compared different cost structures. In plot (a), we plotted the cost function when the cost of a false positive is equal to the cost of a false negative. In plot (b), a false positive costs five times more than a false negative (valid for a loan in a bank).*

Again, the performance object with cost information can be plotted fairly easy. The code below uses the function `par()` to force two plots into one frame, and then resets the frame to one plot. The result is in Figure 22.8 on page 403

```
par(mfrow=c(2,1))
plot(perf_cst1, lwd=2, col='navy', main='(a) cost(FP) = cost(FN)')
plot(perf_cst2, lwd=2, col='navy', main='(b) cost(FP) = 5 x cost(FN)')
```

```
par(mfrow=c(1,1))
```

♣ 23 ♣

Learning Machines

Machine learning is not a new, nor a recent concept. The term "machine learning" (ML) was coined by Artur Samuel 1959 while he worked for IBM. A first book about machine learning for pattern recognition was "Learning Machines" – Nilsson (1965). Machine learning remained an active research topic, and the author was in the 1980s studying perceptron neural networks at the university.

ML
machine learning

Already in 1965, the first framework for "multilayer perceptrons" was published by Alexey Ivakhnenko and Lapa. Since 1986, we call those algorithms "deep learning," thanks to Rina Dechter. However, it was only in 2015, when Google's DeepMind was able to beat the best human Go player in the world with their AlpahGo system, that the term "deep learning" very popular and the old research field got a lot of new interest.

In the early days of machine learning (ML), one focussed on creating algorithms that could mimic human brain function, and achieve a form of artificial intelligence (AI). In the 1990s, computers were fast enough to make practical applications possible and the research became more focussed on wider applications, rather than focusing on reproducing intelligence artificially.

AI
artificial intelligence

Rule based intelligence (as in the computer language Lisp), became less popular as a research subject and in general the focus shifted away from symbolic approaches – inherited from AI, and it became more and more a specific discipline in statistics. This coincided with a period where massive amounts of data became available and computers passed a critical efficiency level. Today, we will continue to see practical results of this machine learning in general and deep learning in particular.

What makes machine learning different as a discipline, is that humans only set the elementary rules of learning and then let the machine learn from data, or even allow the machine to experiment and find out by itself what is a good or a bad result. Usually, one distinguishes the following forms of learning.

- *Supervised learning:* The algorithm will learn from provided results (e.g. we have data of good and bad credit customers) In other words, there is a "teacher" that knows the exact answers.

learning – supervised

- *Unsupervised learning:* The algorithm groups observations according to a given criteria (e.g. the algorithm classifies customers according to profitability without being told what good or bad is).

learning – unsupervised

The Big R-Book: From Data Science to Learning Machines and Big Data, First Edition. Philippe J.S. De Brouwer.

Companion Website: www.wiley.com/go/De Brouwer/The Big R-Book

learning – reinforced
learning – induction

- *Reinforced learning:* The algorithm learns from outcomes: rather than being told what is good or bad, the system will get something like a cost-function (e.g. the result of a treatment, the result of a chess game, or the relative return of a portfolio of investments in a competitive stock market). Another way of defining reinforced learning is that in this case, the environment rather than the teacher provides the right outcomes.

If feedback is available either in the form of correct outcomes (teacher) or can be deduced from the environment, then the task is to fit a function based on example input and output. We distinguish two forms of inductive learning:

classification
regression

- learning a discrete function is called **classification**,
- learning a continuous function is called **regression**.

These problems are equivalent to the problem of trying to approximate an unknown function $f()$. Since we do not know $f()$, we use a hypothesis $h()$ that will approximate $f()$. A first and powerful class of approximations are linear functions and hence linear regressions.[1]

hypothesis

With a regression, we try to fit a curve (line in the linear case) as closely as possible through all data points. However, if there are more observations, then the line will in general not go through all observations. In other words, given the hypothesis space H_L of all linear functions, the problem is *realizable* if all observations are co-linear. Otherwise it is not *realizable*, and we have to try to find a "good fit."

realizable
unrealizable

Of course, we could choose our hypothesis space **H** as large as all Turing machines. However, this will still not be sufficient in case of conflicting observations. Therefore, we need a method that finds a hypothesis that is optimal in a certain sense. The OLS method for linear regression is such example. The line will fit optimally all observations for the L_2-norm in the sense that the sum of the L_2 distances between each point and the fitted line is minimal.

OLS
ordinary least squares

[1] Indeed, we argue here that linear regressions are a form of machine learning.

23.1 Decision Tree

decision tree

A decision tree is another example of inductive machine learning. It is one of the most intuitive methods and in fact many heuristics from bird determination to medical treatment are often summarized in decision trees when humans have to learn. The way of thinking that follows from a decision tree comes natural: if I'm hungry then I check if I have cash on me, if so then I buy a sandwich.

23.1.1 Essential Background

23.1.1.1 The Linear Additive Decision Tree

A decision trees splits the feature space in rectangles and then fits in each rectangle a very simple model such as a constant in each rectangle as follows.

$$\hat{y} = \hat{f}(x) = \sum_{n=1}^{N} \alpha_n I\{x \in R_n\}$$

with $x = (x_1, \ldots, x_m)$ and $I\{b\}$ the identity function so that $I\{b\} := \left\{ \begin{smallmatrix} 1 & \textit{if } b \\ 0 & \textit{if } !b \end{smallmatrix} \right.$

This formula can be illustrated as in Figure 23.1. This illustration maker clear how the partitioning works for only two independent variables x_1 and x_2.

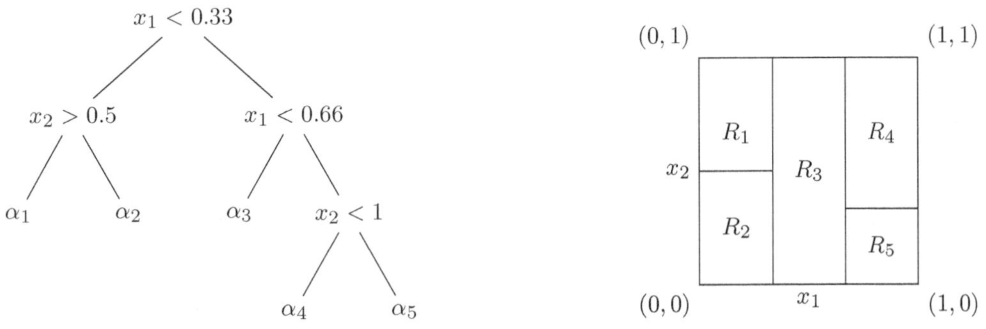

Figure 23.1: *An example of the decision tree on fake data a represented in two ways: on the left the decision tree and on the right the regions R_i that can be identified in the (x_1, x_2)-plane.*

> **Note – Not all sub-sets are possible**
>
> From Figure 23.1, it will be clear that only a certain type of partitioning can be obtained. For example, overlapping rectangles will never be possible with a decision tree, nor will round shapes be obtained.

23.1.1.2 The CART Method

In order to fit the tree (or regression), we need to identify a measure on "how good the fit is." If we adopt the popular sum of squares criterion, $\sum (y_i - f(x_i))^2$, then it can be shown that the average values of observations y_i in each region R_i isthe estimate for y_i:

$$\hat{y_i} = \text{avg}(y_i | x_i \in R_i)$$

Unfortunately, finding the best split of the tree for the minimum sum of squares criterion is in general not practical. One way to solve this problem is working with a "greedy algorithm." Start with all data and considering a splitting variable j (in the previous example we had only two variables x_1 and x_2, so j would be 1 or 2) and a splitting value x_j^s so that we can define the first region R_1 as

$$R_1(j, x_j^s) := \{x | x_j < x_j^s\}$$

and by doing so, we obtain also a second region R_2:

$$R_2(j, x_j^s) := \{x | x \notin R_1\} = \{x | x_j \geq x_j^s\}$$

The best splitting variable is then the one that minimizes the sum of squares between estimates and observations

$$\min_{j,s}\left[\min_{y_1} \sum_{x_i \in R_1(j,s)} (y_i - y_1)^2 + \min_{y_1} \sum_{x_i \in R_2(j,s)} (y_i - y_2)^2\right] \tag{23.1}$$

For any pair (j, x_j^s), we can solve the minimizations with the previously discussed average as estimator:

$$\begin{cases} \hat{y_1} = \text{avg}\left[y_i | x_i \in R_1(j, x_j^s)\right] \\ \hat{y_2} = \text{avg}\left[y_i | x_i \in R_2(j, x_j^s)\right] \end{cases}$$

This is computationally quite fast to solve and therefore, it is feasible to determine the split-point x_j^s for each attribute x_i and hence, completely solve Equation 23.1. This in its turn solves the optimal split for each pair (j, x_j^s), so that the optimal split variable (attribute) can be determined. This is then the first node of the tree. This procedure leads to a partition of the original data in two sets. Each set can be split once more with the same approach. This procedure can then be repeated on each sub-region.

Probably it comes natural to the reader to ponder at this point how large and complex we should grow the tree as well as it probably comes natural to think that stopping for a given minimal improvement in least squares makes sense. This would work if all variables x_i would be independently distributed. However, in reality, it is possible to see complex interactions. Let us consider the imaginary situation where credit risk for men improves with age, but deteriorates with age for women, while on average, there is little difference between creditworthiness between the average man and average woman. In that case, splitting the population on the attribute "gender" does not yield much improvement overall. However, once we only see one of the sexes, the decreasing or increasing trend can yield a significant gain in sum of squares. It might also be clear that first splitting on "age" would not work either if both men and women are equally represented, because then the trends would cancel each other out. Once the split in age group is done, the split according to "gender" will in the next node dramatically improve the sum of squares).

23.1.1.3 Tree Pruning

So the only solution to the problem of optimal tree size is to allow the tree to grow to a given size that is too large and over-fits the data and only then reduce the size. This reducing of size is called "pruning."

pruning

The idea is to minimize the "cost of complexity function" for a given pruning parameter α. The cost function is defined as

$$C_\alpha(T) := \sum_{n=1}^{|E_T|} \mathrm{SE}_n(T) + \alpha|T| \tag{23.2}$$

This is the sum of squares in each end-note plus α times the size of the tree. $|T|$ is the number of terminal nodes in the sub-tree T (T is a subtree to T_0 if T has only nodes of T_0), $|E_T|$ is the number of end-nodes in the tree T and $\mathrm{SE}_n(T)$ is the sum of squares in the end-node n for the tree T. The square errors in node n (or in region R_n) also equals:

$$\begin{aligned}\mathrm{SE}_n(T) &= N_n \ \mathrm{MSE}_n(T)\\ &= N_n \frac{1}{N_n} \sum_{x_i \in R_n}^{N_n} (y_i - \hat{y}_n)^2\\ &= \sum_{x_i \in R_n}^{N_n} (y_i - \hat{y}_n)^2\end{aligned}$$

with $\hat{y}_n$ the average of all y_i in the region n as explained previously.

It appears that for each α there is a unique smallest sub-tree, T_α that minimises the cost function $C_\alpha(T)$. A good way to find T_α is to use "weakest link pruning." This process works as follows: first grow the tree very large, and then collapse that internal node that produces the smallest increase in $\mathrm{SE}_n(T)$ and continue this till we have a single node tree. This produces a finite series of trees and it can be shown that this series includes T_α.

This process leaves the pruning parameter α up to us. Larger values of α lead to smaller trees. With $\alpha = 0$ one will get the full tree.

23.1.1.4 Classification Trees

In case the values y_i do not come from a numerical function but are rather a nominal or ordinal scale,[2] it is no longer possible to use MSE as a measure of fitness for the model. In that case, we can use the average number of matches with class c:

$$\hat{p}_{n,c} := \frac{1}{N_c} \sum_{x_i \in R_n} I\{y_i = c\} \tag{23.3}$$

The class c that has the highest proportion $\hat{p}_{n,c}$, is defined as $argmax_c(\hat{p}_{m,k})$. This is the value that we will assign in that node. The node impurity then can be calculated by one of the following:

$$\text{Gini index} = \sum_{c \neq \tilde{c}} \hat{p}_{n,c}\hat{p}_{n,\tilde{c}} \tag{23.4}$$

$$= \sum_{c=1}^{C} \hat{p}_{n,c}(1 - \hat{p}_{n,c}) \tag{23.5}$$

$$\text{Cross-entropy or deviance} = -\sum_{c=1}^{C} \hat{p}_{n,c} \log_2(\hat{p}_{n,c}) \tag{23.6}$$

$$\text{Misclassification error} = \frac{1}{N_n} \sum_{x_i \in R_n} I\{y_i = c\} \tag{23.7}$$

$$= 1 - \hat{p}_{n,c} \tag{23.8}$$

with C the total number of classes.

[2] For more details about ordinal and nominal scales we refer to Chapter B on page 829.

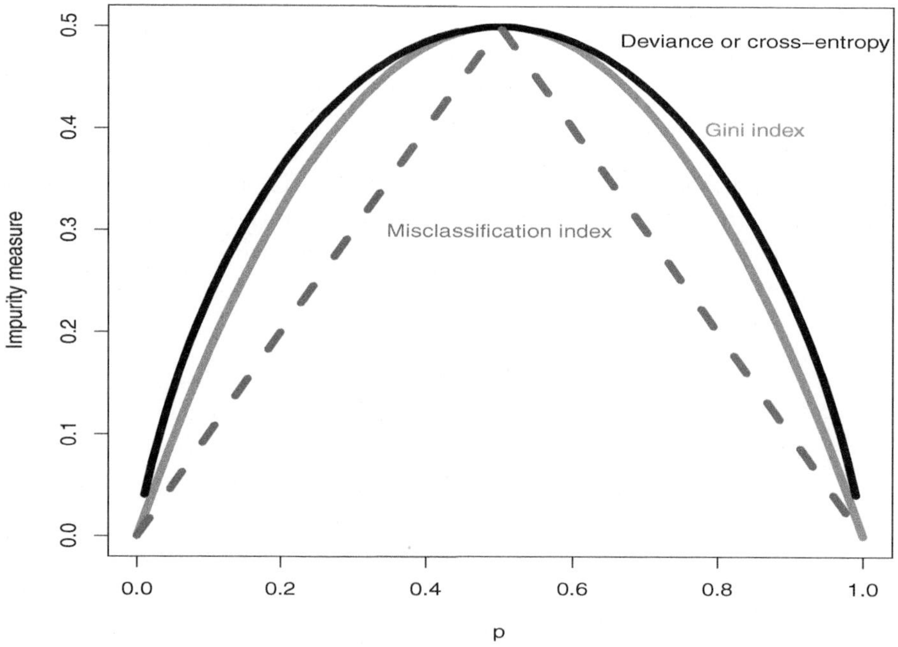

Figure 23.2: *Three alternatives for the impurity measure in the case of classification problems.*

With R we can plot the different measures of impurity. Note that the entropy-based measure has a maximum of 1, while the others are limited to 0.5, so for the representation it has been divided by 2 for the plot. The plot produced by the following code is in Figure 23.2.

Digression – Drawing the misclassification functions

```
# Draw Gini, deviance, and misclassification functions

# Define the functions:
gini <- function(x) 2 * x * (1-x)
entr <- function(x) (-x*log(x) - (1-x)*log(1-x))/log(2) / 2
misc <- function(x) {1 - pmax(x,1-x)}

# Plot the curves:
curve(gini, 0, 1, ylim = c(0,0.5), col = "forestgreen",
      xlab="p", ylab = "Impurity measure", type = "l", lwd = 6)
curve(entr, 0, 1, add = TRUE, col = "black", lwd = 6)

curve(misc, 0, 1, add = TRUE, col = "blue", type = "l",
      lty = 2, lwd = 6)

# Add the text:
text(0.85, 0.4,   "Gini index",                 col = "forestgreen")
text(0.85, 0.485, "Deviance or cross-entropy", col = "black")
text(0.5,  0.3,   "Misclassification index",   col = "blue")
```

The behaviour of the misclassification error around $p = 0.5$ is not differentiable and might lead to abrupt changes in the tree for small differences in data (maybe just one point that is forgotten). Because of that property of differentiability, cross-entropy, and Gini index are better suited for numerical optimisation. It also appears that both Gini index and deviance are more sensitive to changes in probabilities in the nodes. For example, it is possible to find cases where the misclassification error would not prefer a more pure node.

Note also that the Gini index can be interpreted in some other useful ways. For example, if we code the values 1 for observations of class $\tilde{c}$, then the variance over the node of this 0-1 response variable is $\hat{p}_{n,c}(1 - \hat{p}_{n,c})$ and summing over all classes c we find the Gini index.

Also if one would not assign all observations in the node to class c but rather assign them to class c with a probability equal to $\hat{p}_{n,c}$, then the Gini-coefficient is the training error rate in node n: $\sum_{c \neq \tilde{c}} \hat{p}_{n,c}\hat{p}_{n,\tilde{c}}$.

For pruning, all three measures can be trusted. Usually, the misclassification error is not used, because of the aforementioned mathematical properties. The procedure described in this chapter is known as the Classification and Regression Tree (CART) method.

CART
Classification and Regression Tree

23.1.1.5 Binary Classification Trees

While largely covered by the explanation above, it is worth to take a few minutes and study the particular case where the output variable is binary: true or false, good or bad, 0 or 1. This is not only a very important case, but it also allows us to make the parallel with information theory.

Binary classifications are important cases in everyday practice: good or bad credit risk, sick or not, death or alive, etc.

The mechanism to fit the tree works exactly the same. From all attributes, choose that one that classifies the results the best. Split the dataset according to the value that bests separates the goods from the bads.

Now, we need a way to tell what is a good split. This can be done by selecting the attribute that has the most information value. The information – measured in bits – of outcomes x_i with probabilities P_i is

$$I(P_1, \ldots, P_N) = -\sum_{i=1}^{N} P_i \log_2(P_i)$$

Which in the case of two possible outcomes (G the number of "good" observations and B the number of "bad" observations) reduces to

$$I\left(\frac{G}{G+B}, \frac{B}{G+B}\right) = -\frac{G}{G+B}\log_2\left(\frac{G}{G+B}\right) - \frac{B}{G+B}\log_2\left(\frac{B}{G+B}\right)$$

The amount of information provided by each attribute can be found by estimating the amount of information that still is to be gathered after the attribute is applied. Assume an attribute A (for example age of the potential creditor), and assume that we use 4 bins for the attribute age. In that case, the attribute age will separate our set of observations in four sub-sets: S_1, S_2, S_3, S_4. Each subset S_i then has G_i good observations and B_i bad observations. This means that if the attribute A is applied that we still need $I\left(\frac{G_i}{G_i+B_i}, \frac{B_i}{G_i+B_i}\right)$ bits of information to classify all observations correctly.

A random sample from our dataset will be in the ith age group with a probability $P_i := \frac{B_i+B_i}{G+B}$, since $B_i + G_i$ equals the number of observations in the ith age group. Hence, the remaining information after applying the attribute age is:

$$\text{remainder}(A) = \sum_{i=1}^{4} P_i I\left(\frac{G_i}{G_i+B_i}, \frac{B_i}{G_i+B_i}\right)$$

This means that the information gain of applying the attribute A is

$$\text{gain}(A) = I\left(\frac{G}{G+B}, \frac{B}{G+B}\right) - \text{remainder}(A)$$

So, it is sufficient to calculate the information gain for each attribute and split the population according to the attribute that has the highest information gain.

23.1.2 Important Considerations

Decision trees have many advantages: they are intuitive and can be represented in a 2D graph. The nodes in the graph represent an event or choice and the edges of the graph represent the decision rules or conditions. This makes the tool very accessible and an ideal method if a transparent model is needed.

23.1.2.1 Broadening the Scope

Decision trees raise some particular possibilities, we briefly discuss some below.

1. **Loss matrix**: In reality a misclassification in one or the other group does not come at the same cost. If we identify the tumour wrongly as cancer, the health insurance will lose some money for an unnecessary treatment, but if we misclassify the tumour as harmless then the patient will die. That is an order of magnitude worse. A bank might wrongly reject a good customer and fail to earn an interest income of 1000 on a loan of 10 000. However, if the bank accepts the wrong customer then the bank can loose 10 000 or more in recovery costs. This can be mitigated with a loss matrix. Define a $C \times C$ loss matrix $\mathbf{L}$, with L_{kl} the loss incurred by misclassifying a class k as a class l. As a reference, one usually takes a correct classification as a zero cost: $L_{kk} = 0$. Now, it is sufficient to modify the Gini-index as $\sum_{k \neq l}^{C} L_{kl}\hat{p}_{nk}\hat{p}_{nl}$. This works fine when $C > 2$, but unfortunately, the symmetry of a two-class problem this has no effect since the coefficient of $\hat{p}_{nk}\hat{p}_{nl}$ is $(L_{kl} + L_{lk})$. The workaround is weighting the observations with of class k by L_{kl}.

2. **Missing values**: Missing values are a problem for all statistical methods. The classical approach of leaving such observation out or trying to fill them in via some model, can lead to serious issues if the fact that data is missing has a specific reason. For example, males might not fill in the "gender" information if they know that it will increase the price of their car-insurance. In this particular case no gender information can be worse insurance risk than "male" if only the males that already had issues learned this trick. Decision trees allow for another method: assign a specific value "missing" to that predictor value. Alternatively, one can use surrogate splits: first work only with the data that has no missing fields and then try to find alternative variables that provide the same split.

3. **Linear combination splits**: In our approach, each node has a simple decision model of the form $x_i \leq x_j^s$. One can consider decisions of the form $\sum \alpha_i x_i \leq x_j^s$. Depending on the particular case, this might considerably improve the predictability of the model, however, it will be more difficult to understand what happens. In the ideal case, these nodes would lead to some clear attributes such as "risk seeking behaviour" (where the model might create this concept out of a combination of "male," "age group 1" and "marital status unmarried") or affordability to pay ("martial status married," "has a job longer than two years"). In general, this will hardly happen and it becomes unclear what exactly is going on and why the model is refusing the loan. This in its turn makes it more difficult to assess if a model is over-fit or not.

4. **Link with ANOVA:** An alternative way to understand the ideal stopping point is using the ANOVA approach. The impurity in a node can be thought of as the MSE in that node. **anova**

$$\text{MSE} = \sum_{i=1}^{n} (y_i - \bar{y})^2$$

with y_i the value of the ith observation and $\bar{y}$ the average of all observations.

This node impurity can also be thought of as in ANOVA analyses.

$$\frac{\frac{\text{SS}_{\text{between}}}{B-1}}{\frac{\text{SS}_{\text{within}}}{n-B}} \sim F_{n-B,B-1}$$

with

$$\begin{cases} \text{SS}_{\text{between}} & = n_b \sum_{b=1}^{B} (\bar{y}_b - \bar{y})^2 \\ \text{SS}_{\text{within}} & = \sum_{b=1}^{B} \sum_{i=1}^{n_b} (\bar{y}_{bi} - \bar{y})^2 \end{cases}$$

with B the number of branches, n_b the number of observations in branch b, $y_b i$ the value of observation bi.

Now, optimal stopping can be determined by using measures of fit and relevance as in a linear regression model. For example, one can rely on R^2, MAD, etc.

5. **Other tree building procedures**: The method described so far is known as classification and regression tree (CART). Other popular choices are ID3 (and its successors C4.5, C5.0) and MARS.

23.1.2.2 Selected Issues

When working with decision trees, it is essential to be aware of the following issues, and have some plan to mitigate or minimize them.

1. **Over-fitting:** this is one of the most important issues with decision trees. It should never be used without appropriate validation methods such as cross validation or random forest approach before an effort to prune the tree. See e.g. Hastie et al. (2009). **over-fitting**

2. **Categorical predictor values**: Categorical variables that represent a nominal or ordinal scale present a specific challenge. If the variable has κ possible values, then the number of partitions that can be made is $2^{\kappa-1} - 1$. In general, the algorithms tend to favour categorical predictors with many levels with the number of partitions growing exponentially in κ, which will lead to severe over-fitting.

3. **Instability**: Small changes in data can lead to dramatically different tree structures. This is because even a small change on the top of the tree will be cascaded down the tree. This works very different in linear regression models for example where one additional data-point (unless it is an outlier) will only have a small influence on the parameters of the model. Methods such as Random Forest somehow mitigate this instability, but also bagging of data will improve the stability. **bagging**

4. **Difficulties to capture additive relationships**: A decision tree naturally will fit decisions that are not additive. For example, if a person has the affordability to pay and he is honest then he will pay the loan back. This would work fine in a decision tree, if however, the fact

of the customer paying back the loan depends on many other factors that have to be all in place *and* can mitigate each other then a additive relationship might work better.[3]

5. **Stepwise predictions**: The outcome of a decision trees will naturally be approximated by a step-function. Modelling a linear relationship for example is not efficient as it will be approximated by a number of discrete steps. The MARS procedure allows to alleviate this to some extend.

23.1.3 Growing Trees with the Package rpart

23.1.3.1 Getting Started with the Function rpart()

rpart() rpart

The function `rpart()`, of the package with the same name, provides a good implementation of the CART algorithm.

Function use for rpart()

```
rpart(formula, data, weights, subset, na.action = na.rpart,
      method=c(``class'',``anova''), model = FALSE,
      x = FALSE, y = TRUE, parms, control, cost, ...)
```

with the most important parameters:

- `data`: the data-frame containing the columns to be used in `formula`.
- `formula`: am R-formula of the form `y ~ x1 + x1 + ...` – note that the plus signs do not really symbolise the addition here, but only indicate which columns to choose.
- `weights`: optional case weights.
- `subset`: optional expression that indicates which section of the data should be used.
- `na.action`: optional information on what to do with missing values. The default is `na.rpart`, which means that all rows with `y` missing will be deleted, but any x_i can be missing.
- `method`: optional method such as "anova," "poisson," "class" (for classification tree), or "exp". If it is missing, a reasonably guess will be made, based on the nature of `y`.

As usual, more information is in the documentation of the function and the package.

With the package `rpart`, the work-flow becomes:

1. load the package `rpart`

[3] An additional logic to predict payment capacity would be for example having a stable job, having a diploma, having a spouse that can step in, having savings, etc. Indeed, all those things can be added, and if the spouse is gone, the diploma will help to find a next job: these variables compensate. This is not the case for variables such as over-indebted, fraud (no intention to pay), unstable job, etc., in this case there is no compensation: one of those things going wrong will result in a customer that does not pay his loan back.

2. fit the tree with `rpart()` and eventually control the size of the tree with `control = rpart.control(minsplit=20, cp=0.01)`. This example will set the minimum number of observations in a node to 20 and that a split must decrease the overall impurity of fit by a factor of 0.01 (cost complexity factor) before being attempted.

3. investigate and visualize the results with

   - `printcp(t)` to visualize the cp (complexity parameter) table[4]
   - `plotcp(t)` to plot cross validation test
   - `rsq.rpart(t)` to plot R-squared and relative error for different splits (2 plots). Note that labels are only appropriate for the "anova" method.
   - `print(t)` print results
   - `summary(t)` to display the results (with surrogate splits)
   - `plot(t)` to plot the decision tree
   - `text(t)` to add labels to the plot
   - `post(t, file="")` to create create a postscript file of the plot
   - `parms`: optional parameters to the splitting function – below we describe how to correct for prior probabilities with this parameter.

Further information about rpart

More information about the `rpart` package can be found in the documentation of the project: `https://cran.r-project.org/web/packages/rpart/vignettes/longintro.pdf`.

23.1.3.2 Example of a Classification Tree with rpart

In practice, data is usually biased. For example, a bank only gives loans to customers when they are expected to pay back the loan. This means that any dataset of existing customers will show an overwhelming amount of "good" customers compared to very few customers that defaulted on their loans. This means that variance-based methods such as a decision tree will be heavily biased toward reliable customers, the method will try to classify on average each customer as good or bad. The amount of bad customers will never generate the same amount of variance.

Imagine that we have a database of 100 000 customers from which 1 000 defaulted on their loans. There are multiple ways of addressing this problem:

1. copy the existing bad customers 99 times
2. drop 98 000 good customers randomly
3. weight the observations.

The function `rpart` allows to correct these prior probabilities via the variable `parms`. Note that this is what is option 3 in the list above.

[4]The complexity parameter (cp) is used to control the size of the decision tree and to select the optimal tree size. If the cost of adding another variable to the decision tree from the current node is above the value of cp, then tree building does not continue. We could also say that tree construction does not continue unless it would decrease the overall lack of fit by a factor of cp. The complexity parameter as implemented in the package `rpart` is not the same as Equation 23.2 but follows from a similar build-up: $C_{\text{cp}}(T) := C(T) + \text{cp}\,|T|\,C(T_0)$

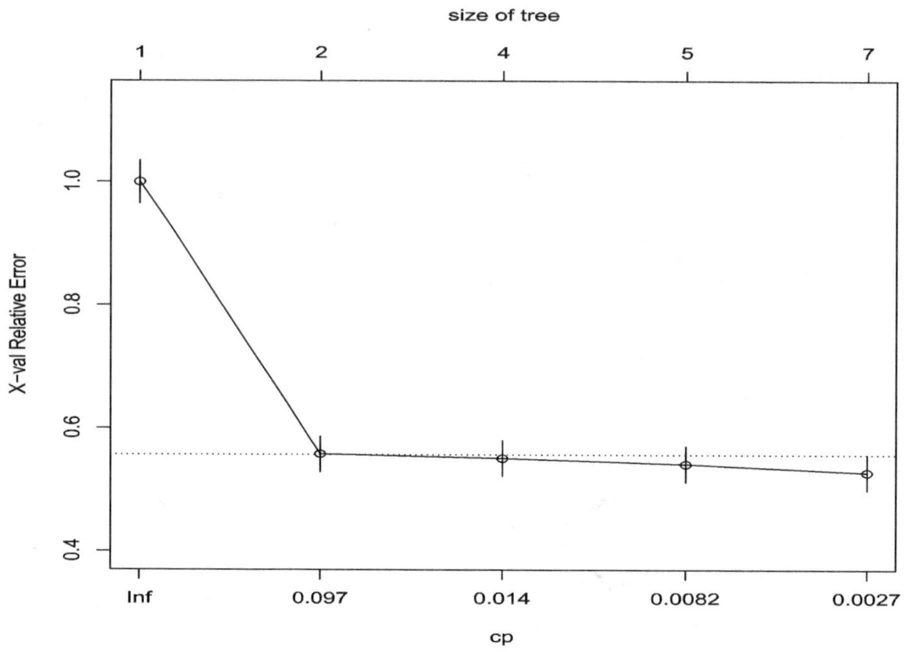

Figure 23.3: *The plot of the complexity parameter (cp) via the function* `plotcp()`.

In this section we will develop an example based on the sinking of the RMS Titanic in 1929. The death toll of the disaster of the RMS Titanic was very high and only 500 of the 1309 passengers survived. This means that the "prior probability" of surviving is about $0.3819 = \frac{500}{1309}$. So, this particular dataset has a similar amount of good and bad results. In this case, one might omit the following parameter:

```
parms = list(prior = c(0.6,0.4))
```

The code below imports the data of the passengers on board of the RMS Titanic, fits a regression tree, prunes the tree, and visualises results. Three plots are produced:

1. A visualisation of the complexity parameter and its impact on the errors in Figure 23.3.
2. The decision tree before pruning in Figure 23.4 on page 417.
3. The decision tree after pruning in Figure 23.5 on page 417.

```
## example of a regression tree with rpart on the dataset of the Titanic
##
library(rpart)
titanic <- read.csv("data/titanic3.csv")
frm     <- survived ~ pclass + sex + sibsp + parch + embarked + age
t0      <- rpart(frm, data=titanic, na.action = na.rpart,
  method="class",
  parms = list(prior = c(0.6,0.4)),
  #weights=c(...), # each observation (row) can be weighted
  control = rpart.control(
  minsplit        = 50,  # minimum nbr. of observations required for split
  minbucket       = 20,  # minimum nbr. of observations in a terminal node
```

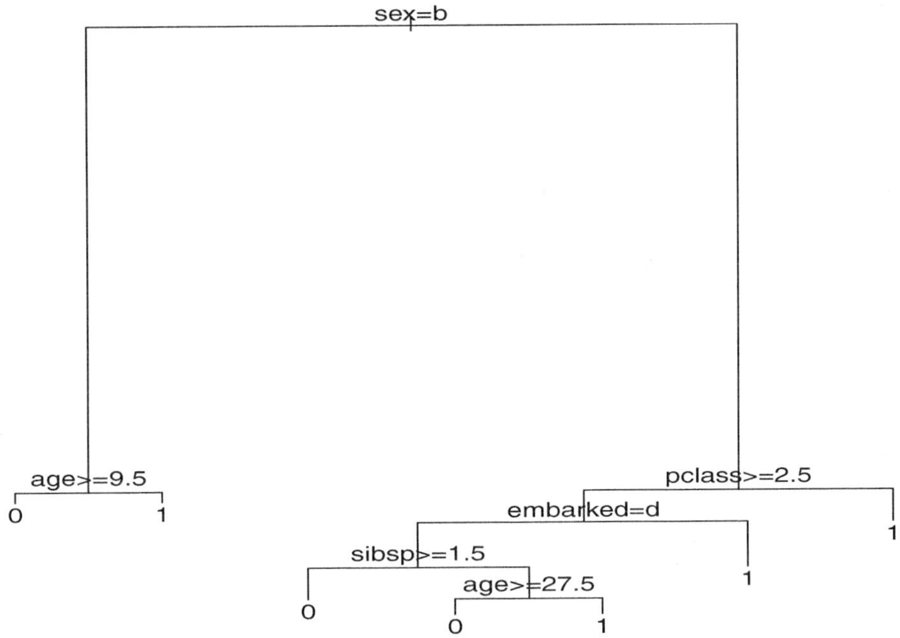

Figure 23.4: *The decision tree, fitted by rpart. This figure helps to visualize what happens in the decision tree that predicts survival in the Titanic disaster.*

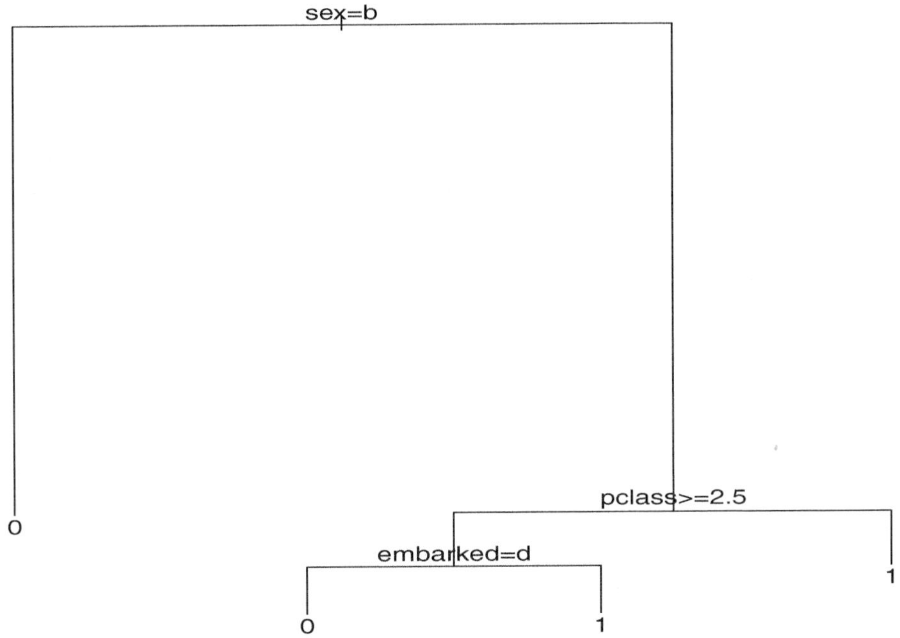

Figure 23.5: *The same tree as in Figure 23.4 but now pruned with a complexity parameter* $\rho = 0.01$. *Note that the tree is .*

```
  cp             = 0.001,# complexity parameter set to a small value
                        # this will grow a large (over-fit) tree
  maxcompete     = 4,   # nbr. of competitor splits retained in output
  maxsurrogate   = 5,   # nbr. of surrogate splits retained in output
  usesurrogate   = 2,   # how to use surrogates in the splitting process
  xval           = 7,   # nbr. of cross validations
  surrogatestyle = 0,   # controls the selection of a best surrogate
  maxdepth       = 6)   # maximum depth of any node of the final tree
  )

# Show details about the tree t0:
printcp(t0)
##
## Classification tree:
## rpart(formula = frm, data = titanic, na.action = na.rpart, method = "class",
##     parms = list(prior = c(0.6, 0.4)), control = rpart.control(minsplit = 50,
##         minbucket = 20, cp = 0.001, maxcompete = 4, maxsurrogate = 5,
##         usesurrogate = 2, xval = 7, surrogatestyle = 0, maxdepth = 6))
##
## Variables actually used in tree construction:
## [1] age      embarked pclass   sex      sibsp
##
## Root node error: 523.6/1309 = 0.4
##
## n= 1309
##
##          CP nsplit rel error  xerror     xstd
## 1 0.4425241      0   1.00000 1.00000 0.035158
## 2 0.0213115      1   0.55748 0.55748 0.029038
## 3 0.0092089      3   0.51485 0.55062 0.029130
## 4 0.0073337      4   0.50564 0.54106 0.028952
## 5 0.0010000      6   0.49098 0.52771 0.028569

# Plot the error in function of the complexity parameter
plotcp(t0)
```

```
# print(t0) # to avoid too long output we commented this out
# summary(t0)

# Plot the original decisions tree
plot(t0)


text(t0)
```

```
# Prune the tree:
t1 <- prune(t0, cp=0.01)
plot(t1); text(t1)
```

23.1.3.3 Visualising a Decision Tree with rpart.plot

When plotting the tree, we used the standard method that is supplied with the library `rplot`: the `plot.rpart` and `text.rpart`. The package `rpart.plot` replaces this functionality and adds useful information and visual pleasing effects. There are options specific for classification trees and others for regression trees.

rpart.plot

prp()

The following code loads the library `rpart.plot` and uses its function `prp()` to produce the plot in Figure 23.6 on page 419

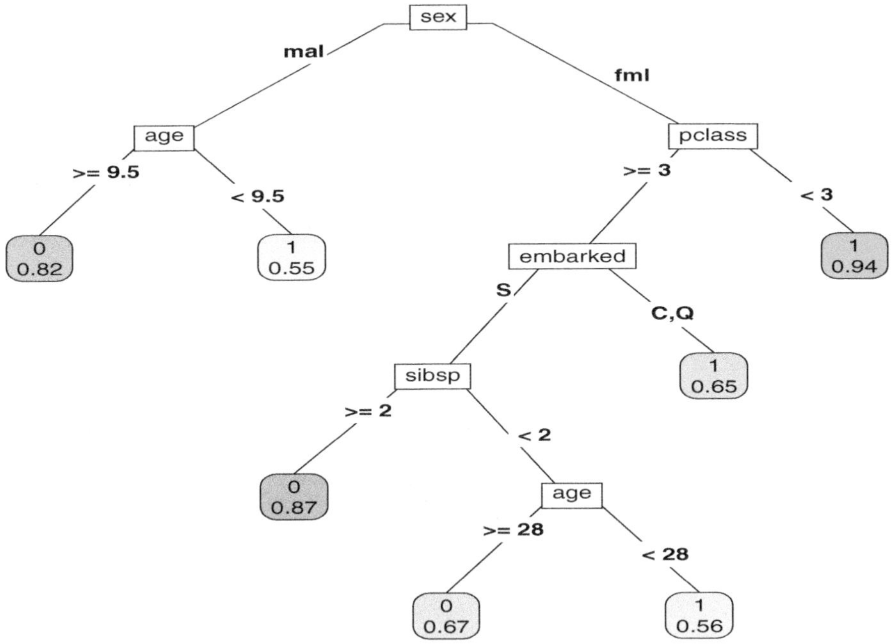

Figure 23.6: *The decision tree represented by the function* `prp()` *from the package* `rpart.plot`. *This plot not only looks more elegant, but it is also more informative and less simplified. For example the top node "sex" has now two clear options in which descriptions we can recognize the words male and female, and the words are on the branches, so there is no confustion possible which is left and which right.*

```
# plot the tree with rpart.plot
library(rpart.plot)
prp(t0, type = 5, extra = 8, box.palette = "auto",
    yesno = 1, yes.text="survived",no.text="dead"
    )
```

The function `prp()` takes many more arguments and allows the user to write functions to obtain exactly the desired result. More information is in the function documentation that is available in R itself by typing `?prp`.

23.1.3.4 Example of a Regression Tree with rpart

A regression tree predicts a continuous variable. We will use in this example the dataset `mtcars`, and predict fuel consumption in miles per gallon. So the dependent variable is the column `mpg` and we use the other columns as independent variables. The work-flow is the same, only details in the function `rpart()` are different: especially the parameter `method` is important. It is via this parameter that we can ensure that the decision tree is a regression tree.

The following code uses `rpart()` with the parameter `method` set to fit the regression tree, and defines appropriate controls to suit the data of `mtcars`. We then print

- the impact of the complexity parameter in Figure 23.7 on page 420,
- the un-pruned tree in Figure 23.8 on page 420, and
- the pruned tree in Figure 23.9 on page 421.

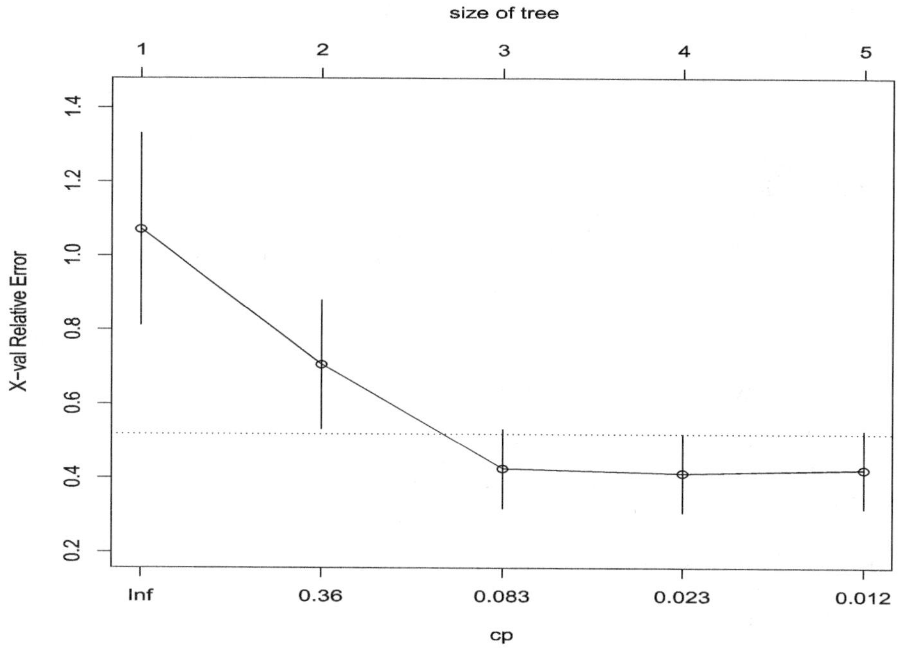

Figure 23.7: *The plot of the complexity parameter (cp) via the function* `plotcp()`

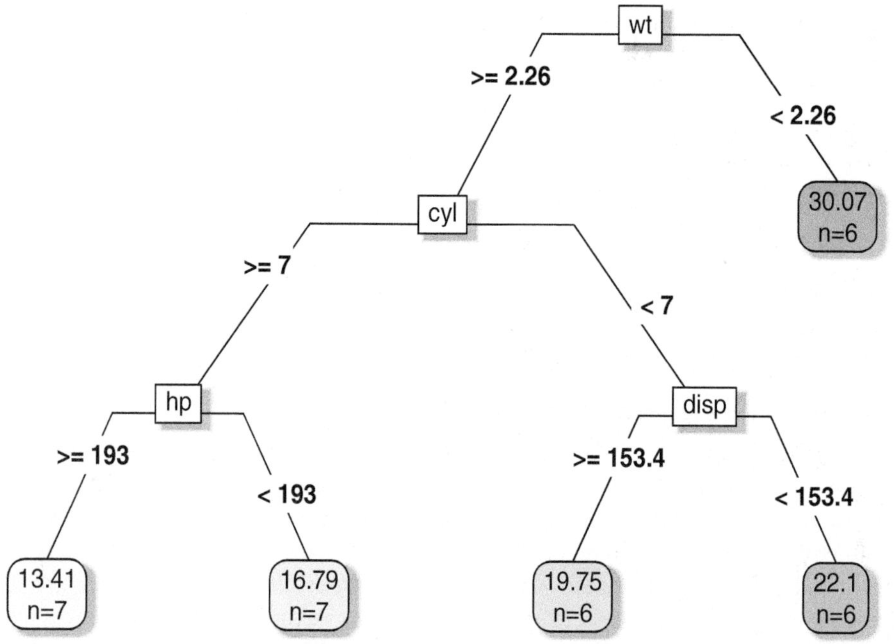

Figure 23.8: *rpart tree on mpg for the dataset mtcars.*

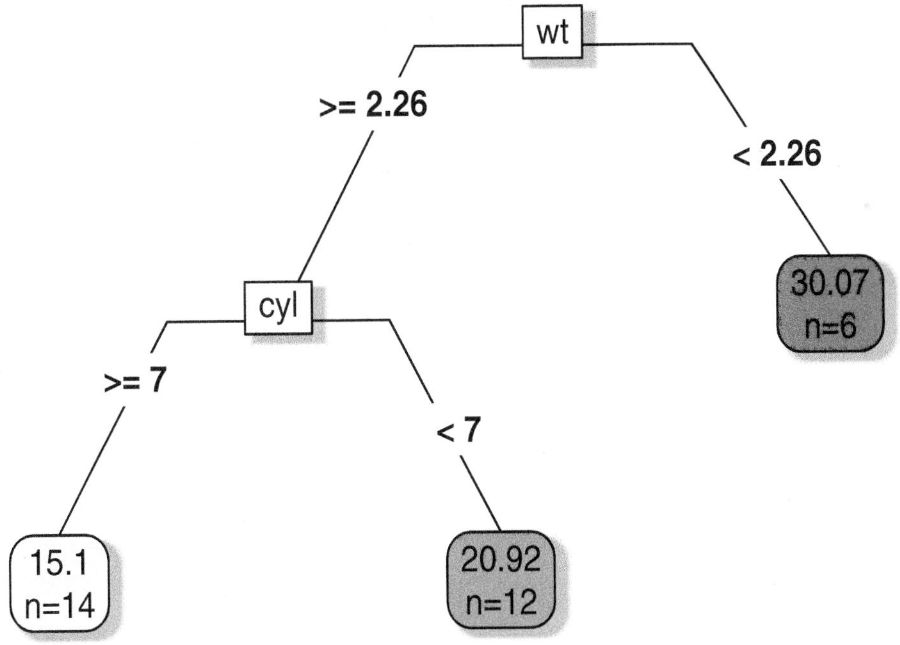

Figure 23.9: *The same tree as in Figure 23.8 but now pruned with a complexity parameter ρ of 0.1. The regression tree is – in this example – too simple.*

```
# Example of a regression tree with rpart on the dataset mtcars

# The libraries should be loaded by now:
library(rpart); library(MASS); library (rpart.plot)

# Fit the tree:
t <- rpart(mpg ~ cyl + disp + hp + drat + wt + qsec + am + gear,
 data=mtcars, na.action = na.rpart,
 method     = "anova",
 control    = rpart.control(
   minsplit       = 10,  # minimum nbr. of observations required for split
   minbucket      = 20/3,# minimum nbr. of observations in a terminal node
                         # the default = minsplit/3
   cp             = 0.01,# complexity parameter set to a very small value
                         # his will grow a large (over-fit) tree
   maxcompete     = 4,   # nbr. of competitor splits retained in output
   maxsurrogate   = 5,   # nbr. of surrogate splits retained in output
   usesurrogate   = 2,   # how to use surrogates in the splitting process
   xval           = 7,   # nbr. of cross validations
   surrogatestyle = 0,   # controls the selection of a best surrogate
   maxdepth       = 30   # maximum depth of any node of the final tree
   )
 )

# Investigate the complexity parameter dependence:
printcp(t)
##
## Regression tree:
## rpart(formula = mpg ~ cyl + disp + hp + drat + wt + qsec + am +
##     gear, data = mtcars, na.action = na.rpart, method = "anova",
##     control = rpart.control(minsplit = 10, minbucket = 20/3,
```

```
##          cp = 0.01, maxcompete = 4, maxsurrogate = 5, usesurrogate = 2,
##          xval = 7, surrogatestyle = 0, maxdepth = 30))
##
## Variables actually used in tree construction:
## [1] cyl  disp hp   wt
##
## Root node error: 1126/32 = 35.189
##
## n= 32
##
##          CP nsplit rel error  xerror    xstd
## 1 0.652661      0   1.00000 1.07140 0.25928
## 2 0.194702      1   0.34734 0.70608 0.17364
## 3 0.035330      2   0.15264 0.42499 0.10673
## 4 0.014713      3   0.11731 0.41249 0.10579
## 5 0.010000      4   0.10259 0.42200 0.10507

plotcp(t)
```

```
# Print the tree:
print(t)
## n= 32
##
## node), split, n, deviance, yval
##       * denotes terminal node
##
##  1) root 32 1126.04700 20.09062
##    2) wt>=2.26 26  346.56650 17.78846
##      4) cyl>=7 14   85.20000 15.10000
##        8) hp>=192.5 7   28.82857 13.41429 *
##        9) hp< 192.5 7   16.58857 16.78571 *
##      5) cyl< 7 12   42.12250 20.92500
##       10) disp>=153.35 6   12.67500 19.75000 *
##       11) disp< 153.35 6   12.88000 22.10000 *
##    3) wt< 2.26 6   44.55333 30.06667 *

summary(t)
## Call:
## rpart(formula = mpg ~ cyl + disp + hp + drat + wt + qsec + am +
##     gear, data = mtcars, na.action = na.rpart, method = "anova",
##     control = rpart.control(minsplit = 10, minbucket = 20/3,
##         cp = 0.01, maxcompete = 4, maxsurrogate = 5, usesurrogate = 2,
##         xval = 7, surrogatestyle = 0, maxdepth = 30))
##   n= 32
##
##           CP nsplit rel error    xerror      xstd
## 1 0.65266121      0 1.0000000 1.0714029 0.2592763
## 2 0.19470235      1 0.3473388 0.7060794 0.1736361
## 3 0.03532965      2 0.1526364 0.4249901 0.1067328
## 4 0.01471297      3 0.1173068 0.4124873 0.1057872
## 5 0.01000000      4 0.1025938 0.4219970 0.1050699
##
## Variable importance
##   wt disp   hp drat  cyl qsec
##   25   24   20   15   10    5
##
## Node number 1: 32 observations,    complexity param=0.6526612
##   mean=20.09062, MSE=35.18897
##   left son=2 (26 obs) right son=3 (6 obs)
##   Primary splits:
##       wt   < 2.26   to the right, improve=0.6526612, (0 missing)
##       cyl  < 5      to the right, improve=0.6431252, (0 missing)
```

```
##       disp < 163.8  to the right, improve=0.6130502, (0 missing)
##       hp   < 118    to the right, improve=0.6010712, (0 missing)
##       drat < 3.75   to the left,  improve=0.4186711, (0 missing)
##   Surrogate splits:
##       disp < 101.55 to the right, agree=0.969, adj=0.833, (0 split)
##       hp   < 92     to the right, agree=0.938, adj=0.667, (0 split)
##       drat < 4      to the left,  agree=0.906, adj=0.500, (0 split)
##       cyl  < 5      to the right, agree=0.844, adj=0.167, (0 split)
##
## Node number 2: 26 observations,    complexity param=0.1947024
##   mean=17.78846, MSE=13.32948
##   left son=4 (14 obs) right son=5 (12 obs)
##   Primary splits:
##       cyl  < 7      to the right, improve=0.6326174, (0 missing)
##       disp < 266.9  to the right, improve=0.6326174, (0 missing)
##       hp   < 136.5  to the right, improve=0.5803554, (0 missing)
##       wt   < 3.325  to the right, improve=0.5393370, (0 missing)
##       qsec < 18.15  to the left,  improve=0.4210605, (0 missing)
##   Surrogate splits:
##       disp < 266.9  to the right, agree=1.000, adj=1.000, (0 split)
##       hp   < 136.5  to the right, agree=0.962, adj=0.917, (0 split)
##       wt   < 3.49   to the right, agree=0.885, adj=0.750, (0 split)
##       qsec < 18.15  to the left,  agree=0.885, adj=0.750, (0 split)
##       drat < 3.58   to the left,  agree=0.846, adj=0.667, (0 split)
##
## Node number 3: 6 observations
##   mean=30.06667, MSE=7.425556
##
## Node number 4: 14 observations,    complexity param=0.03532965
##   mean=15.1, MSE=6.085714
##   left son=8 (7 obs) right son=9 (7 obs)
##   Primary splits:
##       hp   < 192.5  to the right, improve=0.46693490, (0 missing)
##       wt   < 3.81   to the right, improve=0.13159230, (0 missing)
##       qsec < 17.35  to the right, improve=0.13159230, (0 missing)
##       drat < 3.075  to the left,  improve=0.09982394, (0 missing)
##       disp < 334    to the right, improve=0.05477308, (0 missing)
##   Surrogate splits:
##       drat < 3.18   to the right, agree=0.857, adj=0.714, (0 split)
##       disp < 334    to the right, agree=0.786, adj=0.571, (0 split)
##       qsec < 16.355 to the left,  agree=0.786, adj=0.571, (0 split)
##       wt   < 4.66   to the right, agree=0.714, adj=0.429, (0 split)
##       am   < 0.5    to the right, agree=0.643, adj=0.286, (0 split)
##
## Node number 5: 12 observations,    complexity param=0.01471297
##   mean=20.925, MSE=3.510208
##   left son=10 (6 obs) right son=11 (6 obs)
##   Primary splits:
##       disp < 153.35 to the right, improve=0.393317100, (0 missing)
##       hp   < 109.5  to the right, improve=0.235048600, (0 missing)
##       drat < 3.875  to the right, improve=0.043701900, (0 missing)
##       wt   < 3.0125 to the right, improve=0.027083700, (0 missing)
##       qsec < 18.755 to the left,  improve=0.001602469, (0 missing)
##   Surrogate splits:
##       cyl  < 5      to the right, agree=0.917, adj=0.833, (0 split)
##       hp   < 101    to the right, agree=0.833, adj=0.667, (0 split)
##       wt   < 3.2025 to the right, agree=0.833, adj=0.667, (0 split)
##       drat < 3.35   to the left,  agree=0.667, adj=0.333, (0 split)
##       qsec < 18.45  to the left,  agree=0.667, adj=0.333, (0 split)
##
## Node number 8: 7 observations
##   mean=13.41429, MSE=4.118367
```

```
##
## Node number 9: 7 observations
##    mean=16.78571, MSE=2.369796
##
## Node number 10: 6 observations
##    mean=19.75, MSE=2.1125
##
## Node number 11: 6 observations
##    mean=22.1, MSE=2.146667

# plot(t) ; text(t)  # This would produce the standard plot from rpart.
# Instead we use:
prp(t, type = 5, extra = 1, box.palette = "Blues", digits = 4,
    shadow.col = 'darkgray', branch = 0.5)

# Prune the tree:
t1 <- prune(t, cp = 0.05)

# Finally, plot the pruned tree:
prp(t1, type = 5, extra = 1, box.palette = "Reds", digits = 4,
    shadow.col = 'darkgray', branch = 0.5)
```

23.1.4 Evaluating the Performance of a Decision Tree

23.1.4.1 The Performance of the Regression Tree

In order to assess the performance of a regression tree, one can use the same tools as used for linear regression: MSE, MAD, R-squared, etc. These performance measures are explained and discussed in Section 21.3 *"Performance of Regression Models"* on page 384 as basic tools and they are repeated in Chapter 25.1 *"Model Quality Measures"* on page 476, where we also use the `tidyverse` to We will therefore assume that these are commonly known and will focus on the classification tree.

23.1.4.2 The Performance of the Classification Tree

ROCR

The performance of a classification tree can be handled in the same way as a logistic regression and is already discussed in Chapter 22.1 on page 388. using the ROCR package, the code in R is be almost the same.

We continue the example of the RMS Titanic disaster of Section 23.1.3.2 *"Example of a Classification Tree with rpart"* on page 415. The first step is to to make predictions, based on the tree `t0`.

```
# We use the function stats::predict()
predicPerc <- predict(object=t0, newdata=titanic)

# predicPerc is now a matrix with probabilities: in column 1 the
# probability not to survive, in column 2 the probability to survive:
head(predicPerc)
##            0         1
## 1 0.06335498 0.9366450
## 2 0.44928523 0.5507148
## 3 0.06335498 0.9366450
## 4 0.81814936 0.1818506
## 5 0.06335498 0.9366450
## 6 0.81814936 0.1818506

# This is not what we need. We need to specify that it is a
# classification tree. Here we correct this:
```

```
predic <- predict(object=t0, newdata=titanic, type="class")
  # vector with only the fitted class as prediction
head(predic)
## 1 2 3 4 5 6
## 1 1 1 0 1 0
## Levels: 0 1
```

A first and very simple approach to asses performance is to calculate the confusion matrix. This confusion matrix shows the correct classifications and misclassifications.

```
# The confusion matrix:
confusion_matrix <- table(predic, titanic$survived)
rownames(confusion_matrix) <- c("predicted_death",
                                "predicted_survival")
colnames(confusion_matrix) <- c("observed_death",
                                "observed_survival")
confusion_matrix
##
## predic                observed_death observed_survival
##   predicted_death                706               150
##   predicted_survival             103               350

# As a precentage:
confusion_matrixPerc <- sweep(confusion_matrix, 2,
                    margin.table(confusion_matrix,2),"/")

# Here is the confusion matrix:
round(confusion_matrixPerc,2)
##
## predic                observed_death observed_survival
##   predicted_death               0.87              0.30
##   predicted_survival            0.13              0.70
```

Hint – Confusing matrix ready from a package

Above we show the code to produce a confusion matrix. It is also possible to use the function `confusionMatrix()` from the package `caret`. It will not only show the confusion matrix but also some useful statistics.
The package `caret` provides great visualization tools and as well as tools to split data, preprocess data, feature selection, model tuning using resampling and variable importance estimation.
Here is a great introduction: `https://cran.r-project.org/web/packages/caret/vignettes/caret.html`

`confusionMatrix()`

`caret`

The ROC curve can be obtained via the package ROCR as we have seen in Section 22.2.2 *"ROC"* on page 393. To do this, we load the library, create predictions and show the ROC in Figure 23.10 on page 426 with the following code.

```
library(ROCR)
pred <- prediction(predict(t0, type = "prob")[,2],
                                 titanic$survived)

# Visualize the ROC curve:
plot(performance(pred, "tpr", "fpr"), col="blue", lwd=3)
abline(0,1,lty=2)
```

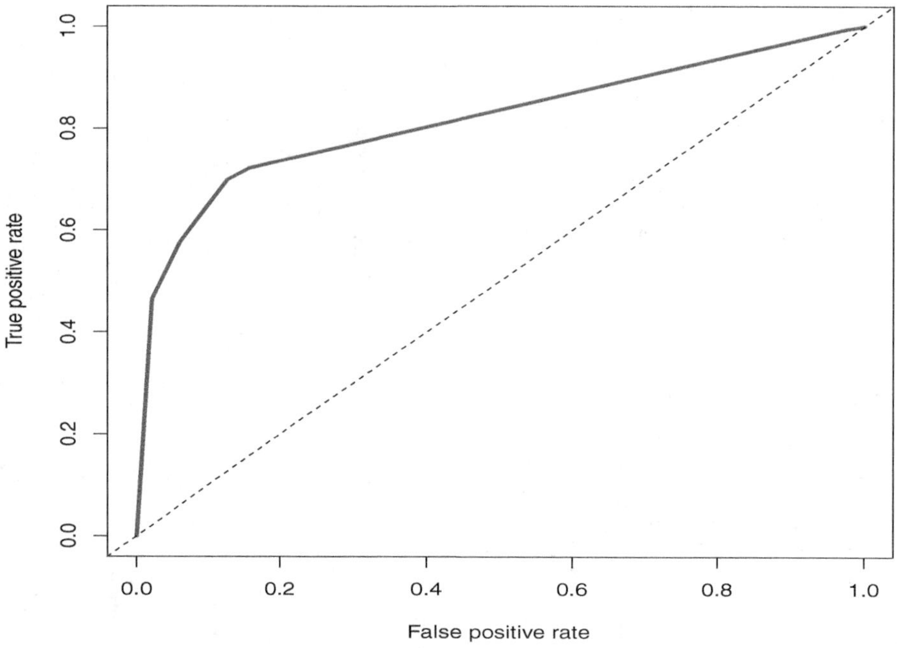

Figure 23.10: *ROC curve of the decision tree.*

The accuracy of the model in function of the cutoff value can be ploted via the `acc` attribute of the performance object. This is done by the following code and the plot is in Figure 23.11.

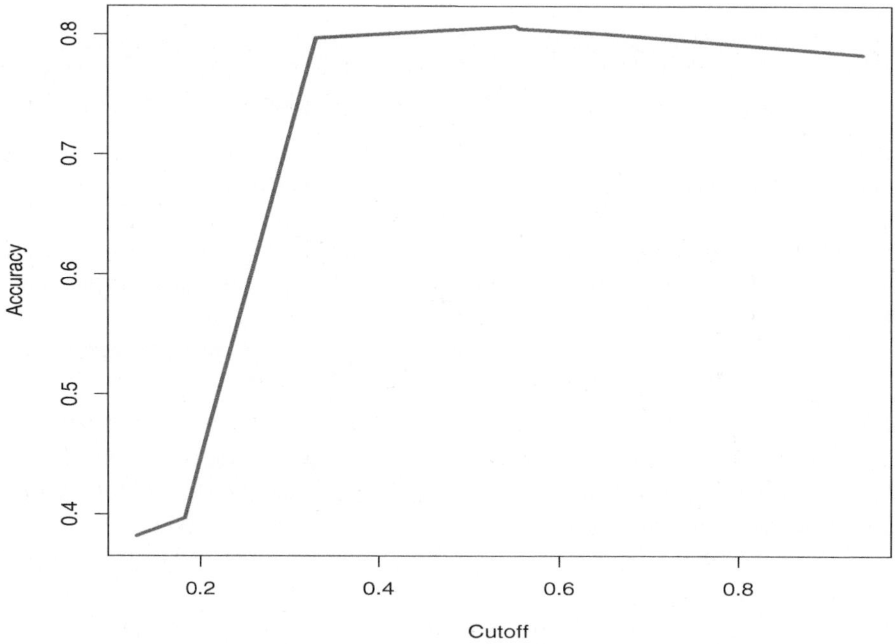

Figure 23.11: *The accuracy for the decision tree on the Titanic data.*

```
plot(performance(pred, "acc"), col="blue", lwd=3)
abline(1,0,lty=2)
```

Finally, we calculate the AUC, GINI and KS in the following code:

```
# AUC:
AUC <- attr(performance(pred, "auc"), "y.values")[[1]]
AUC
## [1] 0.816288

# GINI:
2 * AUC - 1
## [1] 0.632576

# KS:
perf <- performance(pred, "tpr", "fpr")
max(attr(perf,'y.values')[[1]]-attr(perf,'x.values')[[1]])
## [1] 0.5726823
```

23.2 Random Forest

random forest Decision trees are a popular method for various machine learning tasks. Tree learning is invariant under scaling and various other transformations of feature values, is robust to inclusion of irrelevant features, and produces models that are transparent and can be understood by humans.

However, they are seldom robust. In particular, trees that are grown very deep tend to learn coincidental patterns: they "overfit" their training sets, i.e. have low bias, but very high variance.

Maybe the situation would be better if we would have the option to fit multiple decision trees and somehow average the result? It appears that this is not a bad approach. Methods that fit many models and let each model "vote" are called ensemble methods. For example, the random forest method does this.

Random forests are a way of averaging multiple deep decision trees, trained on different parts of the same training set, with the goal of reducing the variance (by averaging the trees). This comes at the expense of a small increase in the bias and some loss of interpretability, but generally greatly boosts the performance and stability of the final model.

Random forests use a combination of techniques to counteract over-fitting. First, many random samples of data are selected, and second variables are put in and out of the formula.

With this process, we obtain a measure of importance of each attribute (input variable), which in its turn can be used to select a model. This can be particularly useful when forward/backward stepwise selection is not appropriate and when working with an extremely high number of candidate variables that needs to be reduced.

In R, we can rely on the library `randomForest` to fit a random forest model. First we install it and then load it:

```
library(randomForest)
```

The process of generating a random forest needs two steps that require a source of randomness, so it is a good idea to set the random seed in R before you begin. Doing so makes your results reproducible next time you run the code. This is done by the function `set.seed(n)`. `set.seed()`

The code below demonstrates how the random forest can be fitted. The package `randomForest` is already loaded in the aforementioned code. So all we have to do is fit the random forest with the function `randomForest()`, and the whole methodology will be executed. The rest of the code is about exploring the model and understanding it. After fitting the model, we show: `randomForest`

- who to get information about the model with the functions `print()`, `plot()` (result in Figure 23.12 on page 429), `summary()`, and `getTree()`; `getTree()`
- how to study the importance of the each independent variable via the function `importance()`; `importance()`
- how to plot a summary of this importance object via the function `plot()` (the result is in Figure 23.13 on page 429);
- and finally we automate the plotting of the partial dependence on each variable and plot those with the function `partialPlot()` – those plots are in Figure 23.14 on page 430, Figure 23.15 on page 430, and Figure 23.16 on page 433. `partialPlot()` `randomForest()`

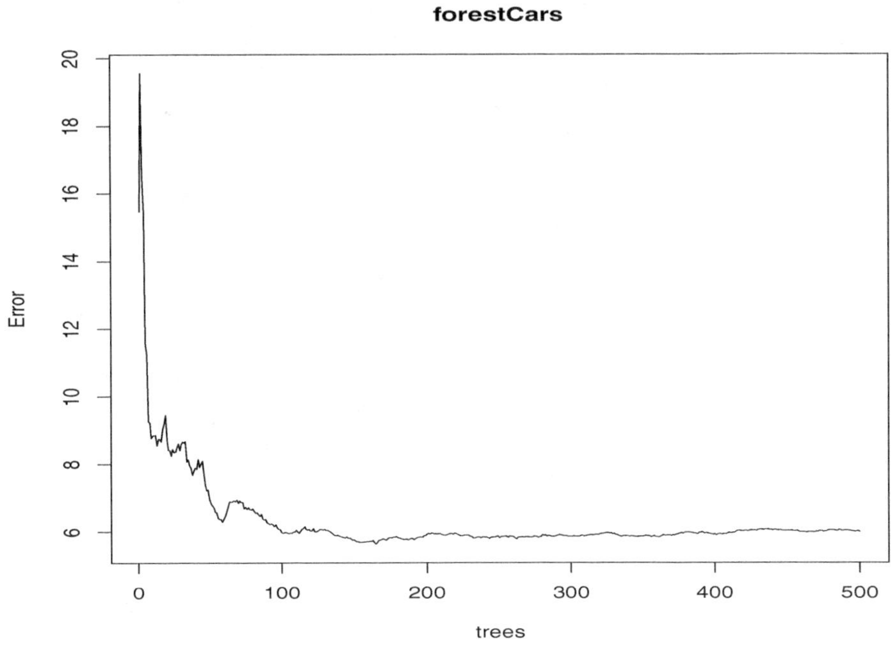

Figure 23.12: *The plot of a randomForest object shows how the model improves in function of the number of trees used.*

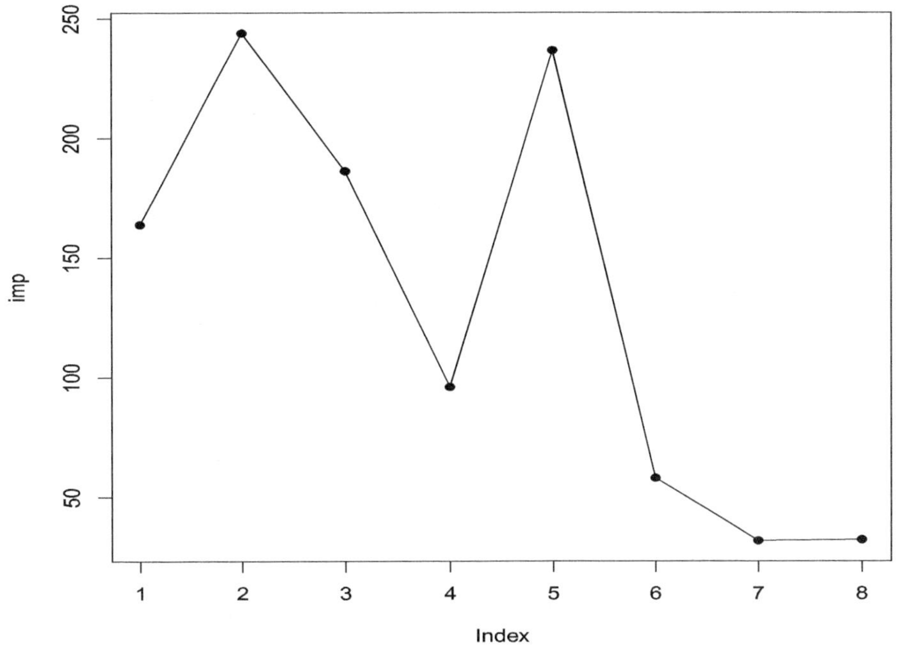

Figure 23.13: *The importance of each variable in the random-forest model.*

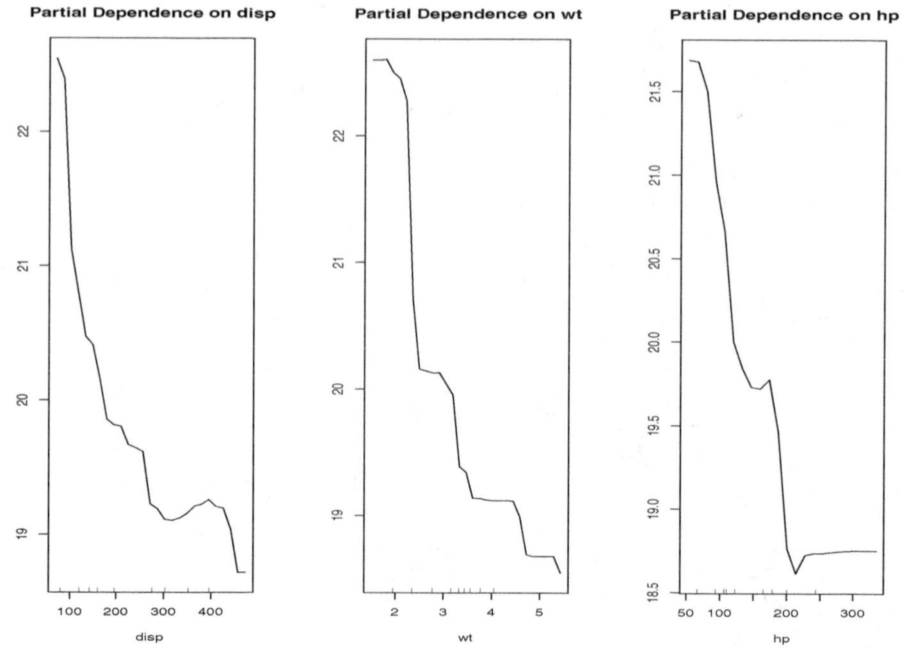

Figure 23.14: *Partial dependence on the variables (1 of 3).*

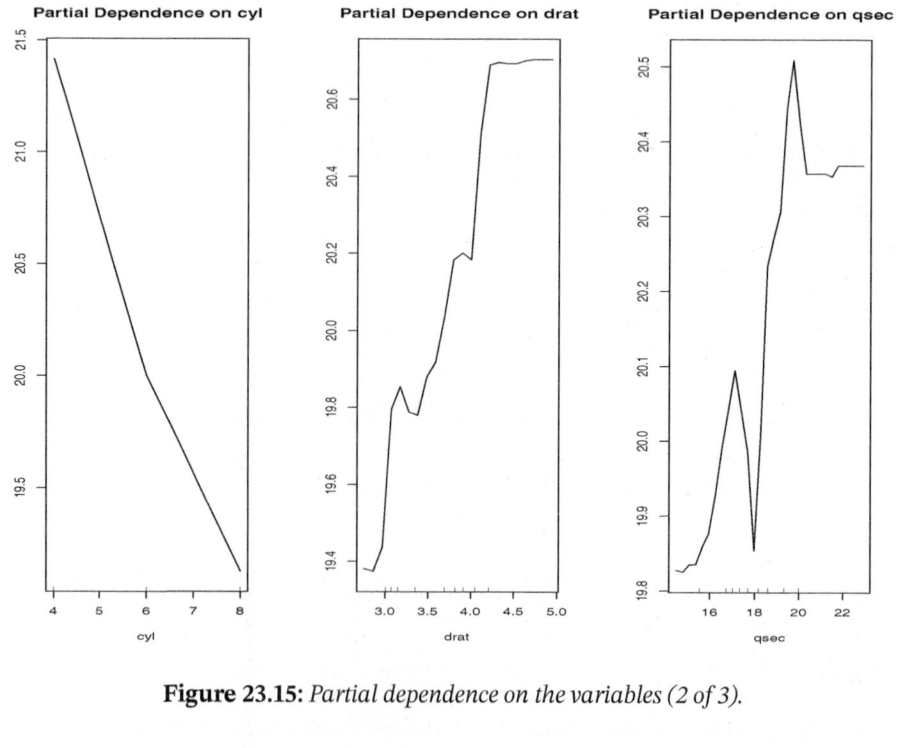

Figure 23.15: *Partial dependence on the variables (2 of 3).*

```
head(mtcars)
##                    mpg cyl disp  hp drat    wt  qsec vs
## Mazda RX4         21.0   6  160 110 3.90 2.620 16.46  0
## Mazda RX4 Wag     21.0   6  160 110 3.90 2.875 17.02  0
## Datsun 710        22.8   4  108  93 3.85 2.320 18.61  1
## Hornet 4 Drive    21.4   6  258 110 3.08 3.215 19.44  1
## Hornet Sportabout 18.7   8  360 175 3.15 3.440 17.02  0
## Valiant           18.1   6  225 105 2.76 3.460 20.22  1
##                   am gear carb
## Mazda RX4          1    4    4
## Mazda RX4 Wag      1    4    4
## Datsun 710         1    4    1
## Hornet 4 Drive     0    3    1
## Hornet Sportabout  0    3    2
## Valiant            0    3    1

mtcars$l <- NULL  # remove our variable
frm      <- mpg ~ cyl + disp + hp + drat + wt + qsec + am + gear
set.seed(1879)

# Fit the random forest:
forestCars   = randomForest(frm, data = mtcars)

# Show an overview:
print(forestCars)
##
## Call:
##  randomForest(formula = frm, data = mtcars)
##                Type of random forest: regression
##                      Number of trees: 500
## No. of variables tried at each split: 2
##
##           Mean of squared residuals: 6.001878
##                     % Var explained: 82.94

# Plot the random forest overview:
plot(forestCars)


# Show the summary of fit:
summary(forestCars)
##                 Length Class  Mode
## call              3    -none- call
## type              1    -none- character
## predicted        32    -none- numeric
## mse             500    -none- numeric
## rsq             500    -none- numeric
## oob.times        32    -none- numeric
## importance        8    -none- numeric
## importanceSD      0    -none- NULL
## localImportance   0    -none- NULL
## proximity         0    -none- NULL
## ntree             1    -none- numeric
## mtry              1    -none- numeric
## forest           11    -none- list
## coefs             0    -none- NULL
## y                32    -none- numeric
## test              0    -none- NULL
## inbag             0    -none- NULL
## terms             3    terms  call

# visualization of the RF:
getTree(forestCars, 1, labelVar=TRUE)
```

```
##    left daughter right daughter split var split point
## 1              2              3      disp     192.500
## 2              4              5       cyl       5.000
## 3              6              7       cyl       7.000
## 4              8              9      gear       3.500
## 5              0              0      <NA>       0.000
## 6              0              0      <NA>       0.000
## 7             10             11      qsec      17.690
## 8              0              0      <NA>       0.000
## 9             12             13      drat       4.000
## 10            14             15      drat       3.440
## 11             0              0      <NA>       0.000
## 12            16             17        am       0.500
## 13            18             19      qsec      19.185
## 14            20             21      drat       3.075
## 15             0              0      <NA>       0.000
## 16             0              0      <NA>       0.000
## 17             0              0      <NA>       0.000
## 18             0              0      <NA>       0.000
## 19             0              0      <NA>       0.000
## 20             0              0      <NA>       0.000
## 21             0              0      <NA>       0.000
##    status prediction
## 1      -3   20.75625
## 2      -3   24.02222
## 3      -3   16.55714
## 4      -3   24.97857
## 5      -1   20.67500
## 6      -1   19.75000
## 7      -3   16.02500
## 8      -1   21.50000
## 9      -3   25.24615
## 10     -3   16.53636
## 11     -1   10.40000
## 12     -3   23.33333
## 13     -3   26.88571
## 14     -3   17.67143
## 15     -1   14.55000
## 16     -1   23.44000
## 17     -1   22.80000
## 18     -1   24.68000
## 19     -1   32.40000
## 20     -1   15.80000
## 21     -1   19.07500

# Show the purity of the nodes:
imp <- importance(forestCars)
imp
##      IncNodePurity
## cyl      163.83222
## disp     243.89957
## hp       186.24274
## drat      96.08086
## wt       236.59343
## qsec      57.99794
## am        31.84926
## gear      32.31675

# This impurity overview can also be plotted:
plot( imp, lty=2, pch=16)
lines(imp)
```

```
# Below we print the partial dependence on each variable.
# We group the plots per 3, to save some space.
impvar = rownames(imp)[order(imp[, 1], decreasing=TRUE)]
op     = par(mfrow=c(1, 3))
for (i in seq_along(impvar)) {
  partialPlot(forestCars, mtcars, impvar[i], xlab=impvar[i],
  main=paste("Partial Dependence on", impvar[i]))
  }
```

randomForest()

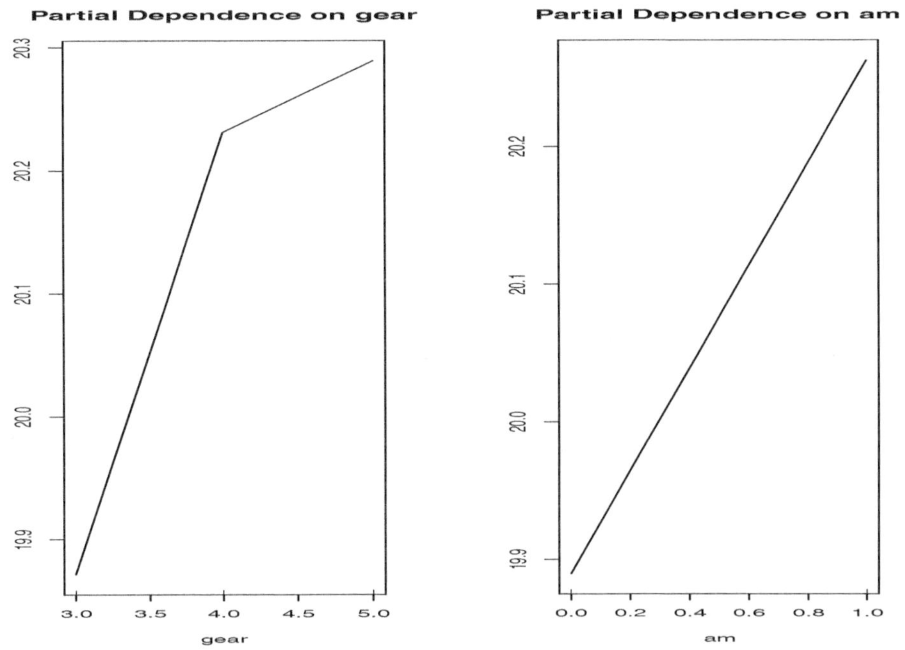

Figure 23.16: *Partial dependence on the variables (3 of 3).*

The random forest method seems to indicate us that in this case one might make a tree fitting on the number of cylinders, displacement, horse power and weight.[5] However, appearances can be deceiving. There is no guarantee that a random forest does not over-fit itself. While the technique of random forest seems to be a good tool against over-fitting, it is no guarantee against it.

Note – Variations for the random forest

The random forest technique (randomly leaving out variables and data) can also be used with other underlying models. Though, most often, it is used with decision trees, and that is also how the function randomForest() implements it.

[5] Of course, in reality one has to be very careful. For example, something can happen that disturbs the dependence of the parameters. For example, once electric or hybrid cars are introduced, this will look quite different.

23.3 Artificial Neural Networks (ANNs)

23.3.1 The Basics of ANNs in R

Neural networks (NNs) are a class of models that are inspired on our understanding of the animal brain. In the brain one has neurons that can be in two states: firing (ie. active) and non-firing (non-active). Those neurons are connected and the "decisions" to fire or not fire are made based on the incoming signals of the surrounding neurons. If the sum of all signals exceeds a certain threshold, the neuron will be in firing mode itself in the next round, if not then it will be inactive in the next round.

NN
neural network

The human brain consist of a neural network, and works at more or less 100 Hz. While the word "neural network" actually refers to the biological brain, for the purpose of this book we use ANN and NN as synonyms.

ANN
artificial neural network
training

Artificial neural networks (ANNs) not only display similarities with our brain in the way they are designed, but also display the ability to "learn" from examples. This is called "training the neural network."

In a first and simple approach, one could consider a two-dimensional neural network of 100 by 100 pixels that can be either black (inactive) or white (firing). One could define an energy function such that it has local minima for the letters A to Z. Then the system will be able to recognize a hand-written letter that is drawn.

This is a very simple neural network with just one layer (no hidden layers and even the output layer is the input layer itself) and the local optima of the energy function are pre-defined. Even in this simple example there are many degrees of freedom to design the neural network.[6]

The potential in each node is something like

$$E_{ij}(t+1) = f\left(\sum_{k=1}^{N}\sum_{l=1}^{M} w_{ij,kl}E_{kl}(t)\right)$$

where f is typically something like

$$f(x) = \begin{cases} 0 & \text{if } g\left(\sum_{k=1}^{N}\sum_{l=1}^{M}) w_{ij,kl}E_{kl}(t)\right) < \theta \\ 1 & \text{if } g\left(\sum_{k=1}^{N}\sum_{l=1}^{M}) w_{ij,kl}E_{kl}(t)\right) \geq \theta \end{cases}$$

The weight $w_{ij,kl}$ is the strength of the connection between neuron ij and neuron kl (two indices since in our example we work from a two-dimensional representation). When the network is being trained, the weights are modified till an optimal fit is achieved. In its turn, there is a possible choice to be made for $g(\alpha)$. Typically, one will take something like the logit, logistic, or hyperbolic tangent function. For example:

$$g(\alpha) = \text{logistic}(\alpha) = \text{logit}^{-1}(\alpha) = \frac{1}{1+e^{(-\alpha)}}$$

or

$$g(\alpha) = \Phi^{-1}\left(\sqrt{\frac{\pi}{8}}\alpha\right)$$

(the scaling of $alpha$ is not necessary but if applied the derivative in 0 will be the same as the logistic function) or

$$g(\alpha) = \tanh(\alpha)$$

The parameter θ (threshold) is a number that should carefully be chosen.

[6] Please note that a one-dimensional, one-layer neural net with a binary output parameter is similar to a logistic regression.

While this approach already has some applications, neural nets become more useful and better models if we allow them to learn by themselves and allow them to create internally hidden layers of neurons.

For example, in image recognition, it is possible to make a NN learn to to identify images that contain gorillas by analysing example images that have been labelled as "has gorilla" or "has no gorilla." Training the NN on a set of sample images will make it capable of recognizing gorillas in new pictures. They do this without any a priori knowledge about gorillas (it is for example not necessary for the neural net to know that gorillas have four limbs, are usually black or grey, have two brown eyes, etc.). The NN will instead create in its hidden layers some abstract concept of a gorilla. The downside of neural networks is that it is for humans very hard to impossible to understand what that concept is or how it can be interpreted.[7]

An ANN is best understood as a set of connected nodes (called artificial neurons, a simplified version of biological neurons in an animal brain). Each connection (a simplified version of a synapse) between artificial neurons can transmit a signal from one to another. The artificial neuron that receives the signals can process it and then signal the other artificial neurons connected to it.

Usually, the signal at a connection between artificial neurons is a real number, and the output of each artificial neuron is calculated by a non-linear function of the sum of its inputs as explained above. Artificial neurons and connections typically have a weight that adjusts as learning proceeds. The weight increases or decreases the strength of the signal at a connection. Artificial neurons may have a threshold such that only if the aggregate signal crosses that threshold is the signal sent. Typically, artificial neurons are organized in layers. Different layers may perform different kinds of transformations on their inputs. Signals travel from the first (input), to the last (output) layer, possibly after traversing the layers multiple times.

So, in order to make an ANN learn it is important to be able to calculate the derivative so that weights can be adjusted in each iteration.

Another way to see ANNs is as an extension of the logistic regression. Actually, the neurons inside the neural network have two possible states: active or inactive (hence 0 or 1). Every neuron in every layer layer in a neural network can be considered as a logistic regression. Figure 23.17 on page 436 shows the scheme for the logistic regression that symbolises this logic. A logistic regression is the same as a neural network with one neuron (and 0 hidden layers).

The interpretation of these neurons in internal layers are quite abstract. In fact they do not correspond necessarily to a real-world aspect (such as age, colour, race, sex, weight, etc.). This means that it is quite hard to understand how the decision process of a neural network works. One can easily observe the weights of the neurons in layer zero, but that does not mean that these are representative for the way a network makes a decision (except in the case where there is only one layer with one neuron) ... hence in the case where there is equivalence with a logistic regression.

[7] This is why banks for example are very slow to adopt neural networks in credit analysis. They will rather rely in less powerful linear regression models. Regressions are extremely transparent and it is easy to explain to regulators or in court why a certain customer was denied a loan and it is always possible to demonstrate that there has been no illegal discrimination (e.g. refusing the loan because a person belongs to a certain racial minority group). With neural networks this is not possible, it might be that the neural net has in its hidden layer something like a racial bias and that the machine derives that racial background via other parameters.

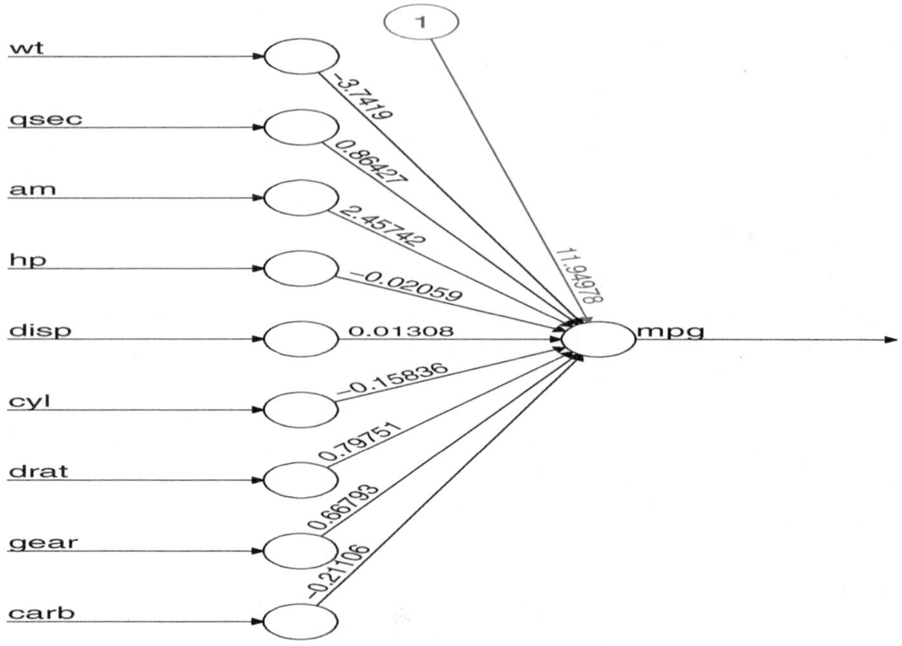

Figure 23.17: *A logistic regression is actually a neural network with one neuron. Each variable contributes to a sigmoid function in one node, and if that one node gets loadings over a critical threshold, then we predict 1, otherwise 0. The intercept is the "1" in a circle. The numbers on the arrows are the loadings for each variable.*

There is a domain of research that tries to extract knowledge from neural networks. This is called "rule extraction," whereby the logic of the neural net is approximated by a decision tree. The process is more or less as follows:

- one will first try to understand the logic of the hidden variables with clustering analysis;
- approximate the ANN's predictions by decisions on the hidden variables;
- approximate the hidden variables by decisions on the input variables;
- collapse the resulting decision tree logically.

See for example Setiono et al. (2008), Hara and Hayashi (2012), Jacobsson (2005), Setiono (1997), or the IEEE survey in Tickle et al. (1998).

As one could expect, there are several packages in R that help to fit neural networks and train them properly. First, we will look at a naive example and then address the issue of over-fitting.

23.3.2 Neural Networks in R

R has a convenient library `neuralnet` that provides a function with the same name. We will use this one to show how NNs can be fitted.

Function use for neuralnet()

```
neuralnet(formula, data, hidden = 1, stepmax = 1e+05
        linear.output = TRUE)
```

The function `neuralnet()` accepts many parameters, we refer to the documentation for more details. The most important are:

- `hidden`: a vector with the number of neurons in each hidden layer,
- `formula`: an R formula that shows all variables to include,
- `data`: the data-frame that holds all columns of the formula,
- `setpmax`: the maximum number iterations to train the NN,
- `linear.output`: set to `TRUE` to fit a regression network, set to `FALSE` to fit a classification network,
- `algorithm`: a string that selects the algorithm for the NN – select from: `backprop` for back-propagation, `rprop+` and `rprop-` for resilient back-propagation, and `sag` and `slr` to use the modified globally convergent algorithm (grprop).

The function `neuralnet()` is very universal and allows for a lot of customisation. However, leaving most options to their default will in many cases lead to good results. Consider the dataset `mtcars` to illustrate how this works.

```
#install.packages("neuralnet") # Do only once.

# Load the library neuralnet:
library(neuralnet)

# Fit the aNN with 2 hidden layers that have resp. 3 and 2 neurons:
# (neuralnet does not accept a formula wit a dot as in 'y ~ .' )
nn1 <- neuralnet(mpg ~ wt + qsec + am + hp + disp + cyl + drat +
                  gear + carb,
            data = mtcars, hidden = c(3,2),
            linear.output = TRUE)
```

An interesting parameter for the function `neuralnet` is the parameter "hidden." This is a vector with the number of neurons in each layer, the number of layers will correspond to the length to the vector provided. For example, `c(10,8,5)` implies three hidden layers (the first has ten neurons, the second eight and the last five).

In fact, it seems that neural networks are naturally good in approaching other functions and already with one layer an ANN is able to approach any continuous function. For discontinuous functions or concepts that cannot be expressed as functions more layers may be required.

It appears that one best chooses a number of neurons that is not too small but does not exceed the number of input layers.

Another important parameter is `linear.output`: set this to `TRUE` in order to estimate a continuous value (regression) and to `FALSE` in order to solve a classification problem.

As expected the function `plot()` applied on an object of the class “nn” will produce a visual that makes sense for that class. This is illustrated in the following code and the plot is in Figure 23.18

plot()

```
plot(nn1, rep = "best", information = FALSE);
```

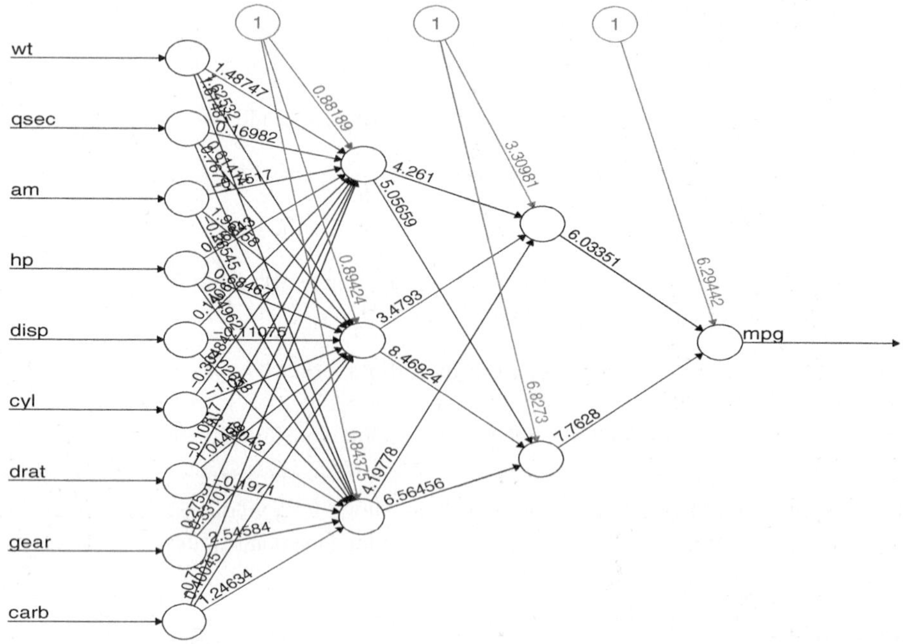

Figure 23.18: *A simple neural net fitted to the dataset of mtcars, predicting the miles per gallon (mpg). In this example we predict the fuel consumption of a car based on some other values in the dataset t mtcars.*

If the parameter `rep` is set to `'best'`, then the repetition with the smallest error will be plotted. If not stated all repetitions will be plotted, each in a separate window.[8]

The black lines in the plot of the ANN represent the connections between each neuron of the previous and the following layer and the number is the weight of that connection. The blue lines show the bias term added in each step. The bias can be thought as the intercept of a linear model.

bias term

23.3.3 The Work-flow to for Fitting a NN

To get a better illustration of the power and potential complexity of neural nets, we will now have a look at a larger dataset. A good starting point is the `Boston` dataset from the `MASS` package. While still very containable, it is a larger dataset containing house values in Boston. We will model the median value of owner-occupied homes (`medv`) using all the other continuous variables available.

knitr
LaTeX

[8]While in interactive mode, this might be confusing, in batch mode, this will create problems. For example, when using an automated workflow with `knitr` and LaTeX, this will cause the plot to fail to render in the document, but not result in any error message.

```
# Get the data about crimes in Boston:
library(MASS)
d <- Boston
```

With this data, we will now step by step prepare the neural network.

Step 1: Missing Data

```
# Inspect if there is missing data:
apply(d,2,function(x) sum(is.na(x)))
##     crim      zn   indus    chas      nox      rm     age
##        0       0       0       0        0       0       0
##      dis     rad     tax ptratio    black   lstat    medv
##        0       0       0       0        0       0       0

# There are no missing values.
```

If there would be missing data, we should first address this issue, but the data is complete. The next step is to split the data in a training and testing subset.

> **Hint – Never delete more data than necessary**
>
> When there are missing values, we might want to remove them first. If you decide to do so, then first, create a dataset that holds only the columns that you need and then use
>
> `d1 <- d[complete.cases(d),]`
>
> That way, we assure that we only delete data that we will not use and we avoid deleting a row where an observation is missing in a variable that we anyhow will not use. More information on treating missing values is in Chapter 18 *"Dealing with Missing Data"* on page 333.

Step 2: Split the Data in Test and Training Set

Neural networks have the tendency to over-fit. So, it is essential to apply some form of cross validation. More information about this subject is in Chapter 25.4 *"Cross-Validation"* on page 483. In our example, we will only use simple cross validation.

```
set.seed(1877) # set the seed for the random generator
idx.train <- sample(1:nrow(d),round(0.75*nrow(d)))
d.train   <- d[idx.train,]
d.test    <- d[-idx.train,]
```

Step 3: Fit a Challenger Model

To be effective, a NN has a certain complexity, but only a few layers lead to so much parameters that it is almost impossible to understand what the NN does and why it does what it does. This makes them effectively black boxes. So, in order to use a neural network, we must be certain that the performance gain is worth the complexity and loss of transparency. Therefore, we will use a challenger model.

```
# Fit the linear model, no default for family, so use 'gaussian':
lm.fit <- glm(medv ~ ., data = d.train)
summary(lm.fit)
##
## Call:
## glm(formula = medv ~ ., data = d.train)
##
## Deviance Residuals:
##      Min       1Q   Median       3Q      Max
## -18.781   -2.632   -0.489    1.838   26.065
##
## Coefficients:
##               Estimate Std. Error t value Pr(>|t|)
## (Intercept)  18.519702   6.032432   3.070  0.00230 **
## crim         -0.126138   0.038328  -3.291  0.00110 **
## zn            0.029671   0.015682   1.892  0.05928 .
## indus         0.026689   0.066923   0.399  0.69027
## chas          2.520510   0.940292   2.681  0.00768 **
## nox         -12.820105   4.135693  -3.100  0.00209 **
## rm            5.351253   0.515365  10.383  < 2e-16 ***
## age          -0.007375   0.014600  -0.505  0.61377
## dis          -1.224021   0.226302  -5.409 1.15e-07 ***
## rad           0.239947   0.072578   3.306  0.00104 **
## tax          -0.010019   0.004077  -2.458  0.01444 *
## ptratio      -0.846958   0.142063  -5.962 5.87e-09 ***
## black         0.014302   0.003022   4.733 3.17e-06 ***
## lstat        -0.432604   0.059364  -7.287 1.96e-12 ***
## ---
## Signif. codes:
## 0 '***' 0.001 '**' 0.01 '*' 0.05 '.' 0.1 ' ' 1
##
## (Dispersion parameter for gaussian family taken to be 20.14508)
##
##     Null deviance: 30439.4  on 379  degrees of freedom
## Residual deviance:  7373.1  on 366  degrees of freedom
## AIC: 2235.3
##
## Number of Fisher Scoring iterations: 2

# Make predictions:
pr.lm  <- predict(lm.fit,d.test)

# Calculate the MSE:
MSE.lm <- sum((pr.lm - d.test$medv)^2)/nrow(d.test)
```

The `sample(x,size)` function outputs a vector of the specified size of randomly selected samples from the vector x. By default the sampling is without replacement, and the variable idx.train is a vector of indices.

We have now a linear model that can be used as a challenger model.

Step 4: Rescale the Data and Split into Training and Testing Set

Now, we would like to start fitting the NN. However, before fitting a neural network it is useful to normalize data between the interval $[0, 1]$ or $[-1, 1]$. This helps the optimisation procedure to converge faster and it makes the results easier to understand. There are many methods available to normalize data (z-normalization, min-max scale, logistic scale, etc.). In this example we will use the min-max method and scale the data to the interval $[0, 1]$.

```
# Store the maxima and minima:
d.maxs <- apply(d, 2, max)
d.mins <- apply(d, 2, min)

# Rescale the data:
d.sc <- as.data.frame(scale(d, center = d.mins,
                               scale  = d.maxs - d.mins))

# Split the data in training and testing set:
d.train.sc <- d.sc[idx.train,]
d.test.sc  <- d.sc[-idx.train,]
```

Note that `scale()` returns a matrix and not a data frame, so we use the function `as.data.frame()` to coerce the results in to a data-frame.

Step 5: Train the ANN on the Training Set

Finally, we are ready to train the ANN. This is straightforward:

```
library(neuralnet)

# Since the shorthand notation y~. does not work in the
# neuralnet() function we have to replicate it:
nm  <- names(d.train.sc)
frm <- as.formula(paste("medv ~", paste(nm[!nm %in% "medv"],
                        collapse = " + ")))

nn2 <- neuralnet(frm, data = d.train.sc, hidden = c(7,5,5),
                 linear.output = T)
```

Visualising the NN is made simple by R's S3 OO model (more in Chapter 6 *"The Implementation of OO"* on page 87): we can use the standard function `plot()`. The output is in Figure 23.19 on page 442.

```
plot(nn2, rep = "best", information = FALSE,
     show.weights = FALSE)
```

Now, we can predict the values for the test dataset based on this model and then calculate the MSE. Since the ANN was trained on scaled data, we need to scale it back in order to make a meaningful comparison.

Step 6: Test the Model on the Test Data

predict.nn()

```
# Our independent variable 'medv' is the 14th column, so:
pr.nn2 <- predict.nn(nn2,d.test.sc[,1:13])

# Rescale back to original span:
pr.nn2 <- pr.nn2$net.result*(max(d$medv)-min(d$medv))+min(d$medv)
test.r <- (d.test.sc$medv)*(max(d$medv)-min(d$medv))+min(d$medv)

# Calculate the MSE:
MSE.nn2 <- sum((test.r - pr.nn2)^2)/nrow(d.test.sc)
print(paste(MSE.lm,MSE.nn2))
## [1] "32.896450960242 14.3230140328093"
```

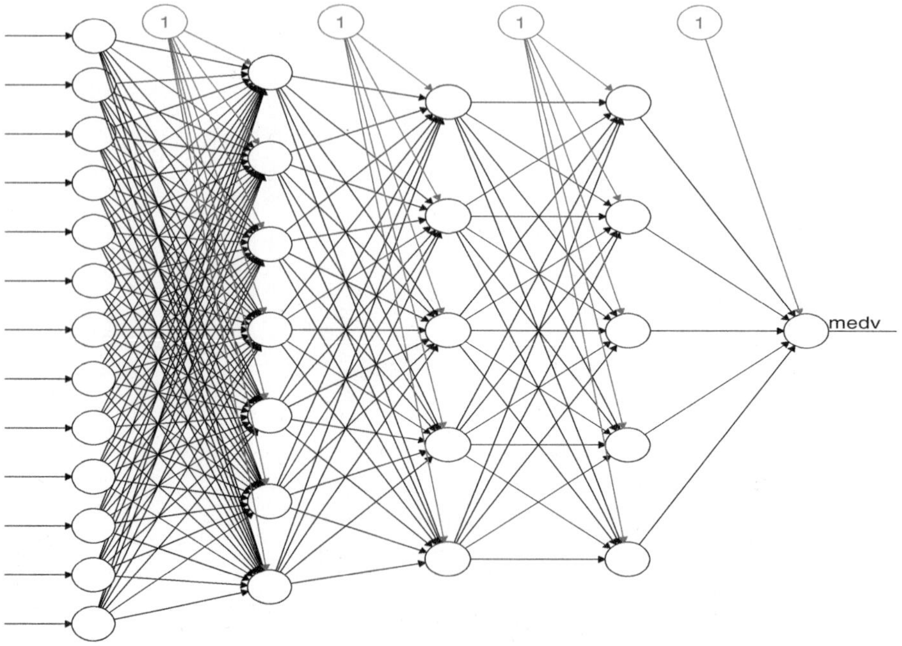

Figure 23.19: *A visualisation of the ANN. Note that we left out the weights, because there would be too many. With 13 variables, and three layers of respectively 7, 5, and 5 neurons, we have* $13 \times 7 + 7 \times 5 + 5 \times 5 + 5 + 7 + 5 + 5 + 1 = 174$ *parameters.*

In terms of RMSE, the NN is doing a better work than the linear model at predicting `medv`. However, this result will depend on

1. our choices for the number of hidden layers and the numbers of neurons in each layer,
2. the selection of the training dataset.

The following code produces a plot to visualize the performance of the two models. We will plot each observed house value on the x-axis and on the y-axis the predicted value. On the same plot we we trace the unit line ($y = x$) that is the reference where we would the points should be. The plot appears in Figure 23.20 on page 443

```
par(mfrow=c(1,2))

plot(d.test$medv, pr.nn2, col='red',
     main='Observed vs predicted NN',
     pch=18,cex=0.7)
abline(0,1,lwd=2)
legend('bottomright', legend='NN', pch=18, col='red', bty='n')

plot(d.test$medv,pr.lm,col='blue',
     main='Observed vs predicted lm',
     pch=18, cex=0.7)
abline(0,1,lwd=2)
legend('bottomright', legend='LM', pch=18,col='blue', bty='n',
       cex=.95)
```

We see that indeed the predictions of the linear model are closer to the observed values. However, again we stress the fact that this picture can be very different for other choices of hidden layers. Since in our dataset there are not too many observations, it is possible to plot both in one graph. The code below does this in and puts the result in Figure 23.21 on page 443.

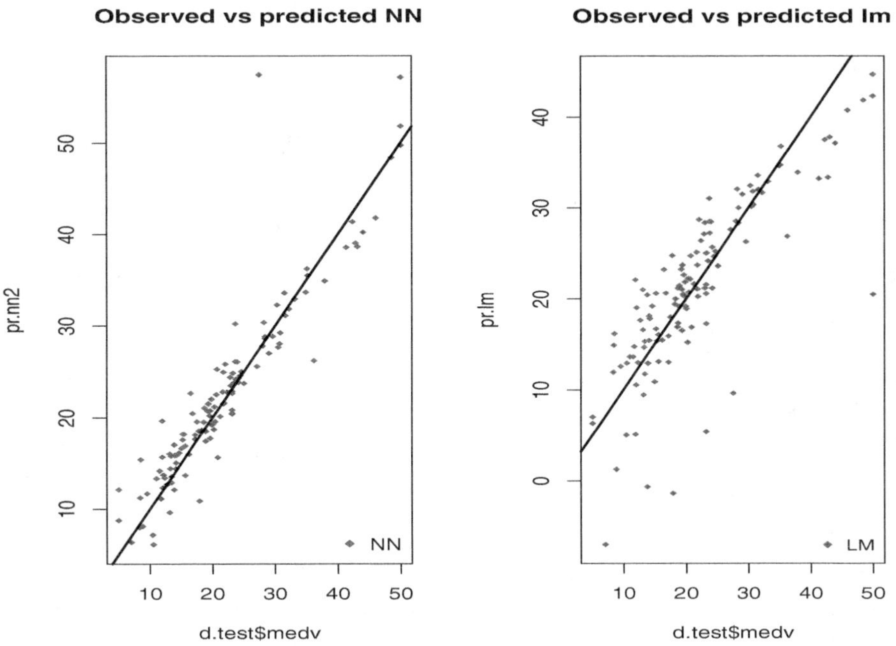

Figure 23.20: *A visualisation of the performance of the ANN (left) compared to the linear regression model (right).*

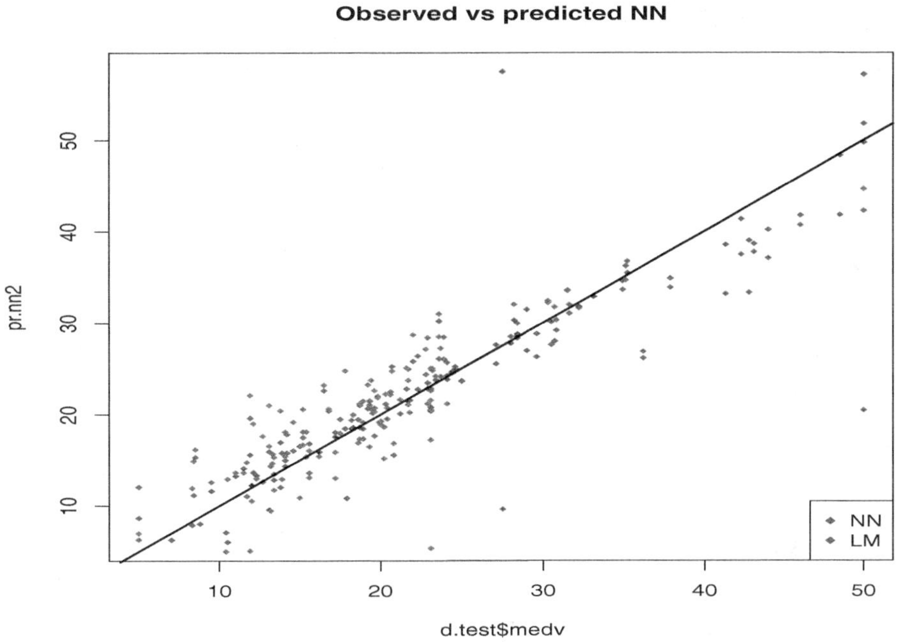

Figure 23.21: *A visualisation of the performance of the ANN compared to the linear regression model with both models in one plot.*

```
plot  (d.test$medv,pr.nn2,col='red',
       main='Observed vs predicted NN',
       pch=18,cex=0.7)
points(d.test$medv,pr.lm,col='blue',pch=18,cex=0.7)
abline(0,1,lwd=2)
legend('bottomright',legend=c('NN','LM'),pch=18,
       col=c('red','blue'))
```

23.3.4 Cross Validate the NN

cross validation

A more profound way to validate a model is to rely on a more advanced form of cross validation. For example we can split K times between training and test data, each time fit the model on another sub-set and compare the results. The average of our chosen error measure will then be indicative for the model.

What we present here is a simple approach with a lot of manual work. For more background and for more efficient code (that also follows the `tidyverse` philosophy), we refer to Chapter 25.4 *"Cross-Validation"* on page 483.

`cv.glm()`

We are going to implement a cross validation using a for- loop for the neural network and the `cv.glm()` function from the `boot` package for the linear model.

To execute the k-fold cross validation for the linear model, we use the function `cv.glm()` from the package `boot`. Below is the code for the 10 fold cross validation MSE for the linear model:

```
library(boot)
set.seed(1875)
lm.fit <- glm(medv ~ ., data = d)

# The estimate of prediction error is now here:
cv.glm(d, lm.fit, K = 10)$delta[1]
## [1] 23.92593
```

Now, we will fit the ANN. Note that we split the data so that we retain 90% of the data in the training set and 10% test set. This we will do randomly for ten times.

`plyr`

It is possible to show a progress bar via `plyr`.

To see this progress bar, uncomment the relevant lines.

```
# Reminders:
d   <- Boston
nm  <- names(d)
frm <- as.formula(paste("medv ~", paste(nm[!nm %in% "medv"],
                        collapse = " + ")))
# Store the maxima and minima:
d.maxs <- apply(d, 2, max)
d.mins <- apply(d, 2, min)

# Rescale the data:
d.sc <- as.data.frame(scale(d, center = d.mins,
                               scale  = d.maxs - d.mins))

# Set parameters:
set.seed(1873)
cv.error <- NULL  # Initiate to append later
k        <- 10    # The number of repetitions

# This code might be slow, so you can add a progress bar as follows:
#library(plyr)
#pbar <- create_progress_bar('text')
#pbar$init(k)
```

```
# In k-fold cross validation, we must take care to select each
# observation just once in the testing set. This is made easy
# with modelr:
library(modelr)
kFoldXval <- crossv_kfold(data = d.sc, k = 10, id = '.id')

# Do the k-fold cross validation:
for(i in 1:k){
    # <see note below>
    train.cv   <- kFoldXval$train[i]
    test.cv    <- kFoldXval$test[i]
    test.cv.df <- as.data.frame(test.cv)

    # Rebuild the formula (names are changed each run):
    nmKfold <- paste0('X', i, '.', nm)
    medvKfld <- paste0('X', i, '.medv')
    frmKfold <- as.formula(paste(medvKfld, "~",
                        paste(nmKfold[!nmKfold %in% medvKfld],
                        collapse = " + ")
                        )
                   )

    # Fit the NN:
    nn2       <- neuralnet(frmKfold, data = train.cv,
                       hidden = c(7, 5, 5),
                       linear.output=TRUE
                       )

    # The explaining variables are in the first 13 rows, so:
    pr.nn2    <- compute(nn2, test.cv.df[,1:13])

    pr.nn2    <- pr.nn2$net.result * (max(d$medv) - min(d$medv)) +
              min(d$medv)
    test.cv.df.r <- test.cv.df[[medvKfld]] *
                 (max(d$medv) - min(d$medv)) + min(d$medv)
    cv.error[i] <- sum((test.cv.df.r - pr.nn2)^2)/nrow(test.cv.df)
    #pbar$step()  #uncomment to see the progress bar
}
```

Note – Cross validation without modelr

Without `modelr`, we have to build the samples as follows:

```
index    <- sample(1:nrow(d),round(0.9*nrow(d)))
train.cv <- d.sc[index,]
test.cv  <- d.sc[-index,]
```

That code would go in the line that reads:

```
# <see note below>
```

While this is short and elegant, it is not exactly the same. This is not a k-fold validation, since it does not ensure that each observations is used in the testing dataset just once. Programming this is a lot longer. That is the reason why we had to used the `tidyverse` here.

Running this code can take a while. Once it is finished, we calculate the average MSE and plot the results as a boxplot in Figure 23.22 on page 446.

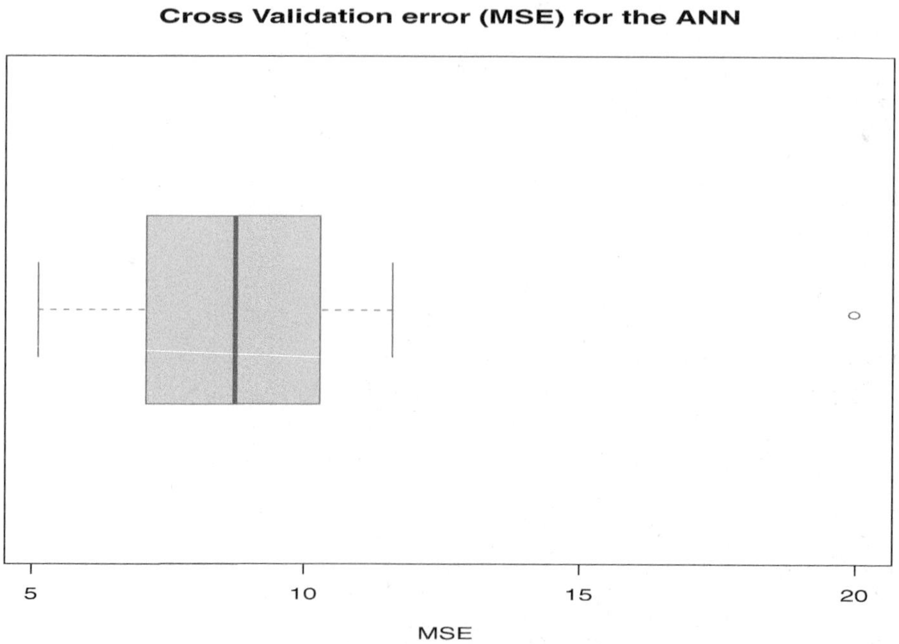

Figure 23.22: *A boxplot for the MSE of the cross validation for the ANN.*

```
# Show the mean of the MSE:
mean(cv.error)
## [1] 9.528274

cv.error
##  [1]  5.190725 11.595832  8.178917 10.288129  9.136706
##  [6] 10.260873  8.326635  7.114097 20.080831  5.109996

# Show the boxplot:
boxplot(cv.error,xlab='MSE',col='gray',
        border='blue',names='CV error (MSE)',
        main='Cross Validation error (MSE) for the ANN',
        horizontal=TRUE)
```

As you can see, the average MSE for the neural network is lower than the one of the linear model, although there is a lot of variation in the MSEs of the cross validation. This may depend on the splitting of the data or the random initialization of the weights in the ANN.

In the left plot of Figure 23.20 on page 443, you can see that there is one outlier. In the MSE of the k-fold cross validation. When that observation was part of the test-data. It also became an outlier in the vector of MSEs.

It is also important to realize that this result, in its turn can be influenced by the choosing seed for the random generator. By running the simulation different times with different seeds, one can get an idea about the sensitivity of the MSE for this seed.

23.4 Support Vector Machine

Support vector machines (SVMs) are a supervised learning technique that is – just as decision trees and neural networks – capable of solving both classification and regression problems. In practice, it is however, best suited for classification problems.

The idea behind support vector machines (SVM) is to find a hyperplane that best separates the data in the known classes. The idea is to find a hyperplane that maximises the distance between the groups.

SVM
support vector machine

The problem is in essence a linear set of equations to be solved, and it will fit a hyperplane, which would be a straight line for two dimensional data.

Obviously, if the separation is not linear, this method will not work well. The solution to this issue is known as the "kernel trick." We add a variable that is a suitable combination of he two variables (for example if one group appears to be centred in the 2D plane, then we could us $z = x^2 + y^2$ as third variable). Then we solve the SVM method as before (but with three variables instead of two), and find a hyperplane (flat surface) in a 3D space span by (x, y, z). This will allow for a much better separation of the data in many cases.

23.4.1 Fitting a SVM in R

Fitting an SVM model in R can be achieved with the function `svm()` from the package `e1071` It allows for the following classification types (parameter `type`):

`e1071`
`svm()`

Function use for svm()

```
svm(formula, data, subset, na.action = na.omit, scale = TRUE,
    type = NULL, kernel = 'radial', degree = 3,
    gamma = if (is.vector(x)) 1 else 1 / ncol(x), coef0 = 0,
    cost = 1, nu = 0.5,  class.weights = NULL, cachesize = 40,
    tolerance = 0.001, epsilon = 0.1, shrinking = TRUE,
    cross = 0, probability = FALSE, fitted = TRUE, ...)
```

Most parameters work very similar to other models such as `lm`, `glm`, etc. For example `data` and `formula` do not need much explanation anymore. The variable `type`, however, is an interesting one and it is quite specific for the SVM model:

1. `C-classification`: The default type of the dependent variable is a factor object;
2. `nu-classification`: Alternative classification – the parameter ν is used to determine the number of support vectors that should be kept in the solution (relative to the size of the dataset), this method will use the parameter ϵ for the optimization, but it is automatically set;
3. `one-classification`: Allows to detect outliers and can be used when only one class is available (say only cars with four cylinders and it allows to detect "unusual cars with four cylinders");
4. `eps-regression`: The default regression type, ϵ regression allows to set the parameter ϵ, the amount of error the model can have so that anything larger than ϵ is penalized in proportion to C, the regularization parameter;

5. `nu-regression`: The regression model that allows to tune the number of support vectors.

Another important parameter is `kernel`. This parameter allows us to select which kernel should be used. The following options are possible:

1. `Linear`: `t(u)*v`
2. `Polynomial`: `(gamma*t(u)*v + coef0)^degree`
3. `Radial` basis: `exp(-gamma*|u-v|^2)`
4. `Sigmoid`: `tanh(gamma*u'*v + coef0)`

When used, the parameters `gamma`, `coef0`, and `degree` can be provided to the function if one wants to over-ride the defaults.

Note – Optimisation types

Excluding the `one-classification`, there are two types of optimization: ν and ϵ and there are two types of target variables and hence we have regression and classification. In the `svm()` function both `C` and `eps` are used to refer to the same mechanism.

Here is a simple example:

```
library(e1071)
svmCars1 <- svm(cyl ~ ., data = mtcars)
summary(svmCars1)
##
## Call:
## svm(formula = cyl ~ ., data = mtcars)
##
##
## Parameters:
##    SVM-Type:  eps-regression
##  SVM-Kernel:  radial
##        cost:  1
##       gamma:  0.1
##     epsilon:  0.1
##
##
## Number of Support Vectors:  17
```

The function `svm` has treated the number of cylinders as numerical and decided that a regression was the best model. However, we are not really interested in fractional cylinders and hence can either round the results or fit a new model but coerce a classification model.

`svm()`

Below we illustrate how classification SVM model can be fitted:

```
# split mtcars in two subsets (not necessary but easier later):
x <- subset(mtcars, select = -cyl)
y <- mtcars$cyl

# fit the model again as a classification model:
```

```
svmCars2 <- svm(cyl ~ ., data = mtcars, type = 'C-classification')

# create predictions
pred <- predict(svmCars2, x)

# show the confusion matrix:
table(pred, y)
##     y
## pred  4  6  8
##    4 11  0  0
##    6  0  7  0
##    8  0  0 14
```

The number of cylinders is highly predictable based on all other data in `mtcars` and the model works just like this. However, in most cases, the user will need to fine tune some parameters. The function `svm` allows for many parameters to be set, and it has multiple kernels that can be used.

23.4.2 Optimizing the SVM

The dataset used (mtcars) resulted from the first try in a confusion matrix that revealed no mistakes. However, in most other datasets this will not be the case. Optimal kernals and values can be found by trial and error. However, there is also the function `tune()` in the package `e1071` that helps us to find optimal parameters.

```
svmTune <- tune(svm, train.x=x, train.y=y, kernel = "radial",
                ranges =  list(cost = 10^(-1:2), gamma = c(.5, 1, 2)))

print(svmTune)
##
## Parameter tuning of 'svm':
##
## - sampling method: 10-fold cross validation
##
## - best parameters:
##  cost gamma
##    10   0.5
##
## - best performance: 0.991777
```

After you have found the optimal parameters, you can run the model again and specify the desired parameters and compare the performance (e.g. with the confusion matrix).

Further information cross validation

In order to evaluate which models works best, it is worth to read Chapter 25 *"Model Validation"* on page 475.

23.5 Unsupervised Learning and Clustering

There are so many possible models and machine learning algorithms that in this book we can only scratch the surface and encourage further study. Till now most models are used to predict the dependant variable. The starting point is usually a clear and well defined segment of the larger pool of possible populations. For example, when we want to predict if a customer is creditworthy, we will make a separate model per credit type and further allocate the customers to groups, such as customers with a current account, customers that are new to the bank.

Would it not be nice if we could ask the machine to tell us what grouping makes sense, or what customers are alike? This type of questions is answered by the branch of machine learning where we do not tell the machine what a good or bad outcome looks like. Therefore, it is also referred to as "unsupervised learning."

unsupervised learning

Typical applications are

- *customer segmentation:* Identify groups of customers with similar traits (so that different segments can get different offers and a more targeted level of service);
- *stock market clustering:* Group stock based so that the resulting group will have similar behaviour under different market conditions;
- Reduce dimensionality of a dataset by grouping observations with similar values;
- Segment data so that for each group a specific and more specialized model can be build.

For example, in Chapter 27.7.6 *"PCA (Gaia)"* on page 553 we ask `ggplot` to tell us which service centre locations are similar and show the groups on the plot in the space of the first two principal components. The underlying algorithm is called "k-means."

Clustering methods identify sets of similar objects – referred to as "clusters" – in a multivariate data set. The most common types of clustering include

1. partitioning methods,
2. hierarchical clustering,
3. fuzzy clustering,
4. density-based clustering, and
5. model-based clustering.

In the next sections, we will explain very briefly what those methods are, show how some of those methods can be executed in R and how great visualizations can be obtained.

23.5.1 k-Means Clustering

Clustering analysis aims to classify data points into clusters so that the observations in the cluster are as similar as possible and differ as much as possible from observations in other clusters. While this premise is simple as such, it is a computationally hard task (actually NP-hard) and it needs the user to define measures of similarity. These similarity measures have to be chosen carefully, because they are determining for the result and can use concepts such as distance, variance, connectivity, or intensity. This means that we should be careful to choose a measure of similarity that suits the data and the application.

Given a set of observations $\mathbf{x} = (x_1, x_2, \ldots, x_n)$ (where each observations x_i is a n-dimensional vector), k-means clustering aims to minimize the variance for k (where $k \leq n$) sets – or clusters, C_i henceforth – between the mean of the set and the members of that group $\mathbf{C} = \{C_1, C_2, \ldots, C_k\}$. So the goal of k-means clustering becomes to find

$$\text{argmin}_{\mathbf{C}} \sum_{i=1}^{k} \sum_{x\ inC_i} ||\mathbf{x} - \mu_{\mathbf{i}}||$$

The standard algorithm start from randomly taking k different observations as initial centre for the k clusters. Each observation is then assigned to the cluster whose centre is the "closest." The distance is usually expressed as the Euclidian distance between that observation and the centroid of the cluster.

Then we calculate again the centre of each cluster[9] and the process is repeated: each observation is now allocated to the cluster that has the centroid closest to the observation. This step is then repeated till there are no changes in the cluster allocations in consecutive steps.

Warning – Hard problems

The clustering problem is actually NP-hard and the algorithm described (or any other known to date) converge fast, but converge to a local minimum. This means that it is possible that better clustering can be found with other initial conditions.

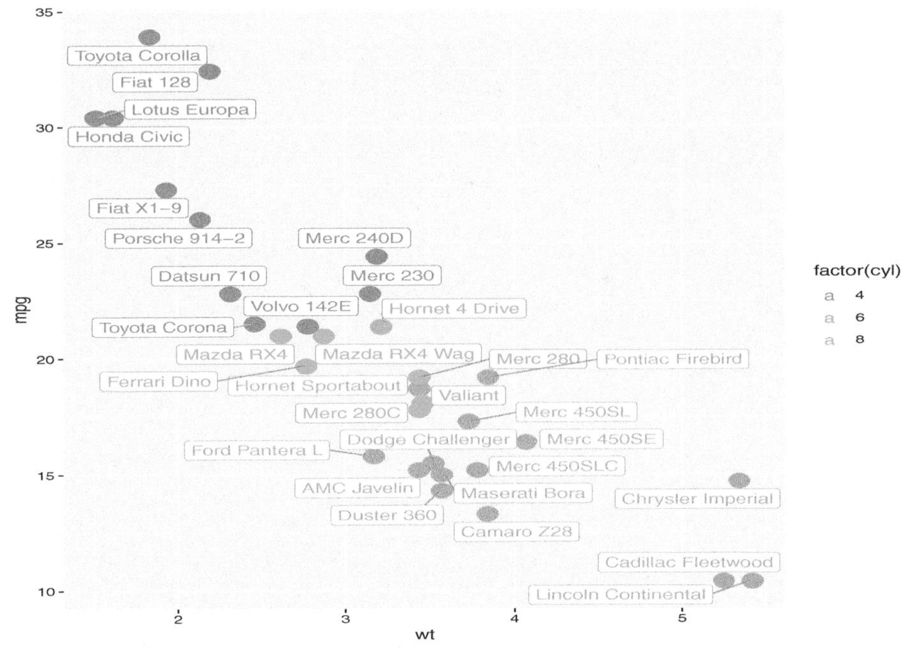

Figure 23.23: *The cars in the dataset mtcars with fuel consumption plotted in function of weight and coloured by the number of cylinders.*

[9]Note that this new centre does not necessarily coincide with an observed point.

23.5.1.1 k-Means Clustering in R

In this section, we will use the dataset `mtcars` that is – by now – well known. The dataset is usually loaded when R starts, and if that is not the case you can find it in the package `datasets`.

ggrepel

First, we have a look at the data `mtcars` and choose weight and fuel consumption as variables of interest for our analysis. Along the way, we introduce you to the package `ggrepel` that is handy to pull labels away from each other. We use this because we want to plot the name of the car next to each dot in order to get some understanding of what is going on.

Most of those things can be obtained with `ggplot2` alone.[10] The output is in Figure 23.23 on page 451.

```
library(ggplot2)
library(ggrepel)  # provides geom_label_repel()
ggplot(mtcars, aes(wt, mpg, color = factor(cyl))) +
      geom_point(size = 5) +
      geom_label_repel(aes(label = rownames(mtcars)),
                 box.padding   = 0.2,
                 point.padding = 0.25,
                 segment.color = 'grey60')
```

Note – Elegant labels

Compare the plot in Figure 23.23 with the result that we could get from adding to our plot the standard `geom_text()`:

```
ggplot(mtcars, aes(wt, mpg, color = factor(cyl))) +
      geom_point(size = 5) +
      geom_text(aes(label = rownames(mtcars)),
               hjust = -0.2, vjust = -0.2)
```

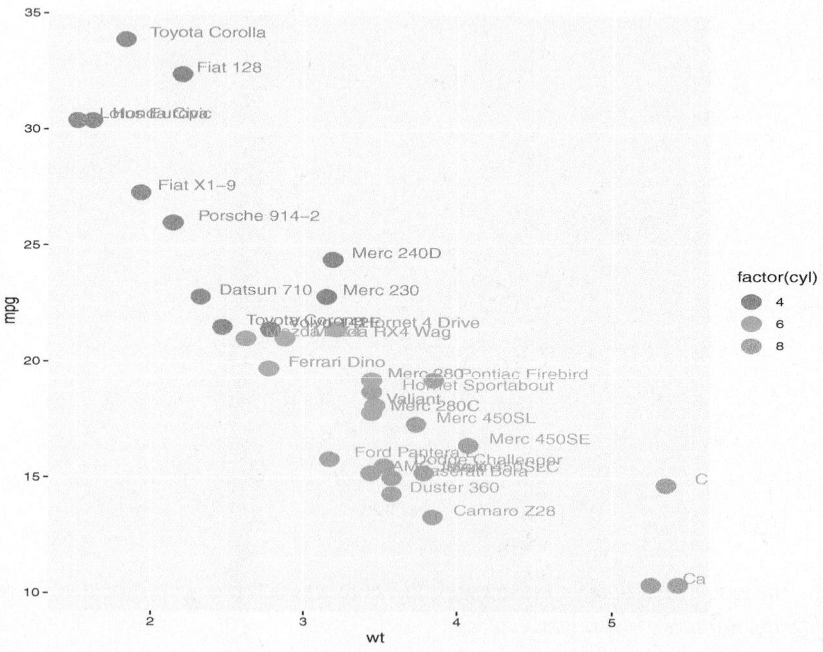

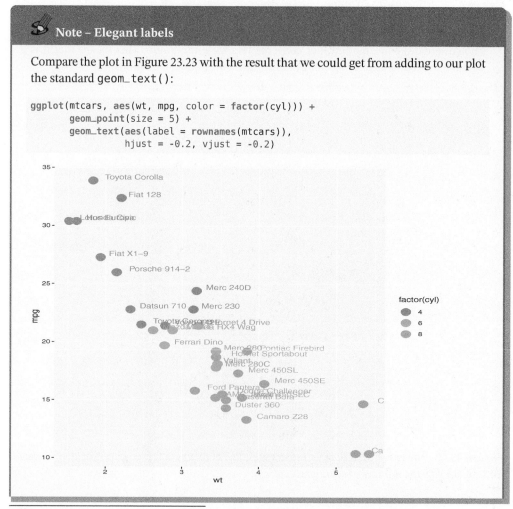

[10] For more information on `ggplot`, see Chapter 31 *"A Grammar of Graphics with ggplot2"* on page 687.

It also works, but geom_label() and geom_label_repel do a lot of heavy lifting: putting a frame around the text, uncluttering the labels, and even adding a small line between the box and the dot if the distance gets too big.

geom_label()

Plotting the cars in the (wt, mpg) plane we notice a certain – almost linear – relation and colouring the dots according to the number of cylinders we might be able to imagine some possible groups.

Before we can use a k-means, however, we need to define a "distance" between cars. This is in essence a process where imagination comes in handy and a deep understanding of the subject is essential. It will also make sense to normalize the data, because any measure would be dominated by the biggest nominal values.

Warning – Distance and units

The Euclidean distance – or any other distance for that matter – assumes a certain equality between dimensions. If we have no idea about what the most important criterion is, then we can use data normalised to one. In case we ask for example users to rate various aspects of user experience and rate their importance, then we have a proxy for such importance. In this case, ratings can be scaled according to importance. This means that we can then multiply the normalised variables with a given weight.

In the next code block, we will cluster cars using weight and fuel consumption.

```
# normalize weight and mpg
d <- data.frame(matrix(NA, nrow = nrow(mtcars), ncol = 1))
d <- d[,-1]  # d is an empty data frame with 32 rows

rngMpg <- range(mtcars$mpg, na.rm = TRUE)
rngWt  <- range(mtcars$wt,  na.rm = TRUE)

d$mpg_n <- (mtcars$mpg - rngMpg[1]) / rngMpg[2]
d$wt_n  <- (mtcars$wt  - rngWt[1])  / rngWt[2]

# Here is the k-means clustering itself.
# Note the nstart parameter (the number of random starting sets)
carCluster <- kmeans(d, 3, nstart = 15)

print(carCluster)
## K-means clustering with 3 clusters of sizes 7, 3, 22
##
## Cluster means:
##        mpg_n       wt_n
## 1 0.54951538 0.07814475
## 2 0.04228122 0.70550639
## 3 0.23518370 0.33595636
##
## Clustering vector:
##  [1] 3 3 1 3 3 3 3 3 3 3 3 3 3 3 2 2 2 1 1 1 3 3 3 3 3 1 1
## [28] 1 3 3 3 3
##
## Within cluster sum of squares by cluster:
## [1] 0.09687975 0.01124221 0.29960626
##  (between_SS / total_SS =  79.5 %)
```

```
##
## Available components:
##
## [1] "cluster"      "centers"      "totss"
## [4] "withinss"     "tot.withinss" "betweenss"
## [7] "size"         "iter"         "ifault"
```

First, we can investigate to what extend the number of cylinders is a good proxy for the clusters as found with the k-means algorithm.

```
table(carCluster$cluster, mtcars$cyl)
##
##      4  6  8
##   1  7  0  0
##   2  0  0  3
##   3  4  7 11

# Note that the rows are the clusters (1, 2, 3) and the number of
# cylinders are the columns (4, 6, 8).
```

Comparing to a known classification is easy, however, if that classification is not to be challenged, then what is the purpose of a clustering algorithm? Typically, a bank or asset manager will classify customers in groups of net worth. Although one will notice that – depending on profit margin structure – that even the most wealthy customers can be loss making for the bank (e.g. customers that make no or too much transactions or FX operations). A more complex customer segmentation, taking into account the customer life time value – see Chapter 29.3.2.3 *"Selected*

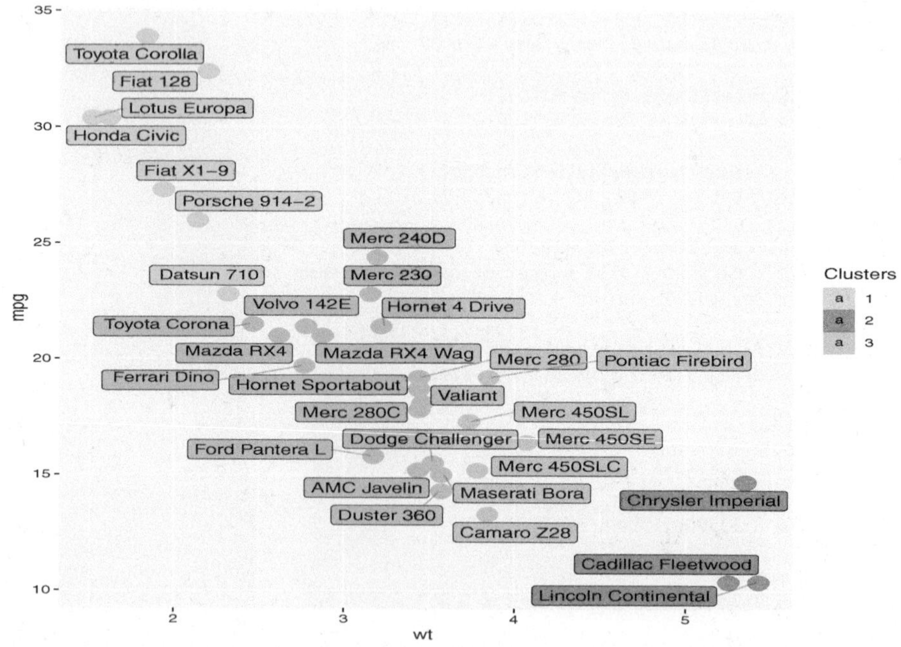

Figure 23.24: *The result of k-means clustering with three clusters on the weight and fuel consumption for the dataset mtcars.*

Useful KPIs" on page 593 – is usually a much better solution that will increase dramatically profitability for the bank and customer satisfaction (since customers will get products and service that suits their needs, not their net worth).

We can also visualize the clustering in the same way as we did before: we only need to colour the observations according to cluster and not according to number of cylinders. The ggplot2 visualisation of the next code block is in Figure 23.24 on page 454.

```
# Additionally, we customize colours and legend title.
# First, we need color names:
my_colors <- if_else(carCluster$cluster == 1, "darkolivegreen3",
                if_else(carCluster$cluster == 2, "coral", "cyan3"))

# We already loaded the libraries as follows:
#library(ggplot2); library(ggrepel)

# Now we can create the plot:
ggplot(mtcars, aes(wt, mpg, fill = factor(carCluster$cluster))) +
       geom_point(size = 5, colour = my_colors) +
       scale_fill_manual('Clusters',
                   values = c("darkolivegreen3","coral", "cyan3"))+
       geom_label_repel(aes(label = rownames(mtcars)),
                  box.padding   = 0.2,
                  point.padding = 0.25,
                  segment.color = 'grey60')
```

Question #21 – k-Means

Try a k-means analysis on the iris dataset. This is an interesting one, because it is a database that holds measurements of three different species of Iris flowers. So, we know in advance which clusters we would like to find. Is it enough to use k-means on Sepal.Length and Sepal.Width in order to classify all flower correctly?

Hint – Adding Voronoi cell borders

Would you like to plot also the Voronoi cell borders? Then we recommend to have a look at the package ggvoronoi.

23.5.1.2 PCA before Clustering

What about normalizing data, making a principal component analysis (PCA) and then use the Euclidean distance in the space that is span by the two major principal components: $(\text{PC}_1, \text{PC}_2)$? This approach takes more information into account and is in most cases preferable, because it makes the model more reliable. The cost is, however, loss of transparency and clarity: it becomes hard to see why a model makes a certain prediction because everything is convoluted. This is a discussion that we will save for later, but we will use the PCA to visualize the situation and make an educated guess about how many clusters there really are.

If you are not familiar with principal component analysis (PCA), then we recommend to read first Chapter 20 *"Factoring Analysis and Principle Components"* on page 363. In this section, we will assume that the reader is more or less familiar with the concept and its tools.

The PCA is executed in R via the function `prcomp` of the package `stats`, which is loaded when R is started.

```
# Normalize the whole mtcars dataset:
d <- data.frame(matrix(NA, nrow = nrow(mtcars), ncol = 1))
d <- d[,-1]  # d is an empty data frame with 32 rows
for (k in 1:ncol(mtcars)) {
rng <- range(mtcars[, k], na.rm = TRUE)
d[, k]  <- (mtcars[, k]  - rng[1])  / rng[2]
}
colnames(d) <- colnames(mtcars)
rownames(d) <- rownames(mtcars)

# The PCA analysis:
pca1 <- prcomp(d)
summary(pca1)
## Importance of components:
##                           PC1    PC2     PC3     PC4
## Standard deviation     0.6960 0.4871 0.20255 0.13916
## Proportion of Variance 0.5993 0.2935 0.05076 0.02396
## Cumulative Proportion  0.5993 0.8929 0.94365 0.96761
##                            PC5     PC6     PC7     PC8
## Standard deviation     0.09207 0.07719 0.06203 0.05801
## Proportion of Variance 0.01049 0.00737 0.00476 0.00416
## Cumulative Proportion  0.97810 0.98547 0.99023 0.99439
##                            PC9    PC10    PC11
## Standard deviation     0.05112 0.03642 0.02432
## Proportion of Variance 0.00323 0.00164 0.00073
## Cumulative Proportion  0.99763 0.99927 1.00000

# Note also:
class(pca1)
## [1] "prcomp"
```

We see that the first two components explain about 90% of the variance. This means that for most applications only two principal components will be sufficient. This is great because the 2D visualizations will be sufficiently clear. The function `plot` on the PCA object (`prcomp`-object in R), will visualize the relative importance of the different principal components (PCs) – in Figure 23.25 on page 457, and the function `biplot()` projects all data in the plane (PC_1, PC_2) and hence should show maximum variance – in Figure 23.26 on page 457:

biplot()

```
# Plot for the prcomp object shows the variance explained by each PC
plot(pca1, type = 'l')
```

```
# biplot shows a projection in the 2D plane (PC1, PC2)
biplot(pca1)
```

The plot in Figure 23.26 on page 457 is cluttered and hardly readable. We can try to make a better impression with `ggplot2`. To make clear what happens, we will do this in two steps. First, we produce the plot (in the following code – plot in Figure 23.27 on page 458) and then we will make sure that the labels are readable.

```
# Same plot with ggplot2:
library(ggplot2)
library(ggfortify)
library(cluster)
```

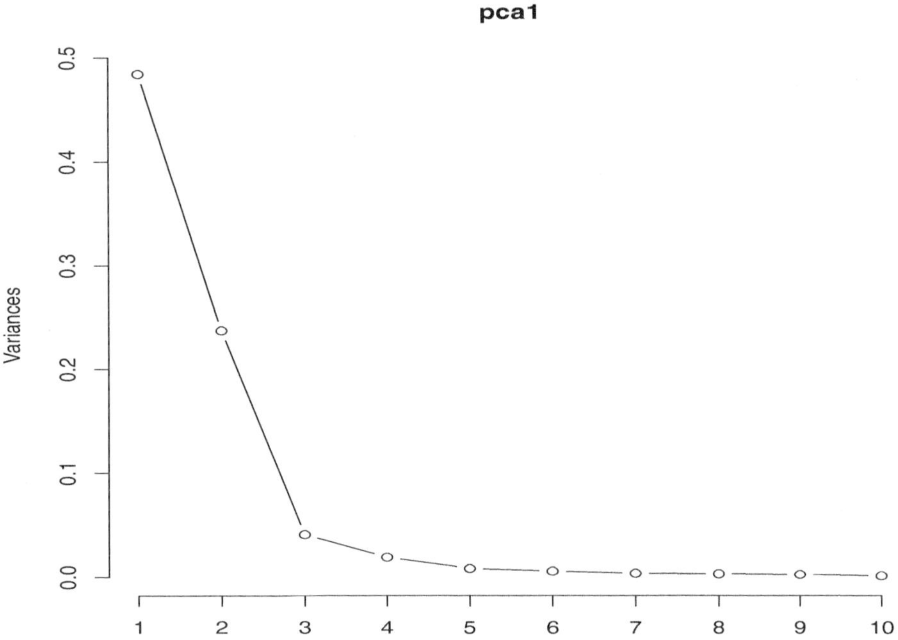

Figure 23.25: *The plot() function applied on a* `prcomp` *object visualises the relative importance of the different principal components.*

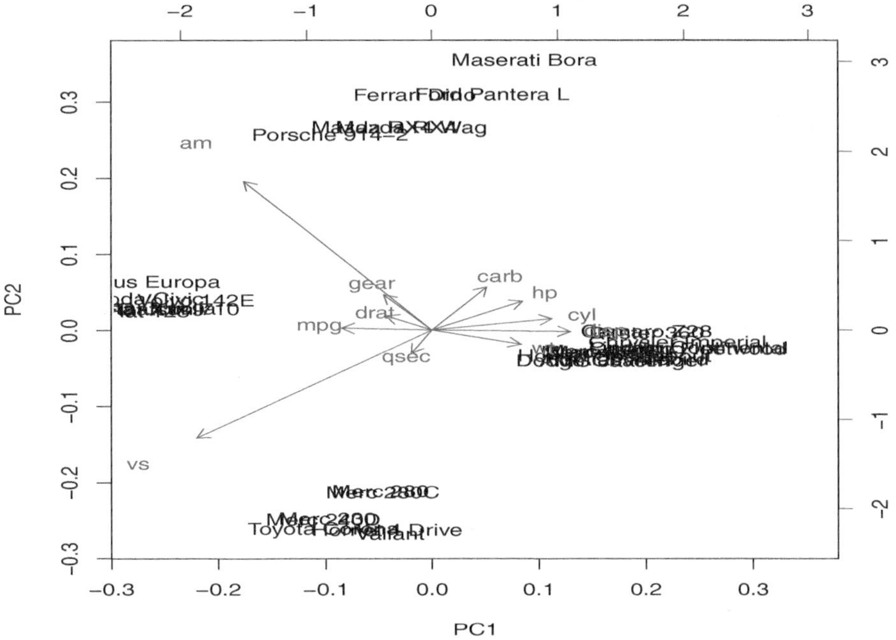

Figure 23.26: *The custom function* `biplot()` *project all data in the plane that is span by the two major PCs.*

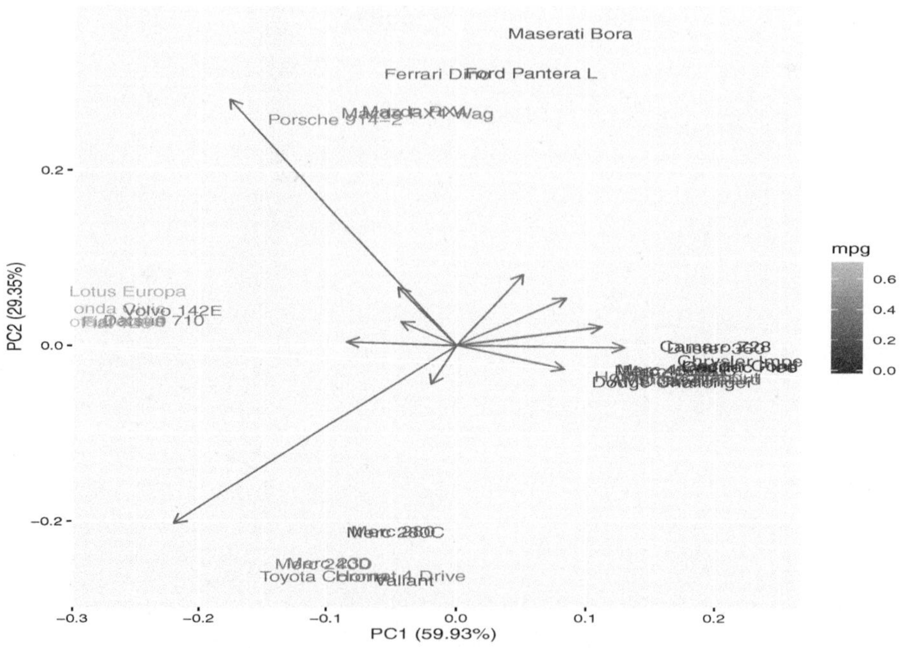

Figure 23.27: *A projection in the plane of the two major principal components via ggplot2. It looks good, but the labels are cluttered.*

```
autoplot(pca1, data=d, label=TRUE, shape=FALSE, colour='mpg',
         loadings = TRUE, loadings.colour = 'blue',
         loadings.label = FALSE, loadings.label.size = 3
         )
```

Both plots, Figure 23.26 on page 457 and Figure 23.27, are not very elegant. However, they server their purpose well: they help us to see that there are actually four types of cars in this database. That means that there are four clusters, and now we can run the k-means algorithm, while requesting four clusters.

We can now run the clustering algorithm on the data as it is or on the data in the orthogonal base (and hence use all the PCs). In this case we could even run the clustering only on the two first PCs, because they explain so much of the variance that the clusters would be the same, when only using these two PCs.

Finally, we are ready for the k-means analysis on the 11-dimensional data (not only two). We choose to run the analysis directly on the normalised data frame, `d`. Of course we decide on four clusters. The last line of the following code also plots the results (with a different colour per cluster), adds labels in a more readable manner in Figure 23.28 on page 459.

```
carCluster <- kmeans(d, 4, nstart = 10)
d <- cbind(d, cluster = factor(carCluster$cluster))
```

We visualize the result with `autoplot` from the library `ggplot2`.

```
library(ggplot2)
library(ggrepel)

autoplot(pca1, data=d, label=FALSE, shape=18, size = 5,
```

```
    alpha = 0.6, colour = 'cluster',
    loadings = TRUE, loadings.colour = 'blue',
    loadings.label = TRUE, loadings.label.size = 5) +
  geom_label_repel(aes(label = rownames(mtcars)),
                  box.padding   = 0.2,
                  point.padding = 0.25,
                  segment.color = 'grey60')
```

From Figure 23.28 we see that the cars indeed seem to group into four clusters. Looking at the "loadings" (projections of the attributes) in the PC_1, PC_2) space we can understand the four groups of cars.

1. *red (West – cluster 1):* the city cars. These cars are best for fuel consumption, have a high rear axle ratio, low displacement, horse power and weight. They rather have a line engine (the 1 in the VS variable) and tend to have a manual gearbox.

2. *green (South – cluster 2):* the light coupe cars (two door sedans). These cars are able to take four passengers and travel longer distances, but motors are not very special (low number of carburators, horse power) and the gearbox has a low number of gears, they are slow to accelerate and have typically cylinders in line. Therefore, they are reasonably fuel efficient.

3. *cyan (East – cluster 3):* the "muscle cars." These are heavy cars with large motors that produce a lot of horse power, hence, need a low rear axle ration and are very fuel inefficient and they tend towards automatic gears. Probably the SUVs would fit also here, but this database precedes the concept SUV.[11]

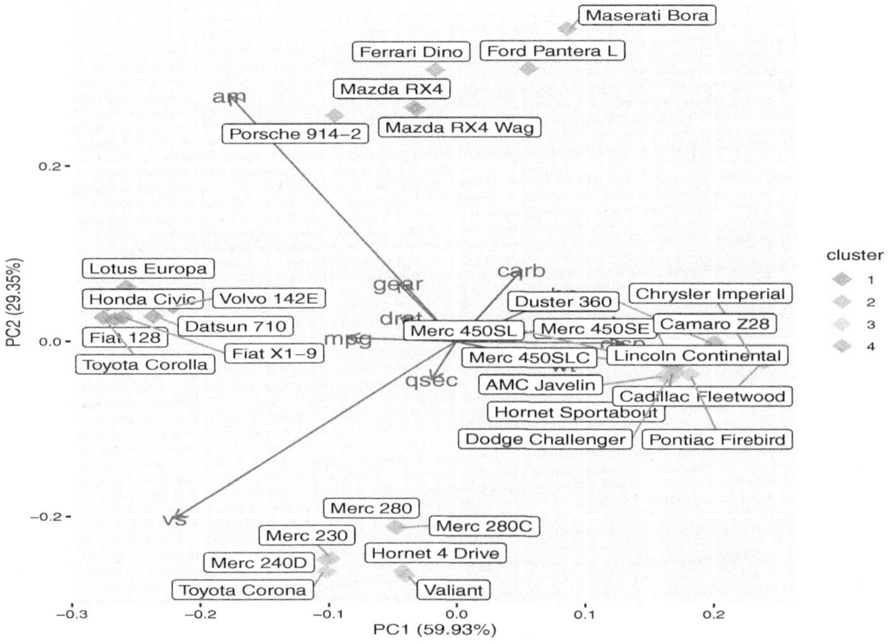

Figure 23.28: *The projection of mtcars in the surface formed by the two first principal components and colored by number of cluster.*

[11] The database mtcars seem to be something that is compiled no later than the early 1980s.

4. *violet (North – cluster 4):* the sportscars. These cars have manual gears, are high in horse power, number of carburators and low in number of seconds to travel 1/4 mile; they are neutral on displacement and fuel consumption (they are fast but light).

Clustering analysis is a method where the algorithm can find by itself which objects should belong to one group. There is no training of the model needed, and hence it is considered as a method of unsupervised learning.

However, after reading so far, it will be clear that the modeller has a major impact on the outcome. First and foremost the selection of variables is key. For example we could leave out axle ratio, weight, etc. and focus on the size of the factory, the country where the car is produced, the number of colours available, etc. This could lead to a different clustering.

Another way to influence results (in absence of PCA) is adding related measures. For example, adding a variable for number of seats would be quite similar to the axle ratio, and hence we would already have two variables that contribute to setting the sports cars apart.

Note – Autoplot or native ggplot?

`autoplot` is a layer of automation over `ggplot2`. It does a great job, but it is not so hard to get nice results with native ggplot2. Note that in the following code most of the lines related to manipulating the colours and the title of the legend.

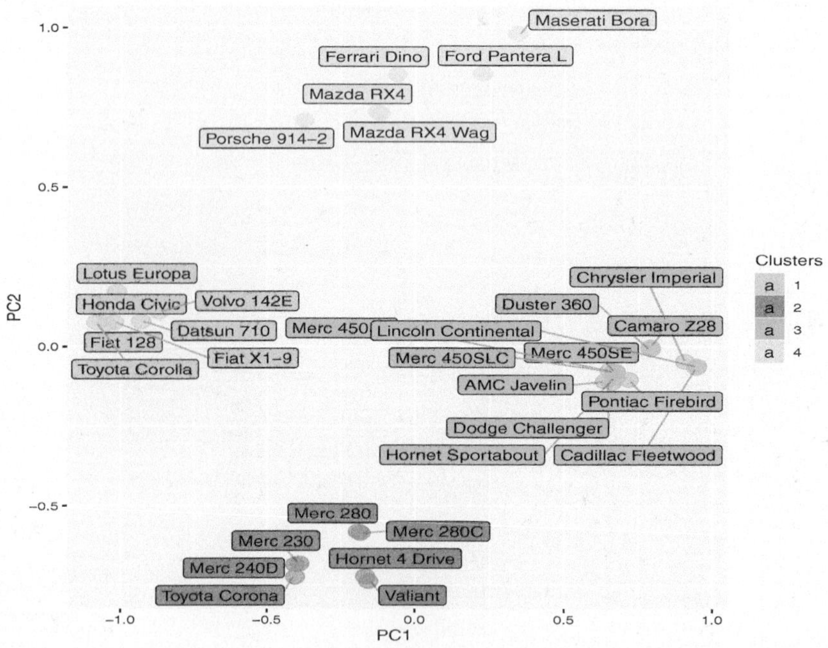

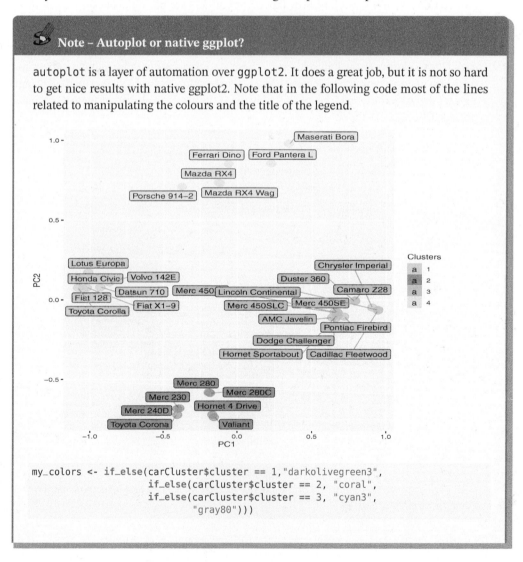

```
my_colors <- if_else(carCluster$cluster == 1,"darkolivegreen3",
                     if_else(carCluster$cluster == 2, "coral",
                     if_else(carCluster$cluster == 3, "cyan3",
                             "gray80")))
```

```
ggplot(pca1$x,
       aes(x=PC1,y=PC2, fill = factor(carCluster$cluster))) +
       geom_point(size = 5, alpha = 0.7, colour = my_colors)+
       scale_fill_manual('Clusters',
                         values =c("darkolivegreen3","coral",
                         "cyan3", "gray80")) +
       geom_label_repel(aes(label = rownames(mtcars)),
                    box.padding    = 0.2,
                    point.padding = 0.25,
                    segment.color = 'grey60')
```

23.5.1.3 On the Relation Between PCA and k-Means

Many practitioners will actually apply the k-means algorithm on the data expressed as coordinates in the orthogonal base formed by the principal components. In many models one could expect a performance and stability boost by first doing the PCA analyses and working from the principal components.

Doing the PCA first, works as follows:

```
pca1 <- prcomp(d)    # unchanged

# Extract the first 3 PCs
PCs <- data.frame(pca1$x[,1:3])

# Then use this in the kmeans function for example.
carCluster <- kmeans(PCs, 4, nstart=10, iter.max=1000)
```

There is a much deeper relations between PCA and k-means clustering. The optimal solution (or even local minima) of the k-means algorithm describe spherical clusters in their Euclidean space. If we only have two clusters, then the line connecting the two centroids is the one-dimensional projection direction that would separate the two clusters most clearly. This is also what the direction of the first PC would indicate. The gravitational centre of that line separates the clusters in the k-means approach.

This reasoning can readily be extended for more clusters. For example, if we consider three clusters, then the two-dimensional plane spanned by three cluster centroids is the two-dimensional projection that shows the most variance in our data. This is similar to the plane spanned by the two first principal components.

So, it appears that the solution of k-means clustering, is given by principal component analysis (PCA).[12] The subspace spanned by the directions of the principal components is identical to the cluster centroid subspace.

In the case where the real clusters are actually not spherical, but for example shaped as the crescent of the moon, this reasoning does not work any more, clusters are not well separated by nor k-means or PCA, and it is possible to construct counterexamples where the cluster centroid subspace is not spanned by the principal directions.[13]

[12] See for example Ding and He (2004).
[13] See for example Cohen et al. (2015).

23.5.2 Visualizing Clusters in Three Dimensions

In our case, the first two principal components explain so much of the variance that it is hardly necessary to consider the other. However, in many cases one will find that there are more than two principal components that explains a significant proportion of the variance. Many analysts will use – as a rule of thumb – that 85% of the variance of the proportion that needs to be captured.

To visualise clusters in more than two components 2D projects are not sufficient. For example, imagine that we decide to work with the three major PCs. Then we can use the standard `plot()` function on the principle components while adding a colour per cluster in order to get insight in how the clusters behave in those three dimensions – this is in Figure 23.29.

```
# Reminder of previous code:
d <- data.frame(matrix(NA, nrow = nrow(mtcars), ncol = 1))
d <- d[,-1]  # d is an empty data frame with 32 rows
for (k in 1:ncol(mtcars)) {
  rng     <- range(mtcars[, k], na.rm = TRUE)
  d[, k]  <- (mtcars[, k]  - rng[1])  / rng[2]
  }
colnames(d) <- colnames(mtcars)
rownames(d) <- rownames(mtcars)
pca1 <- prcomp(d)                 # runs the PCA
PCs <- data.frame(pca1$x[,1:3]) # extracts the first 3 PCs
# Now, PCs holds the three major PCs.

# -- New code below:
# Plot those three PCs:
plot(PCs, col = carCluster$clust, pch = 16, cex = 3)
```

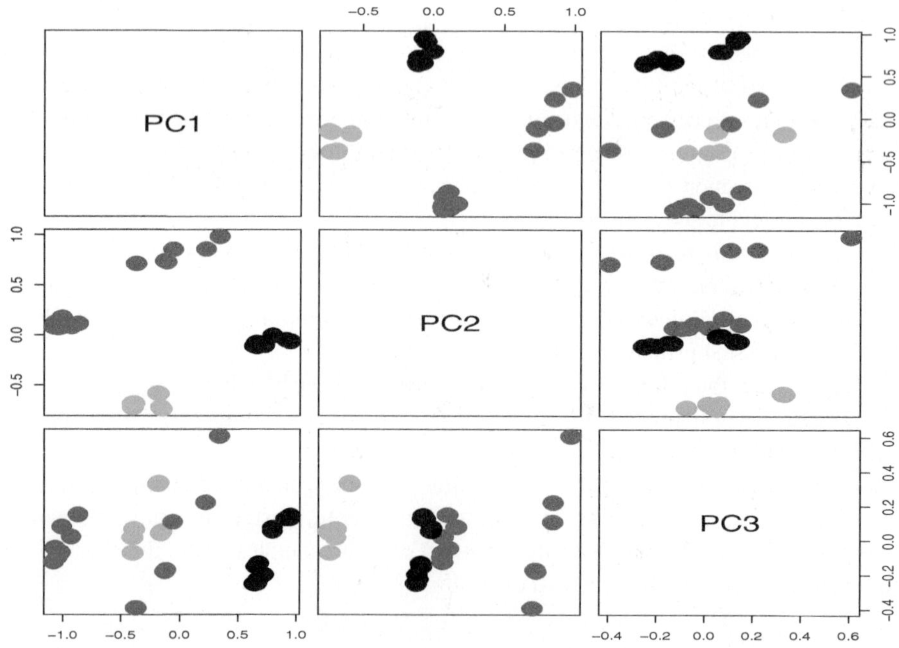

Figure 23.29: *Two dimensional projections of the dependency structure of the data in the first principal components. Note that in this plot, we see different 2D projections of 3D data.*

While, Figure 23.29 provides some insight in how the clusters operate, these are only 2D projections of data that is actually 3D. Seeing the data in three dimensions, is more natural and provides so much more insight.

There are some really great solutions in R to show 3D data. For example, there is the package `plot3D`. The code below shows how to use it and the plot is in Figure 23.30.

```
library(plot3D)
scatter3D(x = PCs$PC1, y = PCs$PC2, z = PCs$PC3,
   phi = 45, theta = 45,
   pch = 16, cex = 1.5, bty = "f",
   clab = "cluster",
   colvar = as.integer(carCluster$cluster),
   col = c("darkolivegreen3", "coral", "cyan3", "gray"),
   colkey = list(at = c(1, 2, 3, 4),
          addlines = TRUE, length = 0.5, width = 0.5,
          labels = c("1", "2", "3", "4"))
   )
text3D(x = PCs$PC1, y = PCs$PC2, z = PCs$PC3,  labels = rownames(d),
       add = TRUE, colkey = FALSE, cex = 1.2)
```

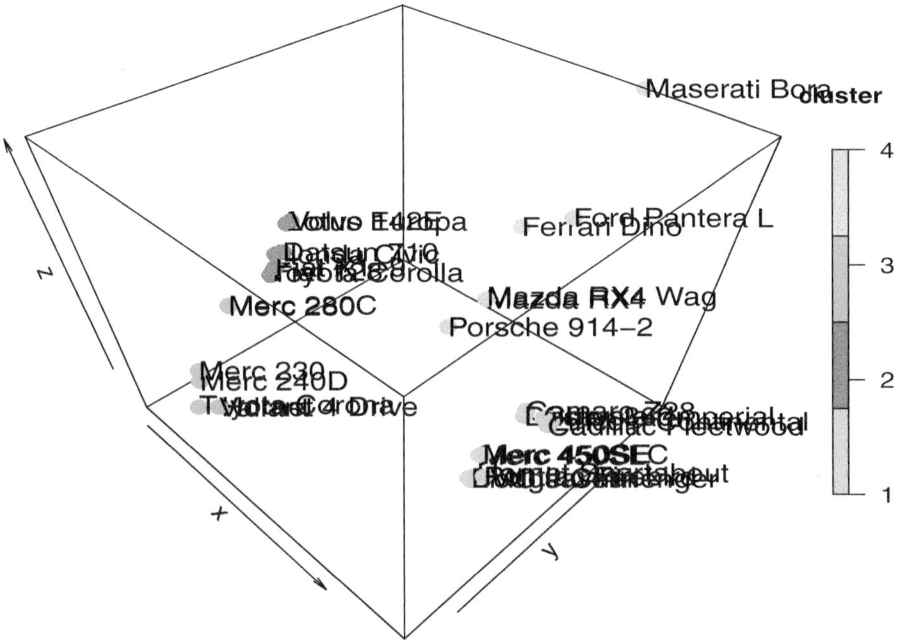

Figure 23.30: *A three dimensional plot of the cars with on the z-axis the first principal component, the second on the y-axis and the third along the z-axis.*

Of course, there is also `ggplot2`, where stunning results can be obtained. We would prefer to present the package `gg3D`, it works like simply by adding a z-axis to ggplot2. Unfortunately, it does not work with the latest versions of R.

A great alternative – for an interactive environment – is `plotly`.[14] The plot of the following code is Figure 23.31 on page 464.

[14] We will also use `plotly` in Chapter 36.3.2 *"A Dashboard with flexdashboard"* on page 731.

```
library(plotly)
plot_ly(x = PCs$PC1, y = PCs$PC2, z = PCs$PC3,
        type = "scatter3d", mode = "markers",
        color = factor(carCluster$cluster))
```

plotly is designed to be used in an interactive environment such as a web-browser. This book is static, hence Figure 23.31 is inserted from a screen-capture. Try it in RStudio to see how it works. You will get automatically tools to turn around the plots, so you can see it from different angles.

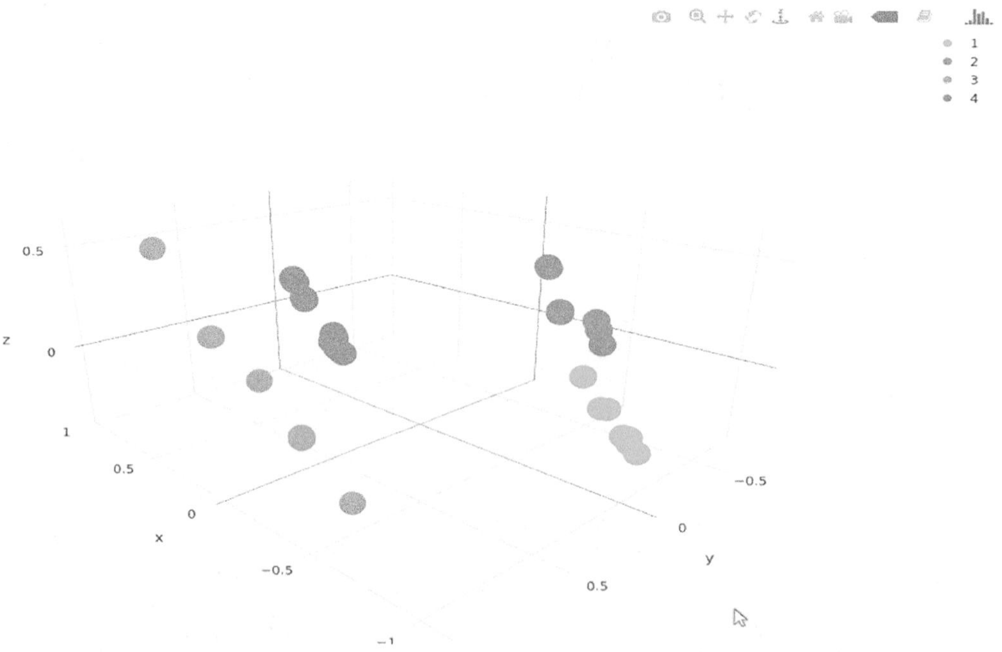

Figure 23.31: *plotly will produce a graph that is not only 3D but is interactive. It can be moved with the mouse to get a better view on how the data is located. It is ideal for an interactive website or RStudio.*

23.5.3 Fuzzy Clustering

Clustering is closely related to Multi Criteria Decision Analysis[15] and as such it also is a field where many algorithms compete and all have their value in creating more insight. Also just like in MCDA, there is not necessarily one unique best solution, and the best solution might depend on interpretation, compromise, and thorough knowledge of the underlying problem.

In k-means each observation belongs to one cluster and the borders between clusters are formed by segments of straight lines (forming the Voronoi diagram). One possible alteration to the algorithm is to come up with a solution where borders between clusters can be blurred and where one observation can belong to more than one cluster. This mimics a situation similar to a citizen having two nationalities, an employee working for more than one department, a person identifying with two genders, etc.

[15]Multi Criteria Decision Analysis (MCDA) is introduced in Chapter 27 *"Multi Criteria Decision Analysis (MCDA)"* on page 511 and it is the field that tries to identify optimal solutions when there are more than one criterion to optimize. This means that there is no unique mathematical solution, but that the best solution depends on compromises.

One such algorithm is called "fuzzy clustering" – also referred to as soft clustering or soft k-means. It works as follows:

1. Decide on the number of clusters, k.
2. Each observation has a coefficient w_{ij} (the degree of x_i being in a cluster j) for each cluster — in the first step assign those coefficients randomly.
3. Calculate the centroid of each cluster as

$$c_j = \frac{\sum_i w_{ij}(x_i)^m x_i}{\sum_i w_{ij}(x_i)^m}$$

   where m is the parameter that controls how fuzzy the cluster will be. Higher m values will result in a more fuzzy cluster. This parameter is also referred to as the "hyper-parameter"

hyper-parameter

4. For each observation calculate again the weights with the updated centroids.

$$w_{ij} = \frac{1}{\sum_l^k \left(\frac{||x_i - c_i||}{x_i - c_l}\right)^{\frac{2}{m-1}}}$$

5. Repeat from step 3, until the algorithm has coefficients that do not change more than a given small value ϵ, the sensitivity threshold

Fuzzy clustering will, given a set of observations $\mathbf{x} = (x_1, x_2, \ldots, x_n)$ (where each observations x_i is a n-dimensional vector), aim to minimize the variance for k (where $k \leq n$) sets – or clusters (C_i henceforth) – between the mean of the set and the members of that group $\mathbf{C} = \{C_1, C_2, \ldots, C_k\}$. So, the goal of k-means clustering becomes to find

$$\text{argmin}_{\mathbf{C}} \sum_{i=1}^{n} \sum_{j=1}^{k} w_{ij} ||\mathbf{x_i} - \mathbf{c_j}||^2$$

With `ggplot2` and `ggfortify` it is easy to obtain nice results with little effort. Below we will use the function `fanny()` from the library `cluster` to execute the fuzzy clustering and plot the results in Figure 23.32 on page 466

```
library(tidyverse)  # provides if_else
library(ggplot2)    # 2D plotting
library(ggfortify)
library(cluster)    # provides fanny (the fuzzy clustering)
library(ggrepel)    # provides geom_label_repel (de-clutter labels)


carCluster <- fanny(d, 4)
my_colors <- if_else(carCluster$cluster == 1, "coral",
               if_else(carCluster$cluster == 2, "darkolivegreen3",
               if_else(carCluster$cluster == 3, "cyan3",
                 "darkorchid1")))

# Autoplot with visualization of 4 clusters:
autoplot(carCluster, label=FALSE, frame=TRUE,  frame.type='norm',
         shape=16,
         loadings=TRUE,  loadings.colour = 'blue',
         loadings.label = TRUE, loadings.label.size = 5,
         loadings.label.vjust = 1.2, loadings.label.hjust = 1.3) +
       geom_point(size = 5, alpha = 0.7, colour = my_colors) +
```

```
geom_label_repel(aes(label = rownames(mtcars)),
          box.padding    = 0.2,
          point.padding = 0.25,
          segment.color = 'grey40') +
          theme_classic()
```

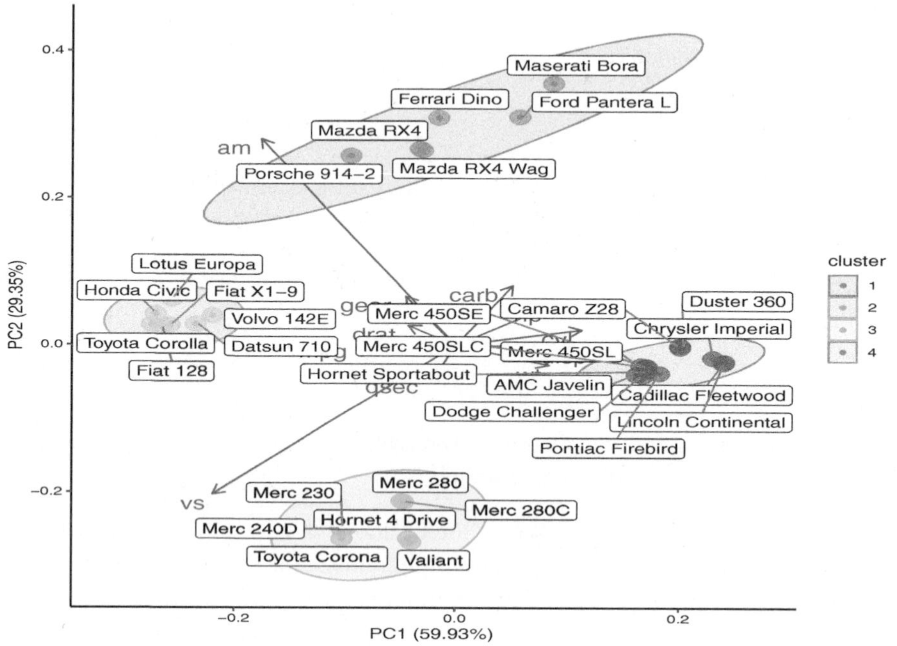

Figure 23.32: *A plot with* `autoplot()`*, enhanced with* `ggrepel` *of the fuzzy clustering for the dataset mtcars.*

23.5.4 Hierarchical Clustering

Hierarchical clustering is a particularly useful approach that provides a lot of insight and does not require to define a number of clusters to be provided by the user. Ultimately, we get a tree-based representation of all observations in our dataset, which is also known as the dendrogram.

This means that we can use the dendrogram itself to make an educated guess on where to separate the dendrogram and hence how much and at what level we make clusters.

The R code to compute and visualize hierarchical clustering is below, and the plot resulting from it is in Figure 23.33 on page 467:

```
# Compute hierarchical clustering
library(tidyverse)
cars_hc <- mtcars                         %>%
           scale                          %>% # scale the data
           dist(method = "euclidean")  %>% # dissimilarity matrix
           hclust(method = "ward.D2")      # hierachical clustering

plot(cars_hc)
```

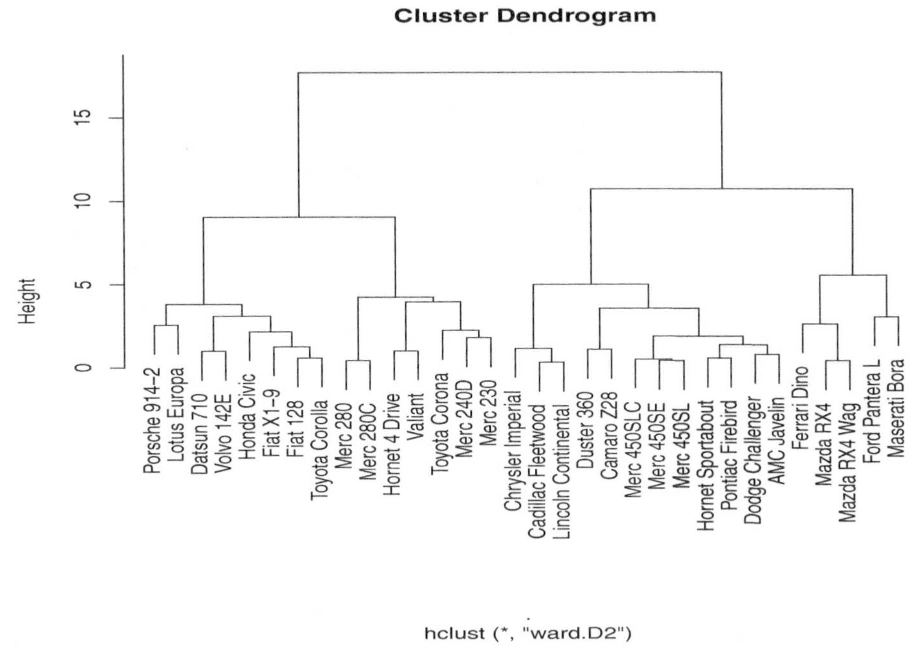

Figure 23.33: *A hierarchical cluster for the dataset mtcars.*

Note that unlike most of the other clustering algorithms used in this section, `hclust` is provided by the package `stats` and not by `cluster`. This results in minor differences in how it functions. The function `hclust()` will not work on the dataset `d` as we did before. This function expects a "dissimilarity structure," this is an object produced by the function `dist()` that calculates a dissimilarity matrix.

`hclust()`
`dist()`

Note – Scaling data

We choose to use the function `scale()` instead of our own simple scaling used before. Scale allows to centre data and scale it at the same time, and it allows different choices per column.

`scale()`

Hierarchical clustering does not require to determine the number of clusters on beforehand and it results in the information-rich dendrogram. It does, however, require us to choose one of the many possible algorithms. `hclust()` alone provides five methods, so it is worth to study the functions and make a careful choice as well as experiment to see what works best for your goal and your data.[16]

[16]We count the `ward.D` and `ward.D2` as just one method. The difference is only that the `ward.D` method expects Euclidean squared distances, while `ward.D2` the not squared ones.

Further information – Ward's method

One of the intersting clustering methods available in the function `hclust()` is "Ward's method."
More about Ward's method can be found in Murtagh and Legendre (2011) and Szekely and Rizzo (2005).

Digression – A nicer dendrogram

factorextra

If you want to have more options and features for the dendrogram (e.g. such as colouring different nodes, control label colours, draw boxes around the clusters, etc.) we recommend to have a look at the library `factorextra` and its function `fviz_dend()`.

23.5.5 Other Clustering Methods

As mentioned before, clustering is more an art than a science and hence there are many more algorithms that can be thought of and each can be used with different scaling and different measures of distance.

For example, one idea could be to replace the centroids by medoids (rather using median instead of average position). The function `pam()` will do this.

pam()

Another idea could be to use correlation as a distance between observations.

Hierarchical clustering algorithms are either top-down or bottom-up. Bottom-up algorithms start from a clustering where each observation has its own cluster and then step by step merge (or agglomerate) pairs of clusters until all clusters have been merged into one single cluster. Bottom-up hierarchical clustering is therefore, called hierarchical agglomerative clustering or HAC. This can be done by `agnes()` from the package `cluster`.

HAC
agnes()

Top-down clustering requires a method for splitting a cluster. It will start from considering all observations to be part of one cluster. Then it will successively split clusters recursively until individual observations are reached.

The "k nearest neighbours" method (kNN) is a simple algorithm that stores all available data points and classifies new cases by a majority vote of its k neighbours. This algorithms segregates unlabelled data points into well defined groups.

kNN

The algorithm is very simple and makes no prior assumption about the underlying data. In R, one can rely on the function `knn()` from the package `class`.

```
library(class)
knn(train, test, cl, k = 1, l = 0, prob = FALSE, use.all = TRUE)
```

One will notice that this function is designed to take a training dataset and a test dataset. This makes it a little impractical for mtcars (which is a really small dataset). It also needs the parameter `cl` which holds the true classifications.

knn()

♣ 24 ♣

Towards a Tidy Modelling Cycle with modelr

The package `modelr` provides a layer around R's base functions that allows not only to work with models using the pipe `%>%` command, but also provides some functions that are more intuitive to work with. `modelr` is not part of the core-tidyverse, so, we need to load it separately. `modelr`

```
library(tidyverse)
library(modelr)
```

In the next sections, we will use the library `modelr` to create predictions, perform cross validations, etc. While it is possible to learn it as you read through the next chapters, it is also useful to have an overview of what `modelr` can do for you. Therefore, we briefly introduce the methods that `modelr` provides and later use them in Chapter 25 *"Model Validation"* on page 475.

To present the functionality, we will focus on a simple model based on the well-known dataset `mtcars` from the package `datasets`. As usual, we will model the miles per gallon of the different car models.

```
d   <- mtcars
lm1 <- lm(mpg ~ wt + cyl, data = d)
```

While we show the functionality on a linear model, you can use virtually any model and `modelr` will take care of the rest.

The Big R-Book: From Data Science to Learning Machines and Big Data, First Edition. Philippe J.S. De Brouwer.

Companion Website: www.wiley.com/go/De Brouwer/The Big R-Book

24.1 Adding Predictions

add_predictions()

Each model leads to predictions and predictions can be used to test the quality of the model. So adding predictions will be the work of every modeller for every model. Therefore it is most important to have a standardised way of doing this. modelr's function add_predictions() provides this.

Function use for add_predictions()

```
add_predictions(data, model, var = "pred", type = NULL)
```

Adds predictions to a dataset for a given model, the predictions are added in a column named by the variable pred.

```
library(modelr)

# Use the data defined above:
d1 <- d %>% add_predictions(lm1)

# d1 has now an extra column "pred"
head(d1)
##                    mpg cyl disp  hp drat    wt  qsec vs
## Mazda RX4         21.0   6  160 110 3.90 2.620 16.46  0
## Mazda RX4 Wag     21.0   6  160 110 3.90 2.875 17.02  0
## Datsun 710        22.8   4  108  93 3.85 2.320 18.61  1
## Hornet 4 Drive    21.4   6  258 110 3.08 3.215 19.44  1
## Hornet Sportabout 18.7   8  360 175 3.15 3.440 17.02  0
## Valiant           18.1   6  225 105 2.76 3.460 20.22  1
##                   am gear carb     pred
## Mazda RX4          1    4    4 22.27914
## Mazda RX4 Wag      1    4    4 21.46545
## Datsun 710         1    4    1 26.25203
## Hornet 4 Drive     0    3    1 20.38052
## Hornet Sportabout  0    3    2 16.64696
## Valiant            0    3    1 19.59873
```

The column pred is easy to understand as the prediction for mpg based on the other variables using the linear model. We can now compare these values with the given mpg, calculate MSE, etc.

24.2 Adding Residuals

The second most used concept is the residual. It is the difference between the observed value and its prediction. To calculate the MSE, for example, one first calculates the residuals, then squares and sums them. Therefore, `modelr` also provides a function to do this: `add_residuals()`.

add_residuals()

Function use for add_residuals()

```
add_residuals(data, model, var = "resid")
```

Adds residuals to a given dataset for a given model. The new column is named by the parameter `var`.

```
d2 <- d1 %>% add_residuals(lm1)

# d2 has now an extra column "resid"
head(d2)
##                    mpg cyl disp  hp drat    wt  qsec vs
## Mazda RX4         21.0   6  160 110 3.90 2.620 16.46  0
## Mazda RX4 Wag     21.0   6  160 110 3.90 2.875 17.02  0
## Datsun 710        22.8   4  108  93 3.85 2.320 18.61  1
## Hornet 4 Drive    21.4   6  258 110 3.08 3.215 19.44  1
## Hornet Sportabout 18.7   8  360 175 3.15 3.440 17.02  0
## Valiant           18.1   6  225 105 2.76 3.460 20.22  1
##                   am gear carb     pred      resid
## Mazda RX4          1    4    4 22.27914 -1.2791447
## Mazda RX4 Wag      1    4    4 21.46545 -0.4654468
## Datsun 710         1    4    1 26.25203 -3.4520262
## Hornet 4 Drive     0    3    1 20.38052  1.0194838
## Hornet Sportabout  0    3    2 16.64696  2.0530424
## Valiant            0    3    1 19.59873 -1.4987281
```

Again we notice the additional column `resid`, that contains the residuals.

24.3 Bootstrapping Data

As we will see in Chapter 25.4 *"Cross-Validation"* on page 483, bootstrapping is an essential element of cross validation and hence an essential tool in assessing the validity of a model.

Function use for bootstrap()

```
bootstrap(data, n, id = ".id")
```

Generates n bootstrap replicates (dataset build from random draws – with replacement – of observations from the source data) of the dataset data.

The following code illustrates how bootstrapping can be used to generate a set of estimates for relevant coefficients for a linear model. The histogram of the estimates is shown in Figure 24.1.

```
set.seed(1872)    # make sure that results can be replicated
library(modelr)   # provides bootstrap
library(purrr)    # provides map, map_df, etc.
library(ggplot2) # provides ggplot
d     <- mtcars
boot <- bootstrap(d, 10)

# Now, we can leverage tidyverse functions such as map to create
# multiple models on the 10 datasets
models <- map(boot$strap, ~ lm(mpg ~ wt + cyl, data = .))

# The function tidy of broom (also tidyverse) allows to create a
```

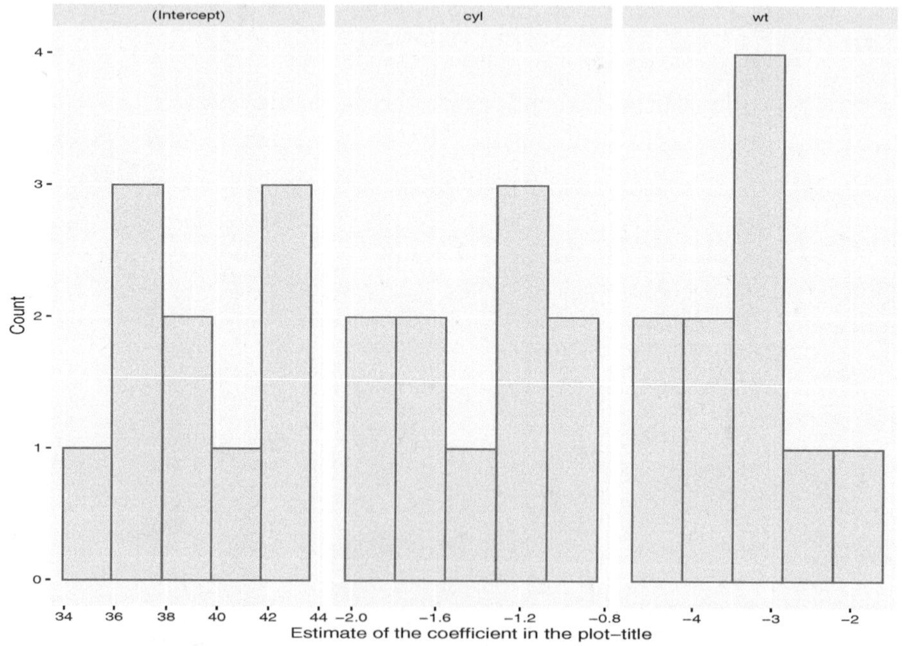

Figure 24.1: *The results of the bootstrap exercise: a set of estimates for each coefficient.*

```
# dataset based on the list of models. Broom is not loaded, because
# it also provides a function bootstrap().
tidied <- map_df(models, broom::tidy, .id = "id")

# Visualize the results with ggplot2:
p <- ggplot(tidied, aes(estimate)) +
     geom_histogram(bins = 5, col = 'red', fill='khaki3',
                    alpha = 0.5) +
     ylab('Count') +
     xlab('Estimate of the coefficient in the plot-title') +
     facet_grid(. ~ term, scales = "free")
p
```

Bootstrapping is explained in more detail in Chapter 25.3 *"Bootstrapping"* on page 479.

24.4 Other Functions of modelr

There are much more functions available in `modelr`, however, they refer to concepts that appear later in the following chapter: Chapter 25 *"Model Validation"* on page 475. At least the following will come in handy:

- `crossv_mc(data, n, test = 0.2, id = ".id")` for Monte Carlo cross validation — see Chapter 25.4.2 *"Monte Carlo Cross Validation"* on page 486
- `crossv_kfold(data, k = 5, id = ".id")` for k-fold cross validation — see Chapter 25.4.3 *"k-Fold Cross Validation"* on page 488

The library `modelr` provides also tools to apply formulae, interact with `ggplot2`, model quality functions (see Chapter 25.1 *"Model Quality Measures"* on page 476), a wrapper around `model.matrix`, tools to warn about missing data, re-sample methods (Chapter 25.4.1 *"Elementary Cross Validation"* on page 483), etc. Some of those will be discussed in the next chapter that deals with model validation: Chapter 25 *"Model Validation"* on page 475.

♣ 25 ♣

Model Validation

The modeller has a vast toolbox of measures to check the power of the model: lift, MSE, AUC, Gini, KS and many more (for more inspiration, see Chapter 25.1 *"Model Quality Measures"* on page 476). However, the intrinsic quality of the model is not the only thing to worry about. Even if it fits really good the data that we have, how will it behave on new data?

First, we need to follow up the power of the model. For example, we choose KS as the key parameter together with a confusion matrix. This means that as new data comes in we build a dashboard (for example see Chapter 36.3 *"Dashboards"* on page 725) that will read in new data and new observations of good and bad outcomes, and as new data becomes available, we can and should calculate the these chosen parameters (KS and confusion matrix) on a regular base.[1]

Secondly, we need an independent opinion. Much like a medical doctor will ask a peer for a second opinion before a risky operation, we need another modeller to look at our model. In a professional setting, this is another team that is specialized in scrutinizing the modelling work of other people. Ideally this team is rather independent and will be a central function, as opposed to the modellers, who should be rather be close to the business.

We do not want to downplay the importance of independent layers (aka "lines of defence"), but for the remainder of this chapter, we will focus on the mathematical aspects of measuring the quality of a model and making sure it is not over-fit.

[1]In fact, we can and should also test the performance of the model on other things that were not so prevalent in our model definition. For example, we work for a bank and made a model on house-loan approvals. We defined our good definition on "not more than two months arrears after 3 years." This means that we can and should monitor things like: never paid anything back, one month arrears after one year, three months arrears any-time, etc.

The Big R-Book: From Data Science to Learning Machines and Big Data, First Edition. Philippe J.S. De Brouwer.

Companion Website: www.wiley.com/go/De Brouwer/The Big R-Book

25.1 Model Quality Measures

modelr also provides a convenient set of functions to access the measures of quality of the model. All functions mentioned in this section are from modelr. Consider a simple linear model that predicts the miles per gallon in the dataset mtcars, and the following risk measures are right away available:

```
# load modelr:
library(modelr)

# Fit a model:
lm1 <- lm(mpg ~ wt + qsec + am, data = mtcars)

# MSE (mean square error):
mse(lm1, mtcars)
## [1] 5.290185

# RMSE (root mean square error):
rmse(lm1, mtcars)
## [1] 2.30004

# MAD (mean absolute error):
mae(lm1, mtcars)
## [1] 1.931954

# Quantiles of absolute error:
qae(lm1, mtcars)
##         5%       25%       50%       75%       95%
## 0.3794271 0.9657082 1.4923568 2.8170045 4.3435305

# R-square (variance of predictions divided by the variance of the
# response variable):
rsquare(lm1, mtcars)
## [1] 0.8496636
```

resample_partition()

A simple cross validation, is also easy and straightforward with modelr. We use the function resample_partition

```
set.seed(1871)

rs  <- mtcars  %>%
       resample_partition(c(train = 0.6, test = 0.4))
lm2 <- lm(mpg ~ wt + qsec + am, data = rs$train)
rmse(lm2, rs$train); rmse(lm2, rs$test)
## [1] 2.18596
## [1] 2.638436

# or with the pipe operator:
lm2 %>% rmse(rs$train)
## [1] 2.18596

lm2 %>% rmse(rs$test)
## [1] 2.638436
```

25.2 Predictions and Residuals

Bespoke functions of `modelr` allow to add predictions and residuals to a dataset. Consider the same linear model as in previous section (predicting `mpg` in the dataset `mtcars`):

```
# Fit the model:
lm1 <- lm(mpg ~ wt + qsec + am, data = mtcars)

# Add the predictions and residuals:
df <- mtcars                %>%
      add_predictions(lm1) %>%
      add_residuals(lm1)

# The predictions are now available in $pred
head(df$pred)
## [1] 22.47046 22.15825 26.28107 20.85744 17.00959 20.85409

# The residuals are now available in $resid
head(df$resid)
## [1] -1.4704610 -1.1582487 -3.4810670  0.5425557  1.6904131
## [6] -2.7540920

# It is now easy to do something with those predictions and
# residuals, e.g. the following 3 lines all do the same:
sum((df$pred - df$mpg)^2) / nrow(mtcars)
## [1] 5.290185

sum((df$resid)^2) / nrow(mtcars)
## [1] 5.290185

mse(lm1, mtcars)
## [1] 5.290185
```

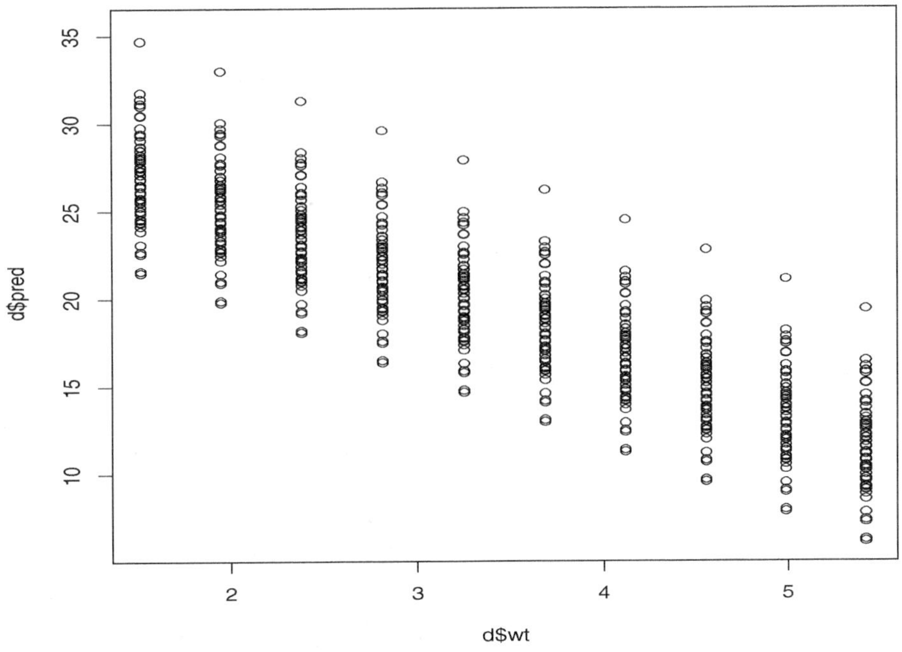

Figure 25.1: *A spacing grid for the predictions of t mpg.*

data_gird() It might be useful to visualize the model with a data grid with even spacing for the independent variables via the function data_grid() – output in Figure 25.1 on page 477:

```
d <- data_grid(mtcars, wt = seq_range(wt, 10), qsec, am) %>%
     add_predictions(lm1)
plot(d$wt, d$pred)
```

25.3 Bootstrapping

Bootstrapping is a technique to enrich the data that we have. It will calculate repeatedly a metric based on a different subset of the data. This subset is drawn randomly and with replacement. It is part of the broader class of resampling methods and it is possible to use bootstrapping to estimate also measures of accuracy (bias, variance, confidence intervals, prediction error, etc.) to the estimates.

bootstrapping
sampling

With bootstrapping, one estimates properties of an estimator (for stochastic variables such as conditional value at risk or variance) by measuring those properties from an approximated distribution based on a sample (e.g. the empirical distribution of the sample data).

The reasons to do this is because

- the whole dataset is too large to work with or
- to test the robustness of the model, we calibrate it on a subset of the data and see how it performs on another set of data.

25.3.1 Bootstrapping in Base R

Base R has the function `sample()`. This function is able to make a random cut of the data, and hence separate it in a training and testing dataset.

`sample()`

Function use for sample()

`sample(x, size, replace = FALSE, prob = NULL)` with

- `x`: either a vector of one or more elements from which to choose, or a positive integer.
- `size`: the number of items to select from x
- `replace`: set to TRUE if sampling is to be done with replacement
- `prob`: a vector of probability weights for obtaining the elements of the vector being sampled

Consider the following example: data from the 500 largest companies on the USA stock exchanges: the Standard and Poors 500 index (S& P500). Let us simply take a sample and compare its histogram with that of the complete dataset. This plot is in Figure 25.2 on page 480.

```
# Create the sample:
SP500_sample <- sample(SP500,size=100)

# Change plotting to 4 plots in one output:
par(mfrow=c(2,2))

# The histogram of the complete dataset:
hist(SP500,main="(a) Histogram of all data",fr=FALSE,
     breaks=c(-9:5),ylim=c(0,0.4))

# The histogram of the sample:
hist(SP500_sample,main="(b) Histogram of the sample",
     fr=FALSE,breaks=c(-9:5),ylim=c(0,0.4))
```

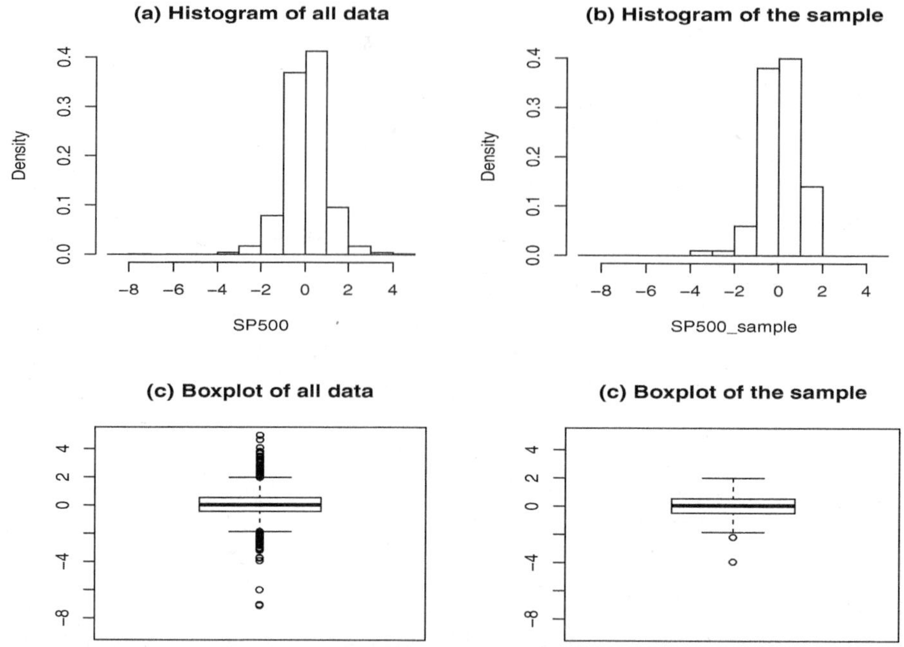

Figure 25.2: *Bootstrapping the returns of the S&P500 index.*

```
# The boxplot of the complete dataset:
boxplot(SP500,main="(c) Boxplot of all data",ylim=c(-9,5))

# The boxplot of the complete sample:
boxplot(SP500_sample,main="(c) Boxplot of the sample",
        ylim=c(-9,5))
```

```
# Reset the plot parameters:
par(mfrow=c(1,1))
```

In base R, the sample is a dataset itself and it can be addressed as any other dataset:

```
mean(SP500)
## [1] 0.04575267

mean(SP500_sample)
## [1] 0.08059197

sd(SP500)
## [1] 0.9477464

sd(SP500_sample)
## [1] 0.9380602
```

25.3.2 Bootstrapping in the tidyverse with modelr

The tidyverse has a new edition of the sampling functions. `modelr` provides the function `bootstrap()`. There are more sampling methods available to suit better modern needs to execute more complex cross validation such as k-fold cross validation.

The main difference, however, is in the details of the functions and objects. In the tidyverse the bootstrapping object is actually only a list of indices. This is much, much lighter than a selection of the dataset and hence can easily be stored for later reference. This is essential in a corporate environment where typically an independent validation team will check the results or in an environment where there is simply a lot of data.

The function `bootstrap()` works as follows:

`bootstrap()`

```
# Bootstrap generates a number of re-ordered datasets
boot <- bootstrap(mtcars, 3)
# The datasets are now in boot$strap[[n]]
# with n between 1 and 3

# e.g. the 3rd set is addressed as follows:
nrow(boot$strap[[3]])
## [1] 32

mean(as.data.frame(boot$strap[[3]])$mpg)
## [1] 20.66875

# It is also possible to coerce the selections into a data-frame:
df <- as.data.frame(boot$strap[[3]])
class(df)
## [1] "data.frame"
```

For example, we can fit linear models on different bootstrapped datasets, collect all coefficients and study then treat the different coefficients and performance parameters as a new dataset and study their distribution. The plots are gathered in Figure 25.3 on page 482.

`tidy()`

```
set.seed(1871)
library(purrr)  # to use the function map()
boot <- bootstrap(mtcars, 150)

lmodels <- map(boot$strap, ~ lm(mpg ~ wt + hp + am:vs, data = .))

# The function tidy of broom turns a model object in a tibble:
df_mods <- map_df(lmodels, broom::tidy, .id = "id")

# Create the plots of histograms of estimates for the coefficients:
par(mfrow=c(2,2))
hist(subset(df_mods, term == "wt")$estimate, col="khaki3",
     main = '(a) wt', xlab = 'estimate for wt')
hist(subset(df_mods, term == "hp")$estimate, col="khaki3",
     main = '(b) hp', xlab = 'estimate for hp')
hist(subset(df_mods, term == "am:vs")$estimate, col="khaki3",
     main = '(c) am:vs', xlab = 'estimate for am:vs')
hist(subset(df_mods, term == "(Intercept)")$estimate, col="khaki3",
     main = '(d) intercept', xlab = 'estimate for the intercept')

par(mfrow=c(1,1))
```

`map()`

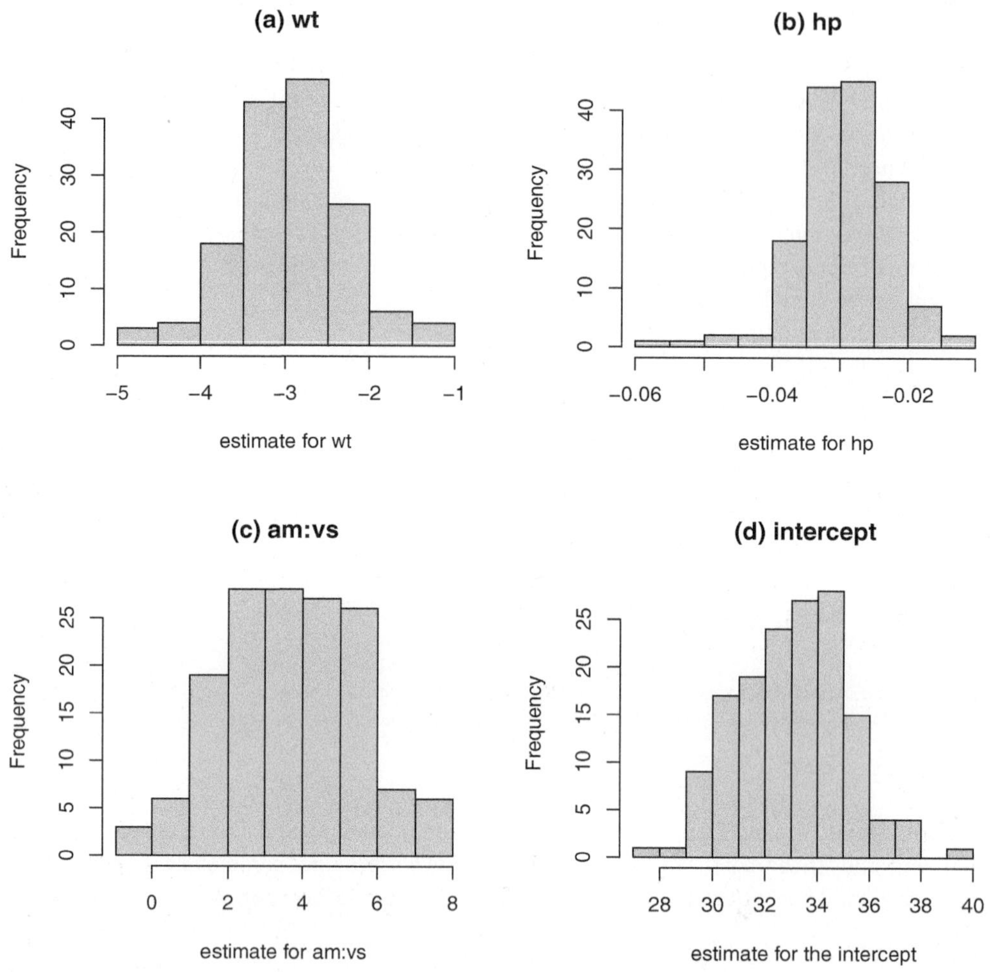

Figure 25.3: *The histograms of the different coefficients of the linear regression model predicting the mpg in the dataset mtcars. We show (a) Estimate for wt., (b) Estimate for hp., (c) Estimate for am:vs., and (d) Estimate for the intercept.*

25.4 Cross-Validation

cross validation

Making a model is actually simple, making a good model is really hard. Therefore, we tend to spend a lot of time validating the model and testing the assumptions. In practice, one of the most common issues is that data is biased: spurious correlations leads us astray and might mask true dependencies, or at least will mislead us to think that the model is really strong, while in reality it is based on coincidental correlations in that particular dataset. Later, when the model is put into production the results will be very different. This effect is called "over-fitting."

25.4.1 Elementary Cross Validation

There are multiple techniques possible to try to separate true dependencies from coincidental ones. The first technique is to separate the dataset in two separate sets: one that will be used to train the model, and another that we keep apart to test the model on data that we did not use to fit it in the first place.

The measure of performance should of course suit the model. For example, we can use MSE for a regression model or AUC for a binary classification model. In any case we need to verify two things:

1. The performance of the model on the training dataset should not be too much lower than that calculated on the training data.

2. The absolute values of the performance calculated on the testing dataset should match the set criteria (not the performance calculated on the training dataset). For example, when we want to compare this model with a challenger model, we use the performance calculated on the testing dataset.

sample()

As usual, there are multiple ways of doing this. The function `sample()` from base R is actually all we need:

```
d <- mtcars                  # get data
set.seed(1871)               # set the seed for the random generator
idx.train <- sample(1:nrow(d),round(0.75*nrow(d)))
d.train <- d[idx.train,]     # positive matches for training set
d.test  <- d[-idx.train,]    # the opposite to the testing set
```

modelr
resample()

If we work with the tidyverse then it makes sense to use the function `resample()` of `modelr` not only because it allows us to use the piping command. That function generates a "resample class object" that is simply a set of pointers to the original data. It can be turned into data by coercing it to a data frame.

```
set.seed(1870)
sample_cars <- mtcars %>%
               resample(sample(1:nrow(mtcars),5)) # random 5 cars

# This is a resample object (indexes shown, not data):
sample_cars
## <resample [5 x 11]> 15, 1, 16, 18, 20

# Turn it into data:
as.data.frame(sample_cars)
```

```
##                      mpg cyl  disp  hp drat    wt  qsec
## Cadillac Fleetwood  10.4   8 472.0 205 2.93 5.250 17.98
## Mazda RX4           21.0   6 160.0 110 3.90 2.620 16.46
## Lincoln Continental 10.4   8 460.0 215 3.00 5.424 17.82
## Fiat 128            32.4   4  78.7  66 4.08 2.200 19.47
## Toyota Corolla      33.9   4  71.1  65 4.22 1.835 19.90
##                     vs am gear carb
## Cadillac Fleetwood   0  0    3    4
## Mazda RX4            0  1    4    4
## Lincoln Continental  0  0    3    4
## Fiat 128             1  1    4    1
## Toyota Corolla       1  1    4    1

# or into a tibble
as_tibble(sample_cars)
## # A tibble: 5 x 11
##     mpg   cyl  disp    hp  drat    wt  qsec    vs    am
##   <dbl> <dbl> <dbl> <dbl> <dbl> <dbl> <dbl> <dbl> <dbl>
## 1  10.4     8 472     205  2.93  5.25  18.0     0     0
## 2  21       6 160     110  3.9   2.62  16.5     0     1
## 3  10.4     8 460     215  3     5.42  17.8     0     0
## 4  32.4     4  78.7    66  4.08  2.2   19.5     1     1
## 5  33.9     4  71.1    65  4.22  1.84  19.9     1     1
## # ... with 2 more variables: gear <dbl>, carb <dbl>

# or use the indices to get to the data:
mtcars[as.integer(sample_cars),]
##                      mpg cyl  disp  hp drat    wt  qsec
## Cadillac Fleetwood  10.4   8 472.0 205 2.93 5.250 17.98
## Mazda RX4           21.0   6 160.0 110 3.90 2.620 16.46
## Lincoln Continental 10.4   8 460.0 215 3.00 5.424 17.82
## Fiat 128            32.4   4  78.7  66 4.08 2.200 19.47
## Toyota Corolla      33.9   4  71.1  65 4.22 1.835 19.90
##                     vs am gear carb
## Cadillac Fleetwood   0  0    3    4
## Mazda RX4            0  1    4    4
## Lincoln Continental  0  0    3    4
## Fiat 128             1  1    4    1
## Toyota Corolla       1  1    4    1
```

resample_partition()

A further advantage of the `resample` family of functions is that there is a concise way to split the data between a training and testing dataset: the function `resample_partition()` of `modelr`.

```
library(modelr)
rs <- mtcars  %>%
      resample_partition(c(train = 0.6, test = 0.4))

# address the datasets with: as.data.frame(rs$train)
#                            as.data.frame(rs$test)

# Check execution:
lapply(rs, nrow)
## $train
## [1] 19
##
## $test
## [1] 13
```

Now, that we have a training and test dataset, we have all the tools necessary. The standard workflow now becomes simply the following:

```
# 0. Store training and test dataset for further use (optional):
d_train  <- as.data.frame(rs$train)
d_test   <- as.data.frame(rs$test)

# 1. Fit the model on the training dataset:
lm1      <- lm(mpg ~ wt + hp + am:vs, data = rs$train)

# 2. Calculate the desired performance measure (e.g.
# root mean square error (rmse)):
rmse_trn <- lm1 %>% rmse(rs$train)
rmse_tst <- lm1 %>% rmse(rs$test)
print(rmse_trn)
## [1] 2.36305

print(rmse_tst)
## [1] 2.097978
```

We were using a performance measure that was readily available via the function `rmse()`, but if we want to calculate another risk measure, we might need the residuals and/or predictions first. Below, we calculate the same risk measure without using the function `rmse()`. Note that step one is the same as in the aforementioned code. `rmse()`

```
# 2. Add predictions and residuals:
x_trn  <- add_predictions(d_train, model = lm1) %>%
          add_residuals(model = lm1)
x_tst  <- add_predictions(d_test,  model = lm1) %>%
          add_residuals(model = lm1)


# 3. Calculate the desired risk metrics (via the residuals):
RMSE_trn  <- sqrt(sum(x_trn$resid^2) / nrow(d_train))
RMSE_tst  <- sqrt(sum(x_tst$resid^2) / nrow(d_test))
print(RMSE_trn)
## [1] 2.36305

print(RMSE_tst)
## [1] 2.097978
```

In general, risk metrics calculated on the test dataset will be worse than those calculated on the training dataset. The difference gives us an indication of the robustness of the model.

This simple test might be misleading on small datasets such as `mtcars`. The results vary from a better result on the training set to significantly worse results. We will discuss solutions in the next sections. If the results are much worse on the testing dataset, then this is an indication that the model is over-fit and should not be trusted for new data.

For most practical purposes, the MSE calculated on the test dataset will be more relevant than the one calculated on the training set.

Note – Which data to use for testing?

In the examples, we selected randomly. For the datasets `mtcars` and `titanic`, this makes sense because all observations have the same time-stamp. If we have an insurance or loan portfolio for example where all customers came in over the last ten years or so, we might consider other options. Assume we have ten years history, then we might test with this method if a model calibrated on the first eight years will still be a good fit for the customers entering the last two years.

Further information – To refit or not to refit?

We split our data in a training and validation set, then calibrate the model on the training data and finally check its validity on the test dataset. This method validates – if the results on the test dataset are satisfactory – the choice of model (linear regression, logistic regression, etc.) as well as the parameters that go in the model (such as `wt` and `hp` for the mtcars data). Along the way we also found an intercept and coefficients.
But which model should we use in production? Should we use the coefficients as produced by the training model or should we refit and use all our data?
There is some controversy on the subject, however, from a statistician point of view it would be hard to defend not using all available data.[a] So, we should indeed do a last effort and recalibrate the model on all relevant data before we put it in production.
However, doing so takes away the ability to validate it again? But wasn't this done already?

[a] There might be a good reason to do so: for example if we assume that old exchange rates or older customers might not be representative for the customers coming in now. In that case, we might choose to calibrate a model on the last five years of data for example and do this every year or month again. Of course, the choice of intervals can then be subject to another cross validation.

Hint – Split Size

The rule of thumb is that both training and validation dataset need enough observations to draw some statistically significant conclusions. Typically, the training dataset is larger than the test dataset. This choice is of course related to the answer to the aforementioned question (in the info-box "to refit or not to refit"). If this is only used for validation purposes, then it does not matter too much. If, however, one wants to use afterwards the model fitted on the training data only, then this set of data must be chosen as large as possible.
In practice – and for larger databases – the training dataset is anything between 15% and 30%.

25.4.2 Monte Carlo Cross Validation

Simply splitting data in a training and a validation set is useful and certainly an important step in the good direction. However, it holds one crucial weakness: the split itself. Whether we choose to split randomly or not, this is just one split of the many possible. Therefore, it makes sense to do this split over and over again, build up a sample of outcomes and consider the outcomes.

Probably, the first idea that comes to mind is to repeat this cross validation process a few times and select each time – randomly – a training dataset. This approach is called Monte Carlo cross validation.

So, the idea is to randomly select – without replacement in that one sample, but replace full over differnt samples – some fraction of the data to form the training set. The rest of the observations go to the validation set (hence forming a partition). This process is then repeated a number of times.

For example, suppose we chose to use 10% of your data as test data. The first rep might then yield rows number 2, 7, 15, and 19. The next run, it might be 8, 10, 15 and 18. Since the partitions are done independently for each run, the same point can appear in the test set multiple times.[2]

While it is not much work to repeat the aforementioned code in a for-loop, `modelr` provides one convenient function to do create the desired number of selections in one step. The following code uses `modelr`'s `crossv_mc()` function and explores the result.

```
# Monte Carlo cross validation
cv_mc <- crossv_mc(data = mtcars, # the dataset to split
            n = 50,          # n random partitions train and test
            test = 0.25, # validation set is 25%
            id = ".id")  # unique identifier for each model

# Example of use:

# Access the 2nd test dataset:
d <- data.frame(cv_mc$test[2])

# Access mpg in that data frame:
data.frame(cv_mc$test[2])$mpg
## [1] 22.8 21.4 18.7 10.4 10.4 13.3 27.3 15.0 21.4

# More cryptic notations are possible to obtain the same:
mtcars[cv_mc[[2]][[2]][2]$idx,1]
## [1] 22.8 21.4 18.7 10.4 10.4 13.3 27.3 15.0 21.4
```

To illustrate how the Monte Carlo cross validation works with in practice, we reuse the example of Chapter 25.3 *"Bootstrapping"* on page 479 fitting a simple linear model on the data of `mtcars` and predicting `mpg`. We use `modelr` and `purrr` and plot the histogram in Figure 25.4 on page 488 in the following code.

```
set.seed(1868)
library(modelr)      # sample functions
library(purrr)       # to use the function map()

cv_mc <- crossv_mc(mtcars, n = 50, test = 0.40)
mods  <- map(cv_mc$train, ~ lm(mpg ~ wt + hp + am:vs, data = .))
RMSE  <- map2_dbl(mods, cv_mc$test, rmse)
hist(RMSE, col="khaki3")
```

Note – Simple cross validation

Note that the results with the simple cross validation of previous section (the RMSE was 2.098) is in the range found by the Monte Carlo cross validation (see Figure 25.4 on page 488. This is normal. The simple cross validation is only one experiment and the Monte Carlo cross validation repeats in some sense this elementary experiment multiple times.

While for the novice a for-loop might be easier to understand, the aforementioned tidy aforementioned code is concise and surprisingly very readable. The way of coding by using the piping

[2]This is the key difference with the method described in the following section.

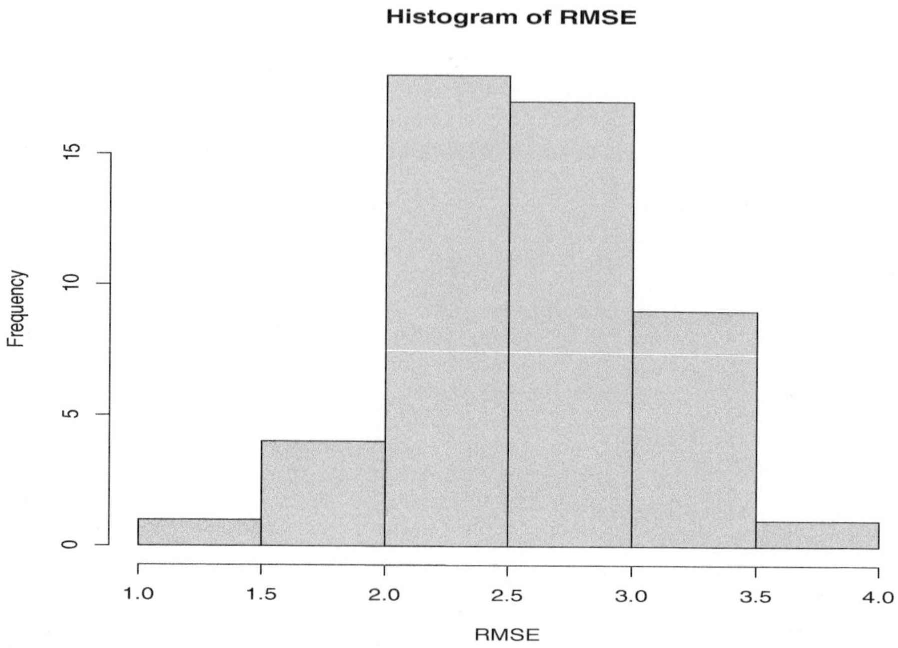

Figure 25.4: *The histogram of the RMSE for a Monte Carlo cross validation on the dataset mtcars.*

command has the advantage of allowing the reader to focus on the logic of what happens instead of what are the intermediary variables in which data is stored. We discuss this programming style and the details on how to use the piping operators in more detail in Section 7.3.2 *"Piping with R"* on page 132.

Digression – Advanced piping

The aforementioned code can also be written to with the pipe operator.

```
library(magrittr)  # to access the %T>% pipe
crossv <- mtcars                                          %>%
        crossv_mc(n = 50, test = 0.40)
RMSE <- crossv                                            %$%
        map(train, ~ lm(mpg ~ wt + hp + am:vs, data = .)) %>%
        map2_dbl(crossv$test, rmse)                       %T>%
        hist(col = "khaki3", main ="Histogram of RMSE",
             xlab = "RMSE")
```

25.4.3 k-Fold Cross Validation

The Monte Carlo cross validation does a good job in validating the assumption that the dynamics as modelled will be valid when the model will be used on new data. One of the possible issues is, however, that it depends on a random selection and this random selection might select the few outliers over and over in our test or train-data. k-fold Cross Validation will address this issue by keeping deterministic control over the sets in between reps.

Suppose that we have a dataset of 100 observations. For k-fold cross validation, these 100 rows are divided into k equal sized and mutually exclusive "folds." Further, assume that we choose

$k = 10$, then we might assign rows 1–10 to fold #1, 11–20 to fold #2, ..., rows 91–100 to fold #10. Now, we will consider one fold as the test dataset and the nine others as the training set. This process is then repeated 10 times till each fold had its turn as test dataset.

The function `crossv_kfold` of `modelr` will prepare the selections as for each run as follows.

```
library(modelr)
# k-fold cross validation
cv_k  <- crossv_kfold(data = mtcars,
           k = 5,        # number of folds
           id = ".id") # unique identifier for each
```

Each observation of the 32 will now appear once in one test dataset:

```
cv_k$test
## $`1`
## <resample [7 x 11]> 4, 10, 11, 14, 17, 18, 25
##
## $`2`
## <resample [7 x 11]> 5, 6, 9, 12, 28, 29, 32
##
## $`3`
## <resample [6 x 11]> 1, 3, 7, 24, 30, 31
##
## $`4`
## <resample [6 x 11]> 2, 15, 20, 21, 22, 26
##
## $`5`
## <resample [6 x 11]> 8, 13, 16, 19, 23, 27
```

The previous example – used in Chapter 25.4.2 *"Monte Carlo Cross Validation"* on page 486 – but with a 5-fold cross validation – becomes now:

```
set.seed(1868)
library(modelr)
library(magrittr)  # to access the %T>% pipe
crossv <- mtcars                                           %>%
          crossv_kfold(k = 5)
RMSE <- crossv                                             %$%
        map(train, ~ lm(mpg ~ wt + hp + am:vs, data = .))  %>%
        map2_dbl(crossv$test, rmse)                        %T>%
        hist(col = "khaki3", main ="Histogram of RMSE",
             xlab = "RMSE")
```

The output of this code segment is in Figure 25.5 on page 490.

25.4.4 Comparing Cross Validation Methods

Is Monte Carlo cross validation better than k-fold cross validation or is it the opposite way around? Each method has its own advantages and disadvantages. Under cross validation, each observation will be exactly once in the test dataset. However, cross validation only uses a few possible ways that the data could have been partitioned. Monte Carlo cross validation allows to explore more possible partitions with another degree of randomness. For example, it would allow that some of the worst returns would be part of the test set simultaneously—but there is no guarantee for this to happen. Even with a few hundred simulations, it is unlikely to have a reasonable sample of all possibilities, indeed there are $\binom{100}{50} \approx 10^{28}$ possible ways to make a 50/50 split of a 100 row dataset. On top of that we had another free parameter: the proportion of the data that is used to fit the model.

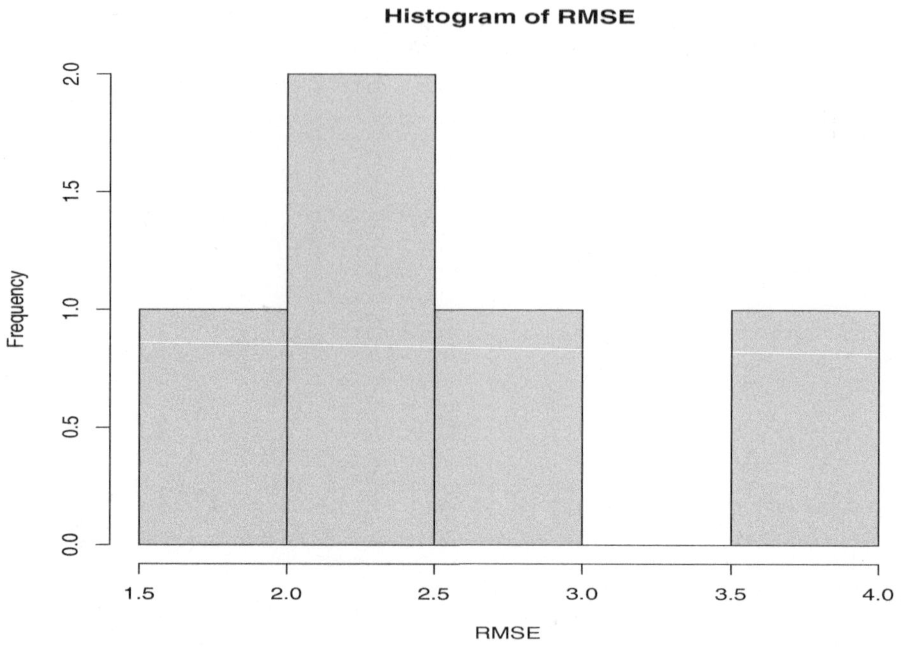

Figure 25.5: *Histogram of the RMSE based on a 5-fold cross validation. The histogram indeed shows that there were 5 observations. Note the significant spread of RMSE: the largest one is about four times the smallest.*

If we use cross validation for inferential purposes – statistically compare two possible algorithms – then averaging the results of a k-fold cross validation run offers a nearly unbiased estimate of the algorithm's performance. This is a strong argument in favour of the k-fold cross validation. Although note that the variance might be rather high, because there are only a k observations. This – of course – becomes more important as the data set gets smaller. Therefore, k-fold validation is mainly useful for larger datasets.

With a Monte Carlo cross validation we will sooner have more observations to average and hence, the result will be less variable. However, the estimator is more biased.

Further information – Hybrid cross validation methods

Taking these observations into account, it is of course possible to design multiple methods that combine advantages of both Monte Carlo and k-fold cross validation. For example, the "5×2 cross validation" (five iterations of a two-fold cross validation), "McNemar's cross validation," etc. (see Dietterich (1998) for a comparison of some methods as well as more details in Bengio and Grandvalet (2004)).

A cross validation is a good idea to check if the model is not over-fit, but it is not a silver bullet. All cross validation methods (even simple splitting data in two sets) are for example vulnerable if the data contains only a limited number of observations.

Also a cross validation method cannot solve problems with the data itself. For example a bank that is making a model to approve credit requests will only have data of customers that were approved in the past. The data does not contain a single customer that was not approved in the past.[3]

In model types that do not directly allow for weighting rare but important observations[4] it might be advisable to force enough such observations to appear in both training and testing set or alternatively multiply the rare observations so that they get more "weight."

Usually we have more observations of one type (e.g. customers that pay back the loans as compared to those defaulting), and usually we have more observations in our training dataset than in our testing dataset. This can lead to the situation that the split in good/bad is very different in our testing dataset as compared to the overall population. This is not a good situation to start from and we need to enrich data or take another sample.

[3] In this particular example the bank can alleviate this issue a to some extend. In most countries there will be a credit bureau that collects credit history of more people than just the customers of this one bank. So, we would also have information about people that got a loan somewhere else. Still, we do not have information of people that got refused in every bank.

[4] For example, the decision tree allows quite directly for such implementation of a weight, the logistic regression on the other hand has no such tool readily available.

25.5 Validation in a Broader Perspective

In the previous sections, we have focussed on validation methods that help us to assess how good a model is, compare models, and via cross validation, we can even make an educated guess how well it will perform when new data comes in. Is this enough? Or is there more that can and should be done?

model validation

Model validation that investigates the power of a model and cross validation alone would be like a medical doctor that only takes your pulse and concludes that you are fine to run the race while he forgets to check if you have legs at all. The model is just one section of the wheel of continuous improvement.

The complexity of our world is increasing fast, more and more data is available. Making a model and just bask in its splendour is not an option. Models are more and more an integrated part of a business cycle and essential to a company's sustainable results. Figure 25.6 shows a simple version of a model cycle.

1. *Formulate a question:* It all starts with a question. For example: can we provide consolidation[5] loans? That question leads directly to the type of data we need. In this case, we could use data from existing products such as credit cards, overdrafts, cash loans, purpose loans, etc.

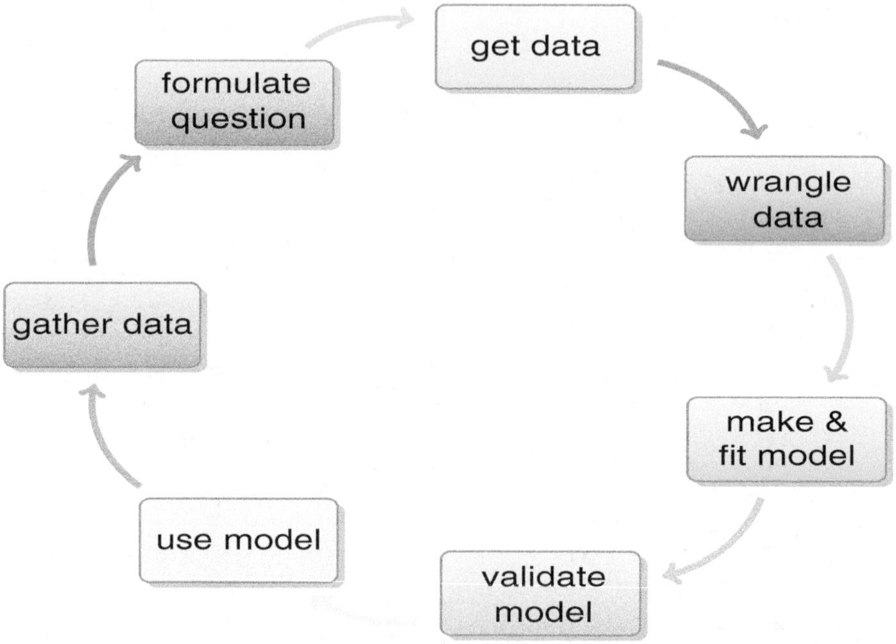

Figure 25.6: *The life cycle of a model: a model is an integrated part of business and focus of continuous improvement. Note how using a model will collect more data and lead to improvement of the model itself.*

[5]These are loans that replace a set of loans that the customer already has. People that have too much loans on short term might struggle to pay the high interests and it might make sense to group all the loans together into one new loan that has a longer horizon. However, this customer already got himself or herself into a difficult situation. Will he or she manage his or her finances better this time?

2. *Get data:* Once we know what data is needed, we can get it from the data-mart or data-warehouse in our company. This is typically a team that resides under the chief data officer. This is the subject of Part III *"Data Import"* on page 213.

3. *Wrangle data:* Once we have the data, it will most probably needs some tweaking before we can use it to fit a model. We might have too much and too sparse information, we might consider to consolidate variables (e.g. the list of 100 account balances at given dates can be reduced to maximal overdraft and maximum time in red). We also might have missing data and we need to decide what to do with the missing variables, and we need to come up with a binning that makes sense and provides good weight of evidence or information value. This is Part IV *"Data Wrangling"* on page 259.

4. *Make and fit the model:* First, we need to decide which model to use and all models come with advantages and disadvantages. It might be worth to make a few, and once we have a model and new model can be considered as a "challenger model." The new model would then need to have significant advantages in terms of performance, stability (measure of how it is not over-fit), simplicity and/or transparency in order to be preferred. For example, we can choose a linear regression, make an appropriate binning of our data, and fit the model. This is the subject of Part V *"Modelling"* on page 373.

5. *Validate the model:* This is the step where we (or preferably and independent person) scrutinize the model. This consists of (i) the mathematical part of fitness and over-fitness of a model (this section: Chapter 25 *"Model Validation"* on page 475) and (ii) an audit of the whole modelling cycle (this is what we are explaining now).

6. *Use the model:* After passing its final exam – the validation – the model can be used. Now, the model will be used for business decisions (e.g. to which customers do we propose our new product, the consolidation loan).

7. *Gather more data:* As the model is being used, it will generate new data. In our example we can already see if customers do at least start to pay back our loan, after a year we can model the customers that are three months in arrears, etc. This data together with the performance monitoring of the model (not visualized in Figure 25.6 on page 492) will open the possibility to ask new questions . . . and the cycle can start again.

Therefore, it is not possible to validate a model by only considering how fit and how over-fit it is. The whole cycle needs to be considered. For example, the validator will want to have a look at the following aspects.

1. *Question:* A model answers a question, but is this the right question to ask? Is there a better question to ask?

2. *Get data:* When pulling data from systems and loading it into other systems, it is important to check if conversions make sense (for example an integer can be encoded in 4-bit and in the other system only two, which means that large numbers will be truncated). More importantly, did we use the right data? Is there data that could be more appropriate?

3. *Wrangle data:* Data wrangling is already preparing directly for the model. Different models will need different type of data and will be more sensitive to certain assumptions. Is there missing data? How much? What did we do with it? Is the selection of target variable a good one? How can it be improved?

4. *Make and fit the model:* The modelling as such is also an art and needs to be in perfect harmony with the data wrangling. Is the model the best that is possible. Is the choice of model good? Does the random forest give enough added value to mitigate it is inherent lack of transparency? What are the weaknesses of data and model. What are the implications for the business? Can we trust that the underlying assumptions that the model makes are valid?

5. *The model maintainance cycle:* Can we trust that the performance, assumptions, data quality, etc. are monitored and revisited at reasonable intervals? Finally, it is essential to have a thorough look at the totality of the business model, question, data, model and the use of the model. Does it all make sense? Is the model used within the field of its purpose or are we using the model for something that it is not designed to do? What are the business risks, and how do those business risk fit into the risk of the whole company (and eventually society)?

This whole process should result in knowledge of the weak points of the model, and it should provide confidence that the model is capable of it is designated task: is it fit for purpose? Finally, this process will result in a document that describes all weaknesses and makes suggestions on how to mitigate risks, enhance the model and make it more reliable. This will ensure that the users of the model understand its limitations. In rare cases, the whole business model, product line, or business logic needs to be challenged.

It is true that it is in everyone's interest to do this and that a good modeller is to be trusted, but it is even more true that an independent investigation will find things that the modeller did not think of.

♣ 26 ♣

Labs

26.1 Financial Analysis with quantmod

The quantmod package for R is directly helpful for financial modelling. It is "designed to assist the quantitative trader in the development, testing, and deployment of statistically based trading models" (see `https://www.quantmod.com`). It allows the user to build financial models, has a simple interface to get data, and has a suite of plots that not only look professional but also provide insight for the trader.

quantmod

Its website is `https:\\www.quantmod.com`.

Also note that this section uses concepts of financial markets. It is best to understand at least the difference between bonds and stocks before going through this section. If these concepts are new, we recommend Chapter 30 *"Asset Valuation Basics"* on page 597.

26.1.1 The Basics of quantmod

The package `quantmod` can be installed as any other package. In the following code, we check if it is already installed and do so if it is not in our local library. An important function in `quantmod` is `getSymbols()`. This function allows to download historic data about financial instruments with just one line of code.

`library()`
`install.packages()`
`if()`
`grepl()`
`getSymbols()`

```
# Install quantmod:
if(!any(grepl("quantmod", installed.packages()))){
    install.packages("quantmod")}

# Load the library:
library(quantmod)
```

Now, we are ready to use `quantmod`. For example, we can start downloading some data with the function `getSymbols()`:

`getSymbols()`

```
# Download historic data of the Google share price:
getSymbols("GOOG", src = "yahoo")         # get Google's history
## [1] "GOOG"

getSymbols(c("GS", "GOOG"), src = "yahoo") # to load more than one
## [1] "GS"   "GOOG"
```

`quantmod` also allows the user to specify lookup parameters, save them for future use, and download more than one symbol in one line of code:

The Big R-Book: From Data Science to Learning Machines and Big Data, First Edition. Philippe J.S. De Brouwer.

Companion Website: www.wiley.com/go/De Brouwer/The Big R-Book

```
setSymbolLookup(HSBC = 'yahoo', GOOG = 'yahoo')
setSymbolLookup(DEXJPUS = 'FRED')
setSymbolLookup(XPTUSD = list(name = "XPT/USD", src = "oanda"))

# Save the settings in a file:
saveSymbolLookup(file = "qmdata.rda")
# Use this in new sessions calling:
loadSymbolLookup(file = "qmdata.rda")

# We can also download a list of symbols as follows:
getSymbols(c("HSBC", "GOOG", "DEXJPUS", "XPTUSD"))
## [1] "HSBC"    "GOOG"    "DEXJPUS" "XPTUSD"
```

26.1.2 Types of Data Available in quantmod

stockSymbols()
New York Stock Exchange

The function stockSymbols() can provide a list of symbols that are quoted on Amex, Nasdaq, and NSYE.

```
stockList <- stockSymbols()  # get all symbols

## Fetching AMEX symbols...
## Fetching NASDAQ symbols...
## Fetching NYSE symbols...

nrow(stockList)     # number of symbols
## [1] 6964

colnames(stockList) # information in this list
## [1] "Symbol"    "Name"      "LastSale"  "MarketCap"
## [5] "IPOyear"   "Sector"    "Industry"  "Exchange"
```

NYSE

Nasdaq

National Association of Securities Dealers Automated Quotations

BatchGetSymbols

If one would like to download all symbols in one data-frame, then one can use the package BatchGetSymbols or simply collect the data with cbind() for example in a data frame.

Hint – Other data sources

For other services it is best to refer to the website of the data provider, e.g. https://fred.stlouisfed.org. Alternatively, one can use a service such as https://www.quandl.com/data/FRED-Federal-Reserve-Economic-Data to locate data from a variety of sources. Macro economic data is available on the website of the World Bank, etc.

foreign exchange

FX
getFX()

For FX data www.oanda.com is a good starting point and the function getFX() will do most of the work.

```
getFX("EUR/PLN", from = "2019-01-01")
## [1] "EUR/PLN"
```

getMetals()

Data about metals can be downloaded with the function getMetals(). More information about all possibilities can be found in the documentation of quantmod: https://cran.r-project.org/web/packages/quantmod/quantmod.pdf.

26.1.3 Plotting with quantmod

We already mentioned that the standard function to access historical market data works as follows:

```
getSymbols("HSBC", src = "yahoo") #get HSBC's data from Yahoo
## [1] "HSBC"
```

This created a data object with the name "HSBC" and it can be used directly in the traditional functions and arithmetic, but quantmod also provides some specific functions that make working with financial data easier.

For example, quantmod offer great and visually attractive plotting capabilities. Below we will plot the standard bar chart (Figure 26.1), line plot (Figure 26.2 on page 498), and candle chart (Figure 26.3 on page 498).

barChart()

lineChart()

candleChart()

```
# 1. The bar chart:
barChart(HSBC)

# 2. The line chart:
lineChart(HSBC)

# Note: the lineChart is also the default that yields the same
#       result as plot(HSBC)

# 3. The candle chart:
candleChart(HSBC, subset = 'last 1 years', theme = "white",
            multi.col = TRUE)
```

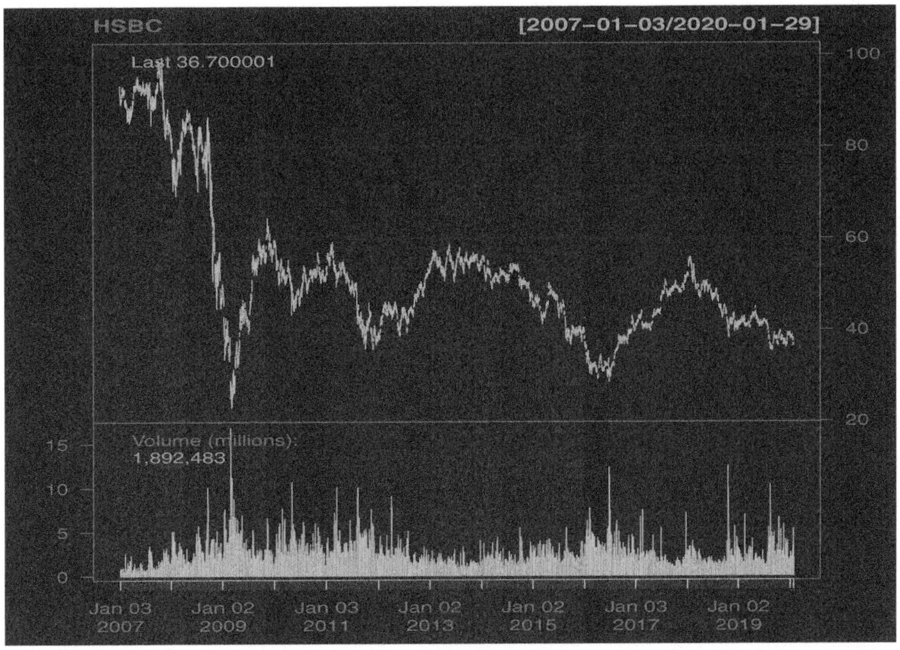

Figure 26.1: *Demonstration of the barChart() function of the package quantmod.*

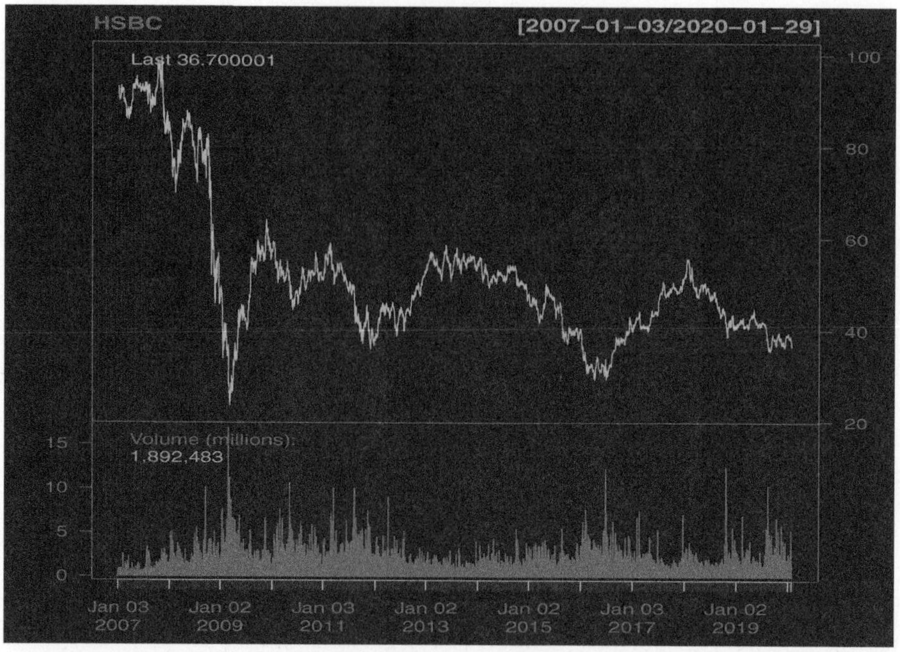

Figure 26.2: *Demonstration of the lineChart() function of the package quandmod.*

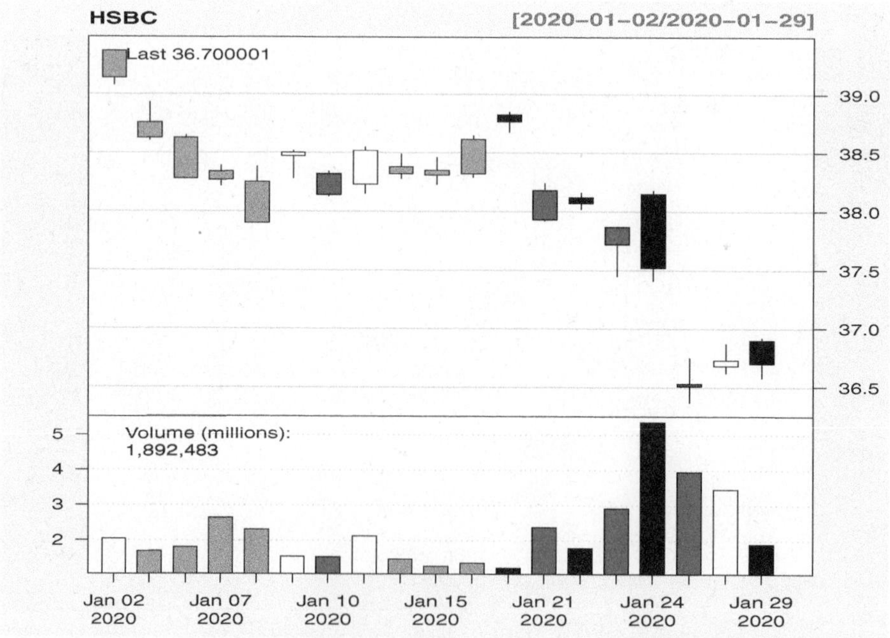

Figure 26.3: *Demonstration of the candleChart() function of the package quantmod.*

> **Hint – short-cuts for dates**
>
> The following two lines of code do exactly the same.
>
> ```
> candleChart(HSBC, subset = '2018::2018-01')
> candleChart(HSBC, subset = 'last 1 months')
> ```

So, there is no need to find a data source, download a portable format (such as CSV), load it in a data-frame, name the columns, etc. The `getSymbols()` function downloads the daily data going 10 years back in the past (if available of course). The plot functions such as `lineChart()`, `barChart()`, and `candleChart()` display the data in a professional and clean fashion. The looks can be customized with a theme-based parameter that uses intuitive conventions. CSV

There is much more and we encourage you to explore by yourself. To give you a taste: let us display a stock chart with many indicators and Bollinger bands with three short lines of code (the plot is in Figure 26.4):

```
getSymbols(c("HSBC"))
## [1] "HSBC"

chartSeries(HSBC, subset = 'last 4 months')
addBBands(n = 20, sd = 2, maType = "SMA", draw = 'bands',
          on = -1)
```

where n is the sampling size of the moving average (of type "maType" – SMA is simple moving average) and sd is the multiplier for the standard deviation to be used as band.

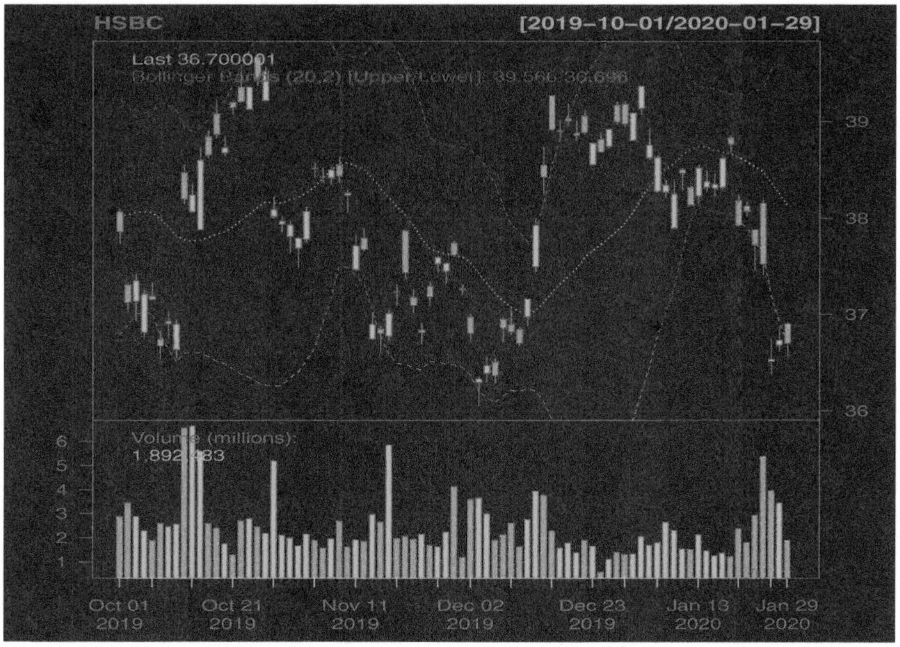

Figure 26.4: *Bollinger bands with the package quandmod.*

Bollinger bands aim to introduce a relative indication of how high a stock is trading relative to its past. Therefore, one calculates a moving average and volatility over a given period (typically $n = 20$) and plots bands with plus and minus a volatility multiplied with a multiplier (typically $\text{sd} = 2$).[1] If the stock trades below the lower band then it is to be considered as low.

The lines drawn correspond to $ma - \text{sd}.\sigma$, ma and $ma + \text{sd}.\sigma$ with σ the standard deviation and where ma and sd take values as define above.

Some traders will then sell it because it is considered as "a buy opportunity," others might consider this as a break in trend and rather short it (or sell off what they had as it is not living up to its expectations).

26.1.4 The quantmod Data Structure

Using quantmod you will find that it is possible to accomplish complex tasks with a very limited number of lines of code. This is made possible thanks to the use of the `xts` package. This means that `quantmod` uses under the hood the extend time series package (short xts). Installing
xts `quantmod` will coerce the installation of `xts`.

The use of `xts` offers many advantages. For example, it makes it more intuitive to manipulate data in function of the time-stamp.

```
myxtsdata["2008-01-01/2010-12-31"]  # between 2 date-stamps

# All data before or after a certain time-stamp:
xtsdata["/2007"]  # from start of data until end of 2007
xtsdata["2009/"]  # from 2009 until the end of the data

# Select the data between different hours:
xtsdata["T07:15/T09:45"]
```

26.1.4.1 Sub-setting by Time and Date

xts The `xts` package offers tools that make it easy to work with time-based series. As such it extends
zoo the `zoo` class, and adds some new methods for sub-setting data.[2]

```
HSBC['2017']     # returns HSBC's OHLC data for 2017
HSBC['2017-08'] # returns HSBC's OHLC data for August 2017
HSBC['2017-06::2018-01-15'] # from June 2017 to Jan 15 2018

HSBC['::']       # returns all data
HSBC['2017::'] # returns all data in HSBC, from 2017 onward
my.selection <- c('2017-01', '2017-03', '2017-11')
HSBC[my.selection]
```

The date format follows the ANSI standard (CCYY-MM-DD HH:MM:SS), the ranges are specified via the "::" operator. The main advantage is that these functions are robust regardless the underlying data: this data can be a quote every minute, every day, or month and the functions will still work.

In order to rerun a particular model one will typically need the data of the last few weeks,
xts months or years. Also here `xts` comes to the rescue with handy shortcuts:

---

[1]The example of such plot with Bolinger bands is in Figure 26.4.

[2]Type `help('[.xts')` to get more information about data extraction of time series.

```
last(HSBC)                # returns the last quotes
last(HSBC, 5)             # returns the last 5 quotes
last(HSBC, '6 weeks')     # the last 6 weeks
last(HSBC, '-1 weeks')    # all but the last week
last(HSBC, '6 months')    # the last 6 months
last(HSBC, '3 years')     # the last 3 years

# these functions can also be combined:
last(first(HSBC, '3 weeks'), '5 days')
#
```

last()

26.1.4.2 Switching Time Scales

One of the interesting aspects in trading is that even while trading on short frequency (such as milliseconds) one will not want to miss bigger trends and the other way around. So it is essential to be able to aggregate data into lower frequency objects. For example, convert daily data to monthly data.

xts provides the tools to do these conversion in a robust and orderly manner with the functions to.weekly() or to.monthly() for example, or the more specific functions such as to.minutes5() and to.minutes10() that will convert to 5- and 10 minute data, respectively.

xts
to.minutes5()
to.minutes10()

```
periodicity(HSBC)
unclass(periodicity(HSBC))
to.weekly(HSBC)
to.monthly(HSBC)
periodicity(to.monthly(HSBC))
ndays(HSBC); nweeks(HSBC); nyears(HSBC)
```

ndays()
nweeks()
nyears()

As these functions are dependent on the upstream xts and not specific to quantmod they can be used also on non-OHLC data:

```
getFX("USD/EUR")
## [1] "USD/EUR"

periodicity(USDEUR)
## Daily periodicity from 2019-08-04 to 2020-01-28

to.monthly(USDEUR)
##          USDEUR.Open USDEUR.High USDEUR.Low USDEUR.Close
## Aug 2019    0.900219    0.909910   0.892214     0.909910
## Sep 2019    0.909908    0.916008   0.902656     0.916008
## Oct 2019    0.916670    0.916670   0.895232     0.896330
## Nov 2019    0.896046    0.908504   0.895620     0.907536
## Dec 2019    0.907528    0.907528   0.891614     0.891614
## Jan 2020    0.891860    0.907806   0.891860     0.907806

periodicity(to.monthly(USDEUR))
## Monthly periodicity from Aug 2019 to Jan 2020
```

periodicity()

to.monthly()
getFX()

26.1.4.3 Apply by Period

Often it will be necessary to identify end points in your data by date with the function endpoints(). Those end points can be used with the functions in the period.apply family. This allows to calculate periodic minimums, maximums, sums, and products as well as the more general user-defined.

```
endpoints(HSBC, on = "years")
## [1]    0  251  504  756 1008 1260 1510 1762 2014 2266
## [11] 2518 2769 3020 3272 3291

# Find the maximum closing price each year:
apply.yearly(HSBC, FUN = function(x) {max(Cl(x)) } )
##             [,1]
## 2007-12-31 99.52
## 2008-12-31 87.67
## 2009-12-31 63.95
## 2010-12-31 59.32
## 2011-12-30 58.99
## 2012-12-31 53.07
## 2013-12-31 58.61
## 2014-12-31 55.96
## 2015-12-31 50.17
## 2016-12-30 42.96
## 2017-12-29 51.66
## 2018-12-31 55.62
## 2019-12-31 44.70
## 2020-01-29 39.37

# The same thing - only more general:
subHSBC <- HSBC['2012::']
period.apply(subHSBC, endpoints(subHSBC, on = 'years'),
             FUN = function(x) {max(Cl(x))} )
##             [,1]
## 2012-12-31 53.07
## 2013-12-31 58.61
## 2014-12-31 55.96
## 2015-12-31 50.17
## 2016-12-30 42.96
## 2017-12-29 51.66
## 2018-12-31 55.62
## 2019-12-31 44.70
## 2020-01-29 39.37

# The following line does the same but is faster:
as.numeric(period.max(Cl(subHSBC), endpoints(subHSBC,
           on = 'years')))
## [1] 53.07 58.61 55.96 50.17 42.96 51.66 55.62 44.70 39.37
```

as.numeric()
period.max()
period.min()

26.1.5 Support Functions Supplied by quantmod

quantmod has some useful features. For example, quantmod dynamically creates data objects creating a model frame internally after going through some of steps to identify the sources of data required (and loading if required).

There are some basic types of data for all financial assets: Op, Hi, Lo, Cl, Vo, and Ad are respectively the Open, High, Low, Close, Volume, and Adjusted price[3] of a given instrument and are stored in the columns of the data object.

close
volume
adjusted
is.OHLC()
has.OHLS()
has.Cl()

There are some functions to test if an object is of the right format and if the instrument has certain data: is.OHLC(), has.OHLC(), has.Op(), has.Cl(), has.Hi(), has.Lo(), has.Ad(), and has.Vo(). Others extract certain data such as Vo(), Cl(), Hi(), Ad(), and Vo(). Still other functions such as seriesHi() and seriesLo() for example return respectively the high and low of a a given dataset.

[3] The adjusted closing price is the closing price where any eventual dividend is added if it was announced after market closure.

```
seriesHi(HSBC)
##            HSBC.Open HSBC.High HSBC.Low HSBC.Close
## 2007-10-31     98.92     99.52    98.05      99.52
##            HSBC.Volume HSBC.Adjusted
## 2007-10-31     1457900      52.73684

has.Cl(HSBC)
## [1] TRUE

tail(Cl(HSBC))
##            HSBC.Close
## 2020-01-22      38.07
## 2020-01-23      37.72
## 2020-01-24      37.52
## 2020-01-27      36.51
## 2020-01-28      36.73
## 2020-01-29      36.70
```

has.Hi()
Hi()
Lo()
Cl()
Vo()
Ad()

has.Cl()
tail()
seriesHi()

There are even functions that will calculate differences, for example:

- OpCl(): daily percent change open to close
- OpOp(): daily open to open change
- HiCl(): the percent change from high to close

OpCl()

HiCl()

These functions rely on the following that are also available to use:

- Lag(): gets the previous value in the series
- Next(): gets the next value in the series
- Delt(): returns the change (delta) from two prices

Delt()

```
Lag(Cl(HSBC))
Lag(Cl(HSBC), c (1, 5, 10)) # One, five and ten period lags
Next(OpCl(HSBC))

# Open to close one, two and three-day lags:
Delt(Op(HSBC), Cl(HSBC), k = 1:3)
```

Lag()
Next()

There are many more wrappers and functions such as period.min(), period.sum(), period.prod(), and period.max.

More often than not it is the return that we are interested in and there is a set of functions that makes it easy and straightforward to do so. There is of course the master-function periodReturn(), that takes a parameter "period' to indicate what periods are designed. Then there is also a suit of derived functions that carry the name of the relevant period.

periodReturn()

The convention used is that the first observation of the period is the first trading time of that period; and the last the last observation is the last trading time of the period, on the last day of the period. xts has adopted the last observation of a given period as the date to record for the larger period. This can be changed via the indexAt argument, we refer to the documentation for more details.

dailyReturn()
weeklyReturn()
monthlyReturn()
quarterlyReturn()
yearlyReturn()

```
dailyReturn(HSBC)
weeklyReturn(HSBC)
monthlyReturn(HSBC)
quarterlyReturn(HSBC)
yearlyReturn(HSBC)
allReturns(HSBC)      # all previous returns
```

26.1.6 Financial Modelling in quantmod

26.1.6.1 Financial Models in quantmod

specifyModel()

specifyModel()

To specify financial models, there is the function `specifyModel()`. Typically, one can specify data within the call to `specifyModel`, and `quantmod` will lookup the relevant data and take care of the data aggregation.

Consider the following naive model:

```
# First, we create a quantmod object.
# At this point, we do not need to load data.

setSymbolLookup(SPY = 'yahoo',
   VXN = list(name = '^VIX', src = 'yahoo'))

qmModel <- specifyModel(Next(OpCl(SPY)) ~ OpCl(SPY) + Cl(VIX))
head(modelData(qmModel))
##              Next.OpCl.SPY      OpCl.SPY  Cl.VIX
## 2014-12-04  0.0006254149  0.0005782548 28447.7
## 2014-12-05 -0.0043851339  0.0006254149 26056.5
## 2014-12-08  0.0102755104 -0.0043851339 23582.8
## 2014-12-09 -0.0133553492  0.0102755104 21274.0
## 2014-12-10  0.0015204875 -0.0133553492 19295.0
## 2014-12-11 -0.0086360048  0.0015204875 17728.3
```

The object `qmModel` is now a `quantmod` object holding the model formula and data structure implying the next (Next) period's open to close of the S&P 500 ETF (`OpCl(SPY)`) is modelled as a function of the current period open to close and the current close of the VIX (`Cl(VIX)`).[4]

modelData()
buildData()

The call to `modelData()` extracts the relevant data set. A more direct function to accomplish the same end is `buildData()`.

26.1.6.2 A Simple Model with quantmod

In this section we will propose a simple (and naive) model to predict the opening price of a stock based on its performance of the day before.

First, we import the data and plot the linechart for the symbol in Figure 26.5 on page 505:

```
getSymbols('HSBC', src = 'yahoo') #google doesn't carry the adjusted price
## [1] "HSBC"

lineChart(HSBC)
```

The line-chart shows that the behaviour of the stock is very different in the period after the crisis. Therefore, we decide to consider only data after 2010.

```
HSBC.tmp   <- HSBC["2010/"]          #see: subsetting for xts objects
```

[4]The VIX is the CBOE Volatility Index, known by its ticker symbol VIX, is a popular measure of the stock market's expectation of volatility implied by S&P 500 index options, calculated and published by the Chicago Board Options Exchange.

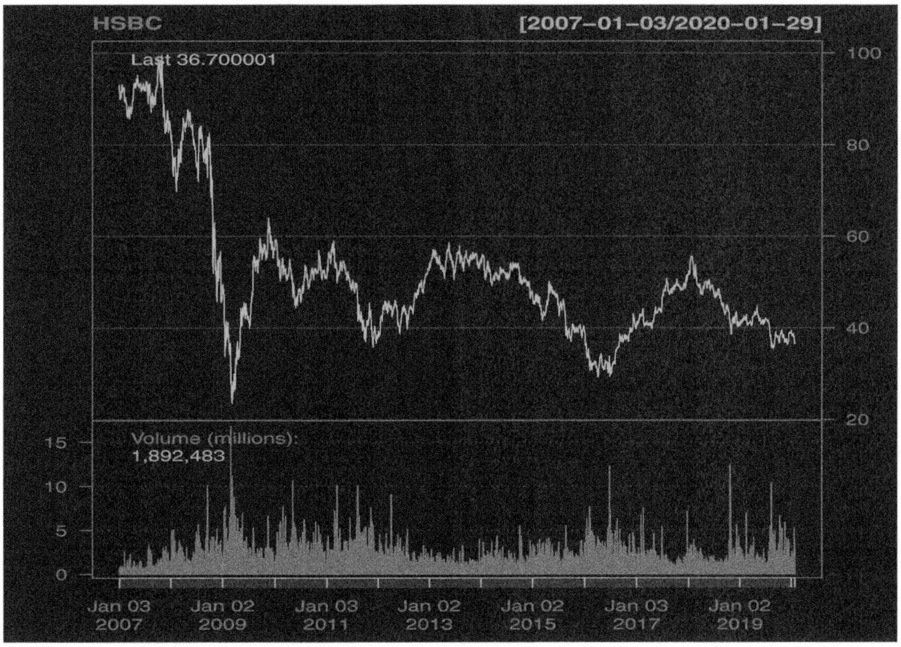

Figure 26.5: *The evolution of the HSBC share for the last ten years.*

The next step is to divide our data in a training dataset and a test-dataset. The training set is the set that we will use to calibrate the model and then we will see how it performs on the test-data. This process will give us a good idea about the robustness of the model.

```
# use 70% of the data to train the model:
n           <- floor(nrow(HSBC.tmp) * 0.7)
HSBC.train <- HSBC.tmp[1:n]                # training data
HSBC.test  <- HSBC[(n+1):nrow(HSBC.tmp)]  # test-data
# head(HSBC.train)
```

Till now we used the functionality of quantmod to pull in data, but the function specifyModel() allows us to prepare automatically the data for modelling: it will align the next opening price with the explaining variables. Further, modelData() allows to make sure the data is up-to-date.

specifyModel()
modelData()

```
# making sure that whenever we re-run this the latest data
# is pulled in:
m.qm.tr <- specifyModel(Next(Op(HSBC.train)) ~ Ad(HSBC.train)
         + Hi(HSBC.train) - Lo(HSBC.train) + Vo(HSBC.train))

D <- modelData(m.qm.tr)
```

We decide to create an additional variable that is the difference between the high and low prices of the previous day.

```
# Add the additional column:
D$diff.HSBC <- D$Hi.HSBC.train - D$Lo.HSBC.train

# Note that the last value is NA:
tail(D, n = 3L)
```

```
##            Next.Op.HSBC.train Ad.HSBC.train Hi.HSBC.train
## 2017-01-17              41.79      34.58311         41.72
## 2017-01-18              41.83      35.19617         41.98
## 2017-01-19                 NA      34.99461         41.92
##            Lo.HSBC.train Vo.HSBC.train diff.HSBC
## 2017-01-17         41.15       1989000  0.569999
## 2017-01-18         41.56       3071400  0.419999
## 2017-01-19         41.49       2159500  0.429996

# Since the last value is NA, let us remove it:
D <- D[-nrow(D),]
```

The column names of the data inherit the full name of the dataset. This is not practical since the names will be different in the training set and in the test-set. So we rename them before making the model.

```
colnames(D) <- c("Next.Op", "Ad", "Hi", "Lo", "Vo", "Diff")
```

Now, we can create the model.

```
m1 <- lm(D$Next.Op ~ D$Ad + D$Diff + D$Vo)
summary(m1)
##
## Call:
## lm(formula = D$Next.Op ~ D$Ad + D$Diff + D$Vo)
##
## Residuals:
##      Min       1Q   Median       3Q      Max
## -22.9066  -2.9761  -0.0378   2.7795  10.7652
##
## Coefficients:
##               Estimate Std. Error t value Pr(>|t|)
## (Intercept) -1.091e+00  9.412e-01  -1.159    0.247
## D$Ad         1.392e+00  2.552e-02  54.535  < 2e-16 ***
## D$Diff       7.559e+00  3.671e-01  20.591  < 2e-16 ***
## D$Vo        -5.705e-07  9.500e-08  -6.005 2.32e-09 ***
## ---
## Signif. codes:
## 0 '***' 0.001 '**' 0.01 '*' 0.05 '.' 0.1 ' ' 1
##
## Residual standard error: 3.973 on 1769 degrees of freedom
## Multiple R-squared:  0.6936,Adjusted R-squared:  0.693
## F-statistic:  1335 on 3 and 1769 DF,  p-value: < 2.2e-16
```

The volume of trading in the stock does not seem to play a significant role, so we leave it out.

```
m2 <- lm(D$Next.Op ~ D$Ad + D$Diff)
summary(m2)
##
## Call:
## lm(formula = D$Next.Op ~ D$Ad + D$Diff)
##
## Residuals:
##      Min       1Q   Median       3Q      Max
## -22.8647  -3.0528   0.0703   2.9178   8.9864
##
## Coefficients:
##             Estimate Std. Error t value Pr(>|t|)
## (Intercept)  -3.8659     0.8280  -4.669 3.25e-06 ***
## D$Ad          1.4558     0.0234  62.215  < 2e-16 ***
```

```
## D$Diff          6.5458     0.3292  19.882  < 2e-16 ***
## ---
## Signif. codes:
## 0 '***' 0.001 '**' 0.01 '*' 0.05 '.' 0.1 ' ' 1
##
## Residual standard error: 4.012 on 1770 degrees of freedom
## Multiple R-squared:  0.6873,Adjusted R-squared:  0.687
## F-statistic:  1945 on 2 and 1770 DF,  p-value: < 2.2e-16
```

From the output of the command `summary(m2)` we learn that all the variables are significant now. The R^2 is slightly down, but in return, one has a much more stable model that is not over-fitted (or at least less over-fitted).

Some more tests can be done. We should also make a Q-Q plot to make sure the residuals are normally distributed. This is done with the function `qqnorm()`.

Q-Q plot
`qqnorm)()`

```
qqnorm(m2$residuals)
qqline(m2$residuals, col = 'blue', lwd = 2)
```

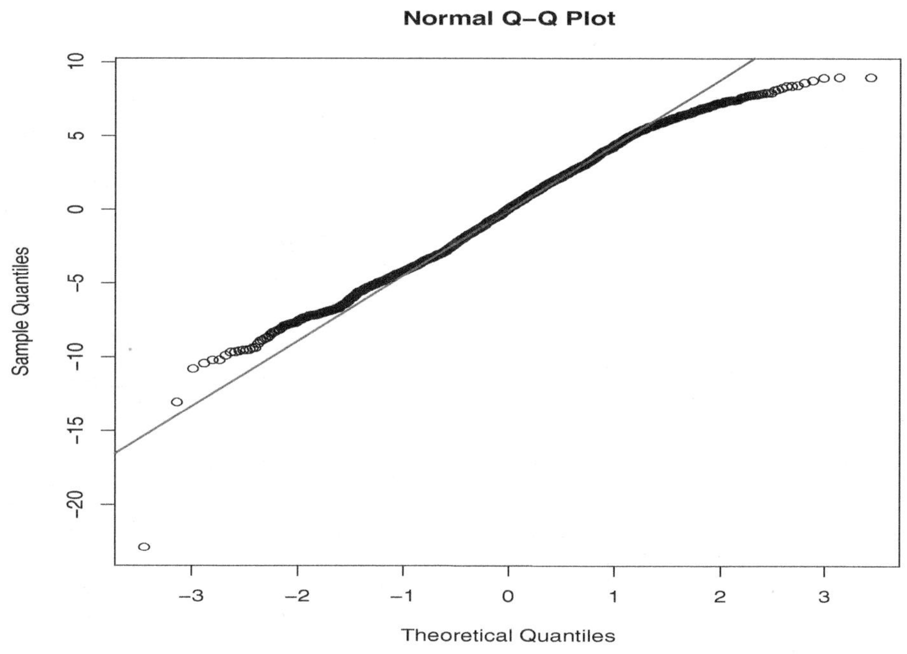

Figure 26.6: *The Q-Q plot of our naive model to forecast the next opening price of the HSBC stock. The results seems to be reasonable.*

Figure 26.6 shows that the model does capture well the tail-behaviour of the forecasted variable. However, the predicting power is not great.

26.1.6.3 Testing the Model Robustness

To check the robustness of our model we should now check how well it fits the test-data. The idea is that since the model was built only on the training data, that we can assess its robustness by checking how well it does on the test-data.

First, we prepare the test data in the same way as the training data:

```
m.qm.tst <- specifyModel(Next(Op(HSBC.test)) ~  Ad(HSBC.test)
            + Hi(HSBC.test) - Lo(HSBC.test) + Vo(HSBC.test))

D.tst <- modelData(m.qm.tst)
D.tst$diff.HSBC.test <- D.tst$Hi.HSBC.test-D.tst$Lo.HSBC.test
#tail(D.tst)                          # the last value is NA
D.tst <- D[-nrow(D.tst),]  # remove the last value that is NA

colnames(D.tst) <- c("Next.Op", "Ad", "Hi", "Lo", "Vo", "Diff")
```

We could of course use the function `predict()` to find the predictions of the model, but here we illustrate how coefficients can be extracted from the model object and used to obtain these predictions. For the ease of reference we will name the coefficients.

```
a   <- coef(m2)['(Intercept)']
bAd <- coef(m2)['D$Ad']
bD  <- coef(m2)['D$Diff']
est <- a + bAd * D.tst$Ad + bD * D.tst$Diff
```

Now, we can calculate all possible measures of model power.

```
# -- Mean squared prediction error (MSPE)
#sqrt(mean(((predict(m2, newdata = D.tst) - D.tst$Next.Op)^2)))
sqrt(mean(((est - D.tst$Next.Op)^2)))
## [1] 4.009343

# -- Mean absolute errors (MAE)
mean((abs(est - D.tst$Next.Op)))
## [1] 3.30732

# -- Mean absolute percentage error (MAPE)
mean((abs(est - D.tst$Next.Op))/D.tst$Next.Op)
## [1] 0.07478334

# -- squared sum of residuals
print(sum(residuals(m2)^2))
## [1] 28493.15

# -- confidence intervals for the model
print(confint(m2))
##                  2.5 %    97.5 %
## (Intercept) -5.489862 -2.241980
## D$Ad          1.409889  1.501675
## D$Diff        5.900040  7.191511
```

residuals()
confint()

These values give us an estimate on what error can be expected by using this simple model.

```
# Compare the coefficients in a refit:
m3 <- lm(D.tst$Next.Op ~ D.tst$Ad + D.tst$Diff)
summary(m3)
##
## Call:
## lm(formula = D.tst$Next.Op ~ D.tst$Ad + D.tst$Diff)
##
## Residuals:
##      Min       1Q   Median       3Q      Max
## -22.8750  -3.0500   0.0732   2.9131   8.9887
##
```

```
## Coefficients:
##              Estimate Std. Error t value Pr(>|t|)
## (Intercept) -3.85764    0.82818  -4.658 3.43e-06 ***
## D.tst$Ad     1.45544    0.02341  62.179  < 2e-16 ***
## D.tst$Diff   6.54810    0.32930  19.885  < 2e-16 ***
## ---
## Signif. codes:
## 0 '***' 0.001 '**' 0.01 '*' 0.05 '.' 0.1 ' ' 1
##
## Residual standard error: 4.013 on 1769 degrees of freedom
## Multiple R-squared:  0.6872,Adjusted R-squared:  0.6868
## F-statistic:  1943 on 2 and 1769 DF,  p-value: < 2.2e-16
```

One will notice that the estimates for the coefficients are close to the values found in model $m2$. Since the last model, $m3$, includes the most recent data it is probably best to use that one and even update it regularly with new data.

Finally, one could compare the models fitted on the training data and on the test-data and consider if on what time horizon the model should be calibrated before use. One can consider the whole dataset, the last five years, the training dataset, etc. The choice will depend on the reality of the environment rather than on naive mathematics. Although one machine-learning approach would consist of using all possible data-horizons and finding the optimal one.

♣ 27 ♣

Multi Criteria Decision Analysis (MCDA)

27.1 What and Why

All models discussed previously use one common assumption: there is just one variable to model. While it might be complicated to find an optimal value for that variable or very complicated to predict that one variable, it does not prepare us well for the situation where there are more than one functions to optimize.

Think for example about the situation where you want to buy something – anything actually: then you certainly want to purchase it cheap but still a good quality. Unfortunately, goods of high quality will have high prices, so there is no such thing as an option that is optimal for both variables. Fortunately our brain is quite well suited to solve such problems: we are able to buy something, get married, decide what to eat, etc. The issue is that there is no clear bad choice but also no clear good choice. People tend hence to decide with gut-feeling. This is great because it helps people to do something: fight or flight in a dangerous situation, start a business, explore a new path, get married, vote, etc.[1] The world would be very different if our brain was unable to solve such questions really efficient: such decisions come natural to us, but they are hard to explain to someone else.

However, in a corporate setting this becomes quite problematic. We will need to be able to explain *why* we recommend a certain solution. That is where Multi Criteria Decision Analysis (MCDA) comes in: it is merely a structured method to think about multi-criteria problems, come to some ranking of alternatives but above all it provides insight in *why* one action or alternative is preferable over another.

MCDA
multi criteria decision analysis

Actually, most most business problems are MCDA problems. Consider for example the levels of decision making proposed by Monsen and Downs (1965). Roughly translated to today's big corporates, this boils down to:

1. *Super-strategic:* Mission statement (typically the founders, supervisory board and/or owners have decided this) — this should not be up to discussion, so nothing to decide here (but

[1] Imagine that people would be bad at naturally solve multi criteria problems. In that case, simply buying something in a shop would be a nightmare, as we would get stuck in a failing optimization over price and quality. Indeed, the cheapest option is seldom the one that has the highest quality, so some trade-off is needed.

The Big R-Book: From Data Science to Learning Machines and Big Data, First Edition. Philippe J.S. De Brouwer.

Companion Website: www.wiley.com/go/De Brouwer/The Big R-Book

note that the company most probably started by a biased vision and a bold move on what was actually a multi-criteria problem).

2. *Managerial Control / strategic:* Typical the executive management (executive committee) — almost all problems will be ideally fit for MCDA analysis.

3. *Operational Control / tactical:* Typical middle management — some multi criteria problems, but most probably other methods – described in Part V *"Modelling"* on page 373 – are more fit.[2]

So, the multi criteria problems that we will study in this chapter typically are more relevant for the higher management.

Digression – Corporate change time scales

Note that these levels roughly correspond to different time frames of change: the super-strategic level should change when the company and/or market changes dramatically, the strategic level every few years, the tactical can range from month to year.

[2] Anyhow, if the problem can be reduced to a uni-criterion problem, then the other methods are always preferable.

27.2 General Work-flow

When more than one function need to be optimized simultaneously there is usually no unique best solution and things start to get messy. As with all things in life, when the situation is messy, a structured approach is helpful.

Actually, all types of modelling profit of a structured approach. While for other models – such as the ones discussed in Part V *"Modelling"* on page 373 – the question to be answered is usually clear, but this is not really the case in multi-criteria problems. More often people mix things that are of very different nature and rather need a strategic choice. For example, if business is good but the company has liquidity issues because of delay between paying for raw materials and selling of finished products, then one might consider being bought by a larger company, get an overdraft facility, increase shares, produce lower quality, save costs, etc.

In the following, we present a – personal – view on how such structured approach should look like.

Step 1: Explore the Big Picture

Make sure that the problem is well understood, that all ideas are on the table, that the environment is taken into account, and that we view the issue at hand through different angles and from different points of view. Use for example exploratory techniques such as:

- SWOT analysis,
- 7Ps of Marketing,
- Business Model Canvas,
- NPV, IRR, cost benefit analysis, etc.,
- time-to-break-even, time to profit, largest cumulated negative, etc.,
- two-parameter criteria (e.g. income/cost) referred,
- make sure that the problem is within one level of decision (strategic / managerial / operational) — see p. 233.

Step 2: Identify the Problem at Hand

Make sure that the question is well formulated and is that it is *the* right questions to ask at this moment in these circumstances.

- Brainstorming techniques or focus groups to
  - get all alternatives
  - get all criteria
  - understand interdependencies
  - etc.
- Make sure you have a clear picture on what the problem is, what the criteria and what the possible alternatives are
- Note: This step is best within one level of decision (strategic/managerial/operational).

Note – MCDA needs a structured approach

An MCDA problem requires identification of alternatives (possible solutions), criteria (what functions need to be optimized) as well as a deep business understanding of all the above.

Step 3: Get Data, Construct and Normalise the Decision Matrix

This step makes the problem quantifiable. At the end of this step, we will have numbers for all criteria for all alternative solutions.

If we miss data, we can sometimes mitigate this by adding a best estimate for that variable, and then using "risk" as an extra parameter.

The work-flow can be summarised as follows:

1. Define how to measure all solutions for all criteria — make sure we have an ordinal scale for all criteria – see Chapter B *"Levels of Measurement"* on page 829.
2. Collect all data so that you can calculate all criteria for all solutions.
3. Put these number is a "decision matrix" – see Section 27.4 *"Step 3: the Decision Matrix"* on page 518.
4. Make sure that the decision matrix is as small as possible: can some criteria be combined into one? For example, it might be useful to fit criteria such as the presence of tram, bus, parking, etc. into one "commuting convenience" criterion.

Normalizing a decision matrix is making sure that

1. All criteria need to be maximized.
2. The lowest alternative for each criterion has a value 0 and the highest equals 1.

Step 4: Leave Out Unacceptable and Inefficient Alternatives

1. Leave out all alternative that do not satisfy the **minimal criteria** – eventually rethink the minimal criteria.
2. Drop the non-optimal solutions (the "dominated ones") – see Chapter 27.5.2 *"Dominance – Inefficient Alternatives"* on page 521.
3. Consider dropping the alternatives that score lowest on some key-criteria.

Step 5: Use a Multi Criteria Decision Method to Get a Ranking

If the problem cannot be reduced to a mono criterion problem then we will necessarily have to make some trade-off when selecting a solution. A – very subjective – top-list of multi criteria decision methods (MCDMs) is the following.

1. Weighted sum method — Chapter 27.7.2 *"The Weighted Sum Method (WSM)"* on page 527
2. ELECTRE (especially I and II) — Chapter 27.7.4 *"ELECTRE"* on page 530
3. PROMethEE (I and II) — Chapter 27.7.5 *"PROMethEE"* on page 540
4. PCA analysis (aka "Gaia" in this context) — Chapter 27.7.6 *"PCA (Gaia)"* on page 553

Step 6: Recommend a Solution

In practice, we never make a model or analysis just out of interest, there is always a goal when we do something. Doing something with our work is the reason why we do it in the first place. The data scientist needs to help the management to make good decisions. Therefore it is necessary to write good reports, make presentations, discuss the results and make sure that the decision maker has understood your assumptions and has a fair idea of what the model does.

This step could also be called "do something with the results." This step is not different from any other model report and hence will be superseded by Part VII *"Reporting"* on page 685. However make sure to take the following into account:

- Connect back to the company, its purpose and strategic goals (steps 1 and 2)
- Provide the rationale
- Provide confidence to decision makers
- Conclude
- Make an initial plan (assuming an Agile approach).

Definition: MCDA wording

- A possible solution for the key-question a_i is called a **alternative**.
- The **set of all alternatives** is $\mathcal{A}$ (in what follows we assume all alternatives to be discrete, and $\mathcal{A}$ is finite (and hence countable - we assume A possible alternatives that are worth to consider) — as opposed to continuous.[a]
- A **criterion** is a measure for success, it is considered to be a function on $\mathcal{A}$ that is indicative of how good an alternative is for on aspect. We consider - without loss of generality - K possible criteria.
- The **decision matrix** $M = (m_{ik})$, is an $A \times K$ matrix for which we choose
  - the alternatives to be headings of the rows (so M has A rows) and
  - the criteria to be headings of the columns (so M has K columns).
- The **normalized decision matrix** is $M = (m_{ik})$, so that $\forall k \in \{1 \dots K\} : \exists i : m_{ik} = 0$ and $\forall k \in \{1 \dots K\} : \exists i : m_{ik} = 1$
- An alternative that cannot be rejected (is not dominated nor preferred under another alternative) is a **solution**.

[a] So, we consider in this chapter problems of *choice* and not problems of *design*.

27.3 Identify the Issue at Hand: Steps 1 and 2

The first step is to get the big picture and explore possible alternatives. There is no conclusive method, no model, no statistical learning possible without a good understanding of the problem, the business, goals and strategy. Multiple methods have been proposed and used in most corporations.

SWOT

Methods such as SWOT analysis for example can be used to investigate ideas, identify new ideas, focus ideas, etc. These methods explore ideas but are no decision models in itself.

Digression – SWOT Analysis

SWOT is a structured method to explore reality. Its letters stand for:

- *Strengths:* Characteristics of the business or project that give it an advantage over others.
- *Weaknesses:* Characteristics that place the business or project at a disadvantage relative to others.
- *Opportunities:* Elements that the business or project could exploit to its advantage.
- *Threats:* Elements in the environment that could cause trouble for the business or project.

While the words "strengths" and "opportunities" are very close to each other (just as "weakness" and "threat") the SWOT method uses the convention that the first two are internal to the team/company and the last two are external. So while a talented and motivated workforce could be seen as an opportunity – in plain English – we will put it in the box of "strengths" and the fact that the competition is unable to attract talent should be considered as an "opportunity."

It really depends on the problem what method to use. Assume for now that we are working in the analytics department of a large multinational, that has a lot of data to take care of and analyse, for example a large multinational bank. Let us call the bank "R-bank", to honour the statistical software that brings this book together. The company has large subsidiaries in many countries and already started to group services in large "service centres" in Asia.

R-bank is UK based and till now it has 10 000 people working in five large service centres in Asia and South America. These centres are in Bangalore, Delhi, Manilla, Hyderabad and São Paulo. These cities also happen to be top destinations for Shared Service Centres (SSC) and Business Process Outsourcing (BPO) – as presented by the Tholons index (see `http://www.tholons.com`).[3]

The bank wants to create a central analytics function to supports its modelling and in one go it will start building one central data warehouse with data scientists to make sense of it for commercial and internal reasons (e.g. risk management).

[3] BPO – Business Process Outsourcing – refers to services delivered by an external (not fully owned) specialized company for third party companies. SSC – Shared Service Center – refers to a company that delivers services by a company that belongs to the same capital group. In our example the SSC in Delhi can provide services to the multiple banks in multiple countries and since both the customers are part of the R-bank group as well as the SSC of R-bank in Delhi are fully owned by R-bank Holding in UK we speak of an "SSC"

As the bank already has SSC centres in low-cost countries it is not a big step to imagine that the centre of gravity of the new risk and analytics centre should be also in a shared service centre. The question that arises is rather: in which SSC? Do we use an existing location or go to a new city? The management of R-bank realizes that this is different from their other services where the processes are well structures, stable, mature and do not require massive communications between the SSC and the headquarters.

This narrative defines the problem at hand and provides understanding of the business question, limitations, strategy and goals. The next step is to identify the criteria to make our selection.

For possible destinations we retain the top ten of Tholons: Bangalore, Delhi, Manilla, Hyderabad and São Paulo, Dublin, Kraków, Chennai, and Buenos Aires.

We use brainstorm sessions, focus groups to evaluate the existing SSCs and have come up with the following list of relevant criteria:

1. *Talent:* Availability of talent and skills (good universities and enough students)
2. *Stability:* Political stability and fiscal stability
3. *Cost:* The of running the centre
4. *Cost inflation:* Salary inflation
5. *Travel:* Cost and convenience of travelling to the centre (important since we expect lots of interaction between the headquarters and the SSC Risk and Analytics)
6. *Time-zone* Time-zone overlap (as alternative to travel)
7. *Infrastructure:* Office space, roads, etc.
8. *Life quality:* Personal risk and quality of life (museums, restaurants, public transport, etc.)
9. An international airport in close proximity.

27.4 Step 3: the Decision Matrix

27.4.1 Construct a Decision Matrix

Further, analysis learns us that the international airport is not a discriminating factor for the Tholons top-10, because all locations have one. So we will drop this criterion.

That leaves us with ten possible solutions and nine criteria. The next step is to get the data and create the decision matrix. After brainstorming and discussions, we came up with the following:

1. *Talent:* Use Tholons' "talent, skill and quality" 2017 index – see `http://www.tholons.com`
2. *Stability:* the 2017 political stability index of the World Bank – see `http://info.worldbank.org/governance/WGI`
3. *Cost:* Use Tholons' "cost" 2017 index – see `http://www.tholons.com`
4. *Cost inflation* "Annualized average growth rate in per capita real survey mean consumption or income, total population (%)" from `https://data.worldbank.org`
5. *Travel:* Cost and convenience of travelling to the centre (important since we expect lots of interaction between the headquarters and the SSC Risk and Analytics) – our assessment of airline ticket price between R-bank's headquarters, the travel time, etc.
6. *Time-zone:* Whether there is a big time-zone differnce – this is roughly one point if in the same time-zone as R-bank's headquarters, zero if more than 6 hours difference.
7. *Infrastructure:* Use Tholons' "infrastructure" 2017 index – see `http://www.tholons.com`
8. *Life quality:* Use Tholons' "risk and quality of life" 2017 index – see `http://www.tholons.com`
9. International airport in close proximity: Not withheld as a criterion, because all cities in the Tholons top-10 have international airports.

These criteria, research, and data result in the raw decision matrix. This matrix is presented in Table 27.1 on page 519

All criteria will at least be evaluated on an ordinal scale[4]. While some are provided by reliable sources, the science behind it is not as exact as measuring a distance. These indexes are usually to be understood as a qualitative KPI. Even when measurable, these quantities will be defined with lots of assumptions underpinning their value.

At this stage we notice that the correlation between travel cost and time-zone is too high to keep the variables separate. So, we need to merge them in to one indicator (we will refer to it as "travel").

The following code characterises this decision matrix.

```
M0 <- matrix(c(
 1.6 , -0.83 , 1.4 , 4.7 , 1 , 0.9 , 1.1 ,
 1.8 , -0.83 , 1.0 , 4.7 , 1 , 0.9 , 0.8 ,
 1.8 , -0.83 , 1.2 , 4.7 , 1 , 0.9 , 0.6 ,
```

[4]See Chapter B *"Levels of Measurement"* on page 829

Location	tlnt	stab	cost	infl	trvl	tm-zn	infr	life
Bangalore	1.6	-0.83	1.4	4.7%	H	1	0.9	1.1
Mumbai	1.8	-0.83	1.0	4.7%	H	1	0.9	0.8
Delhi	1.8	-0.83	1.2	4.7%	H	1	0.9	0.6
Manilla	1.6	-1.24	1.4	2.8%	H	1	0.9	0.8
Hyderabad	0.9	-0.83	1.4	4.7%	H	1	0.7	0.8
Sao Polo	0.9	-0.83	0.8	4.7%	H	1	0.7	0.6
Dublin	0.7	1.02	0.2	2.0%	L	3	1.1	1.3
Krakow	1.1	0.52	1.0	1.3%	L	3	0.6	0.9
Chennai	1.2	-0.83	1.3	4.7%	H	1	0.8	0.5
Buenos Aires	0.9	0.18	0.9	7.3%	H	1	0.8	0.6

Table 27.1: *The decision matrix summarises the information that we have gathered. In this stage the matrix will mix variables in different units, and even qualitative appreciations (e.g. high and low).*

```
 1.6 , -1.24 , 1.4 , 2.8 , 1 , 0.9 , 0.8 ,
 0.9 , -0.83 , 1.4 , 4.7 , 1 , 0.7 , 0.8 ,
 0.9 , -0.83 , 0.8 , 4.7 , 1 , 0.7 , 0.6 ,
 0.7 ,  1.02 , 0.2 , 2.0 , 3 , 1.1 , 1.3 ,
 1.1 ,  0.52 , 1.0 , 1.3 , 3 , 0.6 , 0.9 ,
 1.2 , -0.83 , 1.3 , 4.7 , 1 , 0.8 , 0.5 ,
 0.9,  0.18 , 0.9 , 7.3 ,  1 , 0.8 , 0.6 ),
 byrow = TRUE,
 ncol = 7)
colnames(M0) <- c("tlnt","stab","cost","infl","trvl","infr","life")
# We use the IATA code of a nearby airport as abbreviation,
# so, instead of:
#rownames(M0) <- c("Bangalore", "Mumbai", "Delhi", "Manilla",
#                  "Hyderabad", "Sao Polo", "Dublin", "Krakow",
#                  "Chennai", "Buenos Aires")
# ... we use this:
rownames(M0) <- c("BLR", "BOM", "DEL", "MNL", "HYD", "GRU",
                  "DUB", "KRK", "MAA", "EZE")

M0
##     tlnt  stab cost infl trvl infr life
## BLR  1.6 -0.83  1.4  4.7    1  0.9  1.1
## BOM  1.8 -0.83  1.0  4.7    1  0.9  0.8
## DEL  1.8 -0.83  1.2  4.7    1  0.9  0.6
## MNL  1.6 -1.24  1.4  2.8    1  0.9  0.8
## HYD  0.9 -0.83  1.4  4.7    1  0.7  0.8
## GRU  0.9 -0.83  0.8  4.7    1  0.7  0.6
## DUB  0.7  1.02  0.2  2.0    3  1.1  1.3
## KRK  1.1  0.52  1.0  1.3    3  0.6  0.9
## MAA  1.2 -0.83  1.3  4.7    1  0.8  0.5
## EZE  0.9  0.18  0.9  7.3    1  0.8  0.6
```

One will notice that all data is already numeric. That was not so from the start. In some cases. we need to transform ordinal labels into a numeric scale. As explained in Chapter B *"Levels of Measurement"* on page 829, there is no correct or incorrect way of doing this. The best way is to imagine that there is some quantity as "utility" or "preference" and scale according to this quantity.

27.4.2 Normalize the Decision Matrix

Next, we need to normalize each variable so that it is between 0 and 1 (with 0 worst outcome and 1 the best). The reason to do this is not because it is necessary from a mathematical point of view, but rather because it is useful for the interpretability of the results.

If the scales were different and for example a criterion that takes values between 1 and 10^9 needs much lower coefficients than one that has values between 0 and 10^{-9} to get the same influence. If we normalize all criteria between 0 and 1, and equal weight will mean equal influence in the decision.

Normalizing the decision matrix is easy and base R provides all the tools to do so. We could use the function `range()` or `apply()`. Below is the version that uses the latter.

```
# Political stability is a number between -2.5 and 2.5
# So, we make it all positive by adding 2.5:
M0[,2] <- M0[,2] + 2.5

# Lower wage inflation is better, so invert the data:
M0[,4] <- 1 / M0[,4]

# Then we define a function:

# mcda_rescale_dm
# Rescales a decision matrix M
# Arguments:
#    M -- decision matrix
#         criteria in columns and higher numbers are better.
# Returns
#    M -- normalised decision matrix
mcda_rescale_dm <- function (M) {
  colMaxs <- function(M) apply(M, 2, max, na.rm = TRUE)
  colMins <- function(M) apply(M, 2, min, na.rm = TRUE)
  M <- sweep(M, 2, colMins(M), FUN="-")
  M <- sweep(M, 2, colMaxs(M) - colMins(M), FUN="/")
  M
}

# Use this function:
M <- mcda_rescale_dm(M0)

# Show the new decision matrix elegantly:
knitr::kable(round(M,2))
```

	tlnt	stab	cost	infl	trvl	infr	life
BLR	0.82	0.18	1.00	0.12	0	0.6	0.75
BOM	1.00	0.18	0.67	0.12	0	0.6	0.38
DEL	1.00	0.18	0.83	0.12	0	0.6	0.12
MNL	0.82	0.00	1.00	0.35	0	0.6	0.38
HYD	0.18	0.18	1.00	0.12	0	0.2	0.38
GRU	0.18	0.18	0.50	0.12	0	0.2	0.12
DUB	0.00	1.00	0.00	0.57	1	1.0	1.00
KRK	0.36	0.78	0.67	1.00	1	0.0	0.50
MAA	0.45	0.18	0.92	0.12	0	0.4	0.00
EZE	0.18	0.63	0.58	0.00	0	0.4	0.12

27.5 Step 4: Delete Inefficient and Unacceptable Alternatives

27.5.1 Unacceptable Alternatives

It is worth to make sure that all alternatives meet certain minimal criteria. For example, we need an international airport at less than one hour driving, we need universities in the proximity of the city, etc.

In our case this is most likely to be fine, because we start from the world top-10 of SSC cities. In many other cases, this is the right moment to make sure that we only proceed with solutions that are acceptable when chosen.

It is also worth to consider that some locations might not be desirable regardless the outcome of the MCDA process. It is really unlikely to happen, but not impossible that the most expensive location would appear to be the best. Is that a choice that will be acceptable by the management?

This last argument has to be handled with care: it deals with relative aspects. Would it make sense to leave out a_j because of arguments based on other alternatives?

Applied to our example we might want to consider the following questions. Would we build our SSC in the location that is the most expensive? Would we want to live and hire in the place that offers the lowest quality of life?

This will cast a shadow over Dublin (cost) and both Delhi and Buenos Aires (both for quality of life). Since we know the company philosophy of R-bank we might as well leave those places out, but from a communication point of view it is wiser to leave them in — but take note of the issue and include it in the debriefing. If those places would be chosen, then there is an issue with the model used and we need to carefully rethink. If they do not get chosen, it is wise to point that out in the debriefing.

27.5.2 Dominance – Inefficient Alternatives

A solution a_i is said to be dominated by another solution a_j if a_j scores at least equal on all criteria and outperforms a_i in at least one criterion k.

Below we provide a function that produces a "dominance matrix": a matrix that holds information about which criterion is dominated about which other. If a_{ij} is 1, then criterion i is dominated by criterion j.

```
# mcda_get_dominated
# Finds the alternatives that are dominated by others
# Arguments:
#    M -- normalized decision matrix with alternatives in rows,
#         criteria in columns and higher numbers are better.
# Returns
#    Dom -- prefM -- a preference matrix with 1 in position ij
#                    if alternative i is dominated by alternative j.
mcda_get_dominated <- function(M) {
  Dom <- matrix(data=0, nrow=nrow(M), ncol=nrow(M))
  dominatedOnes <- c()
  for (i in 1:nrow(M)) {
    for (j in 1:nrow(M)) {
      isDom <- TRUE
      for (k in 1:ncol(M)) {
        isDom <- isDom && (M[i,k] >= M[j,k])
      }
      if(isDom && (i != j)) {
        Dom[j,i] <- 1
        dominatedOnes <- c(dominatedOnes,j)
```

```
      }
    }
  }
  colnames(Dom) <- rownames(Dom) <- rownames(M)
  class(Dom) <- "prefM"
  Dom
}
```

The output of this function is a $A \times A$ matrix (with A the number of alternatives). It has zeros and ones. A one in position (i, j) means $a_i \prec a_j$ (or "alternative i is dominated by alternative j). For the rest of this chapter we will look at preference relationships and it might make more sense to flip this around by transposing the matrix.

```
# mcda_get_dominants
# Finds the alternatives that dominate others
# Arguments:
#    M -- normalized decision matrix with alternatives in rows,
#         criteria in columns and higher numbers are better.
# Returns
#    Dom -- prefM -- a preference matrix with 1 in position ij
#                    if alternative i dominates alternative j.
mcda_get_dominants <- function (M) t(mcda_get_dominated(M))
```

Now, a 1 in position (i, j) means $a_i \succ a_j$ (or "alternative i dominates alternative j).
This function provides an easy way to find the dominated alternatives.

```
Dom <- mcda_get_dominants(M)
print(Dom)
##     BLR BOM DEL MNL HYD GRU DUB KRK MAA EZE
## BLR   0   0   0   0   1   1   0   0   1   0
## BOM   0   0   0   0   0   1   0   0   0   0
## DEL   0   0   0   0   0   1   0   0   0   0
## MNL   0   0   0   0   0   0   0   0   0   0
## HYD   0   0   0   0   0   1   0   0   0   0
## GRU   0   0   0   0   0   0   0   0   0   0
## DUB   0   0   0   0   0   0   0   0   0   0
## KRK   0   0   0   0   0   0   0   0   0   0
## MAA   0   0   0   0   0   0   0   0   0   0
## EZE   0   0   0   0   0   0   0   0   0   0
## attr(,"class")
## [1] "prefM"
```

We see that

- Hyderabad (HYD) is dominated by Bangalore: it has a worse talent pool and lower quality of life, while it scores the same for all other criteria.
- São Paulo is dominated by Bangalore, Mumbai, Delhi, and Hyderabad.
- Chennai is dominated by Bangalore.

This should not come as a massive surprise since our list of alternatives is the Tholons top-10 of 2017 (in that order). This helps to leave out some alternatives. In general, it does not make sense to keep alternatives that are worse or equal to another. No decision method would prefer Hyderabad, São Paulo or Chennai over Bangalore.

An easy way to leave out the dominated alternatives could be with the following function.

```
# mcda_del_dominated
# Removes the dominated alternatives from a decision matrix
# Arguments:
#    M -- normalized decision matrix with alternatives in rows,
#         criteria in columns and higher numbers are better.
# Returns
#    A decision matrix without the dominated alternatives
mcda_del_dominated <- function(M) {
  Dom <- mcda_get_dominated(M)
  M[rowSums(Dom) == 0,]
}
```

This function allows us to reduce the decision matrix M to $M1$ that only contains alternatives that are not dominated.

```
M1 <- mcda_del_dominated(M)
round(M1,2)
##     tlnt stab cost infl trvl infr life
## BLR 0.82 0.18 1.00 0.12    0  0.6 0.75
## BOM 1.00 0.18 0.67 0.12    0  0.6 0.38
## DEL 1.00 0.18 0.83 0.12    0  0.6 0.12
## MNL 0.82 0.00 1.00 0.35    0  0.6 0.38
## DUB 0.00 1.00 0.00 0.57    1  1.0 1.00
## KRK 0.36 0.78 0.67 1.00    1  0.0 0.50
## EZE 0.18 0.63 0.58 0.00    0  0.4 0.12
```

Although it makes also clear that dominance is an approach that will leave a lot of alternatives unranked. At this point, we will have to revert to strategies that are on the borderline between science and art. However, using one of the MCDA methods that follow allows to make very clear how you came to the decision and it creates a way to talk about it.

Warning – Rescale the decision matrix

Should we rescale the decision matrix now that we have left out the dominated solutions? In our example the place with the lowest score for "quality of life" has been left out. This means that the range of these criteria is not between 0 and 1 any more. If we do not rescale the results will be hard to interpret. So rescaling is a good idea.

At this point, it makes sense to rescale the decision matrix since the lowest element of some criteria might have been dropped out. We can reuse the function `rescale_dm()` that we have defined previously.

```
M1 <- mcda_rescale_dm(M1)
```

27.6 Plotting Preference Relationships

Before we explore some MCDA methods it is worth to figure out how to visualize preference relationships. We have a simple example in the matrix *Dom* that shows which alternatives are dominated (and hence have others that are preferred over them).

It appears that R has a package called `diagram` that provides a function that is actually designed to visualize a transition matrix, but it is also an ideal fit for our purpose here.

plotmat()

In the following code segment, we create a function `plot.prefM()` that will plot an object of the class `prefM` – as defined in Chapter 27.5.2 *"Dominance – Inefficient Alternatives"* on page 521.

diagram

This function is based on the `plotmat()` function of the library `diagram`.

```
# First, we load diagram:
require(diagram)

## Loading required package: diagram
## Loading required package: shape

# plot.prefM
# Specific function to handle objects of class prefM for the
# generic function plot()
# Arguments:
#    PM  -- prefM -- preference matrix
#    ... -- additional arguments passed to plotmat()
#           of the package diagram.
plot.prefM <- function(PM, ...)
{
  X <- t(PM) # We want arrows to mean '... is better than ...'
             # plotmat uses the opposite convention because it expects flows.
  plotmat(X,
          box.size    = 0.1,
          cex.txt     = 0,
          lwd         = 5 * X,  # lwd proportional to preference
          self.lwd    = 3,
          lcol        = 'blue',
          self.shiftx = c(0.06, -0.06, -0.06, 0.06),
          box.lcol    = 'blue',
          box.col     = 'khaki3',
          box.lwd     = 2,
          relsize     = 0.9,
          box.prop    = 0.5,
          endhead     = FALSE,
          main        = "",
          ...)
}
```

Note – Dots

The three points allow the function `plot.prefM()` to get more arguments passing them on to the function `plotmat()`. This allows us later to pass more arguments to the function `plotmat()` through the function `plot.prefM()`.

Now, it becomes clear why the function `get_dominated()` changes the class argument of the S3 matrix object before returning it. By naming our function `plot.prefM()` we tell R that whenever the dispatcher function `plot()` is used and the class of the object is "prefM" that we expect R to plot it as below and not as a scatter-plot (which is default for a matrix).[5] Hence, plotting the dominance matrix is just one line of code (the output is in Figure 27.1).

```
# We pass the argument 'curve = 0' to the function plotmat,
# since otherwise in this case the arrow from BLR to MAA
# would be hidden after the box of EZE.
plot(Dom, curve = 0)
```

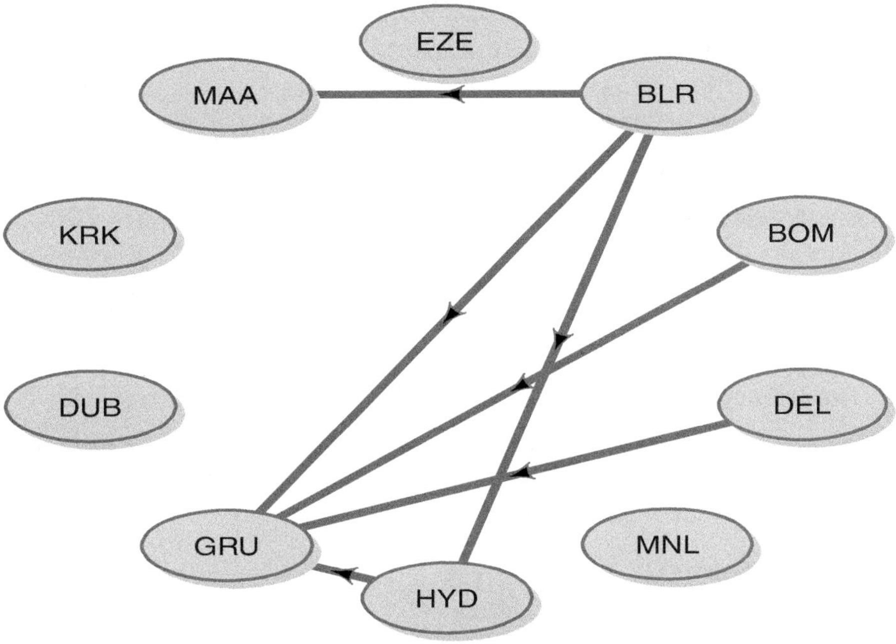

Figure 27.1: *A visualization of the dominance relationship.*

[5]More information about how the object model is implemented in R is in Chapter 6 *"The Implementation of OO"* on page 87 and more in particular Chapter 6.2 *"S3 Objects"* on page 91.

27.7 Step 5: MCDA Methods

At this point, we have left out all dominated solutions and hence are by definition left with a set of "efficient solutions."

Definition: Efficient Solutions

An Efficient Solution is a solution that is not dominated by any other solution

The matrix $M1$ is the decision matrix for all our efficient alternatives. Actually, "dominance" can be considered as the first non-compensatory MCDA method. We recognize two classes of MCDA methods: compensatory methods that allow one strong point to compensate for another weak point, and non-compensatory methods that to not allow for any compensation.

As we have seen, dominance only eliminated one alternative of our list, and most of them could not really be ordered with dominance. That is typical for non-compensatory methods.

Compensatory methods allow weaknesses to be full or partially compensated by strong points. They will typically lead to a richer ranking. In the remainder of this chapter, we will study some of those methods.

27.7.1 Examples of Non-compensatory Methods

Dominance is a useful method that does not allow even one weakness to be offset by many strong points. That is called a non-compensatory method. In this section we list a few other non-compensatory methods that might be useful to consider.

27.7.1.1 The MaxMin Method

1. find the weakest attribute for all solutions
2. select the solution that has the highest weak attribute (0 in a normalized decision matrix)

This method makes sense if

- the attribute values are expressed in the same units, and
- when the "a chain is as weak as the weakest link reasoning" makes sense.

27.7.1.2 The MaxMax Method

1. Find the strongest attribute for all solutions.
2. Select the solution that has the strongest strong attribute.

This method makes sense if

- the attribute values are expressed in the same units, and
- when one knows that the best of the best in one attribute is most important.

In our example, the dimensions (units) are nothing but comparable. Also, if they were all to be expressed in dollar terms, then why not aggregate them and reduce our multi criteria problem to a mono-criterion optimization?

While these methods are of limited practical use, they are important food for thought.

27.7.2 The Weighted Sum Method (WSM)

WSM
weighted sum method

The weighted sum method (WSM) is much akin to a linear regression, or a logistic regression where variables in different dimensions are scaled and then multiplied with a coefficient in order to calculate a "score."[6]

The coefficient is obtained by some choice of optimization criterion such as minimizing the sum of squares of residuals (OLS). This assumes that we have some idea of what is the ideal solution. In MCDA we do not have this, so we will have to make assumptions or choices. One of those choices is the relative importance of criteria, which we will call "weights".

The MCDA is replaced by finding the maximum for:

$$\max_{\mathbf{x} \in \mathcal{A}} \{N(a)\}$$

with $N(.)$ the function $\Re^n \mapsto \Re^n$ so that

$$N(\mathbf{a_i}) = \sum_{k=1}^{K} w_k \, m_{ik} \qquad \text{or}$$

$$\mathbf{N}(\mathbf{a}) = \mathbf{M}.\mathbf{w}$$

where $\mathbf{M}$ is the decision matrix where each element is transformed according to a certain function.

The key thing – especially for the mathematician – is to understand that weights are assigned to "differences in preference" and not to the criteria as such. Since then everything is expressed in the unit "preference," we are indeed allowed to add the scores.

The main advantage of this method is that it is as clear as a logistic regression. Anyone who can count will understand what is happening. So, the WSM method is an ideal tool to stimulate discussion.

In R this can be obtained as follows.

```
# mcda_wsm
# Finds the alternatives that are dominated by others
# Arguments:
#    M -- normalized decision matrix with alternatives in rows,
#         criteria in columns and higher numbers are better.
#    w -- numeric vector of weights for the criteria
# Returns
#    a vector with a score for each alternative
mcda_wsm <- function(M, w) {
  X <- M %*% w
  colnames(X) <- 'pref'
  X
}
```

At this point, we need to assign a "weight" to each criterion. This needs to be done in close collaboration with the decision makers. For example, we might argue that the infrastructure is less important since our SSC will be small compared to the existing ones, etc.

Taking into account that the SSC will not be very large, that we cannot expect employees just to be ready (so we will do a lot of training ourselves and work with universities to fine-tune curricula, etc.), we need a long time to set up such centre of expertise and hence need stability, etc. we came up with the following weights.

[6] The WSM is also known as the "additive method."

```
# The critia: "tlnt" "stab" "cost" "infl" "trvl" "infr" "life"
w <- c(        0.125, 0.2,   0.2,   0.2,  0.175,  0.05,  0.05)
w <- w / sum(w)  # the sum was 1 already, but just to be sure.

# Now we can execute our function mcda_wsm()
mcda_wsm(M1, w)
##          pref
## BLR 0.4282418
## BOM 0.3628739
## DEL 0.3819215
## MNL 0.4162013
## DUB 0.5898333
## KRK 0.7309687
## EZE 0.2850577
```

The WSM produces almost always a complete ranking, but is of course sensitive to the weights that – actually – are arbitrary. So, with this method it is not so difficult to make sure that the winning solution is the one that we preferred from the start anyhow.

The complete ranking can be represented with `plotmat` but it might make more sense to use `ggplot2`. To do so neatly, we take a step back and re-write the function `mcda_wsm()` and make it return a "matrix of scores" ("scoreM" for short).

sweep()

```
# mcda_wsm_score
# Returns the scores for each of the alternative for each of
# the criteria weighted by their weights.
# Arguments:
#    M -- normalized decision matrix with alternatives in rows,
#         criteria in columns and higher numbers are better.
#    w -- numeric vector of weights for the criteria
# Returns
#    a score-matrix of class scoreM
mcda_wsm_score <- function(M, w) {
  X <- sweep(M1, MARGIN = 2, w, `*`)
  class(X) <- 'scoreM'
  X
}

# plot.scoreM
# Specific function for an object of class scoreM for the
# generic function plot().
# Arguments:
#    M -- scoreM -- score matrix
# Returns:
#    plot
plot.scoreM <- function (M) {
  # 1. order the rows according to rowSums
  M <- M[order(rowSums(M), decreasing  = T),]

  # 2. use a bar-plot on the transposed matrix
  barplot(t(M),
     legend = colnames(M),
     xlab   = 'Score',
     col    = rainbow(ncol(M))
     )
}
```

Digression – Colour schemes

The colour scheme is derived from interpolated rainbow colours with the function `rainbow()` from the package `package:grDevices` – which is loaded by default. There are a few more available right out of the box such as `heat.colors(n, alpha = 1)`: have a look at the documentation via `?rainbow`.

This function is now ready to be called automatically via the dispatcher function `plot()`. The code is below and the result is in Figure 27.2.

```
sM <- mcda_wsm_score(M1, w)
plot(sM)
```

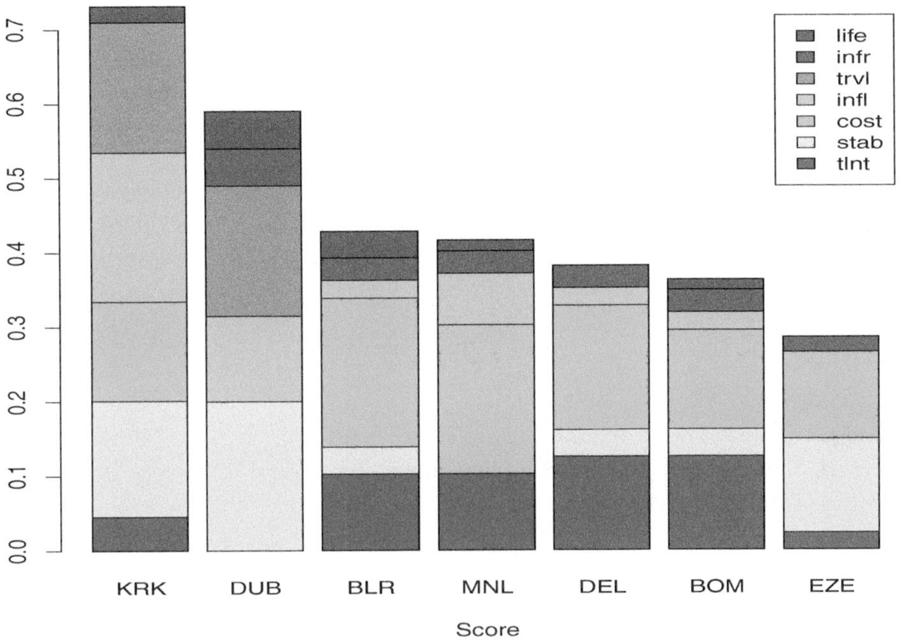

Figure 27.2: *The scores of different cities according to the WSM.*

According to our weighting which attached reasonably similar weights to all criteria, Krakow seems to be a good solution. However, note that this city scores lowest on "infrastructure." Before presenting the solution to the board, we need of course to understand what this really means. Looking deeper in what Tholons is doing, this is due to the fact that it does not have a subway system, public transport relies on trams and buses and both need investment, the ring-road is not finished yet (but there are plans to add three levels of ring-roads, etc.), it misses large office spaces, etc. For our intend, which is building up an SSC with highly skilled workers this is not a major concern. By all standards R-bank's investment in Krakow would be modest and even medium office spaces should suffice.

The next concern is that it scores low at "talent." Again we need to investigate what this really means. Krakow – with its 765 000 inhabitants – is a lot smaller than for example Bangalore that hosts 23.3 million people. At this point, it only employs about 65 000 employees in service centres, but it has great universities and 44 000 students graduate every year. While this might be an issue for a large operation of 10 000 people, it should not be a major concern for our purpose.

Note – MCDA is not exact science

While these considerations are rather particular and not of general interest, it is important to realize that MCDA more than any other method needs a lot of common sense and interpretation.

27.7.3 Weighted Product Method (WPM)

One of the drawbacks to WSM is that really low scores might be compensated by other really high scores. That is why we concluded previous section with a few paragraphs of considerations about the city of preference.

To counteract this problem, it is important to be noted that besides summing scores it is also possible to multiply them. The Weighted Product Method (WPM) works as follows:

Let w_j be the weight of the criterion j, and m_{ij} the score (performance) of alternative i on criterion j then solutions can be ranked according to their total score as follows

$$P(a_i) = \Pi_{j=1}^{n}(m_{ij})^{w_j}$$

This leads to the following definition of preference relation.

Let w_j be the weights of the criteria, and m_{ij} the score (performance) of alternative i on criterion j then a solution a_i is preferred over a solution a_n if the preference $P(a_i, a_j) > 1$, with

$$P(a_i, a_j) := \Pi_{k=1}^{n}\left(\frac{m_{ik}}{m_{jk}}\right)^{w_k}$$

This form of the WPM is often called **dimensionless analysis** because its mathematical structure eliminates any units of measure. Note however, that it requires a ratio scale.

It is less easy to understand what the coefficients of WPM do compared to the WSM. It is also possible to combine both. For example, build a score "human factors" as the product of safety, housing quality, air quality, etc. This "human factor" can then be fed into an additive model such as WSM. This approach can also be useful for scorecards.

27.7.4 ELECTRE

If impossible to find, express, and calculate a meaningful common variable (such as "utility") then we try to find at least a preference structure that can be applied to all criteria.[7] If such a preference structure π exists, and is additive, then we can calculate an inflow and outflow or preference for each alternative.

If the decision matrix M has elements m_{ik}, then we prefer the alternative a_i over the alternative a_j for criterion k if $m_{ik} > m_{jk}$. In other words, we prefer alternative i over alternative j for

[7] If it is possible to define "Utility," then the multi-criteria problem reduces to a mono-criterion problem that is easily solved. In reality however, it is doubtful if "utility" exists and even if it would, it will be different for each decision maker and it might not even be possible to find its expression (if that exists).

criterion k if its score is higher for that criterion. The amount of preference can be captured by a function $\Pi()$.

In ELECTRE the preference function is supposed to be a step-function.

Definition: Preference of one solution over another

The preference of a solution a_i over a solution a_j is

$$\pi^+(a_i, a_j) := \sum_{k=1}^{K} \pi_k(m_{ik} - m_{jk})\, w_k$$

Definition: Anti-preference of one solution over another

The anti-preference of a solution a_i over a solution a_j is

$$\pi^-(a_i, a_j) := \sum_{k=1}^{K} \pi_k(m_{jk} - m_{ik})\, w_k$$

We note that:

$$\begin{aligned}
\pi^+(a_i, a_j) &= \sum_{k=1}^{K} \pi_k(m_{ik} - m_{jk})\, w_k \\
&= -\sum_{k=1}^{K} \pi_k(m_{jk} - m_{ik})\, w_k \\
&= -\pi^+(a_j, a_i) \\
&= -\pi^-(a_i, a_j) \\
&= \pi^-(a_j, a_i)
\end{aligned}$$

Even with a preference function $\pi()$ that is a strictly increasing function of the difference in score, it might be that some solutions have the same score for some criteria and hence are incomparable for these criteria. So, it makes sense to define a degree of "indifference."

Definition: The Weighted Degree of Indifference

The Weighted Degree of Indifference of a solution a and b is

$$\begin{aligned}
\pi^0(a, b) &= \sum_{j=a}^{k} w_j - \pi^+(a_i, a_j) - \pi^-(a_i, a_j) \\
&= 1 - \pi^+(a_i, a_j) - \pi^-(a_i, a_j)
\end{aligned}$$

The last line assumes that the sum of weights is one.

Note – Should the weights be the same as in WSM?

These weights can be the same as in the Weighted Sum Method. However, in general there is no reason why they would be the same.

This way of working makes a lot of sense. Referring to our previous example, we see that Dublin outperforms Krakow for political stability, infrastructure and life quality. We also see that Krakow does better for talent pool, cost, and wage inflation and hence we prefer Krakow for these criteria. They are in the same time-zone and travel costs for almost all relevant stakeholders will be the same. For travel cost, one is indifferent between Krakow and Dublin.

27.7.4.1 ELECTRE I

We have now three preference matrices: Π^+, Π^-, and Π^0. Based on these matrices we can devise multiple preference structures. The first method is "ELECTRE I" and requires to calculate a comparability index.

There are two particularly useful possibilities for this index of comparability. We will call them C_1 and C_2.

Definition: Index of comparability of Type 1

$$C_1(a,b) = \frac{\Pi^+(a,b) + \Pi^0(a,b)}{\Pi^+(a,b) + \Pi^0(a,b) + \Pi^-(a,b)}$$

Note that $C_1(a,b) = 1 \Leftrightarrow aDb$. This, however, should not be the case in our example as we already left out all dominated solutions.

Definition: Index of comparability of Type 2

$$C_2(a,b) = \frac{\Pi^+(a,b)}{\Pi^-(a,b)}$$

Note that $C_2(a,b) = \infty \Leftrightarrow aDb$.

Further, to this index of comparability it makes sense to define a threshold Λ below which we consider the alternatives as "too similar to be discriminated."

For each criterion individually we define:

- for the comparability index a cut-off level and consider the alternatives as equally interesting if $C_i < \Lambda_i$:
  - $\Lambda_1 \in]0,1[$ if one uses C_1
  - $\Lambda_2 \in]0,\infty[$ if one uses C_2
- for each criterion a maximal discrepancy in the "wrong" direction if a preference would be stated: $r_k, k \in \{1 \dots K\}$. This will avoid that a solution a is preferred over b while it is too much worse than b for at least one criterion.

With all those definitions we can define the preference structure as follows:

- for C_1: $\left.\begin{array}{l} \Pi^+(a,b) > \Pi^-(a,b) \\ C_1(a,b) \geq \Lambda_1 \\ \forall j : d_j(a,b) \leq r_j \end{array}\right\} \Rightarrow a \succ b$

- for C_2: $\left.\begin{array}{l} \Pi^+(a,b) > \Pi^-(a,b) \\ C_2(a,b) \geq \Lambda_2 \\ \forall j : d_j(a,b) \leq r_j \end{array}\right\} \Rightarrow a \succ b$

In a last step one can present the results graphically and present the kernel (the best solutions) to the decision makers. The kernel consists of all alternatives that are "efficient" (there is no other alternative that is preferred over the latter).

Definition: Kernel of an MCDA problem

The kernel of a MCDA problem is the set

$$\mathcal{K} = \{a \in \mathcal{A} \mid \nexists b \in \mathcal{A} : b \succ a\}$$

ELECTRE I in R

Below is one way to program the ELECTRE I algorithm in R. One of the major choices that we made was create a function with a side effect. This is not the best solution if we want others to use our code (e.g. if we would like to wrap the functions in a package). The alternative would be to create a list of matrices, that then could be returned by the function.

Since we are only calling the following function within another function this is not toxic, and suits our purpose well.

```
# mcda_electre Type 2
# Push the preference matrixes PI.plus, PI.min and
# PI.indif in the environment that calls this function.
# Arguments:
#    M -- normalized decision matrix with alternatives in rows,
#         criteria in columns and higher numbers are better.
#    w -- numeric vector of weights for the criteria
# Returns nothing but leaves as side effect:
#    PI.plus   -- the matrix of preference
#    PI.min    -- the matrix of non-preference
#    PI.indif  -- the indifference matrix
mcda_electre <- function(M,  w) {
  # initializations
  PI.plus  <<- matrix(data=0, nrow=nrow(M), ncol=nrow(M))
  PI.min   <<- matrix(data=0, nrow=nrow(M), ncol=nrow(M))
  PI.indif <<- matrix(data=0, nrow=nrow(M), ncol=nrow(M))

  # calculate the preference matrix
  for (i in 1:nrow(M)){
    for (j in 1:nrow(M)) {
      for (k in 1:ncol(M)) {
        if (M[i,k] > M[j,k]) {
          PI.plus[i,j] <<- PI.plus[i,j] + w[k]
        }
        if (M[j,k] > M[i,k]) {
          PI.min[i,j] <<- PI.min[i,j] + w[k]
        }
        if (M[j,k] == M[i,k]) {
          PI.indif[j,i] <<- PI.indif[j,i] + w[k]
        }
      }
    }
  }
}
```

This function can now be called in an encapsulating function which calcualtes the ELECTRE preference matrix.

```
# mcda_electre1
# Calculates the preference matrix for the ELECTRE method
# Arguments:
#    M -- decision matrix (colnames are criteria, rownames are alternatives)
#    w -- vector of weights
#    Lambda -- the cutoff for the levels of preference
#    r -- the vector of maximum inverse preferences allowed
#    index -- one of ['C1', 'C2']
# Returns:
#    object of class prefM (preference matrix)
mcda_electre1 <- function(M,  w, Lambda, r, index='C1') {
  # get PI.plus, PI.min and PI.indif
  mcda_electre(M,w)

  # initializations
  CM <- matrix(data=0, nrow=nrow(M), ncol=nrow(M))
  PM <- matrix(data=0, nrow=nrow(M), ncol=nrow(M))
  colnames(PM) <- rownames(PM) <- rownames(M)

  # calcualte the preference matrix
  if (index == 'C1') {
    # for similarity index C1
    for (i in 1:nrow(M)){
      for (j in 1:nrow(M)) {
        CM[i,j] <- (PI.plus[i,j] + PI.indif[i,j]) / (PI.plus[i,j] +
                    PI.indif[i,j] + PI.min[i,j])
        if((CM[i,j] > Lambda) && ((M[j,] - M[i,]) <= r) &&
          (PI.plus[i,j] > PI.min[i,j])) PM[i,j] = 1
      }
    }
  } else {
    # for similarity index C2
    for (i in 1:nrow(M)){
      for (j in 1:nrow(M)) {
        if (PI.min[i,j] != 0)
        {CM[i,j] <- (PI.plus[i,j]) / (PI.min[i,j])}
        else
        {CM[i,j] = 1000 * PI.plus[i,j]} # to avoid dividing by 0
        if((CM[i,j] > Lambda) && ((M[j,] - M[i,]) <= r) &&
           (PI.plus[i,j] > PI.min[i,j])) {PM[i,j] = 1}
      }
    }
  }
  for (i in 1:nrow(PM)) PM[i,i] = 0
  class(PM) <- 'prefM'
  PM
}
```

Digression – Passing on matrices as a list

As mentioned before, we choose to push the matrices `PI.plus`, `PI.min`, and `PI.indif` into the environment of the encapsulating function with the `<<-` operator. Alternatively, we could pass them on in a list of matrices. To do this, we would have to add one last line in the function `mcda_electre()`:

```
list(PI.plus = PI.plus, PI.min = PI.min, PI.indif = PI.indif)
```

Then in the next function, `mcda_electre1()`, we should address the `PI.*` not directly like `Pi.x[i, j]`, but rather as follows:

```
# If we did not push the values PI.plus, PI.min, and
# PI.indif into the environment of this functions, we would write:
X <- mcda_electre(M,w)
# and then address X as in the following code as follows:
X$PI.min[i,j]
```

The function `mcda_electre1()` is now ready for use. We need to provide the decision matrix, weights and the cut-off value and a vector for maximum inverse preferences. The code below does this, prints the preference relations a matrix and finally plots them with our custom method `plot.prefM()` in Figure 27.3 on page 536.

```
# the critia: "tlnt" "stab" "cost" "infl" "trvl" "infr" "life"
w <- c(          0.125, 0.2,   0.2,   0.2,  0.175,  0.05,  0.05)
w <- w / sum(w)  # the sum was 1 already, but just to be sure.
r  <- c(0.3,      0.5,  0.5,    0.5,   1,      0.9,   0.5)

eM <- mcda_electre1(M1, w, Lambda=0.6, r=r)
print(eM)
##     BLR BOM DEL MNL DUB KRK EZE
## BLR   0   1   1   1   0   0   1
## BOM   0   0   0   0   0   0   1
## DEL   0   1   0   0   0   0   1
## MNL   0   1   1   0   0   0   1
## DUB   0   0   0   0   0   0   1
## KRK   0   0   0   0   1   0   1
## EZE   0   0   0   0   0   0   0
## attr(,"class")
## [1] "prefM"

plot(eM)
```

Each arrow in Figure 27.3 on page 536 means "... is preferred over ...." We also notice that the preference relationship in ELECTRE I is transitive, so we might leave out arrows that span over another alternative. Some fiddling would allow us to get rid of those in the preference matrix and we could use argument `pos` for the function `plotmat` in order to fine tune how the cities have to be plotted.

Our plotting functions shows each preference relation. However, in ELECTRE these preference relations are transitive. That means if KRK is preferred over DUB, and DUB is preferred over EZE, than necessarily, KRK is also preferred over EZE. That means that we could reduce the number of arrows. That idea is presented in Figure 27.4 on page 536

Note – Transitiveness of preference

In general preference is transitive for ELECTRE. However, it is not necessarily a universal fact for a preference relationship.

This presentation makes very clear that the ELECTRE method will not always be able to rank alternatives.

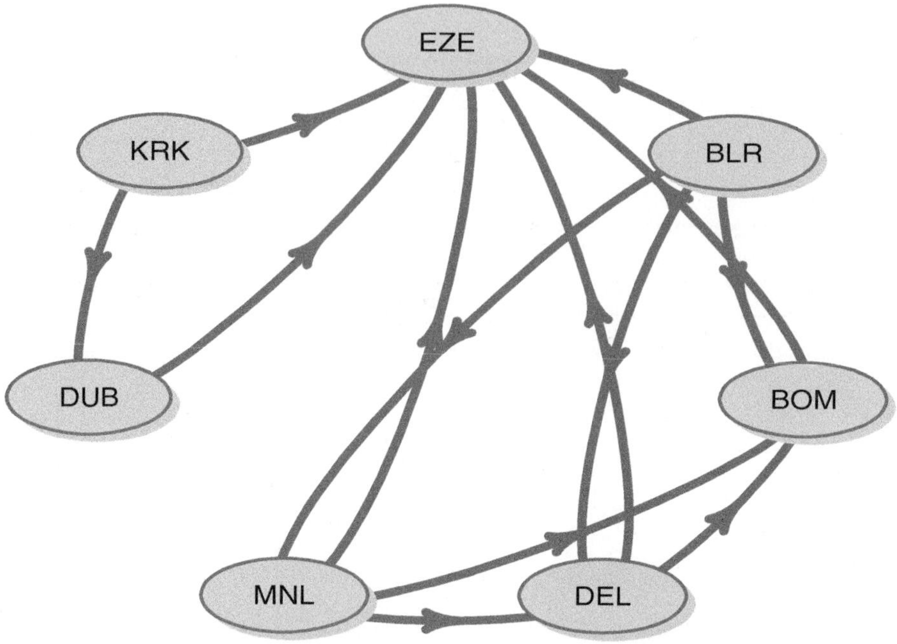

Figure 27.3: *The preference structure as found by the ELECTRE I method given all parameters in the code.*

Besides the plot that we obtain automatically via our functon `plot.prefM()`, it is also possible to create a plot that uses the transitivity to make the image lighter and easier to read. This is presented in Figure 27.4.

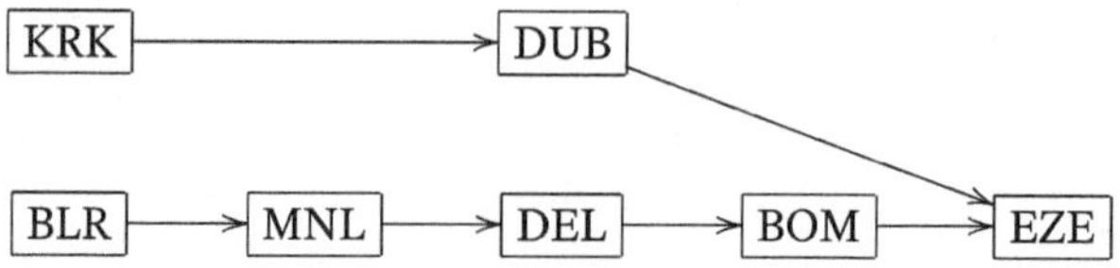

Figure 27.4: *Another representation of Figure 27.3. It is clear that Krakow and Bangalore are quite different places for a SSC. Therefore they are not ranked between each other and choosing between them means making compromises.*

Warning – Handle results with care

All the findings from a method such as ELECTRE I (or II) should be interpreted with the utmost care! All conclusions are made based on many parameters, sometimes moving them a little will yield a slightly different picture. So it is worth to spend a while tweaking so that (a) most are ranked but (b) too dissimilar alternatives are not ranked relative to each other. This situation allows us to better understand what is going on in this particular multi criteria problem.

We can try the same with the $C2$ comparability index. This is done in the following code, and the results are visualised in the last line – output in Figure 27.5 on page 537

```
# the critia: "tlnt" "stab" "cost" "infl" "trvl" "infr" "life"
w <- c(         0.125, 0.2,   0.2,   0.2,  0.175,  0.05,  0.05)
w <- w / sum(w)  # the sum was 1 already, but just to be sure.
r  <- c(0.3,      0.5,  0.5,   0.5,    1,      0.9,    0.5)

eM <- mcda_electre1(M1, w, Lambda=1.25, r=r, index='C2')
plot(eM)
```

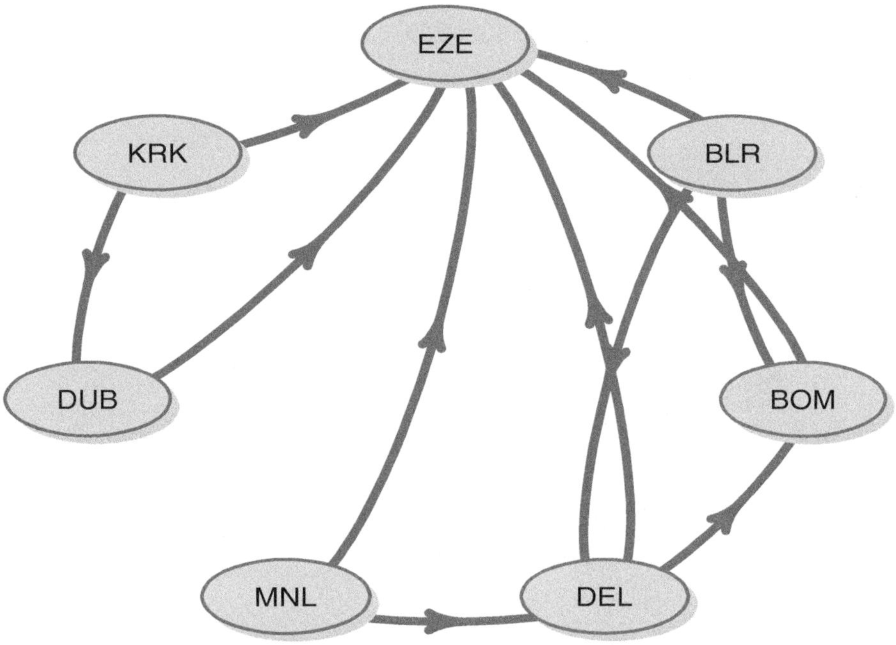

Figure 27.5: *The results of ELECTRE I with comparability index C_2 and parameters as in the aforementioned code.*

Besides the plot that we obtain automatically via our functon `plot.prefM()`, it is also possible to create a plot that uses the transitivity to make the image lighter and easier to read. This is presented in Figure 27.6.

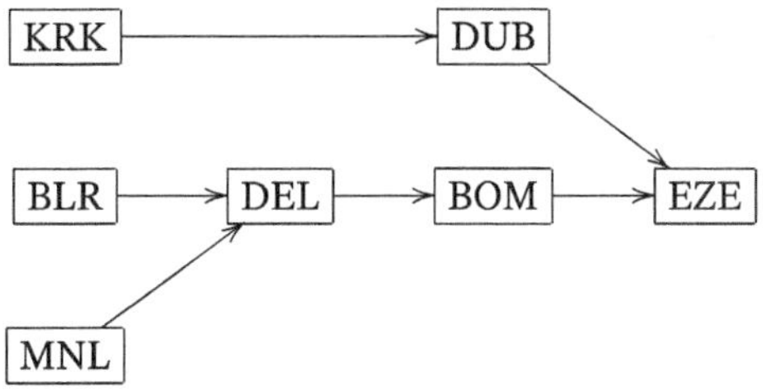

Figure 27.6: *The results for ELECTRE I with comparability index $C2$. The $A \rightarrow B$ means "A is better than B." In this approach and with this visualisation, we find three top-locations: Krakow, Bangalore and Manilla.*

Conclusion for ELECTRE I

The figures and calculations above seem to indicate that given this set of alternatives and our choice of criteria and weights thereof that

1. Krakow and Bangalore are good choices, but for very different reasons;

2. Buenos Aires is not a good choice;
3. The Asian cities and the European cities form two distinct group so that within the group a ranking is possible but it is harder to rank them over groups.

27.7.4.2 ELECTRE II

The strength of ELECTRE I is that it does not force a ranking onto alternatives that are too different. But that is of course its weakness too. In a boardroom it is easier to cope with a simple ranking than with a more complex structure.

Hence, the idea of ELECTRE II was born to force a complete ranking by

- gradually lower the cut-off level Λ_1 and
- increasing the cut-off level for opposite differences in some criteria r_j.

In our example r needs to be equal to the unit vector and Λ can be zero in order to obtain a full ranking. The code below uses these values and plots the preference relations in Figure 27.7 on page 539.

```
# The critia: "tlnt" "stab" "cost" "infl" "trvl" "infr" "life"
w <- c(        0.125, 0.2,    0.2,    0.2,  0.175,  0.05,   0.05)
w <- w / sum(w)  # the sum was 1 already, but just to be sure.
r  <- c(1,      1,  1,     1,     1,      1,    1)

eM <- mcda_electre1(M1, w, Lambda = 0.0, r = r)
print(eM)
##      BLR BOM DEL MNL DUB KRK EZE
## BLR    0   1   1   1   0   0   1
## BOM    0   0   0   0   0   0   1
## DEL    0   1   0   0   0   0   1
## MNL    0   1   1   0   0   0   1
## DUB    1   1   1   1   0   0   1
## KRK    1   1   1   1   1   0   1
## EZE    0   0   0   0   0   0   0
## attr(,"class")
## [1] "prefM"

plot(eM)
```

Note – Alternative

We could define a function `mcda_electre2()` as follows:

```
mcda_electre2 <- function (M1, w) {
  r <- rep(1L, ncol(M))
  mcda_electre1(M1, w, Lambda=0.0, r=r)
  }
```

Note also that in a production environment we should build in further tests (e.g. the number of rows of $M1$ should correspond to the length of w).

The plot shows now an arrow between each of the alternative solutions, so all alternatives are comparable. Since the preference relationship is transitive, we can summarise this complex plot as in Figure 27.8.

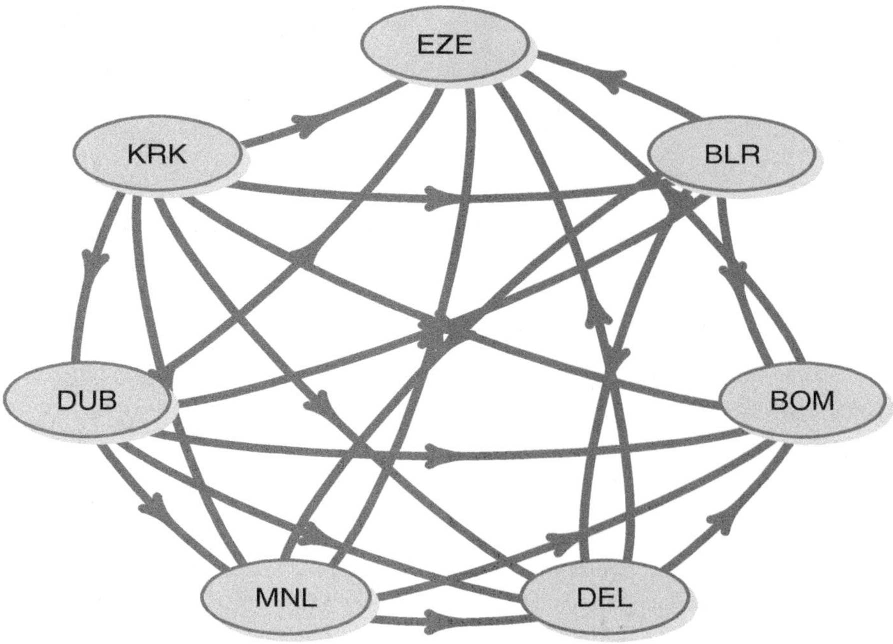

Figure 27.7: *The preference structure as found by the ELECTRE II method given all parameters in the code.*

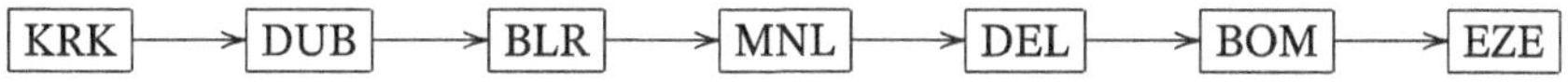

Figure 27.8: *The results for ELECTRE I with comparability index $C2$.*

Note – ELECTRE II

It is not necessary to put r equal to the unit vector and Λ at zero. We could program a goal-seek algorithm using the fact that the associated preference matrix of a full ranking is an upper triangle matrix of all ones (though the rows might not be in order). So, the one will find that in this matrix satisfies the following equation.

```
sum(rowSums(prefM)) == A * A - A
# with prefM the preference matrix and
#       A the number of alternatives.
```

27.7.4.3 Conclusions ELECTRE

The ELECTRE method provides a good balance between allowing full compensation of weaknesses and not allowing any compensation at all. The ELECTRE I provides great insight, and the ELECTRE II usually gets a full ranking. We see the following advantages and disadvantages:
Advantages

- No need to add different variables in different units

- All that is needed is a conversion to "preference" and add this preference
- Richer information than the Weighted Sum Method
- The level of compensation can be controlled

Disadvantages

- There is still an "abstract" concept "preference," which has little meaning and no pure interpretation
- To make matters worse, there are also the cut-off levels
- So to some extend it is still so that concepts that are expressed in different units are compared in a naive way.

27.7.5 PROMethEE

PROMethEE PROMethEE is short for "Preference Ranking Organization METHod for Enrichment of Evaluations".[8] In some way, it is a generalisation of ELECTRE. In ELECTRE the preference relation was binary (0 or 1). So, the obvious generalisation is allowing a preference as a number between 0 and 1. This number can then be calculated in a variety of different ways: that is what PROMethEE does.

The Idea of PROMethEE

- Enrich the preference structure of the ELECTRE method.
- In the ELECTRE Method one prefers essentially a solution a over b for criterion k if and only if $f_k(a) > f_k(b)$.
- This 0-or-1-relation (black or white) can be replaced by a more gradual solution with different shades of grey.
- This preference function will be called $\pi_k(a, b)$ and it can be different for each criterion.

The idea is that the preference for alternative a_i and a_j can be expressed in function of the weighted sum of differences of their scores m_{ik} in the decision matrix.

$$\pi(a_i, a_j) = \sum_{k=1}^{K} P_k(m_{ik} - m_{jk}) w_k \quad (27.1)$$

$$= \sum_{k=1}^{K} P_k\left(d_k(a_i, a_j)\right) w_k \quad (27.2)$$

In which we used the following "distance definition":

Definition: Distance $d_k(a, b)$

$$d_k(a, b) = f_k(a) - f_k(b)$$

[8] PROMethEE is accompanied by a complementary method for "geometrical analysis for interactive aid" – better known as the Gaia – which we will discuss in next section.

The key assumption is that it is überhaupt possible to define a reasonable preference function for each criterion k and that these preferences are additional.

We notice that $d_k(a, b) \in [1, +1]$ and hence for each criterion we need to find a function that over the domain $[1, +1]$ yields a preference that we can also coerce in $[0, +1]$ without loss of generality. Note that these preferences should indeed not become negative, so they cannot compensate.

The preference functions $\pi(a, b)$ is typically considered to be zero between on $[-1, 0]$ (so on the part where b has a better score than a). On the other part it will increase to a maximum of 1 in function of $m_{ak} - m_{bk}$.

Some possible smooth scaling functions are in Figure 27.9. For the use of PROMethEE, we should avoid the preference function to become negative (otherwise we would allow full compensation of weaknesses). So, in practice and for PROMethEE we will use only preference functions that are positive on one side. Some example of that type of functions are in Figure 27.10 on page 542.

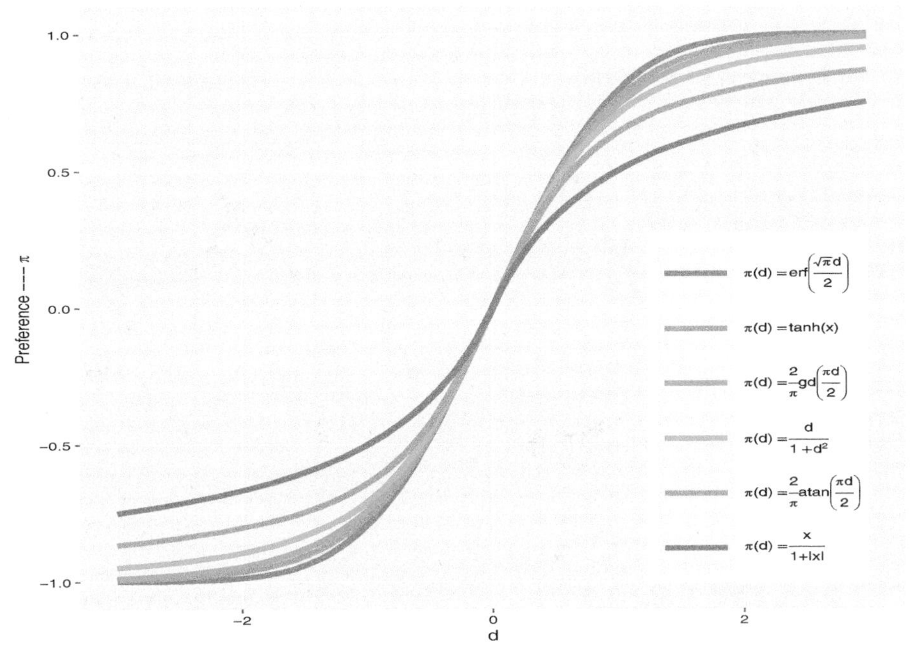

Figure 27.9: *Examples of smooth transition schemes for preference functions $\pi(d)$. The "d" is to be understood as the difference in score for a given criterion.*

Further information – code for plots

The code for the last plots is not included here, but you can find it in the appendix: Chapter D *"Code Not Shown in the Body of the Book"* on page 841. It includes some useful aspects such as fiddling with legends in `ggplot2`, using LaTeX markup in plots, and vectorizing functions.

Figure 27.10 on page 542 shows a few possibilities for the preference functions, but we can of course invent much more. Further, we note – similar to WSM – that not all criteria are

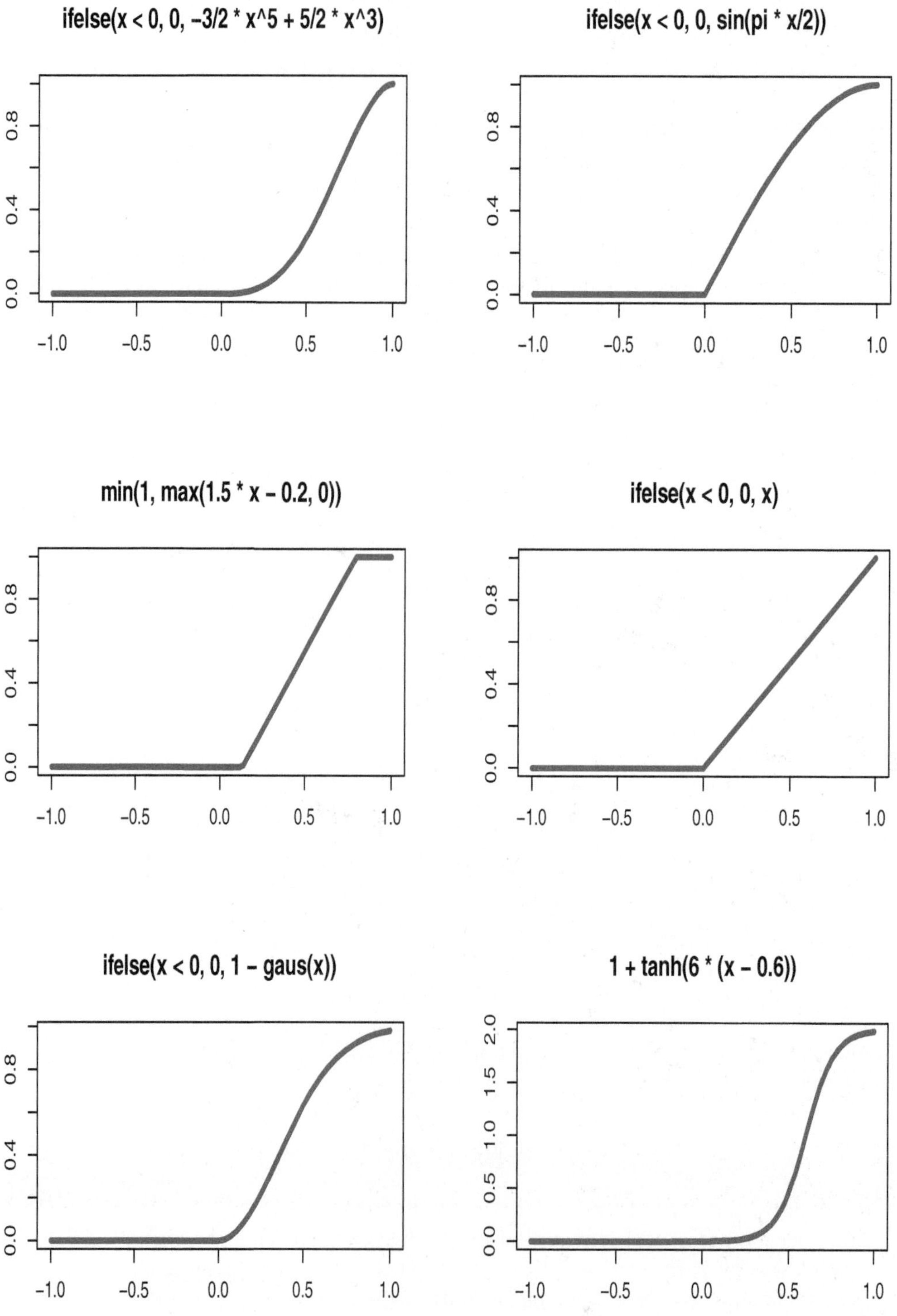

Figure 27.10: *Examples of practically applicable preferences functions $P(d)$. The "x" is to be understood as the difference in score for a given criterion. Note that all scaling factors are optimized for a normalized decision matrix. The function* `gaus()` *refers to $exp(-(x-0)^2/0.5)$.*

equally important in the decision matrix and hence need to be weighted. So for each criterion ($k \in (1 \ldots K)$ we could multiply the preference function $\pi()$ with a weight w_k

Note further that the preference function for each criterion j can be different, so in general we can define the degree of preference as follows.

It is important to understand that there is no exact science for choosing a preference function, nor is there any limit on what one could consider. Here are some more ideas.

Examples:

- step-function with one step (similar to ELECTRE preferences)
- step-function with more than one step
- step-wise linear function
- $\pi(d) = max(0, min(g \times d, d_0))$ (linear, gearing g)
- sigmoid equation: $\pi(d) = \frac{1}{1-\left(\frac{1}{d_0}-1\right)e^{-dt}}$
- $\pi(d) = tanh(d)$
- $\pi(d) = erf\left(\frac{\sqrt{(\pi)}}{2}d\right)$
- $\pi(d) = \frac{d}{\sqrt{1+x^2}}$
- Gaussian: $\pi(d) = \begin{cases} 0 & \text{for } d < 0 \\ 1 - \exp\left(-\frac{(d-d_0)^2}{2s^2}\right) & \text{for } d \geq 0 \end{cases}$
- ...

27.7.5.1 PROMethEE I

Note – Avoiding compensation

Note that in PROMethEE I the preference $\pi(a_i, a_j)$ does not become negative. Otherwise – when added – we would loose information (i.e. differences in opposite directions are *compensated*).

This preference function allows us to define a flow of how much each alternative is preferred, Φ_i^+, as well as a measure of how much other alternatives are preferred over this one: Φ_i^-. The process is as follows.

1. Define preference functions $\pi : \mathcal{A} \times \mathcal{A} \mapsto [0, 1]$
2. They should only depend on the difference between the scores of each alternative as summarized in the decision matrix m_{ik}:

$$\pi_k(a_i, a_j) = \pi_j\left(m_{ik} - m_{jk}\right) = \pi_j\left(d_k(a_i, a_j)\right)$$

3. Define a preference index: $\Pi(a_i, a_j) = \sum_{k=1}^{K} w_k \pi_k(a_i, a_j)$
4. Then sum all those flows for each solution – alternative – to obtain

(a) a positive flow: $\Phi^+(a_i) = \frac{1}{K-1}\sum_{\mathbf{a_j}\in\mathcal{A}}\Pi(a_i,a_j) = \frac{1}{K-1}\sum_{k=1}^{K}\sum_{j=1}^{A}\pi_k(a_i,a_j)$

(b) a negative flow: $\Phi^-(a_i) = \frac{1}{K-1}\sum_{\mathbf{a_j}\in\mathcal{A}}\Pi(a_j,a_i) = \frac{1}{K-1}\sum_{k=1}^{K}\sum_{j=1}^{A}\pi_k(a_j,a_i)$

(c) a net flow: $\Phi(a_i) = \Phi^+(a_i) - \Phi^-(a_i)$

where the w_k are the weights of the preference for each criteria so that $\sum_{k=1}^{K} w_k = 1$ and $\forall k \in \{1\ldots K\} : w_k > 0$

The Preference Relations

Based on these flows, we can define the preference relations for PROMethEE I as follows:

- $$a \succ b \Leftrightarrow \begin{cases} \Phi^+(a) \geq \Phi^+(b) \wedge \Phi^-(a) < \Phi^-(b) \text{ or} \\ \Phi^+(a) > \Phi^+(b) \wedge \Phi^-(a) \leq \Phi^-(b) \end{cases}$$
- indifferent $\Leftrightarrow \Phi^+(a) = \Phi^+(b) \wedge \Phi^-(a) = \Phi^-(b)$
- in all other cases: no preference relation

PROMethEE I in R

We will first define a base function that calculates the flows Φ and pushes the results a in the environment a level higher (similar to the approach for the ELECTRE method).

```
# mcda_promethee
# delivers the preference flow matrices for the Promethee method
# Arguments:
#    M       -- decision matrix
#    w       -- weights
#    piFUNs -- a list of preference functions,
#              if not provided min(1,max(0,d)) is assumed.
# Returns (as side effect)
# phi_plus <<- rowSums(PI.plus)
# phi_min  <<- rowSums(PI.min)
# phi_     <<- phi_plus - phi_min
#
mcda_promethee <- function(M, w, piFUNs='x')
{
  if (piFUNs == 'x') {
       # create a factory function:
       makeFUN <- function(x) {x; function(x) max(0,x) }
       P <- list()
       for (k in 1:ncol(M)) P[[k]] <- makeFUN(k)
       } # in all other cases we assume a vector of functions
# initializations
PI.plus  <<- matrix(data=0, nrow=nrow(M), ncol=nrow(M))
PI.min   <<- matrix(data=0, nrow=nrow(M), ncol=nrow(M))
# calculate the preference matrix
for (i in 1:nrow(M)){
  for (j in 1:nrow(M)) {
    for (k in 1:ncol(M)) {
      if (M[i,k] > M[j,k]) {
        PI.plus[i,j] = PI.plus[i,j] + w[k] * P[[k]](M[i,k] - M[j,k])
      }
      if (M[j,k] > M[i,k]) {
        PI.min[i,j] = PI.min[i,j] + w[k] * P[[k]](M[j,k] - M[i,k])
```

```
      }
    }
  }
}
# note the <<- which pushes the results to the upwards environment
phi_plus <<- rowSums(PI.plus)
phi_min  <<- rowSums(PI.min)
phi_      <<- phi_plus - phi_min
}
```

Question #22 – Avoiding the Side effect in the function

If you want to avoid using side effects to pass on three matrices, how would you re-write this cde so that the three matrices are returned to the environment that call the function? Before you look up the answer, you might want to get Some inpsiration from previous section about the ELECTRE method: read the "digression box" in that section.

Note – Symmetry in the preference functions

In some literature, the preference functions are defined as being symmetrical for rotation along the y-axis. This can be quite confusing and is best avoided, however, the aforementioned code will work with such functions too.

Now, we can define a function mcda_promethee1() that calls the function mcda_promethee() to define the preference flows.

```
# mcda_promethee1
# Calculates the preference matrix for the Promethee1 method
# Arguments:
#    M      -- decision matrix
#    w      -- weights
#    piFUNs -- a list of preference functions,
#              if not provided min(1,max(0,d)) is assumed.
# Returns:
#    prefM object -- the preference matrix
#
mcda_promethee1 <- function(M, w, piFUNs='x') {
  # mcda_promethee adds phi_min, phi_plus & phi_ to this environment:
  mcda_promethee(M, w, piFUNs='x')

  # Now, calculate the preference relations:
  pref     <- matrix(data=0, nrow=nrow(M), ncol=nrow(M))
    for (i in 1:nrow(M)){
      for (j in 1:nrow(M)) {
        if (phi_plus[i] == phi_plus[j] && phi_min[i]==phi_min[j]) {
          pref[i,j] <- 0
          }
        else if ((phi_plus[i] > phi_plus[j] &&
                  phi_min[i] < phi_min[j] ) ||
                (phi_plus[i] >= phi_plus[j] &&
                  phi_min[i] < phi_min[j] )) {
          pref[i,j] <- 1
          }
        else {
```

```
            pref[i,j] = NA
          }
        }
      }
  rownames(pref) <- colnames(pref) <- rownames(M)
  class(pref)    <- 'prefM'
  pref
}
```

All that is left, now is to execute the function that we have created in previous code segment.

```
# We reuse the decision matrix M1 and weights w as defined above.
m <- mcda_promethee1(M1, w)
```

The object m is now the preference matrix of class prefM, and we can plot it as usual – result in Figure 27.11

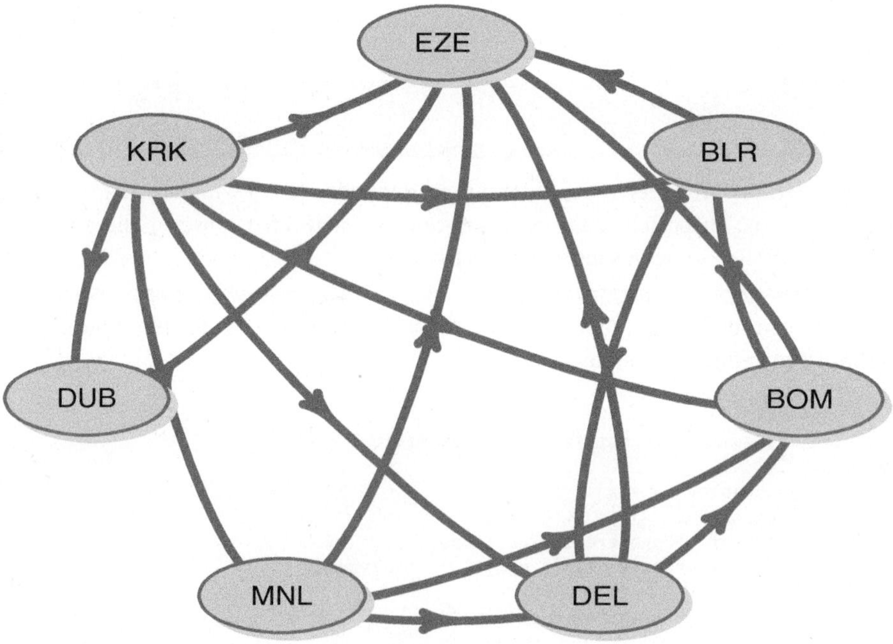

Figure 27.11: *The hierarchy between alternatives as found by PROMethEE I.*

```
# We reuse the decision matrix M1 and weights w as defined above.
m <- mcda_promethee1(M1, w)
plot(m)
```

Again, it is possible to simplify the scheme of Figure 27.11 by leaving out the spurious arrows. This scheme is in Figure 27.12 on page 547.

The function that we have created can also take a list of preference functions via its piFUNs argument. Below, we illustrate how this can work and we plot the results in Figure 27.13 on page 547.

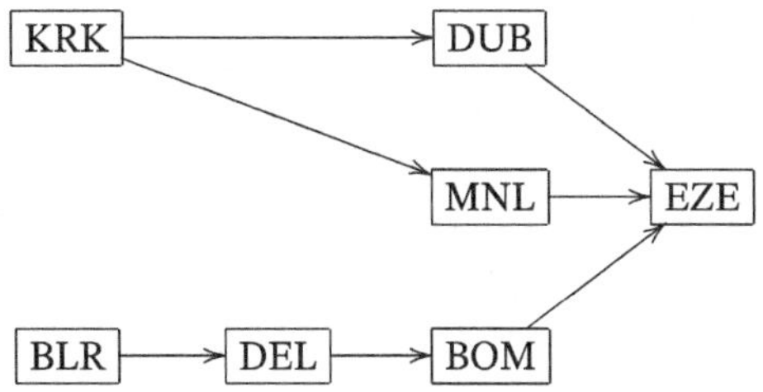

Figure 27.12: *The preference relations resulting from PROMethEE I. For example, this shows that the least suitable city would be Buenos Aires (EZE). It also shows that both Krakow (KRK) and Bangalore (BLR) would be good options, but PROMethEE I is unable to tell us which of both is best, they cannot be ranked based on this method.*

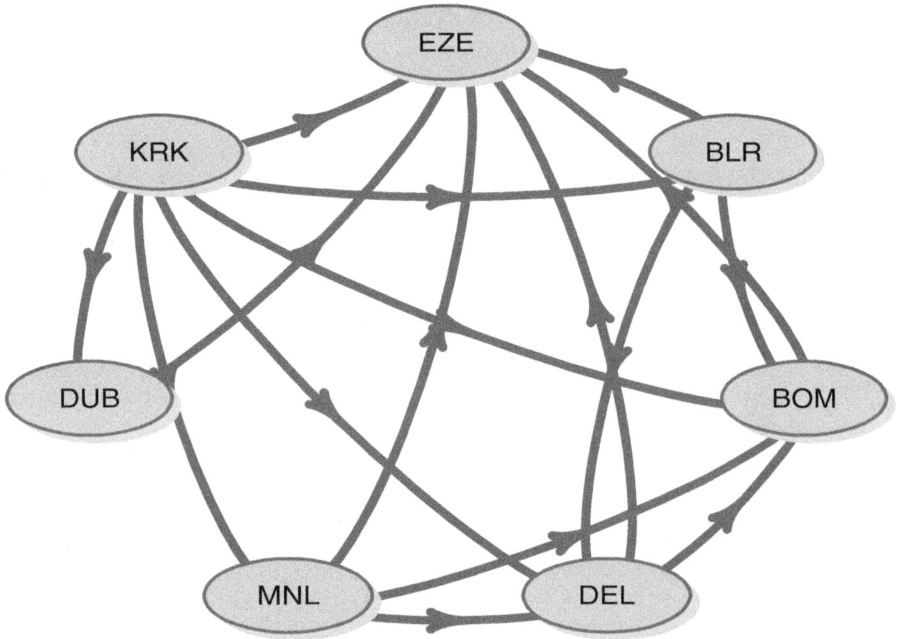

Figure 27.13: *The result for PROMethEE I with different preference functions provided.*

Note that besides the plot that we obtain automatically via our function `plot.prefM()`, it is also possible to create a plot that uses the transitivity to make the image lighter and easier to read. This is presented in Figure 27.14.

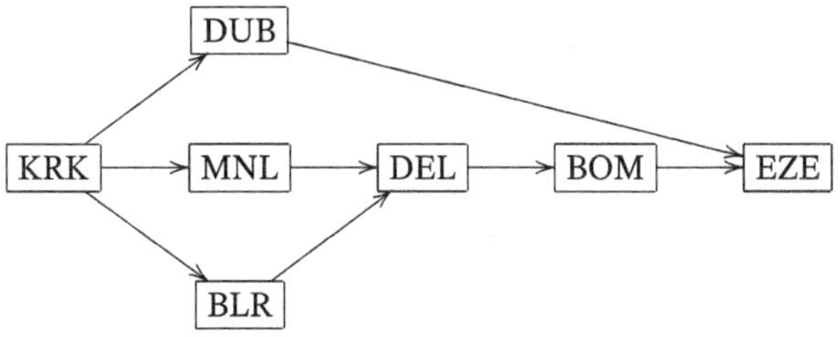

Figure 27.14: *The results for PROMethEE I method with the custom preference functions.*

```
# Make shortcuts for some of the functions that we will use:
gauss_val <- function(d) 1 - exp(-(d - 0.1)^2 / (2 * 0.5^2))
x         <- function(d) max(0,d)
minmax    <- function(d) min(1, max(0,2*(d-0.5)))
step      <- function(d) ifelse(d > 0.5, 1,0)

# Create a list of 7 functions (one per criterion):
f <- list()
f[[1]] <- gauss_val
f[[2]] <- x
f[[3]] <- x
f[[4]] <- gauss_val
f[[5]] <- step
f[[6]] <- x
f[[7]] <- minmax

# Use the functions in mcda_promethee1:
m <- mcda_promethee1(M1, w, f)

# Plot the results:
plot(m)
```

Interestingly, the functions that we have provided, do change the preference structure as found by PROMethEE I, even the main conclusions differ. The main changes are that KRK became comparable to BLR and MNL to DEL.

Advantages and Disadvantages of PROMethEE I

PROMethEE I does not yield a complete ranking (it does not define a total order relation), but along the way we gathered good insight in the problem. Below we list some of the most important advantages and disadvantages of this method.

Advantages:

- It is easier and makes more sense to define a preference function than the parameters Λ_j and **r** in ELECTRE.
- It seems to be stable for addition and deletion of alternatives (the ELECTRE and WPM have been proven inconsistent here).
- No comparison of variables in different units.
- The preference is based on rich information.

Disadvantages:

- Does not readily give too much insight in why a solution is preferred.
- Needs more explanation about how it works than the WSM.
- Some decision makers might not have heard about it.
- There are a lot of arbitrary choices to be made, and those choices can influence the result.

27.7.5.2 PROMethEE II

The idea of PROMethEE II is to allow "full compensation" between positive and negative flows (similar to ELECTRE II) and hence reduces the preference flow calculations to just one flow:

$$\Phi(a,b) = \sum_{j=1}^{k} \pi_j(f_j(a), f_j(b))$$

where here the $\pi_j(a,b)$ can be negative and are symmetrical for mirroring around the axis ($y = -x$). They can also be considered as the concatenation of Φ^+ and Φ^- as follows

$$\Phi = \max(\Phi^+, \Phi^-).$$

The result of this adjusted mechanism will be that we loose some of the insight[9], but gain a total ordering of the alternatives.

We can condense this information further for each alternative:

$$\Phi(a) = \sum_{x \in \mathcal{A}} \sum_{j=1}^{k} \pi_j(f_j(a), f_j(x))$$
$$= \sum_{x \in \mathcal{A}} \pi(a,x)$$

This results in a preference relation that will almost in all cases show a difference (in a small number of cases there is indifference, but all are comparable – there is no "no preference")

- $a \succ b \Leftrightarrow \Phi(a) > \Phi(b)$
- indifferent if $\Phi(a) = \Phi(b)$
- in all other cases: no preference relation

A key element for PROMethEE is that the preference function can be defined by the user. In Figure 27.15 on page 551 we present some functions that can be used to derive a preference from a distance between values of a criterion. While it might be hard to prove the need for other functions, your imagination is the real limit.

We can reuse the function `mcda_promethee()` – that provides the preference flows in the environment that calls it – and build upon this the function `mcda_promethee2()`.

```
# mcda_promethee2
# Calculates the Promethee2 preference matrix
# Arguments:
#    M      -- decision matrix
#    w      -- weights
#    piFUNs -- a list of preference functions,
#              if not provided min(1,max(0,d)) is assumed.
# Returns:
#    prefM object -- the preference matrix
#
mcda_promethee2 <- function(M, w, piFUNs='x')
```

[9] If for example city A is much better on three criteria than city B, but for three other criteria this relation is opposite, we might want to stop and study deeper. PROMethEE II, however will not stop and the small difference on the seventh criterion could determine the ranking.

```
  { # promethee II
   mcda_promethee(M, w, piFUNs='x')
   pref      <- matrix(data=0, nrow=nrow(M), ncol=nrow(M))
     for (i in 1:nrow(M)){
       for (j in 1:nrow(M)) {
           pref[i,j] <- max(phi_[i] - phi_[j],0)
         }
       }
rownames(pref) <- colnames(pref) <- rownames(M)
class(pref) <- 'prefM'
pref
}
```

We can reuse the decision matrix $M1$ with weights w, we can define our own preference functions or use the standard one that is build in $\max(0, d)$.

```
m <- mcda_promethee2(M1, w)
plot(m)
```

The output of this code is the plot in Figure 27.16 on page 552. The preference matrix is now not longer binary and the numbers are the relative flows for each alternative and the preference map will have a preference relation between each data point.

Because of this total relation it is also possible to represent a Promethee II plot as a barchart that represents "preference." The following code does this and the output of this code is in Figure 27.17 on page 552.

```
# We can consider the rowSums as a "score".
rowSums(m)
##         BLR        BOM        DEL        MNL        DUB
##  1.8683911  0.5447135  0.8113801  1.5312562  7.5240937
##         KRK        EZE
## 13.4517776  0.0000000

# So, consider the prefM as a score-matrix (scoreM):
plot.scoreM(m)
```

Advantages and Disadvantages of PROMethEE II

Advantages

- Almost sure to get a full ranking.
- The preference structure is rich and preference quantifiable.
- The preferences are transitive: $a \succ b \wedge b \succ c \Rightarrow a \succ c$.
- No conflicting rankings possible, logically consistent for the decision makers.

Disadvantages

- More condensed information (loss of information, more compensation).
- Might be more challenging to understand for some people.
- A lot of arbitrary functions and parameters relating to preference.

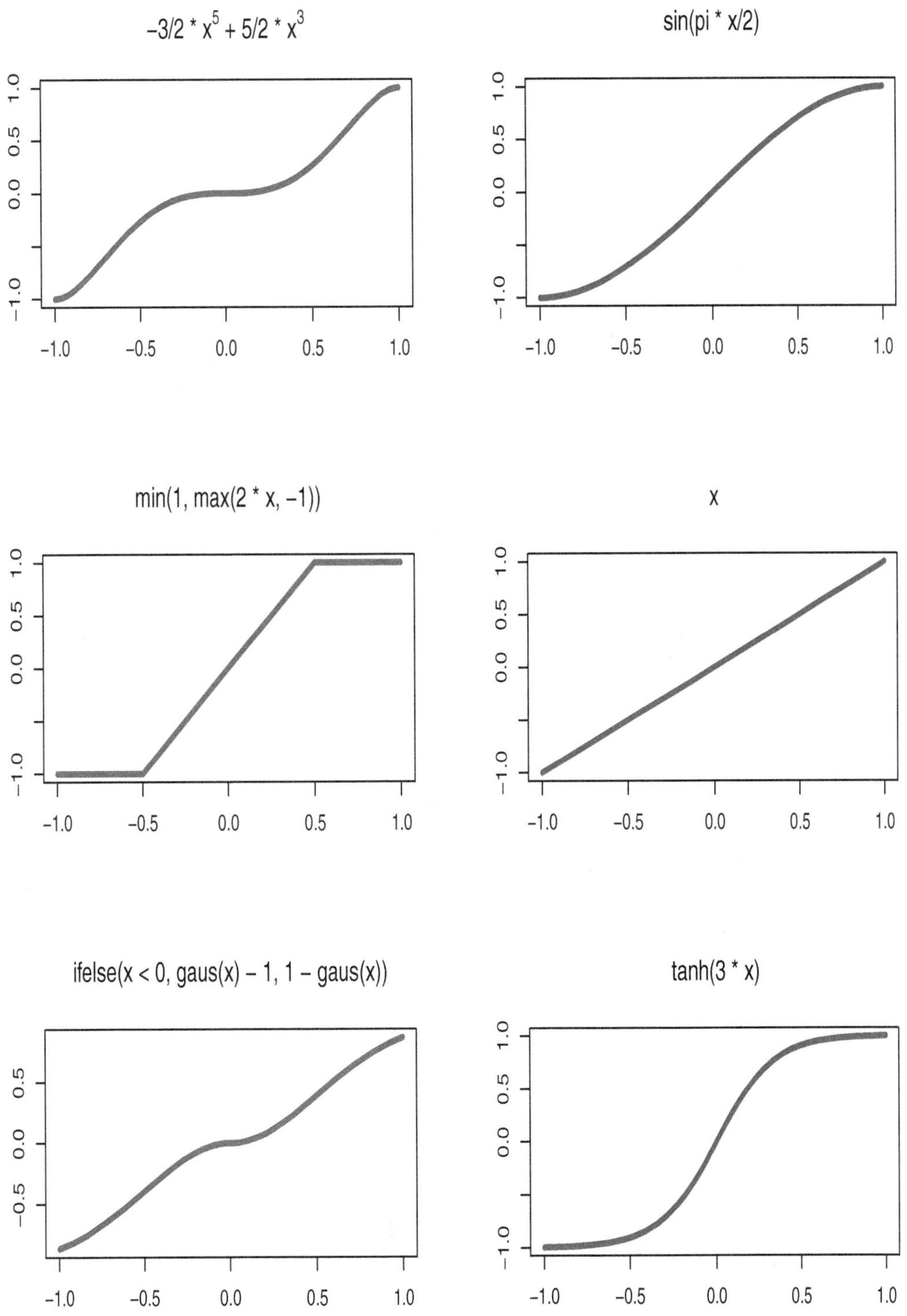

Figure 27.15: *Promethee II can also be seen as using a richer preference structure that can become negative. Here are some examples of practically applicable preferences functions* $P(d)$*. The function* `gaus()` *refers to* $exp(-(x-0)^2/0.5)$.

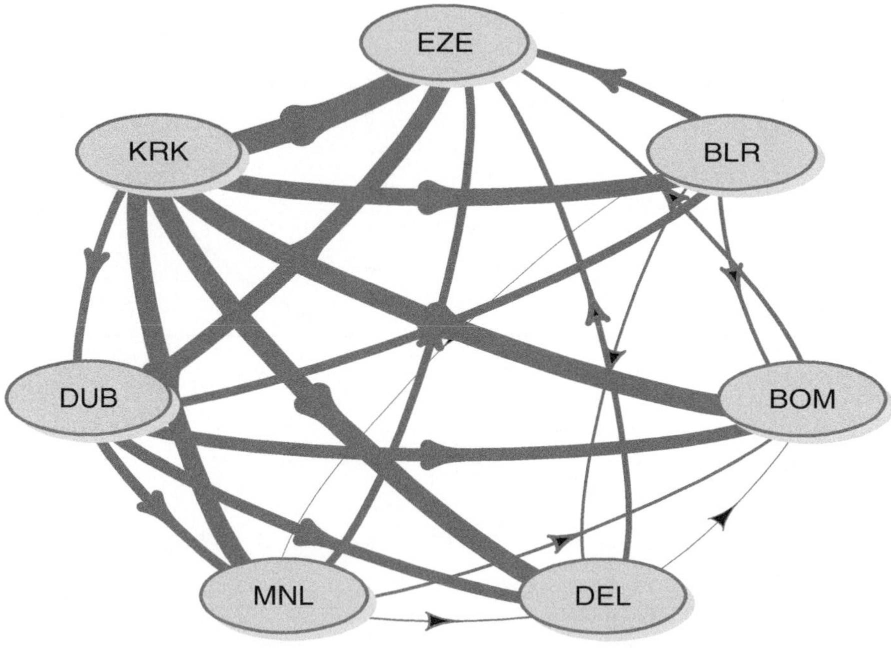

Figure 27.16: *The hierarchy between alternatives as found by PROMethEE II. The thickness of the lines corresponds to the strength of the preference.*

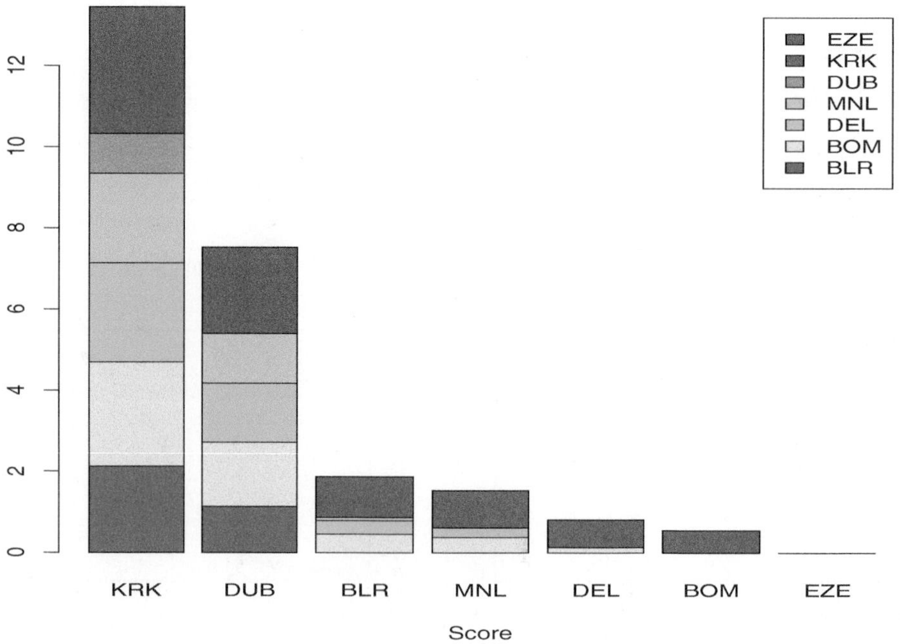

Figure 27.17: *PROMethEE II provides a full ranking. Here we show how much each alternative is preferable over its competitors. The size of the blocks is relative to the amount of preference over the other alternative.*

27.7.6 PCA (Gaia)

Even with the best MCDA we will need to make trade-offs: no solution is perfect. This means that whichever alternative is chosen that this will not be the best on all criteria – if it is, then the problem is solved with the function `del_dominated()`, because that function would then eliminate all alternatives but one.

gaia

Therefore, it is useful to do an effort to understand what we are giving up and what are the main types of alternatives. This analysis is closely related to clustering. In fact, it appears that the first two principal components will already be very insightful because orthogonal projection in the plane (PC_1, PC_2) provides the view with maximized differences between alternatives.

In the context of MCDA, this projection in the (PC_1, PC_2) plane is also referred to as method for "geometrical analysis for interactive aid" (Gaia). It is, however, nothing more than one part of a principal component analysis (PCA).

PCA

Principal component analysis is part of the package `stats` and hence is available by default. We have already demonstrated how to use PCA in R in Section 20 *"Factoring Analysis and Principle Components"* on page 363, here we only repeat the basics. In the following code, we calculate the principle components (PCs), plot the variance explained per principle component in Figure 27.18 and the biplot (projection in the in (PC_1, PC_2) plane) in Figure 27.19 on page 554.

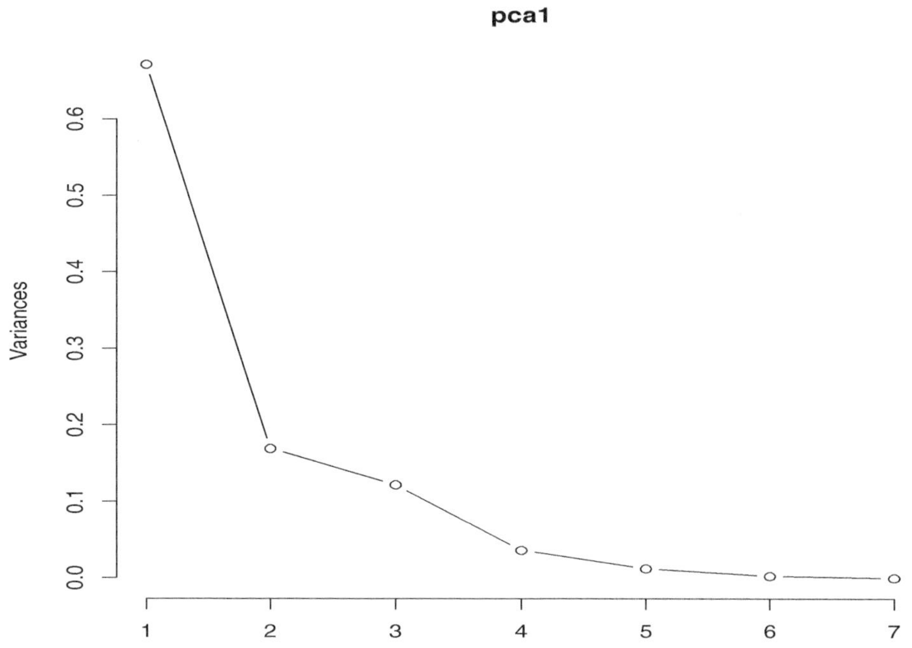

Figure 27.18: *The variance explained by each principal component.*

```
pca1 <- prcomp(M1)
summary(pca1)
## Importance of components:
##                           PC1    PC2    PC3     PC4
## Standard deviation     0.8196 0.4116 0.3492 0.18995
## Proportion of Variance 0.6626 0.1671 0.1203 0.03559
## Cumulative Proportion  0.6626 0.8297 0.9499 0.98555
##                            PC5     PC6       PC7
```

```
## Standard deviation      0.1103 0.04992 4.682e-18
## Proportion of Variance 0.0120 0.00246 0.000e+00
## Cumulative Proportion  0.9975 1.00000 1.000e+00

# plot for the prcomp object shows the variance explained by each PC
plot(pca1, type = 'l')
```

```
# biplot shows a projection in the 2D plane (PC1, PC2)
biplot(pca1)
```

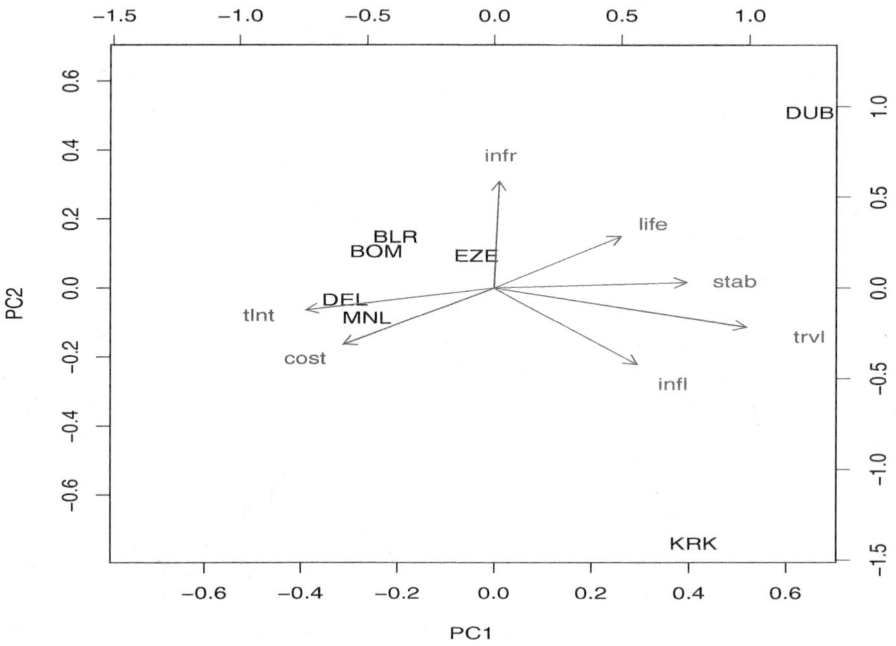

Figure 27.19: *A projection of the space of alternatives in the 2D-plane formed by the two most dominating principal components.*

As mentioned in earlier, also with `ggplot2` and `ggfortify` it is easy to obtain professional results with little effort. The code below does this and shows two versions: first, with the labels coloured according to cost (in Figure 27.20 on page 555), second with the visualisation of two clusters in Figure 27.21 on page 555

```
library(ggplot2)
library(ggfortify)
library(cluster)

# Autoplot with labels colored
autoplot(pca1, data = M1, label = TRUE, shape = FALSE, colour = 'cost', label.size = 6,
         loadings = TRUE, loadings.colour = 'blue',
         loadings.label = TRUE, loadings.label.size = 6
         )
```

```
# Autoplot with visualization of 2 clusters
autoplot(fanny(M1,2), label = TRUE, frame = TRUE, shape = FALSE, label.size = 6,
         loadings = TRUE,  loadings.colour = 'blue',
         loadings.label = TRUE, loadings.label.size = 6)
```

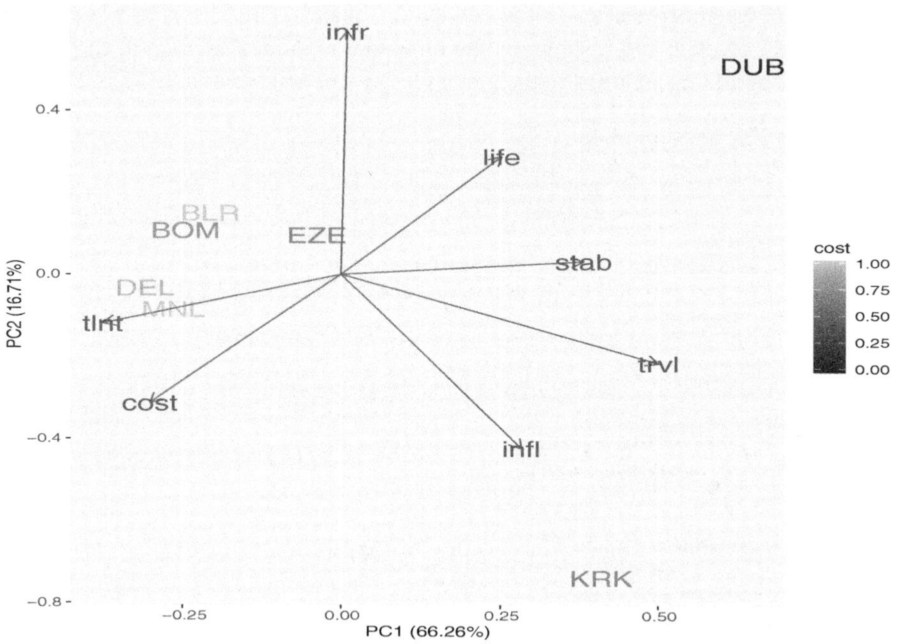

Figure 27.20: *A standard plot with* `autoplot()` *with labels coloured*

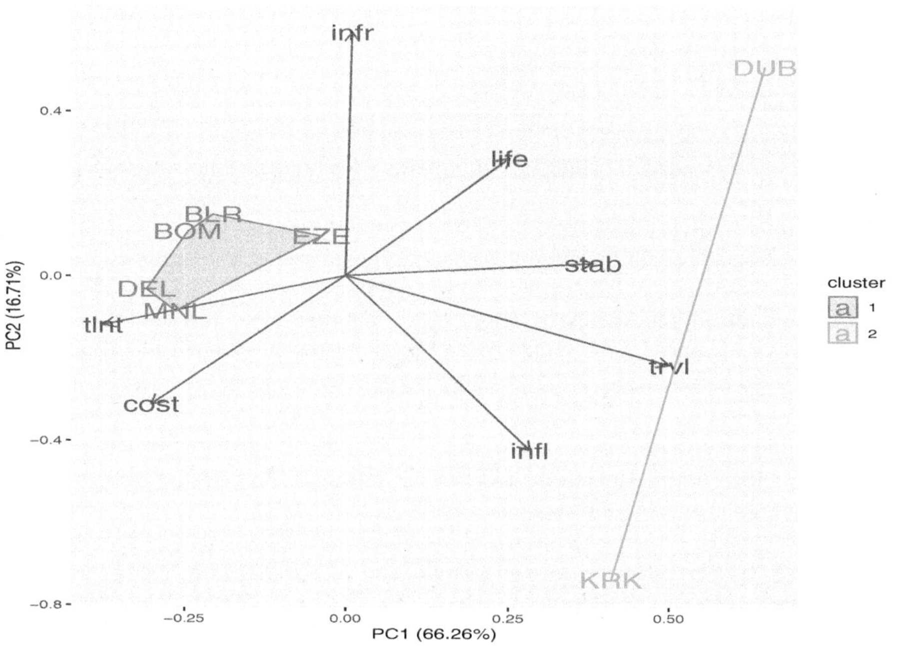

Figure 27.21: *Autoplot with visualization of two clusters*

These visualization show already a lot of information, but we can still add the "decision vector" (the vector of weights projected in the (PC_1, PC_2) plane). This shows us where the main decision weight its located, and it shows us the direction of an ideal soluton in the projection. This can be done by adding an arrow to the plot with the function `annotate()`.

annotate()

```
# Use the weights as defined above:
w
## [1] 0.125 0.200 0.200 0.200 0.175 0.050 0.050

# Calculate coordinates
dv1 <- sum( w * pca1$rotation[,1])  # decision vector PC1 component
dv2 <- sum( w * pca1$rotation[,2])  # decision vector PC2 component

p <- autoplot(pam(M1,2), frame=TRUE, frame.type='norm', label=TRUE,
         shape=FALSE,
         label.colour='blue',label.face='bold', label.size=6,
         loadings=TRUE,  loadings.colour = 'dodgerblue4',
         loadings.label = TRUE, loadings.label.size = 6,
         loadings.label.colour='dodgerblue4',
         loadings.label.vjust = 1.2, loadings.label.hjust = 1.3
         )
p <- p + scale_y_continuous(breaks =
                round(seq(from = -1, to = +1, by = 0.2), 2))
p <- p + scale_x_continuous(breaks =
                round(seq(from = -1, to = +1, by = 0.2), 2))
p <- p + geom_segment(aes(x=0, y=0, xend=dv1, yend=dv2), size = 2,
              arrow = arrow(length = unit(0.5, "cm")))
p <- p + ggplot2::annotate("text", x = dv1+0.2, y = dv2-0.01,
             label = "decision vector",
             colour = "black", fontface =2)
p

## Too few points to calculate an ellipse
```

On plot of Figure 27.22 on page 557 is an orthogonal projection in the (PC_1, PC_2) plane – the plane of the two most important principal components – we find the following information:

1. The name of the alternatives appears centred around the place where they are mapped. The projection coincides with the alternatives being spread out as much as possible.

2. Two clusters are obtained by the function `pam()`: the first cluster has a red ellipsoid around it and the second one generates the error message "Too few points to calculate an ellipse" since there are only two observations in the cluster (KRK and DUB).

3. Each criterion is projected in the same plane. This shows that for example DUB offers great life quality, KRK optimal location and low wage inflation, whereas the group around DEL and MNL have low costs and a big talent pool, etc.

4. A "decision vector," which is the projection of the vector formed by using the weights as coefficients in the base of criteria. This shows the direction of an ideal solution.

When we experiment with the number of clusters and try three clusters, then we see that KRK breaks apart from DUB. Thus we learn that – while both in Europe – Krakow and Dublin are very different places.

This plot shows us how the alternatives are different and what the selection of weights implies. In our example we notice the following.

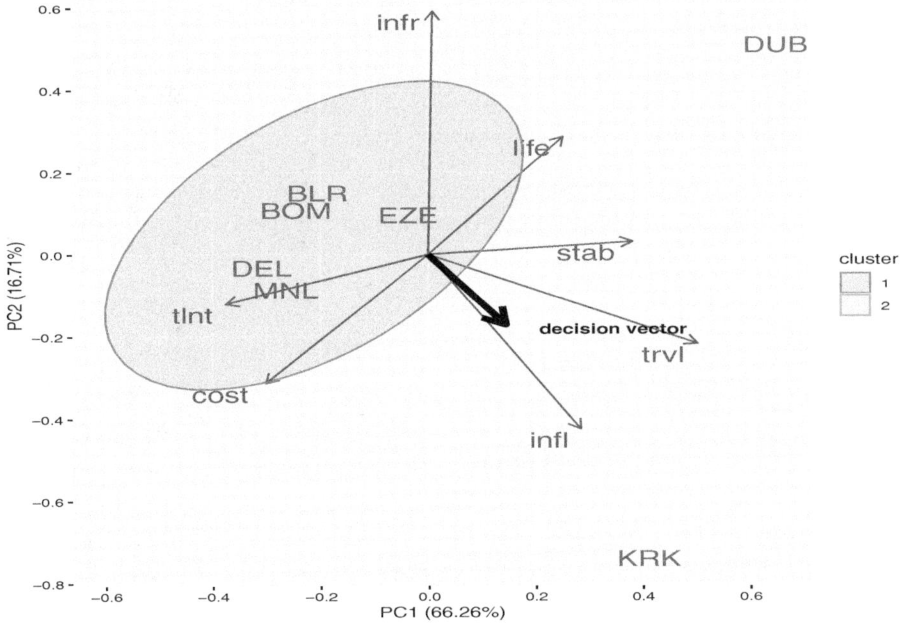

Figure 27.22: *Clustering with elliptoid borders, labels of alternative, projections of the criteria and a "decision vector" – the projection of the weights – constitute a "Gaia-plot."*

- The cities in Asia are clustered together. These cities offer a deep talent pool with hundreds of thousands of already specialized people and are – still – cheap locations: these locations are ideal for large operations where cost is multiplied.
- Dublin offers best life quality and a stable environment. The fact that it has great infrastructure is not so clear in this plot and also note that we left out factors such as "digital enabled" for which again Dublin scores great. Ireland has also as stable low-tax regime. However, we notice that it is opposite to the dimensions "tlnt" and "cost": it is a location with high costs and a really small talent pool. This means that it would be the ideal location for a head-quarter.
- Krakow is – just as Dublin – a class apart. Poland has a stable political environment thanks to the European Union, is close to R-bank's headquarters and further offers reasonable costs and best-in-class wage inflation. However, we note that it sits (almost) opposite to the dimension infrastructure. Krakow is indeed the ideal location for a medium sized operation, where specialization is more important than a talent pool of millions of people. It is also the ideal place for long-term plans (it has low wage inflation and a stable political situation), but still has to invest in its infrastructure. A reality check learns us that this is happening, and hence it would be a safe solution to recommend.

27.7.7 Outranking Methods

The idea of outranking methods is to prefer a solution that does better on more criteria. We can think of the following mechanisms:

- *Direct Ranking:* A solution a is preferred over b if a does better on more criteria than b

- *Inverse Ranking:* A solution a is preferred over b if there are more alternatives that do better than b than there are alternatives that do better than a
- *Median/Average Ranking:* Use the median/average of both previous
- *Weighted Ranking:* Use one of the previous in combination with weights w_j

Actually, outranking methods can be seen as a special case of ELECTRE II – with a step-function as preference function. So, all the code can be reused, though it is also simple to program it directly as the following code fragment shows.

```
### Outrank
# M is the decision matrix (formulated for a maximum problem)
# w the weights to be used for each rank
outrank <- function (M, w)
{
  order      <- matrix(data=0, nrow=nrow(M), ncol=nrow(M))
  order.inv  <- matrix(data=0, nrow=nrow(M), ncol=nrow(M))
  order.pref <- matrix(data=0, nrow=nrow(M), ncol=nrow(M))

  for (i in 1:nrow(M)){
    for (j in 1:nrow(M)) {
      for (k in 1:ncol(M)) {
        if (M[i,k] > M[j,k]) { order[i,j] = order[i,j] + w[k] }
        if (M[j,k] > M[i,k]) { order.inv[i,j] = order.inv[i,j] + w[k] }
      }
    }
  }
  for (i in 1:nrow(M)){
    for (j in 1:nrow(M)) {
      if (order[i,j] > order[j,i]){
        order.pref[i,j] = 1
        order.pref[j,i] = 0
      }
      else if (order[i,j] < order[j,i]) {
        order.pref[i,j] = 0
        order.pref[j,i] = 1
      }
      else {
        order.pref[i,j] = 0
        order.pref[j,i] = 0
      }
    }
  }
 class(order.pref) <- 'prefM'
 order.pref
}
```

27.7.8 Goal Programming

Goal programming brings together linear programming and MCDA. It extends linear programming by defining a target M_i for each criterion. These measures can be considered as given goal or target value to be achieved (e.g. cost lower than x). Deviations from that target y_i should hence be minimised. These deviations can be used directly as a vector or collapsed to a weighted sum (dependent on the goal programming variant used).

Goal programming can – besides identifying the best compromise solution (MCDA) – also

- determine the resources required to achieve a desired set of objectives, and
- determine to what degree the goals goals can be attained with the resources that we have.

Its general formulation is:
Replace $\max\{f_1(x), f_2(x), \ldots, f_n(x)\}$ by

$$\min\left\{y_1 + y_2 + \ldots + y_j + \ldots + y_k \,|\mathbf{x} \in \mathcal{A}\right\}$$

$$\text{with} \begin{cases} f_1(x) & +y_1 & & & & = M_1 \\ f_2(x) & & +y_2 & & & = M_2 \\ \ldots & & & & & = \ldots \\ f_j(x) & & & +y_j & & = M_j \\ \ldots & & & & & = \ldots \\ f_k(x) & & & & +y_k & = M_k \end{cases}$$

- Of course, the y_i have to be additive, so have to be expressed in the same units.
- This forces us to convert them first to the same unit: e.g. introduce factors r_j that eliminate the dimensions, and then minimize $\sum_{j=1}^{k} r_j \, y_j$
- This can be solved by a numerical method.

It should be clear that the r_j play the same role as the $f_j(x)$ in the Weighted Sum Method. This means that the main argument against the Weighted Sum Method (adding things that are expressed in different units) remains valid here.[10].

The target unit that is used will typically be "a unit-less number between zero and one" or "points" (marks) ... as it indeed looses all possible interpretation. To challenge the management, it is worth to try in the first place to present "Euro" or "Dollar" as common unit. This forces a strict reference frame.

Another way of formulating this, is using a target point.

- define a target point, **M** (e.g. the best score on all criteria)
- define a "distance" to the target point: $||\mathbf{F} - \mathbf{x}||$, with $\mathbf{F} = (f_1(x), f_2(x), \ldots, f_k(x))'$ (defined as in the Weighted Sum Method, so reducing all variables to the same units). For the distance measure, be inspired by:
  - the Manhattan Norm: $L_1(\mathbf{x}, \mathbf{y}) = \sum_{j=1}^{k} |x_j - y_j|$
  - the Euler Norm: $L_2(\mathbf{x}, \mathbf{y}) = \left(\sum_{j=1}^{k} (x_j - y_j)^2\right)^{\frac{1}{2}}$
  - the general p-Norm: $L_p(\mathbf{x}, \mathbf{y}) = \left(\sum_{j=1}^{k} (x_j - y_j)^p\right)^{\frac{1}{p}}$
  - the Rawls Norm: $L_\infty(\mathbf{x}, \mathbf{y}) = \max_{j=1 \ldots k} |x_j - y_j|$

The problem was introduced in Page 559 as the Manhattan norm, but we can consider other norms too.

[10] Indeed, we decided to convert differences to preference in order to make this mathematicallly correct, but it remains abstract and arbitrary.

Advantages and Disadvantages

Advantages

- Reasonably intuitive.
- Better adapted to problems of "design" (where $\mathcal{A}$ is infinite).

Disadvantages

- One has to add variables in different units, or at least reduce all different variables to unitless variables via an arbitrary preference function.
- The choice of the weights is arbitrary.
- Even more difficult to gain insight.

27.8 Summary MCDA

It is important to remember that MCDA is not a way to calculate which alternative is best, it is rather a way to understand how alternatives differ and what type of compromise might work best. It is also an ideal way to structure a discussion in a board meeting.

Using MCDA to decide on a multi criteria problem is more an art than a science. All those methods have similar shortcomings:

- MCDA methods used for solving multi-dimensional problems (for which different units of measurement are used to describe the alternatives), are not always accurate in single-dimensional problems (e.g. everything is in Dollar value).
- When one alternative is replaced by a worse one, the ranking of the others can change. This is proven for both ELECTRE and WPM. However, WSM and PROMethEE (most probably) are not subjected to this paradox – see e.g.: Triantaphyllou (2000).
- The methods are also sensitive to all parameters and functions needed to calculate results (e.g. weights and preference functions), and as we have shown, they can also impact the result.

All this means that we should treat them with utmost care, and rather use them to gain insight. May we suggest that the method that gains most insight is the one that does not make a ranking at all: PCA (or Gaia) – see Section 20 *"Factoring Analysis and Principle Components"* on page 363 and Section 27.7.6 *"PCA (Gaia)"* on page 553.

Further information – MCDA

- International Society on Multiple Criteria Decision Making: `http://www.mcdmsociety.org`
- the "Multiple Criteria Decision Aid Bibliography" pages of the "Université Paris Dauphine": `http://www.lamsade.dauphine.fr/mcda/biblio`

Digression – Step 6

You might have noticed that in the chapter about MCDA, we discussed all steps in details, expect the last one: "step 6 – recommend a solution". Recommending a solution is part of reporting and communication, which is the subject of Part VII *"Reporting"* on page 685.

PART VI

Introduction to Companies

A company or a business is an entity that will employ capital with the aim of producing profit. This commercial activity necessarily produces a variety of financial flows. Probably the process that is the most intimately related to all those financial flows. Accounting includes the systematic and comprehensive recording of transactions. The concept "accounting" also refers to the process of summarizing, analysing and reporting these transactions to the management, regulators, and tax offices.

Accounting is one of the quintessential functions in any business of any size. The accounting department will also vary in function of the complexity of the business and fiscal environment. States want to ensure a comfortable inflow of money and the accounting process is used to make sure that private enterprise pays for the state organization. While taxation is still a matter that is very much linked to the relevant country, the European Union lead to harmonisation in legislation. In most countries and companies, accountants use a set of rules that is called "generally accepted accounting principles" (GAAP). GAAP is a set of standards that describe how the assets can be valued and summarized in the balance sheet, how problems like outstanding debt have to be treated, etc. It is based on the double-entry accounting, a method which enters each expense or incoming revenue always mirrored in two places on the balance sheet.

♣ 28 ♣

Financial Accounting (FA)

While financial accounting is not a subject that is part of most programs for mathematicians or data scientists, it does help us to understand how companies work. Accounting is a vast field and is the details tend be dependent on the country of residence and operation. In this chapter, we present only a very brief and general introduction to the subject that will be valid in most countries, and that will help us to understand asset classes and their pricing easier.

The Big R-Book: From Data Science to Learning Machines and Big Data, First Edition. Philippe J.S. De Brouwer.

Companion Website: www.wiley.com/go/De Brouwer/The Big R-Book

28.1 The Statements of Accounts

The process of accounting is essentially registering what happens in the company so that we are able to understand how good the company is doing and to some extent how future looks like. This necessarily implies dealing with both flow variables (cash entering or leaving our books for example) and stock variables (e.g. the reserve of cash, the pile of raw materials, etc.). In order to make sense of this continuously changing reality, one usually employs a standardized approach that consists of daily updating the books and taking regular snapshots. These snapshots are called "statements of accounts" and consist of

- income statement,
- profit and loss statement, and
- balance sheet.

28.1.1 Income Statement

Definition: Income Statement

The Income Statement is all cash income minus all cash expenses.

The income statement informs about cash flow and is therefore, backwards looking only and focuses only on a narrow part of the reality: the cash. Its importance lies of course in the cash management.

28.1.2 Net Income: The P&L statement

The profit and loss statement (P & L) does considers all flows of a certain period and allows to determine how profitable a business was over that period. For example an invoice that has been received or sent will generate soon enough a cash flow that is sufficiently certain to include here.

CoGS
Cost of Goods Sold

R&D
research and development (costs)

EBITDA
earniCompany Value and ngs before interest, taxes and depreciation, amortization

EBIT
earnings before interest and taxes

EBT
earnings before taxes

EAT
earnings after taxes

Definition: P & L

Net sales (= revenue = sales)
- Cost of goods sold
= Gross profit
- SG&A expenses (combined costs of operating the company)
- R&D
= EBITDA
- Depreciation and amortization
= EBIT
- Interest expense (cost of borrowing money)
= EBT
- Tax expense
= Net income (EAT)

The P&L includes all income or loss over a certain period. This defines our tax obligations and determines how much dividend we can pay. However, this might include extra ordinary income or loss. For example, if the business is buying and selling cars, then including the one-off sale of the building will not give a true image on how the core business is doing.

The P&L is already a little more forward looking, but still focuses mainly on the flow variables. It clarifies changes that occur over a certain period.

In order to get that image on how profitable the core-business is, we can use the following.

Definition: NOPAT

NOPAT = Net Operating Income After Taxes (this is EAT minus extra ordinary income)

NOPAT

Net Operating Income After Taxes

This NOPAT is essential when we want to understand how good a company is doing on its core business, and that is important to understand how good it will be doing in the future, because extra ordinary profit or loss is unlikely to be repeated. Instead of defining NOPAT by what it does not contain, it can also be written in terms of its constituents.

$$\begin{aligned}NOPAT =&(\text{Net Income} - \text{after-tax Non-operating Gains}\\&+ \text{after-tax Non-operating Losses} + \text{after-tax Interest Expense})\\&\approx \text{Operating Profit}(1 - \text{tax rate})\end{aligned}$$

With: Operating Profit = EBIT - non operating income

28.1.3 Balance Sheet

The P&L shows the profitability of the company, while this is a good indication of how good things are going on short term, it does not say much of how this short term movement will impact the totality of the company. That is where the balance sheet comes in: the balance sheet highlights the stock variables. That stock of all the company owes and owns is called "balance sheet."

Definition: Balance Sheet

Assets	Liabilities and Owner's Equity
Fixed Assets (Non-current Assets)	Shareholders Equity = Captial Stock + Retained Earnings
Current Assets = *Liquid Assets* + *Stock*	Current Liabilities

The balance sheet is a summary of all what the company owns and all what it owes someone else. This is split in two logical parts: assets and liabilities: The assets are all the company owns and consist of:

Detailed breakdown of assets:

- Current Assets
  - Cash and cash equivalents.
  - Accounts receivable.
  - Prepaid expenses for future services that will be used within a year.

- Non-Current assets (Fixed Assets)
  - Property, plant and equipment.
  - Investment property, such as real estate held for investment purposes.
  - Intangible assets.
  - Financial assets (excluding investments accounted for using the equity method, accounts receivables, and cash and cash equivalents), such as notes receivables.
  - Investments accounted for using the equity method.
  - Biological assets, which are living plants or animals. Bearer biological assets are plants or animals which bear agricultural produce for harvest, such as apple trees grown to produce apples and sheep raised to produce wool.

The liabilities are all the positions that reflect an obligation of the company towards a third parrty.

Detailed Breakdown of Liabilities

- Accounts payable
- Provisions for warranties or court decisions (contingent liabilities that are both probable and measurable)
- Financial liabilities (excluding provisions and accounts payables), such as promissory notes and corporate bonds
- Tax liabilities
- Deferred tax liabilities and deferred tax assets
- Unearned revenue for services paid for by customers but not yet provided
- Shares

The balance sheet of a company is the most comprehensive reflection of the company: it shows all the company's possessions and obligations. The P&L shows how the company is using these assets, and the Income Statement shows how this use of assets reflects in cash-flows.

28.2 The Value Chain

The ultimate goal of a company is to create value and pay back that value to its shareholders. In order to create this value the company can use its assets to generate income, and after paying for costs, the profit will accumulate in the value of the company.

Value Creation

Companies are created to generate profit. The investor will put initial assets in the company (in return the investor gets shares), those assets are used to generate income. Along the way the company incurs also costs and hence we need to deduce those before we can see what the profit is. The value of the company becomes then the discounted sum of all potential profits. This is expressed in Figure 28.1.

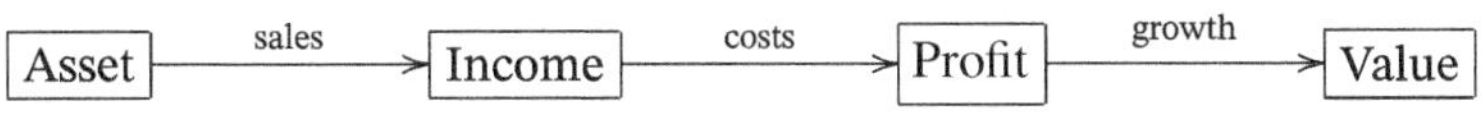

Figure 28.1: *The elements of wealth creation in a company. The company acquires an assets, uses this to generate income, we take into account the costs that were incurred and have the profit. The potential to generate and accumulate profit constitutes the value of a company.*

Note that a company does not need to show profit in order to be valuable. A fast growing company will have to buy more raw materials than it has sales income an hence can show losses–even if each sale in itself is profitable.[1] The value of the company then comes from the "prospect of growing profit" (once the situation stabilizes).

Observation of Value Creation

In order to generate profit, a manager is hired in the company. That person will need to monitor a more detailed value chain in order to create value. The profit will not just happen, there is a conscious effort needed to produce, sell, keep the company safe, etc. This idea is summarised in Figure 28.2.

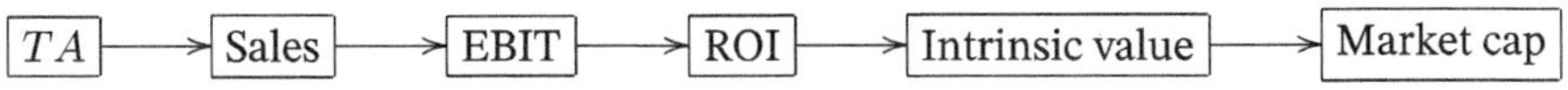

Figure 28.2: *KPIs of the Value Chain that can be used by a manager who wants to increase the value of a company. TA stands for "total assets", "EBIT for "earnings before interest and taxes", and ROI is "return on investments" – we clarify these concepts further in this chapter.*

What the shareholder really should try to obtain is long-term sustainable growth of the share price (and hence the market cap[2]). The idea is that by linking incentives of senior management to sales (annual), EBITDA, growth in share value, etc. that this chain is activated and supported in as many places as possible.

[1] These companies are known as "growth companies".

[2] Market cap is slang for "market capitalisation" and refers to the total amount of stock floated on the stock exchange. We use this concept here as synonym with the total value of the company.

There are a few important points of view. The first is that if one will buy a company then it is essential to have a good idea of it is value. However, once bought the second point of view is managing the value. The owner should make sure that the interest of the management is aligned with the interest of the owners of the company: maximize growth of company value (given the dividend policy).

In the following chapter, we will present management accounting as the tool for the owner to align the interest of the management with that of the owners and as a tool for the management to execute this strategy.

28.3 Further, Terminology

When people interact a lot and have to talk about specific subjects they tend to use specific words more than others, develop specific words or live by acronyms. It is not our aim to list all acronyms or wording used, but we would like to mention a few that will be helpful later on.

Definition: Loans

loans := debt = a sum of money borrowed with the obligation to pay back at pre-agreed terms and conditions.

Note – Wording loans an debt

For the purpose of this section, we will refer to loans (or debt) as the "outstanding amount of debt." So, if for example the initial loan was $10 000 000, but the company already paid back $9 000 000, then we will have only one million of debt.

Definition: Equity

share capital := equity = the value of the shares issued by the company = asset minus cost of liabilities

Example: Equity

A company that has only one asset of $100 000 and a loan against that asset (with outstanding amount of $40 000) has $60 000 equity.

Definition: CapEx

Capital Expenditure (CapEx) is an expense made by a company for which the benefit to the company continues over a long period (multiple accounting cycles), rather than being used and exhausted in a short period (shorter than one accounting cycle). Such expenditure is assumed to be a non-recurring nature and results in acquisition of durable assets.

CapEx

Capital Expenditure

Further information – Capex

CapEx is also referred to as "Capital Expense", both are synonyms.

Capital Expense

In accounting, one will not book CapEx as a cost but rather add them to capital and then depreciate. This allows a regular and durable reduction of taxes. For example, a rail-transport company would depreciate a train over 10 years, because it can typically be used longer.

The counterpart of CapEx is OpEx or recurring expense, which is defined as follows.

OpEx

Operational Expenditure

Definition: OpEx

An Operational Expenditure (OpEx) is an ongoing and/or recurring cost to run a business/
system/product/asset.

OpEx

Operational Expense operational cost

Note – Opex

OpEx is also referred to as "Operational Expense" or "Operational Cost".

In accounting, OpEx is booked as "costs" and will reduce the taxable income of that year (except when local rules force it to be re-added for tax purposes).

For example, the diesel to run a train, its maintenance, salary costs of the driver, oil for the motor, etc. are all OpEx for the train (which would be booked as CapEx).

28.4 Selected Financial Ratios

In many cases, it makes sense to compare the performance on one company to its peers. For example, the potential buyer might be interested to find the most valuable company or the management wants to compare its performance to peers. When comparing companies, it makes little sense to compare just the numbers (such as EAT). Indeed, the bigger company will always have larger numbers. In order to be able to compare the performance of companies we should convert numbers from dollars to percentages. This is done by dividing two quantities that have the same dimensions, and we call the results a "ratio". Note that this procedure makes the ratio a dimensionless quantity.

ratio

Below we list a few common ratios, that can be helpful to understand the strengths and weaknesses of a company relative to another. The management should be interested in monitoring all these margins and make sure they evolve favourably.

Profit Margin

Definition: Profit Margin (PM)

$$\text{PM} = \frac{\text{EBIT}}{\text{Sales}}$$

The profit margin (PM) indicates how much profit the company is able to generate compared to its sales. This is a measure of the power of the sales and can be indicative of strength or weaknesses in the relationship with the customers. If two companies are similar, but company A has a PM much higher, then it indicates that this company is able to gain more with a similar effort.

PM
profit margin

Gross Margin

Definition: Gross Margin (GM)

$$\text{GM} = \frac{\text{Gross Profit}}{\text{Sales}}$$

The gross margin (GM) is related to the profit margin, however, one will notice that while the denominator is the same, the nominator is different. Here, we compare the gross profit with sales. For example if the PM is too low, we should first check if the GM is high enough. If that is so, we can drill down and find out what is reducing the profit. For example, we might find that the cost of sales is too high.

GM
gross margin

If on the other hand, we find that even the GM is too low, then the problem is – predominantly – elsewhere.

Asset Utilisation

Definition: Asset Utilisation (AU)

AU
asset utilisation
TCstE
total cost of equity

$$AU = \frac{\text{Sales}}{\text{net Total Assets}} = \frac{\text{Sales}}{\text{TCstE}}$$

The asset utilisation rate (AU) is hence the quotient of total assets divided by the total cost of equity (TCstE). This indicates of how much net assets the company is able to maintain relative to the cost to raise capital.

Liquid Ratio or Current Ratio

The liquid ratio is defined as follows.

Definition: Liquid Ratio (LR)

LR
liquid ratio

$$LR = \frac{\text{Liquid Assets}}{\text{Liquid Liabilities}} = \frac{\text{Current Assets - Stock}}{\text{Current Liabilities}} = \frac{LA}{LL}$$

where LA is short for liquid assets and LL stands for liquid liabilities.

The liquidity ratio shows us how much liquid assets we have available to cover our liquid liabilities. If that ratio would be small, this means that payments will be due before cash can be raised. It this ratio is high, then the company will soon accumulate excess cash.

This ratio can be used to forecast liquidity squeezes. If it is indeed so that we will soon need more cash to pay for liabilities than we will have income from the realisation of assets, then we should think about an overdraft facility for example.

Sometimes, the LT is also referred to as the current ratio (CR):

Definition: Current Ratio (CR)

CR
current ratio

$$CR = \frac{\text{Current Assets}}{\text{Current Liabilities}} = \frac{CA}{CL} = LR$$

with CA the current assets and CL the current liabilities.

Note – CR or LR

The words "current" and "liquid" are both used, to as synonyms with "short term" in this context so CR and LR are different words for the same concept.

The problem with these ratios is that the current liabilities are certain (the invoice is in our accounting department and we know when we have to pay), but the current assets are composed of cash, cash equivalents, securities, and the stock of finished goods. These goods might or might not be sold. This means that it is uncertain if we will be able to use those assets to cover those liabilities. To alleviate this issue, we can use the quick ratio.

Quick Ratio

Definition: Quick ratio (QR)

$$\begin{aligned} QR &= \frac{\text{Cash + cash equivalents}}{\text{Liquid Liabilities}} \\ &= \frac{\text{Current Assets - Stock}}{\text{Current Liabilities}} \end{aligned}$$

QR
quick ratio

The quick ratio compares the total cash and cash equivalents (marketable securities, and accounts receivable) to the current liabilities. The AR excludes some current assets, such as the inventory of goods. This is hence something like an "acid test": it only takes into account the assets that can quickly be converted to cash.

Note, however that the QR does include accounts receivables. That means we are expecting payment, since we have issued an invoice. In reality, this is not a certainty that we will get the money (in time).

Only the laws of physics are "certain", the laws of men and the customs of business do not only change over time, but cannot be taken for granted.

Operating Assets

Definition: Operating Assets (OA)

$$\begin{aligned} OA &= \text{total assets} - \text{financial assets} \\ &= TA - FA \\ &= TA - \text{cash} \qquad \text{(usually)} \end{aligned}$$

OA
operating assets

The OA are those assets that the company can use for its core business. For most companies (so, excluding banks and investment funds) the operating assets are all the assets minus the financial assets such as shares of other companies.[3] For example the factory, the machines and tools

[3] We exclude financial service companies because their financial assets are their operational assets, in regular companies operational assets are rather buildings, machines, etc.

would qualify as operating assets for a car manufacturer, but not the strategic stake that is held in the competing brand.

Operating Liabilities

OL
operating liabilities

Definition: Operating Assets (OL)

$$\begin{aligned} OL &= \text{total liabilities} - \text{financial liabilities} \\ &= TL - FL \\ &= TL - \text{short term notes} - \text{long term notes} \qquad \text{(usually)} \end{aligned}$$

The operating liabilities are those liabilities that the company should settle in order to pursue its regular business model.

Net Operating Assets

NOA
net operating assets

Definition: Net Operating Assets (NOA)

$$\begin{aligned} NOA &= \text{operating assets} - \text{operating liabilities} \\ &= OA - OL \end{aligned}$$

Digression – non-operating assets

A non-operating asset is an assets that is not essential to the ongoing/usual operations of a business but may still generate income (and hence contributes to the return on investment [ROI]).
Typically, these assets are not listed separately in the balance sheet, one will have to gather the information from the management or on-site analysis in order to separate them.
A suitable acronym for non-operating assets could be NOA, not to be confused with the net operating assets (for which we reserved the acronym NOA in this book).

Working Capital

WC
working capital

Definition: Working Capital (WC)

$$\begin{aligned} WC &= \text{current assets} - \text{current liabilities} \\ &= CA - CL \end{aligned}$$

Hence, WC represents the operating liquidity that is available to the business. A positive WC means that there is enough cash to cover the short term liabilities. If WC is negative this means that the company will need to use more cash to cover short term liabilities than it is getting in via its current assets.

Digression – WC and LR

Compare the WC to the definition of Liquid Ratio on page 576. While both definitions use CL and CA, they do something different and have different unit (WC has monetary unit and CR is dimensionless).

To determine the current assets and current liabilities one needs to use the following three accounts:

- accounts receivable (→ current assets)
- inventory (→ current assets), and
- accounts payable (→ current liabilities) — e.g. short term debts such as bank-loans and lines of credit.

If we observe that the working capital increases, then this means that the current assets grew faster than the current liabilities (for example it is possible that the company has increased its receivables, or other current assets or has decreased current liabilities — or has paid off some short-term creditors, or a combination of both).

Total Capital (Employed)

The total capital employed (TCE) or Total Capital (TC) inform about how much capital the company uses to run the business and invests in future projects. Capital can be raised via equity, bonds or loans, and by accumulating reserves of previous years.

Definition: Total Capital Employed (TC or TCE)

$$\text{TCE} = \text{share capital} + \text{reserves} + \text{loans}$$

TCE
total capital employed

Note that we use TCE and TC interchangeably: TCE := TC.

TC
total capital

Weighted Average Cost of Capital (WACC)

The company can use both equity and debt to finance its business growth. The management should aim to minimize the cost of capital. To some extend the cost of capital is also a measure for the perceived risk of the company.

Definition: Weighted Average Cost of Capital (WACC)

The cost of capital is the cost of a company's funding (debt and equity), or, from an investor's point of view "the required rate of return on a portfolio of the company's existing securities."

WACC

weighted average cost of capital

WACC is used to evaluate new projects of a company: it is the minimum rate of return that a new project should bring. WACC is also the minimum return that investors expect in return for capital they allocate to the company (hereby setting a benchmark that a new project has to meet – so both are equivalent).

V_i
the market value of asset i
R_i
the return of asset i
D
total debt expressed in currency
E
total equity expressed in currency
K_d
K_e
cost of equity
cost of debt

$$\begin{aligned} \text{WACC} &= \frac{\sum_{i=1}^{N} R_i\, V_i}{\sum_{i=1}^{N} V_i} \\ &= \frac{D}{D+E}\, K_d + \frac{E}{D+E}\, K_e \qquad \text{(if only funded by equity and debt)} \end{aligned}$$

With V_i the market value of asset i, R_i the return of asset i, E the total equity, D the total debt, K_e the cost of equity, and K_d the cost of debt.

For our purposes we will have to include tax effects. Assuming a tax rate of τ we get:

$$\text{WACC} = \frac{D}{D+E}\, K_d\,(1-\tau) + \frac{E}{D+E}\, K_e$$

There are factors that make it difficult to calculate the formula for determining WACC (e.g. determining the market value of debt (here one can usually use the book value – in case of a healthy company – and equity (this can be circular), finding a good average tax rate, etc.). Therefore, different stakeholders will make different assumptions and end up with different numbers.

This is the genesis of a healthy market where no one has perfect information and the different players have different price calculations. This will then lead to turnover (buying and selling), and is essential for a healthy economy.

Reinvestment Rate (RIR)

Definition: Reinvestment rate (RIR)

$$\text{RIR} = \text{Reinvestment Rate} = \frac{g}{\text{ROIC}}$$

where g is the growth rate and ROIC is the return on capital.

The RIR measures how much of the profits the company keeps to grow its business. Indeed, when the company pays suppliers, lenders, and authorities it can decide what to do with the profit left. The ordinary shareholders meeting will convene and decide how much of the earnings can remain in the company (this money is called a "reserve", later this can be used for growth purposes g), or how much dividend will be paid (D).

A company that has a lot of confidence from its shareholders, will be allowed to keep more in the company and hence will have a higher growth rate (given the same EAT).

Coverage Ratio

CoverageR
coverage ratio

Definition: Coverage Ratio (CoverageR)

$$\text{CoverageR} = \frac{\text{Operating Income}}{\text{Financial Expenses}}$$

An alternative for the Coverage Ratio is $\frac{debt}{EBIT}$ ratio when one wants to asses the level of leverage in the company.

Another possibility is the Debt-Service Coverage Ratio (DSCR)

Definition: Debt-Service Coverage Ratio (DSCR)

$$\begin{aligned}\text{DSCR} &= \frac{\text{Net Operating Income}}{\text{Debt Services}}\\ &= \frac{\text{Net Income} + \text{Amortization} + \text{Depr.} + \text{Int. Expenses} + \text{Oth. Non Cash}}{\text{Principal Repayments} + \text{Interest Payments} + \text{Lease Payments}}\end{aligned}$$

with Depr. meaning “depreciations”, Int. “interest”, and Oth. meaning “other”.

Gearing

Definition: Gearing Ratio (GR)

GR
gearing ratio

$$\begin{aligned}\text{GR} &= \frac{\text{Loans}}{\text{TCE}}\\ &= \frac{\text{Loans}}{\text{Shareholders Equity} + \text{Reserves} + \text{Loans}}\end{aligned}$$

The gearing ratio expresses how much loans are used to gather capital. A company that is creditworthy will have easier access to the bond market or lending via banks and hence can work with a GR that is higher. This implies that its shareholders will see their stake less diluted. Bond holders or lenders not only earn a fixed percentage, but also get their money first. Only if something is left after paying suppliers, taxes, and debt equity holders earn a dividend.

Debt-to-equity ratio

Definition: Debt-to-equity ratio (DE)

DE
debt to equity ratio

$$DE = \frac{\text{Loans}}{\text{Equity}}$$

The DE is related to the previous ratio and it compares the amount of loans used with the amount of equity.

♣ 29 ♣

Management Accounting

29.1 Introduction

The owner(s) of a company should make sure that the interest of the management is aligned with theirs.[1] In the company form where the owners are most remote from the management (the share company), the owners will have at least once a year an Ordinary General Shareholders Meeting. It is in that meeting that the supervisory board is chosen by a majority of votes – that are allocated in function of the number of shares owned.

In that step, the owners will choose supervisory board members that they can trust to align the executing management's priorities with those of the owners. This supervisory board will typically set goals for the executive management in the form of KPIs (Key Performance Indicators), and the variable pay of the executive manager depends on the results of these KPIs.

For example, the executive management might be pushed to increase share value, market share, and profit. The executive management in it is turn will then be able to set more concrete goals for team leaders who in their turn will put goals for the executing workers. This cascade of goals and their management is "management accounting."

29.1.1 Definition of Management Accounting (MA)

Definition: Management Accounting (MA)

Management Accounting is the provision of financial and non-financial decision-making information to managers.
According to the Institute of Management Accountants (IMA): "Management accounting is a profession that involves partnering in management decision making, devising planning and performance management systems, and providing expertise in financial reporting and control to assist management in the formulation and implementation of an organization's strategy."

Management Accounting (MA) is the section of the company that supports the management to make better informed decisions and planning support. To do so, it will use data to monitor

[1]Of course, if they are themselves the management, then there is no potential conflict of interest and this step becomes trivial.

The Big R-Book: From Data Science to Learning Machines and Big Data, First Edition. Philippe J.S. De Brouwer.

Companion Website: www.wiley.com/go/De Brouwer/The Big R-Book

finances, processes, and people to prepare a decision and after the decision it will help to follow up the impact.

Definition: Management Information — MI

The information presented and consumed in MA is called management information.

The main differences between MI (Management Information) and FI (Financial Information) are as follows.

Financial Accounting	Management Accounting
mainly external use (mainly for shareholders, tax, creditors)	internal use (only for management)
past oriented (shows what happened)	future oriented (supports decisions)
fixed reporting period (annual, quarter, month)	flexible reporting period (period as is relevant)
precise (up to £0.01)	mainly indicates direction
always in one currency	currency but other measures possible
required by law	not required by law
public information	confidential

Managing a company only relying on financial accounting and without good Management Information is similar to driving a car by only using the rear mirrors. While it might work in extremely favourable conditions it is not a method that prepares one for bumpy roads ahead.

29.1.2 Management Information Systems (MIS)

A manager needs to understand the performance of his section, they need to be able to change approaches and solve issues based on the management information or KPIs tha they have. Obviously, it is essential that the information is up to date, but also that it would not slow down operational systems and that it is convenient to collect. For those reasons, people choose to have a separate repository of data that is constantly fed with data from operational system and is ready to create the reports. These systems are called "management information systems" (MIS).

MIS
Management Information System

This MI is ideally kept in a electronic system or systems – separate from transactional systems.

Definition: MIS

A Management Information Systems (MIS) focuses on the management of information to provide guidance for strategic decision making.

The concept may include transaction processing system, decision support systems, expert systems, and executive information systems. The term MIS is often used in the business schools. Some of MIS contents are overlapping with other areas such as information system, information technology, informatics, e-commerce and computer science. Therefore, the MIS term sometimes can be inter-changeable used in above areas.

29.2 Selected Methods in MA

Management accounting needs to support the management in their decision process. In essence, it has to be forward looking, but also it needs to give an accurate image of what is going on in the company. One of the most important additions to the financial accounting is usually a view on what consumes money and what produces money in the company.

Imagine for example that there are two business lines: we produce chairs and tables. Sometimes we sell a set, but most often customers buy only chairs or only the table from us. This means that the price of chair and table should be determined with care. For example, it is possible that the profit margin on the chairs is negative, but on the tables positive. Since we used to sell sets, we did not notice, but now a big contract for chairs comes in.

If we do not have a good mechanism in place to determine what the real cost of the chair production is, then we will make losses on this contract. Such contracts might be decisive for the company.

29.2.1 Cost Accounting

Usually, the income is quite clear: we sell a chair and we get an amount of money in return. Hence, it is very clear what the income is for every product, but the cost to make it is a more delicate issue: while the product might have its dedicated production team and while it might be clear what raw resources it uses, they share the same production hall and sales team for example. The art of giving an educated guess of the cost of each product (eventually in function of its quantity) is called "cost accounting."

Definition: - Cost accounting

Cost accounting is an accounting process that measures and analyses the costs associated with products, production, and projects so that correct amounts are reported on financial statements.

Cost accounting aids in the decision-making processes by allowing a company to evaluate its costs. Some types of costs in cost accounting are direct, indirect, fixed, variable and operating costs.

This idea of cost allocation and understanding is extremely important for any company, and there are many ways of producing a Cost Accounting, for example:

SCA Standard Cost Accounting

- *Standard cost accounting (SCA):* Standard cost accounting (SCA) uses ratios called efficiencies that compare the labour and materials actually used to produce a good with those that the same goods would have required under "standard" conditions – works well if only labour is the main cost driver (as was the case in the 1920s when it was introduced).

ABC activity based costing

- *Activity-based costing (ABC):* Activity-based costing (ABC) is a costing methodology that identifies activities in an organization and assigns the cost of each activity with resources to all products and services according to the actual consumption by each. This model assigns more indirect costs (overhead) into direct costs compared to conventional costing and also allows for the use of activity-based drivers – see for example van der Merwe and Clinton (2006).

- *Lean accounting:* Lean accounting is introduced to support the lean enterprise as a business strategy (the company that strives to follow the principles of Lean Production — see Liker and Convis (2011)). The idea is to promote a system that measures and motivates best business practices in the lean enterprise by measuring those things that matter for the customer and the company.

RCA
Resource Consumption Accounting

- *Resource consumption accounting (RCA):* Resource consumption accounting (RCA) is a management theory describing a dynamic, fully integrated, principle-based, and comprehensive management accounting approach that provides managers with decision support information for enterprise optimization. RCA is a relatively new, flexible, comprehensive management accounting approach based largely on the German management accounting approach Grenzplankostenrechnung (GPK)

- *Grenzplankostenrechnung (GPK)*: is a German costing methodology, developed in the late 1940s and 1950s, designed to provide a consistent and accurate application of how managerial costs are calculated and assigned to a product or service. The term Grenzplankostenrechnung, often referred to as GPK, has been translated as either "marginal planned cost accounting" or "Flexible Analytic Cost Planning and Accounting".[2]

TA
throughput accounting

T
throughput

S
sales

TVC
total variable costs

LCC
life cycle costing

- *Throughput accounting (TA):* Throughput accounting[3] is a principle-based and simplified management accounting approach that aims to maximize throughput[4] (sales reduced with total variable costs). It is not a cost accounting approach as it does not try to allocate all costs (only variable costs) and is only cash focused. Hence, TA tries to maximize Throughput (T): $T = S - TVC$. This throughput is typically expressed as a Throughput Accounting Ratio (TAR), defined as follows: $TAR = \frac{\text{return per factory hour}}{\text{cost per factory hour}}$.

- *Life cycle costing (LCCA):* Life-cycle cost analysis (LCCA) is a tool to determine the most cost-effective option among different competing alternatives to purchase, own, operate, maintain, and finally, dispose of an object or process, when each is equally appropriate to be implemented on technical grounds. Hence, LCCA is ideal to decide what to use and how to do it. For example, it can be used to decide which types of rails to use, which machine to use to put the rails, how to finance the machine, etc.[5]

- *Environmental accounting:* Environmental accounting incorporates both economic and environmental information. It can be conducted at the corporate level, national level or international level (through the System of Integrated Environmental and Economic Accounting, a satellite system to the National Accounts of Countries (those that produce the estimates of Gross Domestic Product (GDP))). Environmental accounting is a field

---

[2] The GPK methodology has become the standard for cost accounting in Germany as a "result of the modern, strong controlling culture in German corporations" (see Sharman, 2004).

[3] TA was introduced by Goldratt et al. (1992).

[4] Where typically cost accounting focuses on reducing all costs, TA focuses on increasing throughput: increasing sales, reducing stocks, … increase the speed at which throughput is generated.

[5] In order to perform an LCCA scoping is critical to understand which aspects are to be included and which not. If the scope becomes too large the tool may become impractical to use and of limited ability to help in decision-making and consideration of alternatives; if the scope is too small then the results may be skewed by the choice of factors considered such that the output becomes unreliable or partisan. Usually, the LCCA term implies that environmental costs are not included, whereas the similar Whole-Life Costing, or just Life Cycle Analysis (LCA), generally has a broader scope, including environmental costs.

that identifies resource use, measures and communicates costs of a company's or national economic impact on the environment. Costs include costs to clean up or contain contaminated sites, environmental fines, penalties and taxes, purchase of pollution prevention technologies and waste management costs.

EA
environmental accounting

- *Target costing (TargetC):* Target costing is a cost management tool for reducing the overall cost of a product over its entire life-cycle with the help of production, engineering, research and design. A target cost is the maximum amount of cost that can be incurred on a product and with it the firm can still earn the required profit margin from that product at a particular selling price.[6]

TargetC
target costing

Further information – Cost accounting

The field of cost accounting is vast and complicated. It is beyond the scope of this book to detail out the subject. Fortunately there are many good books on the subject. For example, a good overview can be found in: Drury (2013) and Killough and Leininger (1977).

29.2.2 Selected Cost Types

The company that produces chairs and tables will need more wood as it sells more furniture, but will not immediately rent a bigger production facility. On short term it is reasonable to see the production hall as a fixed cost (a cost that is always there, regardless the number of chairs sold), whereas the amount of wood used will actually directly depend on the number of chairs and number of tables sold.

It should hence not come as a surprise that people active in cost accounting developed their own language when talking about different cost types. We highlight the most important ones.

Direct Costs

Definition: Direct Cost

A Direct Cost is a cost used to produce a good or service and that can be identified as directly used for the good or service.

For example: Direct Cost can be (raw) materials, labour, expenses, marketing and distribution costs if they can be traced to a product, department or project.

Marginal Cost

Definition: Marginal Cost

The Marginal Cost is the expense to produce one more unit of product. This can also be defined as $C_M = \frac{\partial P}{\partial Q}$ (with P the price, A the quantity produced and C_M the Marginal Cost.

[6]In the traditional cost-plus pricing method, materials, labour, and overhead costs are measured. Further, a desired profit is added to determine the selling price. Target costing works the opposite way around. One starts from the price that can be obtained on the market and then works back what the costs can be and tries to reduce costs as necessary.

Indirect Cost

Definition: Indirect Cost

An Indirect Cost is an expense that is not directly related to producing a good or service, and/or cannot be easily traced to a product, department, activity or project that would be directly related to the good or service considered.

Example: A Computer Assembly Facility

An assembly facility will easily allocate all components and workers to the end-product (e.g. a specific mobile phone or tablet). However, the cost to rent the facility, the electricity and the management are not easily allocated to one type of product, so they can be treated as Indirect Costs.

Fixed Cost

Definition: Fixed Cost

A fixed cost is an expense that does not vary with the number of goods or services produced (at least in medium or short term).

Variable Cost

Definition: Variable Cost

A Variable Cost is an expense that changes directly with the level of production output.

Example: A Computer Assembly Facility

The lease of the facility will be a fixed cost: it will not vary in function of the number of phone and tablets produced. However, the electricity and salaries might be Variable Costs.

Overhead Cost

Definition: Overhead Cost

An Overhead Cost ("operating expense", or "overhead expense") is an on-going expense inherent to operating a business that cannot be easily traced to or identified with any particular cost unit (cost centre).

Overhead expenses can be defined as all costs on the income statement except for direct labour, direct materials, and direct expenses. Overhead expenses may include accounting fees, advertising, insurance, interest, legal fees, labour burden, rent, repairs, supplies, taxes, telephone bills, travel expenditures, and utilities. Note that Overhead Cost can be Variable Overhead (e.g. office supplies, electricity) or Fixed Overhead (lease of a building).

There are of course many other types of costs. Some were already mentioned in the section of financial accounting (e.g. operating costs, cost of goods sold), and others are quite specific (for example "sunk costs" are those costs that have been made on a given project and that will not be recuperated even when the project is stopped).

29.3 Selected Use Cases of MA

While the details of cost accounting are a little beyond the scope of this book, their use is right in the attention scope. Indeed, doing something with the numbers produced and transforming that data in to meaningful insight is the work of the data scientist. Companies gradually discover the importance and value of all the data they harvest and invest more and more in data and analytics. Data scientists will need to make sense of the vast amounts of data available. Besides making models – see Part V *"Modelling"* on page 373 – presenting data is an essential part of the role of the data scientist. This presenting data has a technical part (e.g. choosing the right visualization – see Chapter 9 *"Visualisation Methods"* on page 159, Chapter 31 *"A Grammar of Graphics with ggplot2"* on page 687 and Chapter 36.3 *"Dashboards"* on page 725 – but also has an important content part. That content part is the subject of this section.

29.3.1 Balanced Scorecard

Definition: Balanced scorecard (BS)

BSC
balanced scorecard

The Balanced Scorecard (BSC) is a structured report that helps managers to keep track of the execution of activities, issues, and relevant measures. The critical characteristics that define a balanced scorecard are:

- its focus on the strategic agenda of the organization concerned,
- the selection of a small number of data items to monitor (that are or collections or are expected to monitor a wider concept),
- a mix of financial and non-financial data items,
- a comparison to an expected or hoped for result (closed loop controller)

Example: Diversity dashboard

The dashboard about diversity presented in Chapter 36.3.2 *"A Dashboard with flexdashboard"* on page 731 – based on the explanation of Chapter 36.3.1 *"The Business Case: a Diversity Dashboard"* on page 726 – would be an example of one possible balanced scorecard.

Third Generation Balanced Scorecard

The third-generation version was developed in the late 1990s to address design problems inherent to earlier generations – see Lawrie and Cobbold (2004). Rather than just a card to measure performance, it tries to link into the strategic long-term goals; therefore, it should be composed of the following parts:

- *A destination statement.* This is a one or two page description of the organisation at a defined point in the future, typically three to five years away, assuming the current strategy has been successfully implemented. The descriptions of the successful future are segmented into perspectives for example financial & stakeholder expectations, customer & external relationships, processes & activities, organisation & culture.

- *A strategic linkage model.* This is a version of the traditional "strategy map" that typically contains 12-24 strategic objectives segmented into two perspectives, activities and outcomes, analogous to the logical framework. Linkages indicate hypothesised causal relations between strategic objectives.
- *A set of definitions for each of the strategic objectives.*
- *A set of definitions for each of the measures selected to monitor each of the strategic objectives, including targets.*

Further information – Balanced scorecard

A good overview of the third generation balanced scorecard can be found in: Kaplan and Norton (2001a,b), and Norreklit (2000).

Also essential is that the design process is driven by the management team that will use the balanced scorecard. The managers themselves, not external experts, make all decisions about the balanced scorecard content. The process starts – logically – with the development of the "destination statement" to build management consensus on longer term strategic goals. This result is then used to create the "strategic linkage model" that presents the shorter term management priorities and how they will help to achieve the longer term goals. Then all "strategic objectives" are assigned at least one "owner" in the management team. This owner defines the objective itself, plus the measures and targets associated with the objective. The main difference with the previous generations of BSCs is that the third generation really tries to link in with the strategic objectives, hence improving relevance, buy-in, and comfort that more areas are covered.

29.3.2 Key Performance Indicators (KPIs)

The whole idea of a balanced scorecard is to reduce the complex reality to some parameters that can be measured, that are simple enough to understand and that lead to actionable insight.

Definition: KPI

A Key Performance Indicator (KPI) is a measure used to bring about behavioural change and improve performance.

KPI
key performance indicator

Example: Net Promoter Score (NPS) and customer satisfaction

Our management thinks that customer engagement is key and identifies NPS[a] as a KPI (or even as *the* KPI). Doing so it engages all employees to provide customers with a better experience, better product, sharp price, after sales service, etc. Almost everything the company does will somehow contribute to this KPI.

[a]NPS is formally defined in Section 29.3.2.3 on page 592. It is an indicator that measures customer satisfaction.

A performance indicator or key performance indicator (KPI) measures performance, evaluates the success of an organization or of a particular activity. It can be that success is the repeated, periodic achievement of some levels of operational goal (e.g. zero defects, 10/10 customer satisfaction, etc.), or alternatively it can be the increase or decrease of a measure (e.g. decrease the number of accidents, increase the number of customers in the pipeline, etc.).

29.3.2.1 Lagging Indicators

We can only measure emanations that are in the world now or have existed in the past. Anything that we measure has its cause in the past. For example, when we measure profit, then that is a consequence of income and costs from an anterior period. The costs, in their turn are a result of conditions in that period, but also the results of investments that happened decades ago.

While, we can only measure the past, it is possible to think of indicators that are forward looking. What they measure is in the past, but their impact on the *profit* is rather in the future.

Indicators that have an impact on the profit of this period are called "lagging indicators" (they lag the performance of our capital); while KPIs that will have an impact on future periods are called "leading indicators".

Definition: Lagging Indicator

A Lagging Indicator is an "output" indicator, it is the result of something but and by the time it is measured it is too late for management to intervene. It explains why we have today a given profit.

Example: Lagging Indicator

All financial indicators such as profit, ROI, etc. are lagging indicators, they are the result of many actions, their impact is on the profit that we have today.

29.3.2.2 Leading Indicators

Measuring income will explain the profit in this period. That is a lagging indicator. If, however, we measure the number of prospective customers in the pipleline, the number of hits on our online publicity, etc. then that will not necessarily have impact on this period.[7] Rather, one can expect that these customers become active in next period. Indicators that are predictive for future profit are called "leading indicators."

Leading KPIs

Definition: Leading Indicator

A Leading Indicator is a measure that is indicative for future profit.

[7] Of course it is possible to argue that the impact on this period will be rather adverse if we pay fuel for the cars that are used to visit prospective customers or if we have to pay per hit on the Internet. The point, is however, that these indicators will lead towards income in next period.

Example: Leading Indicator

The textbook example would be that when your strategic goal is to live longer and be physically fitter, then most probably your level one KPI is weight loss. But that is a Lagging Indicator: when you are on the scale and you read out the weight, it is too late to do something about it. Leading indicators that feed into this lagging indicator are for example food intake, hours of workout, etc.

Note – KPIs and corporate organisation

Typically, Leading Indicators feed into Lagging Indicators. The higher up the organization chart, the more KPIs become lagging. One of the finer arts of management is to turn these lagging KPIs into actionable strategies, leading KPIs, leading actions, etc.

29.3.2.3 Selected Useful KPIs

Most good KPIs are very specific to the activity, business, team, etc. Even the experience of a team or person can determine what KPIs are useful. For example a novice employee, might be measured on errors made compared to files submitted, whereas a more experienced employee could be measured on the number of files supervised.

Some indicators, however, are quite universal and have proven their validity of years of practice. In this section, we briefly discuss some particularly useful KPIs that can be applied to any type of business: from a the barber shop to the largest bank in the world.

Customer Value Metric (CVM)

Definition: Customer Value Metric (CVM)

A Customer Value Metric is an estimation of the monetary value that a customer represents.

CVM
customer value metric

Customer Value Metrics can be:

- *Historic:* e.g. the net profit on a customer over the last year.
- *Expected:* also referred to as Customer Lifetime Value or Lifetime Customer Value this is the present value of the total value that be expected to be derived from this customer.
- *Potential:* the maximal obtainable customer value.

CLV
customer lifetime value
LCV
lifetime customer value

Note – The usefullness of past customer profit

The answer is: "it depends". If past income is predictable for the future income (if no cross selling or up-selling is possible), then it is important. In most cases, however, past income on a customer is not the most essential: the future income is the real goal. Banks will for example provide free accounts to youngsters and loss making loans to students in the hope that most of those customers will stay many years and become profitable.

It might seem obvious that a customer value metric is a good idea, but it is also deceivingly difficult to get an estimate that is good enough in the sense that matters. Usually, there are the following hurdles and issues to be considered in CVM calculations.

- Use **Income or Gross Profit** in stead of Net Income.
- **Not allocating costs logically**.
- Allocating costs **politically**.
- Cost allocation is **not detailed enough**.
- Blind use of **existing customer segmentation**.
- Trust **intuition** instead of numbers.
- CVM is an output model (not an input model). If model inputs change then the CVM will change (e.g. better customer service will reduce churn).
- **Correlation** between the CVM of different segments can increase **risk**.

Net Promoter Scores (NPS)

Subjective concepts such as customer satisfaction are best measured on some simple scale – such as "bad," "good", and "excellent." To our experience the best is to use a scale of five options: seven is too much to keep in mind, three is too simple in order to present fine gradation. A "scale from one to five" is almost naturally understood by everyone.[8]

Usually, the scale is more or less as follows:

1. really bad/dissatisfied,
2. acceptable but not good/satisfied,
3. neutral,
4. good/satisfied,
5. really good/extremely satisfied.

It is now possible to define the NPS as follows:

Definition: Net Promoter Score (NPS)

NPS
Net Promoter Score

$$\text{NPS} := \frac{\#\text{Promoters} - \#\text{Detractors}}{\text{Total }\#\text{Customers}}$$

Where Promoters are the people that score highest possible (e.g. 5 out of 5) and Detractors are people that score lowest (1 out of 5 — though it might make sense[a] to make this range wider such as scores 1 and 2). The middle class that is not used in the nominator is called the "Passives": these are the people that do not promote us, nor

[a]In the original work one argues that "customers that give you a 6 or below (on a scale from 1 to 10) are Detractors, a score of 7 or 8 are called Passives, and a 9 or 10 are Promoters."

[8]Note that this "net promoter score" in R would best be described as a factor-object – see Chapter 4.3.7 *"Factors"* on page 45 – and that it is and ordinal scale only – see Chapter B.2 *"Ordinal Scale"* on page 830.

Promoters are *believed* to support the brand and detractors are *believed* to discourage peers to use the brand, while one can expect that the group in between will not actively promote or discourage others. Therefore, it makes logically sense to compare only the two active groups. This does make the ratio "harsh" (in that sense that it can be expected to be low), but it does make the ratio more predictable for what will happen to our customer base.

Digression NPS

The NPS was introduced by Reichheld (2003) and Baine & Co (and still is a registered trademark of Fred Reichheld, Bain & Company and Satmetrix).

If customer satisfaction is rated on a scale from 1 to 10, then the promoter score (PS) is the percentage of users that score 9 or 10, and the brand Detractors are the percentage of clients that score 1 or 2.

An NPS can be between -1 (everybody is a detractor) or as high as $+1$ (everybody is a promoter). So, what is a good and what is a bad NPS? There is of course no rigorous answer, but usually one considers a positive NPS as "good," and an NPS of $+0.5$ or more as "excellent."

It is possible to think of another definition – such as the sum of people that score 1/5 and 2/5 compared to the people that score 4/5 plus those that score 5/5. However, taking less enthusiast people in that promoter or detractor score is not a good idea if we want to measure "who is going to recommend our service or product." This last suggestions seems to be a suitable to measure who is likely to use our product or service again.

We argue that for other uses, it makes sense to use a "Net Satisfaction Score" that can be defined as follows.

Definition: Net Satisfaction Score (NSS)

$$\text{NPS} := \frac{\#\text{Positives} - \#\text{Negatives}}{\text{Total }\#}$$

Where #Positives is the number of people with a positive attitude or appreciation (scores of 4 or 5 on a scale from 1 to 5), and #Negatives are the people with a negative attitude (scores 1 or 2 on a scale from 1 to 5).

NSS
net satisfaction score

For processes that have no immediate competition (for example a self evaluation of a board, the quality of projects run by the project team, etc.) this approach makes sense. However, when customers have a free choice, the NPS is probably a better forward looking measure.

Whatever scale or exact definition of NPS we use, the importance of NPS is that it reduces a complex situation of five or more different values to one number. One number is easier to follow up, present as a KPI and eventually visualize its history over time.

? Question #23 – Using a scale from 1 to 5 as a number

Why is it never a good idea to follow up the average of the satisfaction score (as subjective rating on a scale from 1 to 5)?

♣ 30 ♣

Asset Valuation Basics

In previous chapters – Chapter 28 *"Financial Accounting (FA)"* on page 567 and Chapter 29 *"Management Accounting"* on page 583 – we discussed how a company creates snapshots of the reality with financial accounting and how management accounting is used to gain actionable insight for driving the value of the company. But what is the value of a company? It is easy to reply that the value of a company equals its market capitalization. At least for quoted companies this is a practical definition, but still then someone needs to be able to calculate a fundamental value in order to assess if the market is right or if there is an arbitrage opportunity.

We will build up this chapter towards some ways to calculate the value of a company and along the way introduce some other financial instruments that might be of interest such as cash, bonds, equities, and some derivatives.

The Big R-Book: From Data Science to Learning Machines and Big Data, First Edition. Philippe J.S. De Brouwer.

Companion Website: www.wiley.com/go/De Brouwer/The Big R-Book

30.1 Time Value of Money

Financial instruments typically deliver future cash-flows for the owner, hence before we can start we need some idea how to calculate the value of a future cash flow today.

30.1.1 Interest Basics

Interest payment (in monetary units)

Lending an asset holds risk for the lender and is therefore, compensated by paying interest to the lender. If the asset has a value V_0 today and r is the unit interest rate over a unit period (e.g. one year), then the interest due over one period is

$$I = r\, V_0$$

So, lending an asset over one period and giving it back at the end of that period plus the interest equals paying $(1+r)V_0$ at the end of that period. Therefore, the value of an asset within one year V_1 is:

$$V_1 = (1+r)\, V_0$$

future value

This value in the future is called the "future value" (FV). The FV of an asset after N years becomes then

$$\text{FV} = (1+r)^N\, V_0$$

present value

Note that where we denoted the symbol for the value at this very moment as V_0, in other literature it can be called "present value" (PV). Both concepts are for the purpose of this book the same and will be used interchangeably.

> **? Question #24**
>
> If the interest rate over one year is r_y, then how much interest is due over one month?(While in general that will depend on how many days the month has, for this exercise work with $\frac{1}{12}^{\text{th}}$).

30.1.2 Specific Interest Rate Concepts

It is common knowledge that financial institutions do not always use the same calculations methods and even for commercial reasons it might be possible that for example monthly interest rates are presented in stead of annual ones. For all those reasons financial supervisors and law makers have introduced a concept of Annual Percentage Rate that should be a compound interest rate on an annual base (that takes into account the different costs). This allows easier comparison for the customer.

APR
annual percentage rate
AER
effective annual rate or annual equivalent rate

> **Definition: APR or AER**
>
> the Annual Percentage Rate (APR) or Effective Annual Rate or Annual Equivalent Rate (AER) is the annualized compound interest rate (r_y) that takes into account all costs for the borrower.

Consider now the following example. A loan-shark asks a 5% interest rate for a loan of one month. What is the APR? Well the APR is the annual equivalent interest rate, so we have to see what a 5% interest per week would cost us over a full year.

$$\text{APR} = r_y = (1 + r_w)^{52} - 1$$

This can be formulated in R as follows:

```
r_w <- 0.05                 # the interest rate per week
r_y <- (1 + r_w)^52 - 1  # assume 52 weeks per year
paste0("the APR is: ", round(r_y * 100, 2),"%!")
## [1] "the APR is: 1164.28%!"
```

? Question #25 – Impact of monthly fee on the APR

If a loan shark decides to add an "administration fee" (to be paid immediately after getting the loan) of \$5 on a loan of \$100 for one month. What is the APR?

Nominal vs. Real Interest Rates

Definition: Nominal Interest Rate

nominal interest rate

The nominal interest rate (i_n) is the rate of interest (as shown or calculated) with no adjustment for inflation.

Definition: Real Interest Rate

real interest rate

inflation rate

The real interest rate (i_r) is the growth in real value (purchase power) plus interest corrected for inflation (p).

Example: Real Interest Rate

Assume that the inflation p is 10% and you borrow \$100 for one year and the lender asks you to pay back \$110 after one year. In that case, you pay back the same amount in real terms as the amount that you have borrowed, so the real interest rate is 0%, while the nominal interest rate is 10%.

The relation between the real and nominal interest rate is

$$(1 + i_n) = (1 + i_r)\,(1 + p)$$

and hence

$$i_r = \frac{1 + i_n}{1 + p} - 1$$

So, the relation $i_n = i_r + p$ is only and approximation of the first order (the proof is left as exercise).

30.1.3 Discounting

NPV
Net Present Value

The Net Present Value (NPV) is the Future value discounted to today:

$$\text{PV} = \frac{\text{FV}}{(1+r)^N} \tag{30.1}$$

Cash Flow

Hence, the Net Present Value of a series of cash flows (CF) equals:

$$\text{NPV} = \sum_{t=0}^{N} \frac{\text{CF}_t}{(1+r)^t} \tag{30.2}$$

Example

interest rate

The interest rate is "flat" (this means that it has the same value regardless the time horizon) and equals 10%. We have a project that has no risk and today we need to invest £100, and then it pays £100 in year 5 and 7. Is this a good project?

```
r   <- 0.1
CFs <- c(-100, 100, 100)
t   <- c(   0,   5,   7)
NPV <- sum(CFs / (1 + r)^t)
print(round(NPV, 2))
## [1] 13.41
```

The net present value of the project is £13.41. Since the value is higher than the risk free interest rate (10%), the project is a worthwhile investment and the rational investor should be willing to pay £100 today in order to receive two deferred cash flows of £100 in year 5 and 7.

Note – Element-wise operations in R

The fact that R will treat all operations element per element does make our code really short and neat.[a]

[a] However, be sure to understand vector recycling and read see Chapter 4.3.2 *"Vectors"* on page 29 if necessary.

30.2 Cash

The most simple asset class is cash. This refers to money that is readily available (paper money, current accounts, and asset managers might even use the term for short term government bonds).

Definition: Cash

The strict definition of Cash is money in the physical form of a currency, such as banknotes and coins.
In bookkeeping and finance, cash refers to current assets comprising currency or currency equivalents that can be converted to cash (almost) immediately. Cash is seen either as a reserve for payments, in case of a structural or incidental negative cash flow, or as a way to avoid a downturn on financial markets.

Example: Cash

For example, typically one considers current accounts, savings accounts, short term Treasury notes, etc. also as "cash."

For an asset manager, "cash" is a practical term, used to describe all assets that share similar returns, safety, and liquidity with cash as defined previously. In this wide definition cash can refer to cash held on current accounts, treasury notes of very short duration, etc.

30.3 Bonds

Definition: Bond

In finance, a bond is an instrument of indebtedness of the bond issuer to the holders. It is a debt security, under which the issuer owes the holders a debt and, depending on the terms of the bond, is obliged to pay them interest (the coupon) and/or to repay the principal at a later date, termed the maturity date. Interest is usually payable at fixed intervals (semi-annual, annual, sometimes monthly). A bond is also transferable. That means that the obligor owns money to the holder of he bond and not to the initial holder.

Thus, a bond is a form of loan: the holder of the bond is the lender (creditor), the issuer of the bond is the borrower (debtor), and the coupon is the interest. Bonds provide the borrower with external funds to finance long-term investments, or, in the case of government bonds, to finance current expenditure. Certificates of deposit (CDs) or short term commercial paper are considered to be money market instruments and not bonds for investment purposes: the main difference is in the length of the term of the instrument.

Bonds and stocks are both securities, but the major difference between the two is that (capital) stockholders have an equity stake in the company (i.e. they are investors, they own part of the company), whereas bondholders have a creditor stake in the company (i.e. they are lenders). Being a creditor, bondholders have priority over stockholders. This means they will be repaid in advance of stockholders, but will rank behind secured creditors in the event of bankruptcy. Another difference is that bonds usually have a defined term, or maturity, after which the bond is redeemed, whereas stocks are typically outstanding indefinitely. An exception is an irredeemable bond (perpetual bond), ie. a bond with no maturity.

30.3.1 Features of a Bond

Bonds are characterised by the obligation for the debtor to pay back at set moments. This implies that there is a fixed amount, investment horizon, interest rate, and other conditions. These conditions bear specific names in financial markets. Below, we list some of the most important concepts that you will encounter on a bond market.

Definition: Principal

Nominal, principal, par, or face amount is the amount on which the issuer pays interest, and which – usually – has to be repaid at the end of the term. Some structured bonds can have a redemption amount which is different from the face amount and can be linked to performance of particular assets.

Definition: Maturity

The issuer has to repay the nominal amount on the maturity date. As long as all due payments have been made, the issuer has no further obligations to the bond holders after the maturity date. The length of time until the maturity date is often referred to as the term or tenor or maturity of a bond. The maturity can be any length of time. Most bonds have a term of up to 30 years, however, some issues have no maturity date ("irredeemables" or "eternal bonds").

In the market for United States Treasury securities, there are three categories of bond maturities:

- *Short term (bills):* Maturities between one to five year; (instruments with maturities less than one year are called Money Market Instruments).
- *Medium term (notes):* Maturities between six to twelve years.
- *Long term (bonds):* Maturities greater than twelve years.

Definition: Coupon

The coupon is the interest rate that the issuer pays to the holder. Usually, this rate is fixed throughout the life of the bond. It can also vary with a money market index, such as LIBOR.

The name "coupon" arose because in the past, paper bond certificates were issued which had coupons attached to them, one for each interest payment. On the due dates the bondholder would hand in the coupon to a bank in exchange for the interest payment. Interest can be paid at different frequencies: generally semi-annual, i.e. every six months or annual.

Definition: Yield

The yield is the rate of return received from investing in the bond. It usually refers either to

- the current yield, or running yield, which is simply the annual interest payment divided by the current market price of the bond (often the clean price), or to
- the yield to maturity or redemption yield, which is a more useful measure of the return of the bond, taking into account the current market price, and the amount and timing of all remaining coupon payments and of the repayment due on maturity. It is equivalent to the internal rate of return of a bond.

Definition: Credit quality

The quality of the issue refers to the probability that the bondholders will receive the amounts promised at the due dates. This will depend on a wide range of factors. High-yield bonds are bonds that are rated below investment grade by the credit rating agencies. As these bonds are more risky than investment grade bonds, investors expect to earn a higher yield. These bonds are also called junk bonds.

Definition: Market price

The market price of a trade-able bond will be influenced amongst other things by the amounts, currency and timing of the interest payments and capital repayment due, the quality of the bond, and the available redemption yield of other comparable bonds which can be traded in the markets.

On bond markets, the price can be quoted as "clean" or "dirty". (a "dirty" price includes the present value of all future cash flows including accrued interest for the actual period. In Europe, the dirty price is most commonly used. The "clean" price does not include accrued interest, this price is most often used in the U.S.A.)

The issue price – at which investors buy the bonds when they are first issued – will in many cases be approximately equal to the nominal amount. This is because the interest rate should be chosen so that it matches the risk profile of the debtor. This means that the issuer will receive the issue price, minus all costs related to issuing the bonds. From that moment, the market price of the bond will change: it may trade "at a premium" (this is a price higher than the issue price, also referred to as "above par", usually because market interest rates have fallen since issue), or at a discount (lower than the issue price, also referred to as "below par", usually due to higher interest rates or a deterioration of the credit risk of the issuer).

30.3.2 Valuation of Bonds

In a first, naive approach, we will assume that the value of a bond is simply given by the discounted cash flows that it generates. This can be expressed as follows.

$$P_{\text{bond}} = \sum_{i=0}^{N} \frac{\text{CF}_i}{(1+r_i)^{t_i}} \tag{30.3}$$

$$= \sum_{t=1}^{N} \frac{\text{coupon}_i}{(1+r_i)^t} + \frac{\text{nominal}}{(1+r_N)^N} \tag{30.4}$$

for a bond that pays annual coupon. where P_{bond} is the price of the bond (this can also be understood as the actual price, or in other words, the net present value – see Chapter 30.1.3 *"Discounting"* on page 600, and more in particular Equation 30.2 on page 600). CF_t is the cash flow at moment t, and r_i the relevant interest rate for an investment of time t. Note that in the second line, we assume that each moment t correspond exactly with one year. This allows to simplify the notations.

Digression – Required interest rate

In this formula, we use an interest rate, r. For now, we will assume that this is the risk free interest rate for the relevant investment horizon for the relevant currency. This has the – mathematical – advantage that it is a known number (we can find it as the interest rate on the government bonds). In fact, we should include a risk premium:

$$r = R_{\text{RF}} + \text{RP}$$

where r_{RF} is the risk free interest rate and RP is the risk premium. This is further clarified in Chapter 30.4 *"The Capital Asset Pricing Model (CAPM)"* on page 610.

Note – Bond prices change every day, even every second

In the aforementioned formulae, we make the simplifying assumption that we are at the beginning of the year. In reality the price of a bond needs to be adapted as interest is accrued and when interest rates are changed. This means that the price of a bond will differ every minute of the day.

The nominal is also called face value, it is the amount that will be paid back at the end.

Example: – Bond value

Consider the following example. A bond with nominal value of \$100 pays annually a coupon of \$5 (with the first payment in exactly one year from now), during 4 years and in the fifth year the debtor will pay the last \$5 and the nominal value of \$100. What is its fair price given that the risk free interest rate is 3%?

```
# bond_value
# Calculates the fair value of a bond
# Arguments:
#    time_to_mat -- time to maturity in years
#    coupon      -- annual coupon in $
#    disc_rate   -- discount rate (risk free + risk premium)
#    nominal     -- face value of the bond in $
# Returns:
#    the value of the bond in $
bond_value <- function(time_to_mat, coupon, disc_rate, nominal){
  value  <- 0
  # 1/ all coupons
  for (t in 1:time_to_mat) {
     value <- value + coupon * (1 + disc_rate)^(-t)
     }
  # 2/ end payment of face value
  value <- value + nominal * (1 + disc_rate)^(-time_to_mat)
  value
}

# We assume that the required interest rate is the
# risk free interest rate of 3%.

# The fair value of the bond is then:
bond_value(time_to_mat = 5, coupon = 5, disc_rate = 0.03,
           nominal = 100)
## [1] 109.1594
```

Example: – Higher rates increase

Assume now that the interest rates increase to 3.5%. What is the value now?

```
bond_value(time_to_mat = 5, coupon = 5, disc_rate = 0.035,
           nominal = 100)
## [1] 106.7726
```

As one could expect, the value of the bond is decreased as the interest rates went up. Most bonds are reasonably safe investments. The emitter has the legal obligation to pay back the bonds. In case of adverse economic climate that company will not pay dividend but still will have to pay its bonds.

Of course, companies and governments can default on their debt. When the income is not sufficient to pay back the dividends and/or nominal values, then the bond issuer defaults on its bonds. In this section we assume that there is no default risk.

? Question #26 - Calculate a bond value

Assume a bond that has pays for the next five years each year one coupon of 5% while the interest rate is 5% (and the first coupon is due in exactly one year). What is the value of a bond emission of PLN 1 000? This means that buyer of the bond will see the following cash flows:

Year	Cash flow
0	-V_{bond}
1	50 PLN
2	50 PLN
3	50 PLN
4	50 PLN
5	1050 PLN

Find V_{bond} assuming a discount rate of 5%.

? Question #27 - Lower interest rates

You have just bought the bond and the interest rate drops to 3%. How much do you loose or win that day?

? Question #28 - Higher interest rates

You have just bought the bond and the interest rate goes up to 7% in stead of going down. How much do you loose or win that day?

duration

30.3.3 Duration

In general, the word "duration" is the average of times when payments are received, weighted with the amount received. For bonds there are two concepts that are referred to as "duration": the Macaulay duration and the Modified duration. Both concepts are different though numeric values are close to each other.

Macaulay duration

30.3.3.1 Macaulay Duration

The Macaulay duration – named after Frederick Macaulay – is the weighted average maturity of the cash flows.

$$
\begin{aligned}
\text{MacD} &:= \frac{\sum_{i=0}^{N} t_i \text{PV}_i}{\sum_{i=0}^{N} \text{PV}_i} \\
&= \frac{\sum_{i=0}^{N} t_i \text{PV}_i}{V_{\text{bond}}} \\
&= \frac{1}{V_{\text{bond}}} \sum_{i=0}^{N} \frac{t_i \text{CF}_i}{(1+r_i)^{t_i}}
\end{aligned}
$$

Using the same example as aforementioned: a bond with nominal value of \$100, an annual coupon of \$5 (with the first payment in exactly one year from now), with a maturity of five years so that in the fifth year the debtor will pay the last \$5 and the nominal value of \$100. What is its Macaulay Duration given that the risk free interest rate is 3%?

```
V <- bond_value(time_to_mat = 5, coupon = 5, disc_rate = 0.03,
           nominal = 100)
CFs <- c(seq(5, 5, length.out=4), 105)
t   <- c(1:5)
r   <- 0.03
MacD <- 1/V * sum(t * CFs / (1 + r)^t)
print(MacD)
## [1] 4.56806
```

In many calculations one will rather use yield to maturity (y) to calculate the PV_i, with $\text{PV}_i = \sum_{i=1}^{N} \text{CF}_i \exp\{-y\, t_i\}$, and hence:

$$\text{MacD} = \frac{1}{V} \sum_{i=1}^{N} t_i \text{CF}_i \, \exp\{-y\, t_i\}$$

where we have used V instead of V_{bond} to make the notations lighter.

Further information – Yield to maturity

yield to maturity

The yield to maturity is the internal rate of return (IRR) of the bond, earned by an investor who buys the bond now at market price and keeps the bond till maturity. Yield to maturity is the same as the IRR; hence it is the discount rate at which the sum of all future cash flows from the bond (coupons and principal) is equal to the current price of the bond. The y is often given in terms of Annual Percentage Rate (APR), but more often market convention is followed. In a number of major markets (such as gilts) the convention is to quote annualized yields with semi-annual compounding (see compound interest)

$$y := \sqrt[T]{\frac{\text{nominal}}{\text{PV}}} - 1$$

30.3.3.2 Modified Duration

The name modified duration is actually somehow misleading in this case. Modified duration is actually the price elasticity with respect to changes in the yield. It is defined as the percentage derivative of price with respect to yield, which is the same as the logarithmic derivative of bond price with respect to yield.

$$\text{ModD} := \frac{1}{V}\frac{\partial V}{\partial y} = -\frac{\partial \log(V)}{\partial y}$$

When the yield is expressed as a continuously compounded rate, the Macaulay duration and modified duration are numerically equal. This can be verified by calculating the derivative of the value to the yield:

$$\frac{\partial V}{\partial y} = \sum_{i=1}^{N} CF_i \, \exp(-y\, t_i) = -\sum_{i=1}^{N} t_i . CF_i \, \exp(-y\, t_i) = -\text{MacD}\, V$$

Inserting this in the definition of the modified duration shows that $MacD = ModD$.

However, in most financial markets interest rates are not presented as continuously compounded interest rates but rather as a periodically compounded interest rate (usually annually compounded). If we write k as the compounding frequency (1 for annual, 2 for semi-annual, 12 for monthly, etc.) and y_k the yield to maturity expressed as periodically compounded, then it can be shown that

$$\text{MacD} = \left(1 + \frac{y_k}{k}\right) \text{ModD}$$

To see this express the value of a bond in function of the periodically compounded interest rates.

$$V(y_k) = \sum_{i=1}^{N} \text{PV}_i = \sum_{i=1}^{N} \frac{\text{CF}_i}{\left(1 + \frac{y_k}{k}\right)^{k.t_i}}$$

Its Macaulay duration becomes then

$$\sum_{i=1}^{N} \frac{t_i}{V(y_k)} \frac{\text{CF}_i}{1 + \frac{y_k}{k}}$$

And its Modified duration is then

$$\begin{aligned}
\text{ModD} &= -\frac{1}{V(y_k)} \frac{\partial V(y_k)}{\partial y_k} && (30.5)\\
&= \frac{1}{V(y_k)} \frac{1}{\left(1 + \frac{y_k}{k}\right)^{k\, t_i}} \sum_{i=1}^{N} \frac{t_i\, \text{CF}_i}{\left(1 + \frac{y_k}{k(y_k)}\right)^{k\, t_i}} && (30.6)\\
&= -\frac{1}{V(y_k)} \frac{\text{MacD}\; V(y_k)}{1 + \frac{y_k}{k}} && (30.7)\\
&= \frac{\text{MacD}}{1 + \frac{y_k}{k}} && (30.8)
\end{aligned}$$

This calculation shows that the Macaulay duration and the modified duration numerically will be reasonably close to each other and you will understand why there is sometimes confusion between the two concepts.

Note – First order estimate of price change

Since the modified duration is a derivative it provides us with a first order estimate of the price change when the yield changes a small amount. Hence

$$\frac{\Delta V}{\Delta y} \approx -\text{ModD}\; V$$

So for a bond with a modified duration of 4% and given a 0.5% interest rate increase, we can estimate that the bond price will decrease with 2%.

Digression – DV01

In professional markets it is common to use the concept "dollar duration" (`DV01`). Sometimes, it is also referred to as the "Bloomberg Risk". It is defined as negative of the derivative of the value with respect to yield:

$$D_{\$} = \text{DV01} = \frac{\partial V}{\partial y}$$

Using, Equation 30.5, we find:

$$D_{\$} = \text{DV01} = V\,\frac{\text{ModD}}{100}$$

which is makes clear that it is expressed in Dollar per percentage point change in yield. Alternatively, it is not divided by 100, but rather by 10 000 and to express the change per "base point" (sometimes called "bips", the base point is simply one hundredth of a percentage point).

The DV01 is analogous to the delta in derivative pricing (see Chapter 30.7.8 *"The Greeks"* on page 664). The DV01 is the ratio of a price change in output (dollars) to unit change in input (a basis point of yield).

Note that it is the change in price in Dollars, not in percentage. Usually, it is measured per 1 basis point[a] Sometimes, the DV01 is also referred to as the BPV (basis point value) or "Bloomberg Risk"

$$\text{BPV} = \text{DV01}$$

[a]That explains its name. The DV01 is short for "dollar value of a 01 percentage point change".

CAPM
Capital Asset Pricing Model

30.4 The Capital Asset Pricing Model (CAPM)

In our naive presentation of the valuation of a bond, we mentioned that the lender will ask an interest rate that compensates for the risk that the borrower will fail to pay back. At that point, we did not elaborate on how to calculate that "risk premium". In fact the interest rate that the lender will ask is the interest rate of the risk free investment (government bonds) plus a risk premium adequate for that particular borrower.

A simple and useful framework to estimate a risk premium – as defined in Equation 30.14 on page 616 – is the Capital Asset Pricing Model (CAPM). It uses the "the market" as reference and relies on some simplifications that create a workable framework.

30.4.1 The CAPM Framework

The CAPM was introduced by Treynor (1961, 1962), Sharpe (1964), and Mossin (1968).[1] It is therefore, also called the "Sharpe-Lintner-Mossin mean-variance equilibrium model of exchange" – the Capital Asset Pricing Model (CAPM) – is used to determine a theoretically appropriate required rate of return of an asset in function of that asset's non-diversifiable risk and the inherent risk to the market. The model takes into account the asset's sensitivity to non-diversifiable risk (also known as systemic risk or market risk), often represented by the quantity beta (β) in the financial industry, as well as the expected return of the market, and the expected return of a theoretical risk free asset.

The CAPM is expressed as follows:

$$\frac{E[R_k] - R_{\text{RF}}}{\beta_k} = E[R_M] - R_{\text{RF}} \tag{30.9}$$

The market reward-to-risk ratio is effectively the market risk premium. Rearranging the aforementioned equation and solving for $E(R_k)$, we obtain the expected return from the asset k via the CAPM:

$$E[R_k] = R_{\text{RF}} + \beta_k \left(E[R_M] - R_{\text{RF}}\right) \tag{30.10}$$

where:

- $E[R_k]$ is the expected return on the capital asset.
- R_{RF} is the risk free rate of interest, such as interest arising from government bonds.
- β_k (the beta coefficient) is the sensitivity of the asset returns to market returns, or also $\beta_k = \frac{\text{Cov}(R_k, R_M)}{\text{VAR}(R_M)}$.
- $E[R_M]$ is the expected return of the market.
- $E[R_M] - R_{\text{RF}}$ the market premium or risk premium.
- $\text{VAR}(R_M)$ is the variance of the market return.

VAR
variance

SML
security market line

The CAPM is a model for pricing an individual security or portfolio. For individual securities, we make use of the security market line (SML) and its relation to expected return and systemic risk (β), in order to show how the market must price individual securities in relation to their security risk class. The SML enables us to calculate the reward-to-risk ratio for any security in

[1] All these authors were building on the earlier work of Harry Markowitz on diversification and his Mean Variance Theory – see Markowitz (1952). Sharpe received the Nobel Memorial Prize in Economics (jointly with Markowitz and Merton Miller) for this contribution to the field of financial economics.

relation to the reward-to-risk ration of the overall market. Therefore, when the expected rate of return for any security is deflated by its beta coefficient, the reward-to-risk ratio for any individual security in the market is equal to the market reward-to-risk ratio. For any security k:

Restated in terms of risk premium:

$$E[R_k] - R_{\mathrm{RF}} = \beta_k \left(E[R_M] - R_{\mathrm{RF}}\right) \tag{30.11}$$

which states that the individual risk premium equals the market premium times beta.

The CAPM provides a framework that can be used to calculate the value of a company. The caveat is that the framework is self recursive. The riskiness of the company determines the beta, that in its turn influences the required rate of return, that is then used to calculate the value of the company ... using the beta.

Below we provide some examples, to illustrate how the CAPM can be used to calculate the value of a company.

Example: Company A

The company "A Plc." has a β of 1.25, the market return is 10% and the risk free return is 2%.
What is the expected return for that company?

Since beta and R_{RF} are given, we can use the CAPM to calculate the required rate of return R_A as follows:

$$E[R_A] = R_{\mathrm{RF}} + \beta_A \ \left(E[R_M] - R_{\mathrm{RF}}\right)$$

```
R_A <- 0.02 + 1.25 * (0.10 - 0.02)
print(paste0('The RR for company A is: ',
             round(R_A, 2) * 100, '%'))
## [1] "The RR for company A is: 12%"
```

Example: Company B

The company "B" has a β of 0.75 and all other parameters are the same (the market return is 10% and the risk free return is 2%).
What is the expected return for that company?

```
R_B <- 0.02 + 0.75 * (0.10 - 0.02)
print(paste0('The RR for B is: ', round(R_B, 2) * 100, '%'))
## [1] "The RR for B is: 8%"

print(paste0('The beta changed by ',
             round((0.75 / 1.25 - 1) * 100, 2),
             '% and the RR by ',
             round((R_B / R_A - 1) * 100, 2), '%.'))
## [1] "The beta changed by -40% and the RR by -33.33%."
```

30.4.2 The CAPM and Risk

There are many studies that show that diversification should decrease risk. If one invests his total wealth in one asset A, and that asset defaults, then the invetor loses everything. If we diversify over two assets (A and B) then we only loose everything if both assets A and B together disappear. Unless they are 100% correlated, that probability should be lower. This idea goes back to Bernoulli (1738) and is so fundamental to our thinking about risk that it became one of the axioma's in "Thinking Coherently" by Artzner et al. (1997).[2]

Thinking of a portfolio of assets that are quoted on a stock exchange (or available on a given market), it makes sense to break down the risk of the portfolio in the two following components.

1. *Systematic risk or undiversifiable risk:* cannot be diversified away — it is inherent to the market under consideration ("market risk").
2. *Unsystematic risk, idiosyncratic risk or diversifiable risk:* the risk of individual assets. Unsystematic risk can be reduced by diversifying the portfolio (specific risks "average out").

The CAPM implies the following.

- *A rational investor should not take on any diversifiable risks:* — therefore the required return on an asset (i.e. the return that compensates for risk taken), must be linked to its riskiness in a portfolio context – i.e. its contribution to the portfolio's overall riskiness — as opposed to its "stand-alone riskiness."
- In CAPM, *portfolio risk is represented by variance.* — therefore the beta of the portfolio is the defining factor in rewarding the systematic exposure taken by an investor.
- The CAPM assumes that the volatility-return profile of a *portfolio can be optimized as in Mean Variance Theory.*
- Because the unsystematic risk is diversifiable, the *total risk of a portfolio can be viewed as beta.*

30.4.3 Limitations and Shortcomings of the CAPM

The CAPM makes some bold assumptions about financial markets. In particular the following assumptions are not always or never really true in reality. The CAPM assumes that all investors

1. try to maximize utility that is a function of only return and volatility,
2. have a stable utility function (does not depend on the level of wealth),
3. are rational and volatility-averse,
4. consider all assets in one portfolio,
5. do not care about other live goals apart from money (investments are a life goal in their own right and do not serve to cover other liabilities or goals),
6. are price takers, i.e. they cannot influence prices,
7. are able to lend and borrow under the risk free rate of interest with no limitations,

[2] The original paper of Artzner et al. (1997) is quite good, but we can also refer to De Brouwer (2012) for a complete and slower paced introduction to risk measures and coherent risk measures in particular.

8. trade without transaction costs,
9. are not taxed in any way on their investments or transactions,
10. deal with securities that are all highly divisible into small units, and
11. assume all information is at the same time available to all investors.

Further information – CAPM

The CAPM methods also holds important conclusions for the construction of investment portfolios and there is a lot more to be said about the limitations. For a more complete treatment we refer to De Brouwer (2012).

30.5 Equities

Equity (or share) refers to a title of ownership in a company. The equity holder (owner) typically gets a fair share in the dividend and voting rights in the shareholders meeting. Shares represent ownership in a company.

30.5.1 Definition

Definition: Stock, shares and equity

A share in a company a title of ownership. The capital stock of an incorporated business constitutes the equity stake of its owners. It represents the residual assets of the company that would be due to stockholders after discharge of all senior claims such as secured and unsecured debt.

There are some different classes of shares that are quite different.

1. **Common stock** usually entitles the owner to vote at shareholders' meetings and to receive dividends.
2. **Preferred stock** generally does not have voting rights, but has a higher claim on assets and earnings than the common shares. For example, owners of preferred stock receive dividends before common shareholders and have priority in the event that a company goes bankrupt and is liquidated.

Digression – Local use of definitions

In some jurisdictions such as the United Kingdom, Republic of Ireland, South Africa, and Australia, stock can also refer to other financial instruments such as government bonds.

30.5.2 Short History

The joint stock company is omnipresent in our world and most probably your car, building, street, food etc.is produced or processed by such company. A joint stock company transcends the level of one investor and allows to pool resources and share risks. Therefore, one will notice that societies that embraced this way of sharing risk and pooling resources were able to bring development to another level.

Some early examples of the use of shares include.

- *Roman Republic*, the state outsourced many of its services to private companies. These government contractors were called *publicani*, or *societas publicanorum* (as individual company). These companies issued shares called *partes* (for large cooperatives) and *particulae* for the smaller ones.[3]
- *ca. 1250:* 96 shares of the *Société des Moulins du Bazacle* were traded (with varying price) in Toulouse

[3]Sources: Polybius (ca. 200—118 BC) mentions that "almost every citizen" participated in the government leases. Marcus Tullius Cicero (03/01/-106 — 07/12/-43) mentions "partes illo tempore carissimae" (this translates to "share that had a very high price at that time," and these words provide early evidence for price fluctuations)

- *31/12/1600:* the East India Company was granted the Royal Charter by Elizabeth I and became the earliest recognized joint-stock company in modern times.[4]

- *1602:* saw the birth of the stock exchange: the "Vereenigde Oostindische Compagnie" issued shares that were traded on the Amsterdam Stock Exchange. The invention of the joint stock company and the stock exchanges allowed to pool risk and resources much more efficiently and capital could be gathered faster for more expensive enterprises. The trade with the Indies could really start and bring wealth to the most successful countries. Soon England and Holland would become superpowers as the sea.

- Dutch stock market of the **17th century** had

  - stock futures,
  - stock options,
  - short selling,
  - credit to purchase stock (margin trading or "trading on a margin"),
  - ...and famously gave rise to the first market crash "the Tulipomania" in 1637 – Mackay (1841)

30.5.3 Valuation of Equities

Where a bond entitles the owner to a flow of coupons and payback of the face value, an equity entitles the holder only to dividends. So the value of a company must equate the present value of all future dividends. Future dividends are subject to much uncertainty. Many aspects are uncertain: will that market still be attractive, will this company do well in that market, how evolves the potential customer base, etc. There are largely two groups of approaches to estimating the value of a company.

1. *Absolute value models* try to predict future cash flows and then discount them back to the present value. The Dividend Discount Model – see Chapter 30.5.4.1 *"Dividend Discount Model (DDM)"* on page 616 – and the Free Cash Flow Method – see Chapter 30.5.4.2 *"Free Cash Flow (FCF)"* on page 620 are examples of absolute value models.

2. *Relative value models* rely on the collective wisdom of the financial markets and determine the value based on the observation of market prices of similar assets. Relative value models are discussed in Chapter 30.5.5 *"Relative Value Models"* on page 625

It also makes sense to distinguish some different types of prices.

- *market value:* The value that "the market" is willing to pay (the price on the stock exchange).

- *fair value:* If there is no market price for this asset, but we can determine its value based on other market prices then we call this the fair value.

- *intrinsic value:* This is the underlying or true value.

[4]The Royal Charter effectively gave the newly created Honourable East India Company a 15-year monopoly on all trade in the East Indies. This allowed it to acquire auxiliary governmental and military functions and virtually rule the East Indies.

For instance, when an analyst believes a stock's intrinsic value is greater (less) than its market price, an analyst makes a "buy" ("sell") recommendation. Moreover, an asset's intrinsic value may be subject to personal opinion and vary among analysts.

Since a share will only yield dividend, it stands to reason that the value of a share today is the discounted value of all those dividends. Hence, a simple model for the price of a share can be just this: the present value of all cash-flows.

$$P_{\text{equity}} = \sum_{t=0}^{N} \frac{\text{CF}_t}{(1+r)^t} \tag{30.12}$$

$$= \sum_{t=0}^{\infty} \frac{D_t}{(1+r)^t} \tag{30.13}$$

with $D = dividend$.

This model is the Dividend Discount Model, and we will come back to it later, because although the model is concise and logical, it has two variables that will be hard to quantify.

RP
risk premium

1. *The discount rate* should also include the risk. It is easy to write that the discount rate in Equation 30.13 should be the risk free interest rate plus a risk premium

$$r = R_{\text{RF}} + \text{RP} \tag{30.14}$$

R_{RF}
risk free interest rate

with R_{RF} the risk free interest rate and RP the risk premium. The real problem is only deferred: finding the risk premium that is appropriate for the assets being valued.

2. *The dividends* themselves are unknown. In fact, we should estimate an infinite series of future dividends.

In order to make sense of this simple model, we need first to find methods to find good estimates for the risk premium and the dividends.

30.5.4 Absolute Value Models

Thanks to the CAPM we have a method to determine a reasonable discount rate – risk premium or required rate of return, so it is time to revisit Equation 30.13 and use that to construct the absolute value models. Absolute value models – or intrinsic Value models – do not compare the value of a company with its price on the financial markets but rather estimate it from fundamental, underlying factors such as the expected dividends.

30.5.4.1 Dividend Discount Model (DDM)

Since the value of a company equates its future potential to generate dividend, it would be logical to try to calculate the value of a company as the discounted flow of future dividends. The problem is that the future flow of dividends is not known. It is, however, possible to make educated guesses about the future potential to generate dividends, by creating a model for different variables that influence these dividends.

DDM
dividend discount model

Theorem 30.5.1 (DDM). *The value of a stock is given by the discounted stream of dividends:*

$$V_0 = \sum_{t=1}^{\infty} \frac{D_t}{(1+r)^t}$$

Capital gains appear as expected sales value and are derived from expected dividend income.

V_0 the intrinsic value of the stock now

D_t the dividend paid in year t

r is the capitalization rate and is the same as $E[R_k]$ in the CAPM, see Equation 30.10 on page 610

In this section, we will develop this idea further. We start from the most simple case, where we assume that the dividend will increase at a constant and given rate g. This model is called the constant growth DDM.

Constant Growth DDM (CGDDM)

CGDDM
constant growth dividend discount model

A particularly simple yet powerful idea is to assume that the dividends grow at a constant rate. It might not be the most sophisticated method, but is a good zero-hypothesis and it leads to elegant results that can be used as a rule of thumb when a quick price estimate is needed or when one tries to assess if the result of a more complex model makes sense.

If every year the dividend increases with the same percentage, $100g\%$ (with g the growth rate), then each dividend can be written in function of the previous one.

$$\begin{aligned} D_1 &= D_0(1+g) \\ D_2 &= D_1(1+g) &= D_0(1+g)^2 \\ &\ldots \\ D_n &= D_{n-1}(1+g) &= D_0(1+g)^n \end{aligned}$$

Theorem 30.5.2 (constant-growth DDM). *Assume that* $\forall t : D_t = D_0(1+g)^t$, *then the DDM collapses to*

$$V_0 = \frac{D_0(1+g)}{r-g} = \frac{D_1}{r-g}$$

This model is very simple and the following examples illustrate how to calculate a reasonable approximation of a the value of a private company. In those examples, we use a fictional company that has the name "ABCD."

Note – The required rate of return

Note, that in all those examples, we need to calculate first the required rate of return for that given company (denoted R_{ABCD}). To approximate this, we use the CAPM (see Chapter 30.4 *"The Capital Asset Pricing Model (CAPM)"* on page 610). This, requires of course the assumptions that the beta is endogenous (which is not entirely correct, but a good approximation in many cases).

Example: ABCD with $g = 0\%$

Consider the company, ABCD. It pays now a dividend of €10 and we believe that the dividend will grow at 0% per year. The risk free rate (on any horizon) is 1%, and the market risk premium is 5% and the β is 1. What is the intrinsic value of the company?

Using the CGDDM, $V_0 = \frac{D_0(1+g)}{R_{ABCD}-g}$, and the CAPM $R_{ABCD} = R_{\text{RF}} + \beta.RP_M$ we get:

```
V_0 <- 10 * (1 + 0.00) / (0.01 + 0.05 - 0.00)
print(round(V_0,2))
## [1] 166.67
```

The company value that assumes a zero growth rate is called the "no-growth-value." Unless when the outlook is really bleak, this is seldom a valid assumption. The concept of investing in a company and hence accepting its increased risk, depends on the fact that it should – on average – be a better investment than investing in bonds or safer interest bearing products.

Example: ABCD with $g = 2\%$

Expected growth rate of the dividend is 2%, ceteris paribus.

```
V_0 <- 10 * (1 + 0.02) / (0.01 + 0.05 -0.02)
print(round(V_0,2))
## [1] 255
```

The difference in value compared to the previous example is called the PVGO (present value of growth opportunities). So,

$$P_{\text{equity}} = \text{no-growth-value} + \text{PVGO} \tag{30.15}$$

$$V_0 = \frac{D_0}{r} + \text{PVGO} \tag{30.16}$$

Example: ABCD with $g = 2\%$ and $\beta = 1.5$

The β is assumed to be 1.5, ceteris paribus. What is the value of ABCD?

```
V_0 <- 10 * (1 + 0.02) / (0.01 + 1.5 * 0.05 - 0.02)
print(round(V_0,2))
## [1] 156.92
```

Since the company is more risky and all other things remained the same, the price must be lower. For the price to be the same, investors would expect higher returns.

Example: ABCD — extreme growth

The dividend growth rate is now expected to be 10%, ceteris paribus.

Simply adding that growth rate in the formula leads to impossible results:

```
V_0 <- 10 * (1 + 0.02) / (0.01 + 1.5 * 0.05 - 0.10)
print(round(V_0,2))
## [1] -680
```

A companies equity can never become negative, because the equity holder is only liable up to the invested amount. Therefore this result is not possible: it indicates a situation where the DDM fails (the growth rate cannot be larger than the required rate of return).

This example illustrates that the DDM is only valid for dividend growth rates smaller than the required rate of return. A company that would grow faster, would be deemed to be more risky and hence would have a higher beta, leading to a higher discount rate. The model states that anything above that is unsustainable and will lead to a correction.

When buying a company, it is not realistic to expect it to grow eternaly at the same rate. The growth rate can be assumed to follow certain patterns to match economic cycles or the investor might even assume zero growth after 20 years – simply as a rule of thumb to avoid overpaying.

Relationship between growth rate and ROE

To better understand how the growth rate related to certain accounting values, we introduce the following concepts.

Definition: earnings

E := net income

E

Definition: dividend payout ratio

$\text{DPR} := \frac{D}{E}$

DPR

Definition: plow-back ratio (earnings retention ratio)

PBR := 1 − DPR

PBR

Definition: Return on Equity

$\text{ROE} := \frac{E}{P}$

ROE

Note

Note that all definitions work as well per share as for the company as a whole! **We will use all concepts per share** unless otherwise stated.

The growth rate of the dividend is the amount of ROE that is not paid out as dividend, hence

$$g = \text{ROE} \times \text{PBR} \tag{30.17}$$

This is because if the company retains $x\%$ earnings, then the next dividend will be $x\%$ higher. More generally:

$$g = \frac{\text{reinvested earnings}}{\text{BV}} = \frac{\text{reinvested earnings}}{\text{TE}} \frac{\text{TE}}{\text{BV}}$$

BV
book value

where BV stands for "book value", and TE is the "total earnings" (note that this is similar to the concept of total earnings in the context of personal income).

Hence, – as mentioned earlier –

$$Price = \text{no-growth-value} + \text{PVGO} \tag{30.18}$$

$$V_0 = \frac{D_0}{r} + \text{PVGO} \tag{30.19}$$

If the stock trades at its intrinsic value (i.e. $P_0 = V_0$) and if we assume the CGDDM then

$$V_0 = \frac{D_1}{r - g} \tag{30.20}$$

$$= \frac{D_1}{r - ROE \times PBR} \tag{30.21}$$

Conclusions for the DDM Method

The DDM has certain distinct *advantages*; the DDM is

- logical and complete in a liquid market,
- easy to understand, and
- it only makes assumptions about the outcome (dividend) and not the thousands of variables that influence this variable.

However, in order to calculate the model one needs

- to forecast an infinite amount of dividends,
- to find a good discount rate (which is complex and actually circular), and hence
- it is incomplete without stress testing the result.

30.5.4.2 Free Cash Flow (FCF)

FCF
free cash flow

The DDM takes an outsider's view: it assumes that the dividends are "black box," ie. we have no insight in their dynamics and we can only observe the end result. However, if one buys a significant stake in a company this comes with influence on the management and insight in the inner workings of the company. This means that it becomes possible to make a simulation that builds the estimation of the dividend from a simulation of the balance sheet. This approach is called the free cash flow method (FCF method).

The first thing to do is to identify the cash that is available to pay dividends. It is reasonable to assume that the only value a company brings to the shareholder is the financial income. This means that the value of the company is the present value of all future cash flows to the equity holder paid up-front (now and at once). Therefore, it is important to use only cash flows that go to the shareholder and exclude salaries, bonuses, taxes, etc. ... so we use free cash flow (FCF).

Free Cash Flow (FCF)

Definition: Free Cash Flow (FCF)

Free Cash Flow is the cash flow available for distribution to equity holders of the company.

$$\text{FCF} = \text{EBIT}(1-\tau) + \text{Depreciation} + \text{Amortization} - \Delta\text{WC} - \text{CapEx}$$

FCFF
free cash flow to firm

In other words, starting from EBIT one will want to

- take out the tax paid since that is not going to the shareholders,
- add again depreciations and amortizations because these are no cash outflows,[5]
- reduce by changes in working capital (if the working capital increased, this means that the company needed more cash to operate and this will reduce the owner earnings), and
- reduce by changes in capital expenses, because these costs really reduce liquidity (these are of course linked to the amortizations and depreciations).

Digression – FCF or FCFF?

FCF is also referred to as FCFF (Free Cash Flow to Firm). These are essentially the same concepts.

Alternative Ways to Calculate FCF

There are multiple ways to calculate the FCF, depending on the data available one could choose for example the following format.

$$\begin{aligned}
\text{FCF} &= \text{EBIT}\ (1-\tau) + \text{Depreciations} + \text{Amortisations} - \Delta\text{WC} - \text{CapEx} \\
&= \text{NOPAT} + \text{Interest Expense} + \text{Depreciations} + \text{Amortisations} \\
&\quad - \Delta\text{WC} - \text{CapEx} - \text{Tax Shield On Interest Expense} \\
&= \text{PAT} - (1-d)\ (\text{Depreciations} + \text{Amortisations} - \Delta\text{WC} - \text{CapEx})
\end{aligned}$$

where NOPAT stands for "net operational profit after tax", PAT for "profit after tax", WC for "working capital", CapEx for "capital expenses" (expenses that get booked as capital), and τ is the tax rate.

NOPAT
WC
CapEx

For dividend estimations, it is useful to use the concept "Net free cash Flow". This is the FCF available for the company to maintain operations without making more debt, its definition also allows for cash available to pay off the company's short term debt and should also take into account any dividends.

Net Free Cash Flow

$$\begin{aligned}
\text{Net Free Cash Flow} = {} & \text{Operation Cash flow} \\
& - \text{Capital Expenses to keep current level of operation} \\
& - \text{Dividends} - \text{Current Portion of long term debt (LTD)} \\
& - \text{Depreciations}
\end{aligned}$$

[5] The defining difference between depreciation and amortization is that amortization charges off the cost of an intangible asset, where depreciation charges of cost of a tangible asset

Here, the Capex definition should not include additional investment on new equipment. However, maintenance cost can be added. Further, we should consider the following.

- *Dividends:* This will be base dividend that the company intends to distribute to its share holders.
- *Current portion of LTD:* This will be minimum debt that the company needs to pay in order to not default.
- *Depreciation:* This should be taken out since this will account for future investment for replacing the current PPE.

Net Free Cash Flow is a useful measure for the management of a company but we will not need it for company valuation.

30.5.4.3 Discounted Cash Flow Model

DCFM
discounted cash flow model

In the previous section, we explained the concept Free Cash Flow (FC). Since these free cash flows should be equal to the potential dividend payments, we can now use the FCF to estimate the value of a company. To do this we need to model the balance sheet and P&L for the future and from this building up estimates for the money that can be used reasonably to pay dividend. Then we discount all those FCF estimates to today to obtain the present value of the company. This is called the Discounted Cash Flow Model (DCF Model). In essence this method is an application of the basic formula for net present value.

Discounted Cash Flow

DCF

Definition: Discounted Cash Flow

Discounted cash flow (DCF) is a method of valuing a company, project, or any other asset by discounting future cash flow to today's value (in other words: using the time value of money) and then summing them.

In practice, future cash flows are first estimated and then discounted by using the relevant cost of capital to give their present values (PVs). The sum of all future cash flows, is the then called NPV, which is taken as the value or price of the cash flows in question. We remind the definition of NPV

NPV

Definition: NPV

The Net Present Value (NPV) is then the sum of all present values and represents today's value of the asset.

The DCF model in company valuation is simply calculating the NPV of the companies FCF.

Essentially, the DCF model is the sum of all future cash flows discounted for each moment t, and can hence be written as follows:

$$\text{PV} := \frac{\text{CF}_t}{(1 + r_t)^t}$$

In that model we will substitute both cash flow CF_t and required rate of return r_t as follows.

- CF by FCF, because that is the relevant cash flow for the potential buyer of the company.
- r by WACC, because the company should at least make good for compensating its capital. needs

Just as for the DDM it is possible to make some simplifying assumptions:

$$\begin{aligned}
V &= \sum_{t=1}^{\infty} \frac{\text{FCF}_t}{(1+\text{WACC})^t} \\
&= \sum_{t=1}^{\infty} \frac{\text{FCF}_{t-1}(1+g_t)}{(1+r)^t} && \text{(with } FCF_0 \text{ known)} \\
&= \text{FCF}_0 \sum_{t=1}^{\infty} \frac{(1+g)^t}{(1+\text{WACC})^t} && \text{(assuming } \forall g_t : g_t = g) \\
&= \text{FCF}_0 \frac{1+g}{\text{WACC}-g} && \text{(assuming } g < \text{WACC and } g \neq -1) \\
&= \frac{\text{FCF}_0}{\text{WACC}} && \text{(assuming } g = 0)
\end{aligned}$$

Of course, it will not always be appropriate to simplify too much, but – also here – these simplification might be a good rule of thumb.

Advantages and Disadvantages of the DCF method

The DCF method for company valuation has the following distinct *advantages*.

- Always applicable to all companies.
- Logical and complete.
- Easy to understand.
- DCF allows for the most detailed view on the company's business model.
- It can be used to model synergies and/or influence on the company's strategy.

However, there are a few *caveats*:

- Determining the FCF is not too easy.
- One needs to forecast an infinite amount of FCFs (and therefore, one needs to model the whole balance sheet).
- Therefore, one needs many assumptions (costs, inflation, labour costs, sales, etc.)
- One needs to find a good discount rate (which is complex and actually circular).
- It is incomplete without stress testing the result.

30.5.4.4 Discounted Abnormal Operating Earnings Model

Calculating the Net operating asset value (NOA) is necessary for applying the Discounted Abnormal Operating Earnings valuation model (DAOE). DAOE is one of the most widely accepted valuation models because it is considered the least sensitive to forecast errors. NOA can also be used in the calculation of Free cash flow (FCF) and therefore, the Discounted cash flow model. However, it is not necessary to calculate FCF. **NOA** **DAOE**

$$\begin{aligned}
\text{DAOE} &= \frac{\text{NOPAT}(t) - \text{WACC} \times \text{NOA}(t-1)}{\text{WACC}} + \text{NOA} - \text{BVD} \\
\text{FCF} &= \text{NOPAT} - \Delta\text{NOA} \\
\text{DCF} &= \frac{\text{FCF}}{\text{WACC}} - \text{BVD} \qquad \text{(in case of zero growth)}
\end{aligned}$$

30.5.4.5 Net Asset Value Method or Cost Method

The Net Asset Value Method (hereafter NAV method) is probably the simplest method to value assets but it also has the most narrow field of application. To value a company it will simply take the value of the assets of the company.

NAV
net asset value

> **Definition: Net Asset Value Method**
>
> The Net Asset Value Method is also known as the Liquidation Method. The idea is to use the liquidation value as a proxy for the company's value. The question one answers is: "if the company stopped trading now, what would be got from all assets (reduced with all liquidation costs)."

This method only considers the assets and liabilities of the business. At a minimum, a solvent company could shut down operations, sell off the assets, and pay the creditors. The money that is left can then be distributed to the shareholders and hence can be considered as the value of the company.

Of course, companies are supposed to grow and create value, hence this method is rather a good floor value for the company. In general, the discounted cash flows of a well-performing company exceeds this floor value.

Zombie companies that needs subsidiaries to survive for example that own many tangible assets might be worth more when liquidated than when operations are continued.

This method is probably a good alternative for valuing non-profit organisations, because generating profit (cash flow) is not the main purpose of these companies.

liquidation cost

Further, it is essential to consider the purpose of this type of valuation. If it is really the idea to stop trading and liquidate the company, in that case one will have to add the liquidation cost. For example, selling an asset might involve costs to market it, have it valued, maintain it till it sold, store it (e.g. keep a boat in a harbour), etc.

The time scale becomes also relevant here. If one needs urgently the cash then the expected price will be lower, but also the cost of storing and maintaining the asset might be lower.

Depending on the purpose of the valuation one will also have to choose what value to consider: book value or market value? For example, a used car might be worth 0 in the books but still can be sold for good money.

Investment Funds

Investment funds are a very specific type of companies that are created with the sole purpose to invest in other financial assets.[6] The most common type of investment funds will invest in liquid assets and not try to influence the management of the company.

Investment funds can invest in all other financial assets that are explained in this chapter, and can do much more. For example, an investment fund can invest in real estate, labour ground, or eventually actively play a role in infrastructure works.

UCITS
Undertaking for Collective Investments in Transferable Securities

The investment funds that are the most relevant for the investor that wants to save money for retirement – for example – are the liquid investment funds that are investing in the assets that are explained in this chapter. In Europe they are known as UCITS (Undertaking for Collective Investments in Transferable Securities) and regulated by the UCITS IV regulations.

holding company

[6] In that sense investment funds are comparable to holding companies, which have as sole purpose to invest in other companies and participate in their management.

UCITS will never invest in their own shares. They have a variable capital and buying shares is considered as "redeeming" shares. This means that in stead of buying and holding its own shares these shares stop to exist.

In the same way the fund can create new shares when more people want to buy the fund. Typically, there will be a market maker to facilitate this process.

Market Maker

Advantages and Disadvantages of the NAV Method

While for investment funds and bankrupted companies the NAV method is the only method necessary, this method will not give an appropriate picture of normal operating companies.

Again, there are some *advantages* that stand out.

- Simple and straightforward.
- Easy to understand.
- No assumption needed about a discount rate.
- It is all we need for companies in receivership and investment funds.

For normal operating companies that are not in receivership:

- The NAV is only the lower limit of the real value (and hence merely a reality check) and for normal companies it misses the point of a valuation.
- It is irrelevant for growth companies (e.g. Google, Uber, Facebook).

30.5.4.6 Excess Earnings Method

In this method, first the tangible assets are estimated and an appropriate return on those tangible assets. Then one subtracts that return from the total return for the business, leaving the "excess return." This excess return is presumed to come from the intangible assets. An capitalization rate is applied to this excess return, resulting in the value of those intangible assets. That value is added to the value of the tangible assets and any non-operating assets, and the total is the value estimate for the business as a whole.

30.5.5 Relative Value Models

30.5.5.1 The Concept of Relative Value Models

Relative Value methods are also known as "Guideline Companies Method," "Comparative Value Models," "Comparable Companies Analysis" etc.

The idea is to determine the value of a firm by adjusting known prices of similar companies. The comparison is for example done via one or more indices such as the price-earnings or price-to-book ratios.

Market Value vs Instrinsic Value

intrinsic value
market value

The intrinsic value is the true fair value of a company. However, this is not always equal to what someone else is willing to pay for it. The amount that others are willing to pay is the market value.

Definition: Price or market value

P_0 = the price paid for a company on the market (market price).

Definition: Value

V_0 = the real value of a company (intrinsic value).

A short-cut: the price is the consensus of the market about the value.

Definition: Market capitalization

The market capitalisation (often shortened to "market cap") is the total value of all outstanding stocks at the market price. It is the value of the company as fixed by the market.

If you have ever bought or sold a house or an apartment, you will know how this process works. Typically, when one plans to buy a property, one will scan the market for suitable properties. Suitable properties are those that are in a certain location, have the required number of rooms, etc.

Comparing the price per square meter makes a lot of sense, but it is not the whole story. The neighbourhood, quality of the property, age, willingness to sell, etc. will all play an important role.

Similarly, we can compare companies via the price earnings ratio, but many other elements will need to be taken into account.

30.5.5.2 The Price Earnings Ratio (PE)

Relative value models try to estimate the relative price of a company by comparing it with others in a meaningful way. This can not be done by considering the dollar value of the profit for example, since then the bigger company always will have a higher number. When using ratios instead, we compare variables that can be compared.

For example we can compare the price of the share on the stock exchange with its profit. Both are expressed in the same currency, and hence the ratio is devoid from dimensions and can be compared to other companies.

The price earnings ratio (PE), compares the price of the stock with the profit of the company, and hence is quite directly a measure for future income for the equity holder.

PE
price earnings ratio

Definition: – Price earnings ratio (PE ratio)

The price earnings ratio is the market price of a company divided by its last annual earnings.

$$\text{PE} := \frac{V_0}{E_0}$$

Using this definition and rearranging Equation 30.19 on page 620 shows that:

$$\frac{V_0}{E_1} = \frac{1}{r} + \frac{\text{PVGO}}{E} \tag{30.22}$$

$$= \frac{\text{DPR}}{r - \text{ROE} \times \text{PBR}} \tag{30.23}$$

$$= \frac{\text{DPR}}{r - g} \tag{30.24}$$

$$= \frac{1 - \text{PBR}}{r - g} \tag{30.25}$$

where we remind that DPR is the dividend payout ratio, PBR the plow-back ratio, PVGO is the present value of growth opportunities, r is the required discount rate and g is the growth.

These formulae make clear that PE is lower for more risky firms; and that when ROE increases that then PE decreases.

Digression – How bonds and shares are (dis-)similar

If PVGO = 0, then Equation 30.22 shows that $V_0 = \frac{E_1}{r}$: the stock is then valuated as a perpetual bond with coupon E_1; and the PE ratio is then $\frac{1}{r}$. Further, one will remark that if $g = 0$, that then $E_1 = E_0$ and hence

$$\text{PE} = V_0/E_0.$$

30.5.5.3 Pitfalls when using PE Analysis

When using a relative value method to calculate the value of a company, it is usually not a good idea to use only one ratio. While PE is has good prediction power, it needs to be handled with care and it is worth to consider the following issues.

- *Accounting details:* The earnings are sourced from the accounting system and it is worth to investigate how this particular company is using the various accounting rules and which rules apply.
- *Earnings management:* the management has some freedom within the guidelines of the accounting rules to show more or less profit on short term.
- *Economic cycles:* Economic cycles might influence the earnings of companies in various ways and patterns can change when the economic environment changes (e.g. company A can do relatively better than B in economic downturns, while otherwise B does better in the same market).
- *Estimation of the first future earnings:* The formula tells us to use E_1, but that value is not known yet. In practice, one uses the earnings of previous accounting year, E_{-1}.
- *Circular reasoning in value and riskiness:* PE ratio includes the future growth potential and the riskiness in one measure, hence it is extremely important to compare only with companies that share similar potential and risk (e.g. companies of the same sector in the same country).
- *Short term fluctuations for a long term estimate:* For the same reason – that PE ratios include the future growth potential and the riskiness in one measure – they will jump up when the economic cycle is on its low in short term.

30.5.5.4 Other Company Value Ratios

Besides the PE ratio discussed above there are a lot more ratios that can be used. Usually, it is a good idea to use a few ratios and compare results.

The price-to-book ratio is similar, but uses the book value instead of the market value of the share.

Definition: Price-to-book ratio (PTB)

$$\text{PTB} := \frac{P}{\text{BV}} \qquad \text{with BV} = \text{book value}$$

PTB
price to book ratio

The book value (BV), is the value of the company as per accounting standards – in other words the size of the balance sheet. The advantage of using the book value is of course its stability, the downside is that accounting rules are designed to collect taxes. That makes accounting rules inherently backwards looking, while company valuation is essentially forward looking.

The list of possible ratios to use is only limited by your imagination and can differ for different companies. For example a steel producer can build up a stock of completed products, in a bank that works differently: they need assets to compensate risk.

PTCF

price-to-cash-flow ratio

Definition: Price-to-cash-flow ratio (PTCF)

$\text{PTCF} := \frac{P}{\text{CF}}$ with CF = (free) cash flow

Sales is important for any company and easy to isolate in the balance sheet.

PTS

price to sales ratio

Definition: Price-to-sales ratio (PTS)

$\text{PTS} := \frac{P}{S}$ with S = sales

These ratios are necessarily a small sample of what is used in practice. It is not really possible to provide an exhaustive list. For each sector, country or special situation some other measures might be useful. Actually all measures that we have presented in Chapter 28 *"Financial Accounting (FA)"* on page 567 can also be used to calculate the value of a company. It is even possible to define your own ratio, that makes most sense to you and the special situation that you're dealing with.

Already in Section 28.4 *"Selected Financial Ratios"* on page 575, we discussed some ratios that help the managers of the company to manage value. They are different from the ratios discussed in Section 30.5.5.2 *"The Price Earnings Ratio (PE)"* on page 626, because they cannot directly be used to calculate the value of a company. However, they are indirectly linked to company value and can be used to amend the value of a company or to gain more insight in how the company is doing compared to its competitors.

As data scientist in a commercial company, you are very likely to encounter these ratios or you might be in a position to include them in a report.

ROIC
return on invested capital

ROI
return on invested capital

Return on Invested Capital (ROIC)

Most important for the investor is the return that is earned on the capital that is invested. Therefore both management and shareholder will every year carefully monitor the return on invested capital (ROIC).

Definition: Return on Invested Capital (ROIC)

$$\begin{aligned}\text{ROIC} &= \frac{\text{Operating Profit}(1-\text{tax rate})}{\text{Book value of Invested Capital}_{t-1}}\\ &= \frac{\text{Operating Profit} - \text{Adjusted Taxes}}{\text{Invested Capital}}\\ &= \frac{\text{EBIT}}{\text{Fixed Assets} + \text{Intangible Assets} + \text{CA} - \text{CL} - \text{cash}}\\ &= \frac{\text{EBIT}}{\text{debt + equity - cash (- goodwill)}}\\ &= \frac{\text{EBIT}}{\text{TCE}}\end{aligned}$$

we remind that CA stands for current assets and CL for current liabilities.

Return on Capital Employed (ROCE)

The ROIC only includes the equity capital employed. So, while for the investors that is the most important, the management can also raise capital via debt. Therefore, it is customary to use also ROCE. ROCE uses the total capital employed by the company (this is the sum of Debt and equity). Another difference is that ROCE is a pre-tax measure, whereas ROIC is an after-tax measure. So ROCE measures the effectiveness of a company as the profit exceeding the cost of capital.

Definition: Return on Captial Employed (ROCE)

$$\text{ROCE} = \frac{\text{net operating profit}}{\text{total capital employed}} = \frac{\text{NOP}}{\text{TCE}}$$

where NOP is the net operating profit.

ROCE
return on capital employed

NOP
net operating profit

Return on Equity (ROE)

Definition: Return on Equity (ROE)

$$\begin{aligned} \text{ROE} &= \frac{\text{Net Income}_t}{\text{equity}} \\ &= \frac{\text{Net Income}}{S} \times \frac{S}{\text{TA}} \times \frac{\text{TA}}{\text{equity}} \qquad \text{(DuPont Formula)} \\ &= \frac{\text{NOPAT}}{\text{equity}} \end{aligned}$$

where S is the sales and TA are the total assets

Note – Difference ROIC and ROE

The main difference with ROIC is that ROE does not include debt (including loans, bonds and overdue taxes) in the de-numerator. The second difference is that ROE uses in the numerator earnings after taxes (but before dividends), where RoC uses earnings before interest and taxes (EBIT).

ROE
Return on Equity

ROE shows how profitable a business is for the investor/shareholder/owner, because the denominator is simply shareholders' equity. ROIC and ROCE show the overall profitability of the business (and for the business) because the denominator includes debt in addition to equity (which is also capital employed, but not necessarily provided by the owner).

ROE and ROCE will differ widely in businesses that employ a lot of leverage. Banks for example, earn a very low return on assets because they earn a small spread (e.g.: borrow at 0.5%, lend at 3.5%). Regular saving banks have the majority of their capital structure in depositors' money (i.e. low-interest bearing debt) and this leverage magnifies their returns compared to equity. It is typical for banks to have low ROCE but a high ROE.

An important footnote is that in the denominator of ROE one will find the book value of the equity (of course one might make the calculation with the market value). However, that is not necessarily the most important reference for the investor. The investor might have his/her own book value, purchase price or other price as a reference.

Economic Value Added (EVA)

economic value added

EVA

> **Definition: Economic Value Added (EVA)**
>
> Economic Value Added (EVA) is an estimate of the company's economic profit (the value created in excess of the required return of the company's shareholders). In other words, EVA is the net profit less the opportunity cost for the firm's capital.

$$\begin{aligned} \text{EVA} &= (\text{ROIC} - \text{WACC})\,(\text{TA} - \text{CL}) &\quad (30.26)\\ &= \text{NOPAT} - \text{WACC}\,(\text{TA} - \text{CL}) &\quad (30.27) \end{aligned}$$

Market Value Added (MVA)

MVA
market value added

Market Value (of a company)

> **Definition: Market Value Added (MVA)**
>
> Market value added (MVA) is the difference between the company's current market value and the capital contributed by investors.

$$\text{MVA} = V_{\text{market}} - K \quad (30.28)$$

with V_{market} the market value and K the capital paid by investors.

If a company has a positive MVA this means that it has created value (in case of a negative MVA it has destroyed value). However, to determine if the company has been a good investment one has to compare the return on the invested capital with the return of the market (r_M), adjusted for the relative risk of that company (its β).

The MVA is the present value of the series EVA values. MVA is economically equivalent to the traditional NPV measure of worth for evaluating an after-tax cash flow profile of a company if the cost of capital is used for discounting.

$$\text{MVA} = \sum_{t=0}^{\infty} \frac{\text{EVA}_t}{(1 + \text{WACC})^t}$$

There is no best measure, and all measures have to be used with care. It is important to consider what one wants to obtain before making a choice. It is very different to make a comparative analysis within one sector or compare different sectors for example.

30.5.6 Selection of Valuation Methods

Selecting a valuation method is not an exact science, there are many methods with each their strong and weak points, point of view, and assumptions endogenous and exogenous to the model itself. In general, it is not a bad idea to use multiple methods and compare results.

The basic idea could be as follows.

1. Is it my purpose to buy the company and stop its activities or did it already stop trading or is it an investment fund? – If yes, use NAV method (in all other cases this should be the lower limit). If not, then continue to next question.

2. Will you be an important share holder and can you make a business plan? – If yes, try to use DCF, if not continue.

3. Do you have the option not to invest? – If yes, use DDM otherwise continue.

4. If you ended up here, this means that you have to invest anyhow in similar stocks (e.g. you are an equity fund manager and need to follow your benchmark). – In this case, you might want to use a relative value method.

30.5.7 Pitfalls in Company Valuation

None of the methods shown in this book, no statistical method, nothing that is in our hands is able to predict the future. Ask the same valuator to calculate the value of the same company after a year and most probably the outcome will be different. As the future reveals itself we have more information and our assumptions about the future will change.

No method is able to solve this problem and there are many more issues to discuss and be aware of. For example, calculating FCF makes a lot of assumptions and even if we perfectly forecast the sales, inflation, labour costs, etc. then still some variable can be influenced by decisions of the management or higher organs.

In this section we will have a closer look at some of those problems.

30.5.7.1 Forecasting Performance

Valuating a company is a delicate exercise in forecasting the future while one only has the past and the present to one's disposal. It compares to driving a car by using rear mirrors and side windows. Most notably, we will have to consider the following issues.

- *relevance of history:* is the past data relevant for the future?

- *short history:* it might be easier to forecast mature companies with long, stable history. Buying a company with only a few years history is a leap of confidence.

- *management differences:* will you attribute cash differently? Is the salary that the owner (not) took relevant for your case? etc.

There are many other factors to consider. For example: when identifying the non-operating assets in the accountancy of a company one needs to be careful to avoid double counting in valuation.

30.5.7.2 Results and Sensitivity

Since a company valuation is about forecasting the future, it is not a well determined value. Different valuators working on the same valuation will find different prices. The key is to understand how you got your results, what assumptions you made an to what extend variations in price can be accepted.

Elaboration on sensitivity analysis can be found for example in De Brouwer (2016). In this section we will explain the basic tools that should help the reader to get started in practice.

Stress Test

Stress Testing

In order to gain some insight in how robust a certain result of our valuation is or what bad cases can be expected a simple stress test can answer that question.

A simple example could be: allow the price of certain raw materials to fluctuate (simply test a few possibilities), then do the same with labour prices, allow the effect of a strike, an earthquake, fluctuations in exchange rates, one of the lenders that gets into problems, we have to halt digging because we stumbled upon a site of historic importance, etc.

Soon, one of the problems with stress testing becomes obvious: it becomes bewildering how much possibilities there are, it is impossible to say which is more probable that the other, etc. The answer to that shortcoming is simply to restrict stress testing to what it does best: explore extreme risks – without knowing how likely it is. So, for example assume that we are building an airport and an earthquake destroys a lot of the half-build site, killed a few people, causes a strike of our crew, and creates a negative climate that makes the currency plunge, this in return pushes and the domestic bank in the syndicate into problems, etc. Then we have just one scenario, something that we can calculate with your spreadsheet and that gives us a "worst case scenario."

The relevance for each investor is that he should ask the question "can I afford to loose that much." If the answer is "no," then the investor should seek another partner in the syndicate in order to diversify risks. Failing to do so, is planning for disaster.

In order to do that in practice, a spreadsheet might still be sufficient, however, it might be advisable to follow a few simple rules to keep it organized. For example:

P&L
Profit and Loss

- Use different tabs (sheets) for (i) assumptions, (ii) costs, (iii) income, (iv) P&L, and (v) ratios.
- Make sure that each sheet has the same columns (they are your time axis).
- Use different colours to make the different function of each cell clear: for example pale yellow for an input cell, no background for the result of a calculation, etc.
- Avoid – where possible – obscure formulae that are difficult to read for humans.

VB
Visual Basic

- Do use as much as possible underlying programming language (Visual Basic for example) and never ever use macros (macros are very difficult to read by other humans, not reuseable, slow, and confusing).
- Keep different versions, have frequent team-meetings when working on one file, and agree who will modify what.

Following these simple rules will help you to make rather complex models in the simple spreadsheet that a modern computer offers. If you find that the spreadsheet becomes difficult to read or slow we suggest to have a look at the alternatives presented on page 634.

Monte Carlo Simulation

Monte Carlo Simulations

A Monte-Carlo simulation can be understood as hundreds of thousands of valuations run by an automated machine so that it becomes possible to get an idea about how probable certain outcomes are. This is of course only possible if we are able to say something sensible about the underlying risk factors.

With "something sensible" we mean that we know something about the likelihood of something to happen. We might not know the exact distribution, but at least some probability. For example, we might expect an earthquake of force 4 to happen once in 1 000 years. This simple number is far less than knowing the probability density function, but it can already work.

In that case, we would have a 0.000 083 probability each month that such earthquake would occur. However, if it occurs, then the knock-on effects will be significant for the project: damage, delays, other problems in the region needing attention, etc. It is here that the limitations of a spreadsheet become all too clear. It becomes impossible to model correctly the effect of such events, not only because of the interdependence with other parameters, but also in time. If such event occurred, then is it more or less likely to happen again? Some effects will be immediate (such as if the currency drops 20% with respect to the currency that we use to pay a certain material or service, then that service or material is immediately more expensive). This can still be modelled in a spreadsheet, but in the realistic case with the earthquake one must take into account a whole different scenario for the rest of the project and that becomes almost impossible and at least very convoluted.

The alternative is to use a programming language that allows us to model anything. Best suited for large projects are languages that allow for some object oriented code. We can use the features of an object oriented programming language to represent actors and input in our project. For example, the engineering company can be one "object" and it will decide to hedge currency risk if the exchange rate hits a certain barrier, etc.

This allows us to model dependencies such as in our example with the earthquake. If the earthquake happened, then other objects can "see" that and react accordingly, the exchange rate (also an object) will switch regime (ie. draw its result from a different distribution), the workers can see the impact of the safety conditions and consider a strike with a given probability, etc. This way of working is not so far removed from the way modern computer games work.

Good examples of programming languages that allow vast amounts of complex calculations are C++ and R. The high level of abstraction offered by object oriented programming languages allows the programmer to create objects that can interact with each other and their environment. For example, the Engineering Company can be such object. That object can be instructed to employ more workers when a delay threatens to happen but up to the limit that the extra costs are offset by the potential penalties. As the simulation then runs, market parameters change and events happen according to their probability of occurrence and each object will then interact in a pre-programmed or stochastic way.

This allows very complex behaviour and dependencies to be modelled, yet everything will be in a logical place and any other programmer can read it as a book. On top of that there are good free solutions to create a professional documentation set with little effort. For example, Doxygen (see `http://www.doxygen.org`) is free and able to create both an interactive website as well as a LaTeX[7] book for the documentation, that details each class, function, handle, property, etc. Code written in such way and documented properly is not only easy to maintain, but also straightforward to audit.

[7]LaTeX is a high-quality typesetting system; it includes a large set of features designed for the production of technical and scientific documentation. LaTeX is the de facto standard for the communication and publication of scientific documents. LaTeX is available as free software in the repositories of your distribution and at `http://www.latex-project.org`. Information on how to link it with R, is in Chapter 33 *"knitr and LaTeX"* on page 703.

Beyond the Monte Carlo Simulation

Now, that we have a good idea how the distribution of the results will look like, we can use this distribution to calculate the relevant risk parameters. In many cases, the "historic" distribution that we got by our Monte Carlo simulation will be usable, however, for large and complex projects the distribution might not be very smooth. If we believe that this is a sign of the limited number of simulations, then we can try to apply a kernel estimation in order to obtain a smoother results that yield more robust risk parameters.

KDE
kernel density estimation

The technique of kernel density estimation (KDE) could be helpful for all distributions that are estimated from a histogram. As an alternative to parametric estimation where one infers a certain distribution it avoids the strong assumption that the data indeed follows that given distribution. Note a KDE can be used also for any input parameter where the distribution used is based on observations.

Of course, one can choose a standard distribution if we have reasons to assume that this would be a good approximation. However, choosing a non-parametric KDE, has the advantage of avoiding any assumptions about the distribution, and on top of that:

- It is well documented in the case of expected shortfall – see e.g. Scaillet (2004), Chen (2008), Scaillet (2005), and Bertsimas et al. (2004).
- There is research on its sensitivity with respect to the portfolio composition, $\boldsymbol{w}$ – see e.g. Scaillet (2004), Fermanian and Scaillet (2005).

Using a non-parametric KDE, however, requires one arbitrary parameter: "the bandwidth." The bandwidth is a parameter that is related to the level to which the data sample is representative of the real underlying distribution. If one makes a choice of this parameter that is too small, one forces the estimated distribution function, f_{est}, to stick too much to the data, and there is too little of a smoothing effect. If, on the other hand, the parameter is insufficiently restrictive, then f_{est} will be smeared out over an area that is too large.[8] More information on bandwidth selection can be found in Jones et al. (1996b).

Of course, one can ask if it is necessary at all to use a kernel estimation instead of working with the histogram obtained from the data. Using the histogram as pdf has a few disadvantages:

- It is not smooth (this observation tells us that the use of histograms is similar to noticing that The dataset is imperfect and not doing anything about it).
- it depends on the end points of the bins that are used (changing the end points can dramatically change the shape of the histograms).
- It depends on the width of the bins (this parameter can also change the shape of the histogram).
- It introduces two arbitrary parameters: the start point of the first bin, and the width of the bins.

An answer to the first two points (and half of the last point) is to use a kernel density estimation (KDE). In that procedure, a certain function is centred around each data point (for example, an indicator function, a Gaussian distribution, the top of a cosine, etc.), these functions then are

[8] Note that we do not use the usual notation for the estimated distribution density function, $\hat{f}$, because we have reserved that notation for the Fourier transform.

summed to form the estimator of the density function. The KDE is currently the most popular method for non-parametric density estimation – see e.g. in the following books: Scott (2015), Wand and Jones (1994), and Simonoff (2012)

This method consists in estimating the real (but unknown) density function $f(x)$ with

$$f_{est}(x;h) = \frac{1}{N}\sum_{n=1}^{N} K_h(x - x_n) = \frac{1}{Nh}\sum_{n=1}^{N} K\left(\frac{x - x_n}{h}\right) \tag{30.29}$$

where K is the kernel

Definition: Kernel

A kernel is a function $K(x) : \mathbb{R} \mapsto \mathbb{R}^+$ that satisfies the following conditions:

$$\begin{cases} \int_{-\infty}^{+\infty} K(u)\,\mathrm{d}u = 1 \\ \forall u \in \mathbb{R} : K(u) = K(-u) \end{cases}$$

If K is a kernel, then also $K^*(u) := \frac{1}{h}K\left(\frac{u}{h}\right)$ (with $h > 0$) is a kernel. This introduces an elegant way to use h as a smoothing parameter, often called "the bandwidth."

This method was hinted by Rosenblatt et al. (1956) and further developed in its actual form by Parzen (1962). The method is thus also known as the "Parzen-Rozenblatt window method."

The Epachenikov kernel (see Epanechnikov (1969)) is optimal in a minimum variance sense. However, it has been shown by Wand and Jones (1994) that the loss of efficiency is minimal for the Gaussian, triangular, biweight, triweight, and uniform kernels.

Two of those kernels are illustrated in Figure 30.1 on page 636.

If an underlying pdf exists, kernel density estimations have a some distinct advantages over histograms: they can offer a smooth density function for an appropriate kernel and bandwidth, and the end points of the bins are no longer an arbitrary parameter (and hence we have one arbitrary parameter less, but still the bandwidth remains an arbitrary parameter).

We also note that Scott (1979) proves the statistical inferiority of histograms compared to a Gaussian kernel with the aid of Monte Carlo simulations. This inferiority of histograms is measured in the L^2 norm, usually referred to as the "mean integrated squared error" (MISE), which is defined as follows.

$$MISE(h) = E\left[\int_{-\infty}^{+\infty} \{f_{est}(x;h) - f(x)\}^2\,\mathrm{d}x\right] \tag{30.30}$$

A variant of this, the AMISE (asymptotic version), can also be defined, and this allows us to write an explicit form of the optimal bandwidth, h. Both measures have their relevance in testing a specific bandwidth selection method. However, for our purpose these formulae cannot be used since they contain the unknown density function $f(x)$. Many alternatives have been proposed and many comparative studies have been carried out. A first heuristic was called "cross validation selectors" – see Rudemo (1982), Bowman (1984), and Hall et al. (1992). Sheather and Jones (1991) developed "plug-in selectors" and showed their theoretical and practical advantages over existing methods, as well as their reliable performance. A good overview is in Jones et al. (1996a).

In Figure 30.2 on page 636, we show how the histogram and Epachinov KDE are different.

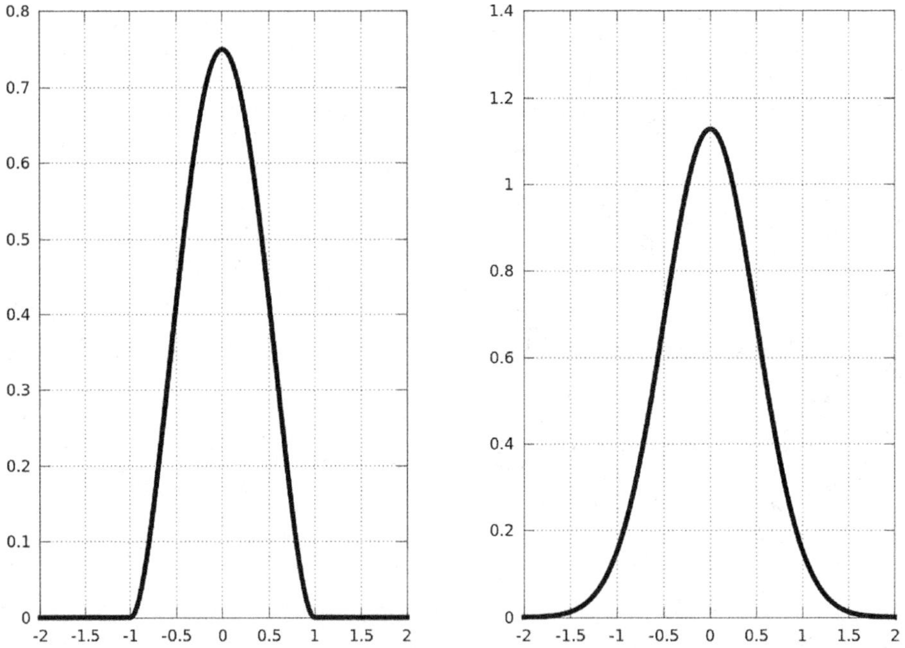

Figure 30.1: *The Epachenikov kernel (left),* $K_h^E(x) = \frac{3}{4h}\left(1 - \left(\frac{u}{h}\right)^2\right)\mathbf{1}_{\{|u/h| \leq 1\}}$ *for* $h = 1$*; and the Gaussian kernel (right),* $K_h^G(u) = \frac{1}{\sqrt{2\Pi h}} e^{-\frac{u^2}{h^2}}$ *for* $h = 0.5$.

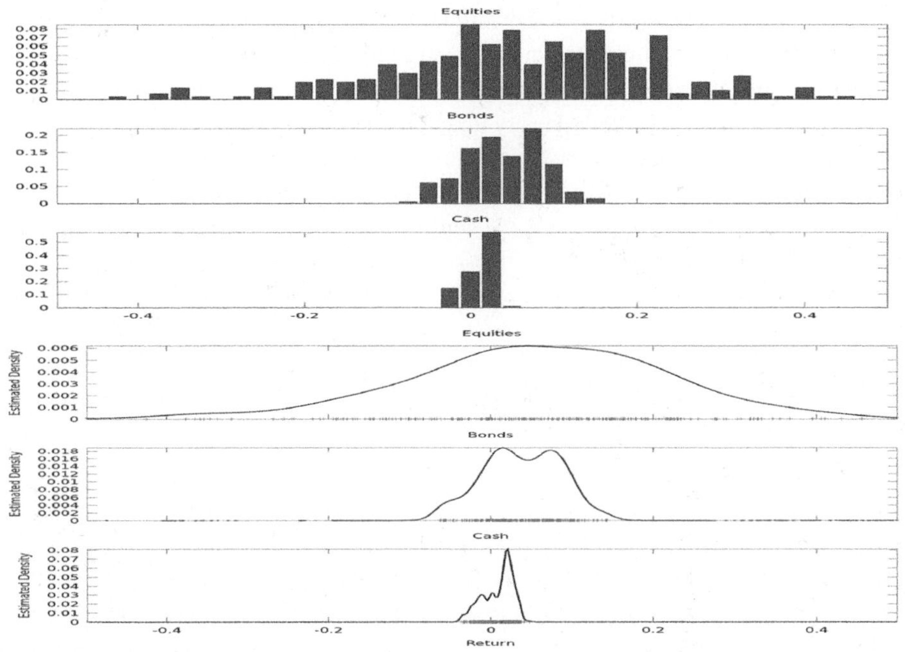

Figure 30.2: *As illustration on how the Epachenikov Kernel Estimation works, we present in the upper graph the histogram of the annual inflation corrected returns of standard asset classes. The lower graph offers a view on what a non-parametric kernel density estimation on those data can do.*

Conclusion

Kernel estimation is a widely accepted and used method that has many advantages. However, it introduces the arbitrary choice of bandwidth and of type of kernel. However, we note a novel method that automates this selection without the use of arbitrary normal reference rules Botev et al. (2010) see. We also note that it is blind to specific aspects, such as the boundedness of the domain of values (e.g. prices cannot become negative). Therefore, the method has to be used with care, and preferably on non-bounded data (e.g. log-returns).

30.6 Forwards and Futures

Forwards and futures give exposure to the underlying asset without actually buying it right now. Both forwards and futures are agreements to sell or buy at a future date at market price at that moment. They are defined as follows.

Definition: Future

A future is an agreements to sell or buy at a future date at market price at that moment, when the agreement is quoted on a regulated stock exchange.

Definition: Forward

A forward an agreements to sell or buy at a future date at market price at that moment, when the agreement is an OTC agreement.

The difference between a forward and future is only about the form of the agreement.

The value of a forward or future at maturity is the difference between the delivery price (K) and the spot price of the underlying as at maturity (S_T)

- For a long position, the payoff is: $F_T = S_T - K$.
- For a short position, the payoff is: $F_T = K - S_T$.

Reasoning that there should be no rational difference between buying an asset today and keeping it compared to buying it later and investing the money at the risk free interest rate till we do so, this relationship becomes – for assets that yield no income:

$$F_0 = S_0 \, e^{rT}$$

with T the time to maturity and r the continuously compounded risk free interest rate.

This formula can be modified to include income. If an asset pays income then this advantage is for the holder of the assets and hence we subtract it for the cash portfolio where one does not have the asset. For example, if the income is known to be a a discrete series of income I_t then the formula becomes

$$F_0 = \left(S_0 - \sum_{t=1}^{N} \mathrm{PV}(I_t) \right) \exp(r\,T)$$

A continuous stream of income ι can be modelled as follows:

$$F_0 = S_0 \exp\left[(r - \iota)T\right]$$

Note – Commodities

For commodities such as for example gold or silver there is no income from the asset, but rather a storage cost. This cost can be modelled in the same way as the income, but of course with the opposite sign.

This means that the value of a future will evolve quite similar to the value of the underlying asset. For example, to gain exposure to the S&P500 one would have to buy 500 shares in exact proportions, which is complicated, costly, and only possible for large portfolios, but it is possible to gain that exposure with just one future.

It is possible to manage a portfolio by only using futures, so that it looks like we have the underlying asset in portfolio. When the future is close to maturity, it is sufficient to "roll the contract": sell the old one and buy a new contract.

Forwards and futures play an important role on the stock exchange, they are efficient for exposure or hedging purposes and are usually very liquid.

One particular forward is the Forward Rate Agreement (FRA) that fixes an interest rate for a future transaction such as borrowing or lending. They are used to hedge against interest rate changes.

FRA
forward rate agreement

More specifically, the FRA is a cash settlement that compensates the counterpart for promising a transaction at a given interest rate in the future.

30.7 Options

30.7.1 Definitions

Options as financial derivative instruments are very much what one would expect from the English word "option." It is an agreement that allows the owner of the contract to do or not do something at his/her own discretion. Usually, it is something along the lines of "the bearer of this agreement has the right to purchase N shares of company ABC for the price of X at a given date T from the writer of this option, DEF".

There are two base types of options:

Definition: Call Option

A Call Option is the right to buy the underlying asset at a given price (the Strike) at some point in the future (the maturity date).

Definition: Put Option

A Put Option is the right to sell the underlying asset at a given price (the Strike) at some point in the future (the maturity date).

The option market is a very specialised market and has developed an elaborate vocabulary to communicate about the options. For example,

- The **strike** or **execution price** is the price at which an option can be executed (e.g. for a call the price at which the underlying can be sold when executing the option)
  The strike price is denoted as X.
- The **maturity date** is the expiry date of an option.

So, the owner of the option can at maturity date execute the right that a particular option provides. This act is called "exercising an option."

Example: buying an option

For example, you have bought a call option on HSBC Holdings plc with strike price of £600, the price today is £625. This means that you have the right to buy the share from the writer of the contract for £600, but you can sell it right away on the market for £625. This transaction leaves you with £25 profit.

Actually, option traders will seldom use the words "option buyer," they rather say that you have **long position**. The option buyer has the long position, he or she has **right** to sell or buy at the pre-agreed price.

The world is balanced and for each person or group that has a right, there is an other person or group that has an obligation to match that right. For each long position, there is a **short position**. The option writer has the **obligation** to sell or buy at the pre-agreed price. He/she has a short position.

Note – long and short

The words long and short are also used with relation to futures and shares. For example, having a short position in Citybank means that you have sold shares of Citybank without having them in portfolio. This means that at a later and pre-defined moment in time you will need to buy them at market price do deliver. This means that you earn money on this transaction when the price would go down.
The long position in shares – or any other symmetric asset – means that you have the contract in portfolio. As in "I have two shares of HSBC Holdings Plc."

The example shows that having a long position in a call option for example gives you the right to buy a certain underlying asset. This is all nice, but it means that you should not forget to use the right to buy the option, then sell it, etc. Maybe there is a better option and ask for "cash settlement." So in general, options provide two ways of closing the contract:

- *Delivering of the underlying:* Providing or accepting the underlying from the option buyer who exercises his/her option.
- *Cash settlement:* The option writer will pay out the profit of the option to the buyer in cash in stead of delivering the asset.

Still, we are not finished listing the slang of an option trader. Today's price of the underlying asset is usually referred to as the **Spot Price**. This language use is in line with its use in the forwards and futures market. The spot price is traditionally denoted as S.

So, at maturity the value of the options equates the profit that the long position will provide. For a call, that profit will equal the difference between the spot price at that moment and the strike price. Even before maturity one can observe if the spot price is higher or lower than the strike price. This difference is called the **Intrinsic Value**. The Intrinsic Value is the payoff that the option would yield if the spot price would remain unchanged till maturity (not discounted, just nominal value). For example,

- $IV_{call} = max(S - X, 0)$
- $IV_{put} = max(X - S, 0)$.

This leads to another set of bespoke vocabulary.

- *ITM:* an option is **in-the-money** if its Intrinsic Value is positive. So if the price of the underlying would be the same at maturity date, the option buyer would get some payoff.
- *ATM:* An option is **at-the-money** if $S = X$
- *OTM:* An option is **out-of-the-money** if the spot price is not equal to the strike and the intrinsic value of the option is zero. For a call, this means that $S < X$. This would mean that if at maturity the spot price would be the same as now, then the buyer would get no payoff.

While the concept "marked-to-market" is not specific for the option market, it is worth to notice that **MTM** or **Marked-to-Market** is the value of a financial instrument that market participants would pay for it (regardless if you believe this price to be fair or correct).

These concepts are visualised in Figure 30.3 on page 642.

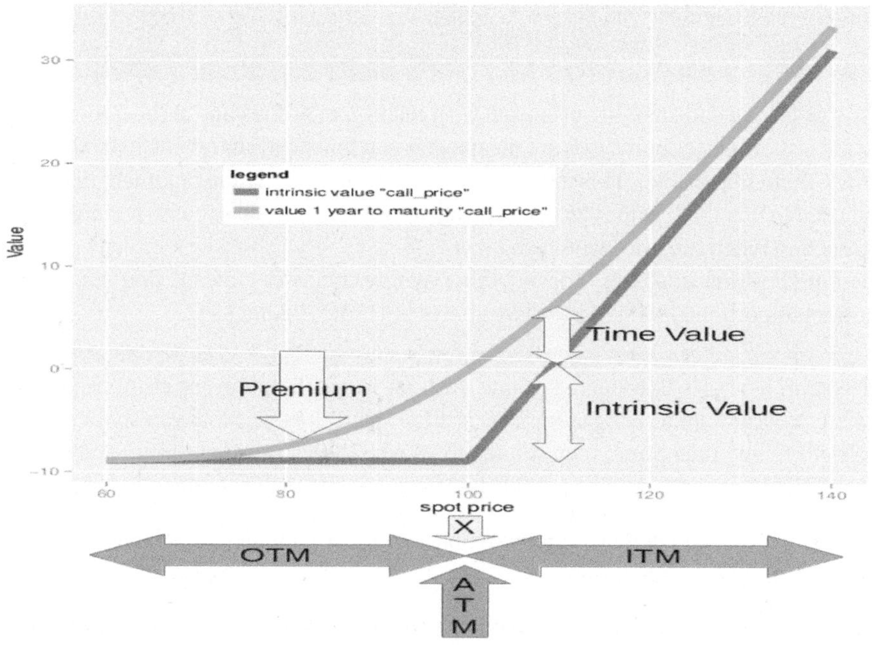

Figure 30.3: *Some concepts illustrated on the example of a call option with on the z-axis the sport price S and on the y-axis the payoff of the structure. The length of the blue arrows illustrates the concept, while the grey arrows indicate the position of the concept on the z-axis.*

Finally, there are two types of these basic options that are worth distinguishing.

- *European Option:* A European Option is an option that can be executed by the buyer at the maturity date and only at the maturity date.
- *American Option:* An American Option is an option that can be executed by the buyer from the moment it is bought and till the at maturity date.

Even the price of an option has it specific name: **the option premium**. The premium includes obviously the present value of the intrinsic value, but usually will be higher. There is indeed an extra value hidden in the optionality of the contract: the right of being allowed do do something – but not having the obligation – as such has a positive value. We come back to this idea in Chapter 30.7.5 *"The Black and Scholes Model"* on page 649.

30.7.2 Commercial Aspects

The option market is quite specific and not only uses its own vocabulary, it is also specific in how it works. Options can be bought on highly regulated markets, but can also be bought "over the counter" (OTC). OTC options are very customizable, they are bespoke options that are not quoted on any exchange. Negotiations happen between professional parties.

Definition: OTC

A financial instrument is said to be bought/sold Over the counter if it is bought/sold **outside** a regulated stock exchange.

The main differences between OTC and Exchange Traded options are:

1. *The counter-party risk:* On the exchange those risks are covered by the clearing house;
2. *Settlement and clearing* have to be specified in the OTC agreement, while on the stock exchange the clearing house will do this; and
3. *The regulatory oversight* is different.

Options are available on most stock exchanges. Already during the Tulipomania in 1637 there were options on tulip onions available both on the London Stock Exchange (LSE) and the Amsterdam exchange. **LSE**

While the amount of options traded on regulated exchanges is huge, the amounts traded OTC are much higher and typically amount in hundreds of trillions of dollars per year. This is because after those OTC trades are usually professional counterparts such as investment funds or pension funds that collect savings of thousands of investors and then build a structure with capital protection based on options.

Digression – The International Swap and Derivatives Organization (ISDA)

The ISDA aims to make OTC transactions more structured and safer. To achieve this it developed the ISDA Master Agreement and engages with policy makers and legislators around the world. The result is reduced counterparty risk, increased transparency, and improved operational infrastructure.

It has over 800 member institutions in 64 countries. These members include:

- market participants corporations, investment managers, government and supranational entities, insurance companies, energy and commodities firms, and international and regional banks.
- others: exchanges, clearing-houses and repositories, law firms, accounting firms and other service providers.

30.7.3 Short History

While it might seem that options are highly complex and modern financial instruments, they are actually older than most modern states. For illustrative purposes we list a few examples of how options were used in the past.

1. Supposedly, the first option buyer in the world was the ancient Greek mathematician and philosopher **Thales of Miletus** (ca. 624 – ca. 546 BCE). On a certain occasion, it was predicted that the season's olive harvest would be larger than usual, and during the off-season he acquired the right to use a number of olive presses the following spring. When spring came and the olive harvest was larger than expected he exercised his options and then rented the presses out at much higher price than he paid for his "option." – see Kraut (2002)
2. **Tulipomania** (March 1637): On February 24, 1637, the self-regulating guild of Dutch florists, in a decision that was later ratified by the Dutch Parliament, announced that all futures contracts written after November 30, 1636 and before the re-opening of the cash market in the early Spring, were to be interpreted as option contracts. See for example: Mackay (1841)

3. In **London**, puts and "refusals" (calls) first became well-known trading instruments in the 1690s during the reign of William and Mary. See: Smith (2004)

4. **Privileges** were options sold OTC in nineteenth century America, with both puts and calls on shares offered by specialized dealers. Their exercise price was fixed at a rounded-off market price on the day or week that the option was bought, and the expiry date was generally three months after purchase. They were not traded in secondary markets.

5. In real **estate market**, call options have long been used to assemble large parcels of land from separate owners; e.g., a developer pays for the right to buy several adjacent plots, but is not obligated to buy these plots and might not unless he can buy all the plots in the entire parcel.

6. **Film or theatrical producers** often buy the right – but not the obligation – to dramatize a specific book or script.

7. **Lines of credit** give the potential borrower the right – but not the obligation – to borrow within a specified time period and up to a certain amount.

8. Many choices, or embedded options, have traditionally been included in **bond contracts**. For example, many bonds are convertible into common stock at the buyer's discretion, or may be called (bought back) at specified prices at the issuer's option.

9. Mortgage borrowers have long had the option to repay the loan early, which corresponds to a callable bond option.

As you will have noticed from this list, options are not limited to financial markets. For example a lease contract for a car has usually an option to buy the car at the end of the contract.

30.7.4 Valuation of Options at Maturity

To start with, assume that we have bought a call option on HSBC with a strike price of £600. Today is the maturity of the option and today's price of HSBC is £650.

Executing the option means that we will use our right to "call" the share of HSBC at the strike price of 600. When we do so, we have on share for which we have paid £600. We can sell it immediately for £650 and hence make a profit of £50, therefore, the value of the option at the day of the maturity is the difference between the market price and the strike price – if the strike is below the market price. If the market price is lower than the strike price, then we will simply not execute the option and hence will not incur any loss.

So, the value of a call at maturity is

$$C = \max(0, X)$$

with X the strike price and C the value of the call.

long call

30.7.4.1 A Long Call at Maturity

A "long" position is the position of the option buyer (the party that has the right to exercise the option). Let us dive in with programming an example of the aforementioned formula in R, and plotting the results in Figure 30.4 on page 645.

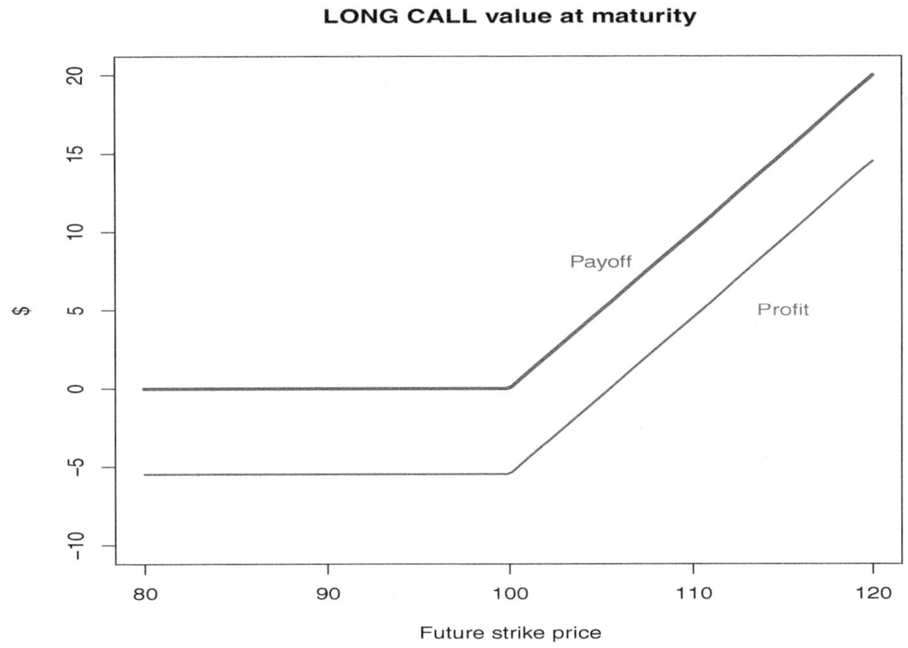

Figure 30.4: *The intrinsic value of a long call illustrated with its payoff and profit. The profit is lower, since it takes into account that the option buyer has paid a fixed premium for the option.*

```
# Let us plot the value of the call in function of the strike
FS <- seq(80, 120, length.out=150) # future spot price
X  <- 100                          # strike
P  <- 5                            # option premium
T  <- 3                            # time to maturity
r  <- 0.03                         # discount rate
payoff <- mapply(max, FS-X, 0)
profit <- payoff - P * (1 + r)^T

# Plot the results:
plot(FS, payoff,
     col='red', lwd=3, type='l',
     main='LONG CALL value at maturity',
     xlab='Future strike price',
     ylab='$',
     ylim=c(-10,20)
     )
lines(FS, profit,
      col='blue', lwd=2)
text(105,8, 'Payoff', col='red')
text(115,5, 'Profit', col='blue')
```

30.7.4.2 A Short Call at Maturity

short call

A "short" position means that we have sold a call. This will of course reverse the position (mirror symmetry along the x-axis). The results are plotted in Figure 30.5 on page 646

```
FS <- seq(80, 120, length.out=150) # future spot price
X  <- 100                          # strike
P  <- 5                            # option premium
```

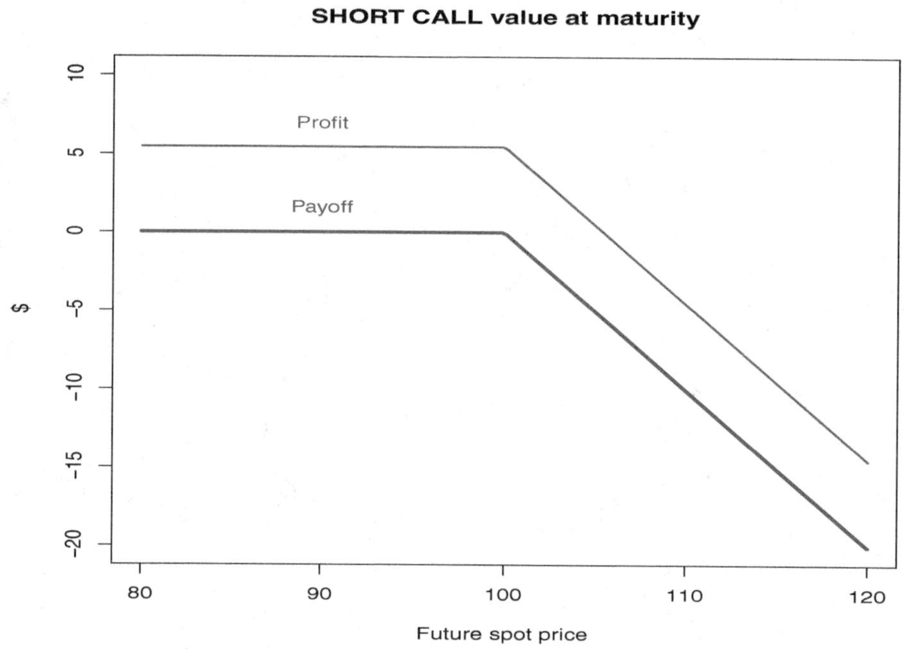

Figure 30.5: *The intrinsic value of a short call illustrated with its payoff and profit. The profit is higher, since this the position of the option-wirter, so this party has got the premium at the start of the contract. Note that the loss is unlimited.*

```
T  <- 3                              # time to maturity
r  <- 0.03                           # discount rate
payoff <- - mapply(max, FS-X, 0)
profit <- P * (1 + r)^T + payoff

# Plot the results:
plot(FS, payoff,
     col='red', lwd=3, type='l',
     main='SHORT CALL value at maturity',
     xlab='Future spot price',
     ylab='$',
     ylim=c(-20,10)
     )
lines(FS, profit,
      col='blue', lwd=2)
text(90,1.5, 'Payoff', col='red')
text(90,7, 'Profit', col='blue')
```

While the option buyer (the long position) can loose maximum the premium paid, the option writer can loose an unlimited amount while his profit will maximum be equal to the option premium.

30.7.4.3 Long and Short Put

A put option is an option that allocates the right to sell something to the option buyer – where a call is the right to buy something. A key difference is that in order to execute a put, the option buyer needs to own the underlying asset. This implies that this party needs to buy this asset first and hold it till the option expires.[9] The cost to hold the asset is known as the "cost of carry".

The code below, calculates the intrinsic values of both a long and a short position and then plots the results in Figure 30.6.

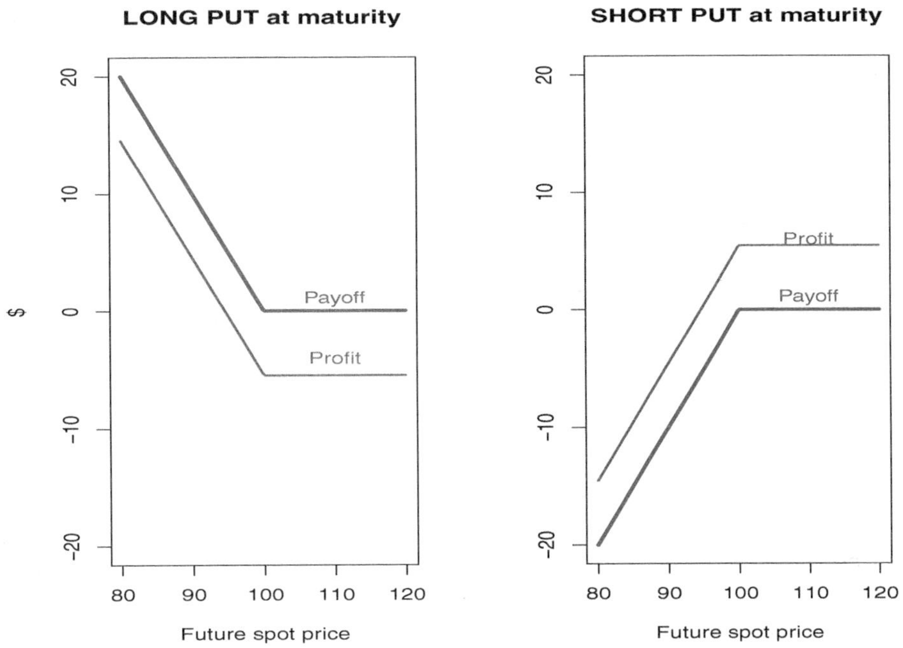

Figure 30.6: *The payoff and profit for a long put (left) and a short put (right).*

```
FS <- seq(80, 120, length.out=150) # future spot price
X  <- 100                          # strike
P  <- 5                            # option premium
T  <- 3                            # time to maturity
r  <- 0.03                         # discount rate

# the long put:
payoff <- mapply(max, X - FS, 0)
profit <- payoff - P * (1 + r)^T

par(mfrow=c(1,2))

plot(FS, payoff,
     col='red', lwd=3, type='l',
     main='LONG PUT at maturity',
     xlab='Future spot price',
     ylab='$',
     ylim=c(-20,20)
     )
lines(FS, profit,
      col='blue', lwd=2)
text(110,1,  'Payoff', col='red')
text(110,-4, 'Profit', col='blue')
```

[9]If the option ends favourably, but the option buyer does not have the underlying asset, then he needs to buy it at that moment at the market price in order to sell it to the option buyer. This does not make sense, as it cancels out the potential profit on the option.

```
# the short put:
payoff <- - mapply(max, X - FS, 0)
profit <- payoff + P * (1 + r)^T

plot(FS, payoff,
     col='red', lwd=3, type='l',
     main='SHORT PUT at maturity',
     xlab='Future spot price',
     ylab='',
     ylim=c(-20,20)
     )
lines(FS, profit,
      col='blue', lwd=2)
text(110,1, 'Payoff', col='red')
text(110,6, 'Profit', col='blue')
```

```
par(mfrow=c(1,1))  # reset the plot interface
```

30.7.4.4 The Put-Call Parity

Before diving into the Black and Scholes model it is worth to reflect first on the relationship between the premium for a call and a put option. This relationship will later help us to derive the price for the latter from the price of the first – calculated with the Black and Scholes model.

Consider Put and a Call that

- are both European,
- have the same strike,
- have the same maturity, and
- have the same underlying.

First, consider a long call and a short put position. This will provide us – at maturity – with profit if the future spot price S_T is higher than the strike price X, but it will also expose us to any price decreases because of the short put position. The result is that the payoff of the structure is $S(T) - X$. This means that at maturity the following holds:

$$C(T) - P(T) = S(T) - X$$

If those two portfolios are equal at maturity, then also today their market value must be the same. The present value of the left-hand side is the long call and short put position today: $C(0) - P(0)$ and the second portfolio is the underlying as it is today $S(0)$ minus the present value of the strike price. If we denote the discounting operation as $D(.)$ then this becomes.

$$C - P = D(F - X)$$

with

- C = the price of the Call
- P = the price of the Put

- F = the future price of the underlying
- X = the strike price
- D = the discount factor so that $S = D \times F$.

This can be rewritten as

$$C - P = S - D \times X$$

the right hand side is the same as buying a forward contract on the underlying with the strike as delivery price. So, a portfolio that is long a call and short a put is the same as being long a forward.

The Put-Call Parity can be rewritten as

$$C + D \times X = P + S$$

In this case, the left-hand side is a fiduciary call, which is long a call and enough cash (or bonds) to pay the strike price if the call is exercised, while the right-hand side is a protective put, which is long a put and the asset, so the asset can be sold for the strike price if the spot is below strike at expiry. Both sides have payoff $\max(S(T), K)$ at expiry (i.e., at least the strike price, or the value of the asset if more), which gives another way of proving or interpreting put-call parity.

30.7.5 The Black and Scholes Model

While the payoff of an option is obvious at maturity it is a non-trivial problem before maturity because while the intrinsic value gives some indication, the time value is essential too. This time value is a reflection of the rights that one acquires with the option. For any given intrinsic value, there is always a certain non-zero probability that the option ends up in-the-money.

30.7.5.1 Pricing of Options Before Maturity

This problem was first solved by Black and Scholes (1973) and in essence they assume that log-returns are normally distributed (so compound returns are log-normally distributed) and calculate the expected value of the option. This approach gives us for example the price of a European call and through some smart reasoning – see below – we can use this result to find the premium for the put option.

The main assumptions that allow us to derive the Black and Scholes model (BS Model) are as follows: BS

1. Log-returns follow a Gaussian distribution on each time interval
2. The returns of one period are statistically independent of the return in other periods
3. Volatility and expected return exist and are and stable
4. The continuous time assumption:
   - interest rates are continuous: so that $e^{rt} = (1 + i)^t$, implying that the compounded continuous rate can be calculated from the annual compound interest rate: $r = \log(1 + i)$
   - also returns can be split infinitesimally and be expressed as a continuous rate.

These assumptions allow to calculate the fair value for a call option as in Black and Scholes (1973) and via the call-put parity derive the value of the put option — see Chapter 30.7.4.4 *"The Put-Call Parity"* on page 648. The result is the following:

- Call Price: $C(S, X, \tau, r, \sigma) = N(d_1)S - N(d_2)Xe^{-r\tau}$
- Put Price: $P(S, X, \tau, r, \sigma) = Xe^{-r\tau} - S + C(S, X, \tau, r, \sigma)$
- with:

  $N(\cdot)$ the cumulative distribution function of the standard normal distribution

  τ the time to maturity

  S the spot price of the underlying asset

  X the strike price

  r the risk free rate (annual rate, expressed in terms of continuous compounding)

  σ the volatility of the returns of the underlying asset

  $$d_1 := \frac{\log\left(\frac{S}{X}\right)+\left(r+\frac{\sigma^2}{2}\right)(\tau)}{\sigma\sqrt{\tau}}$$

  $$d_2 := \frac{\log\left(\frac{S}{X}\right)+\left(r-\frac{\sigma^2}{2}\right)(\tau)}{\sigma\sqrt{\tau}} = d_1 - \sigma\sqrt{\tau}$$

30.7.5.2 Apply the Black and Scholes Formula

The Black and Scholes formula provides us an analytic expression for the value of basic call and put options. This is easy to program in R and experimenting with the formula allows us to gain insight in what drives the price of an option.

We will work with the following example in mind (unless stated otherwise).

- $S = 100$
- $X = 100$ (when $S = X$ one says that the "option is at-the-money")
- $\sigma = 20\%$
- $r = 2\%$
- $\tau = 1$ year

We will convert the Black and Scholes formula into functions, so we can easily reuse the code later. First, we also make functions for the intrinsic value. In these functions we leave out the time value of the premium, because want to use them later to compare portfolios at the moment of the purchase of the option. The code below defines these functions.

```
# call_intrinsicVal
# Calculates the intrinsic value for a call option
# Arguments:
#    Spot   -- numeric -- spot price
#    Strike -- numeric -- the strike price of the option
# Returns
#    numeric -- intrinsic value of the call option.
call_intrinsicVal <- function(Spot, Strike) {max(Spot - Strike, 0)}
```

```
# put_intrinsicVal
# Calculates the intrinsic value for a put option
# Arguments:
#    Spot   -- numeric -- spot price
#    Strike -- numeric -- the strike price of the option
# Returns
#    numeric -- intrinsic value of the put option.
put_intrinsicVal <- function(Spot, Strike) {max(-Spot + Strike, 0)}


# call_price
# The B&S price of a call option before maturity
# Arguments:
#    Spot   -- numeric -- spot price in $ or %
#    Strike -- numeric -- the strike price of the option  in $ or %
#    T      -- numeric -- time to maturity in years
#    r      -- numeric -- interest rates (e.g. 0.02 = 2%)
#    vol    -- numeric -- standard deviation of underlying in $ or %
# Returns
#    numeric -- value of the call option in $ or %
#
call_price <- function (Spot, Strike, T, r, vol)
 {
  d1 <- (log(Spot / Strike) + (r + vol ^ 2/2) * T) / (vol * sqrt(T))
  d2 <- (log(Spot / Strike) + (r - vol ^ 2/2) * T) / (vol * sqrt(T))
  pnorm(d1) * Spot - pnorm(d2) * Strike * exp(-r * T)
  }

# put_price
# The B&S price of a put option before maturity
# Arguments:
#    Spot   -- numeric -- spot price in $ or %
#    Strike -- numeric -- the strike price of the option  in $ or %
#    T      -- numeric -- time to maturity in years
#    r      -- numeric -- interest rates (e.g. 0.02 = 2%)
#    vol    -- numeric -- standard deviation of underlying in $ or %
# Returns
#    numeric -- value of the put option in $ or %
#
put_price <- function(Spot, Strike, T, r, vol)
 {
 Strike * exp(-r * T) - Spot + call_price(Spot, Strike, T, r, vol)
 }
```

These functions can now be deployed as in the following examples:

```
# Examples:
call_price (Spot = 100, Strike = 100, T = 1, r = 0.02, vol = 0.2)
## [1] 8.916037

put_price  (Spot = 100, Strike = 100, T = 1, r = 0.02, vol = 0.2)
## [1] 6.935905
```

It is even possible to use our functions to plot the market value of different options. First we plot an example of the market value of a long call and compare this to its intrinsic value. The results are in Figure 30.7 on page 652.

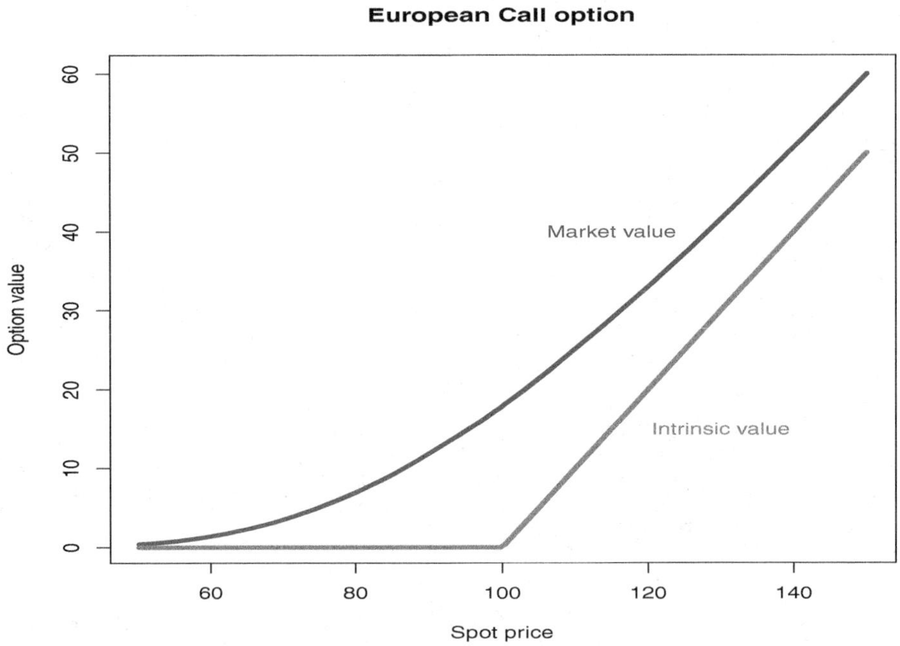

Figure 30.7: *The price of a long call compared to its intrinsic value. The market value is always positive.*

```
# Long call
spot <- seq(50,150, length.out=150)
intrinsic_value_call <- apply(as.data.frame(spot),
                              MARGIN=1,
                              FUN=call_intrinsicVal,
                              Strike=100)
market_value_call    <- call_price(Spot = spot, Strike = 100,
                              T = 3, r = 0.03, vol = 0.2)
plot(spot, market_value_call,
     type = 'l', col= 'red', lwd = 4,
     main = 'European Call option',
     xlab = 'Spot price',
     ylab = 'Option value')
text(115, 40, 'Market value', col='red')
lines(spot, intrinsic_value_call,
      col= 'forestgreen', lwd = 4)
text(130,15, 'Intrinsic value', col='forestgreen')
```

Let us try the same for a long put position. The code below does this and plots the results in Figure 30.8.

```
# Long put
spot <- seq(50,150, length.out=150)
intrinsic_value_put <- apply(as.data.frame(spot),
                             MARGIN=1,
                             FUN=put_intrinsicVal,
                             Strike=100)
market_value_put    <- put_price(Spot = spot, Strike = 100,
```

```
                           T = 3, r = 0.03, vol = 0.2)
plot(spot, market_value_put,
     type = 'l', col= 'red', lwd = 4,
     main = 'European Put option',
     xlab = 'Spot price',
     ylab = 'Option value')
text(120, 8, 'market value', col='red')
lines(spot, intrinsic_value_put,
      col= 'forestgreen', lwd = 4)
text(75,10, 'intrinsic value', col='forestgreen')
```

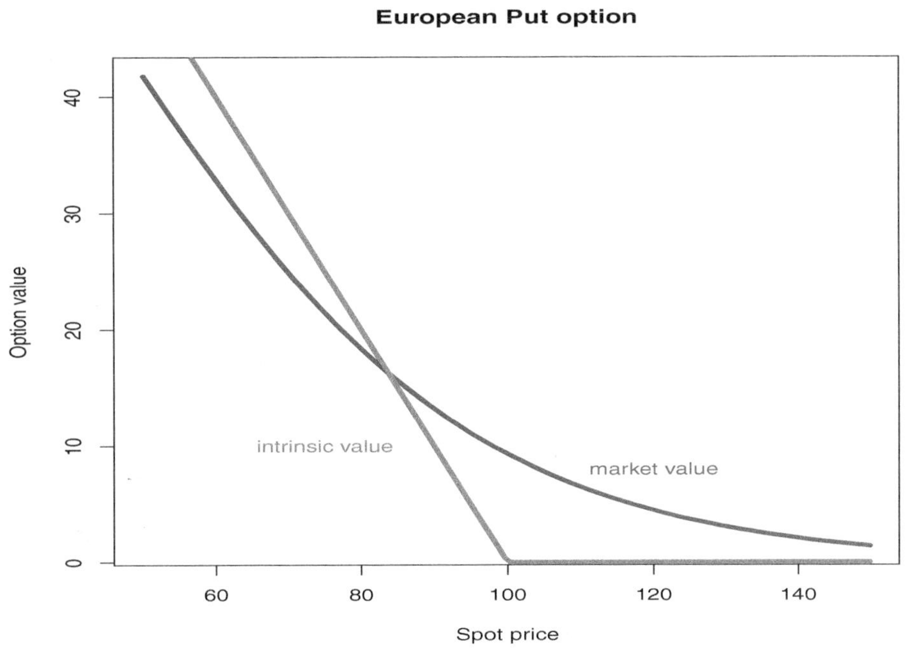

Figure 30.8: *The price of a long put compared to its intrinsic value. Note that the market price of a put can be lower than its intrinsic value. This is because of the cost of carry.*

It is noteworthy that the time value of a put can be negative (ie. the market value minus the intrinsic value). This is because it is somehow costly to exercise a put. In order to exercise a put we first need to have the asset. So, the equivalent portfolio is lending money in order to buy the asset so that we would be able to sell the assets at maturity. Lending costs money and hence when the probability that the put will end in the money, this costs will push down the value of the put. We call this the cost of carry.

30.7.5.3 The Limits of the Black and Scholes Model

Academics love the BS model for it is a beautiful application of basic statistics that lead to insightful and practical results. However, when applying the BS formula to calculate the price of options that there is something like a volatility skew. A trader will apply a higher volatility for an option that is out of the money, resulting in a price that seems higher than expected.

Maybe the problem is that the BS model assumes log-normal returns and that in reality distributions in financial markets have "heavy tails". Big losses appear much more frequent than the normal distribution would imply. This is for example illustrated in Chapter 8.4.1 *"Normal Distribution"* on page 150 with the S&P500 index.

The BS model needs a few strong assumptions that are not really reflecting the reality on financial markets. For example, in order to derive the model, we need to assume that:

- markets are efficient (follow a Wiener process, we can always buy and sell, etc.),
- the underlying is not paying any dividend (and if it is, this is a continuous flow),
- the log-returns at maturity are normally distributed (Gaussian distribution), and
- that the volatility is stable.

The BS model can be modified to accommodate many of these shortcomings. However, there are other limitations for the BS Model. Some options can simply not be priced with the model. While a simple knock-in or knock-out might still fit within the model's possibilities, more complex options require a complete different approach. In the next section (Section 30.7.6) we describe another model that can more naturally cope with any underlying distribution: the binomial model.

30.7.6 The Binomial Model

A "lookback" option is an option that will use the best possible strike price over a certain period. This exceeds the possibilities of the Black and Scholes model and hence we need another pricing method. One such approaches that is successful is the binomial model. The logic of this model is well-adapted to path-dependent options.

The idea is as simple as it is powerful. The underlying spot price at moment 0 is S_0 and then consider the next (small) time period. Obviously, the price S_1 will be governed by a stochastic distribution – more or less-centred around S_0. Simplify this idea to leave just two possibilities: one positive scenario (the spot is now $S_0.u$ with $u > 1$) and one negative scenario (the spot is now $S_0.d$ with $0 < d < 1$. Then repeat this process many times.

This will result in a tree of possibilities, but in each path it is possible to determine exactly what the payoff of the option exactly is. If we made our splits in a smart way we know the probability of each possible price and it is sufficient to discount all the results back to this moment (weighted with their probabilities) and we have the price of the option.

Step One

The first step of this approach can be visualized as in Figure 30.9, where we use the following definitions:

$$\begin{cases} S_0 & = stock\ price\ now \\ p & = probability\ of\ stock\ increase \\ u & = price\ increase\ (when\ it\ increases) \\ d & = price\ decrease\ (when\ it\ decreases) \end{cases}$$

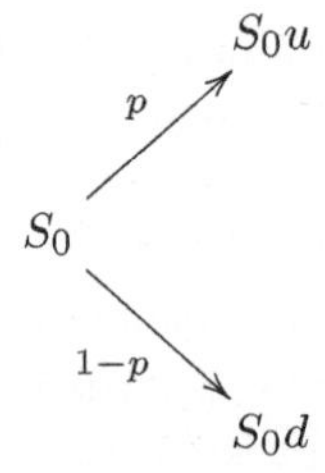

Figure 30.9: *Step 1 in the binomial model.*

Hence, the future option price is:

$$P_1 = p \text{ Payoff}(S_0 u) + (1 - p) \text{ Payoff}(S_0 d)$$

Today's price would then be:

$$P = \frac{P_1}{1 + r_1}$$

(with r_1 the interest rate over one period)

Example: One step binomial pricing model

Calculate the price of a long ATM European call option; using one step in the binomial model and the following assumptions: $S_0 = \$100, p = 0.70, u = 1.03\ (\equiv 3\%$ increase), $d = 0.95\ (\equiv 5\%$ decrease), $r = 0.1$. Assume zero interest rates.

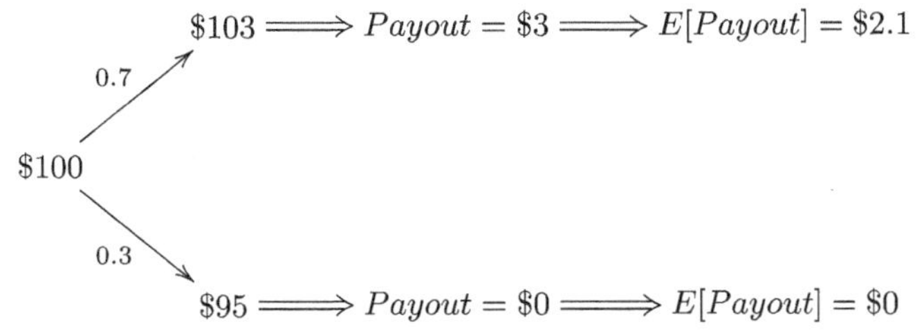

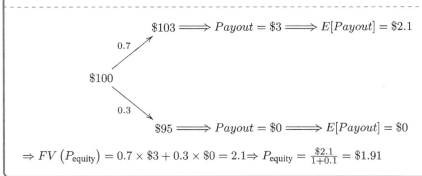

$$\Rightarrow FV\left(P_{\text{equity}}\right) = 0.7 \times \$3 + 0.3 \times \$0 = 2.1 \Rightarrow P_{\text{equity}} = \frac{\$2.1}{1+0.1} = \$1.91$$

The Second Step of the Binomial Model

binomial model

In the next step, we will split each node again, as illustrated in Figure 30.10 on page 656

Each ending node of the second step will be the split in step 3 and so on. The total number of end-nodes is hence 2^{K-1} with K the number of steps.

This approach is called the binomial model, since in each node we consider two scenarios.[10] The key issue is to find good parameters so that the probabilities and up and down moves all together create a realistic picture. The first approach is the risk neutral approach.

30.7.6.1 Risk Neutral Method

The first method to choose the parameters u, d, and p consistently is the risk neutral method or also the Cox–Ross–Rubinstein model.

The binomial model then can be summarized in a few simple steps.

1. Choose (u, d, p) consistent with some other theory of observation. For example, the Cox–Ross–Rubinstein model:
   - $u = e^{\sigma\sqrt{\delta t}}$
   - $d = e^{-\sigma\sqrt{\delta t}}$
   - $p = \frac{e^{R_{RF}\delta t} - d}{u - d}$

[10] The reason why we only consider two branches for each node is mainly that it is the most simple system that works. Adding more branches makes everything a lot more complex and does not really add any precision. More precision rather can be obtained by taking smaller steps and hence a larger tree.

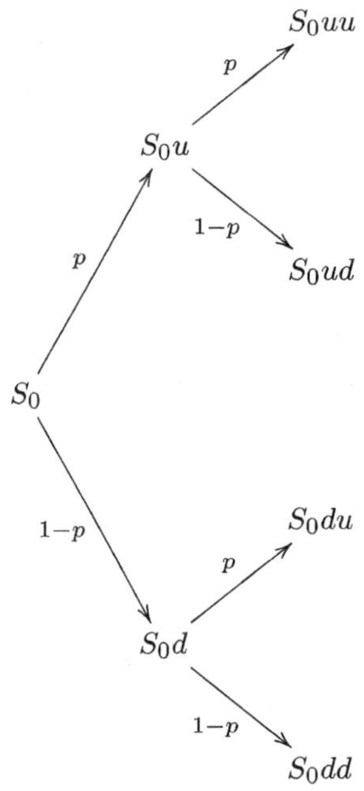

Figure 30.10: *The first 2 steps of the binomial model.*

2. Iterate till some convergence is satisfactory.
3. Discount the expected value of the option back to today.

Cox–Ross–Rubinstein (option pricing method)

CRR

To illustrate how this can work, we provide here an implementation of the Cox–Ross–Rubinstein model (CRR).

```
# CRR_price
# Calculates the CRR binomial model for an option
# Arguments:
#         S0 -- numeric -- spot price today (start value)
#         SX -- numeric -- strike, e.g. 100
#      sigma -- numeric -- the volatility over the maturity period,
#                          e.g. ca. 0.2 for shares on 1 yr
#        Rrf -- numeric -- the risk free interest rate (log-return)
# optionType -- character -- 'lookback' for lookback option,
#                            otherwise vanilla call is assumed
#    maxIter -- numeric -- number of iterations
# Returns:
#    numeric -- the value of the option given parameters above
CRR_price <- function(S0, SX, sigma, Rrf, optionType, maxIter)
  {
  Svals <- mat.or.vec(2^(maxIter), maxIter+1)
  probs <- mat.or.vec(2^(maxIter), maxIter+1)
  Smax  <- mat.or.vec(2^(maxIter), maxIter+1)
  Svals[1,1] <- S0
  probs[1,1] <- 1
  Smax[1,1]  <- S0
  dt <-  1 / maxIter
  u <- exp(sigma * sqrt(dt))
  d <- exp(-sigma * sqrt(dt))
  p = (exp(Rrf * dt) - d) / (u - d)
  for (n in 1:(maxIter))
```

```
    {
    for (m in 1:2^(n-1))
     {
     Svals[2*m-1,n+1] <- Svals[m,n] * u
     Svals[2*m,n+1]   <- Svals[m,n] * d
     probs[2*m-1,n+1] <- probs[m,n] * p
     probs[2*m,n+1]   <- probs[m,n] * (1 - p)
     Smax[2*m-1,n+1]  <- max(Smax[m,n], Svals[2*m-1,n+1])
     Smax[2*m,n+1]    <- max(Smax[m,n], Svals[2*m,n+1])
     }
    }
  if (optionType == 'lookback')
   {
     exp.payoff <- (Smax - SX)[,maxIter + 1]  * probs[,maxIter + 1]
   } # lookback call option
   else
    {
     optVal <- sapply(Svals[,maxIter + 1] - SX,max,0)
     exp.payoff <- optVal * probs[,maxIter + 1]
    } # vanilla call option
  sum(exp.payoff) / (1 + Rrf)
  }
```

Now, we still add another function that can be used as a wrapper function for the previous one in order to visualize the results:

```
# plot_CRR
# This function will call the CRR function iteratively for
# number of iterations increasing from 1 to maxIter and
# plot the results on screen (or if desired uncomment the
# relevant lines to save to disk).
# Arguments:
# optionType -- character -- 'lookback' for lookback option,
#                            otherwise vanilla call is assumed
#    maxIter -- numeric -- maximal number of iterations
#   saveFile -- boolean -- TRUE to save the plot as pdf
# Returns:
#    numeric -- the value of the option given parameters above

plot_CRR <- function(optionType, maxIter, saveFile = FALSE)
  {
  x <- seq(1,maxIter)
  y <- mat.or.vec(maxIter,1)
  for (k in 1:maxIter) {y[k] <- CRR_price(100,
                                          100,
                                          sigma      = 0.2,
                                          Rrf        = 0.02,
                                          optionType = optionType,
                                          maxIter    = k)}
  d <- data.frame(x, y)
  colnames(d) <- c('maxIter', 'value')
  p <- qplot(maxIter, value, data=d, geom = "point",size=I(3) )
  p <- p + geom_line() + ggtitle(optionType)
  p <- p + xlab('Number of iterations' ) +
       ylab('Estimated value of the option')

  if(saveFile) {
    p
    ggsave(paste('img/binomial_CRR_',optionType,'.pdf',sep=''))
    }
  # Return the plot:
  p
  }
```

This code allows us to plot the CRR model and show how it converges. We will plot it for two options: a in Figure 30.11 on page 658.

```
library(ggplot2)
# Plot the convergence of the CRR algorithm for a call option.
plot_CRR("Call", maxIter = 20)
```

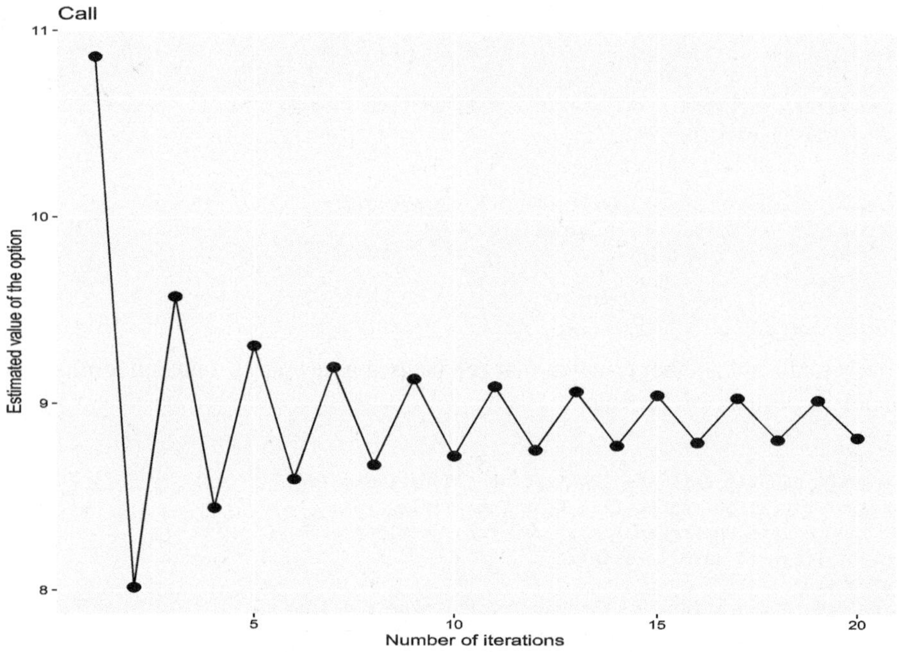

Figure 30.11: *The Cox–Ross–Rubinstein model for the binomial model applied to a call option. Note how the process converges smooth and quick.*

To illustrate this further, we will do the same for an unlimited lookback option ("Russian" option). This option allows the buyer to set the strike price at the lowest quoted price during the lifetime of the option. The code of that example is below and the plot that results from it in Figure 30.12 on page 659.

```
# Plot the convergence of the CRR algorithm for a call option.
plot_CRR("lookback", maxIter = 15)
```

> **Warning – Memory use and running time**
>
> Note that the binomial model has an exponentially growing number of nodes in every step. This can be seen in the line: `for (m in 1:2^(n-1))`. This might become a limiting factor to see the result. Best is to experiment with small numbers and once the code is de bugged check what your computer can do. Changing the number of steps can be achieved by fine-tuning the `maxIter` in the aforementioned code.

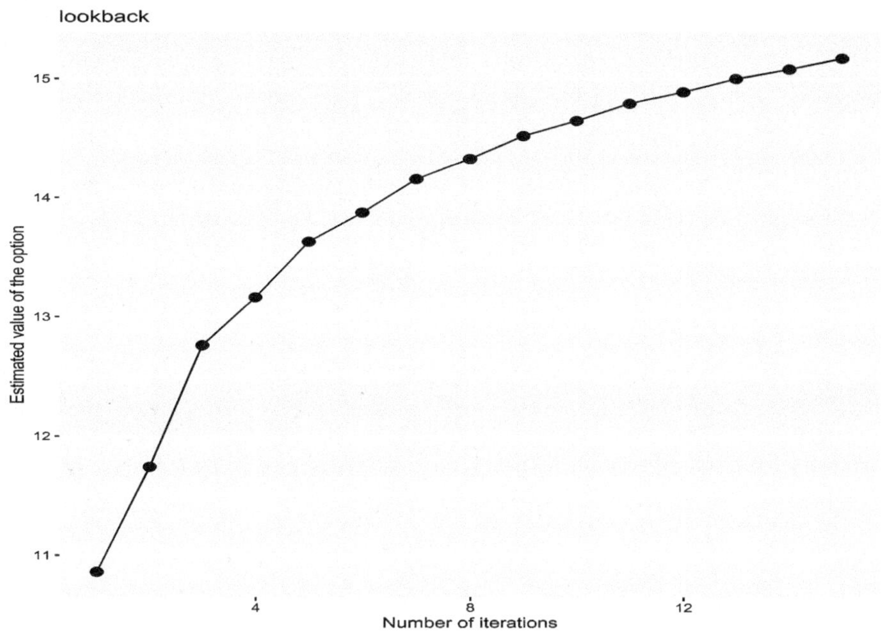

Figure 30.12: *The Cox–Ross–Rubinstein model for the binomial model applied to an unlimited look-back option (aka Russian Option). For the Russian option, the convergence process is not obvious. In fact the more steps we allow, the more expensive the option becomes, because the more there are moments that can have a lower price.*

30.7.6.2 The Equivalent Portfolio Binomial Model

Another approach is to reason along the lines of delta hedging – see Chapter 30.7.9 *"Delta Hedging"* on page 665 – and argue that the portfolio that has the option must equate the portfolio that is being managed by the delta hedging. The latter portfolio consists of $\delta.S$ financed by borrowed money.

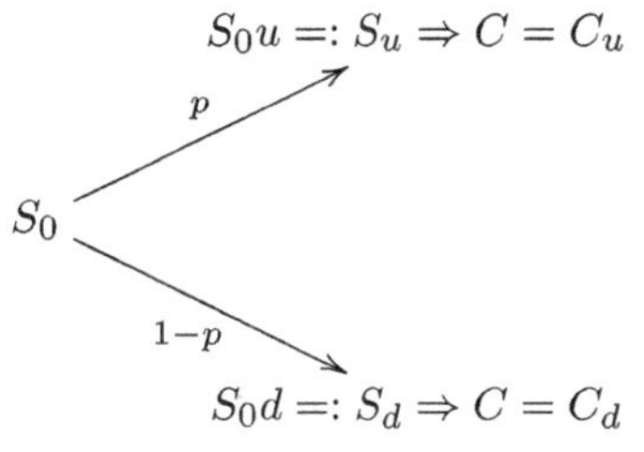

The tree is built in the same way, but notice that

- If two portfolios have the same pay-off at time T, then they have the same price at time T-1.
- Choose the Equivalent Portfolio as "invest delta dollar in the underlying and borrow B cash," then

$$C = \delta S - B$$

If both portfolios have the same price now, then

$$C = \delta S - B$$

If both portfolios have the same price at T + 1, then

$$\begin{cases} C_u & = \delta S_u - (1+r)B \\ C_d & = \delta S_d - (1+r)B \end{cases}$$

We have now two equations with two unknown (δ and B), and hence easily find that:

$$\begin{cases} \delta & = \frac{C_u - C_d}{S_u - S_d} \\ B & = \frac{\delta S_u - C_u}{1+r} = \frac{\delta S_d - C_d}{1+r} \end{cases}$$

1. $\delta = \frac{C_u - C_d}{S_u - S_d}$
2. $B = \frac{\delta S_u - C_u}{1+r}$ or $B = \frac{\delta S_d - C_d}{1+r}$
3. $C = \delta S - B$

Example: – First order binomial model

What is the value of a call option, assuming that: the strike price, X is \$100; the spot price, S is \$100, S_d = \$98, S_u = \$105, and the interest is 2%.
$\delta = \frac{5\$ - 0\$}{105\$ - 98\$} = 0.714 \Rightarrow B = \frac{0.714 \times 105\$ - 5\$}{1+0.02} = 68.60\$ \Rightarrow C = 0.714 \times 100\$ - 68.60\$ =$ \$2.88

30.7.6.3 Summary Binomial Model

The binomial model is more flexible than the BS Model, it naturally copes with different distributions (and eliminates the need for a volatility smile or smirk), dividends, path dependent options, etc. This method is one of the widely used option valuation strategies.

In order to choose between the Risk Neutral and the Equivalent Portfolio Model, consider the following.

- Both yield mathematically the same option value
- Risk neutral method
  - Is easier to calculate
  - Does not use the economic probability of the stock going up or down
- Equivalent portfolio method
  - Is more challenging computationally
  - Draws upon sound economic principle of arbitrage
  - Provides insight in option delta

30.7.7 Dependencies of the Option Price

In this section, we will illustrate how the price of an option changes in function of the different market variables on which it depends. For example, the option price depends on the volatility of the underlying, but does the option price increase or decrease when the volatility increases?

To study these relationships, we will use the Black and Scholes formula, as presented in Chapter 30.7.5 *"The Black and Scholes Model"* on page 649.

The formulae `call_price()` and `put_price()` – as set up in Chapter 30.7.5.2 *"Apply the Black and Scholes Formula"* on page 650 – can also be used to study how the price of and option changes in function of the different parameters such as time to maturity, volatility, interest rates, etc. So, we will use these as a starting point to create the illustrations.

To visualise the results, we will use the library ggplot2, and hence we need to load it first:

ggplot2

```
# We still use ggplot2
library(ggplot2)
```

Then we create a generic function that we will use to plot the price dependencies. This function will take an argument varName, that is the variable that we will study. The function takes even the option name as an argument.

```
# plot_price_evol
# Plots the evolution of Call price in function of a given variable
# Arguments:
#     var       -- numeric   -- vector of values of the variable
#     varName   -- character -- name of the variable to be studied
#     price     -- numeric   -- vector of prices of the option
#     priceName -- character -- the name of the option
#     reverseX  -- boolean   -- TRUE to plot x-axis from high to low
# Returns
#     ggplot2 plot
#
plot_price_evol <- function(var, varName, price, priceName,
                            reverseX = FALSE)
{
  d <- data.frame(var, price)
  colnames(d) <- c('x', 'y')
  p <- qplot(x, y, data = d, geom = "line", size = I(2) )
  p <- p + geom_line()
  if (reverseX) {p <- p + xlim(max(var), min(var))} # reverse axis
  p <- p + xlab(varName ) + ylab(priceName)
  p   # return the plot
}
```

30.7.7.1 Dependencies in a Long Call Option

Now, we will use this function plot_price_evol() in order to study the behaviour of the option value in function of its different parameters. First we will study the call option and take its long position.

The code below produces the plot Figure 30.13 on page 662:

```
# Define the default values:
t      <- 1
Spot   <- 100
Strike <- 100
r      <- log(1 + 0.03)
vol    <- 0.2


## ... time
T <- seq(5, 0.0001, -0.01)
Call <- c(call_price (Spot, Strike, T, r, vol))
p1 <- plot_price_evol(T, "Time to maturity (years)", Call, "Call",
                      TRUE)

## ... interest
R <- seq(0.001, 0.3, 0.001)
Call <- c(call_price (Spot, Strike, t, R, vol))
p2 <- plot_price_evol(R, "Interest rate", Call, "Call")
```

```
## ... volatility
vol <- seq(0.00, 0.2, 0.001)
Call <- c(call_price (Spot, Strike, t, r, vol))
p3 <- plot_price_evol(vol, "Volatility", Call, "Call")

## ... strike
X <- seq(0, 200, 1)
Call <- c(call_price (Spot, X, t, r, vol))
p4 <- plot_price_evol(X, "Strike", Call, "Call")

## ... Spot
spot <- seq(0, 200, 1)
Call <- c(call_price (spot, Strike, t, r, vol))
p5 <- plot_price_evol(spot, "Spot price", Call, "Call")

# In the next line we use the function grid.arrange()
# from the gridExtra package
library(gridExtra)
grid.arrange(p1, p2, p3, p4, p5, nrow = 3)
```

gridExtra
grid.arrange()

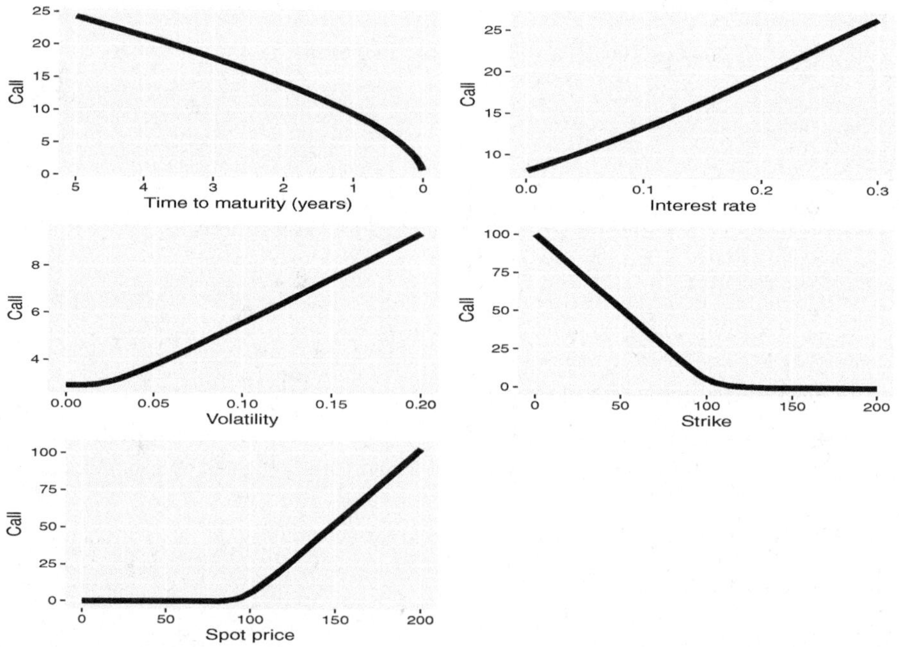

Figure 30.13: *The value of a call option depends on many variables. Some are illustrated in these plots.*

30.7.7.2 Dependencies in a Long Put Option

It is now easy to do the same for a put-option. The code below creates different plots of how the value of a put option depends on its parameters, and then plots the results in Figure 30.14 on page 663:

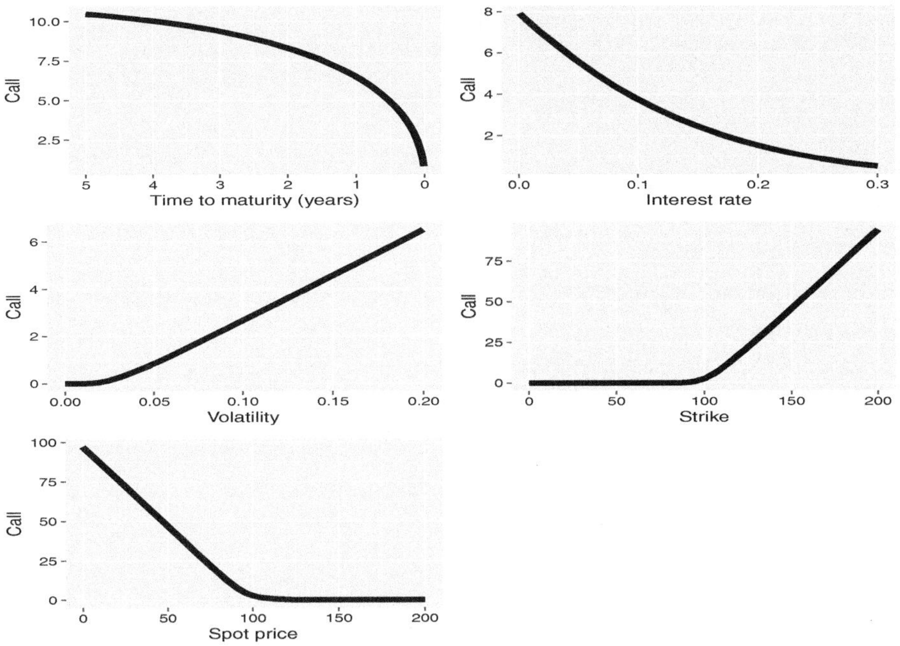

Figure 30.14: *The value of a put option depends on many variables. Some are illustrated in these plots.*

```
# Define the default values:
t      <- 1
Spot   <- 100
Strike <- 100
r      <- log(1 + 0.03)
vol    <- 0.2


## ... time
T <- seq(5, 0.0001, -0.01)
Call <- c(put_price (Spot, Strike, T, r, vol))
p1 <- plot_price_evol(T, "Time to maturity (years)",
                      Call, "Call", TRUE)

## ... interest
R <- seq(0.001, 0.3, 0.001)
Call <- c(put_price (Spot, Strike, t, R, vol))
p2 <- plot_price_evol(R, "Interest rate", Call, "Call")

## ... volatility
vol <- seq(0.00, 0.2, 0.001)
Call <- c(put_price (Spot, Strike, t, r, vol))
p3 <- plot_price_evol(vol, "Volatility", Call, "Call")

## ... strike
X <- seq(0, 200, 1)
Call <- c(put_price (Spot, X, t, r, vol))
p4 <- plot_price_evol(X, "Strike", Call, "Call")

## ... Spot
spot <- seq(0, 200, 1)
```

```
Call <- c(put_price (spot, Strike, t, r, vol))
p5 <- plot_price_evol(spot, "Spot price", Call, "Call")

# In the next line we use the function grid.arrange()
# from the gridExtra package
library(gridExtra)
grid.arrange(p1, p2, p3, p4, p5, nrow = 3)
```

30.7.7.3 Summary of Findings

The findings of the plots Figure 30.14 on page 663 and Figure 30.13 on page 662 can be summarised as per Table 30.1

Dependency on ...	Call	Put
Spot (value of underlying)	+	-
Volatility of underlying	+	+
Time to maturity	+	+
Interest rate	+	-
Dividend	-	+

Table 30.1: *An overview of the price dependency for call and put options. A plus sign indicates that if the variable goes up, then the option premium goes up. The minus sign indicates that in the same case the option premium goes down.*

30.7.8 The Greeks

Option books can be huge and the risks are enormous. Hence, it is of paramount importance to follow up the value and predict potential moves. It should hence not come as a surprise that the derivatives of the option-price got their own name.

The names of the derivatives got a name that (in most cases resembles) a Greek letter. Hence, they are collectively referred to as "Greeks." The definitions for the Black and Scholes model are given in Table 30.2.

The derivatives of the option price are key in managing an option book. For example an option book with positive delta will increase in value as the underlying asset increases in price. So, in order to reduce – and even eliminate if at all possible – the risk related to the price of the underlying, the option trader will keep a delta neutral position.

This can be done by adding other options with negative delta. The total position of an option book needs to be managed so that all the Greeks are zero at any given time. This is only made more

	What	Call	Put
delta	$\frac{\partial C}{\partial S}$	$N(d_1)$	$-N(-d_1) = N(d_1) - 1$
gamma	$\frac{\partial^2 C}{\partial S^2}$	$\frac{N'(d_1)}{S\sigma\sqrt{\tau}}$	
vega	$\frac{\partial C}{\partial \sigma}$	$SN'(d_1)\sqrt{\tau}$	
theta	$\frac{\partial C}{\partial t}$	$-\frac{SN'(d_1)\sigma}{2\sqrt{\tau}} - rKe^{-r(\tau)}N(d_2)$	$-\frac{SN'(d_1)\sigma}{2\sqrt{\tau}} + rKe^{-r(\tau)}N(-d_2)$
rho	$\frac{\partial C}{\partial r}$	$K(\tau)e^{-r(\tau)}N(d_2)$	$-K(\tau)e^{-r(\tau)}N(-d_2)$

Table 30.2: *An overview of "the Greeks:" the most relevant derivatives of the option price.*

complicated by the fact that transactions come at a cost. This means that one will have to make a trade-off between constantly trading but keeping risks low and incurring too much transaction costs.

So, the Greeks are very important to the option trader. Therefore, we will illustrate how they in their turn depend on the spot price of the underlying asset.

We can now visualize the value of the delta of a call and a put as follows – plot in Figure 30.15 on page 666:

```
# Define the functions to calculate the price of the delta

# call_delta
# Calculates the delta of a call option
# Arguments:
#    S      -- numeric -- spot price
#    Strike -- numeric -- strike price
#    T      -- numeric -- time to maturity
#    r      -- numeric -- interest rate
#    vol    -- numeric -- standard deviation of underlying
call_delta  <- function (S, Strike, T, r, vol)
 {
  d1 <- (log (S / Strike)+(r + vol ^2 / 2) * T) / (vol * sqrt(T))
  pnorm(d1)
  }


# put_delta
# Calculates the delta of a put option
# Arguments:
#    S      -- numeric -- spot price
#    Strike -- numeric -- strike price
#    T      -- numeric -- time to maturity
#    r      -- numeric -- interest rate
#    vol    -- numeric -- standard deviation of underlying
put_delta  <- function (S, Strike, T, r, vol)
 {
  d1 <- (log (S / Strike)+(r + vol ^2 / 2) * T) / (vol * sqrt(T))
  pnorm(d1) - 1
  }


## DELTA CALL
spot <- seq(0,200, 1)
delta <- c(call_delta(spot, Strike, t, r, vol))
p1 <- plot_price_evol(spot, "Spot price", delta, "Call delta")

## DELTA PUT
spot <- seq(0,200, 1)
delta <- c(put_delta(spot, Strike, t, r, vol))
p2 <- plot_price_evol(spot, "Spot price", delta, "Put delta")

# plot the two visualizations:
grid.arrange(p1, p2, nrow = 2)
```

30.7.9 Delta Hedging

While options put the option holder in a very comfortable position, the option writer can have extremely high losses. Simply buying the underlying would offset all profits and is hence not a good option. Rather, we need some system that works like an insurance for the portfolio of options.

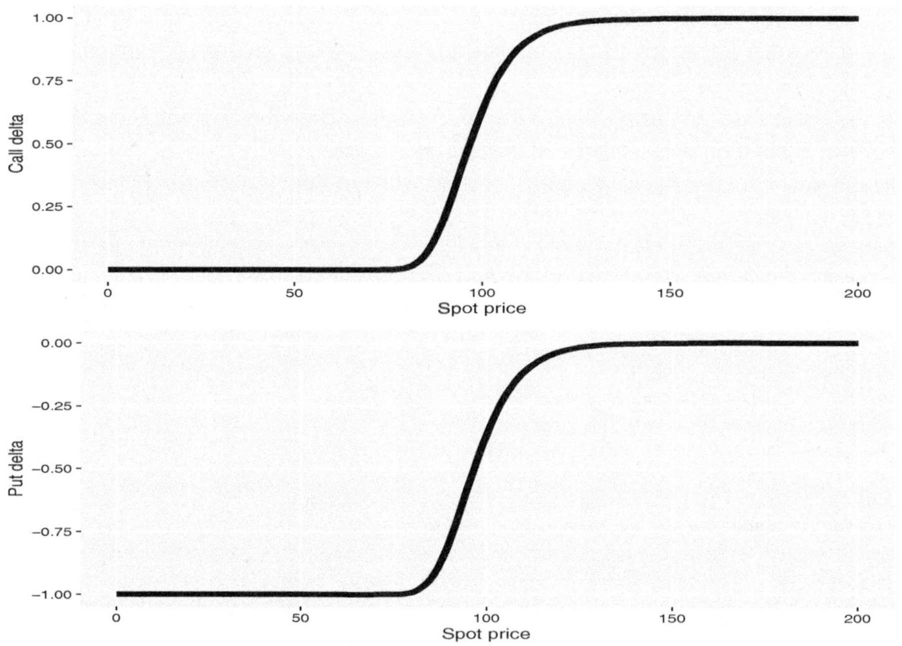

Figure 30.15: *An illustration of how the delta of a call and put compare in function of the spot price. Note the difference in scale on the y-axis.*

So, if we have a short position in a call for example and the price of the underlying increases, then we might have more losses because it becomes more probable that the option will end in the money. So, if that happens we need to buy more of the underlying asset.

The first derivative of the option price is a linear approximation of how the option price will move with the underlying asset. It appears that this thinking works, always adjusting our position of the underlying to the delta of the option would result in a risk-less strategy. Although it is of course not cost-less as it will require many transactions.

Of course, we need to adjust our strategy to the type of option that is in portfolio. This is the essence of "delta hedging." Summarized delta hedging is:

delta hedging

- Buy or sell as much of the underlying as indicated by the delta.
- Repeat this step as often as possible.

Note that the delta of a call is positive and that of a put is negative.

Delta Hedging Example

Assume an asset with $strike = \$100$, $time\ to\ maturity = 5\ years$, $\sigma = 0.2$, $r = 0.02$ (as continuous, so as percent: 1.98%).
In previous example – where we adapt our position only once a year – the option writer has to buy and sell during the hedge. Note that:

- the payoff of the option is $\$15$, that is what the option writer pays the option buyer;
- our option writer has spent $\$107$ (non-discounted) and can sell this portfolio for $\$115$ (difference is [non-discounted] $\$8$), this is $\$7$ short for paying his customer, but he did get the premium;

tm 2 mat.	Spot	Delta	To hedge	Portf.	To buy	Cash
5.00	100.00	0.67	67.26	0.00	67.26	67.26
4.00	110.00	0.74	81.22	73.99	7.23	74.49
3.00	95.00	0.58	54.97	70.14	-15.18	59.32
2.00	110.00	0.73	80.55	63.65	16.91	76.22
1.00	105.00	0.67	70.50	76.89	-6.39	69.83
0.00	115.00	1.00	115.00	77.22	37.78	107.62

Table 30.3: *Delta hedging of a hypothetical example where we only hedge our position once per year.*

- compared to the option price: $22.02\% \Longrightarrow \$15.02$ profit (non-discounted);
- The difference at the end is $\$37.78$ (additional shares to buy), this is a big risk and results from leaving the position open for one year.

So the option writer should

- hedge more often ... as long as it is efficient for transaction costs, and
- keep all Greeks at zero, it is not sufficient to hedge only delta.

Warning – Option hedging

To safely hedge a portfolio of options, it is not sufficient that the position remains delta neutral. We would for example not be covered if the volatility would suddenly increase. Hence, we need to manager *all* greeks (all derivatives and even some second derivatives) and keep them all zero.

30.7.10 Linear Option Strategies

Options are extremely versatile instruments, they can be used to reduce risk or even to enhance risk and potential profit dramatically. In this section we will show and name some of the more popular option strategies.

30.7.10.1 Plotting a Portfolio of Options

We want to plot a chart using the functions from the section above. However, for ease of future use we will first make a function that could be given a portfolio of options and then will aggregate them all together and show us the total effect.

Furthermore in these plots we will not use the standard `plot()` function, but rely on `ggplot2`'s functionality. `ggplot2` is discussed in Chapter 31 *"A Grammar of Graphics with ggplot2"* on page 687.

```
# load the plotting library
library(ggplot2)
```

The code for the function `portfolio_plot()` is below. This function produces the plots of intrinsic value and market value. It is quite long because if will have to work for different option types.

```
# portfolio_plot
# Produces a plot of a portfolio of the value in function of the
# spot price of the underlying asset.
# Arguments:
#   portf              - data.frame - composition of the portfolio
#                    with one row per option, structured as follows:
#                        - ['long', 'short'] - position
#                        - ['call', 'put']   - option type
#                        - numeric           - strike
#                        - numeric           - gearing (1 = 100%)
#   structureName="" - character - label of the portfolio
#   T = 1            - numeric   - time to maturity (in years)
#   r = log(1 + 0.02) - numeric   - interest rate (per year)
#                                  as log-return
#   vol = 0.2        - numeric   - annual volatility of underlying
#   spot.min = NULL  - NULL for automatic scaling x-axis, value for min
#   spot.max = NULL  - NULL for automatic scaling x-axis, value for max
#   legendPos=c(.25,0.6) - numeric vector - set to 'none' to turn off
#   yLims = NULL     - numeric vector - limits y-axis, e.g. c(80, 120)
#   fileName = NULL  - character - filename, NULL for no saving
#   xlab = "default" - character - x axis label, NULL to turn off
#   ylab = "default" - character - y axis label, NULL to turn off
# Returns (as side effect)
#   ggplot plot
#   pdf file of this plot (in subdirectory ./img/)

portfolio_plot <- function(portf,
           structureName="", # name of the option strategy
           T = 1,            # time to maturity (in years)
           r = log(1 + 0.02),# interest rate (per year)
           vol = 0.2,        # annual volatility of the underlying
           spot.min = NULL,  # NULL for automatic scaling x-axis
           spot.max = NULL,  # NULL for automatic scaling x-axis
           legendPos=c(.25,0.6), # set to 'none' to turn off
           yLims = NULL,     # limits of y-axis, e.g. c(80, 120)
           fileName = NULL,  # NULL for no saving plot into file
           xlab = "default", # set to NULL to turn off (or string)
           ylab = "default"  # set to NULL to turn off (or string)
           ) {
# portf = data frame with: long/short, call/put, strike, gearing

the_S = 100       # The spot price today is always 100
#
  strikes <- as.numeric(portf[,3])
  strike.min <- min(strikes)
  strike.max <- max(strikes)
  if (is.null(spot.min)) {
    spot.min <- min(0.8*strike.min, max(0,2*strike.min - strike.max))
    }
  if (is.null(spot.max)) {
    spot.max <- max(1.2 * strike.max, 2 * strike.max - strike.min)}
  if (structureName == ""){
    structureName<- paste(deparse(substitute(fileName)),
                          collapse = "", sep="")
    }
  nbrObs   <- 200
  spot     <- seq(spot.min,spot.max,len=nbrObs)
  val.now  <- seq(0,0,len=nbrObs)
  val.end  <- seq(0,0,len=nbrObs)
  for (k in 1:nrow(portf))
    {
     Strike  <- as.numeric(portf[k,3])
```

```
    gearing <- as.numeric(portf[k,4])
    if (portf[k,1] == 'long'){theSign <- 1}else{theSign = -1}
    if (portf[k,2] == 'call')
      {
       purchasePrice <- call_price(the_S, Strike, T, r, vol)
       callVal  <- sapply(spot, call_price, Strike=Strike, T=T,
                          r=r, vol=vol)
       val.now.incr <- callVal - purchasePrice
       val.end.incr <- sapply(spot, call_intrinsicVal,
                        Strike = Strike) - purchasePrice
      }
      else
      {
       if (portf[k,2] == 'put')
         {
         purchasePrice <- put_price(the_S, Strike, T, r, vol)
         callVal  <- sapply(spot, put_price, Strike=Strike, T=T,
                            r=r, vol=vol)
         val.now.incr <- callVal - purchasePrice
         val.end.incr <- sapply(spot, put_intrinsicVal,
                          Strike = Strike) - purchasePrice
         }
         else # then it is 'underlying'
         {
         val.now.incr <- spot - Strike
         val.end.incr <- spot - Strike
         }
      }
    val.now <- val.now + val.now.incr * gearing * theSign
    val.end <- val.end + val.end.incr * gearing * theSign
    }
  d1 <- data.frame(spot, val.end,
         paste('intrinsic value',structureName,sep=" "), 3)
  d2 <- data.frame(spot, val.now,
         paste('value 1 year to maturity',structureName,sep=" "), 2)
  colnames(d1) <- c('spot', 'value', 'legend')
  colnames(d2) <- c('spot', 'value', 'legend')
  dd <- rbind(d1,d2)
  p <- qplot(spot, value, data=dd, color = legend,
             geom = "line",size=I(2) )
  if(is.null(xlab)) {
      p <- p + theme(axis.title.x = element_blank())
      } else {
        if(xlab == "default") {p <- p + xlab('spot price')
           } else {p <- p + xlab(xlab)}}
  if(is.null(ylab)) {
      p <- p + theme(axis.title.y = element_blank())
      } else {
        if(ylab == "default") {p <- p + ylab('Value')
           } else {p <- p + ylab(ylab)}}
  p <- p + ylab('Value')
  p <- p + theme(legend.position=legendPos)
  if(legendPos == "none") {p <- p + ggtitle(structureName)}
  if (!is.null(yLims)) {p <- p + scale_y_continuous(limits=yLims)}
  if(!is.null(fileName)) {
    # remove punctuation:
    fileName <- str_replace_all(fileName, "[[:punct:]]", "")
    # remove spaces:
    fileName <- str_replace_all(fileName, " ", "")
    # save file in sub-directory img
    ggsave(paste('img/',fileName,'.pdf',sep=''),
           width = 6, height = 3)
```

```
    }
  # return the plot:
  p
}
```

The function above caters for various visual effects. While not too complicated, this tends to be quite verbose.

Digression – Plotting with base R

We could have used base R's plotting facility (as we have done in Figure 30.7 on page 652 and Figure 30.8 on page 653), but instead we chose to use `ggplot2`. This only changes the last lines of the function where the plot is parsed. In this particular case the main advantage is that a plot is an object in R that can be passed on and later parsed into one grid of plots. More about `ggplot2` is in Chapter 31 *"A Grammar of Graphics with ggplot2"* on page 687.

30.7.10.2 Single Option Strategies

First, we can use the functionality created in previous section to plot simple, single option strategies.

```
# long call
portfolio <- rbind(c('long','call',100,1))
p1 <- portfolio_plot(portfolio, 'Long call',
                     legendPos="none", xlab = NULL)

# short call
portfolio <- rbind(c('short','call',100,1))
p2 <- portfolio_plot(portfolio, 'Short call', legendPos="none",
                     xlab = NULL, ylab = NULL)

# long put
portfolio <- rbind(c('long','put',100,1))
p3 <- portfolio_plot(portfolio, 'Long put', legendPos="none",
                     xlab=NULL)

# short put
portfolio <- rbind(c('short','put',100,1))
p4 <- portfolio_plot(portfolio, 'Short put', legendPos="none",
                     xlab = NULL, , ylab = NULL)

# -- long call and short put
portfolio <- rbind(c('long','call',100,1))
portfolio <- rbind(portfolio, c('short','put',100,1))
p5 <- portfolio_plot(portfolio, 'Long call + short put',
                     legendPos="none", xlab = NULL)

# -- call
portfolio <- rbind(c('long','call',100,1))
p6 <- portfolio_plot(portfolio, 'Call', legendPos="none",
                     xlab = NULL, ylab = NULL)

# -- put
portfolio <- rbind(c('long','put',100,1))
p7 <- portfolio_plot(portfolio, 'Put', legendPos="none")
```

```
# -- callput
portfolio <- rbind(c('short','put',100,1))
portfolio <- rbind(portfolio, c('long','call',100,1))
p8 <- portfolio_plot(portfolio, 'Call + Put', legendPos="none",
                     ylab = NULL)

# show all visualizations:
grid.arrange(p1, p2, p3, p4, p5, p6, p7, p8, nrow = 4)
```

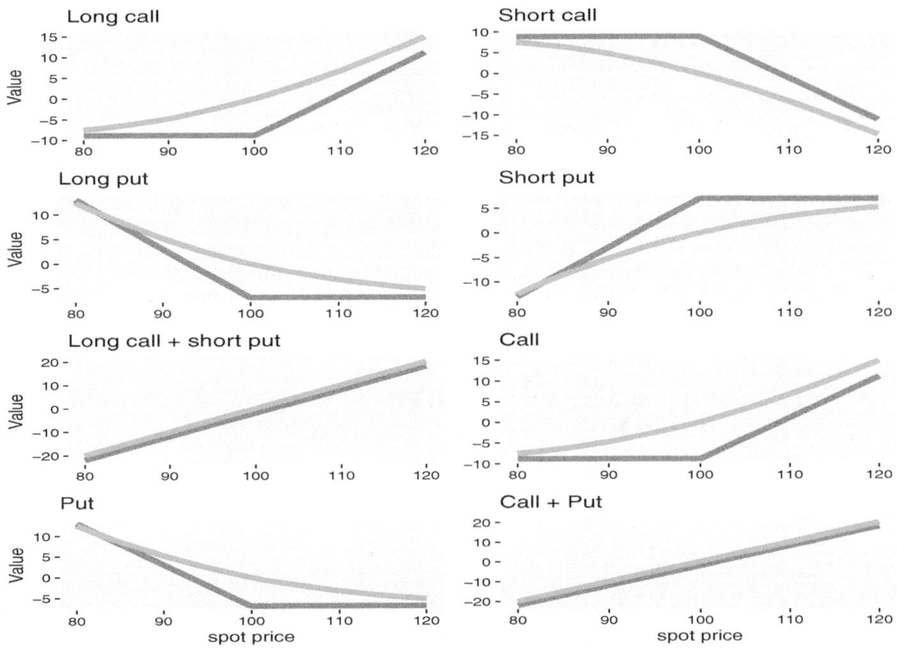

Figure 30.16: *Linear option strategies illustrated. The red line is the intrinsic value and the green line is the value of today if the spot price would move away from 100. Part 1 (basic strategies).*

The output of the aforementioned code is in Figure 30.16.

30.7.10.3 Composite Option Strategies

It becomes more interesting when we combine more options in one portfolio. In the following code we produce a few plots of popular option structures that generally can be used to reduce risk.

The first in the list – for example – is the callspread. This is a popular structure that allows to construct a portfolio where upside and downside are limited. This means that the worst outcome is known in advance and so is the best outcome.

In the second batch of option strategies that follow below, the effects on the portfolio are more directly useful in portfolio management.

```
# -- callspread
portfolio <- rbind(c('short','call',120,1))
portfolio <- rbind(portfolio, c('long','call',100,1))
p1 <- portfolio_plot(portfolio, 'CallSpread',
                     legendPos="none", xlab = NULL)

# -- short callspread
portfolio <- rbind(c('long','call',120,1))
portfolio <- rbind(portfolio, c('short','call',100,1))
p2 <- portfolio_plot(portfolio, 'Short allSpread',
                     legendPos="none", xlab = NULL, ylab = NULL)

# -- callspread differently
portfolio <- rbind(c('short','put',120,1))
portfolio <- rbind(portfolio, c('long','put',100,1))
p3 <- portfolio_plot(portfolio, 'Short putSpread',
                     legendPos="none", xlab = NULL)

# -- putspread
portfolio <- rbind(c('short','put',80,1))
portfolio <- rbind(portfolio, c('long','put',100,1))
p4 <- portfolio_plot(portfolio, 'PutSpread',
                     legendPos="none", xlab = NULL, ylab = NULL)

# -- straddle
portfolio <- rbind(c('long','call',100,1))
portfolio <- rbind(portfolio, c('long','put',100,1))
p5 <- portfolio_plot(portfolio, 'Straddle', spot.min = 50,
                     spot.max = 150,legendPos="none", xlab = NULL)
# Note that our default choices for x-axis range are not suitable
# for this structure. Hence, we add spot.min and spot.max

# -- short straddle
portfolio <- rbind(c('short','call',100,1))
portfolio <- rbind(portfolio, c('short','put',100,1))
p6 <- portfolio_plot(portfolio, 'Short straddle',spot.min = 50,
                     spot.max = 150, legendPos="none",
                     xlab = NULL, ylab = NULL)

# -- strangle
portfolio <- rbind(c('long','call',110,1))
portfolio <- rbind(portfolio, c('long','put',90,1))
p7 <- portfolio_plot(portfolio, 'Strangle',
                     spot.min = 50, spot.max = 150,
                     legendPos="none", xlab = NULL)

# -- butterfly
portfolio <- rbind(c('long','call',120,1))
portfolio <- rbind(portfolio, c('short','call',100,1))
portfolio <- rbind(portfolio, c('long','put',80,1))
portfolio <- rbind(portfolio, c('short','put',100,1))
p8 <- portfolio_plot(portfolio, 'Butterfly',
                     spot.min = 50, spot.max = 150,
                     legendPos="none", xlab = NULL, ylab = NULL)


# show all visualizations:
grid.arrange(p1, p2, p3, p4, p5, p6, p7, p8, nrow = 4)
```

The output of this second batch of linear option strategies is in Figure 30.17 on page 673.

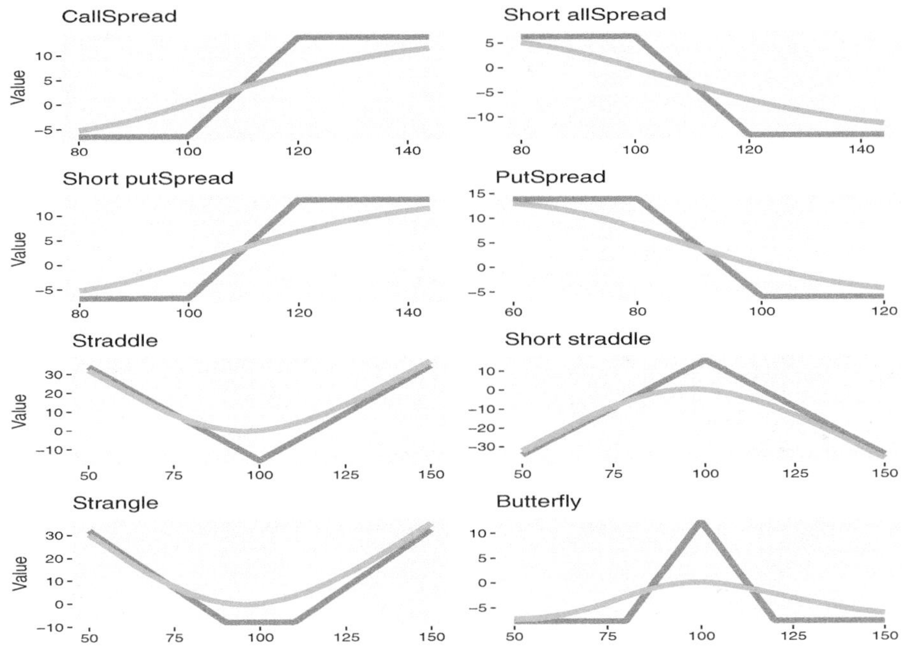

Figure 30.17: *Linear option strategies illustrated. Part 2 (basic composite structures).*

Note – Callspread and putspread

What stands out in Figure 30.17 is that a callspread can be equivalent to a putspread. The payoffs will be the same in every market scenario and hence the market price must also be the same.

Finally, we complete our list of popular structures and add one that is maybe not very practical but illustrates the fact that options can be used as building blocks to build any strategy imaginable – this structure is called "a complex structure" in the code and it is the last portfolio defined in the following code block. The plot is in Figure 30.18 on page 675

```
# -- condor
portfolio <- rbind(c('long','call',140,1))
portfolio <- rbind(portfolio, c('short','call',120,1))
portfolio <- rbind(portfolio, c('long','put',60,1))
portfolio <- rbind(portfolio, c('short','put',80,1))
p1 <- portfolio_plot(portfolio, 'Condor',spot.min = 40,
                     spot.max = 160, legendPos="none",
                     xlab = NULL)

# -- short condor
portfolio <- rbind(c('short','call',140,1))
portfolio <- rbind(portfolio, c('long','call',120,1))
portfolio <- rbind(portfolio, c('short','put',60,1))
portfolio <- rbind(portfolio, c('long','put',80,1))
p2 <- portfolio_plot(portfolio, 'Short Condor',spot.min = 40,
                     spot.max = 160, legendPos="none",
                     xlab = NULL, ylab = NULL)
```

```
# -- geared call
portfolio <- rbind(c('long','call',100.0,2))
p3 <- portfolio_plot(portfolio,
                     structureName="Call with a gearing of 2",
                     legendPos="none", xlab = NULL)

# -- nearDigital (approximate a digital option with a geared call)
portfolio <- rbind(c('short','call',100.1,10))
portfolio <- rbind(portfolio, c('long','call',100,10))
p4 <- portfolio_plot(portfolio, 'Near digital',
                     legendPos="none", xlab = NULL, ylab = NULL)

# -- a complex structure:
portfolio <- rbind(c('long','call',110,1))
portfolio <- rbind(portfolio, c('short','call',105,1))
portfolio <- rbind(portfolio, c('short','put',95,1))
portfolio <- rbind(portfolio, c('long','put',90,1))
portfolio <- rbind(portfolio, c('long','put',80,1))
portfolio <- rbind(portfolio, c('long','call',120,1))
portfolio <- rbind(portfolio, c('short','call',125,1))
portfolio <- rbind(portfolio, c('short','put',70,10))
portfolio <- rbind(portfolio, c('short','put',75,1))
portfolio <- rbind(portfolio, c('short','call',130,10))
portfolio <- rbind(portfolio, c('long','call',99,10))
portfolio <- rbind(portfolio, c('short','call',100,10))
portfolio <- rbind(portfolio, c('short','put',100,10))
portfolio <- rbind(portfolio, c('long','put',101,10))
p5 <- portfolio_plot(portfolio, 'Fun',legendPos='none',
                     spot.min=60, spot.max=140,
                     yLims=c(-0,25))

# show all visualizations:
# Pasing a layout_matrix to the function grid.arrange()
# allows to make the lower plot bigger:
layoutM <- rbind(c(1,2),
                 c(3,4),
                 c(5,5))
grid.arrange(p1, p2, p3, p4, p5, nrow = 3, layout_matrix = layoutM)

## Warning: Removed 201 rows containing missing values (geom_path).
```

By now it should be clear that options can be used as building block to create payoff structures as desired for portfolio management.

Note also that each structure has its counterpart as a short position. To visualise this "short" structure, it is sufficient to switch "long" with "short" in each definition.

30.7.11 Integrated Option Strategies

Options can be used in investment portfolios to limit downside risk or to set up for maximal gain in certain market conditions. Options are also ideal assets to mix with regular portfolios. In this section, we will present three such structures that can be useful when managing a long only portfolio of assets.

The advantage of these structures is that they only use vanilla options, and hence the portfolio manager can use options that are traded on the stock exchange and enjoy a liquid market and the protection of a clearing house.

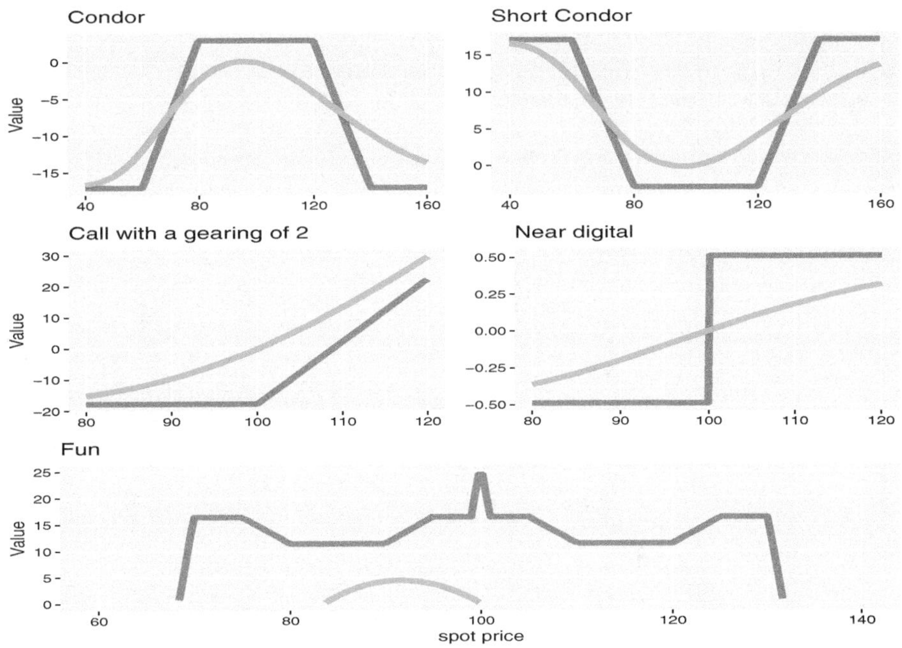

Figure 30.18: *Linear option strategies illustrated. Part 3 (some more complex structures and extra one structure that is entirely made up).*

30.7.11.1 The Covered Call

In the following code, we will study a few of those possibilities with the aid of the functions that we created in Section 30.7.10 *"Linear Option Strategies"* on page 667. The first example is the "covered call". This is a portfolio that contains both a short call and the underlying (for the same nominal amounts). The visualisation is in Figure 30.19 on page 676.

```
## --- covered call ----

nbrObs      <- 100
the.S       <- 100
the.Strike <- 100
the.r       <- log (1 + 0.03)
the.T       <- 1
the.vol     <- 0.2
Spot.min    = 80
Spot.max    = 120
LegendPos   = c(.5,0.2)
Spot    <- seq(Spot.min,Spot.max,len=nbrObs)
val.end.call <-   - sapply(Spot, call_intrinsicVal, Strike = 100)
call.value <- call_price(the.S, the.Strike, the.T, the.r, the.vol)

d.underlying <- data.frame(Spot, Spot - 100,  'Underlying',   1)
d.shortcall  <- data.frame(Spot, val.end.call,  'Short call', 1)
d.portfolio  <- data.frame(Spot,
                           Spot + val.end.call + call.value - 100,
                           'portfolio', 1.1)
colnames(d.underlying) <- c('Spot', 'value', 'Legend','size')
colnames(d.shortcall)  <- c('Spot', 'value', 'Legend','size')
colnames(d.portfolio)  <- c('Spot', 'value', 'Legend','size')
dd <- rbind(d.underlying,d.shortcall,d.portfolio)
```

```
p <- qplot(Spot, value, data = dd, color = Legend, geom = "line",
           size=size )
p <- p + xlab('Value of the underlying' ) + ylab('Profit at maturity')
p <- p + theme(legend.position = LegendPos)
p <- p + scale_size(guide = 'none')
print(p)
```

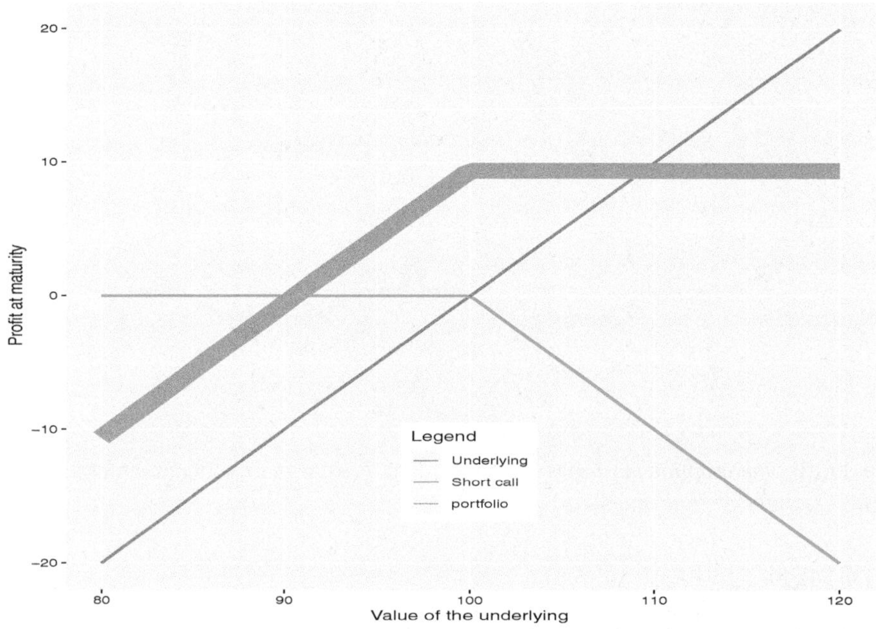

Figure 30.19: *A covered call is a short call where the losses are protected by having the underlying asset in the same portfolio.*

30.7.11.2 The Married Put

In case we want to protect our investments from large adverse market movements, we can add a put to the underlying asset. The effect is that we pay the premium for the option, and hence move profit down by a fixed amount regardless the market. However, when the markets are down we can execute our put option (since we have the underlying in portfolio) and compensate potential losses.

This structure is called "a married put". It is constructed in the following code and visualised in Figure 30.20 on page 677.

```
## --- married put ----

LegendPos = c(.8,0.2)
Spot        <- seq(Spot.min,Spot.max,len=nbrObs)
val.end.put <-  sapply(Spot, put_intrinsicVal, Strike = 100)
put.value   <- - put_price(the.S, the.Strike, the.T, the.r, the.vol)

d.underlying <- data.frame(Spot, Spot - 100,  'Underlying', 1)
d.shortput   <- data.frame(Spot, val.end.put,  'Long put',  1)
d.portfolio  <- data.frame(Spot,
                           Spot + val.end.put + put.value - 100,
```

```
                        'portfolio',
                        1.1)
colnames(d.underlying) <- c('Spot', 'value', 'Legend','size')
colnames(d.shortput)   <- c('Spot', 'value', 'Legend','size')
colnames(d.portfolio)  <- c('Spot', 'value', 'Legend','size')
dd <- rbind(d.underlying,d.shortput,d.portfolio)
p  <- qplot(Spot, value, data = dd, color = Legend, geom = "line",
          size = size )
p <- p + xlab('Value of the underlying' ) + ylab('Profit at maturity')
p <- p + theme(legend.position = LegendPos)
p <- p + scale_size(guide = 'none')
print(p)
```

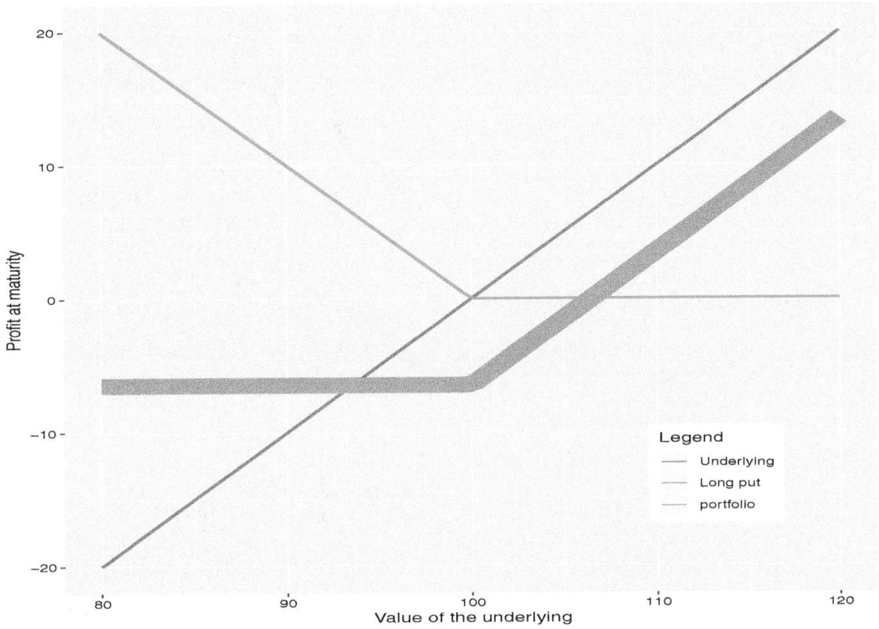

Figure 30.20: *A married put is a put option combined with the underlying asset.*

30.7.11.3 The Collar

If we want to limit the cost of the married put, then we can construct a collar option by means of put options or call options. We can use a long and a short call, a long and a short put or even mix the structures. Below, we use a long put and a short call and visualise the result in Figure 30.21 on page 678.

```
## --- collar ----

# Using the same default values as for the previous code block:
LegendPos = c(.6,0.25)
Spot          <-   seq(Spot.min,Spot.max,len=nbrObs)
val.end.call <- - sapply(Spot, call_intrinsicVal, Strike = 110)
val.end.put  <- + sapply(Spot, put_intrinsicVal, Strike = 95)
call.value   <- call_price(the.S, the.Strike, the.T, the.r, the.vol)
put.value    <- put_price(the.S, the.Strike, the.T, the.r, the.vol)
```

```
d.underlying <- data.frame(Spot, Spot - 100,     'Underlying', 1)
d.shortcall  <- data.frame(Spot, val.end.call, 'Short call', 1)
d.longput    <- data.frame(Spot, val.end.put,  'Long call',  1)
d.portfolio  <- data.frame(Spot, Spot + val.end.call + call.value +
                    val.end.put - put.value - 100, 'portfolio',1.1)
colnames(d.underlying) <- c('Spot', 'value', 'Legend','size')
colnames(d.shortcall)  <- c('Spot', 'value', 'Legend','size')
colnames(d.longput)    <- c('Spot', 'value', 'Legend','size')
colnames(d.portfolio)  <- c('Spot', 'value', 'Legend','size')
dd <- rbind(d.underlying,d.shortcall,d.longput,d.portfolio)
p  <- qplot(Spot, value, data=dd, color = Legend, geom = "line",
           size=size )
p <- p + xlab('Value of the underlying' ) + ylab('Profit at maturity')
p <- p + theme(legend.position = LegendPos)
p <- p + scale_size(guide = 'none')
print(p)
```

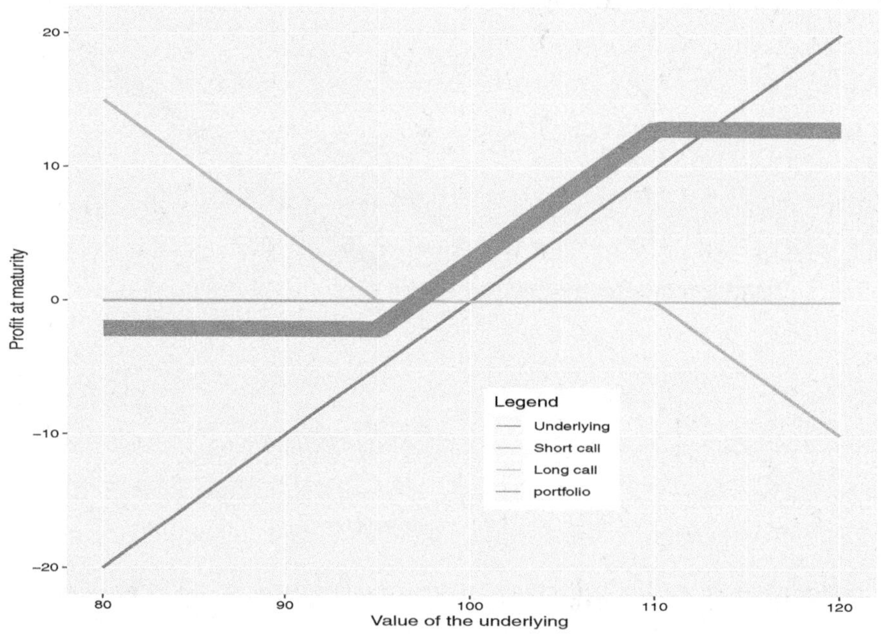

Figure 30.21: *A collar is a structure that protects us from strong downwards movements at the cost of a limited upside potential.*

option – exotic

30.7.12 Exotic Options

So far, we have studied linear combinations from elementary options such as the put and the call. It is possible to extend the concept "option" beyond the call and put as described before. Actually, we can take any parameter such as strike or fixing of spot price, and modify that definition. Doing so, we can create a whole set op new options that answer a particular need. Those modified option types are called "exotic options."

This might seem rather specialised, however, you might already have these exotic options in portfolio. These types of options are a popular way to design capital protected investment funds or insurance linked investments. Studying those options, will give you insight in how those investment products work. These are actually products that you might use for your own retirement in the package as explained in the following section: Chapter 30.7.13 *"Capital Protected Structures"* on page 680.

Digression – Investment advice

There is a lot to say about who the investment portfolio of a private investor should look like. The author's opinion is published in De Brouwer (2012). In general, it is worth to ask oneself if the complex structure really adds value. Most important is to keep in mind what one wants to do with this investment. Investments should answer a life goal, and not tailored to someone's risk appetite.

Before listing the more complex types it is probably worth to understand some of the basic building blocks for Exotic Options.

- *Knock-in:* This option only becomes active when a certain level (up or down) is reached.
- *Knock-out:* This option will have a fixed (zero or more) return when a certain level (up or down is reached); if the level is not reached it remains an option.
- *Barrier Option:* Another word for Knock-in or Knock-Out options.
- Options on "stock baskets" as opposed to options on stocks or indices, eventually accompanied by algorithms that chagne the basket during the lifetime of the option.

In the context of exotic options, the basic options such as call and put are referred to as "(plain) vanilla options." The vanilla options and the aforementioned advanced options are used as building blocks to construct even more customised option types. It almost looks as if it is done on purpose, but those options come in different groups that are roughly named after counties or mountain ranges.

- *Asian options:* The strike or spot is determined by the average price of the underlying taken at different moments.
- *Full Asianing:* The average of the Asian feature is taken over some moments over the whole lifetime of the option.
- *Look-back options:* The spot is determined as the best price of different moments
- *Russian Look-back options:* The look-back feature is unlimited (the best price over the whole lifetime of the option.
- *Callable or Israeli options:* The writer has the opportunity to cancel the option, but must pay the payoff at that point plus a penalty fee.
- *Himalayan:* Payoff based on the performance of the best asset in the portfolio.
- *Everest:* Payoff based on the worst-performing securities in the basket.
- *Annapurna:* In which the option holder is rewarded if all securities in the basket never fall below a certain price during the relevant time period. **Annapurna**
- *Atlas:* In which the best and worst-performing securities are removed from the basket prior to execution of the option. **Atlas**

- *Altiplano:* A vanilla option is combined with a compensatory coupon payment if the underlying security never reaches its strike price during a given period.
- *Bermuda Option:* An option where the buyer has the right to exercise at a set (always discretely spaced) number of times, so this structure is between an American and a European option.
- *Canary Option:* Can be exercised at quarterly dates, but not before a set time period has elapsed.
- *Verde Option:* Can be exercised at incremental dates(typically annually), but not before a set time period has elapsed.

30.7.13 Capital Protected Structures

People do not like losses when profits are involved – in other words: people are loss averse.[11] Hence, it makes a lot of sense to offer to investors the possibility to invest in a structure that would never show losses. It would for example after three years pay at least the initial capital back and – if a certain market or stock is up – it might pay additional profit.

The structures were popular in Spain and Belgium since the 1990s and quickly conquered the world in the aftermath of the financial meltdown of 2008. It is quite likely that you or a family member has investments in such "capital protected" investment fund or insurance product.

As long as interest rates are positive, it is possible to think of such structure as a portfolio that contains

- a zero bond – or deposit – that will increase to exactly the initial amount invested after a certain period,
- an option that is bought with the money left over from the first operation. This option will then provide the additional profit if the market evolves positively. In adverse market conditions it expires out-of-the money, but the initial capital is reconstructed via the zero bond.

Example: capital protected structure

Using our standard example parameters (see page 650), and assuming that

1. we can place a deposit at $r = 2\%$,
2. that we build a five year capital protected structure,
3. that we have no transaction, nor spreads, nor holding costs, and
4. that our nominal amount is €1 000;

then what what gearing can we give to a capital protected structure that only uses a fixed term deposit (or custom zero bond) and a vanilla call option?

We need to invest € $1000\frac{1}{(1+0.02)^5} =$ €905.73 in the fixed term deposit in order to make it increase to 1 000 in five year.
This leaves us €94.27 for buying an option. A call option on 5 years costs us €95.35 on a nominal of €1 000, so we can make a structure with a gearing of ca. 99%.

[11] The aspect of being loss averse as opposed to being risk averse is for example described in De Brouwer (2012), Thaler (2016), or Kahneman (2011). In traditional economic theory actors on markets are "rational" and always "risk-averse," however, people tend to be rather loss-averse. This implies that for profits people are risk-averse, but for losses they are risk-seeking.

The previous example refers to a structure that has no costs. That is of course not realistic. We will need to procure the services of an investment manager, transfer agent, deposit bank, and we need to pay costs related to the prospectus of the fund. Assume now that we have a cost of (on average) 1% per year.

In R, this example can be programmed as follows (note that we use our function `call_price()` – which is defined in Chapter 30.7.5.2 *"Apply the Black and Scholes Formula"* on page 650):

```
##--------- the example of the capital protected structure
N           <- 5
nominal     <- 1000
inDeposit   <- nominal * (1.02)^(-N)
cst         <- 0.01  *0          # in PERCENT
pvCosts     <- N * cst * nominal # one should rather use the present value here
rest4option <- 1000 - inDeposit - pvCosts
callPrice   <- call_price (100, Strike = 100, T = 5, r = 0.02, vol = 0.02)
# reformulate this price as a percentage and then adjust to nominal
callPrice   <- callPrice / 100 * 1000
gearing     <- rest4option / callPrice
paste('The gearing is:', round(gearing * 100, 2))
## [1] "The gearing is: 46.43"
```

So, when adding 1% of annual costs to this structure (and not investing the provisions for these costs), our gearing decreases to 46.43%. That is roughly half of the original gearing.

PART VII

Reporting

In this book we already studied data and building mathematical models on that data. The next logical step is to communicate clearly and concisely, bring our opinion across and point the management to potential opportunities and risks. The first step is generally building a good presentation.

Again, R can be our workhorse and be the cornerstone of a free work-flow that produces professional results. While certain applications might have an edge in environments where data is not measured in Gigabytes but in Terrabytes, or where the inflow of information is extremely fast and one has to make sense of the flow of data rather than a given snapshot. The reality is that in almost all applications in almost all organisations from the sole trader to the giant that employs three hundred thousand people and processes hundred thousand payments per second one will find that a free set of tools based around R, C++, MySQL and presented in a pdf or html5 web-page is more than enough.

The difference is in the mindset and in the price tag. The large corporate can save millions of dollars switching to a free workflow. However, that entails taking responsibility as there is no company to blame in the unexpected case that something could go wrong. Should it also be mentioned that the relevant manager's bonus is linked to tangible micro-results – which are easier to obtain with the support of a commercial company – rather than to the money that is not spend (or overall company profit).

In this section we provide the elements that will allow the reader to build very professional reports, dashboard, and other documentation, using only free and open source software.

For example, this book is written in LaTeX markup language and the R code is compiled with the `knitr` library.

First, we will have a look at the library `ggplot2`, that we already encountered a few times in this book.

♣ 31 ♣

A Grammar of Graphics with ggplot2

We already used `ggplot2` in Chapter 9.6 *"Violin Plots"* on page 173, Chapter 22.2.3 *"The AUC"* on page 396 and in Chapter 30.7.7 *"Dependencies of the Option Price"* on page 660 (and following). We were confident in doing so, because `ggplot2` is rather intuitive and it is not so difficult to understand what a given code segment does. In this section, we will use little words, but allow the code to speak for itself and provide tactical examples that learn how to use `ggplot2` in practice. `ggplot2`

With `ggplot2` comes the notion of a "grammar of graphics." Just as the grammar for the English language, it is a set of rules that allow different words to interact and produce something that makes sense. That something is in the case of `ggplot2` is a professional and clear chart.

Further information – Extensions

Another important advantage of `ggplot2` is that it is designed to be extendible. For example, the website `https://www.ggplot2-exts.org` hosts a set of jaw dropping extension. Before starting to write our own extension it is worth to have a look at this gallery, and most probably, you can start from something that is already there.

The Big R-Book: From Data Science to Learning Machines and Big Data, First Edition. Philippe J.S. De Brouwer.

Companion Website: www.wiley.com/go/De Brouwer/The Big R-Book

31.1 The Basics of ggplot2

To explore `ggplot2`, we will use the dataset `mtcars` from the library `datasets`. It is usually loaded as the start of R, and it is already known from other sections in this book.

```
# install once: install.packages('ggplot2')
library(ggplot2)
```

Most of the functionality is obtained by producing a plot object. This is done by the function

ggplot()

`ggplot`. It takes first a data-frame.[1] Then, it will still need an aesthetics parameter and a geometry one.

This plot object with only dots to be printed is created in the following code. The plot is visualised in Figure 31.1.

```
p <- ggplot(mtcars, aes(x=wt, y=mpg))
# So far printing p would result in an empty plot.
# We need to add a geom to tell ggplot how to plot.
p <- p + geom_point()

# Now, print the plot:
p
```

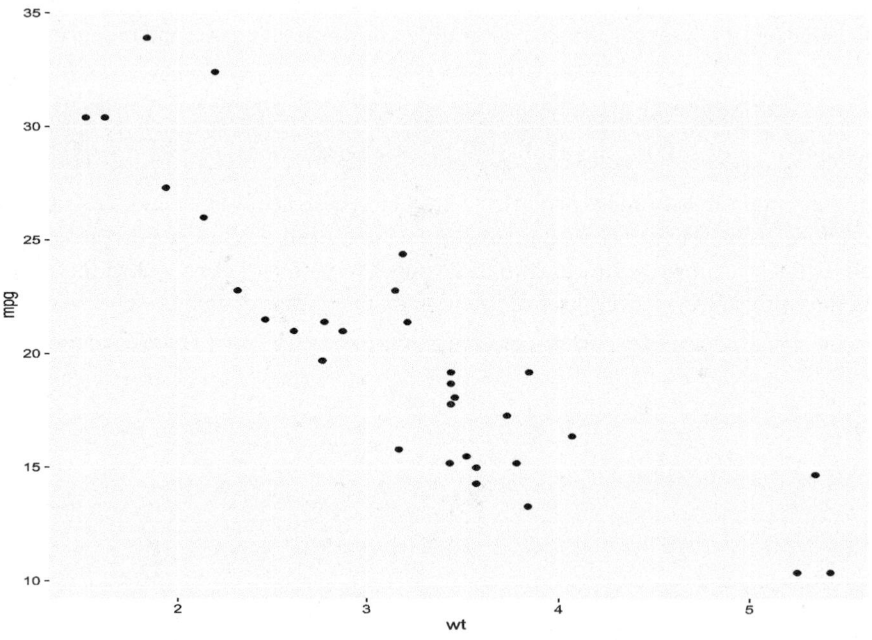

ggplot()

Figure 31.1: *A basic and simple scatter-plot generated with* `ggplot2`.

This is a basic plot. From here, there are an endless series of possibilities. For example, we notice that there is indeed some relationship between the fuel consumption `mpg` and the weight (`wt`) of a car. It seems as we are on to something.

With the following code segment, we investigate this assumption with a Loess estimate and improve the layout of the plot (add axis names, change font face and size, and add a clear title). The result is shown in Figure 31.2 on page 689.

[1] This is in line with the tidyverse philosophy, however, `ggplot2` does not allow to work with the pipe operator.

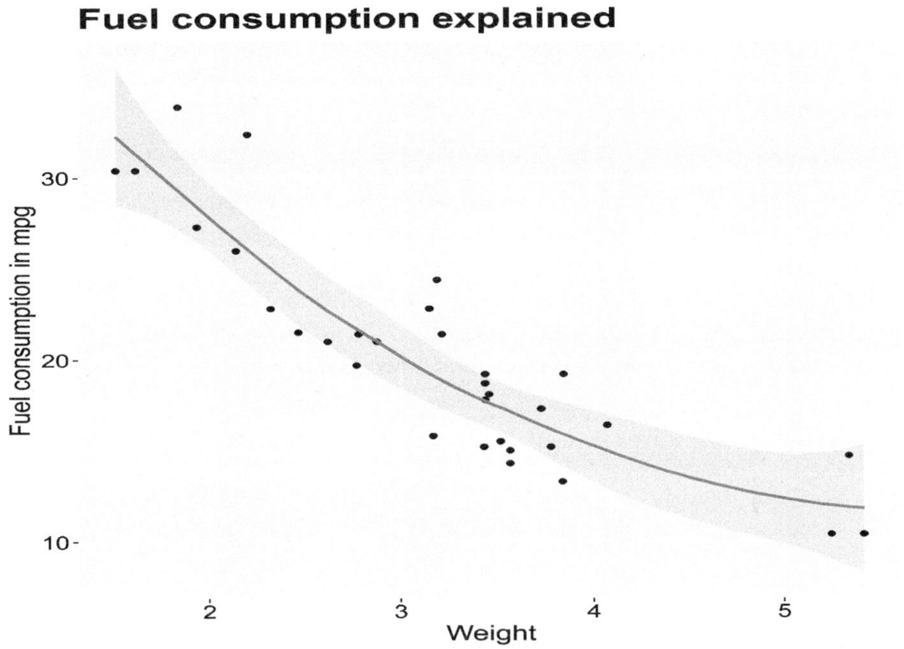

Figure 31.2: *The same plot as in previous figure, but now enhanced with Loess estimates complete with their confidence interval, custom title, axis labels, and improved font properties.*

```
# Note that we can reuse the object p
p <- p + geom_smooth(method = "loess", span = 0.9, alpha = 0.3)
p <- p + xlab('Weight') + ggtitle('Fuel consumption explained')
p <- p + ylab('Fuel consumption in mpg')
p <- p + theme(axis.text  = element_text(size = 14),
               axis.title = element_text(size = 16),
               plot.title = element_text(size = 22, face = 'bold'))
p
```

The shaded area is by default the 95% confidence interval. This confidence interval level can be changed by passing the argument `level` to the function `geom_smooth`. The type of estimation can also be changed. For example, it is possible to change the loess estimation by a linear model by passing "lm" to the method argument. For example: `geom_smooth(method = "lm", level = 0.90)`.

Hint – Themes

`ggplot2` allows the user to specify a "theme," which is a pre-defined list of options that give the plot a certain look and feel. Here are some options:

```
# Try *one* of the following:
p <- p + theme_minimal(base_size = 16)  # change also font size
p <- p + theme_light()                  # light theme
p <- p + theme_dark()                   # dark theme
p <- p + theme_classic()                # classic style
p <- p + theme_minimal()                # minimalistic style
# Much more themes are available right after loading ggplot2.

# Change the theme for the entire session:
theme_set(theme_classic(base_size = 12))
```

> Would you like even more eye-candy? Or you want to make a book according to the famous Tufte style? Or rather have a layout as in the Economist? then you might want to have a look at the package `ggthemes`.

It is now easy to add layers and information to the plot. For example, we can investigate the effect of the gearbox (in the parameter `am`: this is zero for an automatic gearbox and one for a manual transmission). Further, we will change the size of the dot in function of the acceleration of the car (in the variable `qsec`), and show the new plot in Figure 31.3.

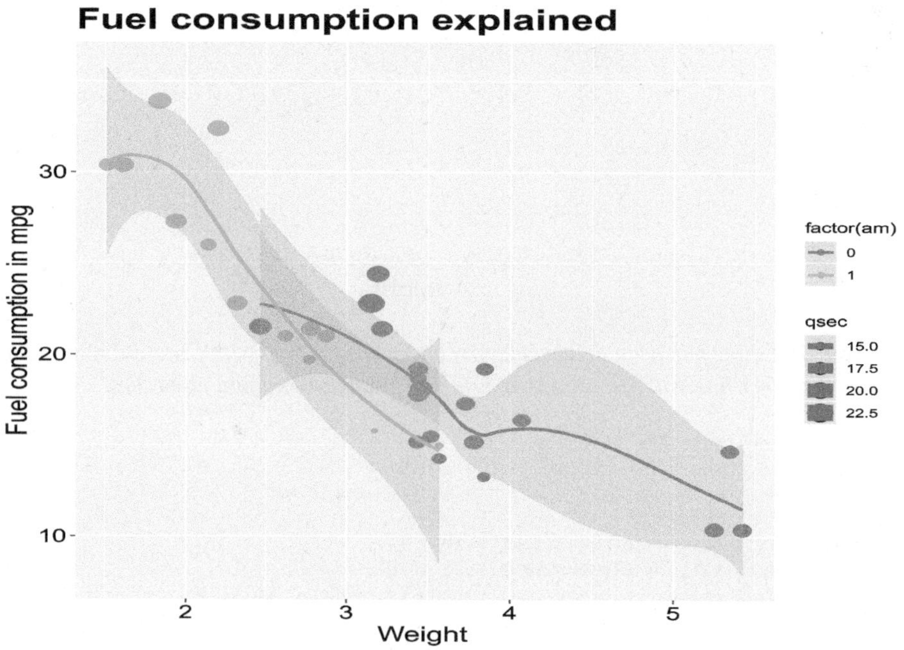

Figure 31.3: *The same plot as in previous Figure, but now enhanced with different colours that depend on the gearbox (the parameter* `am`*, and the size of the dot corresponds to the time the car needs to get from standstill to the distance of 0.2 Miles (ca. 322 m).*

```
# We start from the previously generated plot p:
p <- p + aes(colour = factor(am))
p <- p + aes(size   = qsec)
p
```

Note that it is not necessary to factorize the variable `am`. However, the result will be different. If the variable is not factorised, ggplot will scale the colour between two colours in a linear way (in other words, it will genearate a smooth transition, which does not make sense for a binary variable).

Another powerful feature of `ggplot2` is its ability to split plots and produce a "facet grid." A facet grid is executed in function of one or more variables. This can be passed on to `ggplot2` in the familiar way of passing on an expression (note the `~` in the following code). The output is in Figure 31.4 on page 691 and we also show how to use the pipe operator.

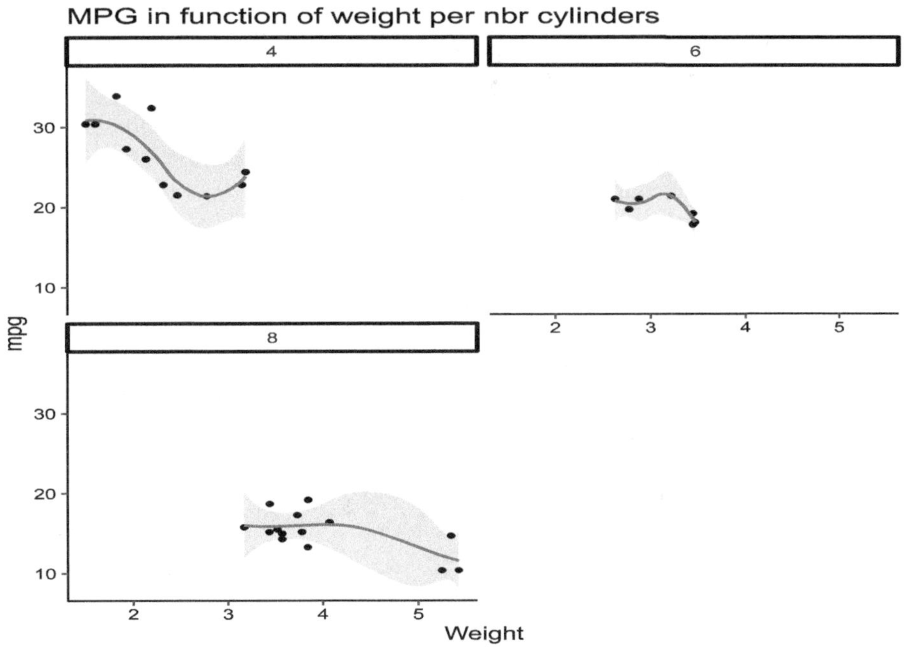

Figure 31.4: *A facet plot will create sub-plots per discrete value of one or more variables. In this case we used the number of cylinders, so each sub-plot is for the number of cylinders that is in its title.*

```
library(tidyverse)  # provides the pipe: %>%
mtcars %>%
  ggplot(aes(x = wt, y = mpg)) +
        geom_point() +
        geom_smooth(method = "loess", span = 0.99, alpha = 0.3) +
        xlab('Weight') +
        ggtitle('MPG in function of weight per nbr cylinders') +
        facet_wrap( ~ cyl, nrow = 2) +
        theme_classic(base_size = 14) # helps the grey to stand out
```

? Question #29 – Explore data with ggplot

The package ggplot2 comes with a dataset mpg that has more rows and different information. Try to find out more about the cars in the dataset by using ggplot2. Better still go to https://www.fueleconomy.gov and download a recent dataset.

31.2 Over-plotting

The dataset mtcars has only 32 rows, and the aforementioned plots are not cluttered and clear. Generally, we can expect more than 32 rows in a dataset and soon plots will be so busy that it is not possible to spot any trend or pattern. For example, a loan portfolio can easily have a few million customers, high frequency market data can have multiple thousands of observations per second, etc.

For the larger dataset, a simple scatterplot will soon become hard to read. To illustrate this, we will generate our own dataset with the following code and plot the result in Figure 31.5. In this example it might be useful to think of the variables to be defined as: "LTI" meaning "loans to installments" (the amount of loan repayments due very month divided by the income over that period), and "DPD" as "days past due" (the number of days that the payments were due but not paid).

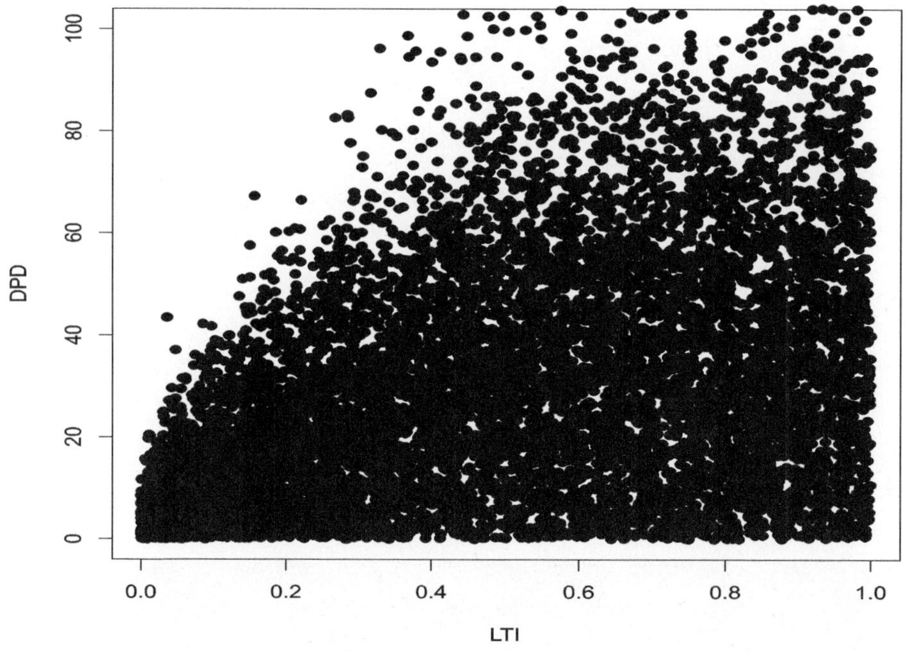

Figure 31.5: *The standard functionality for scatterplots is not optimal for large datasets. In this case, it is not clear what the relation is between LTI and DPD. ggplot2 will provide us some more tools to handle this situation.*

```
# Set the seed to allow results to be replicated:
set.seed(1868)

# Generate the data:
LTI <- runif(10000, min = 0, max = 1)
DPD <- abs(rnorm(10000,
                 mean = 70 * ifelse(LTI < 0.5, LTI, 0.5),
                 sd = 30 * sqrt(LTI) + 5))

# Plot the newly generated data and try to make it not cluttered:
```

```
plot(LTI, DPD,
     pch=19,            # small dot
     ylim =c(0, 100)) # not show outliers, hence zooming in
```

We can think of some tools to solve this problem in base R: use a plot character that is small (like pch = 1, 3, or 5) or use boxplots. However, ggplot2 has a larger set of tools to tackle this situation.

ggplot2 has all the tools of the traditional plot() function, but it has a few more possibilities. We can choose one or more of the following:

- add additional information with geom_smooth(), geom_quantile(), or geom_density_2d(); geom_density()
- use boxplots via geom_boxplot(); geom_boxplot()
- use violin plots: see Chapter 9.6 *"Violin Plots"* on page 173;
- it is possible to summarise the density of points at each location and display that in some way, using geom_count(), geom_hex(), geom_bin2d() or geom_density2d(). The latter will first make a a 2D kernel density estimation using MASS::kde2d() and display the results with contours; geom_count()
- add transparency to the points by adding geom_point(alpha = 0.05); and
- use small dots: geom_point(shape = ".").

There is no unique best solution that is the best for each dataset. Usually, the solutions described previously will allow to visualize the dependencies. If the database is really too big then the effect might still not be optimal or simply it might take too long to produce the plots. In that case, it is of course possible to use a random sample of the dataset to produce the plot.

A fist solution that ggplot2 offers is a contour plot. Using the contour plot on the same dataset, we can work as follows and plot the results in Figure 31.6 on page 694:

```
# We add also the colour schemes of viridisLite:
library(viridisLite)

d <- data.frame(LTI = LTI, DPD = DPD)
p <- ggplot(d, aes(x = LTI, y = DPD)) +
     stat_density_2d(geom = "raster", aes(fill = ..density..),
                     contour = FALSE) +
     geom_density_2d() +
     scale_fill_gradientn(colours = viridis(256, option = "D")) +
     ylim(0,100)
p

## Warning: Removed 91 rows containing non-finite values
## (stat_density2d).
## Warning: Removed 91 rows containing non-finite values
## (stat_density2d).


# Note that ggplot will warn us about the data that is not shown
# due to the cut-off on the y-axis -- via the function ylim(0, 100).
```

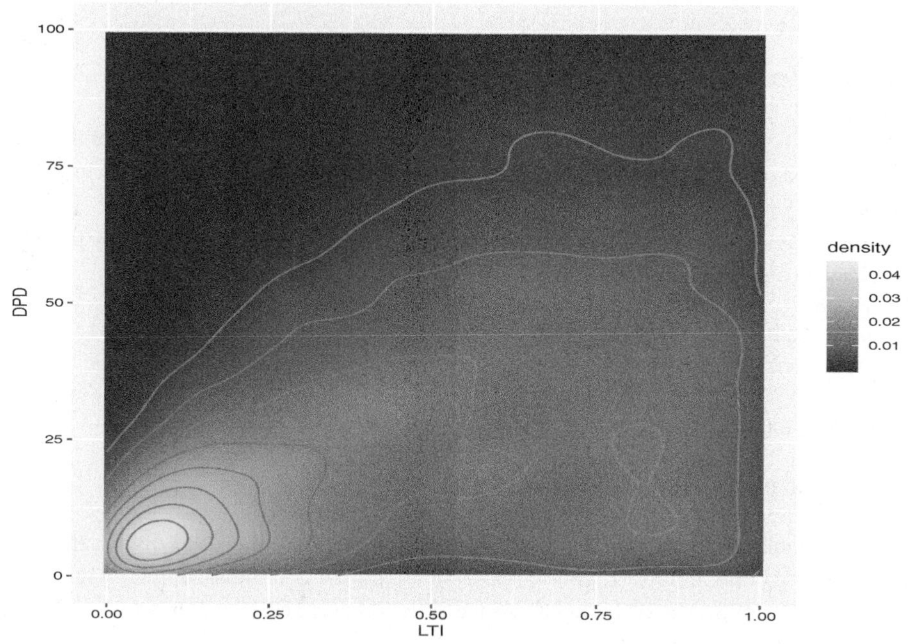

Figure 31.6: *The contour plot is able to show where the density of points is highest by a visually attractive gradient in colour.*

Another approach could be based on a scatter plot to which we add transparency for the dots and (if the dataset is not too large[2]) a Loess fit via `geom_smooth()` – the results is shown in Figure 31.7 on page 695:

```
# Loess smoothing with ggplot2:
# This will generate warnings, because the observations that are not
# plotted will not be used for our smoothing algorithm.
ggplot(d, aes(x = LTI, y = DPD)) + geom_point(alpha=0.25) +
      geom_smooth(method='loess') +
      ylim(0,100)

## Warning: Removed 91 rows containing non-finite values
## (stat_smooth).
## Warning: Removed 91 rows containing missing values (geom_point).
```

Warning – Not plotting data is not including it at all

Any method passed to `geom_smooth()` such as `lm`, `gam`, and `loess` will only take into account the observations that are not excluded from being plotted. In our case we have a cut-off on the y-axiswith `ylim(0, 100)`, so any outliers that have a DPD higher than 100 do not contribute in the model being fitted.

[2]For a Loess estimation the time to calculate and memory use are quadratic in the number of observations, and hence with actual computers a dataset of about thousand observations is maximum. For larger datasets, one can then resort to taking a random sample of the data or using another model fit, for example, a linear model can be used with the option `method='lm'`.

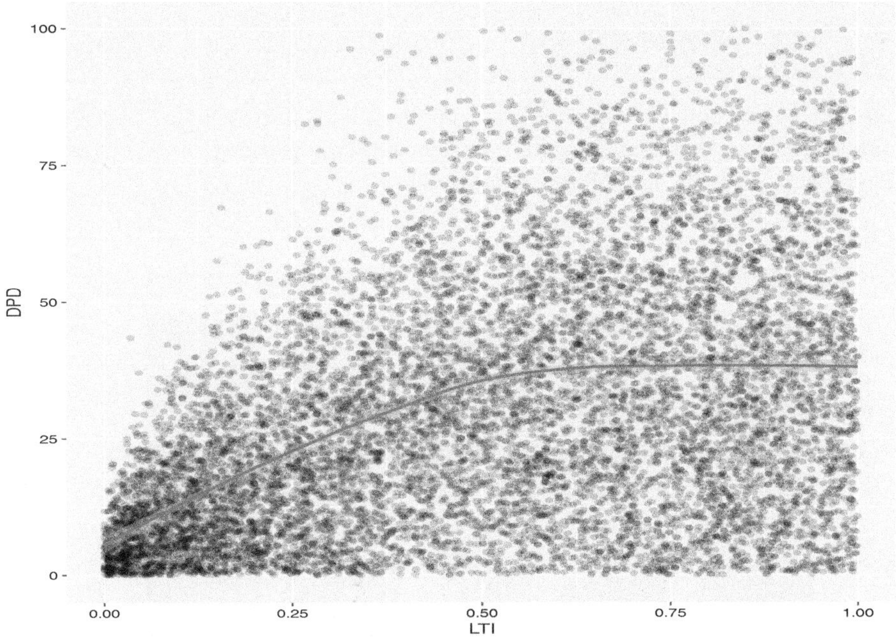

Figure 31.7: *Adding a Loess estimate is a good idea to visualize the general trend in the data. The algorithm, however needs a time that increases as the square of the number of observations and will be too slow for larger datasets.*

Note – Selection of the smoothing method

The geom "smooth" takes an argument `method`. This refers to the model that is fitted. For example we can use `lm`, `glm`, `gam`, `loess`, or `rlm`. The default is `method = "auto"`. In that case, R will select the smoothing method is chosen based on the size of the largest group (across all panels) that is being plotted. It will select "loess" for sample sizes up to 1 000 observations and in all other cases it will use "gam" (Generalized Additive Model) with the formula `formula = y ~ s(x, bs = "cs")`. The reason for that process is that usually "loess" is preferable because it makes less assumptions and hence can reveal other trends than only linear. However, Loess estimation will need $O(n^2)$ memory, which can become fast insurmountable. This is the reason why we in our example have reduced the number of data-points.

31.3 Case Study for ggplot2

To summarize the section about `ggplot2`, we will show one more example on another dataset. We choose the dataset "diamonds." This dataset is provided by the package `ggplot2` and is ideally suited to show the excellent capabilities of `ggplot2`.

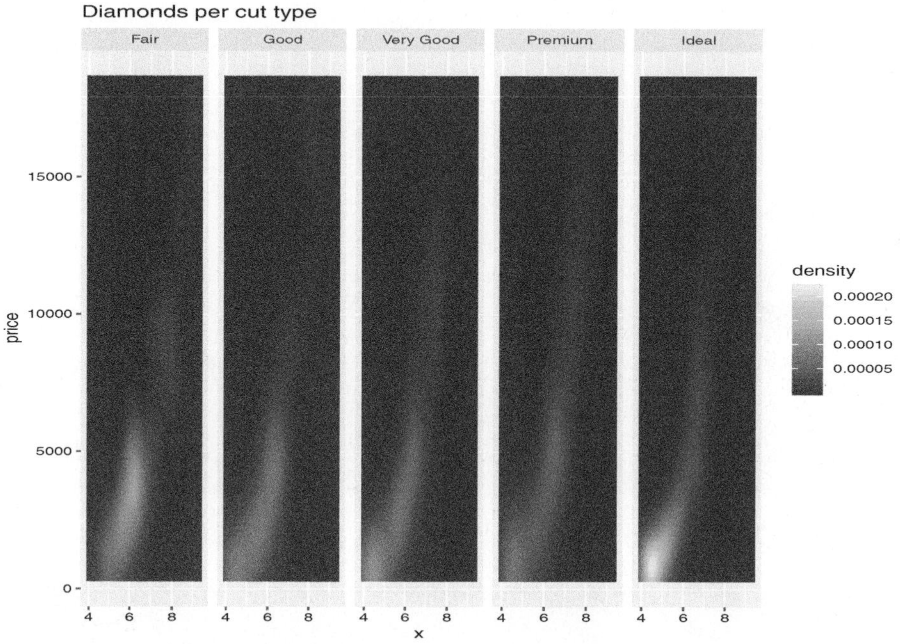

Figure 31.8: *This plot shows a facet plot of a contour plot with customised colour scheme.*

The code following code segment will generate the plot in Figure 31.8.

```
library(ggplot2)
library(viridisLite)

# take a subset of the dataset diamonds
set.seed(1867)
d <- diamonds[sample(nrow(diamonds), 1500),]
p <- ggplot(d, aes(x, price)) +
     stat_density_2d(geom = "raster", aes(fill = ..density..),
                     contour = FALSE) +
#     geom_density_2d() +
     facet_grid(. ~ cut) +
     scale_fill_gradientn(colours = viridis(256, option = "D")) +
     ggtitle('Diamonds per cut type')
p
```

Note – ggplot and the pipe

While `ggplot` is not really compatible with the piping operator (`%>%`),[a] it is possible to use different notation styles. In this section we changed from creating an object `p` and adding to this, over moving the `+` to the end of the line so that R expects further input, to omitting the object and allowing the function `ggplot()` to send the output directly.

[a] While we can pass data to `ggplot2` with the pipe opeartor, later on we cannot use it to add a geom for example. The workaround is to use the package `wrapr` and its `%.>%` operator.

Question #30 – Predicting days past due

Given the data in the example above. Assume that LTI stands for "loans to income" (or rather loan instalments to income ratio) and DPD stands for "days past due" (the arrears on the payments of the loans). Further, assume that you would be requested to build a model that predicts the DPD in function of LTI. How would you proceed and what model would you build? Do think this data is realistic in such case? What does it mean?

♣ 32 ♣

R Markdown

Once an analysis is finished and conclusions can be drawn, it is important to write up a report. Whether that report will look like a slideshow, scientific paper, or a book, some tasks will be quite laborious. For example, saving all plots, then adding them to the document, and – even worse – do the whole cycle again when something changes (new data, a great idea, etc.).

R Markdown is a great solution. R Markdown is a variant to markdown (that is – despite its name – a "markup language") that is designed to mix text, presentation, and R-code.

Hint – RStudio

It is entirely possible to do everything from the command prompt, but working with RStudio will be most efficient.[a]

[a]It is worth to note that `rmarkdown` is produced and maintained by the RStudio team. RStudio has produced and is maintaining `rmarkdown`, yet made it freely available for everyone. The website of RStudio is `http://www.rstudio.com`.

RStudio shortens the learning curve by facilitating the creation of a new document (see Figure 32.1 on page 700) and inserting some essential example text in a new document. Most of the content is self-explanatory and hence one can soon enjoy the great results instead of spending time to learn how R Markdown works.

Further information – More about R Markdown

R Markdown is a product of RStudio Inc. Its official homepage is `https://rmarkdown.rstudio.com`, and we especially recommend to browse through the galleries of examples.

For example, when creating a new file in RStudio we can select that we want PDF slides – and a popup message tells us to, make sure we have LaTeX on our system installed. The new file that we create this way will not be empty, but contains a basic structure for PDF slides.

An example content of such R-markdown file is shown below.

```
---
title: "R Markdown"
author: "Philippe De Brouwer"
date: "January 1, 2020"
```

The Big R-Book: From Data Science to Learning Machines and Big Data, First Edition. Philippe J.S. De Brouwer.

Companion Website: www.wiley.com/go/De Brouwer/The Big R-Book

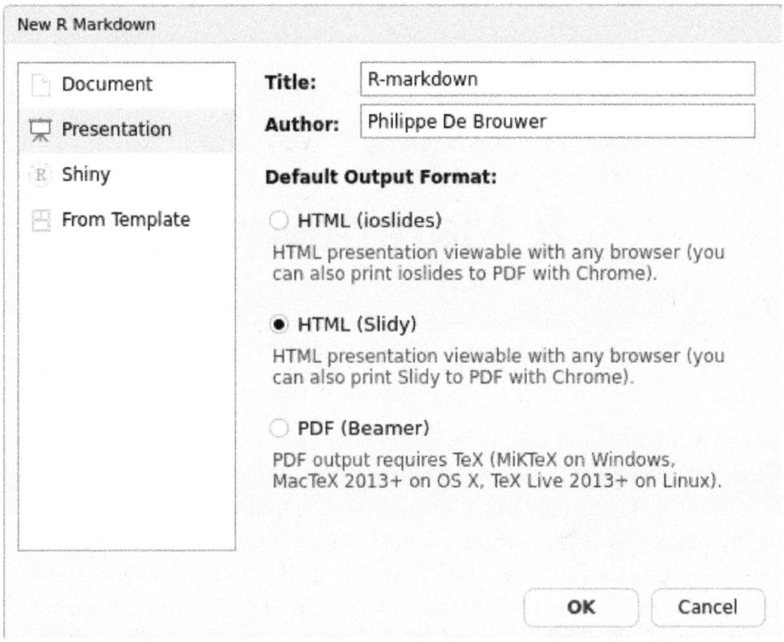

Figure 32.1: *Selecting* `File` → `New File` → `R Markdown...` *in RStudio will open this window that allows us to select which R Markdown file type we would like to start with.*

```
output: beamer_presentation
---

```{r setup, include=FALSE}
knitr::opts_chunk$set(echo = FALSE)
```

R Markdown

This is an R Markdown presentation. Markdown is a simple formatting
syntax for authoring HTML, PDF, and MS Word documents. For more
details on using R Markdown see <http://rmarkdown.rstudio.com>.

When you click the **Knit** button a document will be generated
that includes both content as well as the output of any embedded R
code chunks within the document.

Slide with Bullets

- Bullet 1
- Bullet 2
- Bullet 3

Slide with R Output

```{r cars, echo = TRUE}
summary(cars)
```

Slide with Plot
```

````
```{r pressure}
plot(pressure)
```
````

Warning – Be careful with specific symbols

Some aspects that might seem cosmetic are actually very important. For example, the backticks, pound-symbols, and hyphens have to be at the beginning of the line. Put them somewhere else and R will not recognize them any more.

The first section – between the `---` signs – are the YAML headers. They tell what type of file we want to create. Do not alter these `---` signs, nor move them, change the content between quotes as desired. The last entry in the YAML headers (`output: beamer_presentation`) is the one that will be responsible for creating a PDF slideshow (based on the package "beamer" for LaTeX) – hence the need for a working version of LaTeX. **YAML** **LaTeX**

RStudio will now show automatically a `Knit` button. This button will run all the R code, add the output to the R-Markdown file, convert the R-Markdown to LaTeX, and finally compile this to PDF.

Then following the R-Markdown content, a new header slide will be created for each level one title.

```
# Level 1: title slide

## level 2: title or new slide
This line goes on slide the slide with title 'level 2: title
or new slide'

### level 3: box on slide
This line goes in the body of the box that will have the title
'level 3: box on slide'
```

Executable code is placed between the three backtics marker (```` ``` ````), we tell that it is R-code by putting the letter "r" between curly brackets. That is also the place where we can name the code chunk and eventually override the default options `knitr::opts_chunk$set(echo = FALSE)`. knitr

For example, the following code will create a histogram on the slide with title "50 random numbers".

````
## 50 random numbers
```{r showPlot}
hist(runif(50))
```
````

Hint – Change document format

Do you need to convert your presentation to a handout text or online documentation? Simply change the YAML headers. For example, change the output option to `html_document`, `pdf_document` or `word_document`. The document should right away look professional and much more readable than printed slides.

Digression – Free book on R Markdown

A really great resource on R Markdown is the free online book of Yihui Xie, J. J. Allaire, and Garrett Grolemund: "R Markdown: The Definitive Guide". It is not only free, but it is also regularly updated and improved. You can find it here: `https://bookdown.org/yihui/rmarkdown`

Digression – Notebook format

Would you like to see your code and text in an interactive notebook format, where you can execute code and see the output with a click of a mouse (instead of seeing the full document already compiled)? That is possible with `R Notebook`. More information is here for example: `https://bookdown.org/yihui/rmarkdown/notebook.html`

Digression – R Bookdown

There is a flavour of R Markdown that is especially designed to facilitate the creation of books. It is called "R Bookdown" and its homepage is here: `https://bookdown.org`. As a bonus, you will find some free books on that page.

knitr and LaTeX

It is hard to beat R-Markdown when the document needs limited customization and when R-output is omnipresent. When the balance is more toward the text, it might be a good option to do the opposite: instead of working from R-Markdown and R and calling LaTeX to create the finished product, it might make more sense to work in LaTeX and call R from there.

This gives us the unrivalled power of LaTeX to typeset articles, presentations, and books and neatly include R-code and/or output generated from R. This book is compiled this way.

LaTeX is a markup language (just as R-Markdown itself for that matter) that is extremely versatile – and Turing complete. It is often said that LaTeX produces neat and professional looking output without any effort but that producing a confusing document requires effort in LaTeX. This is quite the opposite of a regular WYSIWYG text editor.

This is only one of the many advantages of LaTeX. There are many more advantages, but probably the most compelling advantages are that it is the de facto standard for scientific writing, allows for high automation and – since it is Turing complete – will always be more capable than any text editor.[1] For example, in this book all references to title, figures, and tables are done automatically; and we can really customize how the reference looks like. That experience is more close to a programming language than to a document editor.

Explaining the finer details of working with LaTeX is a book in itself and beyond the scope of this book. In the remainder of this chapter we want to give a flavour to the reader of how easy LaTeX can be and how straightforward it is to make professional documents that include R-code and the results of that code. We will build a minimal viable example. To get started you will need a working version of R and a LaTeX compiler.

Already in section Chapter 4 *"The Basics of R"* on page 21 we explained how to install R. LaTeX is installed on the CLI as follows.[2]

```
sudo apt-get install texlive-full
```

On Windows you want to download and install MitTex – see `https://miktex.org` and follow the instructions there.

LaTeX uses normal text files that follow a specific syntax and grammar. So, it is not necessary to use a specific text editor. However, a text editor that understands the mark-up language is easier to work with as it will help by auto-completing commands and highlighting keywords. Similar to what RStudio is to R, there are also IDEs available for LaTeX. For example,you might want to use `texmaker` or `kile`?

[1]While these objective arguments in favour of LaTeX are very convincing, for the author the most important one is that it is intellectually satisfying, while working with a WYSIWIG text editor is intellectually frustrating and condescending.

[2]We assume here that you know to updated all your packages before installing anything on your system.

The Big R-Book: From Data Science to Learning Machines and Big Data, First Edition. Philippe J.S. De Brouwer.

Companion Website: www.wiley.com/go/De Brouwer/The Big R-Book

```
sudo apt-get install kile
# or:
sudo apt-get install texmaker
```

Now, create a text-file (by using one of the IDE or your favourite text-file editor) with the following content and name it `latex_article.Rnw`. The name is of course not important, but if you use another name you will have to substitute this in the following code.

```
\documentclass[a4paper,12pt]{article}
\usepackage[utf8]{inputenc}

\title{About \LaTeX and knitr}
\author{Philippe J.S. De Brouwer}
\date{2019-02-21}

\begin{document}
<<echo=FALSE,include=FALSE>>=
library(knitr)              # load knitr
opts_chunk$set(echo=TRUE,
               warning=TRUE,
          message=TRUE,
          out.width='0.5\\textwidth',
               fig.align='center',
               fig.pos='h'
       )
@

\maketitle

\begin{abstract}
A story about FOSS that resulted in software that rivals any --even very
    expensive-- commercial software \ldots and has the best user support
    possible: a community of people that help each other.
\end{abstract}

\section{This is how it works}
The next code will include the R-code and its output.
<<fig.cap='a plot in LaTeX via knitr'>>=
N <- 50
hist(runif(N), col = 'deepskyblue3', border = 'white')
@
In this plot we used \Sexpr{N} random numbers and then plotted the histogram of
     the result.
\end{document}
```

If not already done, save this in a text-file and give it the name `latex_article.Rnw`. Now, we are ready to compile this article. It is sufficient to execute two lines in the Linux CLI:

```
R -e 'library(knitr);knit("latex\_article.Rnw")'
pdflatex latex\_article.tex
```

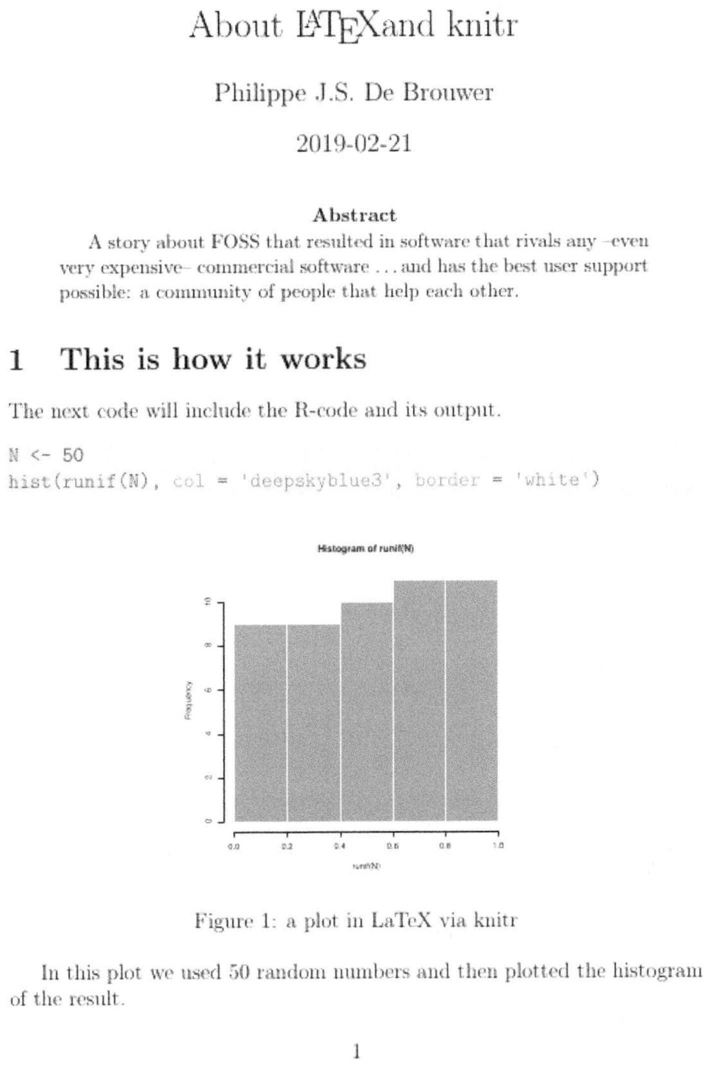

About LaTeXand knitr

Philippe J.S. De Brouwer

2019-02-21

Abstract

A story about FOSS that resulted in software that rivals any –even very expensive– commercial software . . . and has the best user support possible: a community of people that help each other.

1 This is how it works

The next code will include the R-code and its output.

```
N <- 50
hist(runif(N), col = 'deepskyblue3', border = 'white')
```


Figure 1: a plot in LaTeX via knitr

In this plot we used 50 random numbers and then plotted the histogram of the result.

1

Figure 33.1: *The LaTeX article looks like this. Note that this is a cropped image so it would be readable in this book. The result will be centred on an A4 page.*

Alternatively the textfile can be compiled from the R-promt as follows:

```
library(knitr);                          # load the knitr package
setpwd('/path/to/my/file')               # move to the appropriate directory
knit("latex_article.Rnw")                # creat a .tex file by executing all
                                         # R-code
system('pdflatex latex_article.tex')     # compile the .tex file to a .pdf
```

The first line calls the R software and executes two commands: it first loads the `knitr` library and then "knits" the document. This means that it takes the Rnw file, and extracts all expressions that are between `<<>>=` (eventually with optional arguments between the double brackets) and the `@` as well as all the code wrapped in the macro `\Sexpr{}` and executes that in R, then "knits" all this code, its output and plots together with the remainder of the LaTeX code. The result of this process is then stored in the file `latex_article.tex`.

knitr

The second line uses the command `pdflatex` to compile the tex file into a pdf file.

Note – Short and long code

Note that knitr provides two methods to include R-code in the LaTeX file.

1. long code goes between `<<>>=` and `@`
2. short code that appears in the normal text can be wrapped in the function `\Sexpr{}`

The result will be a pdf file that is called `latex_article.pdf`. It will look as in Figure 33.1 on page 705.

Further information – LaTeX

Latex is a mature FOSS project. A good place to start looking is `https://www.latex-project.org`. There are many online articles, user guides, and books available – for example see Lamport (1994)

Digression – RStudio

While it is easy enough to work in the CLI and often preferable to use our most loved IDE or text editor, it is also possible to use RStudio. With RStudio it is not necessary to use any command line and a click of the button will compile the R-markdown to LaTeX code, then compile that and present the result in a separate screen. Maybe you prefer this workflow?

♣ 34 ♣

An Automated Development Cycle

R allows for a completely automated development cycle, from data extraction over data-wrangling, data exploration, modelling, and reporting. There are so many possible ways to automate a model development cycle, and there is no clear best solution. When building an automated modelling cycle it is, however, worth to consider the following hints.

- R has a function `source()` that allows to read in a file of commands – for example the file that loads in the new dataset or provides some frequently used functions. `source()`
- R can be invoked from the command line and from the command line we can ask R to execute some commands. We demonstrated this in previous section (Chapter 33 *"knitr and LaTeX"* on page 703). This implies that together with the `source()` function it is possible to automate about everything.
- Further adding the governing commands to the `crontab` file of your computer will cause them to be executed at any regular or irregular time interval that you might choose. For example, we can run every week a file that extracts data from the servers and updates our data-mart or we can run a weekly dashboard that checks the performance of our models.
- While database servers have their own variations of the `source()` function and can usually also be controlled from the command line, it is also possible to access those database servers from R – as demonstrated in Chapter 15 *"Connecting R to an SQL Database"* on page 253. For example **crontab**
- The object oriented structure in R allows to build objects that hold everything together. For example we can build an object that knows where to find data, how to wrangle that data, what to model and how to present it, we can even add to this nice package the documentation of the model and contact details. The objects can for example have methods that update data, run analysis, check model performance, and feed that information to a model-dashboard – see Chapter 36.3 *"Dashboards"* on page 725. More information about the OO model is in Chapter 6 *"The Implementation of OO"* on page 87 – we recommend especially to have a look at the RC objects for this purpose.
- It is also not too difficult to write one's own packages for R. For example, we could build a package for our company that facilitates connection to our servers, provides functions that

The Big R-Book: From Data Science to Learning Machines and Big Data, First Edition. Philippe J.S. De Brouwer.

Companion Website: www.wiley.com/go/De Brouwer/The Big R-Book

can be used in `shiny` to use the look and feel of our company, etc. This package can hold whatever is relevant to our model and provides a neat way to store the logic, the data, and the documentation on one place. More about building packages is in Chapter A *"Create your own R Package"* on page 821.

- It is possible that the model will be implemented in a specific production system, but in many cases the goal is to present the analysis. R has great tools to produce professional documents with a gentle learning curve – see for example Chapter 32 *"R Markdown"* on page 699, Chapter 36.3 *"Dashboards"* on page 725, etc. This allows us even to publish results on a corporate server, which eliminates the need to send spreadsheets around and will be more useful to the recipient as he/she can interact with the data, while the report owner can follow up who is using the report.

This toolbox is very powerful and allows for almost any size corporate to increase efficiency, reduce risks, and produce more sustainable results ... while spending nothing on expensive software. Not only can the aforementioned software be used free of charge – even for commercial use – but also the stack on which it is build (such as operating system, windows manager, etc.) can be completely free.

♣ 35 ♣

Writing and Communication Skills

Writing up conclusions is often an essential part of the work of the data scientist. Indeed, the best analysis is of no use if it does not convince decision makers. In order to be effective, the data-scientist needs also to be a skilled communicator.

The exact style and the attention to things such as abstract, summary, conclusions, executive summary, and many more aspects will depend on the tradition in your company. However, some things are bound to be universal. For example, we live in an age where information is abundant and it is not abnormal to communicate on different social platforms and still receive hundreds of mails every day while the normal pattern is meetings back-to-back. The average manager will hence have little time to read your document. While you want of course the document to be readable (avoid slang and acronyms for example), you also will want to keep the message as short and clear as possible. Indeed, in this time and age of information overload, concise communication is of paramount importance.

A first thing to do is re-read your text and make sure it is logical, clear, and that every word really has to be there. If something can be written shorter then that is usually the way to go.

However, even before we start writing it is worth to check some basic rules. When asked to look into a problem, make sure to ask yourself the following ideas.

1. Identify the main question or task at hand. Usually, there is a situation and a complication in some form that leads directly to one main question. Focus that single question.
2. Formulate the questions so that it addresses the concerns of the audience.
3. Break down the issue into smaller (solvable) problems.
4. Then for each of the small problems follow the logic of this book: get the relevant data, wrangle and explore, model, gather evidence, and draw conclusions.
5. Convince others by a written report and/or a convincing presentation.

The Big R-Book: From Data Science to Learning Machines and Big Data, First Edition. Philippe J.S. De Brouwer.

Companion Website: www.wiley.com/go/De Brouwer/The Big R-Book

Example: Customer profitability

Consider the situation where you have been tasked to investigate customer profitability. You have set up a team that for one year has been building a model to calculate a customer value metric (CVM), you have not only made good proxies of actual costs and incomes streams but also used machine learning to forecast the expected customer lifetime value (CLTV) for each customer. During that year your team has grown from yourself only to 15 people, it was hard work but you managed to keep the staff motivated and are very proud of the hard work.

CLTV

While the initial idea was to "do something extra for very profitable customers," your conclusions are that

1. there are also a lot of non-profitable customers that only produce losses for us;
2. the actual customer segmentation – based on wealth (for a bank) – does not make sense.

Therefore, you believe it is best to review actual customer segmentation, and find solutions for the loss-producing customers as well as focussing more on profitable customers.

The analysis has revealed that there are more problems that need addressing. What is now the best way forward?

Suppose that we are faced with a situation as in the aforementioned example. How would we best communicate and to whom? Given the request, there is certainly an interest in profitability and most probably people are open to discuss what drives customer profitability. Challenging the customer segmentation is an entirely different task. This would involve restructuring teams, re-allocating customers and most probably will shift P&L from one senior manager to the other. This is not something that can be handled overnight and probably will need board sign-off.

Therefore, we might want to focus – for each existing customer segment – on making the customer portfolio more profitable. We might want to propose to offboard some customers, cross-sell to others and provide a better service to a third category that is already profitable.

When compiling the presentation it is a good idea to follow certain simple guidelines. The whole presentation can best be compared with a pyramid where at each level we start with the conclusions and provides strong evidence. At each level we find similar things (arguments, actions, etc.). For example, we could structure our presentation as follows.

1. We need a simple matrix of three actions for each customer segment. This is about segment A, and we need to get rid of some customers, cross-sell to others, and make sure we keep the third category.
2. Introduction: Remind the reader what he/she already knows and hence what the key question is to be answered. Then for each segment elaborate the following.
   (a) There is a group of customers that are unprofitable, and never will be profitable. It is unfair towards other customers to keep them.
      - demonstrate how unprofitable these customers are
      - show how even cross selling will not help
      - show how much we can save by focussing on profitable customers

   (b) The second group of customers can be made more profitable by cross-selling (and we have suggestions on how to do that)
      - show the sub-optimal nature of these customer accounts
      - show how suitable cross-selling helps
      - estimate of the gains
   (c) The last group of customers are the ones that are very profitable, we can help you identify them and we suggest to do all you can to keep them as customers.
      - show how profitable these customers are
      - show why they are profitable
      - forecast their contribution to future profits

3. conclusions: repeat the three actions per client segment and propose first steps

4. appendix
   (a) our approach to CVM
   (b) our approach to CLTV
   (c) histogram of CLTV

The point that stands out most in the presentation structure above is probably that what matters most to us (how smart the team was, how hard it worked, what concepts it developed, etc.) is banned to the appendix and is only used in case someone asks about it.

Hint – Think from the audience's point of view

Making a presentation, always start thinking from the point of view of the audience: start from their needs and goals, ask yourself what makes them tick, what is acceptable and what not for them . . . and what is important to convince them (seldom the overtime that you have made, the genius breakthrough, or cute mathematical formula will fit that category).

There are a few golden rules to keep in mind about the structure of a presentation.

1. The ideas – at any level in the structure – are summaries of the ideas below – in other words, start with the conclusion on the title slide, the title of the next slide is the summary of the things on the slide, etc. On a smaller scale, an enumeration groups things of the same type (e.g. does not mix actions and findings) and the title of that slide will be a summary of that list.

2. Ideas in each grouping should be of the same kind – never mix things like arguments and conclusions, observations and deductions, etc.

3. Ideas in each grouping should be logically ordered – ask someone else to challenge your presentation before sending it to the boss.[1]

[1] If you fail to do this step, your boss will play the role as challenger, but that is not best use of his/her time, nor will it be helpful for your reputation.

Hint – How R can help

Of course, it is possible to make a slide-show presentation. In that case, we recommend to have a look at the section about R-Markdown (see Chapter 32 *"R Markdown"* on page 699), but it might have more impact to prepare a website with interactive content, in that case you might want to have a look at the "storyboard" layout of Chapter 36.3.2 *"A Dashboard with flexdashboard"* on page 731.

Note – Two very different slide decks

Making slides is an art, and it really depends on what you want to achieve, for what audience, what size of audience, etc. Are you making a presentation for thousand people or are you having a one-to-one discussion and will you leave the slides with the decision maker? It is obvious that while the structure can be largely the same that the layout will vary dramatically. In the first case, we might focus on a few visuals and eliminate text at all; in the second case text is of paramount importance and titles can even be multiple lines long. For each level in the presentation, the mantra "the ideas at any level in the structure should be summaries of the ideas below" holds, this means that you should put summaries of the ideas on the slide in the titles.

Finally, one will want to pay attention to the layout of the slides. When working with R-Markdown (see Chapter 32 *"R Markdown"* on page 699) or with the Beamer class in LaTeX, good layout is the standard and it will be hard to make an unprofessional layout. Working with WYSIWYG slide producing software it is easy to have a different font, font-size and colour on each page. That will not look professional, nor will it be conducive for the reader to focus on the content. Most companies will have a style-guide and templates that will always look professional.

♣ 36 ♣

Interactive Apps

A static, well written report, a good book, or a scientific paper have their use and value, but some things look better on screen. More importantly it also allows the user to interact with the data. For example, putting the data on a website, can with the right tools allow the user to visualize different cuts of the data, zoom in or drill down to issues of interest or move around 3D plots on the screen.

In the larger company, too many people spend their time on manually processing the same data in an electronic spreadsheet over and over again, then mail it to too many people. Typically, no one reads those weekly reports, but when asked if the report is still needed the answer is usually a confirming "yes." Be warned, though, this asking is dangerous. That usually triggers someone to open the file and spawn out a mail with requests: add a plot here, a summary there and of course some other breakdown of data.

Putting the dashboard online, not only allows people to "play" with the data and get more insight from the same data, but also it allows the publisher to follow up the use of the dashboard. It is also much faster: it can be updated as the data flows in and the user is not bound to weekly or monthly updates. It will take a little more time to produce it the first time, but then it will not take any human time at all to make every second an update.

Unfortunately, companies tend to use old technology (such as spreadsheet and slides) and then make the jump to super-powerful engines that are able to make sense of ultra-fast changing data flows that flows into Hadoop infrastructure. Usually, that it overkill, not needed and way too expensive.

Knowledge is power and that is even more true in this age of digitization: "data is the new oil." Many companies have data that can be monetized (for example by enhancing cross-sales). Business intelligence tools hold the promise to unlock the value of that data.

BI stands for "business intelligence", and it is a powerful buzzword. So, powerful that all the large IT companies have their solution: SAP, IBM Cognos, Oracle BI, Dundas BI, and Microsoft Power BI. Also some specialized players offer masterpieces of software: QlikView, Tableau, PowerBI and Looker, just to name a few. These things can do wonders, they all offer great data visualization libraries, OLAP[1] capacity, analytical tools to make simple or more advanced models, offer support for different document formats as well as online interactive display, allow thresholds to be defined, and can send alarm emails when a threshold is breached, they offer great integration with almost all database systems and big data platforms. **OLAP**

[1]OLAP is the buzzword for "online analytical processing," it means that data can come in raw format and that at the moment the user requests the plot, that only then a snapshot is taken to calculate the plot.

The Big R-Book: From Data Science to Learning Machines and Big Data, First Edition. Philippe J.S. De Brouwer.

Companion Website: www.wiley.com/go/De Brouwer/The Big R-Book

SLA
service level agreement

These systems are truly great solutions, but did we not just describe R? Indeed, R, with some libraries R can do all of that.[2] Maybe it looks a little less flashy for an executive demo, and there is a friendly online community to help you out instead of a helpline with and SLA[3] based on a contract.

Using R it is also of crucial importance to manage the updates and the dependencies of packages carefully. It can happen that you use a package X and that after updating R this package will not load anymore. In this case, you might want to wait to update R before the update of the package X is available or alternatively update the package yourself. Because R is FOSS, there are so many packages and it is not realistic that they are all tested and updated as a new release of R comes available. Further, it might be that package X depends on package Y and it might be that Y is not yet updated. Commercial software usually has better backwards compatibility and the management of dependencies is less crucial. However, most R-code will usually continue to work fine for many years.

However, the difference that stands out is the price tag. The commercial tools typically will set you back millions of dollars and R is free. Not only free, but also open source so you know exactly what you get, and if you need to change something, then you do that and recompile instead of adding your request to a feature list.

The reader might also appreciate the fact that free and open source is the best warranty against back-doors in the software or even data-leakages. If you are in doubt, it is possible to check the source code, compile that code and only this version.

[2] R is not the only free and open source solution for data manipulation, model building and visualization. There are other possible solutions such as Python, C++, etc.

[3] SLA stands for "service level agreement," it is an agreement that details KPIs for the service. For example, the response time or metrics such as not more than ten percent of the cases can take more than one business day to solve.

36.1 Shiny

The father (or mother?) of R's fantastic online capacity is the library `shiny`. It is made available by the company around RStudio. Shiny provides a straightforward interface that facilitates building interactive web apps directly from R. Apps can be hosted as standalone apps on a web-page or embedded in R Markdown documents. Because it is built for online processing and visualization it is ideal for dashboards. It will even generate HTML 4 and use Bootstrap style CSS styling. So, the website or app can be extended and modified with CSS themes,[4] htmlwidgets, and JavaScript interaction.

Hint – Online gallery for Shiny

Have a look at the online gallery `https://shiny.rstudio.com/gallery` to get an idea of what `Shiny` can do for you. Get more inspiration in the user gallery: `https://www.rstudio.com/products/shiny/shiny-user-showcase`

In order to display the app online, it has to be hosted on a webserver. It will work on the usual LAMP stack (Linux, Apache, MySQL, and PHP/Python/Perl), however, it is designed for the newer `gdebi`. In theory, it can co-exist with Apache, though it is bound to raise issues. So, if you already have other webservers it might be wise to use a dedicated server for `Shiny` apps. There are free and commercial versions available. More information can be found on `https://www.rstudio.com`

Hint – Run apps in your browser

In order to show `Shiny` apps on your webserver, it is also possible to publish them on `http://www.shinyapps.io` and then include them in the html code of your page via an `<iframe>` tag and the app will seamlessly integrate with the page that is calling it.

Using `Shiny` is easy enough thanks to the excellent help that is available online. For example, RStudio provides an excellent tutorial – available at `http://shiny.rstudio.com/tutorial` – that will get you started in no time.

The first thing to do is – as usual – installing the package then load it. Then we can immediately look at an example.

```
library(shiny)
runExample("01_hello")
```

Pressing enter after the line `runExample("01_hello")` will produce the output `Listening on http://127.0.0.1:7352` and open a web-page with the content in Figure 36.1 on page 716. In the meanwhile, the R-terminal is not accepting any further commands. `[CTRL]+C` (or `[ESC]` in RStudio) will allow you to stop the webserver and make the command prompt responsive again.

[4]The standard layout that Shiny provides looks very professional and nicely integrates with Twitter's Bootstrap. In order to access more of the Bootstrap widgets directly, use the package `shinyBS`.

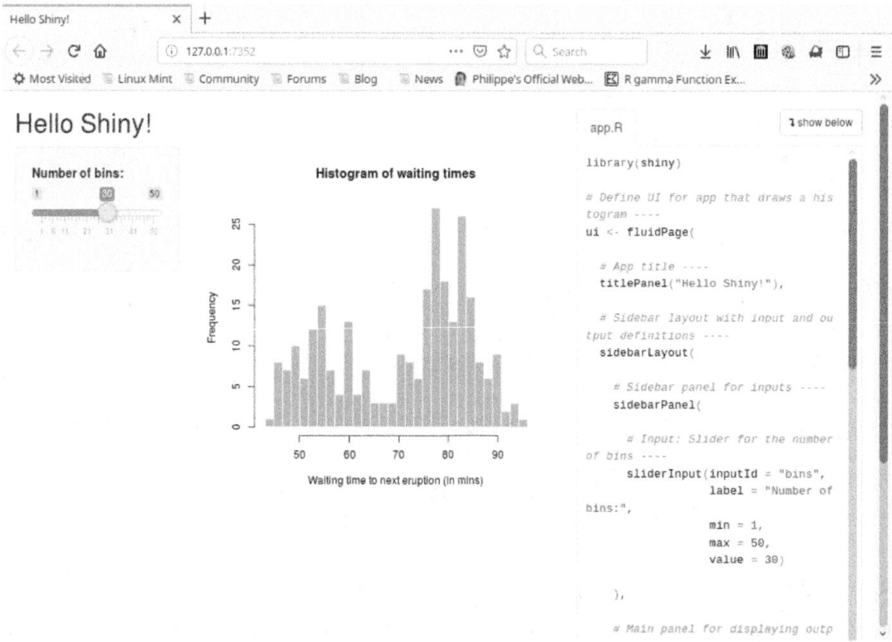

Figure 36.1: *The output of one of the examples supplied by the Shiny package. Note the slider bar on the left and the code on the right.*

The examples provided show the code and are so well documented that it is hardly necessary to explain here too many details. Rather we will explain the main work-flow, concepts and focus on a few caveats.

Hint – See all available examples

The command `runExample()` will list all examples provided. Look at their code and start modifying the one that is closest to your needs.

There are a few things to be noted, though. First, is that there are two main ways to code an app. First, – the older method – is to provide two files: `server.R` and `ui.R` and the – newer – method to fit all code into one file `app.R`

Here is a simple example of the second method that has all code in one file:

```
# The filename must be app.R to execute it on a server.

# The name of the server function must match the argument in the
# shinyApp function also it must take the arguments input and output.
server <- function(input, output) {
  output$distPlot <- renderPlot({
    # any plot must be passed through renderPlot()
    hist(rnorm(input$nbr_obs), col = 'khaki3', border = 'white',
         breaks = input$breaks,
         main = input$title,
         xlab = "random observations")
    })
```

```
  output$nbr_txt <- renderText({
    paste("You have selected", input$nbr_obs, "observations.")
  })  # note the use of brackets ({ })
}

# The name of the ui object must match the argument in the shinyApp
# function and we must provide an object ui (that holds the html
# code for our page).
ui <- fluidPage(
  titlePanel("Our random simulator"),
  sidebarLayout(
    sidebarPanel(
      sliderInput("nbr_obs", "Number of observations:",
                  min = 10, max = 500, value = 100),
      sliderInput("breaks", "Number of bins:",
                  min = 4, max = 50, value = 10),
      textInput("title", "Title of the plot",
                value = "title goes here")
    ),
    mainPanel(plotOutput("distPlot"),
              h4("Conclusion"),
              p("Small sample sizes combined with a high number of
              bins might provide a visual image of the
              distribution that does not resemble the underlying
              dynamics."),
              "Note that we can provide text, but not
              <b>html code</b> directly.",
              textOutput("nbr_txt")  # object name in quotes
              )
  )
)

# finally we call the shinyApp function
shinyApp(ui = ui, server = server)
```

This code can be executed in an R terminal, and the function `shinyApp()` will start a webserver on the IP loopback address of your computer (usually 127.0.0.1) and broadcast to port 7352 the webpage. Then, it opens a browser and directs it to that ip address and port. The browser will then show an image as in Figure 36.2 on page 718.

Feel free to test the app in the browser. The app is interactive: changing an input will modify the plot and text below it immediately. What you're looking at now, is a simple `shiny` app, but you probably get an idea about the possibilities of this amazing platform.

It is of course possible to include the app on a web page, so that it it is accessible from anywhere on the Internet. RStudio provides a free version of "ShinyServer" that can be put on your (dedicated) webserver of choice. There are also non-free versions available.

They also provide a website `ShinyApps.io` on which you can put your apps for free and then include them on your web pages via an `iframe`. The process is simple, and after signing in on that website the user is guided through the process.

It tells us to install a package `rsconnect`, then authorize the account with a function provided by the package. Finally, the app can be deployed as follows:

```
# Load the library:
library(rsconnect)

# Upload the app (default filename is app.R)
rsconnect::deployApp('path/to/your/app',
                     server="shinyapps.io", account="xxx")
```

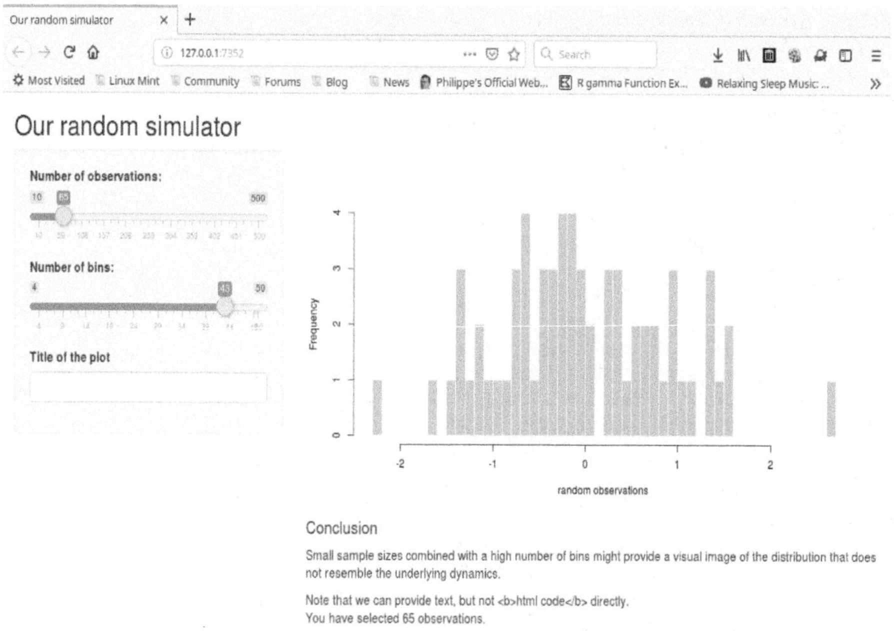

Figure 36.2: *The web-page produced by the code fragment above.*

Then it becomes available via an url like `https://xxx.shinyapps.io/normsimulator`, where the letters "xxx" are your account name on the platform. Now, any html-page can be configured to display the app via an iframe.

```
<iframe src="https://xxx.shinyapps.io/normsimulator/"
      frameborder="0"
      allowfullscreen
      style="width:100%;height:800px;"
      >
      </iframe>
```

Listing 36.1: *The html code to include our Shiny app on a live web-page.*

The great thing is that when people visit your website, they will only see the app that sits on one of your web pages. This means that your visitors will not even notice that the app itself and R behind it are actually hosted on another server.

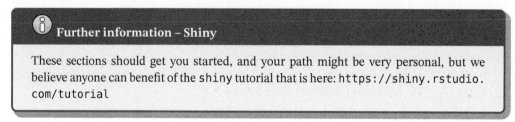

Further information – Shiny

These sections should get you started, and your path might be very personal, but we believe anyone can benefit of the `shiny` tutorial that is here: `https://shiny.rstudio.com/tutorial`

36.2 Browser Born Data Visualization

At this point, it is quite natural to be impressed by the capabilities of R to produce professional online visualizations and the question might arise on how to make it even more impressive. Sure, `ggplot2` is a really good option and works nicely together with `shiny`. However, there are two minor points that could be corrected. Firstly, `ggplot2` does not work with the piping operator `%>%` and second it is actually designed to produce offline plots.

It does make sense to opt for a library that is designed to produce rich, interactive visualizations and still allows data graphics to be declaratively described as in `ggplot2`. Below we discuss a limited and opinionated selection of libraries and tools that can spice up your `shiny` apps.

36.2.1 HTML-widgets

Adding more interactivity is made easy with many free libraries that are called "HTML-widgets." The concept "HTML-widgets" is a little misleading in that sense that HTML is a passive markup language to present content on a s screen. The interactivity usually comes from Javascript, but it is enough to have some idea how HTML works in order to use those widgets.

javascript

Here is a selection of some personal favourites of the author:

- `ggvis`: Interactive plots – see Chapter 36.2.3 *"Interactive Data Visualisation with ggvis"* on page 721;
- `dygraphs`: Interactive time series visualization – see `http://dygraphs.com`;
- `leaflet`: Interactive maps from all over the world – see `https://rstudio.github.io/leaflet`;
- `DiagrammeR`: A variety of diagrams (org-chars, flowcharts, etc.) – see `http://rich-iannone.github.io/DiagrammeR`;
- `3dheadmap`: Heat-maps – see `https://github.com/rstudio/d3heatmap`
- `DT`: interactive tables – see `https://rstudio.github.io/DT`;
- `network3D`: Network graphs – see `https://christophergandrud.github.io/networkD3`;
- `threeJS`: 3D scatterplots and globes – see `https://bwlewis.github.io/rthreejs`

Further information – Stunning visualisations

Need even more eye-candy? An ever growing list of widgets can be found online at `http://www.htmlwidgets.org`.

36.2.2 Interactive Maps with leaflet

Here is a basic example with `leaflet`[5] that shows how to visualize geographical information.

```
library(leaflet)
content <- paste(sep = "<br/>",
 "<b><a href='http://www.de-brouwer.com/honcon/'> Honorary Consulate of Belgium</a></b>",
 "ul. Marii Grzegorzewskiej 33A",
 "30-394 Krakow"
 )
map <- leaflet()                                                  %>%
 addProviderTiles(providers$OpenStreetMap)                        %>%
 addMarkers(lng = 19.870188, lat = 50.009159)                     %>%
 addPopups(lng = 19.870188, lat = 50.009159, content,
   options = popupOptions(closeButton = TRUE))                    %>%
 setView(lat = 50.009159,lng = 19.870188, zoom = 12)
map
```

When the aforementioned code is run, it will display an interactive map in a browser that looks like the figure in Figure 36.3.

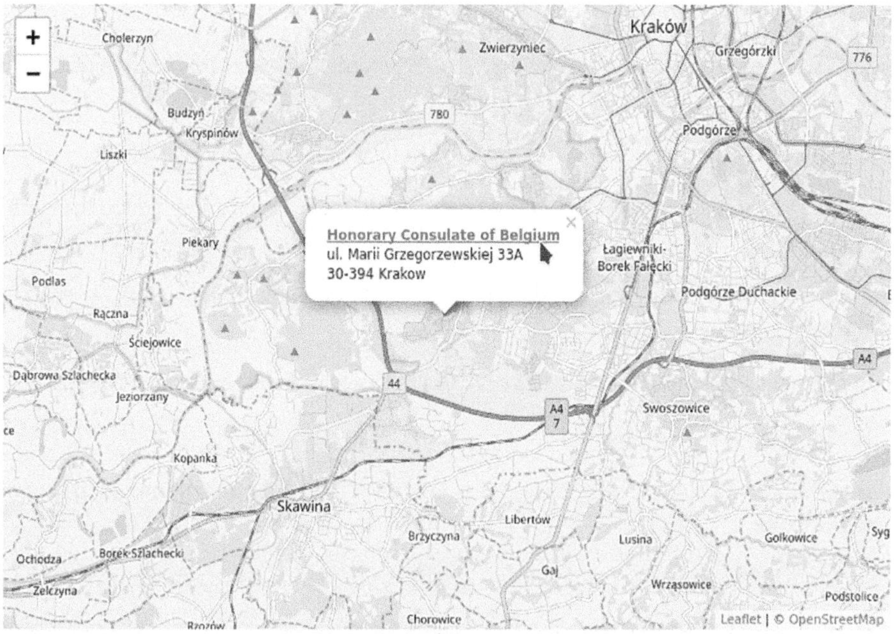

Figure 36.3: *A map created by* `leaflet` *based on the famous* `OpenStreetMap` *maps. The ability to zoom the map is standard, the marker and popup (with clicking-enabled link) are added in the code.*

Note that `leaflet` can also be integrated with `shiny`, in order to create interactive maps that use your own data and display it for the user according to his or her preferences and interests.

[5] To experience the interactivity it must be included in an interactive application, for example based on the `shiny` framework.

Further information – leaflet

This simple example should help you to get started with `leaflet`. We refer to its website for deeper information, more customization and other examples: `https://rstudio.github.io/leaflet`

Digression – References for leaflet

The homepage[a] of `leaflet` introduces the package as follows:

> "Leaflet is one of the most popular open-source JavaScript libraries for interactive maps. It's used by websites ranging from The New York Times and The Washington Post to GitHub and Flickr, as well as GIS specialists like OpenStreetMap, Mapbox, and CartoDB."

[a] The homepage of `leaflet` is `https://rstudio.github.io/leaflet`.

36.2.3 Interactive Data Visualisation with ggvis

An excellent solution to improve the data visualisation in your interactive app is `ggvis`, a library that is made available by the company behind RStudio. The `ggvis` data visualization package has the following advantages.

- It uses a declarative description of data graphics with a syntax that has kept all the best of `ggplot2`, though does deviate in some key areas.
- With little effort it is possible to create rich interactive graphics that are designed to be viewed in a web-browser.
- It works smoothly together with `shiny`.

The package `ggvis` is provided by RStudio and hence – as expected – it integrates seamlessly with RStudio. Just as for the standard `shiny` apps, it is possible to test how it works on your own computer, and later deploy it to a webserver. If you are using RStudio, the graphics display of RStudio will also display the plots and allow you to use the interactivity.

Hint – Interactive data for data scientists

Interactive apps are not only a good way for high ranking managers to consume their custom MI and allow them to zoom in on the relevant information for them. It is also a great tool that allows the data scientist to get used to new data before making a model or analysis.

36.2.3.1 Getting Started in R with ggvis

Below a simple example.

```
library(titanic)    # for the data
library(tidyverse)  # for the tibble
library(ggvis)      # for the plot
```

```
titanic_train$Age      %>%
    as_tibble          %>%
    na.omit            %>%
    ggvis(x = ~value) %>%
    layer_densities(
      adjust = input_slider(.1, 2, value = 1, step = .1,
                            label = "Bandwidth"),
      kernel = input_select(
        c("Gaussian"      = "gaussian",
          "Epanechnikov" = "epanechnikov",
          "Rectangular"  = "rectangular",
          "Triangular"   = "triangular",
          "Biweight"     = "biweight",
          "Cosine"       = "cosine",
          "Optcosine"    = "optcosine"),
        label = "Kernel")
    )
```

Executing this code fragment will open a browser with the content shown in Figure 36.4.

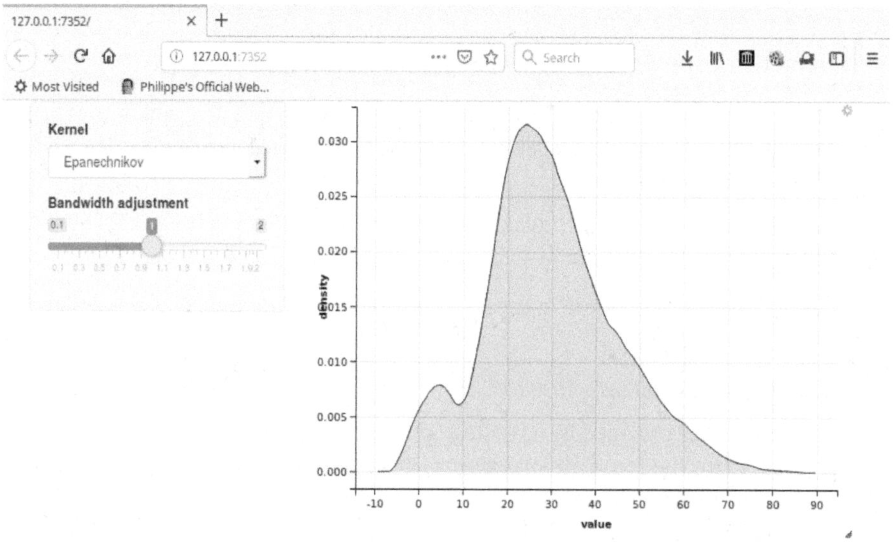

Figure 36.4: *A useful tool to explore new data and/or get an intuitive understanding of what the different kernels and bandwidth actually do for a kernel density estimation.*

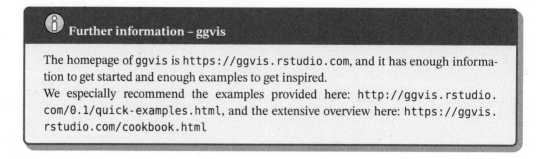

> **Further information – ggvis**
>
> The homepage of `ggvis` is `https://ggvis.rstudio.com`, and it has enough information to get started and enough examples to get inspired.
> We especially recommend the examples provided here: `http://ggvis.rstudio.com/0.1/quick-examples.html`, and the extensive overview here: `https://ggvis.rstudio.com/cookbook.html`

36.2.3.2 Combining the Power of ggvis and Shiny

ggvis has a lot to offer, but it really shines when combined with Shiny. To do this, we can use the traditional construct of a Shiny app. We choose to provide two files.

Below we show an example based on the well known dataset mtcars. The server.R file contains the following code:

```
library(ggvis)

function(input, output, session) {
  # A reactive subset of mtcars:
  reacCars <- reactive({ mtcars[1:input$n, ] })

  # Register observers and place the controls:
  reacCars %>%
     ggvis(~wt, ~mpg, fill=(~cyl)) %>%
     layer_points() %>%
     layer_smooths(span = input_slider(0.5, 1, value = 1,
        label = 'smoothing span:')) %>%
     bind_shiny("plot1", "plot_ui_div")

  output$carsData <- renderTable({ reacCars()[, c("wt", "mpg")] })
}
```

Next, we need the ui.R-file that provides the code regulating how the items are placed on the screen:

```
library(ggvis)

fluidPage(sidebarLayout(
  sidebarPanel(
    # Explicit code for a slider-bar:
    sliderInput("n", "Number of points", min = 1, max = nrow(mtcars),
                value = 10, step = 1),
    # No code needed for the smoothing span, ggvis does this:
    uiOutput("plot_ui_div") # produces a <div> with id corresponding
                            # to argument in bind_shiny
  ),
  mainPanel(
    # Place the plot "plot1" here:
    ggvisOutput("plot1"),    # matches argument to bind_shiny()
    # Under this the table of selected card models:
    tableOutput("carsData")  # parses the result of renderTable()
  )
))
```

The application can now be placed on a server – such as ShinyApps.io for example – or opened in a web-browser directly via the following command.

```
shiny::runApp("/path/to/my/app")
```

Executing this line will open a browser with the content shown in Figure 36.5 on page 724. This allows us to inspect how the app works before publishing it.

36.2.4 googleVis

If we are thinking about a professional looking dashboard or a graphical library that can make fancy plots but also gauges, maps, organization charts, sankey charts on top of all the normal plots such as scatter-plots, line charts, and bar charts that is designed to be interactive then GooglerVis imposes itself the natural choice.

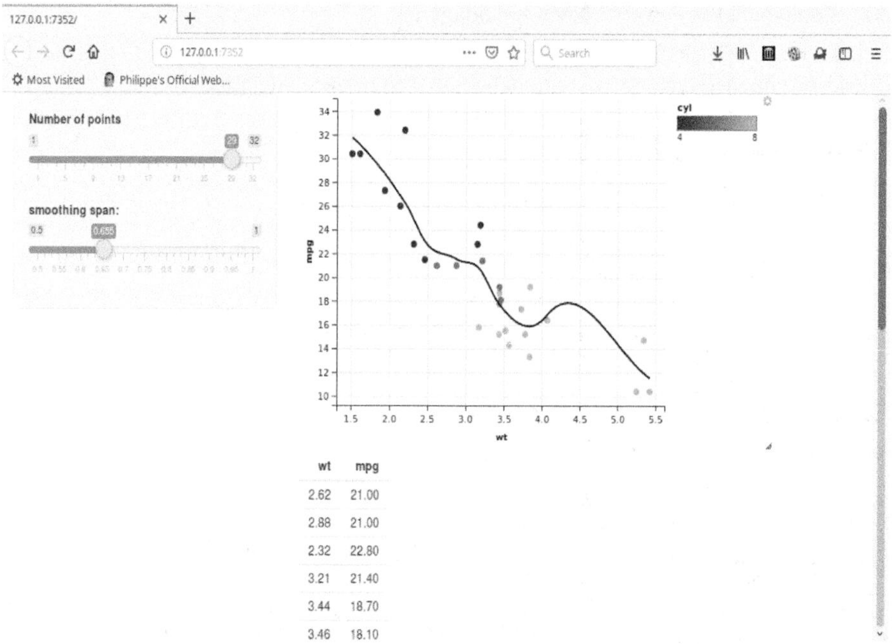

Figure 36.5: *An interactive app with* `ggvis`. *This particular example uses the data from the dataset mtcars, to experiment with two parameters: the size of the sample and the span of the Loess smoothing. This is most useful to get a better intuitive understanding of the impact of sample size and span.*

googleVis

The `googleVis` package provides an interface between R and the Google Charts API. As all other packages, it is not only free but also open source. Its code is maintained on github, just as all others packages. Its github home is: `https://github.com/mages/googleVis`.

Note that the interactivity and animations for `googleVis` will generally require Flash. This means that in order to see the special effects, your browser will have to support Flash and have it allowed in its settings. Especially plots with film-like animation will require Flash.

In order to get an idea of the possibilities, please execute the following code.

```
library(googleVis)
demo(package='googleVis')
```

36.3 Dashboards

Dashboards merit a special place in this book and in the heart of every data scientist. They are the ideal way to visualize data and call for appropriate action and are in many aspects similar to the function of a periscope in a submarine. Most companies spend massive amounts of working days on producing "reports": static visualizations and tables produced, that are usually produced by Microsoft Excel, but that always require a lot of manual work to maintain. Usually, there is good interaction with the management for whom the MI ("management information") is intended when the report is created. Then, the report is produced every week, eating up time that could be used wiser, and unfortunately the report will typically see an ever decreasing amount of people interested in it.

MI
Management Information

The answer is – partly – to have dynamic dashboards, that allow the user to change selections, drill down in the data, and find answers to questions that were never asked before. Unfortunately, companies are spending every increasing amounts to visualize data on expensive systems, while R can do this for free ... and with quite great results.

It is possible to use `Shiny` and build a dashboard with `Shiny` alone. However, good dashboards will do some the same things over and over again. For example, most dashboards will use colours to visualize importance of the numbers (for example a simple gauge or a RAG[6] status; the famous red-amber-green can already be very helpful). So, it does make sense to build that common infrastructure once and do it thorough. This is what the packages `flexdashboard` and `shinydashboard` do for you.

`flexdashboard`
`shinydashboard`

If it is your task to make a dashboard and you want to use R, then we recommend to have a thorough look at those two options. The two packages compare as in Table 36.1.

Aspect	Flexdashboard	Shinydashboard
Learning curve	Easy	Harder
Coding	R Markdown	Shiny UI code
Objective	Static or dynamic	Dynamic
Layout via	`flexbox` CSS	Bootstrap grid
Homepage	`https://rmarkdown.rstudio.com/flexdashboard`	`http://rstudio.github.io/shinydashboard`

Table 36.1: *A comparision of* `flexdashboard` *and* `shinydashboard`.

Both solutions will provide you with a flexible and powerful environment that is easily customized to the style guide of your company, that allows certain central management of things that are always the same (such the company logo). Both solutions have an active online community to provide advice and examples.

Maybe you want to choose `shinydashboard` because your company website anyhow uses Bootstrap and your team has also experience with `Shiny`? Maybe you want to choose `flexdashboard` because making a dashboard is just an ancillary activity and you do not have large teams specializing only in dashboards? Either way you will end up with a flexible and

[6]RAG is corporate slang for "red–amber–green", and it usually means that an item market green does not require attention, and item market amber should be monitored as it could go wrong soon, and an item market red is requires immediate attention.

professional solution that can rival the most expensive systems available, while incurring no direct software cost.

For any business, for any company, for any manager, for any team, and for any individual employee having a "scorecard" or "a set of KPIs" this is a powerful idea.[7] In the first place, it helps to create clarity about what is to be done, but most importantly the author's adagio "what you measure is what you get" always seem to hold. Measuring KPIs and discussing them aligns minds and creates a common focus. If the manager measure sales per quarter, then your staff will pursue sales at any cost and converge towards a short time perspective. If the manager has a place on the scorecard for "feedback from the customer," then employees will converge towards entirely different values and value sustainable, long-term results.

A scorecard should be simple (say each employee should focus on maximum seven personal KPIs), but also offer enough detail to find out when things go wrong and ideally offer enough detail to get us started when investigating mitigating measures.

A scorecard that has SMART goals (goals that are specific, measurable, attainable but realistic and limited in time) also creates a moment to celebrate success. This is another powerful motivator for any team!

The underlying mechanics of dashboards also apply to any other aspect of any business: from showing what happens in a production line, over what customers are profitable to what competitors are doing. Using a scorecard will create clarity about what is done and people will strive for it.

In the remainder of this chapter we will show how to build a simple dashboard. To focus the ideas we will create a dashboard that is about diversity of employees.

36.3.1 The Business Case: a Diversity Dashboard

Whatever the task at hand is, one will need a diverse, motivated and focussed team. This book is not about how to get this motivated team; instead, we warmly recommend Lencioni (2002) and Marquet (2015) to get started. We will focus on the measurable mechanics in order to create a simple example of a dashboard.

In business, so many aspects are important. Indeed, it is essential to follow up on the cash flows and be sure that we have no cash squeeze anywhere soon. It is equally important to monitor sales, production, errors, etc. But looking more into the future, survival will be determined by our ability to adapt to the changing environment and not to miss important and structural trends. This is what will prevent any company to avoid the fate of Kodak. That went from absolute world-market leader over inventing the digital camera and finally deciding not to use it to – today – almost extinct.

Just as the in biological evolution "it is not the most intellectual of the species that survives; it is not the strongest that survives; but the species that survives is the one that is able best to adapt and adjust to the changing environment in which it finds itself"[8]; the same holds about the survival of companies.

There is also some research on correlation between diversity and profit. For example, McKinsey has a strong reputation in researching correlation between gender diversity and

[7] More information about scorecards is in Section 29.3.1 *"Balanced Scorecard"* on page 590 and about KPIs is in Section 29.3.2 *"Key Performance Indicators (KPIs)"* on page 591.

[8] While usually attributed to Charles Darwin, and most certainly inspired by his work, we could not find any trace of this citation in his works. The earliest reference appears to be from Leon C. Megginson: June 1963, Southwestern Social Science Quarterly, Volume 44, Number 1, "Lessons from Europe for American Business by Leon C. Megginson", (Presidential address delivered at the Southwestern Social Science Association convention in San Antonio, Texas, 12 April 1963), Published jointly by The Southwestern Social Science Association and the University of Texas Press.

financial performance, and they find that "companies with more diverse top teams were also top financial performers" — see Barta et al. (2012). However, these kind of studies fail to prove any causal relationship. The paper phrases it as follows: "We acknowledge that these findings, though consistent, aren't proof of a direct relationship between diversity and financial success."

Indeed, most of those studies only study correlation. We argue that it would be possible to study causation by for example taking a few years lag between board composition and relative financial results and carefully filter industry effect and use more control variables. A rare example is found in Badal and Harter (2014) where one controlling variable "employee engagement" is used. They find that both gender diversity and this employee engagement independently are able to "explain" the financial success of a company.

In summary those studies generally prove that financial performance and diversity can be found together, but fail to answer what is the cause. Therefore, we prefer – at this point – our logical argument based on biological evolution as previously presented. More importantly, striving to equal chances and more inclusiveness is also the right thing to do. It seems also to go well with the evolutionary argument of capturing the diversity of ideas.

Now, that we have convinced ourselves that a more diverse and inclusive company is the way to go, we still need a simple quantification of "diversity" that is universal enough so it can be applied on multiple dimensions.

What could be more natural measure for diversity than Boltzmann's definition of entropy? In 1877, he defined entropy to be proportional to the natural logarithm of the number of micro-states a system could potentially occupy. While this definition was proposed to describe a probabilistic system such as a gas and aims to measure the entropy of an ensemble of ideal gas particles, it is remarkably universal and perfectly suited to quantify diversity.

Under the assumption that each microstate is improbable in itself but possible, the entropy S is the natural logarithm of the number of microstates Ω, multiplied by the Boltzmann constant k_B:

$$S = k_B \log \Omega$$

When those states are not equally probable, the definition becomes:

$$S = -k_B \sum_i^N p_i \log p_i$$

Where there are N possible and mutually exclusive states i. This definition shows that the entropy is a logarithmic measure of the number of states and their probability of being occupied.[9]

If we choose the constant k_B so that equal probabilities yield a maximal entropy of 1 (or in other words $k_B := \frac{1}{\log(N)}$) then we can program in R a simple function.

```
# diversity
# Calculates the entropy of a system with equiprobable states.
# Arguments:
#    x -- numeric vector -- observed probabilities of classes
# Returns:
#    numeric -- the entropy / diversity measure
diversity <- function(x) {
  f <- function(x) x * log(x)
  x1 <- mapply(FUN = f, x)
  - sum(x1) / log(length(x))
  }
```

[9]Note also how similar entropy is to "information" as defined in Chapter 23.1.1.5 *"Binary Classification Trees"* on page 411.

In the context of diversity, entropy works fine as a measure for diversity. Many authors use a similar definition.[10] In this section we will take a practical approach.

If there is a relevant prior probability (e.g. we know that the working population in our area consist of 20% Hispanic people and 80% Caucasian people) then we might want to show the maximum diversity for that prior probability (e.g. 20% Hispanic people and not 50%).

In such case, it makes sense to rescale the diversity function so that a maximum is attained at the natural levels (the expected proportions in a random draw). This can be done by scaling $s(x)$ so that the scaled prior probability of each sub-group becomes $\frac{1}{N}$. So, we want for each group i to find a scaling so that

$$\begin{cases} s(0) & = & 0 \\ s(P_i) & = & \frac{1}{N} \\ s(1) & = & 1 \end{cases}$$

with P_i the prior probability of sub-group i. For example, we could fit a quadratic function through these three data-points. A broken line would also work, but the quadratic function will be smooth in P_i and has a continuous derivative.

Solving the simple set of aforementioned equations, we find that $s(x)$ can be written as:

$$s(x) = ax^2 + bx + c,$$

where

$$\begin{cases} a & = & \frac{1-\frac{1}{NP_i}}{1-P_i} \\ b & = & \frac{1-NP_i^2}{NP_i(1-P_i)} \\ c & = & 0 \end{cases}$$

with N the number of sub-groups and P_i the prior probability of the group i.

To add this as a possibility but not make it obligatory to suppy these prior probabilities, we re-write the function `diversity()` so that it takes an optional argument of prior probabilities; and if that argument is not given, it will use the probabilities as they are.

```
# diversity
# Calculates the entropy of a system with discrete states.
# Arguments:
#    x     -- numeric vector -- observed probabilities of classes
#    prior -- numeric vector -- prior probabilities of the classes
# Returns:
#     numeric -- the entropy / diversity measure
diversity <- function(x, prior = NULL) {
  if (min(x) <= 0) {return(0);} # the log will fail for 0
  # If the numbers are higher than 1, then not probabilities but
  # populations are given, so we rescale to probabilities:
  if (sum(x) != 1) {x <- x / sum(x)}
  N <- length(x)
  if(!is.null(prior)) {
    for (i in (1:N)) {
      a <- (1 - 1 / (N * prior[i])) / (1 - prior[i])
      b <- (1 - N * prior[i]^2) / (N * prior[i] * (1 - prior[i]))
      x[i] <- a * x[i]^2 + b * x[i]
    }
   }
  f <- function(x) x * log(x)
  x1 <- mapply(FUN = f, x)
  - sum(x1) / log(N)
}
```

[10] See for example Jost (2006), Keylock (2005), Botta-Dukát (2005), or Kumar Nayak (1985).

For example, if we have prior probabilities of three subgroups of 10%, 50% and 40% then we consider our population as optimally diverse when these probabilities are obtained:

```
# Consider the following prior probabilities:
pri <- c(0.1,0.5,0.4)

# No prior priorities supplied, so 1/N is most diverse:
diversity(c(0.1,0.5,0.4))
## [1] 0.8586727

# The population matches prior probabilities, so index should be 1:
diversity(c(0.1,0.5,0.4), prior = pri)
## [1] 1

# Very non-diverse population:
diversity(c(0.999,0.0005,0.0005), prior = pri)
## [1] 0.002478312

# Only one sub-group is represented (no diversity):
diversity(c(1,0,0), prior = pri)
## [1] 0

# Numbers instead of probabilities provided, also this works:
diversity(c(100,150,200))
## [1] 0.9656336
```

We can also visualize what this function does. For example, assume prior population of men and women equal and consider gender as binary, then we can visualize the evolution of our index as follows – the plot is in Figure 36.6 on page 730:

```
females <- seq(from = 0,to = 1, length.out = 100)
div <- numeric(0)
for (i in (1:length(females))) {
  div[i] <- diversity (c(females[i], 1 - females[i]))
  }

d <- as.data.frame(cbind(females, div)  )
colnames(d) <- c('percentage females', 'diversity index')
library(ggplot2)
p <- ggplot(data = d,
        aes(x = `percentage females`, y = `diversity index`)) +
     geom_line(color = 'red', lwd = 3) +
     ggtitle('Diversity Index') +
     xlab('percentage females') + ylab('diversity index')
p
```

Note – Special characters in variable names

Note the back-ticks around the variable names. That is necessary because the variable names contain spaces. Alternatively, it is possible to refer to them using the function `get('percentage females')`. In either case, `ggplot2` will insist in adding the function `get()` or back-ticks to the labels, so it is best to re-define them with the functions `xlab()` and `ylab()`.

Now, that we have some concept of what we can use as a diversity index, we still need some data. This should of course come from your HR system, but for the purpose of this book, we will produce data. The advantage of this approach is that you can do this on your own computer

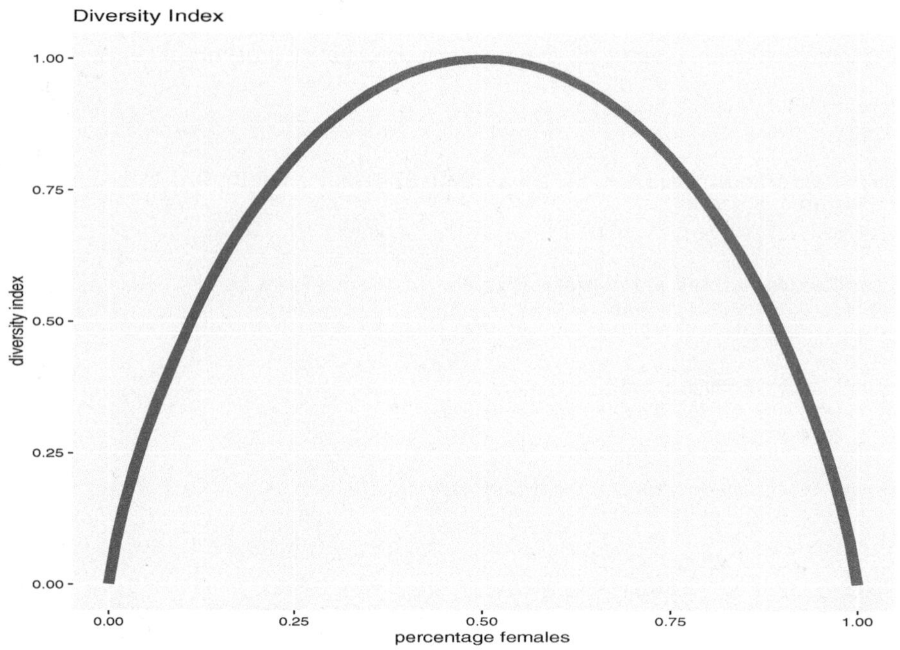

Figure 36.6: *The evolution of the gender-diversity-index in function of one of the representation of one of the genders in our population.*

and play with the sample size and learn about the impact of sample size on diversity. Another advantage is that we can build in a bias into the data, and that known bias will make the concepts more clear.

```
library(tidyverse)
N <- 200
set.seed(1866)

d0 <- data.frame("ID"          = 1:N,

          # Log-normal age distribution
          "age"          = round(rlnorm(N, log(30), log(1.25))),
          # A significant bias towards the old continent:
          "continent"    = ifelse(runif(N) < 0.3, "America",
                              ifelse(runif(N) < 0.7,"Europe","Other")),
          # A mild bias towards males:
          "gender"       = ifelse(runif(N) < 0.45, "F", "M"),
          # Grade will be filled in later:
          "grade"        = 0,
          # Three teams of different sizes:
          "team"         = ifelse(runif(N) < 0.6, "bigTeam",
                               ifelse(runif(N) < 0.6,
                               "mediumTeam",
                               ifelse(runif(N) < 0.8, "smallTeam",
                                      "XsmallTeam"))),
          # Most people have little people depending on them:
          "dependents"   = round(rlnorm(N,log(0.75),log(2.5))),
          # Random performance (no bias linked, but different group sizes):
          "performance" = ifelse(runif(N) < 0.1, "L",
                               ifelse(runif(N) < 0.6, "M",
                               ifelse(runif(N) < 0.7, "H", "XH"))),
```

```
        # Salary will be filled in later:
        "salary"       = 0,
        # We make just a snapshot dashboard, so we do not need this now,
        # but we could use this later to show evolution:
        "timestamp"    = as.Date("2020-01-01")
        )

# Now we clean up age and fill in grade, salary and lastPromoted without
# any bias for gender, origin -- but with a bias for age.
d1 <- d0                                                    %>%
  mutate(age     = ifelse((age < 18), age + 10, age))       %>%
  mutate(grade   = ifelse(runif(N) * age < 20, 0,
                   ifelse(runif(N) * age < 25, 1,
                   ifelse(runif(N) * age < 30, 2, 3))))     %>%
  mutate(salary = round(exp(0.75*grade)*4000 +
                   rnorm(N,0,1500)))                        %>%
  mutate(lastPromoted = round(exp(0.05*(3-grade))*1 +
                   abs(rnorm(N,0,5))) -1)
```

While we are usually very comfortably using R in the CLI, it is – at this point – a good idea to use RStudio. RStudio will make working with dynamic content a lot easier – just as working with Chapter 36.1 *"Shiny"* on page 715.

36.3.2 A Dashboard with flexdashboard

Now, we have the data and want to build a dashboard to show a snapshot of the diversity in our company. We start with building a static dashboard with the package `flexdashboard`.

36.3.2.1 A Static Dashboard

`flexdashboard` allows to create professional dashboards with the ease of R Markown. First, we install the package `flexdashboard`; and once that is done, we load the package.

```
# If not done yet, install the package:
# install.packages('flexdashboard')

# Then load the package:
library(flexdashboard)
```

Then in RStudio, chose "new file " in the file menu and select then "R Markdown," then "from template" and finally select "Flex Dashboard" from the list. The screen will look like Figure 36.7 on page 732, and the document will now look similar the following code. Although, note that this is only the framework (sections without the actual code).

```
---
title: "Untitled"
output:
  flexdashboard::flex_dashboard:
    orientation: columns
    vertical_layout: fill
---
```{r setup, include=FALSE}
library(flexdashboard)
```
```

```
Column {data-width=650}
-----------------------------------------------------

### Chart A

```{r}

```

Column {data-width=350}

Chart B
```{r}

```

Chart C
```{r}

```
```

This prepares the framework for a dashboard that contains a wide column on the left, and a narrower one at the right (the `{data-width=xx}` arguments make the columns). Similar to R Markdown, sections are created by the third-level titles (the words following three pound signs ###).

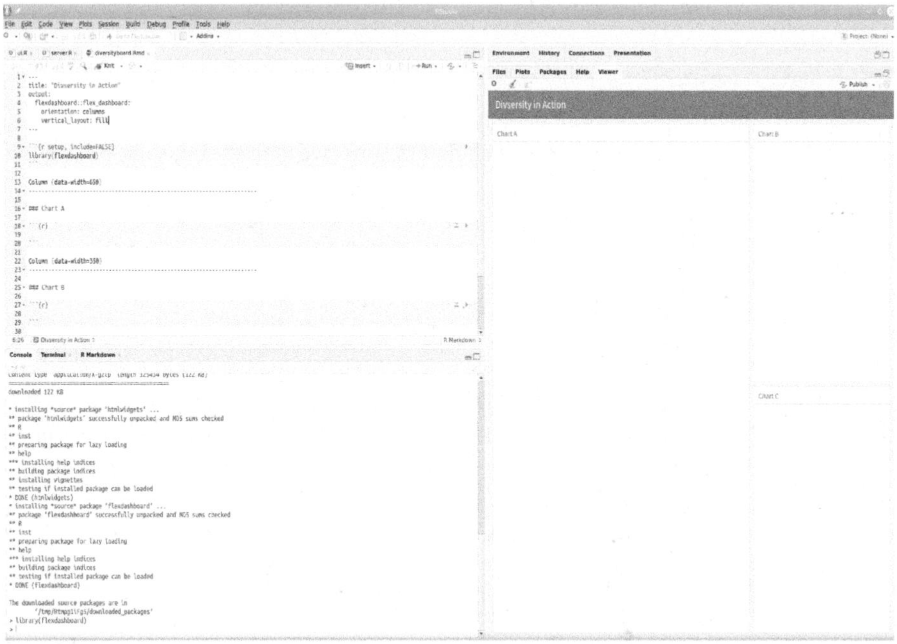

Figure 36.7: *Creating a flexdashboard from the template provides a useful base-structure for three plots. It can readily be extended to add one's own content.*

The dashboard that we intend to make will have one front-page that provides an overview of the diversity in our population based on the indices as defined above.

The structure is as follows.

````
---
title: "Diversity in Action"
output:
  flexdashboard::flex_dashboard:
    theme: cosmo
    orientation: rows
    vertical_layout: fill
    social: menu
    source: embed
---

```{r setup, include=FALSE}
# load packages and functions
```
Overview
========
Row
-------------------------------------
### Gender
```{r}
code to generate the first gauge
```

Row
-------------------------------------

Gender
=====================================

Row {.tabset}
-------------------------------------

Age
=====================================
Row {.tabset}
-------------------------------------
### first tab
### second tab

Roots
=====================================
Row {.tabset}
-------------------------------------

Dependants
=====================================
Row {.tabset}
-------------------------------------
````

Hint – Find the full code of the dashboard

Here we present only the structure of the dashboard. The complete working demo-dashboard can be seen at the website of this book, and the code can be downloaded from that website (click on the `</> Source Code` button in the right upper corner), and of course the code is also in the code-document that goes with this book in the code-box named "flexdash."

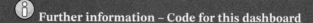

Further information – Code for this dashboard

The code listing is too long to print here, but it is available in the code that goes with this book: `www.de-brouwer.com/publications/r-book`.

The dashboard looks now as represented in Figure 36.8.

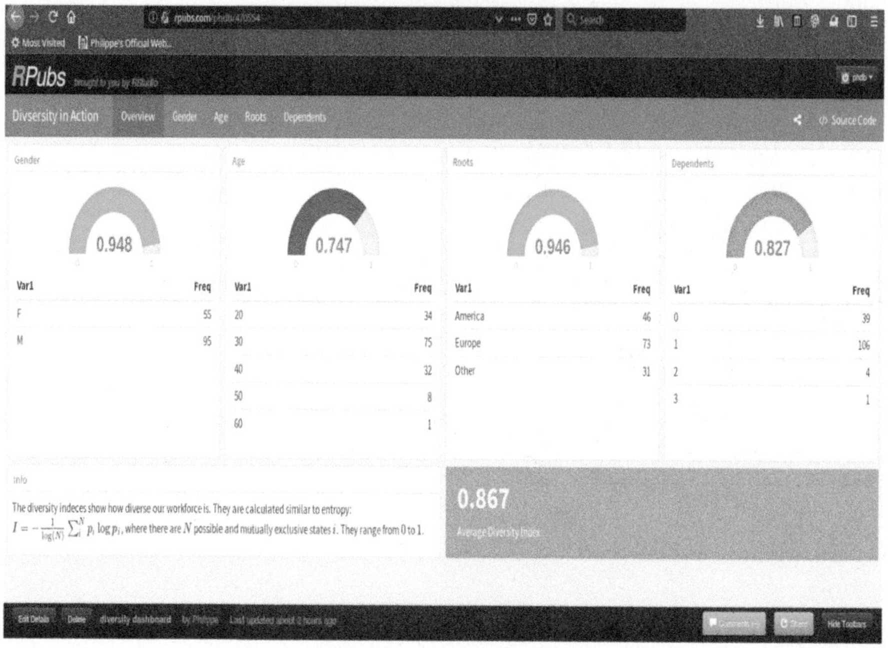

Figure 36.8: *The welcome page of the dashboard provides the overview (menu) at the top and in the body an overview of some diversity indices.*

The structure of the dashboard starts with the header of the rmarkdown document. In the header we tell R what the title is, what theme to use, whether to include buttons to share on social media, whether to show the source code but also how the layout will look like. We have chosen "orientation: rows." This means that sections – characterized by three pound signs: `###` – will be will be shown as rows.

The rmarkdown document then consists of the first level titles – underlined by equal signs – and second level titles – underlined by at least three dash signs. The first level titles get a separate page (they appear in the blue header of all pages and eventually collapse on smaller screens).

That brings us to the level of a page. Each page can still contain more than one plot or more than one row. For example, the first page – the overview – contains four columns in one row and then two columns in the next row.

The code for the gauges works as follows:

```
Overview
========

Row
-------------------------------------

```{r}
# here goes the R-code to get data and calculate diversity indices.
```
```

```
### Gender
```{r genderGauge}
# ranges:
rGreen   <- c(0.900001, 1)
rAmber   <- c(0.800001, 0.9)
rRed     <- c(0, 0.8)
iGender  <- round(diversity(table(d1$gender)),3)
gauge(iGender, min = 0, max = 1, gaugeSectors(
      success = rGreen, warning = rAmber, danger = rRed
      ))
kable(table(d1$gender))
```
This is only the first gauge, the next one can be described below.
```

The widget "gauge" is provided by `flexdashboard` and takes the obvious parameters such as minimum, maximum, etc. After the gauge, we simply output a table that sums up the data. This table is then formatted with the function `kable()` from `knitr`.

knitr
kable()

Further information – More eye candy

Also `gvis` provides a long list of widgets such as gauges for example:

```
gauge <-  gvisGauge(iGender,
             options=list(min = 0, max = 1,
                     greenFrom = 0.9,greenTo = 1,
                     yellowFrom = 0.8, yellowTo = 0.8,
                     redFrom = 0, redTo = 0.8,
                     width = 400, height = 300))
```

Of course, the package `plot_ly` has also out-of-the box the possiblity to produce amazing gauges.

It is of course possible to use a similar layout on every page (main titles) that simply relies on columns and rows. However, it is also possible to make more interesting structures such as "tabsets" and "storyboards." A tabset will layout the page in different sections that are hidden one after the other and that can be revealed by clicking on the relevant tab.

The storyboard is a great tool to present conclusions of a particular bespoke analysis. Each "tab" is now a little text-field that is big enough to contain an observation, finding, or conclusion. Clicking on this information reveals then the data-visualization that leads to that conclusion or observation.

For example, the page "Gender" has a structure with tabs. That is made clear to R by adding `{.tabset}`

```
Gender
======================================

Row {.tabset}
-------------------------------------

### Composition
```{r}
p2 <- ggplot(data = d1, aes(x=gender, fill=gender)) +
  geom_bar(stat="count", width=0.7) +
  facet_grid(rows=d1$grade) +
  ggtitle('workforce composition i.f.o. salary grade')
ggplotly(p2)
```
```

```
### Tab2
```{r RCodeForTab2}
# etc.
```
```

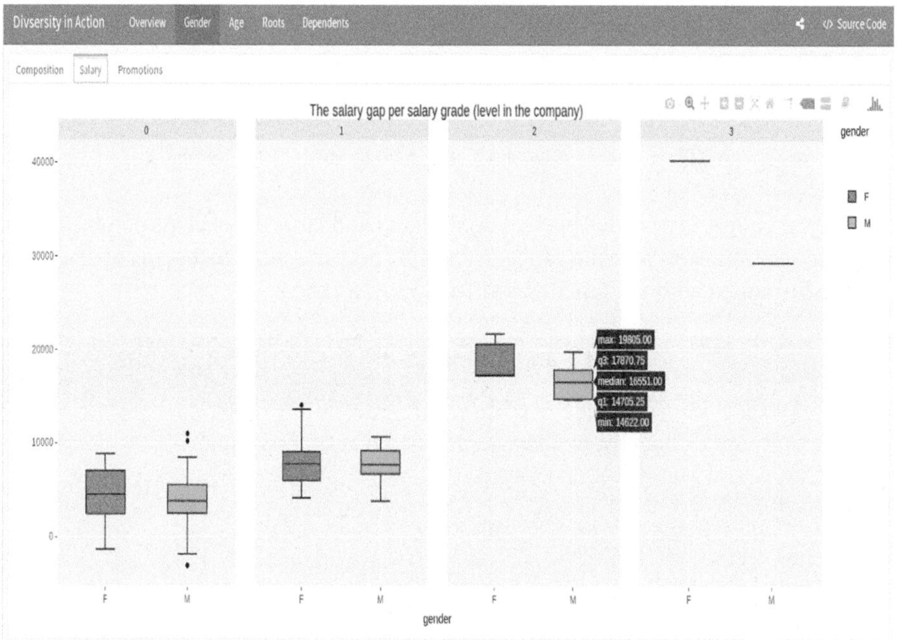

Figure 36.9: *The page "Gender" of the dashboard provides multiple views of the data that are each made visible by clicking on the relevant tab. The effect of* `ggplotly()` *is visible by the toolbar that appears just right above the plot and the mouse-over effect on the boxplot for the salary grade 2/male population.*

Note – Visualising is not via printing the plot

Note that we use the function `ggplotly()`, from the package `plotly` instead of printing the plot as we are used to in `ggplot2`. This function will print the plot but on top of that add interactive functionality to show underlying data, zoom, download, etc. Some of that interaction and the interface that it creates is visible in Figure 36.9.

36.3.2.2 Interactive Dashboards with flexdashboard

To make the dashboard interactive there are some elegant solutions available. The first and most standardized – but already very powerful – interaction can be obtained by using `ggvis()` for the plots or using `ggplotly()` as shown in previous section.

If we need other interaction than only the dynamics provided by `ggplotly`, then we can integrate the dashboard with `Shiny`. `shiny` is more flexible and allows to select different data, upload new data, etc.

> **Further information – flexdashboard**
>
> Read more about flexdashboard at https://rmarkdown.rstudio.com/flexdashboard/index.html

Adding interactive functionality to flexdashboard is rather straightforward. Consider the following steps:

1. Add `runtime: shiny` to the header of the file (in the YAML front matter).
2. Add the input controls. For example, add the `{.sidebar}` attribute to the first column of and it will create a box for the Shiny input fields.
3. Add Shiny outputs on the dashboard and when including plots that need to react to the input controls, make sure to to wrap them renderPlot().

36.3.3 A Dashboard with shinydashboard

An equally exciting solution to create a dashboard is the package shinydashboard. If you are used to Shiny, you will be familiar with how to get an interactive app together – if not we recommend to read Chapter 36.1 *"Shiny"* on page 715. A good way of thinking about shinydashboard is that it acts like an extension of Shiny. All you learned in order to create a great app with Shiny is still applicable, but there are some specific functionalities that will come in handy to create a dashboard.

shinydashboard

To get the most out of shinydashboard, it is useful to master building interactive sites with Shiny. This means that the learning curve will typically be a lot steeper than with flexdashboard. However, once over that initial effort, one will enjoy the flexibility and power of Shiny to produce interactive and highly customized dashboards.

Just as with any other package, working with shinydashboard requires us to download the package first with `install.packages('shinydashboard')` and then load it in the R-headers of our file.

In this section we will not build a fully fledged diversity dashboard, but rather opt for showing how a very simple dashboard can be build, but in return show all the code necessary. Such simple dashboard could look like the one in Figure 36.10 on page 738.

The code to obtain this dashboard is as follows:

```
# this file should be called app.R
library(shinydashboard)

# general code (not part of server or ui function)
my_seed <- 1865
set.seed(my_seed)
N <- 150

# user interface ui
ui <- dashboardPage(
  dashboardHeader(title = "ShinyDashBoard"),
  dashboardSidebar(
      title = "Choose ...",
      numericInput('N', 'Number of data points:', N),
      numericInput('my_seed', 'seed:', my_seed),
      sliderInput("bins", "Number of bins:",
                  min = 1, max = 50, value = 30)
  ),
```

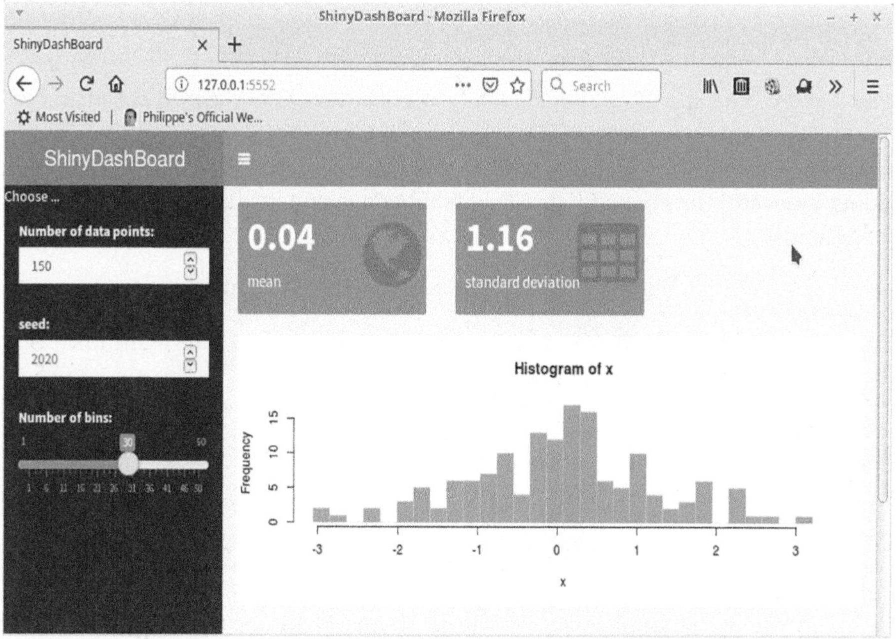

Figure 36.10: *A simple dashboard with* `shinydashboard`.

```
  dashboardBody(
    fluidRow(
      valueBoxOutput("box1"),
      valueBoxOutput("box2")
      ),
    plotOutput("p1", height = 250)
    )
  )

# server function
server <- function(input, output) {
  d <- reactive({
    set.seed(input$my_seed)
    rnorm(input$N)
    })
  output$p1 <- renderPlot({
    x    <- d()
    bins <- seq(min(x), max(x), length.out = input$bins + 1)
    hist(x, breaks = bins, col = 'deepskyblue3', border = 'gray')
    })
  output$box1 <- renderValueBox({
    valueBox(
      value    = formatC(mean(d()), digits = 2, format = "f"),
      subtitle = "mean", icon     = icon("globe"),
      color    = "light-blue")
    })
  output$box2 <- renderValueBox({
      valueBox(
        value    = formatC(sd(d()), digits = 2, format = "f"),
        subtitle = "standard deviation",  icon     = icon("table"),
        color    = "light-blue")
      })
```

```
}

# load the app
shinyApp(ui, server)
```

Note – Make your website reactive

Note how in the server function, the data is dynamically determined and stored in a variable `d`. This is useful because the variable is used in more than one function later on. However, to make clear that this is dynamic and will use methods of the `input` object, it has to be wrapped in the function `reactive()`. This will create a method to find the value of `d` rather than a dataset. Therefore, any future references can only be done in a reactive function and by referring to it as `d()`.

`reactive()`

Also the diversity-dashboard will look good with `shinydashboard` and – if one can work with `Shiny` in general – will be fast and elegant. It can look for example as in Figure 36.11.[11]

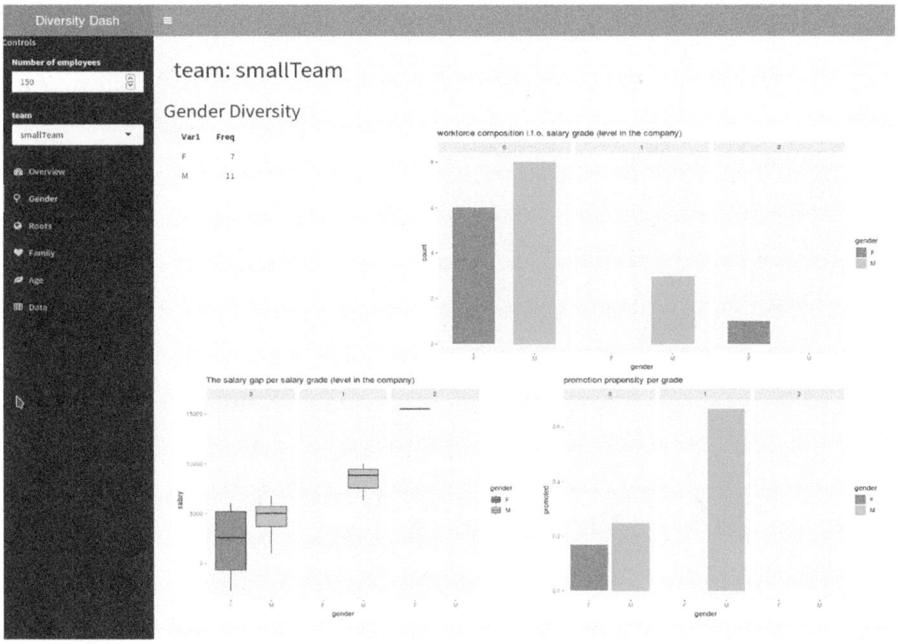

Figure 36.11: *Another take on the diversity dashboard, rendered with the help of* `shinydashboard` *and sporting dynamic content with* `Shiny`.

Also here we can choose from an abundance of free html-widgets to spice up our content, and make the presentation more interactive, useful, attractive, and insightful. See Chapter 36.2.1 *"HTML-widgets"* on page 719 for more information and suggestions.

[11] The code for this dashboard will be shared completely on the website of this book: http://www.de-brouwer.com/r-book.

PART VIII

Bigger and Faster R

♣ 37 ♣

Parallel Computing

It may happen that running your code takes an impractical[1] amount of time. If the calculations do not depend on each other's outcome then the calculations can be run in parallel.

In its standard mode, R is single threaded and all operations go through one processor. The package `parallel` allows us to split the runtime of a loop over different cores available on our system. Note that the library `parallel` is part of core-R and hence does not need to be installed, so we can load it immediately.

`parallel`

```
# Load the library:
library(parallel)

# The function detectCores finds the number of cores available:
numCores <- detectCores()
numCores
## [1] 12
```

So, the computer that was used to compile this book has 12 cores at its disposal. Larger computers have vast arrays of CPUs to which calculations can be dispatched and the library `parallel` provides exactly the tools to use that multitude of cores to our advantage. The function `mclapply()` that is an alternative to `lapply()` that can use multiple cores.

The following code illustrates how simple we can use multiple cores with the function `mcapply()`, and we will also time the gains with the function `system.time()`.

`mclapply()`
`lapply()`

```
# Set the sample size:
N <- 50

# Set the starting points for the k-means algorithm:
starts <- rep(100, N)   # 50 times the value 100

# Prepare data as numeric only:
t <- as.data.frame(titanic_train)
t <- t[complete.cases(t),]
t <- t[,c(2,3,5,6,7,8)]
t$Sex <- ifelse(t$Sex == 'male', 1, 0)
```

[1] An impractical amount of time is somehow a personal concept. For the author it is fine if code runs less than a few minutes or longer than an hour up to a few days. In the first case, one can wait for the result to appear and in the second case we can do something else while the code is running. Anything in between is annoying enough to do an effort to speed it up. Anything that runs longer than a good night rest is probably also a good candidate to speed up.

The Big R-Book: From Data Science to Learning Machines and Big Data, First Edition. Philippe J.S. De Brouwer.

Companion Website: www.wiley.com/go/De Brouwer/The Big R-Book

```
# Prepare the functions to be executed:
f <- function(nstart) kmeans(t, 4, nstart = nstart)

# Now, we are ready to go, we try two approaches:

# 1. With the standard function base::lapply
system.time(results <- lapply(starts, f))
##    user  system elapsed
##   1.311   0.186   1.444

# 2. With parallel::mclapply
system.time(results <- mclapply(starts, f, mc.cores = numCores))
##    user  system elapsed
##   1.124   0.536   0.287
```

The total elapsed time is about *five times* shorter when we use all cores.[2] This comes at some overhead cost (the System time is longer) and the gain is not exactly the number of cores. The gain of using more cores is significant. Problems that can be programmed as independent blocks on different cores should use this way of parallel programming.

Our system had 12 cores and the gain was roughly fivefold. That is not bad, and can make a serious difference. Remember, however, that this book is compiled on a laptop that is designed to read email and play an occasional game. Large computers that are equipped with an array of CPUs are called "Supercomputers." On such computers, the gain can be much, much more. Note also that modern supercomputers not only have arrays of CPUs, but that each of those CPUs can be equipped with a GPU.[3]

supercomputer

Computers that are designed for massive parallel computing are still quite big, and usually are the sole tenants of a large building. A list of supercomputers can be found at `https://www.top500.org`. The rankings are updated each six months and in June 2019 the fastest computer is IBM's "Summit." Summit has 2,414,592 cores available for calculations. This is a lot more than our laptop and the performance gain will be accordingly.

Indeed, it is that simple: we can just install R on a supercomputer and use its massive amount of cores. Almost all supercomputers run Linux,[4] and hence, can run R very much in the same way as your computer can do that. The command line interface can be accessed via remote login (ssh) or eventually we can install RStudio server on the supercomputer to provide a more user-friendly experience for the end-user.

[2]The Elapsed Time is the time charged to the CPU(s) for the expression, the User Time is the time that you as user experience, and System Time is the time spent by the kernel on behalf of the process. While usually elapsed time and system time are quite close. If the elapsed time is longer than the user time this means that the CPU was spending tiime on other operations (maybe external to our process). Especially in the case of `mcapplyl()` the elapsed time is a lot less than the user time. This means means that our computer has multiple cores and is using them.

[3]The CPU is the central processing unit, and the GPU is the graphical processing unit. These additions became popular on the back of the popularity of computer games, but the modern GPU has thousands of processors on boards and has its own RAM. This makes them an ideal platform for speeding up calculations. More about GPU programming is in Chapter 37.3 *"Using the GPU"* on page 752.

[4]At the time of the top-list at `https://www.top500.org`, 100% of the 500 fastest supercomputers run Linux.

37.1 Combine foreach and doParallel

If you rather work with for-loops then the library `foreach` will help you to extend the for-loop to the `foreach` loop. This construct combines the power of the for-loop with the power of the `lapply()` function. The difference with `lapply` is, that it evaluates an expression and not a function and hence returns a value (and avoids side effects).

foreach()

Here is a short illustration on how it works:

```
# 1. The regular for loop:
for (n in 1:4) print(gamma(n))
## [1] 1
## [1] 1
## [1] 2
## [1] 6

# 2. The foreach loop:
library(foreach)
foreach (n = 1:4) %do% print(gamma(n))
## [1] 1
## [1] 1
## [1] 2
## [1] 6
## [[1]]
## [1] 1
##
## [[2]]
## [1] 1
##
## [[3]]
## [1] 2
##
## [[4]]
## [1] 6
```

Note – foreach

Note that `foreach` returns a an object of class `list`.

```
class(foreach (n = 1:4))
## [1] "foreach"
```

The operation is best understood by noticing that the structure of the `foreach` loop is based on the definition of the operator `%do%`. This operator, `%do%`, acts on a `foreach` object (created by the function `foreach()`). This operator controls how the expression that follows is evaluated.

While the operator `%do%` will sequentially run each iteration over the same core, `%dopar%` will utilise the power of parallel processing by using more cores. The operator can be initialised with the library `doParallel`. Note that this package most probably needs to be installed first via the command
`install.packages('doParallel')`.

doParallel

```
# Once installed, the package must be loaded once per session:
library(doParallel)

# Find the number of available cores (as in previous section):
numCores <- parallel::detectCores()

# Register doParallel:
```

```
registerDoParallel(numCores)

# Now, we are ready to put it in action:
foreach (n = 1:4) %dopar% print(gamma(n))
## [[1]]
## [1] 1
##
## [[2]]
## [1] 1
##
## [[3]]
## [1] 2
##
## [[4]]
## [1] 6
```

Hint – Expressions

The expression – that is passed to `foreach` – can span multiple lines provided that it is encapsulated by curly brackets. This is similar to how other functions such as `if`, `for`, etc. work in R.

Note also that the function `foreach` allows to simplify the result (i.e. collapse the result to a simpler data type). To achieve this, we need to provide the creator function to such simple data type. For example, we can supply `c()`, to simplify the result to a vector.

`c()`

In the following code we provide a few examples to that collapse the results to a more elementary data type with the `.combine` argument.

```
# Collapse to wide data.frame:
foreach (n = 1:4, .combine = cbind) %dopar% print(gamma(n))
##      result.1 result.2 result.3 result.4
## [1,]        1        1        2        6

# Collapse to long data.frame:
foreach (n = 1:4, .combine = rbind) %dopar% print(gamma(n))
##          [,1]
## result.1    1
## result.2    1
## result.3    2
## result.4    6

# Collapse to vector:
foreach (n = 1:4, .combine = c)     %dopar% print(gamma(n))
## [1] 1 1 2 6
```

We can rewrite our example of calculating 50 times k-means with 100 starting points with the `foreach` syntax as follows.

```
# We reuse parameters and data as defined in the previous section:
system.time(
foreach (n = 1:N) %dopar% {
    kmeans(t, 4, nstart=100)
    }
  )
##    user  system elapsed
##   1.707   0.644   0.344
```

As can be seen, the time is similar to `mclapply`. The difference is in the formalism that can be more intuitive for some people. This might be up to personal taste, but as many other open-source solutions, R always offers choice.

Further information – Alternatives for multi-core processing

R has a multitude of solutions available. If you need multi-core processing extensively, you might want to have a look at

- `Multicore` in case the nodes do not need to communicate; and
- when inter-node communication is needed, you might want to consider `foreach` and `parallel` or, alternatively, explore `doMC` or `doSNOW`.

37.2 Distribute Calculations over LAN with Snow

In order use more cores it is not always necessary to buy time on a supercomputer. It is also possible to use the workstations that are already connected to the network to dispatch calculations so that they can be executed in parallel. The library `snow` (an acronym for Simple Network Of Workstations) provides such framework that allows computers to collaborate over a network while using R. The model is that of Master/Slave communication in which one node that has the Master role controls and steers the calculations on the slave nodes.

snow

Master
slave

To achieve this, `snow` implements an application programming interface (API) to different mechanisms that allow the Slave and Master to connect and exchange information and commands. The following protocols can be used:

API

- Socket
- PVM (Parallel Virtual Machine), and
- MPI (Message Passing Interface).

Hint – Random numbers

In bootstrapping and many other applications, it is important that the random numbers that are used in all nodes are independent. This can be assured via the `rlecuyer` or the `rsprng` library.

First, we install the library `snow` and when that is done we can load it.

```
library(snow)

##
## Attaching package: 'snow'
## The following objects are masked from 'package:parallel':
##
##    clusterApply, clusterApplyLB, clusterCall,
##    clusterEvalQ, clusterExport, clusterMap,
##    clusterSplit, makeCluster, parApply, parCapply,
##    parLapply, parRapply, parSapply, splitIndices,
##    stopCluster
```

Note – snow overwrites many functions

We had loaded `parallel` in previous section, and the output shows how many of the functions overlap in both packages. This means two things. First, they should not be used together, they do the same but in different ways. Second, the similarity between both will shorten the total learning time.
The better approach is to unload `parallel` first:

```
detach("package:parallel", unload = TRUE)
library(snow)
```

Now, the slave process needs to be started on each slave node and finally the cluster can be started on the Master node. The socket method allows to connect a node via the `ssh` protocol. This protocol is always available – on *nix systems – and easy to set up.

ssh

***nix**

It allows also to create a simulated slave on the same machine by referring to "localhost."[5] This will create a cluster on the same machine.

localhost

```
cl <- makeCluster(c("localhost", "localhost"), type = "SOCK")
```

In this particular case, it is possible to use the function's defaults and `makeCluster(2)` will have the same effect. However, remember that this functionality only makes sense when you use other machines than "localhost."

Now, that the cluster is defined, we can run operations over the cluster. We will use the same example as we used in previous section.

```
library(titanic)  # provides the dataset: titanic_train
N <- 50
starts <- rep(100, N)   # 50 times the value 100

# Prepare data as numeric only:
t <- as.data.frame(titanic_train)
t <- t[complete.cases(t),]
t <- t[,c(2,3,5,6,7,8)]
t$Sex <- ifelse(t$Sex == 'male', 1, 0)

# Prepare the functions to be executed:
f <- function(nstart) kmeans(t, 4, nstart=nstart)

# 1. with the standard function
system.time(results <- lapply (starts, f))
##    user  system elapsed
##   1.314   0.203   1.463

# 2. with the cluster
# First, we must export the object t,  so that it can
# be used by the cluster:
clusterExport(cl, "t")
#clusterExport(cl, "starts") # Not needed since it is in the function f
system.time(
  result2 <- parLapply(cl, starts, f)
  )
##    user  system elapsed
##   0.011   0.008   0.591
```

In summary, one defines a cluster object, and this allows the Master node to dispatch part of parallel calculations to the slave nodes by using functions such as `parLapply()` as an alternative to `lapply()` or `parSapply()` as an alternative to `sapply()`, etc. Those functions will then take the cluster as the first argument and work for the rest similar to their base R counterparts and are shown in Table 37.1 on page 750.

`parLapply()`
`parSapply()`

[5]Creating a cluster on your own computer will of course not be faster than the solutions described earlier in this chapter (e.g. the library `parallel`). The reason why we do this here, is that the whole book is using life connections and calculations and therefore this is the most efficient way to show how it works.

Base R	snow
lapply	parLapply
sapply	parSapply
vapply	NA
apply (row-wise)	parRapply or parApply(,1)
apply (column-wise)	parCapply or parApply(,2)

Table 37.1: *Base R functions and their alternative in* `snow`.

Hint – Cost efficiency

Snow is the ideal solution to use existing hardware before buying an expensive supercluster or renting time on an existing supercomputer. It allows all computers to collaborate and use computing power that sits idle.

`clusterCall()`

Some applications will not suit the `apply` family of functions. The function `clusterCall()` will call a given function with identical arguments on each slave in the cluster.

```
f <- function (x, y) x + y + rnorm(1)
clusterCall(cl, function (x, y) {x + y + rnorm(1)}, 0, pi)
## [[1]]
## [1] 3.046134
##
## [[2]]
## [1] 4.187535

clusterCall(cl, f, 0, pi)
## [[1]]
## [1] 1.608131
##
## [[2]]
## [1] 5.024353

# Both forms are semantically similar to:
clusterCall(cl, eval, f(0,pi))
## [[1]]
## [1] 3.182824
##
## [[2]]
## [1] 3.182824

# However, note that in the last example the random numbers are exactly the same
# on both clusters.
```

`clusterCallEvalQ()`

The cluster version of the function `evalq()` is `clusterCallEvalQ()` and it is called as follows.

```
# Note that ...
clusterEvalQ(cl, rnorm(1))
## [[1]]
## [1] -0.1861496
##
## [[2]]
## [1] 1.014375
```

```
# ... is almost the same as
clusterCall(cl, evalq, rnorm(1))
## [[1]]
## [1] -0.5018691
##
## [[2]]
## [1] -0.5018691

# ... but that the random numbers on both slaves nodes are the same
```

Note that the evalq function in base R is equivalent to eval(quote(expr), ...). This function, eval(), evaluates its first argument in the current scope before passing it to the evaluator, whereas evalq avoids evaluating the argument first and hence passes it on before evaluating.

eval()
evalq()

When not all arguments of the functions should be the same, the function clusterApply() allows to run a function with a list of provided arguments, so that each cluster will run the same function but with its unique given argument. The function will therefore, take a cluster, a sequence of arguments (in the form of a vector or a list), and a function, and calls the function with the first element of the list on the first node, with the second element of the list on the second node, etc. Obviously, the list of arguments must have at most as many elements as there are nodes in the cluster. If it is shorter, then it will be recycled.

```
clusterApply(cl, c(0, 10), sum, pi)
## [[1]]
## [1] 3.141593
##
## [[2]]
## [1] 13.14159
```

Hint – Load balancing with clusterApplyLB(

The package snow allows for automatic load-balancing with the function clusterApplyLB(). The function definition is:

clusterApplyLB()

```
clusterApplyLB(cl, seq, fun, ...)
```

It takes the same arguments and behaves the same as clusterApply(). It will dispatch a balanced work load to slave nodes when the length of seq is greater than the number of cluster nodes. Note that it does not work on a cluster that is defined with Type=Socket as we did.

Note also that there is a function clusterSplit(cl, seq) that allows the sequence seq to be split over the different clusters so that each cluster gets an equal amount of data. There are also specific functions to dispatch working with matrices over the cluster: parMM(cl, A, B) will multiply the matrices A and B over the cluster cl.

clusterSplit()

Finally, after all our code if finished running, we terminate the cluster and clean up all connections between master and slave nodes with the function stopCluster():

stopCluster()

```
stopCluster(cl)
```

37.3 Using the GPU

GPU

In all previous sections of this book we used the CPU (the "central processing unit") to handle all our requests. Modern computers often have a dedicated a graphics card on board. That card usually has many cores that can work simultaneously to compute and render complex landscapes, shadows and artefacts in a computer game. The processors on such graphics card (GPU) are usually not as fast as the CPU, but there are many of them on one card, and even have their own memory. While the GPU is designed to improve experience for playing computer games, it is also a superb source for parallel computing. Indeed, a high-end graphics card has multiple thousands processors on board. Compare this to the CPU that nowadays has around ten cores.

C
C++
Fortran

Unfortunately, programming a GPU is a rather advanced topic and the interfaces depend on the brand and make of the card itself. Most graphic cards producers provide libraries that can be used. While the interfaces provided are typically geared towards C, C++, or Fortran, the R-user can rely, on R-packages that simplify the interface to using functions in R.

Example: Nvidia

CUDA
rpud

One of the major players on the graphics cards market is Nvidia. This company provides a programming platform for accessing the multiple processors on their cards: CUDA (Compute Unified Device Architecture). The interface that accesses the processors is a C or C++ interface, but the library `rpud` – for example – takes much of the complexity away for the R-user. This means that in R we can simply access the cores of the GPU.
To give you an idea: today, a decent graphics card has thousands[a] cores that run at a clock-speed that is typically one third to half of a good CPU. This means that when the code benefits of parallelism, that speed gains of many hundreds of times are very realistic.

[a] For example, one of the later models of Nvidia is the Titan. It offers 3840 CUDA cores that have a clock-speed of about 1.5GHz.

PU
processing unit

To take this concept a step further, it is also possible to use the GPUs in a cluster of computers. Some supercomputers have GPU cards for each of the CPU cards that constitute the cluster. If the array has 1 000 CPUs, and each CPU has 1 000 GPU cores on board, then the total number of processing units (PUs) exceeds a million.

Using the GPU depends critically on the card in your computer. So, a first thing to do, is to determine what graphics card you have on board. If you have invested in a good GPU, you will know what model you have, otherwise, you can use the command below (not in the R-console, but in the CLI of the OS).

```
sudo lshw -numeric -C display
```

The output can look as follows:

```
*-display
       description: VGA compatible controller
       product: Crystal Well Integrated Graphics Controller [8086:D26]
       vendor: Intel Corporation [8086]
       physical id: 2
       bus info: pci@0000:00:02.0
       version: 08
```

```
      width: 64 bits
      clock: 33MHz
      capabilities: msi pm vga_controller bus_master cap_list rom
      configuration: driver=i915 latency=0
      resources: irq:30 memory:f7800000-f7bfffff memory:e0000000-efffffff
   ioport:f000(size=64) memory:c0000-dffff
```

This result means that there is actually no GPU other than the one that Intel has added to the CPU. This means that for common rendering tasks the CPU will be able offload some tasks to this card, however, we cannot expect much gain. No dedicated GPU means that on this computer it is not an alternative to program the GPU.

If you have a dedicated GPU, the output will provide information about the GPU:

```
  *-display
description: VGA compatible controller
product: NVIDIA Corporation [10DE:2191]
vendor: NVIDIA Corporation [10DE]
physical id: 0
bus info: pci@0000:01:00.0
version: a1
width: 64 bits
clock: 33MHz
capabilities: pm msi pciexpress vga_controller bus_master cap_list rom
configuration: driver=nvidia latency=0
resources: irq:154 memory:b3000000-b3ffffff memory:a0000000-afffffff memory:
    b0000000-b1ffffff ioport:4000(size=128) memory:c0000-dffff
```

If you have a dedicated GPU, then chances are that it is an AMD or an Nvidia device, these are the two market leaders. The AMD GPU will understand the open source OpenCL framework, and a modern Nvidia GPU will use the proprietary CUDA framework (but will also understand OpenCL)

Digression – CUDA or OpenCL?

Most papers will find a performance advantage in CUDA (e.g. Karimi et al. (2010)). However, Fang et al. (2011) find that "for most applications, CUDA performs at most 30% better than OpenCL. We also show that this difference is due to unfair comparisons: in fact, OpenCL can achieve similar performance to CUDA under a fair comparison."

Both systems have strong and weak points. Usually, Nvidia will give a better performance, the CUDA platform is readily integrated in C, C++, and Fortran; and there are a handful of libraries for R available that take care of the complexity of programming the GPU.[6]

Hint – Nvidia

If you have an Nvidia GPU that understands the CUDA framework, then you might want to have a look at the packages `gputools`, `cudaBayesreg`, `HiPLARM`, `HiPLARb`, and `gmatrix` that are specifically designed with the CUDA framework in mind.

[6]CUDA also supports programming frameworks such as OpenACC and OpenCL, which are a little more user friendly to use than the older platforms.

37.3.1 Getting Started with gpuR

OpenCL

ViennaCL

OEM and AMD cards will typically use the OpenCL framework – which is also understood by a Nvidia GPU. A good place to start is the package gpuR. This library uses `ViennaCL library`[7] This library has been made available in the `RViennaCL` package (which is rather designed to be used in other R-packages). The package gpuR allows the user to access the GPU via OpenCL on – virtually – any GPU.

The package comes actually in two variants. The twin brother `gpuRcuda` is the bridge for gpuR to use the CUDA, which can work better if you have an Nvidia GPU. This is important because it will work significantly faster.

> **Warning – Use the memory of the GPU**
>
> Be sure to read the documentation of the package of your choice. Most packages will not use the memory of the GPU, and this increases transfer times. This means that copying variables will not be so time and memory friendly as in base R.

Once installed, the packages are usually very user friendly. For example, the following code is enough to use the GPU to calculate a matrix product.

```
# Install.packages("gpuR")  # do only once
library(gpuR)               # load gpuR
## Number of platforms: 1
## - platform: NVIDIA Corporation: OpenCL 1.2 CUDA 10.1.0
##   - context device index: 0
##     - GeForce GTX 1660 Ti
## checked all devices
## completed initialization

## gpuR 2.0.3
## Attaching package: 'gpuR'
## The following objects are masked from 'package:base':
## colnames, pmax, pmin, svd

detectGPUs()                # check if it all works
## [1] 1

# Prepare an example:
N <- 516
A <- matrix(rnorm(N^2), nrow = N, ncol = N)
gpuA <- gpuMatrix(A)        # prepare the matrix to be used on GPU


# In base R we could do this:
B <- A %*% A

# gpuR works as one would expect:
gpuB <- gpuA %*% gpuA

# note its structure:
gpuB
## An object of class "fgpuMatrix"
## Slot "address":
## <pointer: 0x564551a5d190>
##
## Slot ".context_index":
## [1] 1
##
## Slot ".platform_index":
## [1] 1
```

[7] See: `http://viennacl.sourceforge.net`.

```
##
## Slot ".platform":
## [1] "NVIDIA CUDA"
##
## Slot ".device_index":
## [1] 1
##
## Slot ".device":
## [1] "GeForce GTX 1660 Ti"
```

Note – Milage may vary

Unlike other sections in this book, do not expect output to be exactly the same when you execute the code on your computer. This chapter tests the hardware and not numeric logic, and hence the results will vary depending on the hardware that you're using.

We refer to the Chapter 40.1 *"Benchmarking"* on page 794 to compare the speed difference. Here we will simply use `system.time()` to compare the performance.

```
# base R:
system.time(B <- A %*% A)
##    user  system elapsed
##   0.052   0.000   0.051

# gpuR works exactly as one would expect:
system.time(gpuB <- gpuA %*% gpuA)
##    user  system elapsed
##   0.022   0.000   0.023
```

The time gain, when using the GPU, depends of course on the size of the matrix. This is illustrated with the following code. We run an experiment,[8] where we test different matrix sizes and then plot the times. The results are in Figure 37.1 on page 756.

```
set.seed(1890)
NN <- seq(from = 500, to = 4500,by = 1000)
t  <- data.frame(N = numeric(), CPU = numeric(), GPU = numeric())
i <- 1
for(k in NN) {
  A <- matrix(rnorm(k^2), nrow = k, ncol = k)
  gpuA <- gpuMatrix(A)
  t[i,1] <- k
  t[i,2] <- system.time(B <- A %*% A)[3]
  t[i,3] <- system.time(gpuB <- gpuA %*% gpuA)[3]
  i <- i + 1
}
# Print the results
t
##      N    CPU   GPU
## 1  500  0.047 0.006
## 2 1500  1.597 0.057
## 3 2500  8.634 0.285
## 4 3500 22.995 0.686
## 5 4500 48.545 1.094

# Tidy up the data-frame:
library(tidyr)
tms <- gather(t, 'PU', 'time', 2:3)
```

[8]Please refer to the section Chapter 40.1 *"Benchmarking"* on page 794 for more details on how to run more reliable tests. For now, it is sufficient to realize that the same code can run at different times. This depends for example on other processes running. Therefore, our experiment can show minor deviations.

```
# Plot the results:
library(ggplot2)
p <- ggplot(tms, aes(x = N, y = time, colour = PU)) +
     geom_point(size=5) +
     geom_line() +
     xlab('Matrix size (number of rows and columns)') +
     ylab('Time in seconds') +
     theme(axis.text = element_text(size=12),
           axis.title = element_text(size=14)
           )
print(p)
```

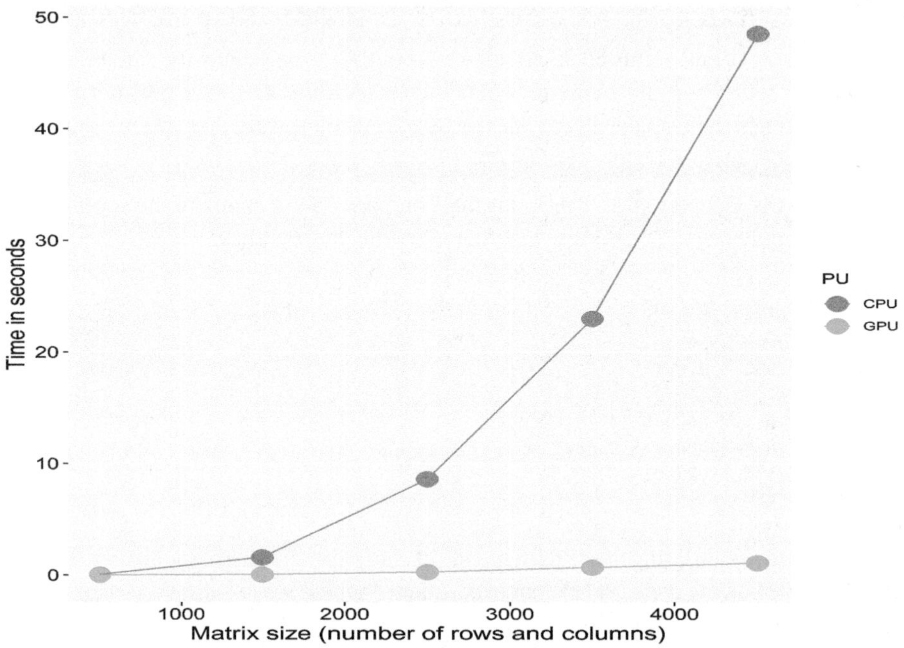

Figure 37.1: *The runtimes for matrix multiplication compared on the CPU versus the GPU.*

Hint – Cleaning up

After such experiment, we have not only large matrices in the memory, but also cluttered memory in the RAM of the GPU (at least this is the case with gpuR version 2.0.3). Therefore, it is a good idea to clean up:

```
# Remove a list of selected variables:
rm (list = c('A','B', 'gpuA', 'gpuB', 'vlcA', 'vlcB'))

# Alternatively remove *all* variables previously defined (be careful!):
rm(list = ls())

# Run the garbage collector:
gc()

# If not longer needed, unload gpuR:
detach('package:gpuR', unload = TRUE)
```

The usability of a GPU programming library depends on the calculations that one can spread over the cores of the graphics card. The library gpuR also provides methods for basic arithmetic and calculus such as: `%*%`, `+`, `-`, `*`,`/`, `t`, `crossprod`, `tcrossprod`, `colMeans`, `colSums`, `rowMeans`, `rowSums`, `sin`, `asin`, `sinh`, `cos`, `acos`, `cosh`, `tan`, `atan`, `tanh`, `exp`, `log`, `exp`, `abs`, `max`, `min`, `cov`, `eigen`. There are many other operations available such as Euclidian distance, etc.

With all those functions gpuR is one of the most complete, but it is also one of the most universal GPU programming options available for the R programmer.

Further information – Machine learning libraries

Most of the packages and interfaces focus on matrix algebra or at least will include this. That makes sense, since it is close to the primordial task of GPUs. The power of a GPU can be extended to much more functions such as deep learning, adiabatic quantum annealing, clusters, distances, optimisation problems, cross validations, and many more. Nvidia, for example, is now also offering frameworks for pre-fitted artificial intelligent solutions (such as for interpolating pixels). To learn more, we refer to their website: `https://www.nvidia.com/en-us/gpu-cloud`.

matrix

These simple experiments show that working with a GPU is not too difficult in R, but more importantly that significant performance gains can be obtained in some cases.

37.3.2 On the Importance of Memory use

The advantage of doing the calculations on the GPU is a balance between the time lost to transfer the object to the GPU and back to the CPU RAM, and the time gained by executing the calculations distributed over the many cores of the GPU. The package gpuR provides another type of matrix – created with `vclMatrix()` – that keeps the matrix on the RAM of the GPU. This means that we only transfer the matrix once: when we use the function `vclMatrix()`, we push the matrix in the memory of the GPU and it will stay there, ready to be used on the GPU.

We illustrate how this works, by modifying previous example. The results are in: Figure 37.2 on page 758.

```
require(gpuR)
require(tidyr)
require(ggplot2)
set.seed(1890)
NN <- seq(from = 500, to = 5500,by = 1000)
t  <- data.frame(N = numeric(), `CPU/CPU` = numeric(),
                 `CPU/GPU` = numeric(), `GPU/GPU` = numeric())
i <- 1

# Run the experiment
for(k in NN) {
A <- matrix(rnorm(k^2), nrow = k, ncol = k)
                    # storage in CPU-RAM, calculations in CPU
gpuA <- gpuMatrix(A) # storage in CPU-RAM, calculations in GPU
vclA <- vclMatrix(A) # storage in GPU-RAM, calculations in GPU
t[i,1] <- k
t[i,2] <- system.time(B <- A %*% A)[3]
t[i,3] <- system.time(gpuB <- gpuA %*% gpuA)[3]
t[i,4] <- system.time(vclB <- vclA %*% vclA)[3]
```

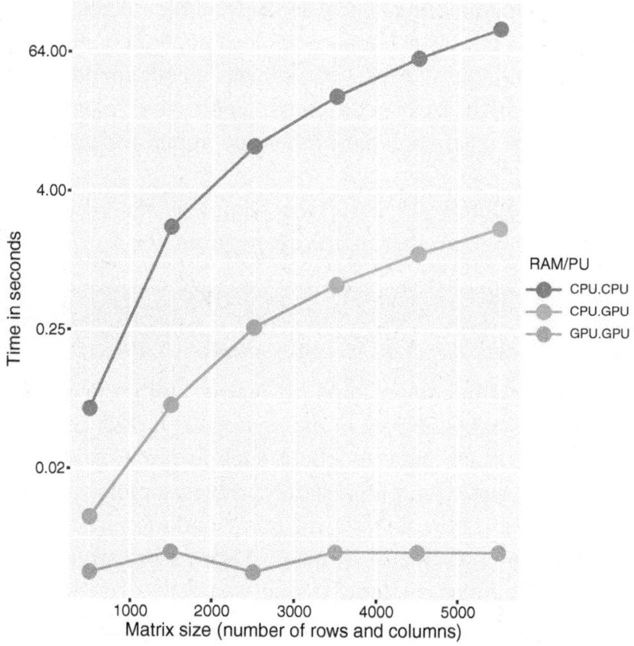

Figure 37.2: *The runtimes for matrix multiplication for storage/calculation pairs: CPU/CPU, CPU/GPU, GPU/GPU. Note that the scale of the y-axisis logarithmic.*

```
i <- i + 1
}

# Print the results
t
##       N CPU.CPU CPU.GPU GPU.GPU
## 1  500   0.047   0.005   0.004
## 2 1500   1.635   0.054   0.018
## 3 2500   8.088   0.244   0.003
## 4 3500  22.172   0.577   0.003
## 5 4500  47.235   1.111   0.003
## 6 5500  86.114   1.843   0.003

# Tidy up the data-frame:
tms <- gather(t, 'RAM/PU', 'time', 2:4)

# Plot the results:
scaleFUN <- function(x) sprintf("%.2f", x)

p <- ggplot(tms, aes(x = N, y = time, colour = `RAM/PU`)) +
  geom_point(size=5) +
  geom_line() +
  scale_y_continuous(trans = 'log2', labels = scaleFUN) +  # NEW!!
  xlab('Matrix size (number of rows and columns)') +
  ylab('Time in seconds') +
  theme(axis.text = element_text(size=12),
  axis.title = element_text(size=14)
)
print(p)
```

```
# unload gpuR:
detach('package:gpuR', unload = TRUE)
```

The performance gain in this simple example is massive. For a multiplication of two matrices of 5 500 by 5 500, the `vclMatrix` is about 50 times faster than the `gpuMatrix`. If we compare the saem `vclMatrix` with the CPU alternative, then it is about 30 000 times faster. Taking into account the low effort that is needed to learn the package `gpuR`, we must conclude that it really pays of to use the GPU.

Digression – Deviations in execution time

The use of a seed – with `set.seed()` – is not sufficient to make the speed-experiment reproducible. All the measures of speed in this section are just one experiment. The run-time of any given code is actually a stochastic variable. Each time we run it, the time that it runs will be different. It depends on many small and big things, such as the memory used, the temperature of the CPU, the dynamic response of the clock speed, the time used by other code, memory fragmentation, etc. More about benchmarking in Section 40.1 *"Benchmarking"* on page 794.

37.3.3 Conclusions for GPU Programming

Both the mathematics behind rendering a landscape and fitting a neural network include matrix calculus. The GPU that was originally designed to aid stunning visuals in computer games offers a relatively cheap cluster of PUs that can be programmed for any distributed calculation task. We have shown that even for relatively small matrices the gain can be truly massive, for very little effort with the dedicated R libraries.

There are many other algorithms, such as various forms of clustering, deep learning, cross validation, Monte Carlo simulations, quantum annealing, that can benefit to various extends of a GPU. Other algorithms, such as networking analysis or searching in unstructured data might easier get bogged down by the data transfer times. To facilitate problems, that include really large data, we need another solution: massive, resilient distributed data with distributed (or local) PUs. That is the subject of Chapter 38 *"R and Big Data"* on page 761.

There are also processes, that require one thread, and hence, cannot be ran on distributed PUs. However, many algorithms allow distribution, and whenever we can use the GPU, it is low hanging fruit to speed up code. If your code is running slow, it is a good excuse to invest in a GPU ...and then you could also allow yourself to be seduced to play an occasional computer game.

Further information – Advanced GPU programming

In this section, we have only presented the option to use an R package in order to program the GPU. Knowledge of R is sufficient. The package `gpuR` has a lot of many functions available, but if you need something that is not there, you might want to use the library `OpenCL` and create your own functions. A good introduction is here: `https://cnugteren.github.io/tutorial`. However note that knowledge of C or C++ is required.

Hint – Multiple GPUs

A good GPU offers thousands of PUs, but did you know that it is possible to have more than one GPU in your desktop computer? For example two identical video cards can be bridged together. Nvidia calls their solution SLI, while the AMD version is called Crossfire.

Further information – Nvidia's CUDA

If you are using an Nvidia GPU, then you should read `https://devblogs.nvidia.com/accelerate-r-applications-cuda`. This tutorial shows how to accelerating R with the CUDA libraries, call bespoke parallel algorithms written in CUDA C/C++ or CUDA Fortran from R, and adds information on profiling GPU-accelerated R applications via the CUDA Profiler.

♣ 38 ♣

R and Big Data

In the rest of this book we implicitly assume that the data can be loaded in memory. However, imagine that we have a dataset of all the data that you have transmitted over your mobile phone, including messages, images, complete with timestamps and location that results from being connected. Imagine now that we have this dataset for 50 million customers. This dataset will be so large that this becomes impractical, or even impossible to store on one computer. In that case, it is fair to speak of "big data."

Usually, the academic definition of big data implies that the data has to be big in terms of **big data**

- velocity: the speed at which the data comes in,
- variety: the number of columns and formats,
- veracity: the reliability or data quality, and
- volume: the amount of data.

Commercial institutions will add "value" as a fourth word that starts with the letter "v." While this definition of "big data" has its merits, we rather opt for a very practical approach. We consider our data to be "big" if it is no longer practically possible to store all data on one machine and/or use all processing units (PUs)[1] of that one machine to do all calculations (such as calculating a mean or fitting a neural network).

memory limit

Note – Memory limits in R

R is in the first place limited to the amount of memory that the OS allocates to R. A 32-bit system has a limit of something like 3GB, a 64-bit system can allocate a lot more. However, – by default – it is not really possible to manage data that has the number of rows multiplied with the number of columns more than $2^{31} - 1 = 2\,147\,483\,647$.

When our data is too big to be handled elegantly in the RAM of our computer, we can follow the following steps to get the analysis done.

[1]With "all PUs", we mean "all the cores of the CPU(s) and all the cores on the GPU(s) connected to that CPU array". In this sentence it is equivalent to say that you have tried to make your computer as powerful as possible and pushed the limits of what you can do on one machine. The thing is that if the data is too large, no array of processors will be sufficient, that's what we intend to call "big data."

The Big R-Book: From Data Science to Learning Machines and Big Data, First Edition. Philippe J.S. De Brouwer.

Companion Website: www.wiley.com/go/De Brouwer/The Big R-Book

1. Can the data be reduced at read-in. Maybe you do not need to copy the whole table, but only some columns? To obtain this, do as much as possible data wrangling on the RDBMS (or NoSQL server). If this does not solve the problem, go to next point — see Part III *"Data Import"* on page 213.

2. Is it possible to take a sub-set of the data, do the modelling and then check if the model can be generalized to the whole database? Take a sample of the data that can be handled, make the model and then calculate the model performance on the whole table. If the results are not acceptable, move to the next step — see e.g. Chapter 21.3 *"Performance of Regression Models"* on page 384 and Chapter 25 *"Model Validation"* on page 475.

3. Would the data fit on a more powerful computer? This can be the computer of a friend, a server at the workplace of a supercomputer. Chapter 38.1 *"Use a Powerful Server"* on page 763 is a reference about this solution.

4. Force R to work a little different, and do not try to load all data in memory, but rather work with a moving window on the data and leave long tables on the hard-drive. This will – usually – slow down the calculations, but it might just be possible to get the calculations done. This can be done by packages like `ff`, `Biglm`, `RODBC`, `snow`, etc. We will discuss this in Chapter 38.2 *"Using more Memory than we have RAM"* on page 765. This is typically a solution for a few Gigabytes of data. It your data is rather petabytes, then move on to the next step.

5. Use parallelism in data-storage and calculations, and then distribute data and calculations over different nodes with an appropriate ecosystem to support this massive parallel processing and distributed, resilient file-system. This will be explained in Chapter 39 *"Parallelism for Big Data"* on page 767.

38.1 Use a Powerful Server

38.1.1 Use R on a Server

When your personal computer fails to get the task done, it might be possible for a larger computer to handle the task. Depending on the exact problem, it might be necessary to install more RAM, more cores, more disk space or a combination of all of the above.

The word "server" can refer to a computer dedicated to provide a specific service in the network (ie. serve other working nodes). We are aware that this is not necessarily a powerful computer. For example, it is possible to use a Raspberry computer as file server or mail server. In this section, we reserve the word "server" for a powerful machine.

It is also possible to use R as a service that runs on a server. R can be installed on the server and you can login remotely or use `ssh` to open a terminal that is visible on your screen but only reflects the CLI of the server. With R installed there, you can access the data that sits there and potentially benefit of the large array of disks that is probably expressed in giga-bytes and maybe even in peta-bytes.

Hint – RStudio

If you do not like the command line too much, then you might want to check "RStudio Server." It provides a browser based interface to a version of R running on a remote Linux server. So, it will look as if you use the RStudio IDE locally, but actually you are using the power and storage capacity of the server.

RStudio Server

38.1.2 Let the Database Server do the Heavy Lifting

In a first approach of taming big data, we can try if the big-data problem can be reduced to a small data problem by leaving all unnecessary data on a database-server and doing the data wrangling as much as possible on the database server. The techniques, packages and work-flow related to the data importing and wrangling are explained in Part III *"Data Import"* on page 213 and Part IV *"Data Wrangling"* on page 259. If this can be done, then the problem is solved, but when that data wrangling becomes too slow in R, then we might want to be more selective in what we download and eventually prepare the data better.

Most of the data-wrangling can be done directly on the server or via R. If we choose to do this via R, then there are largely two approaches.

1. Connect to the server – as explained in Chapter 15 *"Connecting R to an SQL Database"* on page 253 – and run commands native to the SQL-server.

2. Using the library `dplyr` to generate the SQL code for you and execute the query only when needed. This is the subject of Chapter 17 *"Data Wrangling in the tidyverse"* on page 265.

Since the "how to" part is explained earlier in this book, we will not elaborate on the subject further in this chapter.

Hint – A general package to talk ODBC

RODBC
ODBC

Some databases have their own drivers for R, others not. It might be a good idea to have a look at the package `RODBC`. It is a package that allows to connect to many databases via the `ODBC` protocol.[a]

[a] ODBC stands for Open Database Connectivity and is the standard API for many DBMS.

RStudio
dplyr

Further information – A nice intro from RStudio

Did you know that `dplyr` is made available by RStudio and that they have further great documentation on the subject? More information is on their website: `https://www.rstudio.com/resources/webinars/working-with-big-data-in-r`.

38.2 Using more Memory than we have RAM

If the problem is that the data is too large to keep in memory, but calculation times are (or would be) still acceptable, then we can choose one of the libraries to change the default behaviour or R so that data is not – by default – all loaded into RAM.[2]

To the library `ff` provides data-structures that are stored on the disk rather than in RAM. When the data is needed, it will be loaded chunk after chunk. `ff`

There are also a few other packages that do the same, while targeting specific use. For example:

- `bigmemory` for large matrices, `bigmemory`
- `biglm` for building generalised linear models. `biglm`

For the purpose of this book, we will not discuss further those packages, but look into the next step: what if your hard-disk is no longer sufficient to store the date in the first place?

[2]Competing software such as the commercial SAS engine has by default the opposite behaviour. Hence, SAS is out of the box capable of handling really large data that would not fit in the RAM.

♣ 39 ♣

Parallelism for Big Data

Since the 1990s providers such as Terradata specialize in solutions that store data and then allow to operate parallel on massive amounts of data. CERN[1] on the other hand is a textbook example of an institution that handles big datasets, but they rather rely on super computers (high performance computers) instead of resilient distributed computing. In general – and for most applications – there are two solutions: one is resilient distributed computing where data and processing units (PUs) are commodity hardware and redundant, and the second is high performance computing with racks of high grade CPUs.[2]

One particularly successful solution is breaking up the data in parts and storing each part (with some redundancy built in) on a computer its own CPU and data storage. This allows to bring the calculations to the multiple parts of data instead of bringing all the data to the single CPU.

MapReduce

In 2004 Google published the seminal paper in which they described the process "MapReduce" that allows for a parallel processing model, to process huge amounts of data in parallel nodes. MapReduce will split queries and distributes them over the parallel nodes, where they are processed simultaneously (in parallel). This phase is called the "Map step." Once a node has obtained its results, it will pass them on to the layer higher. This is the "Reduce step," where the results are stitched together for the user.

This approach has many advantages:

- it is able to handle massive amounts of data in incredible speeds,
- because of the built-in redundancy it is very reliant and resilient,
- therefore, it is also possible to use cheaper hardware in the nodes (if one node is busy or not online, then the task will be dispatched to the alternate CPU that also has the same part of the data), and it becomes also a cost efficient solution.

The downside is that parallelism comes at an overhead cost of complexity. So, as long as the other methods work, big data solutions based on massive redundant parallel systems might not be the best choice. However, when data reaches sizes that are best expressed in peta-bytes, then this becomes the only viable solution that is fault tolerant, resilient, fast enough and affordable. At that point you will need to get specialist knowledge in the company to build and maintain the big data solution.

[1] CERN is the European Organization for Nuclear Research, it's website is here `https://home.cern` and the letters stand for "centre Européen de recherche nucléaire."

[2] This is also referred to as "supercomputer", as understood by `www.top500.org`.

The Big R-Book: From Data Science to Learning Machines and Big Data, First Edition. Philippe J.S. De Brouwer.

Companion Website: www.wiley.com/go/De Brouwer/The Big R-Book

However, it is worth noticing that R offers an intermediate solution with `snow`.[3] However, at some point that protocol will not be sufficient to provide the performance boost and storage capacity needed to finish the calculations in a practical time. Then it is time to think about massive resilient dataset and why not try to free and open source ecosystem of "Hadoop"?

[3]For a quick introduction to `snow` and references see Chapter 37.2 on page 748.

39.1 Apache Hadoop

The world of parallelism and big data becomes fast very complex. Fortunately there is the Apache Software Foundation. In some way, Apache is for the big-data community similar to what the RStudio community is for R-users. The structure is different, but both provide a lot of tools and software that can be used for free and in some way both entities produce tools that are so useful and efficient that they become unavoidable in their ecosystem.

Apache

Further information – Apache Sotware Foundation

The Apache Software Foundation is probably best known and loved for their http webserver "Apache." Today, the Apache Software Foundation is – with more than 350 active software projects – arguably the largest open software foundation. More information about the Apache Software Foundation is here: `https://www.apache.org`

MapReduce

Implementing MapReduce in a high volume, high speed environment requires to re-invent how a computer works and all components (that in a traditional operating system (OS) – such as Windows or Linux – are all on one computer) should now be distributed over many systems. Therefore, parallelism becomes fast very complex and the need for some standard imposes itself. Under the umbrella "Hadoop," the Apache Software Foundation has a host of software solutions that allows to build and operate a solution for manipulating and storing massive amounts of data.

Hadoop Common

The core of Hadoop is known as "Hadoop Common" and consists of:

- `HDFS` is the Hadoop Distributed File System is the lowest level of the ecosystem: it manages the physical storage of the distributed files in a redundant way.
- `Hadoop YARN` takes care of the cluster resource management.
- `Hadoop MapReduce` is the mapping and reducing engine.

The family of software that populate the Hadoop ecosystem is much larger than Hadoop Common. Here are some of the solutions that usually are found together.

- `Zookeeper:` is a high availability coordination and configuration service for distributed systems.
- `Apache Spark:` is an in-memory data-flow engine: it is a cluster computing framework for massive distributed data and provides an interface that makes abstraction of the data parallelism with built-in fault tolerance.
- `Apache Storm:` is the computation framework for the distributed stream processing.
- `Oozie:` is the workflow scheduling system that manages different Hadoop jobs for the client.
- *Query and data intelligence interfaces* that are designed to allow for massively parallel processing.
  - `Apache Pig:` allows to create programs that run on Hadoop via the high-level language "Pig Latin."

   - `HBase`: the non relational distributed database.
   - `Apache Phoenix`: the relational database engine (using HBase underneath).
   - `Apache Impala`: the SQL query engine.
   - `Mahout` and `Spark MLib`: machine learning libraries.
   - `Drill`: for interactive analysing data.

- `Sqoop`: allows to import data from relational databases
- `Apache Flume` takes care of the massive amounts of log-data.

In some way this ecosystem is the equivalent of a computer that allows to store data, query it, and run calculations on it. However, the data is not stored on one disk but distributed over a massive cluster of commodity hardware, the data can come in at high speeds, the amounts of data gathered are massive. Hadoop does all of that and is surprisingly fast, scalable, and reliable.

While R allows us to interface with different components of the Hadoop ecosystem, we will focus on Apache Spark in the next section because it offers – for R-users – probably the ideal combination of similar concepts and the right level of API.

39.2 Apache Spark

The Apache foundation succeeded to implement the MapReduce framework framework in an open-soruce solution "Hadoop." This soon became the industry standard because on top of all the advantages of the MapReduce algorithm one could rely on free software that was backed by a large and friendly community. However, Hadoop had limitations and implementation issues that pushed Apache to produce in 2012 "Spark." In essence, Spark allows for more complex steps than just the map and reduce steps. It is also free and open source.

Apache – Spark

Within the Hadoop project and the wider Hadoop ecosystem, Spark takes a special place with the resilient distributed dataset concept (RDD). It is a read-only equivalent of the data-frame in R, but it is designed to keep the dataset distributed over a cluster of machines.

RDD

> **Further information – Spark**
>
> More about spark can be found in its website: `https://spark.apache.org`

Apache Spark runs also a lot faster than Hadoop MapReduce.[4] Spark can be run stand alone or on a variety of platforms including Hadoop, Cassandra, and HBase. It has APIs for R, but also for other languages such as Python, SQL, and Scala.

In the remainder of this section we will illustrate how one can harvest the power of Spark via R. To test this, one will first need to install Spark. Of course, you will argue that it is of limited use to install Spark just on your own laptop computer. You're right, but we will do this here so that we have a self-contained solution that can be tested by everyone, and as a bonus when you need to connect to a real Hadoop cluster, then you will find that this works in a very similar way.

39.2.1 Installing Spark

Spark will require Java in order to install. Java is probably already installed on your system and you can test its presence via the command:

Java

```
java -version
```

If this produces some output, then Java is installed. If you do not have Java installed yet, install it for example on a Debian system as follows:

```
sudo apt install default-jdk
```

> **Warning – Installing Spark**
>
> It appears that the installation of Spark is still changing, so it might make sense to refer to the website of Spark before going ahead.

For now, it is not part of most packaging systems of most distributions. So we need to download it first. Best is to refer to `https://www.apache.org/dyn/closer.lua/spark/spark-2.4.3/spark-2.4.3-bin-hadoop2.7.tgz` and choose a mirror that will work best for you. For example we can use `wget` to download the files:

[4] The Apache website reports a speed-up of about 100 times.

```
mkdir ~/tmp
cd ~/tmp
wget https://www-eu.apache.org/dist/spark/spark-2.4.3/spark-2.4.3-bin-hadoop2
    .7.tgz
```

This will result in the file `park-2.4.2-bin-hadoop2.7.tgz` in the directory `~/tmp`. This is a compressed file, also known as "tarball." It can be decompressed as follows:

```
tar -xvf spark-2.4.3-bin-hadoop2.7.tgz
```

Hint – Installing via the Apache website

CLI

If you prefer not using the command line interface (CLI), then you might want to refer to the website of Spark and download it from there: `https://spark.apache.org/downloads.html` is your place to start then. Once the file is on your computer, you can access it via a file manager (such as Dolphin or Thunar), double click on it and you will be presented with a window that is a graphical user interface to decompress the file.

tarball

Now, you should have a subdirectory with the code that will be able to run Spark – in our case the directory is called `spark-2.4.3-bin-hadoop2.7`, and it is a sub-directory to the place where we have extracted the tarball (our `~/tmp`. This directory needs to be moved to the `/opt` directory. Note that in order to do that you will need administrator rights. Chances are that you do not have the directory `/opt` on your system yet and in this case you might want to opt to create it or put the binaries elsewhere (maybe in `/bin`)[5].

```
# Create the /opt directory:
sudo mkdir /opt

# Move Spark to that directory:
sudo mv spark-2.4.3-bin-hadoop2.7/ /opt/spark
```

Now, you should have the binaries of Spark in the `/opt/spark` directory. Test this as follows.

```
$ ls /opt/spark
bin      data       jars  LICENSE   NOTICE     R
RELEASE  yarn       conf  examples  kubernetes licenses
python   README.md  sbin
```

Take some time to go through the `READMDE.md` file. As the extension suggests, this is a MarkDown file. While it is readable as a text-file, it is better viewed via a markdown reader such as Grip, ReText, etc. For example:

```
retext /opt/spark/README.md
```

ReText

will open the ReText[6], but it still will not look good. Activate the live-preview by pressing the keys `CTRL + L`.

Spark will still not work: some manual configuration is still necessary. We need to add the directory of Spark to the system path (the directories where the system will look for executable

[5] Keeping Spark in a separate directory has the advantage that it will be easy to identify, remove, and update it if necessary. The documentation on the Apache website recommends to use a directory called `/opt`.

[6] If necessary, install it first. On a Debian OS, this is don by the usual `sudo apt install retex` in the CLI.

files). This can be done by editing the ~/.bashrc file. Any text-editor will do, we like vi, but you can use any other text editor.

vi

```
vi ~/.bashrc
```

Working in vi requires some practice (it pre-dates GUI interfaces), so we describe here how to get things done. At the very end of the file we need to add the two following lines. To do this: copy the text below, then press the following keys [CTRL]+[END], A, [ENTER], CTR+V, [ESC], :wq and finally [ENTER].

```
export SPARK_HOME=/opt/spark
export PATH=$PATH:$SPARK_HOME/bin:$SPARK_HOME/sbin
```

Save and quit (in vi this would be ESC, :wq, ENTER). Activate the changes by opening a new terminal or executing

```
source ~/.bashrc
```

Digression – Spark installer

We thank RStudio and Microsoft to have created and made available a useful cross-platform installer for Apache Spark. This installer is designed to use system resources efficiently under a common API. It will support as well R as Python.

```
# install from github
devtools::install_github(repo = "rstudio/spark-install",
                         subdir = "R")
library(sparkinstall)

# lists the versions available to install
spark_available_versions()

# installs an specific version
spark_install(version = "2.4.3")

# uninstalls an specific version
spark_uninstall(version = "2.4.3", hadoop_version = "2.6")
```

sparkinstall

39.2.2 Running Spark

Spark is a high-end, high-capacity engine and not an interactive tool as most of the other software that we use on a daily basis. Just as for a file-server we need to activate the “daemon”.[7] This can be done with the command:

daemon

```
start-master.sh
```

The system will reply with the location of the log-file and return to the command prompt. It might seem that not too much happened, but Spark is really active on the system now and ready to reply to instructions. We can use the CLI to interrogate the system status or use a web-browser and point it to http://localhost:8080. The browser-interface will look similar to the one in Figure 39.1 on page 774.

[7] A daemon a software that runs on a computer but is rather invisible to the normal user. Daemons provide services, and other software can communicate with them.

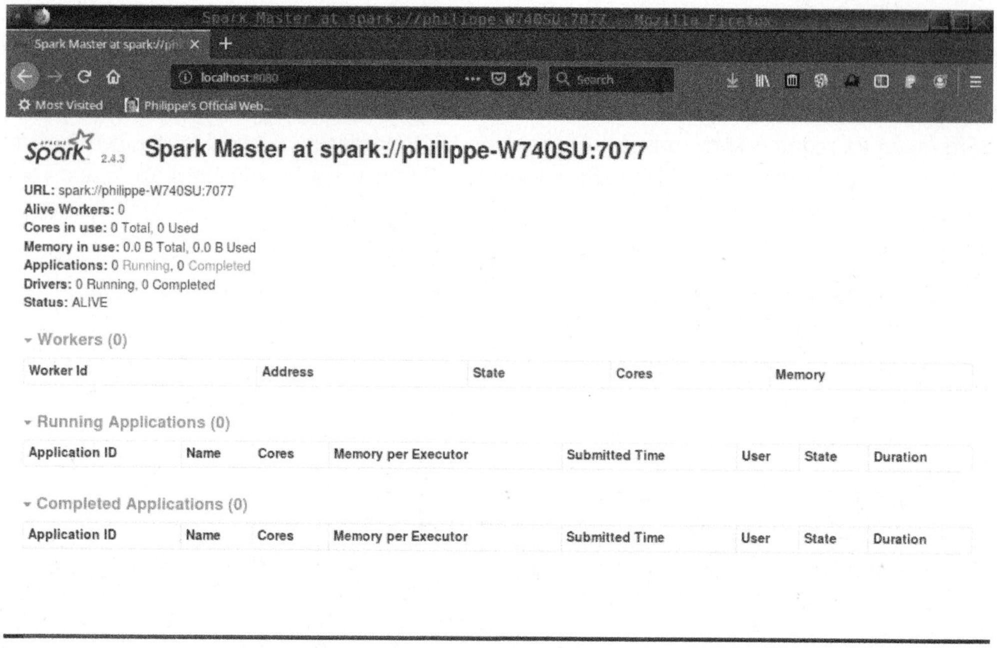

Figure 39.1: *The status of Spark can be controlled via a regular web-browser, that is directed to* `http://localhost:8080`.

There is – of course – no need to use a web-browser nor even have a windows manager active: the status of Spark can also be checked via the CLI. This can be done via the command `ss`.

```
$ ss -tunelp | grep 8080
tcp    LISTEN     0      1         :::8080                 :::*
      users:(("java",pid=26295,fd=351)) uid:1000 ino:5982022 sk:1d v6only:0
    <->
```

Now, we are sure that the Spark-master is up and running and can be used on our system. So, we can start a Spark-slave via the shell command `start-slave.sh` or via the Spark-shell. This shell can be invoked via the command `spark-shell` and will look as follows.

```
$ spark-shell
19/07/14 21:16:43 WARN Utils: Your hostname, philippe-W740SU resolves to a
    loopback address: 127.0.1.1; using 192.168.100.120 instead (on interface
    enp0s25)
19/07/14 21:16:43 WARN Utils: Set SPARK_LOCAL_IP if you need to bind to another
     address
19/07/14 21:16:44 WARN NativeCodeLoader: Unable to load native-hadoop library
    for your platform... using builtin-java classes where applicable
Using Spark's default log4j profile: org/apache/spark/log4j-defaults.properties
Setting default log level to "WARN".
To adjust logging level use sc.setLogLevel(newLevel). For SparkR, use
    setLogLevel(newLevel).
Spark context Web UI available at http://192.168.100.120:4040
Spark context available as 'sc' (master = local[*], app id = local
    -1563131811133).
Spark session available as 'spark'.
```

```
Welcome to
      ____              __
     / __/__  ___ _____/ /__
    _\ \/ _ \/ _ `/ __/  '_/
   /___/ .__/\_,_/_/ /_/\_\   version 2.4.3
      /_/

Using Scala version 2.11.12 (IBM J9 VM, Java 1.8.0_211)
Type in expressions to have them evaluated.
Type :help for more information.

scala>
```

Read this welcome message, because it is packed with useful information. For example, it refers you to a web user interface (in our case at `http://192.168.100.120:4040`). Note that when you work on the computer that has the Spark-master installed and running, that this is always equivalent to `http://localhost:4040`. The screen of this page is shown in Figure 39.2

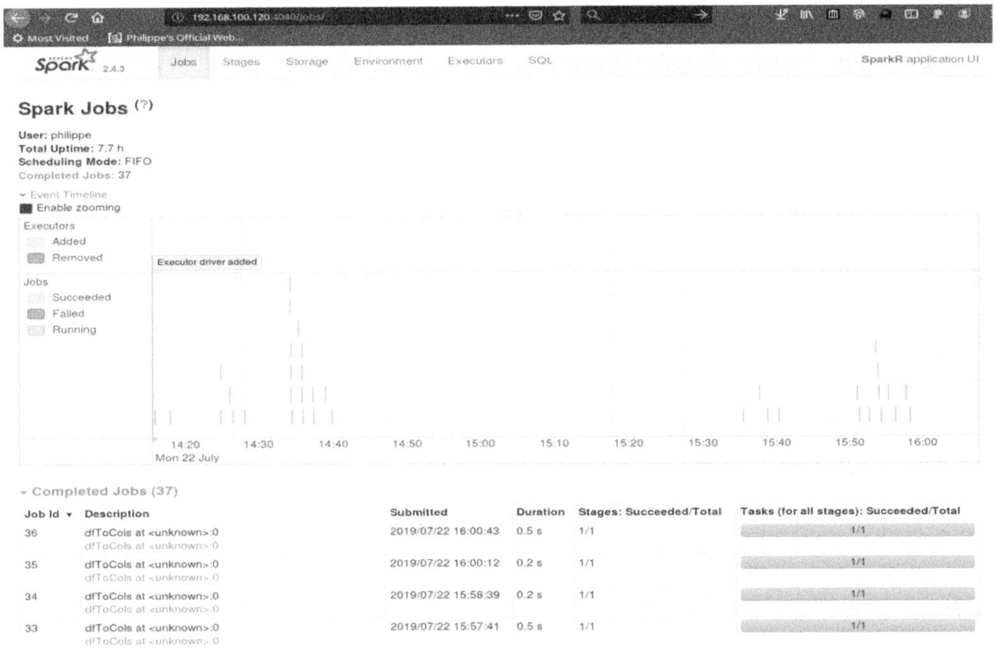

Figure 39.2: *Each user can check information about his own Spark connection via a web-browser directed to* `http://localhost:4040`*. Here we show the "jobs" page after running some short jobs in the Spark environment.*

Digression – Scala

Note that the spark command prompt is `scala>`. Scala is a high level, general purpose language. It saw first light in 2014 as an alternative to Java, and it addressed some of the criticism on Java. Scala code can be compiled to Java bitmode and hence can run in the Java Virtual Machine, and hence the same code can be executed on most modern operating systems.

In order to test this shell environment, we can execute the "Hello world" program as follows.

```
scala> println("Hello World!")
Hello World!

scala>
```

To leave the Spark-shell, type CTRL+C.

Instead of using the spark-shell, we will now keep the Spark-master running and go back to R.

Note – Stopping Spark

To stop respectively the Spark master and slave, use the following commands:

```
stop-slave.sh
stop-master.sh
```

Finally, note that we can also start Spark from the R-prompt. The following command will start the daemon for Spark and then return the R-command prompt to normal status ready for further input.

```
system('start-master.sh')
```

39.2.3 SparkR

As usual in R, there is more than one way to get something done. To address Apache Spark, there are two main options: SparkR from the Apache Software Foundation and sparklyr from RStdio. For a comparison between those two solutions, we refer to Chapter 39.2.5 *"SparkR or sparklyr"* on page 791. In this first section, we will describe the basics of SparkR.

SparkR

SparkR provides a new data type: the DataFrame. As the data.frame in base R, it holds rectangular data, but it is designed so that the data frame resides on a resilient and redundant cluster of computers and not be read in memory at once.

The Spark data frame is, however, also designed to be user friendly and most functions of dplyr have an equivalent in SparkR. This means that what we have learned in Chapter 17 *"Data Wrangling in the tidyverse"* on page 265 can be used with minor modifications and all the additional complexity will be handled in the background.

To start, load all libraries that we will use, connect to the Spark-Master and define our first SparkDataFrame, with the function as.DataFrame().

SparkDataFrame

```
library(tidyverse)
library(dplyr)
library(SparkR)
# Note that loading SparkR will generate many warning messages,
# because it overrrides many functions such as summary, first,
# last, corr, ceil, rbind, expr, cov, sd and many more.

sc <- sparkR.session(master = "local", appName = 'first test',
                     sparkConfig = list(spark.driver.memory = '2g'))
## Launching java with spark-submit command /home/philippe/.cache/spark/spark-2.4.4-
## bin-hadoop2.7/bin/spark-submit   --driver-memory "2g" sparkr-shell /tmp/Rtmpc5L8yL/
## backend_port5f87d94f2ef
```

```
# Show the session:
sc
## Java ref type org.apache.spark.sql.SparkSession id 1

DF <- as.DataFrame(mtcars)

# The DataFrame is for big data, so the attempt to print all data,
# might surprise us a little:
DF
## SparkDataFrame[mpg:double, cyl:double, disp:double, hp:double, drat:double, wt:double,
## qsec:double, vs:double, am:double, gear:double, carb:double]

# R assumes that the data-frame is big data and does not even
# start printing all data.

# head() will collapse the first lines to a data.frame:
head(DF)
##    mpg cyl disp  hp drat    wt  qsec vs am gear carb
## 1 21.0   6  160 110 3.90 2.620 16.46  0  1    4    4
## 2 21.0   6  160 110 3.90 2.875 17.02  0  1    4    4
## 3 22.8   4  108  93 3.85 2.320 18.61  1  1    4    1
## 4 21.4   6  258 110 3.08 3.215 19.44  1  0    3    1
## 5 18.7   8  360 175 3.15 3.440 17.02  0  0    3    2
## 6 18.1   6  225 105 2.76 3.460 20.22  1  0    3    1
```

Hint – Adding checkpoints

The central data-structure in Spark is called RDD ("Resilient Distributed Dataset"). The resilience is obtained by the fact that the dataset can at any point be restored to a previous state via the "RDD lineage information." The lineage information refers to the additional information that RDD stores about the origin of each dataset and each number.
This lineage of RDD can grow fast and can become problematic if we use the same data multiple time (e.g. in a loop). In that case, it is useful to add "checkpoints."
The function `checkpoint()` (from `SparkR`) will clear all the lineage information and just keep the data.

```
# If DDF is a distributed DataFrame defined by SparkR,
# we can add a checkpoint as follows:
DDF <- SparkR::checkpoint(DDF)
```

lineage
RDD

checkpoint

Note – Our convention

We will use capitals to represent a SparkDataFrame and small letters to represent a data.frame from base-R.

To develop an example, we will use the well known database `titanic`. This data was introduced in Chapter 22.2 *"Performance of Binary Classification Models"* on page 390. In the following code, we create the `DataFrame` and show some of its properties.

```
library(titanic)
library(tidyverse)
```

```
# This provides a.o. two datasets titanic_train and titanic_test.
# We will work further with the training-dataset:
T <- as.DataFrame(titanic_train)

# The SparkDataFrame inherits from data.frame, so most functions
# work as expected on a DataFrame:
colnames(T)
##  [1] "PassengerId" "Survived"    "Pclass"
##  [4] "Name"        "Sex"         "Age"
##  [7] "SibSp"       "Parch"       "Ticket"
## [10] "Fare"        "Cabin"       "Embarked"

str(T)
## 'SparkDataFrame': 12 variables:
##  $ PassengerId: int 1 2 3 4 5 6
##  $ Survived   : int 0 1 1 1 0 0
##  $ Pclass     : int 3 1 3 1 3 3
##  $ Name       : chr "Braund, Mr. Owen Harris" "Cumings, Mrs. John Bradley
##  (Florence Briggs Thayer)" "Heikkinen, Miss. La
##  $ Sex        : chr "male" "female" "female" "female" "male" "male"
##  $ Age        : num 22 38 26 35 35 NA
##  $ SibSp      : int 1 1 0 1 0 0
##  $ Parch      : int 0 0 0 0 0 0
##  $ Ticket     : chr "A/5 21171" "PC 17599" "STON/O2. 3101282" "113803"
##  "373450" "330877"
##  $ Fare       : num 7.25 71.2833 7.925 53.1 8.05 8.4583
##  $ Cabin      : chr "" "C85" "" "C123" "" ""
##  $ Embarked   : chr "S" "C" "S" "S" "S" "Q"

summary(T)
## SparkDataFrame[summary:string, PassengerId:string, Survived:string,
## Pclass:string, Name:string, Sex:string, Age:string, SibSp:string, Parch:string,
## Ticket:string, Fare:string, Cabin:string, Embarked:string]

class(T)
## [1] "SparkDataFrame"
## attr(,"package")
## [1] "SparkR"

# The scheme is a declaration of the structure:
printSchema(T)
## root
##  |-- PassengerId: integer (nullable = true)
##  |-- Survived: integer (nullable = true)
##  |-- Pclass: integer (nullable = true)
##  |-- Name: string (nullable = true)
##  |-- Sex: string (nullable = true)
##  |-- Age: double (nullable = true)
##  |-- SibSp: integer (nullable = true)
##  |-- Parch: integer (nullable = true)
##  |-- Ticket: string (nullable = true)
##  |-- Fare: double (nullable = true)
##  |-- Cabin: string (nullable = true)
##  |-- Embarked: string (nullable = true)

# Truncated information collapses to data.frame:
T %>% head %>% class
## [1] "data.frame"
```

Note – SparkDataFrame or DataFrame

The documentation of Spark, `https://spark.apache.org/docs/latest/index.html` only mentions "DataFrame", however many other people will use the wording SparkDataFrame, the function to create one is `as.DataFrame()`, and it reports on itself as a `SparkDataFrame`. Both terms are used and we will also use them interchangeably.

The data manipulation capacities of `SparkR` are modelled on `dplyr` and `SparkR` over-rides most of `dplyr`'s functions so that they will work on a SparkDataFrame. This means that for the R-user, all functions from `dplyr` work as expected.[8]

`dplyr`

For example, selecting a row can be done in a very similar way as in base-R. The following lines of code do all the same: select one column of the Titanic data and then show the first ones.

```
X <- T %>% SparkR::select(T$Age)          %>% head
Y <- T %>% SparkR::select(column('Age')) %>% head
Z <- T %>% SparkR::select(expr('Age'))    %>% head
cbind(X, Y, Z)
##   Age Age Age
## 1  22  22  22
## 2  38  38  38
## 3  26  26  26
## 4  35  35  35
## 5  35  35  35
## 6  NA  NA  NA
```

The package `dplyr` offers SQL-like functions to manipulate and select data, and this functionality works also on a DataFrame using `SparkR`. In many cases, this will allow us to reduce the big data problem to a small data problem that can be used for analysis.

For example, we can select all young males that survived the Titanic disaster as follows:

```
T %>%
  SparkR::filter("Age < 20 AND Sex == 'male' AND Survived == 1") %>%
  SparkR::select(expr('PassengerId'), expr('Pclass'), expr('Age'),
                 expr('Survived'), expr('Embarked'))              %>%
  head
##   PassengerId Pclass   Age Survived Embarked
## 1          79      2  0.83        1        S
## 2         126      3 12.00        1        C
## 3         166      3  9.00        1        S
## 4         184      2  1.00        1        S
## 5         194      2  3.00        1        S
## 6         205      3 18.00        1        S

# The following is another approach. The end-result is the same,
# however, we bring the data first to the R's working memory and then
# use dplyr. Note the subtle differences in syntax.
SparkR::collect(T)                                              %>%
      dplyr::filter(Age < 20 & Sex == 'male' & Survived == 1)   %>%
      dplyr::select(PassengerId, Pclass, Age, Survived,
                    Embarked)                                   %>%
      head
```

[8]The package `dplyr` is introduced in Chapter 17 *"Data Wrangling in the tidyverse"* on page 265.

```
##   PassengerId Pclass   Age Survived Embarked
## 1          79      2  0.83        1        S
## 2         126      3 12.00        1        C
## 3         166      3  9.00        1        S
## 4         184      2  1.00        1        S
## 5         194      2  3.00        1        S
## 6         205      3 18.00        1        S
```

SparkR has its functions modelled to dplyr and most functions will work as expected. For example, grouping, summarizing, changing and arranging data works as in dplyr.

```
# Extract the survival percentage per class for each gender:
TMP <- T                                          %>%
       SparkR::group_by(expr('Pclass'), expr('Sex'))  %>%
             summarize(countS = sum(expr('Survived')), count = n(expr('PassengerId')))
N   <- nrow(T)

TMP                                                              %>%
    mutate(PctAll   = expr('count') / N  * 100)                  %>%
    mutate(PctGroup = expr('countS') / expr('count')  * 100) %>%
    arrange('Pclass', 'Sex')                                     %>%
    SparkR::collect()
##   Pclass    Sex countS count    PctAll PctGroup
## 1      1 female     91    94 10.549944 96.80851
## 2      1   male     45   122 13.692480 36.88525
## 3      2 female     70    76  8.529742 92.10526
## 4      2   male     17   108 12.121212 15.74074
## 5      3 female     72   144 16.161616 50.00000
## 6      3   male     47   347 38.945006 13.54467
```

Note – Statements that are not repeated further

In the later code fragments in this section we assume that the following statements are run and we will not repeat them.

```
library(tidyverse)
library(SparkR)
library(titanic)
sc <- sparkR.session(master = "local", appName = 'first test',
                     sparkConfig = list(spark.driver.memory = '2g'))
T <- as.DataFrame(titanic_train)
```

39.2.3.1 A User Defined Function on Spark

The fact that the dplyr functions can be used on a SparkDataFrame without taking care of the complexity behind is fantastic. But it is equally important to be able to run a user-defined function on a Resilient Distributed Dataset (RDD). SparkR provides the function dapply() as workhorse to send any function to a resilient distributed data set and execute this function on the different fragments of the data and then collect the results back.

RDD

UDF

dapply

dapply()

Running a user defined function (UDF) on a SparkDataFrame (a resilient distributed dataset) is remarkably simple with the function dapply(). The UDF is passed as a parameter to dapply and it should have only use one parameter: an R data.frame. The output must also be an R-data.frame and nothing else.

```
# The data:
T <- as.DataFrame(titanic_train)

# The schema can be a structType:
schema <- SparkR::structType(SparkR::structField("Age", "double"),
                             SparkR::structField("ageGroup", "string"))

# Or (since Spark 2.3) it can also be a DDL-formatted string
schema <- "age DOUBLE, ageGroup STRING"

# The function to be applied:
f <- function(x) {
  data.frame(x$Age, if_else(x$Age < 30, "youth", "mature"))
  }

# Run the function f on the Spark cluster:
T2 <- SparkR::dapply(T, f, schema)

# Inspect the results:
head(SparkR::collect(T2))

##      age          ageGroup
## 1     22             youth
## 2     38            mature
## 3     26             youth
## 4     35            mature
## 5     35            mature
## 6     NA              <NA>
```

Hint – Extracting the schema

The schema can be extracted via the function `schema()`. Below a trivial example.

```
DFcars  <- createDataFrame(mtcars)
DFcars2 <- dapply(DFcars, function(x) {x}, schema(DFcars))
head(collect(DFcars2))
##    mpg cyl disp  hp drat    wt  qsec vs am gear carb
## 1 21.0   6  160 110 3.90 2.620 16.46  0  1    4    4
## 2 21.0   6  160 110 3.90 2.875 17.02  0  1    4    4
## 3 22.8   4  108  93 3.85 2.320 18.61  1  1    4    1
## 4 21.4   6  258 110 3.08 3.215 19.44  1  0    3    1
## 5 18.7   8  360 175 3.15 3.440 17.02  0  0    3    2
## 6 18.1   6  225 105 2.76 3.460 20.22  1  0    3    1
```

The function `apply` has a sister function `dapplyCollect` that does basically the same plus "collecting" the data back to a data.frame on the client computer (whereas `dapply` will leave the results as a resilient distributed dataset (SparkDataFrame) on the Hadoop cluster.

dapplyCollect

The "collect" version, `dapplyCollect()`, will apply the provided user defined function to each partition of the SparkDataFrame and collect the results back into a familiar R data.frame. Note also that the schema does not have to be provided to the function `dapplyCollect()`, because it returns a `data.frame` and not a `DataFrame`.

`dapplyCollect()`

```
# The data:
T <- as.DataFrame(titanic_train)

# The function to be applied:
f <- function(x) {
  y <- data.frame(x$Age, ifelse(x$Age < 30, "youth", "mature"))

  # We specify now column names in the data.frame to be returned:
    colnames(y) <- c("age", "ageGroup")

  # and we return the data.frame (base R type):
  y
  }

# Run the function f on the Spark cluster:
T2_DF <- dapplyCollect(T, f)

# Inspect the results (T2_DF is now a data.frame, no collect needed):
head(T2_DF)
##    age ageGroup
## 1  22     youth
## 2  38    mature
## 3  26     youth
## 4  35    mature
## 5  35    mature
## 6  NA      <NA>
```

The Group Apply Variant: gapply

In many cases, we need something different than just working with the partitions of the data as they are on the distributed file system. Maybe – referring to previous example – we want to know if in the different classes of passengers on the Titanic the age had a different dynamic. For example, we need to understand if richer people (in class one) were also on average older than those of class three. This will not work with the `dapply`, we need to be able to specify ourselves the grouping parameter, so that we can calculate the age average per group (class).

gapply() The function `gapply()` and `gaplyCollect()` will connect to spark and instruct it to run a user defined function on a given dataset. They both work similar, with the major difference that `gapplyCollect` will "collect" the data (ie. coerce the result into a non-distributed `data.frame`.

The function that is applied to the Spark data-frame need to comply with the following rules. It should only have two parameters: a grouping key and a standard data.frame as we are used to in R. The function will also return just a R `data.frame`.

`gapply` takes as arguments the SparkDataFrame, the grouping key (as a string), the user defined function and the schema of the output. The `schema` specifies the format of each row of the resulting a SparkDataFrame..

```
# Define the function to be used:
f <- function (key, x) {
  data.frame(key, min(x$Age,  na.rm = TRUE),
                  mean(x$Age, na.rm = TRUE),
                  max(x$Age,  na.rm = TRUE))
  }

# The schema also can be specified via a DDL-formatted string:
schema <- "class INT, min_age DOUBLE, avg_age DOUBLE, max_age DOUBLE"

maxAge <- gapply(T, "Pclass", f, schema)
```

```
head(collect(arrange(maxAge, "class", decreasing = TRUE)))
##   class min_age  avg_age max_age
## 1     3    0.42 25.14062      74
## 2     2    0.67 29.87763      70
## 3     1    0.92 38.23344      80
```

gapplyCollect

The function gapply has a variant that "collects" the result to an R data.frame. The main difference in usage is that the schema does not need to be supplied.

```
# Define the function to be used:
f <- function (key, x) {
  y <- data.frame(key, min(x$Age,  na.rm = TRUE),
                      mean(x$Age, na.rm = TRUE),
                      max(x$Age,  na.rm = TRUE))
  colnames(y) <- c("class", "min_age", "avg_age", "max_age")
  y
  }

maxAge <- gapplyCollect(T, "Pclass", f)

head(maxAge[order(maxAge$class, decreasing = TRUE), ])
##   class min_age  avg_age max_age
## 2     3    0.42 25.14062      74
## 3     2    0.67 29.87763      70
## 1     1    0.92 38.23344      80
```

spark.lapply

The equivalent of lapply from R is in SparkR the function spark.lapply. This function allows to run a function over a list of elements and execute the work over the cluster and its distributed dataset.[9] Generally it is necessary that the results of the calculations fit on one machine, however, workarounds via lists of data.frames are possible. spark.lapply()

```
# First a trivial example to show how spark.lapply works:
surfaces <- spark.lapply(1:3, function(x){pi * x^2})
print(surfaces)
## [[1]]
## [1] 3.141593
##
## [[2]]
## [1] 12.56637
##
## [[3]]
## [1] 28.27433
```

The power of spark.lapply resides of course not in calculating a list of squares, but rather in executing more complex model fitting over large distributed datasets. Below we show how it can fit a model.

[9]The function spark.lapply works similar to doParallel and lapply.

```
mFamilies   <- c("binomial", "gaussian")
trainModels <- function(fam) {
  m <- SparkR::glm(Survived ~ Age + Pclass,
                   data = T,
                   family = fam)
  summary(m)
  }
mSummaries <- spark.lapply(mFamilies, trainModels)
```

39.2.3.2 Some Other Functions of SparkR

Hadoop, Spark and any other big data solution are massive projects and the list of functions continues to grow. In this section, we list a short selection of other useful functions in SparkR. This list is not intended to be exhaustive, but rather to get you started.

Back and Forth Between R and Spark

It is essential to be able to move data.frames to the distributed filesystem as a resilient distributed dataset (or SparkDataFrame) and it is also necessary to be able to collect the data back to R. The following code shows conversions in both directions.

collect()
createDataFrame()

```
# df is a data.frame (R-data.frame)
# DF is a DataFrame (distributed Spark data-frame)

# We already saw how to get data from Spark to R:
df <- collect(DF)

# From R to Spark:
DF <- createDataFrame(df)
```

Reading CSV files

Data can reside in SQL or NoSQL databases, but sometimes it is necessary to input plain textfiles such as CSV-files. This can be done as follows.

CSV

```
loadDF(fileName,
       source = "csv",
       header = "true",
       sep = ",")
```

Changing Columns and Adding New Ones

Apart from the functions mutate() and select, there are much more functions from dplyr re-engineered to work on a SparkDataFrame. For example, we can change and add columns via the function withColumn.

withColumn()

```
T %>% withColumn("AgeGroup", column("Age") / lit(10))        %>%
      SparkR::select(expr('PassengerId'), expr('Age'),
                     expr('AgeGroup'))                        %>%
      head
##   PassengerId Age AgeGroup
## 1           1  22      2.2
## 2           2  38      3.8
```

```
## 3           3  26      2.6
## 4           4  35      3.5
## 5           5  35      3.5
## 6           6  NA       NA
```

Note that in the code fragment above, the function `lit()` returns the literal value of 10. `lit()`

Aggregation

In addition to the `group_by()` function used before, SparkR also provides the aggregation function `agg()` as well as the OLAP cube operator: `cube()`. `agg()` `cube()`

```
T %>% cube("Pclass", "Sex")  %>%
      agg(avg(T$Age))        %>%
      collect()
##    Pclass    Sex avg(Age)
## 1       3   <NA> 25.14062
## 2       1   male 41.28139
## 3       2   male 30.74071
## 4       1   <NA> 38.23344
## 5       2 female 28.72297
## 6      NA   <NA> 29.69912
## 7       1 female 34.61176
## 8       2   <NA> 29.87763
## 9       3   male 26.50759
## 10     NA female 27.91571
## 11     NA   male 30.72664
## 12      3 female 21.75000
```

The aggregation function can also work with sums (via the function `sum()`) and counts (via the functions `count()` or `n()`).

Also the functions `merge()`, `join()`, etc. will work as expected and there is a whole set of additional functions available that for example allow to work with certain "windows on data." For example, we can partition data per class with the function `windowPartitionBy("Pclass")`. `SparkR` also foresees a complete range of functions to works with strings, dates and cast variables to different types, etc. In this book we only want to give a flavour of what R can do for you when working with really big data.

39.2.3.3 Machine learning with SparkR

One of the strengths of using Spark is that machine learning techniques can be applied on the resilient distributed dataset via "MLlib." MLlib is the machine learning component of the Hadoop ecosystem. **MLlib**

While `SparkR` is still developing and the list below will still grow, we give you some oversight of the machine learning possibilities. The following list is a name of the relevant function in `SparkR`.

1. **Classification**
   - `spark.logit`: Logistic Regression
   - `spark.mlp`: Multilayer Perceptron (MLP)
   - `spark.naiveBayes`: Naive Bayes
   - `spark.svmLinear`: Linear Support Vector Machine

2. **Regression**
   - spark.glm or glm: Generalized Linear Model (GLM)
   - spark.survreg: Accelerated Failure Time (AFT) Survival Model
   - spark.isoreg: Isotonic Regression
3. **Tree**
   - spark.gbt: Gradient Boosted Trees for Regression and Classification
   - spark.randomForest: Random Forest for Regression and Classification
4. **Clustering**
   - spark.kmeans: K-Means
   - spark.bisectingKmeans: Bisecting k-means
   - spark.gaussianMixture: Gaussian Mixture Model (GMM)
   - spark.lda: Latent Dirichlet Allocation (LDA)
5. **Collaborative Filtering**
   - spark.als: Alternating Least Squares (ALS)
6. **Frequent Pattern Mining**
   - spark.fpGrowth : FP-growth
7. **Statistics**
   - spark.kstest: Kolmogorov-Smirnov Test

Note – Familiar MLliB

Behind the scenes SparkR will dispatch the fitting of the model to MLlib. That does not mean that we need to change old R-habits too much. SparkR supports a subset of the available R formula operators for model fitting (for example, we can use the familar operators ~, ., :, + and -).

Model Persistence and Sampling

write.ml()
read.ml()

SparkR makes it possible to store results of models and retrieve them via the functions write.ml() and read.ml().

```
# Prepare training and testing data:
T_split <- randomSplit(T, c(8,2), 2)
T_train <- T_split[[1]]
T_test  <- T_split[[2]]

# Fit the model:
M1 <- spark.glm(T_train,
                Survived ~ Pclass + Sex,
                family = "binomial")
```

```
# Save the model:
path1 <- tempfile(pattern = 'ml', fileext = '.tmp')
write.ml(M1, path1)

# Retrieve the model
M2 <- read.ml(path1)

# Do something with M2
summary(M2)
##
## Saved-loaded model does not support output 'Deviance Residuals'.
##
## Coefficients:
##              Estimate  Std. Error   t value    Pr(>|t|)
## (Intercept)   3.30997     0.33860    9.7755  0.0000e+00
## Pclass       -0.98022     0.11989   -8.1758  4.4409e-16
## Sex_male     -2.62236     0.20723  -12.6542  0.0000e+00
##
## (Dispersion parameter for binomial family taken to be 1)
##
##     Null deviance: 932.42  on 704  degrees of freedom
## Residual deviance: 654.45  on 702  degrees of freedom
## AIC: 660.5
##
## Number of Fisher Scoring iterations: 5

# Add predictions to the model for the test data:
PRED1 <- predict(M2, T_test)

# Show the results:
x <- head(collect(PRED1))
head(cbind(x$Survived, x$prediction))
##      [,1]      [,2]
## [1,]    0 0.4273654
## [2,]    1 0.9113117
## [3,]    0 0.2187741
## [4,]    0 0.4273654
## [5,]    0 0.4273654
## [6,]    0 0.4273654

# Close the connection:
unlink(path1)
```

Further information – SparkR

Spark and SparkR are cutting edge technologies and are still changing. While the library of functions is already impressive, good documentation is a little sparse. Maybe you find Ott Toomet's introduction useful? It is here: `https://otoomet.github.io/sparkr_notes.html`. Also do not forget that the Apache Foundation offers a website with the official documentation, that is well written and easy to read. See `https://spark.apache.org/docs/latest/index.html`.

39.2.4 sparklyr

RStudio sparklyr dplyr

Another way to connect R to Spark is the package `sparklyr`. It is provided by RStudio and hence – as you can expect – it makes sense to use the IDE RStudio, it encourages the tidyverse phylosophy and has a gentle learning curve if you already know the other RStudio tools such as `dplyr`.

`sparklyr` provides a complete `dplyr` back-end. So, via this library, it is a breeze to filter and use an RDD in Spark and then bring the results into R for what R does best: analysis, modelling and visualization. Of course, it is possible to use the MLlib machine learning library that comes with Spark all from your comfortable environment in R. The advanced user will also appreciate the possibility to create extensions to use the Spark API and provide interfaces to Spark packages.

Note – Short intro to sparklyr

Note that in this section, we merely demonstrate how easy it is to use Spark from the R with `sparklyr` and we do not try to provide a complete overview of the package.

Before we start we will make sure that we work with a clean environment. First, we could stop the Spark Master, but that is not really necessary.[10] It makes sense, however, to unload SparkR before loading `sparklyr`.[11] We will also de-connect the tidyverse libraries and then reconnect them on the desired order.

```
# We unload the package SparkR.
detach("package:SparkR",     unload = TRUE)
detach("package:tidyverse", unload = TRUE)
detach("package:dplyr",      unload = TRUE)
```

If the package `sparklyr` is not yet on your system, install it via:

```
install.packages('sparklyr')
```

After this, the package is downloaded, compiled and kept on our hard disk, ready to be loaded in each session where we need to use it and connect to Spark.

```
# First, load the tidyverse packages:
library(tidyverse)
library(dplyr)

# Load the package sparklyr:
library(sparklyr)

# Load the data:
library(titanic)

# Our spark-master is already running, so no need for:
```

[10] To stop the Spark-Master from R, we can use the command `system('stop-master.sh')`.

[11] Failing to unload `SparkR` will result in strange behaviour. For example, the function `group_by()` is provided by both packages and R will try to use the one from SparkR (because it was loaded earlier) and this will result in cryptic errors, such as `unable to find an inherited method for function 'group_by' for signature '"tbl_spark"'`. The workaround is then to specify which `group_by` function you want to use: `sparklyr::group_by(...)`.

```
# system('start-master.sh')

# Connect to the local Spark master
sc <- spark_connect(master = "local")
```

This connection object, `sc`, allows us to connect tot Spark as follows:

```
Titanic_tbl <- copy_to(sc, titanic_train)
Titanic_tbl

## # Source: spark<titanic_train> [?? x 12]
##    PassengerId Survived Pclass Name    Sex     Age SibSp Parch ...
##          <int>    <int>  <int> <chr>   <chr> <dbl> <int> <int> ...
##  1           1        0      3 Brau... male     22     1     0 ...
##  2           2        1      1 Cumi... female   38     1     0 ...
##  3           3        1      3 Heik... female   26     0     0 ...
##  4           4        1      1 Futr... female   35     1     0 ...
##  5           5        0      3 Alle... male     35     0     0 ...
##  6           6        0      3 Mora... male    NaN     0     0 ...
##  7           7        0      1 McCa... male     54     0     0 ...
##  8           8        0      3 Pals... male      2     3     1 ...
##  9           9        1      3 John... female   27     0     2 ...
## 10          10        1      2 Nass... female   14     1     0 ...
## # ... with more rows, and 1 more variable: Embarked <chr>


# More datasets can be stored in the same connection:
cars_tbl <- copy_to(sc, mtcars)

# List the available tables:
src_tbls(sc)
## [1] "mtcars"        "titanic_train"
```

The functionality that we are familiar with from `dplyr` can also be applied on the Spark table via `sparklyr`. The following code demonstrates this by summarizing some aspects of the data.

```
Titanic_tbl %>% summarise(n = n())
## # Source: spark<?> [?? x 1]
##       n
##   <dbl>
## 1   891

# Alternatively:
Titanic_tbl %>% spark_dataframe() %>% invoke("count")
## [1] 891

Titanic_tbl                                  %>%
  dplyr::group_by(Sex, Embarked)             %>%
  summarise(count = n(), AgeMean = mean(Age)) %>%
  collect
## # A tibble: 7 x 4
##   Sex    Embarked count AgeMean
##   <chr>  <chr>    <dbl>   <dbl>
## 1 male   C           95    33.0
## 2 male   S          441    30.3
## 3 female C           73    28.3
## 4 female ""           2    50
## 5 female S          203    27.8
## 6 male   Q           41    30.9
## 7 female Q           36    24.3
```

Hint – Use SQL from R on RDD

DBI

It is also possible to use the power of SQL directly on the spark array via the library DBI.

```
library(DBI)
sSQL <- "SELECT Name, Age, Sex, Embarked FROM titanic_train
        WHERE Embarked = 'Q' LIMIT 10"
dbGetQuery(sc, sSQL)
##                             Name Age    Sex Embarked
## 1               Moran, Mr. James NaN   male        Q
## 2           Rice, Master. Eugene   2   male        Q
## 3    McGowan, Miss. Anna "Annie"  15 female        Q
## 4 O'Dwyer, Miss. Ellen "Nellie" NaN female        Q
## 5       Glynn, Miss. Mary Agatha NaN female        Q
```

Note that dbGetQuery() does not like a semicolon at the end of the SQL statement.

Spark_apply

spark_apply()

Of course, sparklyr has also a workhorse function to apply user defined functions over the distributed resilient dataset. This is the function spark_apply(). This allows us to write our own functions – that might not be available in Apache Spark – to run fast on the distributed data.

Typically, one uses spark_apply() to apply an R function on a SparkDataFrame (or more general "Spark object"). Spark objects are by default partitioned so they can be distributed over a cluster of working nodes. To some extend spark_apply is a combination of different functions in SparkR: it can be used over the default partitions or it can be run over a chosen partitioning. This can be obtained by simply specifying the group_by argument.

Note that the R function – that is supplied to spark_apply – takes a Spark DataFrame as argument and must return another Spark DataFrame. As expected, spark_apply will apply the given R function on each partition and then aggregate the result to a single Spark DataFrame.

```
# sdf_len creates a DataFrame of a given length (5 in the example):
x <- sdf_len(sc, 5, repartition = 1) %>%
     spark_apply(function(x) pi * x^2)
print(x)
## # Source: spark<?> [?? x 1]
##       id
##    <dbl>
## 1  3.14
## 2 12.6
## 3 28.3
## 4 50.3
## 5 78.5
```

Machine Learning

The library sparklyr comes with a wide range of machine learning functions as well as with transforming functions.

- **feature transformers** are the functions that create buckets, binary values, element-wise products, binned categorical values based on quantiles, etc. The name of these functions starts with ft_. Note especially the function sql_transformer() – an exception on the naming convention – it is the function that allows to transform data based on an SQL statement.

- **machine learning algorithms** are the functions that perform linear regression, PCA, logistic regression, decision trees, random forest, etc. The name of these functions starts with ml_.

machine learning

Below is an example of how the functions all work together.

```
# Transform data and create buckets for Age:
t <- Titanic_tbl %>%
  ft_bucketizer(input_col  = "Age",
                output_col = "AgeBucket",
                splits     = c(0, 10, 30, 90))

# Split data in training and testing set:
partitions <- t %>%
  sdf_random_split(training = 0.7, test = 0.3, seed = 1942)
t_training <- partitions$training
t_test     <- partitions$test

# Fit a logistic regression model
M <- t_training %>%
    ml_logistic_regression(Survived ~ AgeBucket + Pclass + Sex)

# Add predictions:
pred <- sdf_predict(t_test, M)

# Show details of the quality of the fit of the model:
ml_binary_classification_evaluator(pred)
```

Further information – sparklyr

As one came to expect from RStudio, all their contributions to the R-community come with great documentation on their website. The pages of sparklyr are here: https://spark.rstudio.com.

39.2.5 SparkR or sparklyr

Both libraries are very powerful, but unlike in most other sections of this book we decided to present the two options. For most applications, both will work fine and the major concern is to remain in line with a previously made corporate choice.

However, the philosophy of both packages is different, and hence, it is worth to compare them. First, we will have a look at the similarities between the two tools.

Both tools

- mimic RStudio's dplyr,
- should feel reasonably familiar (knowing both base R and the tidyverse),
- allow to filter data and aggregate Spark datasets,

- have a substantial machine learning library that links through to MLlib,
- allow the results to be brought back to R for further analysis, visualization and reporting.

So both libraries achieve roughly something similar, however, we believe that it is fair to say that

- `SparkR` is a tool to use Spark via R – it is a Spark API – whereas
- `sparklyr` is a tool to use Apache Spark as a back-end for `dplyr`.

Reasons to choose SparkR

Python

- `SparkR` has a Python equivalent, so if you use both languages you will appreciate the shortened learning curve;
- `sparkR` follows closer the scala-spark API and so also here are learning synergies.

Those are strong arguments if you are the data scientist that also works with Python or directly on Spark. However, if you are the person that has an analyst profile and work usually with R, then `sparklyr` has some noteworthy advantages.

Reasons to choose sparklyr

- smoother integration with `dpylr`,
- function naming conventions that are in line with the tidyverse conventions, and
- probably easier to learn for the modern R-user.

There is no conclusive winner, it is, however, worth to spend some time considering which interface to use and then stick with that one.

♣ 40 ♣

The Need for Speed

In previous section, we focused on the issue where data becomes too large to be read into memory or causes the code to run too slowly. The first aspect – large data – is covered in the previous section, however, the reason why our code is slow, is not always due to data being too big. It might be related to the algorithm, the programming style, or we might have hit limits of what R naturally can do. In this section we will have a look on how to optimize code for speed – and assume that data-size is not the main blocking factor.

In this section we will show how we can evaluate and recognize efficient code and study various ways to speed up code. Most of those ways to reduce runtime are related to how R is implemented and what type of language it is. R is a high level interpreted language and provides complex data types with loads of functionalities and hides away much complexity. Using simpler data types, pre-allocating memory, and compiling code will therefore, be part of our basic toolbox.

The Big R-Book: From Data Science to Learning Machines and Big Data, First Edition. Philippe J.S. De Brouwer.

Companion Website: www.wiley.com/go/De Brouwer/The Big R-Book

40.1 Benchmarking

Before we start to look into details on optimizing for speed, we will need an objective way to tell if code is really faster or not. R has built-in tools to measure how long a function or code block runs via the function `system.time()` of `base` R, which we used before (for example in Chapter 37.3 *"Using the GPU"* on page 752).

`system.time()`

We supply an expression to be timed to the function `system.time()` as its argument.

```
x <- 1:1e4

# Timing the function mean:
system.time(mean(x))
##    user  system elapsed
##       0       0       0
```

The code block above illustrates immediately some issues with timing code. The function `mean()` is too fast to be timed and hence we need to repeat it a few times to get a meaningful measure.

```
N <- 2500

# Repeating it to gather longer time:
system.time({for (i in 1:N) mean(x)})
##    user  system elapsed
##   0.038   0.000   0.037
```

This result is meaningful. However, when we execute the function once more, the result is usually different.

```
system.time({for (i in 1:N) mean(x)})
##    user  system elapsed
##   0.037   0.000   0.037

system.time({for (i in 1:N) mean(x)})
##    user  system elapsed
##   0.039   0.000   0.039
```

We realize that the timing is a stochastic result and know how to measure it, but as long as we have no alternative algorithm; this measurement is not so useful. So, let us come up with a challenger model and calculate mean as the sum divided by the length of the vector. This will only work on vectors, but since our `x` is a vector this is a valid alternative.

Digression – Expressions and curly brackets

Notice the use of the curly brackets. The function `system.time()` expects one expression and if we want to use multiple lines of code, we have to wrap it in curly brackets.

```
# Timing a challenger model:
system.time({for (i in 1:N) sum(x) / length(x)})
##    user  system elapsed
##   0.029   0.000   0.029
```

The difference in runtime is clear: our alternative to the function `mean` is faster: it takes about half of the time to run. However, the difference will not always so clear and run-times are not always exactly the same each time we run the code. Measuring the same block of code on the same computer for a few times will typically yield different results. So, in essence the run-time is a stochastic variable. To compare stochastic variables, we need to get some insight in the shape of the distribution, study the mean, median, minimum, quartiles, maximum, etc. This can be done manually or by using the library `microbenchmark`.

microbenchmark

This lightweight package provides us with all those tools that we need. It even provides automated visualization of the results. After installing the packages, we can load and use it.

```
N <- 1500

# Load microbenchmark:
library(microbenchmark)

# Create a microbenchmark object:
comp <- microbenchmark(mean(x),                # 1st code block
                       {sum(x) / length(x)}, # 2nd code block
                       times = N)              # number times to run
```

The object `comp` is an object of class `microbenchmark` and – because of R's useful implementation of the S3 OO system – we can print it or ask for a summary with the known functions `print()` and `plot()` for example.

microbenchmark()

```
summary(comp)
##                         expr    min      lq     mean
## 1                    mean(x) 12.849 13.1315 13.95813
## 2 {      sum(x)/length(x) } 10.018 10.1490 10.55872
##    median     uq    max neval
## 1 13.2295 13.374 66.145  1500
## 2 10.2060 10.267 45.464  1500
```

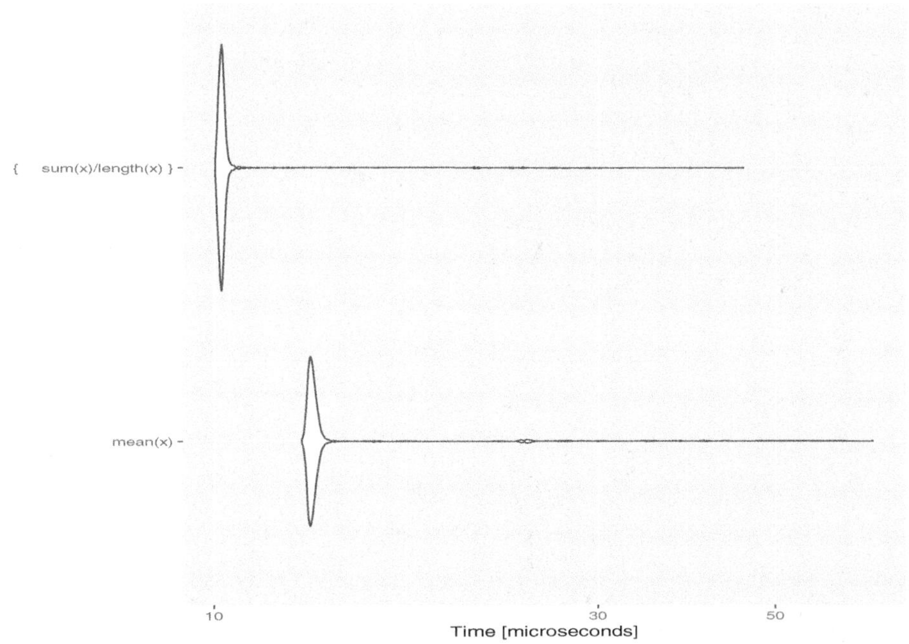

Figure 40.1: *The package* `microbenchmark` *also defines a suible visualisation via* `autoplot`. *This shows us the violin plots of the different competing methods. It also makes clear how heavy the right tails of the distribution of run-times are.*

`microbenchmark` allows us to confirm with more certainty that our own code is faster than the native function `mean()`. This is because the function `mean()` is a dispatcher function that first will find out what object `x` is, then dispatch it to the relevant function, where still some logic about handling NAs, error handling, and more functionality slows down the calculation.

The `microbenchmark` object holds information about the different times that were recorded when the code was run and it has a suitable method for visualizing the histograms via `autoplot()` – a wrapper around `ggplot2`. The function `autoplot()` is a dispatcher function for wrappers around `ggplot2` – see Chapter 31 *"A Grammar of Graphics with ggplot2"* on page 687. The plot methods that `microbenchmark` provides is a violin plot: see Chapter 9.6 on page 173. As demonstrated via the following code, which creates the plot in Figure 40.1 on page 795.

```
# Load ggplot2:
library(ggplot2)

# Use autoplot():
autoplot(comp)
```

Thanks to the visualisation, such as in Chapter 31 *"A Grammar of Graphics with ggplot2"* on page 687, we get an idea how robust the results are. In this case, we can safely conclude that – in the given use – our function will almost always run faster than `mean()`. When we will use numeric vectors that have no missing values, no infinite numbers, etc. then we can replace `mean()` with our challenger function.

40.2 Optimize Code

The place to start looking for improvement is always the programming style. Good code is logically structured, easy to maintain, readable and fast. However, usually some compromise is needed: we need to make a choice and optimize for readability or speed.

Below we look at some heuristics that can be used to optimize code for speed. This list might not be complete, but in our experience the selected subjects make a significant difference in how efficient the code is.

40.2.1 Avoid Repeating the Same

Minimize repeating the same might sound obvious, but it will surprisingly often appear in early versions of code. For example, if you have a for-loop that needs in the length of a vector, then it is usually faster to do this once – before the loop – and store the result to be used in the for-loop.

```
x <- 1:1e+4
y <- 0

# Here we recalculate the sum multiple times.
system.time(for(i in x) {y <- y + sum(x)})
##    user  system elapsed
##   0.106   0.000   0.106

# Here we calculate it once and store it.
system.time({sum_x <- sum(x); for(i in x) {y <- y + sum_x}})
##    user  system elapsed
##   0.004   0.000   0.004
```

40.2.2 Use Vectorisation where Appropriate

In R, vectors (and the family of `*apply`-functions) are usually faster than `for`-loops. This is true, but many tutorials on speed will make the bold statement that it is always better to use vectors than scalars. The following example, however, illustrates that there is a nuance to this.

```
# Define some numbers.
x1 <- 3.00e8
x2 <- 663e-34
x3 <- 2.718282
x4 <- 6.64e-11

y1 <- pi
y2 <- 2.718282
y3 <- 9.869604
y4 <- 1.772454

N <- 1e5

# 1. Adding some of them directly:
f1 <- function () {
  for (i in 1:N) {
    x1 + y1
    x2 + y2
    x3 + y3
    x4 + y4
```

```
    }
  }
system.time(f1())
##     user  system elapsed
##    0.031   0.005   0.036

# 2. Converting first to a vector and then adding the vectors:
f2 <- function () {
  x <- c(x1, x2, x3, x4)
  y <- c(y1, y2, y3, y4)
  for (i in 1:N) x + y
}
system.time(f2())
##     user  system elapsed
##    0.015   0.000   0.015

# 3. Working with the elements of the vectors:
f3 <- function () {
  x <- c(x1, x2, x3, x4)
  y <- c(y1, y2, y3, y4)
  for (i in 1:N) {
    x[1] + y[1]
    x[2] + y[2]
    x[3] + y[3]
    x[4] + y[4]
  }
  }
system.time(f3())
##     user  system elapsed
##    0.046   0.000   0.045

# 4. Working with the elements of the vectors and code shorter:
f4 <- function () {
  x <- c(x1, x2, x3, x4)
  y <- c(y1, y2, y3, y4)
  for (i in 1:N) for(n in 1:4) x[n] + y[n]
  }
system.time(f4())
##     user  system elapsed
##    0.026   0.011   0.036
```

Note – Overhead created by the for-loops

The for-loops in the aforementioned code and in most of the rest of the section only repeats the same action in order to create a longer time that is measurable by `system.time()`. The for-loops themselves create indeed also some overhead, but they are the same in each of the methods and hence the differences are to be allocated to the methods discussed.

The direct sums (`f1()`) is slower than first creating a vector and then summing the vectors (`f2()`). The overhead seems to come from accessing the vector rather than using the scalars directly (`f3()` and `f4()`). However, considering that option 2 is not only the fastest option, but also the most readable, we strongly recommend to use this.

While our experiment shows a a clear winner (method two), it is not a bad idea to use `microbenchmark`, and make sure that we make the right choice. The results of this approach are in Figure 40.2 on page 799, which is generated by the code below.

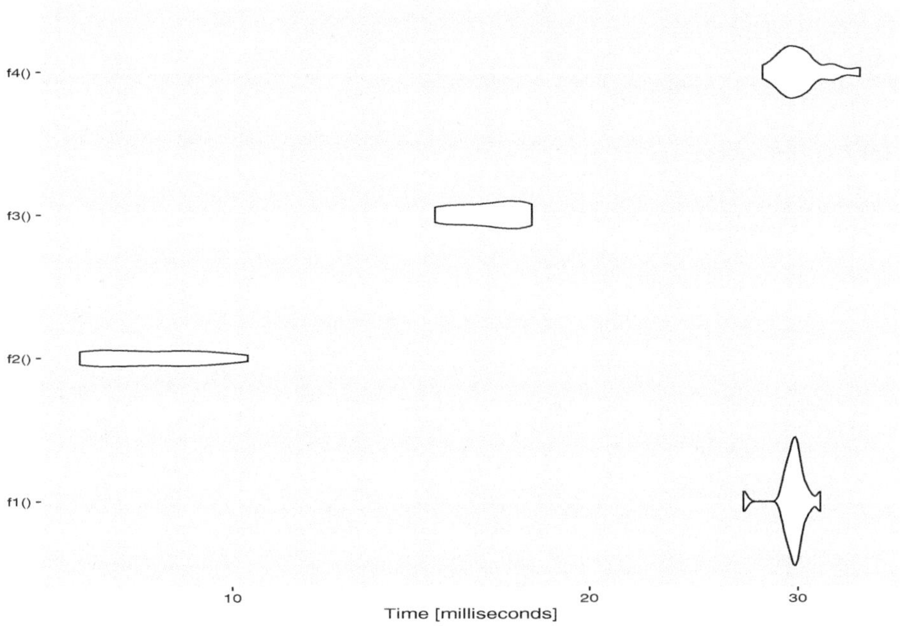

Figure 40.2: *This figure shows that working with vectors is fastest (f2()).*

```
# Remind the packages:
library(microbenchmark)
library(ggplot2)

# Compare all four functions:
comp <- microbenchmark(f1(), f2(), f3(), f4(), times = 15)

# Create the violin plots:
autoplot(comp)
```

40.2.3 Pre-allocating Memory

Allocating memory takes a small time. While this is of the order of nanoseconds, it becomes significant if that is repeated enough. This is especially true for more complex objects. The preferable way to go is allocating the empty object first and then adding elements.

```
N <- 2e4

# Method 1: using append():
system.time ({
    lst <- list()
    for(i in 1:N) {lst <- append(lst, pi)}
   })
##    user  system elapsed
##   1.215   0.006   1.225

# Method 2: increasing length while counting with length():
system.time ({
    lst <- list()
    for(i in 1:N) {
        lst[[length(lst) + 1]] <- pi}
   })
```

```
##    user  system elapsed
##   0.026   0.000   0.026

# Method 3: increasing length using a counter:
system.time ({
    lst <- list()
    for(i in 1:N) {lst[[i]] <- pi}
  })
##    user  system elapsed
##   0.023   0.000   0.024

# Method 4: pre-allocate memory:
system.time({
    lst <- vector("list", N)
    for(i in 1:N) {lst[[i]] <- pi}
    })
##    user  system elapsed
##   0.017   0.003   0.020
```

Using the function `append()` is not very efficient compared to the other methods, because the function is designed to cater for much more than what we are doing.

The methods 2 and 3 are very close and yield good results in other programming languages. However, in R they need – just as method 1 – quadratic time ($O(N)$), and are therefore bound to fail for large enough N. Apart from the obvious inefficiency in method 2 (actually related to the first principle discussed) these methods require that R allocates more memory for the list object in each iteration. Only method 4, where the size of the list is defined only once, is efficient.

Whenever you have the choice between a complex data-structure (RC object, list, etc.) and there is a more simple data-structure available – that can be used in the particular code – then choose for the most simple one. When an object provides more flexibility (e.g. a data-frame can also hold character types, where a matrix can only have elements of the same type), then this additional flexibility comes at a cost.

For example, let us investigate the difference between using a matrix and a data frame. Below we will compare both for simple arithmetic.

```
N <- 500
# simple operations on a matrix
M <- matrix(1:36, nrow = 6)
system.time({for (i in 1:N) {x1 <- t(M); x2 <- M + pi}})
##    user  system elapsed
##   0.009   0.000   0.009

# simple operations on a data-frame
D <- as.data.frame(M)
system.time({for (i in 1:N) {x1 <- t(D); x2 <- D + pi}})
##    user  system elapsed
##   0.274   0.012   0.284
```

Obviously, the aforementioned code does nothing useful: it only has unnecessary repetitions, but it illustrate that the performance gain of a matrix over a data-frame is massive (the matrix is more than 30 times faster in this example).

40.2.4 Use the Fastest Function

Functions in base R and larger projects such as the tidyverse are very fast and it will for the casual user not be possible to create a function with the same functionality that runs faster. Often these functions are written in C or C++ and are complied for speed.

There is, however, a caveat. That caveat is in the words "same functionality." Base functions such as `mean()` do a lot more than just `sum(x)/length(x)`. These base-functions are dispatcher functions: they will check the data-type of x and then dispatch to the correct function to calculate the mean.

We refer to the example from the beginning of this chapter:

```
x <- 1:1e4
N <- 1000
system.time({for (i in 1:N) mean(x)})
##    user  system elapsed
##   0.017   0.000   0.017

system.time({for (i in 1:N) sum(x) / length(x)})
##    user  system elapsed
##   0.015   0.000   0.014
```

40.2.5 Use the Fastest Package

R is free software, has a stable core, and it has a large community of people, who contribute their personal work to be used by everyone. Some of the libraries, however, are more focussed on getting a job done than on speed. If you are in the need for speed, then you might want to search for he best optimized package.

For example, when handling time series there is a lot of choice. It is likely that `xts`, `zoo`, `timeSeries`, `ts` or `tseries` get the work done for you.

xts
zoo

However, they are not created equal. Consider the following example:

```
N <- 732
# Use ts from stats to create the a time series object:
t1 <- stats::ts(rnorm(N), start = c(2000,1), end = c(2060,12),
                frequency = 12)
t2 <- stats::ts(rnorm(N), start = c(2010,1), end = c(2050,12),
                frequency = 12)

# Create matching zoo and xts objects:
zoo_t1 <- zoo::zoo(t1)
zoo_t2 <- zoo::zoo(t2)
xts_t1 <- xts::as.xts(t1)
xts_t2 <- xts::as.xts(t2)

# Run a merge on them:
# Note that base::merge() is a dispatcher function.
system.time({zoo_t <- merge(zoo_t1, zoo_t2)})
##    user  system elapsed
##   0.042   0.003   0.045

system.time({xts_t <- merge(xts_t1, xts_t2)})
##    user  system elapsed
##       0       0       0

# Calculate the lags:
system.time({for(i in 1:100) lag(zoo_t1)})
##    user  system elapsed
##   0.028   0.000   0.028

system.time({for(i in 1:100) lag(xts_t1)})
##    user  system elapsed
##   0.009   0.000   0.009
```

This example makes clear that the library `xts` is much faster than the `zoo` library. Similar differences can be found among many other libraries. For almost any task, R offers choice on how to achieve a certain goal. It is up to you to make a wise choice that is intuitive, fast and covers your needs.

40.2.6 Be Mindful about Details

Did you think that curly brackets are as fast as round brackets? Or did you think that raising something to the power -1 is the same as division? Well, mathematically, these are the same things, but in a programming language such as R, they are typically executed in a different way and display differences in performance.

Note – Cold code ahead

Till now the differences in performance were both obvious and significant. For the remainder of the chapter, we study some effects that can be smaller and more subtle. Therefore, we have chosen to not to run the calculations and performance tests separate from compiling the book. In other words, while in the rest of the book – almost everywhere – the code that you see is directly generating the output and plots, below the code is run separately and results were manually added. The reason is that while generating this book, the code will be wrapped in other functions such as `knitr()`, `tryCatch()`, etc. and that this has a less predictable impact.
You can recognize the static code on the response of R: the output lines – starting with ## – appear in bold green and not in black text.

```
# standard function:
f1 <- function(n, x = pi) for(i in 1:n) x = 1 / (1+x)

# using curly brackets:
f2 <- function(n, x = pi) for(i in 1:n) x = 1 / {1+x}

# adding unnecessary round brackets:
f3 <- function(n, x = pi) for(i in 1:n) x = (1 / (1+x))

# adding unnecessary curly brackets:
f4 <- function(n, x = pi) for(i in 1:n) x = {1 / {1+x}}

# achieving the same result by raising to a power
f5 <- function(n, x = pi) for(i in 1:n) x = (1+x)^(-1)

# performing the power with curly brackets
f6 <- function(n, x = pi) for(i in 1:n) x = {1+x}^{-1}

N <- 1e6
library(microbenchmark)
comp <- microbenchmark(f1(N), f2(N), f3(N), f4(N), f5(N), f6(N),
                       times = 150)

comp
## Unit: milliseconds
##   expr      min       lq     mean   median       uq      max ...
##   f1(N) 37.37476 37.49228 37.76950 37.57212 37.79876 39.99120 ...
##   f2(N) 37.29297 37.50435 37.79612 37.63191 37.81497 41.09414 ...
##   f3(N) 37.96886 38.18751 38.59619 38.28713 38.68162 47.66612 ...
```

```
##    f4(N) 37.88111 38.06787 38.41134 38.16297 38.36706 42.53103 ...
##    f5(N) 45.12742 45.31632 45.67364 45.45465 45.69882 49.65297 ...
##    f6(N) 45.93406 46.03159 46.51151 46.15287 46.64509 52.95426 ...

# Plot the results:
library(ggplot2)
autoplot(comp)
```

The plot, generated in the last line of the aforementioned code is in Figure 40.3, and shows how the two first methods are the most efficient. It seems that while the curly brackets perform a little more consistent, there is little to no reason to prefer either of them based on speed. However, not overusing brackets and using the simplest operation does pay off.

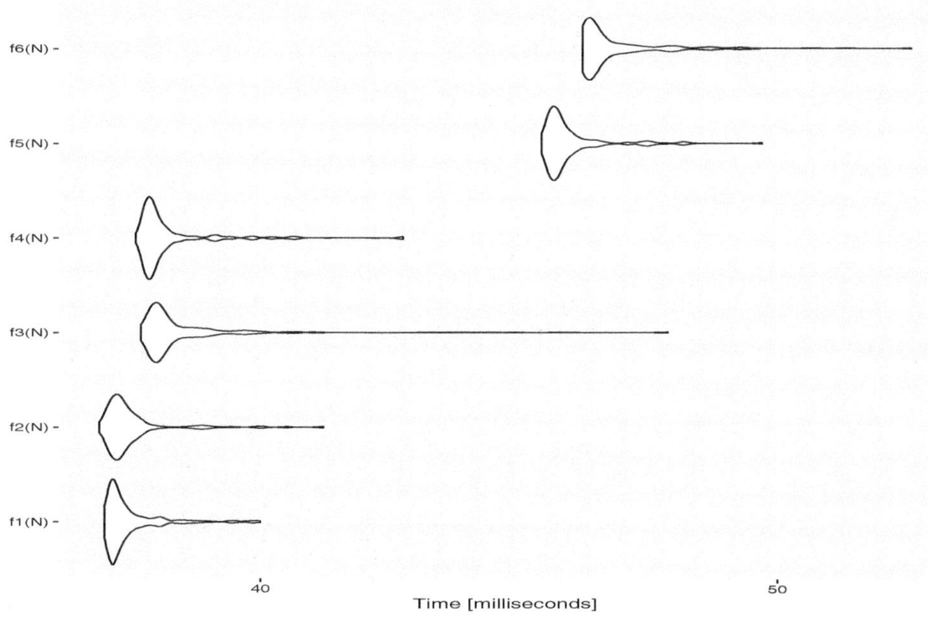

Figure 40.3: *Different ways to calculate $\frac{1}{1+x}$. The definition of the functions $f_i()$ is in the aforementioned code.*

This simple example allows us to note that:

- Curly brackets are typically a little faster than the round ones, but the difference is really small, and (in this case) much smaller than the variations between different runs.[1]
- More brackets slow down the process,
- A power is a more general functions than $\frac{1}{x}$ and hence comes at more overhead.

[1]The differences are so small that answers of the statistics are not always univocal. For example, the difference between function $f_1()$ and $f_2()$ is really small. The median of f_2 is better, but the mean of f_1 shows a slight performance advantage in the opposite direction.

40.2.7 Compile Functions

R is an interpreted language and hence necessarily comes with a lot of overhead at each run of code. While the core team of R has spent a great deal of time optimizing the code and making sure R runs as fast as possible, there is still a gain to be obtained by simply compiling code.

compiler

Depending on the particular code, compiling R-code can be a major time gain. Core-R offers the library `compiler` to achieve this in a surprisingly easy and straightforward way.

```
library(compiler)
N <- as.double(1:1e7)

# Create a *bad* function to calculate mean:
f <- function(x) {
  xx = 0
  l = length(x)
  for(i in 1:l)
    xx = xx + x[i]/l
  xx
}

# Time the function:
system.time(f(N))
##    user  system elapsed
##  0.61    0.000    0.61

# Compile the function:
cmp_f <- cmpfun(f)

# Time the compiled version
system.time(cmp_f(N))
##  user   system elapsed
##  0.596  0.00    0.596

# The difference is small, so we use microbenchmark
library(microbenchmark)
comp <- microbenchmark(f(N), cmp_f(N), times = 150)

# See the results:
comp
## Unit: milliseconds
##      expr      min       lq     mean   median       uq     ...
##      f(N) 552.2785 553.9911 559.6025 556.1511 562.7207 601.5...
##  cmp_f(N) 552.2152 553.9453 558.1029 555.8457 560.0771 588.4...


# Plot the results.
library(ggplot2)
autoplot(comp)
```

The plot, generated by the aforementioned code is in Figure 40.4 on page 805. The compiled function is just a little faster, but shows a more consistent result.

It seems that in this particular case (meaning, the determining factors such as function, implementation of R, operating system, processor, etc.), we can get little to no gain of compiling the function. In general, one can expect little gain if there are a lot of base-R functions in your code, or conversions of to different data-types. Compiling functions typically performs better when there are more numeric calculations, selections, etc.

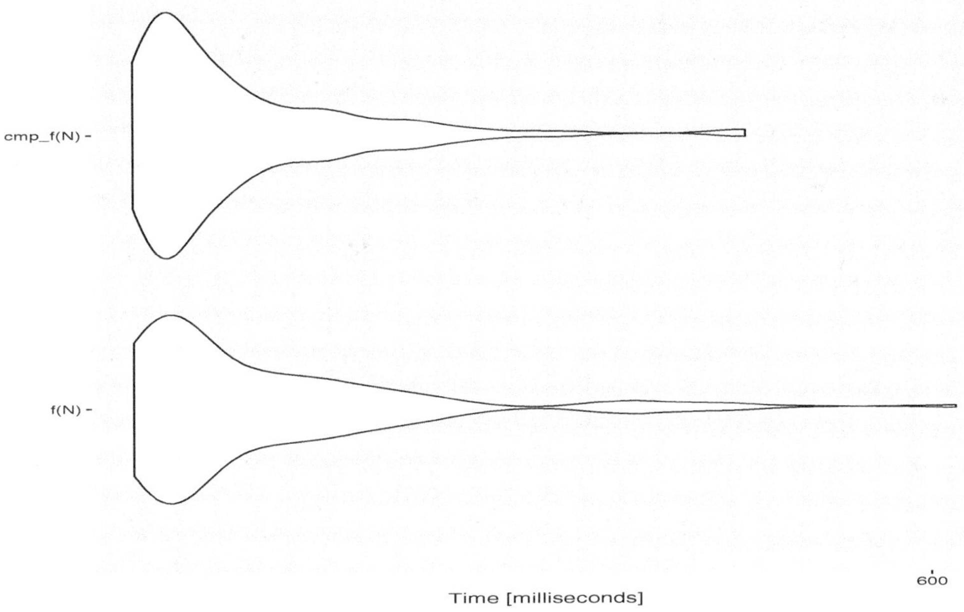

Figure 40.4: *Autoplot generates nice violin plots to compare the speed of the function defined in the aforementioned code (f) versus its compiled version (cmp_f).*

Note that there is more than one way to compile code. `cmpfun()` will go a long way, but – for example – when creating packages, we can force the code to be compiled at installation by adding: `ByteCompile: true` in the `DESCRIPTION` file of the package.[2]

Actually, when using `install.packages()`, and the programmer did not specify the `ByteCompile` variable, then the package is not compiled. It is possible to force R to compile every package that gets installed by setting the `R_COMPILE_PKGS` variable set to a positive integer value by executing:

```
options(R_COMPILE_PKGS = 1)
```

A final option to use just-in-time (JIT) compilation. This means that the overhead of compiling a function only comes when needed and this might – depending on how you are using R – deliver a better experience. The function `enableJIT()` – from the package `compiler` – will set the level of JIT compilation for the rest of the session. `enableJIT()`

It disables JIT compilation if the argument is 0, and for the arguments 1, 2, or 3 it implements different levels of optimisation. **JIT**

JIT can also be enabled by setting the environment variable R_ENABLE_JIT, to one of the values mentioned above. For example:

```
options(R_ENABLE_JIT = 0)
```

[2]See the section Chapter A *"Create your own R Package"* on page 821 to learn more about creating R-packages and the files that are delivered with a package.

40.2.8 Use C or C++ Code in R

C++ To push this a little further, we can write code in C or C++ and call it via our R-code. C++ will in general have a performance advantage over R. However, it is important to understand that C++ (and C of course) are lower level languages, with a lot less overhead. This means that they have no automated variable initialization, type checking, etc. and that code will run really fast. The cost is in the programming itself: you can expect a steeper learning curve, more errors, debugging can be more challenging, and (depending on the libraries that you use) longer code that is more difficult to read.

For example, when functions are called, the small overhead of calling the function is less in C++ than in R. The difference is naturally really small, but when that calling of a function happens a lot all these small times will accumulate and can become significant. C++ will even allow you to suggest to the compiler that small functions should be repeated in the binary code where they are needed, this decission can be left to the compiler and/or we can specify this per function. Other typical examples are loop structures, also here C++ has a measurable performance advantage. C++ allows even to decide which variables are stored in the RAM and which stay in the registers of the processor for faster access. R will never have such fine grained control over how the code is excecuted, nor should it have.

So, where R is a high level interpreted language, C++ is a medium-to-low level compiled language. Notably, its implementation of OO is extremely flexible and efficient and hence will OO also allows more complex structures than R. The OO implementation in C++ is similar to the RC – see Chapter 6.4 *"The Reference Class, refclass, RC or R5 Model"* on page 113 – in R, but a lot RC more flexible, and feature-rich.

There is a way to combine the best of both worlds. The `Rcpp` package allows to use C or C++ Rcpp code without having to worry too much about compiling everything first.

Learning C or C++ has – initially – a relatively steep learning curve and so it is not possible to squeeze this in a book about R. We recommend, however, to start with C, it has only 20 keywords and is – though a little abstract – easy enough to learn. Knowledge of C will be helpful to understand a lot of other modern computer languages: in many ways it is similar to learning Italian (e.g. Java) when you already know Latin (C). C++ was first written as an optional extension on C by Bjarne Stroustrup, and hence keeps all advantages of C. To push the analogy with languages a little further, we could say that C is classic Latin and C++ is Medieval Latin in this analogy.

To illustrate how simple is is to use C++ code via `Rcpp` we will create a function to calculate the Fibonacci numbers. The Fibonacci numbers are so that $F_n = F_{n-2} + F_{n-1}$ and $F_0 = F_1 = 1$. This inspires us to write the following naive and inefficient function:

```
# Naive implementation of the Fibonacci numbers in R:
Fib_R <- function (n) {
  if ((n == 0) | (n == 1)) return(1)
  return (Fib_R(n - 1) + Fib_R(n - 2))
  }

# The R-function compiled via cmpfun():
library(compiler)
Fib_R_cmp <- cmpfun(Fib_R)

# Using native C++ via cppFunction():
Rcpp::cppFunction('int Fib_Cpp(int n) {
  if ((n == 0) || (n == 1)) return(1);
  return (Fib_Cpp(n - 1) + Fib_Cpp(n - 2));
}')

library(microbenchmark)
N <- 30
comp <- microbenchmark(Fib_R(N), Fib_R_cmp(N),
                       Fib_Cpp(N), times = 25)
```

```
comp
## Unit: milliseconds
##          expr         min          lq        mean      median          uq
##       Fib_R(N) 1449.755022 1453.320560 1474.679341 1456.202559 1472.447928
##   Fib_R_cmp(N) 1444.145773 1454.127022 1489.742750 1459.170600 1554.450501
##     Fib_Cpp(N)    2.678766    2.694425    2.729571    2.711567    2.749208
##          max neval cld
##  1596.226483    25   b
##  1569.764246    25   b
##     2.858784    25  a

library(ggplot2)
autoplot(comp)
```

The last line of the aforementioned code generates a plot, that is in Figure 40.5.

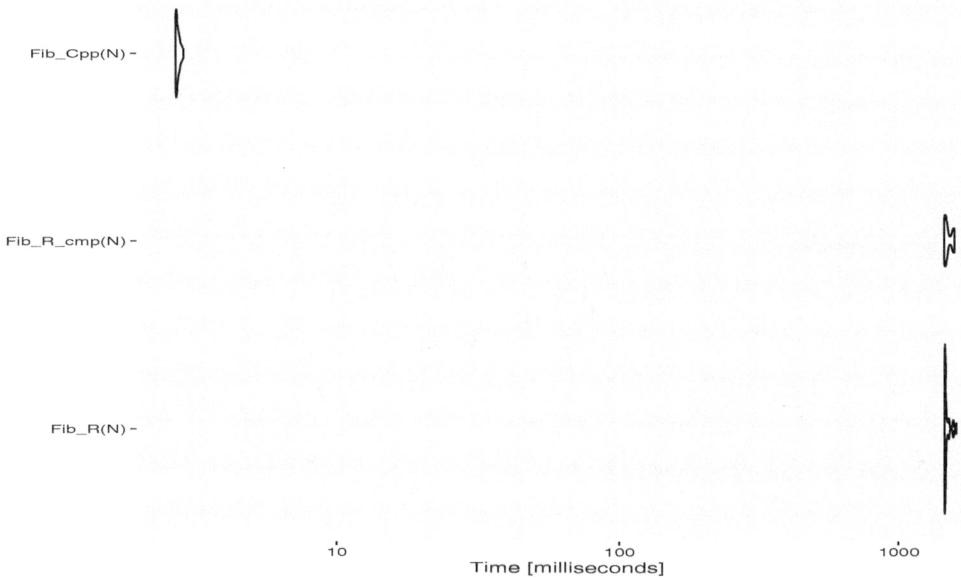

Figure 40.5: *A violin plot visualizing the advantage of using C++ code in R via* Rcpp*. This picture makes very clear that the "compilation" (via the function* cmpfun()*) has only a minor advantage over a native R-function. Using C++ and compiling via a regular compiler (in the background* cppFunction() *uses* g++*), however, is really a game changer.*

Digression – Background operation

While the function is compiling, there is quite some text that is printed in the R console. From that code, we see that the C++ function will be compiled with the regular compiler g++. This means that this compiler must be installed earlier.

The performance advantage of the compilation of native C++ code with Rcpp() is massive. The C++ function is about 500 times faster than the R function, while the compiled function in R seems to have almost no advantage.

The reader will by now probably notice that the code can still be improved further. Better results can be obtained by realizing that the way this is programmed we need to do twice the same calculations. In other words, we can still tune the algorithm.

```
# Efficient function to calculate the Fibonacci numbers in R:
Fib_R2 <- function (n) {
 x = 1
 x_prev = 1
 for (i in 2:n) {
   x <- x + x_prev
   x_prev = x
   }
 x
 }

# Efficient function to calculate the Fibonacci numbers in C++:
Rcpp::cppFunction('int Fib_Cpp2(int n) {
  int x = 1, x_prev = 1, i;
  for (i = 2; i <= n; i++) {
    x += x_prev;
    x_prev = x;
    }
  return x;
  }')

# Test the performance of all the functions:
N <- 30
comp <- microbenchmark(Fib_R(N), Fib_R2(N),
                       Fib_Cpp(N), Fib_Cpp2(N),
                       times = 20)

comp
## Unit: microseconds
##         expr         min           lq         mean      median ...
##     Fib_R(N) 1453850.637 1460021.5865 1.495407e+06 1471455.852...
##    Fib_R2(N)       2.057       2.4185 1.508404e+02       4.792...
##   Fib_Cpp(N)    2677.347    2691.5255 2.757781e+03    2697.519...
##  Fib_Cpp2(N)       1.067       1.4405 5.209175e+01       2.622...
##          max neval cld
##  1603991.462    20   b
##     2925.070    20  a
##     3322.077    20  a
##      964.378    20  a

library(ggplot2)
autoplot(comp)
```

The violin plots generated by the last line of this code are in Figure 40.6 on page 809 and show the supremacy of C++, only to be rivalled by he intelligence of the brain of the programmer. Finally, this version of the function `Fib_Cpp2()` is almost a million times faster than the native and naive approach in R (`Fib_R()`).

When writing code, it is really worth to

1. write code not only with readability in mind, but also write intelligently with speed in mind
2. and – if you are fluent in C++ – use it to your advantage via the package `Rcpp`.

Summarising, we can state `Rcpp` is an amazing tool, but it cannot rival smart coding.

Further information – Rcpp

Using C++ code in R not only requires knowledge of both C++ and R, but it also requires you to deal with different data definitions in the two environments, but it is really worth learning. We recommend to start with Hadley Wickham's introduction: `http://adv-r.had.co.nz/Rcpp.html` and read Eddelbuettel and Balamuta (2018).

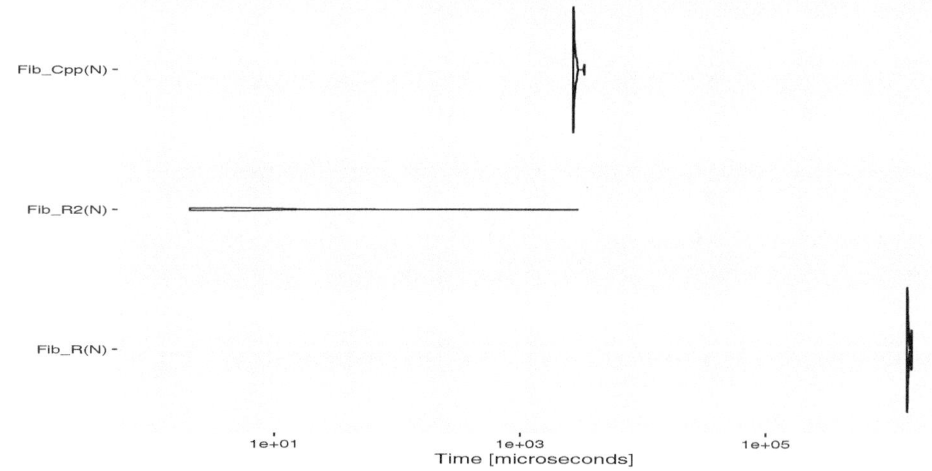

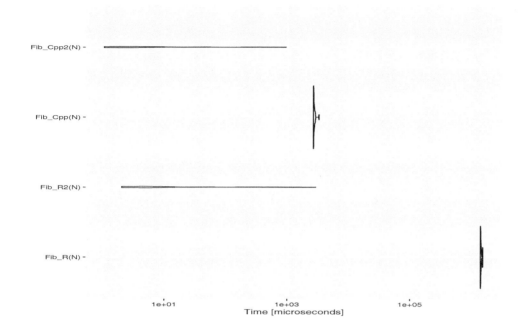

Figure 40.6: *A violin plot visualizing the advantage of using C++ code in R via* Rcpp. *Note that the scale of the x-axis defaulted to a logarithmic one. This image makes very clear that using native C++ via* Rcpp *is a good choice, but that it is equally important to be smart.*

Digression – Efficient R

Notice how close the performance of the functions Fib_R2 and Fib_Cpp2 is. R has come a long way and in simple routines (so e.g. no complex data structures, no overload for calling functions, etc.) R approximates the speed of C++ in the smart implementation of the example.

40.2.9 Using a C++ Source File in R

cppFunction() uses a string as argument and while this works fine for short functions, it becomes unwieldy for large chunks of code. In fact, we might want to include header files or simply define more than one function. In such case, we can include C++ source files and via sourceCpp(). This allows us also to use a regular C++ IDE or text editor, with C++ syntax highlighting and line numbers, save time programming and debugging, while it becomes also a smoother experience.

To achieve the integration of the C++ code in R, it is sufficient to abide to the following rules. First, the C++ source file should have the usual .cpp extension, and needs to start with: **.cpp**

```
#include <Rcpp.h>
using namespace Rcpp;
```

Further, the functions that needs to be exported to R should be preceded with exactly the following comment (nothing more and nothing less in the line preceding the function – not even an empty line):

```
// [[Rcpp::export]]
```

Hint – R code within the C++ code within R

Note also that it is possible to embed R code in special C++ comment blocks. This is really convenient if you want to run some test code, or want to keep functions logically in the same place (regardless the programming language used):

```
/*** R
# This is R code
*/
```

The R code is run with `source(echo = TRUE)` so you don't need to explicitly print output.

Once this source file is created and saved on the hard disk we can compile it in R via `sourceCpp()`.

sourceCpp()

For further reference we will assume that the file is saved under the name `/path/cppSource.cpp`.

```
sourceCpp("/path/cppSource.cpp")
```

The function `sourceCpp()` will create all the selected functions and make them available to R in the current environment.

Warning – C++ function not saved in .RData

As you know R is able to remember the current environment via saving all variables and functions in the `.Rdata` file, and then load all that information when a new session is opened. Note that the C++ functions will not be saved in and must be created again in a future session.

In the following example we use the example from previous section: the faster implementations of the Fibonacci series in both C++ and R. To import this function from C++ source, we need to create a file with the following content:

```
#include <Rcpp.h>
using namespace Rcpp;

// [[Rcpp::export]]
int Fib_Cpp2(int n) {
  int x = 1, x_prev = 1, i;
  for (i = 2; i <= n; i++) {
    x += x_prev;
    x_prev = x;
    }
  return x;
  }

/*** R
Fib_R2 <- function (n) {
 x = 1
 x_prev = 1
 for (i in 2:n) {
   x <- x + x_prev
   x_prev = x
```

```
  }
 x
 }

N <- 30
comp <- microbenchmark(Fib_R2(N), Fib_Cpp2(N), times = 20)

comp
*/
```

This source file will not only make the C++ function available but also run some tests, so that we can immediately assess if all works fine. When this file is imported via the function `sourceCpp()`, the functions `Fib_Cpp2()` and `Fib_R()` will both be available to be used in R.

Further information – Using C in R

To understand this subject better, we recommend Hadley Wickham's blog "High performance functions with Rcpp": `http://adv-r.had.co.nz/Rcpp.html`.

The code above will most probably work for you just like that, but in case you don't have the `Rcpp.h` header, it is possible to download it from `https://github.com/RcppCore/Rcpp`. It will also be part of the packages of your linux distribution. For example on Debian and its derivatives one can install it via:

```
sudo apt install r-cran-rcpp  # this prepares the use of the package Rcpp
sudo apt install r-base-dev   # this is more general and most probably what you
    need for this section
```

40.2.10 Call Compiled C++ Functions in R

Existing (compiled C and C++) functions – that sit on the hard-drive of the computer as binaries – can be called from the R environment via `.Call()` function.

R objects passed to these routines have type `SEXP`. These objects are pointers to a structure that holds the object's type, value, and other attributes. The R application programming interface (API) provides a limited set of functions to call these.

Further information – Calling C functions

More information about the subject of calling C routines in R, is for example here: `http://adv-r.had.co.nz/C-interface.html`

Digression – R in C++

It is also possible to call R functions from within a C++ program. Apart from solid knowledge of C++, it also requires that R on your computer is compiled to allow linking, and you must have a shared or static R library. We consider this beyond the scope of this book, but wanted to make sure that you know it exists and refer you to the "RInside project": `https://github.com/eddelbuettel/rinside` and its documentation: `http://dirk.eddelbuettel.com/code/rinside.html`.

40.3 Profiling Code

benchmarking

We already discussed in Chapter 40.1 *"Benchmarking"* on page 794 how benchmarking helps to provide a good insight in how fast a function really is. This is very useful, however, in a typically work-flow, one will first make a larger piece of code that "does the job." Once we have a piece of code that works, we can start to optimize it and make sure that it will run as fast as possible. We could recursively work our way through the many lines and functions and via benchmarking and keep track of which function is fastest, yet this process is not efficient nor does it provide insight in what parts of the code are called most often. The process of finding out what part of the code takes most time -and hence is the low-hanging fruit for optimization – is called "profiling."

Note – Cold code ahead

While in the rest of the book the code shown generates directly the output and plots that appear below or near to it, the code in this section is run separately and results were manually collated. The reason is that while generating this book, the code will be wrapped in other functions such as `knitr()`, `tryCatch()`, etc. and that this has a less predictable impact and – most importantly – those functions might also appear in the output that we present here and this would be confusing for the reader.

`utils`

`Rprof()`

The package `utils` that is part of base-R provides the function `Rpof()` that is the profiling tool in R. The function is called when the logging should start with the name of the file where the results should be stored. Then follows the code to be profiled and finally we call `Rprof()` again to stop the logging.

So, the general work-flow is as follows.

```
Rprof("/path/to/my/logfile")
... code goes here
Rprof(NULL)
```

As a first example we will consider a simple construction of functions that mainly call each other, in such way that some parts should display a predictable performance difference (added complexity or double amount of repetitions).

```
f0 <- function() x <- pi^2
f1 <- function() x <- pi^2 + exp(pi)
f2 <- function(n) for (i in 1:n) {f0(); f1()}
f3 <- function(n) for (i in 1:(2*n)) {f0(); f1()}
f4 <- function(n) for (i in 1:n) {f2(n); f3(n)}
```

In order to do the profiling we have to mark the starting point of the profiling operation by calling the function `Rprof()`. This indicates to R that it should start monitoring. The function takes one argument: the file to keep the results. The profiling process will register information till the process is stopped via another call to `Rprof()` with the argument `NULL`.

```
# Start the profiling:
Rprof("prof_f4.txt")

# Run our functions:
```

```
N <- 500
f4(N)

# Stop the profiling process:
Rprof(NULL)
```

Now, the file prof_f4.txt is created and can be read and analysed. The function SummaryRprof() from the package utils is a first port of call and usually provides very good insight.

```
# show the summary:
summaryRprof("prof_f4.txt")
## $by.self
##      self.time self.pct total.time total.pct
## "f0"      0.18    37.50       0.18     37.50
## "f3"      0.16    33.33       0.34     70.83
## "f1"      0.08    16.67       0.08     16.67
## "f2"      0.06    12.50       0.14     29.17
##
## $by.total
##      total.time total.pct self.time self.pct
## "f4"       0.48    100.00      0.00     0.00
## "f3"       0.34     70.83      0.16    33.33
## "f0"       0.18     37.50      0.18    37.50
## "f2"       0.14     29.17      0.06    12.50
## "f1"       0.08     16.67      0.08    16.67
##
## $sample.interval
## [1] 0.02
##
## $sampling.time
## [1] 0.48
```

The times shown are:

- by.self: the time spent at the level of a certain function;
- by.total: the time spent in a function, including all the functions that this function calls.

40.3.1 The Package profr

If you prefer a more visual presentation and other summary, then the library profr will be of interest to you. This is a wrapper around Rprof() and it creates a profr object that is a data-frame. The package provides also a function to visualize the data via the standard plot() command. profr

```
N <- 1000
require(profr)
pr <- profr({f4(N)}, 0.01)
plot(pr)
```

40.3.2 The Package proftools

If you are really serious about performance and want even more useful visualisation methods, then the package proftools will be a better choice. It offers a range of information summaries, allows a high level of customization, and has more advanced visualisations.

For example, the package allows to produce a structured table with the paths of the code that uses most time, complete with the percentage of time spent in the function and its sub-routines

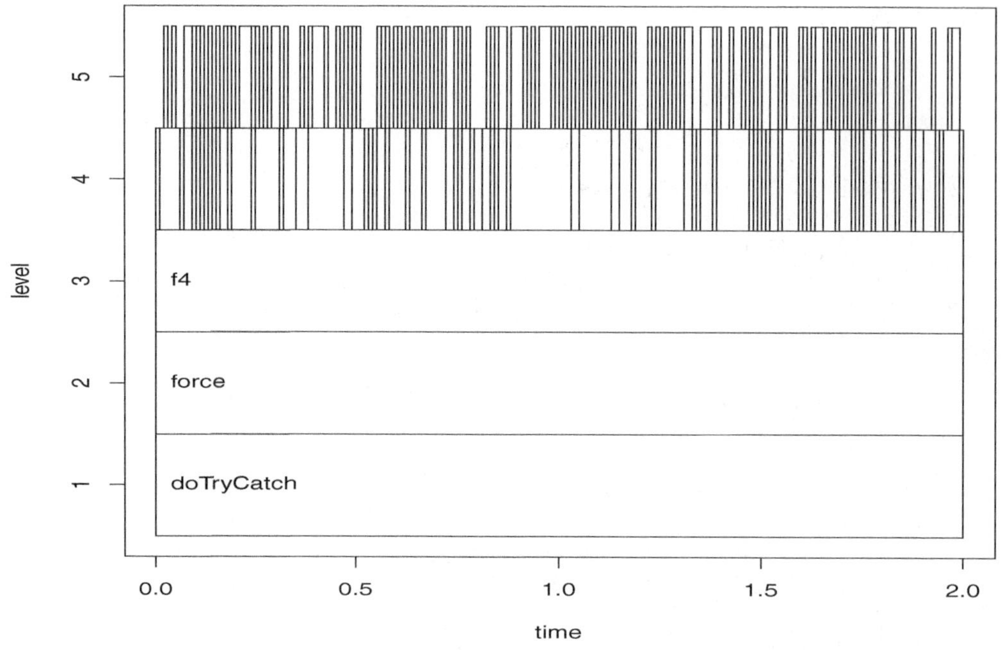

Figure 40.7: *A visualization of the profiling of our functions.*

hotPaths()
flameGraph()
calleeTreeMap()

via the function `hotPaths()` – illustrated in the following code. The function `flameGraph()` produces a colourful flame graph or a "callee tree map" for the call tree in the profile stack trace as produced by `Rprof()` – illustrated in Figure 40.8 on page 815.

This same plot can also be produced via the function `calleeTreeMap()` in order to get another appealing visual effect – – illustrated in Figure 40.9 on page 815.

```
library(proftools)

# Read in the existing profile data from Rprof:
pd <- readProfileData("prof_f4.txt")

# Print the hot-path:
hotPaths(pd, total.pct = 10.0)
## path    total.pct self.pct
## f4      100.00      0.00
##  . f3     70.83    33.33
##  . . f0  29.17    29.17
##  . f2     29.17    12.50

# A flame-graph (stacked time-bars: first following plot)
flameGraph(pd)
```

```
# Callee tree-map (intersecting boxes with area related to time
# spent in a function: see plot below)
calleeTreeMap(pd)
```

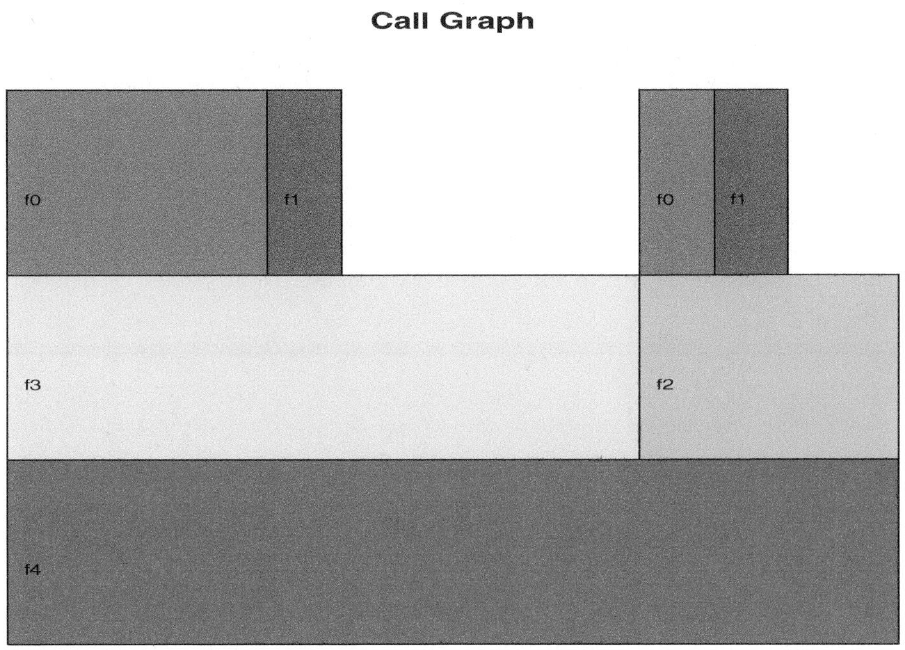

Figure 40.8: *A flame-plot produced by the function* `flameGraph()` *from the package* `proftools`.

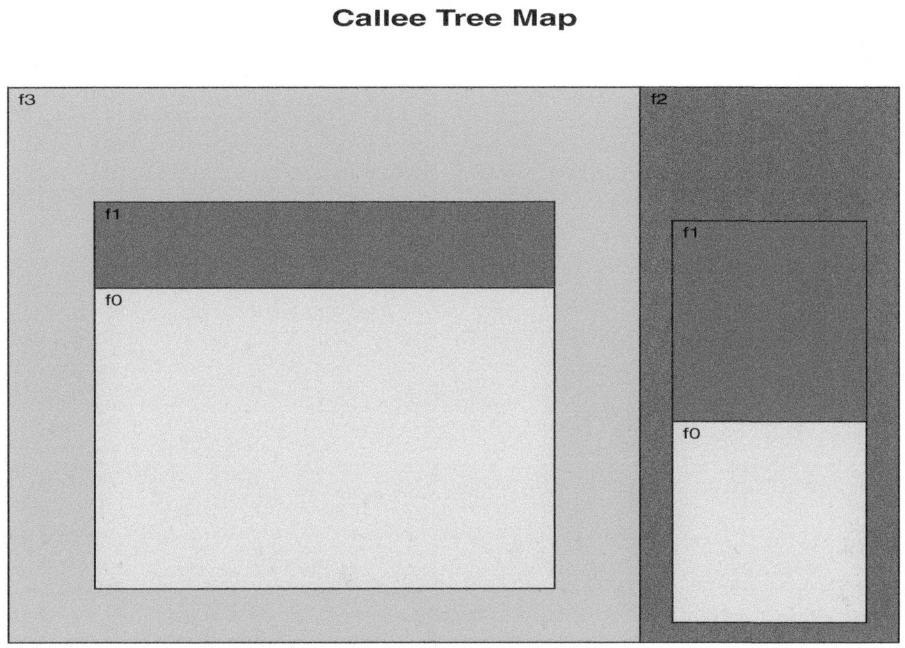

Figure 40.9: *A Callee Tee Map produced by the function* `calleeTreeMap()` *from the package* `proftools`. *This visualization shows boxes with surfaces that are relative to the time that the function takes.*

unlink()

Hint – Delete file

In order to delete the log-file via R, you can use the `unlink()` function.

```
unlink("prof_f4.txt")
```

This is useful when automating since `Rprof()` expects a not-existing file. However, note that `Rprof()` also allows you to append to an existing file via the parameter `append = TRUE`.

40.4 Optimize Your Computer

Look back at Figure 40.4 on page 805 and notice how all the performance distributions have heavy tails to the right. This means that code "usually runs take a given time to run, but in some rare cases it takes a lot more time." This is mainly due to other – high priority code – running on the same computer.

To compile this book, we use a general purpose Linux distribution, and such systems by default do not allocate the highest performance to the user processes, and if they do they will make sure that for example music in the background will not get interrupted when R starts calculating.

One way to improve this aspect is allocate a higher priority to your R process. In Linux each process gets a "niceness factor" when it is started. Niceness ranges from -20 to 19, with -20 the "meanest process" that will not be "nice" to other processes and not share processor time, and hence can be expected to be a little faster. **niceness**

Warning – Be nice

Below we show how it is possible to allocate a higher priority to R. Note that

1. in general this will make very little to no difference,
2. in the cases where it does make a difference it will interup other processes and the computer will be not responsive to user input (e.g. switch to anohter window).

Much more speed gain is to be expected from a clean programming style, optimised, and eventually C++ code.

To get some insight in what is going on, use the command `top` **top**

```
top
```

```
top - 16:18:34 up 5 days,  1:36,  4 users,  load average: 1,71, 1,41, 1,13
Tasks: 261 total,   2 running, 169 sleeping,   0 stopped,   0 zombie
%Cpu(s): 12,1 us,  1,1 sy,  0,0 ni, 86,7 id,  0,1 wa,  0,0 hi,  0,0 si,  0,0 st
KiB Mem :  8086288 total,  1551740 free,  3683512 used,  2851036 buff/cache
KiB Swap: 17406972 total, 17406972 free,        0 used.  3590084 avail Mem

  PID USER      PR  NI    VIRT    RES    SHR S  %CPU %MEM     TIME+ COMMAND
24401 philippe  20   0  144080  32172   5620 R 100,0  0,4   0:11.53 R
19962 philippe  20   0 2350004 350280 117596 S  49,7  4,3 524:02.78 thunderbird
19710 philippe  20   0 3293888 542864 191312 S   6,7  6,7  40:55.48 Web Content
20202 philippe  20   0 2974192 157296  65004 S   6,7  1,9   7:42.16 kile
19654 philippe  20   0 4092076 638440 248148 S   3,3  7,9  63:23.38 firefox
19165 root      20   0  562188 153696 138864 S   1,7  1,9   5:54.70 Xorg
 9594 philippe  20   0  579892  50712  42140 S   1,3  0,6   0:12.59 x-terminal-
    emul
 2051 philippe  20   0  578240  11732   8160 S   1,0  0,1  11:39.49 pulseaudio
    8 root      20   0       0      0      0 I   0,3  0,0   0:55.84 rcu_sched
...
```

As you can see, the user processes are started with a niceness ("NI") of 0 by default and R is not the only process running and claiming processor time. To change the priority you need the command `nice`.[3]

It is possible to change the niceness of processes while it is running with the command `renice`, but probably the most straightforward is to set the niceness at launch as follows.

```
# Launch R with niceness of +10
nice -n 10 R
```

However, this will run R with a positive niceness, which is probably not what you need. Setting a negative niceness needs administrator rights. So, you can use:

```
# Launch R with niceness of -15
sudo nice -n -15 R
```

Now, the R process will be less willing to share processor time with other processes and this will force R to run at a higher priority:

```
  PID USER      PR  NI    VIRT    RES    SHR S  %CPU %MEM     TIME+ COMMAND
11975 root       5 -15  549320 243072  22964 R  99,7  3,0   0:26.59 R
19962 philippe  20   0 2359076 197480  46964 S  49,5  2,4 594:28.54 thunderbird
19710 philippe  20   0 3328052 388672 119256 S   7,3  4,8  45:32.56 Web Content
19654 philippe  20   0 3973576 521712 177524 S   4,0  6,5  67:34.24 firefox
[...]
```

Digression – Batch R code

If you have all the code in one file that can be executed, then you can run that whole file as follows:

```
sudo nice -n -15 R -e "source('path/to/my/file')"
```

Note that setting negative niceness factors requires root privileges – hence the `sudo` command. There is a good reason why this is needed: "it is not nice to do so." Other processes such as communication, user interaction, streaming, etc. do need priority. Expect your computer to behave less nice if you do this.

A second way to improve performance via software is trying different distributions, like "Scientific Linux," "Arch," "MX Linux" or – with a little more patience but more control – "Gentoo" or "LFS." Have a look at `www.distrowatch.com` and explore some of the possible free options.

Also the hardware can be tuned and these days modern motherboards will make it easy to over-clock the CPU. This is, however, not without risk. It is important to monitor the temperature of the CPU and other elements, eventually improve cooling. In any case, this is not without danger and can physically damage the computer.

We really add this section only for completeness and mainly to allow us to make the point that all the other methods mentioned in earlier sections are more helpful and most likely the way to go. Especially read Chapters 37 to 39 and the first part of this one (40).

[3] You will also notice that there is a column with "PR" (for priority), for non-real-time processes niceness and priority are related as $PR = 20 + NI$ (where for real-time processes this is $PR = -1 - real\_time\_priority$ (real_time_priority ranges from 1 to 99, priorities for normal processes are from 100 to 300)).

PART IX

Appendices

♣ A ♣

Create your own R Package

package creation

While this book focuses on the work-flow of the data scientist and it is not a programming book per se; it still makes a lot of sense to introducing you to building your own package. A package is an ideal solution to keep together the functions and data that you need regularly for specific tasks. Of course, it is possible to keep these functions in a file that can be read in via the function `source()`, but having it all in a package has the advantage of being more standardized, structured and easier to document and so it becomes more portable so that also others can use your code. Eventually, if you have built something great and unique then it might be worth to share it via the "Comprehensive R Archive Network" (CRAN).[1]

package

source()

> **Further information – More on creating packages**
>
> This chapter provides only a quick introduction. More information can be found online, for example `http://r-pkgs.had.co.nz`, which is the website version of Hadley Wickham's book "R-Packages" — see Wickham (2015).

Before we can get started, we need to install two packages first: `devtools` – that provides the essentials to build the package – and `roxygen` – that facilitates the documentation. This is done as follows:

devtools
roxygen

```
install.packages("devtools")
install.packages("roxygen2")
```

`devtools` is the package that is the workhorse to develop packages, it is built and designed to make the process of building packages easier. For readers that use C++, it might suffice to say that `roxygen2` is the equivalent in R for Doxygen.[2] This means that `roxygen2` will be able to create documentation for R-source code if that code is documented according to a specific syntax.

[1] The link to submit a package is: `https://cran.r-project.org/submit.html`.

[2] Doxygen is the de facto standard tool to generate documentation directly form C++ source code. This is achieved by annotating the code in a specific way. `doxygen` also supports other languages such as C, C#, PHP, Java, Python, IDL, Fortran, etc. Its website is here: `http://www.doxygen.nl`.

The Big R-Book: From Data Science to Learning Machines and Big Data, First Edition. Philippe J.S. De Brouwer.

Companion Website: www.wiley.com/go/De Brouwer/The Big R-Book

There are multiple work-flows possible to create a package. One can start from an existing .R file or first create the empty skeleton of a package and then fill in the code. In any case, the two packages are needed and hence, the first step is loading the packages.

```
library(devtools)

## Loading required package: usethis
```

```
library(roxygen2)
```

A.1 Creating the Package in the R Console

We could select the all the functions used in this book to create a package and name it after the book, but it makes more sense to use a narrower scope, such as the functionality around asset pricing (see Chapter 30 *"Asset Valuation Basics"* on page 597), diversity (see Chapter 36.3.1 *"The Business Case: a Diversity Dashboard"* on page 726) or multi criteria decision analysis (see Chapter 27 *"Multi Criteria Decision Analysis (MCDA)"* on page 511). The functions for multi criteria decision analysis in this book follow a neat convention where each function name starts with `mcda_`, this would be a first step of good house-holding that is necessary for a great package in R.

We choose the functions around the diversity dashboard to illustrate how a package in R is built. The code below walks through the different steps. First, we have the function `setwd()`, that sets the working directory on the file system of the computer to the place where we want to create the package. Then, in step one, it defines the function – as it was done before, not using the `roxygen` formatting commenting style for now.

```
setwd(".")  # replace this with your desired directory

# Step 1: define the function(s)
# -- below we repeat the function as defined earlier.
# diversity
# Calculates the entropy of a system with discrete states.
# Arguments:
#    x      -- numeric vector -- observed probabilities of classes
#    prior -- numeric vector -- prior probabilities of the classes
# Returns:
#     numeric -- the entropy / diversity measure
diversity <- function(x, prior = NULL) {
  if (min(x) <= 0) {return(0);} # the log will fail for 0
  # If the numbers are higher than 1, then not probabilities but
  # populations are given, so we rescale to probabilities:
  if (sum(x) != 1) {x <- x / sum(x)}
  N <- length(x)
  if(!is.null(prior)) {
    for (i in (1:N)) {
      a <- (1 - 1 / (N * prior[i])) / (1 - prior[i])
      b <- (1 - N * prior[i]^2) / (N * prior[i] * (1 - prior[i]))
      x[i] <- a * x[i]^2 + b * x[i]
    }
   }
  f <- function(x) x * log(x)
  x1 <- mapply(FUN = f, x)
  - sum(x1) / log(N)
  }
```

Now, we are working from the correct working directory and we have the function `diversity()` in the working memory of R. This is the moment where we can create the package with the function `package.skeleton()` of `devtools`.

```
# Step 2: create the empty package directory:
package.skeleton(list = c("diversity"), # list all the functions
                                        # from the current
                                        # environment that we want
                                        # in the package.
                name = "div"            # name of the package
                )
```

package.skeleton()

The function parameter `list` is a vector of the names of all functions that we want to be included in the package. The argument `name` supplied to the function `package.skeleton()` is the name of the package and it will become a subdirectory in the current directory of our file-system. In that directory there you will find now the following files and directories:

```
list.files(path = "div", all.files = TRUE)
## [1] "."                    ".."
## [3] "DESCRIPTION"          "man"
## [5] "NAMESPACE"            "R"
## [7] "Read-and-delete-me"
```

If you find these files, it means that you have created your first package. Those files are the package and can be used as any other package.

RStudio

Digression – Creating packages in RStudio

If you prefer a GUI and you work with RStudio, then it is possible create packages via the menu structure. Click on "File," then "New project ...," then "New Directory" (or use an existing one), choose "R Package" and fill out the details. After confirming, RStudio will create the skeleton for the package, and you will find almost exactly the same files as directly using the function `package.skeleton()` as we did in the aforementioned code. If you have followed the RStudio work-flow, you should have now a new menu item: `build`. This menu allows to load the package and test it. Just as in our aforementioned approach, RStudio has created the following files for you:

- `DESCRIPTION` provides meta-data about your package. We edit this shortly.
- `NAMESPACE` declares the functions your package exports for external use and the external functions your package imports from other packages. At the moment, it holds temporary-yet-functional place-holder content.
- `R` contains the R-code: ideally a .R file for each function.
- `man` contains the documentation for the functions and the package.

Compared to the aforementioned example with no options in `package.skeleton()`, RStudio, has created some extra files:

- `.gitignore` anticipates Git usage and ignores some standard, behind-the-scenes files created by R and RStudio. If you do not plan to use Git, this will will have no impact on anything else.
- `.Rbuildignore` lists files that we need to have around but that should not be included when building the R package from this source.
- `foofactors.Rproj` is the file that makes this directory an RStudio Project. Even if you do not use RStudio, this file is harmless. Or you can suppress its creation with create(..., rstudio = FALSE).

A.2 Update the Package Description

Anyone, who will use your package, needs a short description to understand what the package does and why he or she would need it. Adding this documentation is done by editing the `./div/DESCRIPTION` file. For example, replace its content by:

```
Package: div
Type: Package
Title: Provides functionality for reporting about diversity
Version: 0.8
Date: 2019-07-10
Author: Philippe J.S. De Brouwer
Maintainer: Who to complain to <philippe@de-brouwer.com>
Description: This package provides functions to calculate diversity of discrete
     observations (e.g. males and females in a team) and functions that allow
    a reporting to see if there is no discrimination going on towards some
 of those categories (e.g. by showing salary distributions for the groups of
    observations per salary band).
License: LGPL
RoxygenNote: 6.1.1
```

The last line is used at by R to document the functions, so it is essential to leave it exactly as it was.

Now, that the package has a high-level explanation, it is time to document each function of the package.

A.3 Documenting the Functions

You want, of course, that the functions are neatly documented in the usual way that R recognizes so that all methods such as asking for documentation should also work for your package. For example the following line of code should lead to the documentation of the function `diversity()`, and display standard help-file formatted in the usual way.

```
?diversity
```

Each function will have its own file with documentation. R will look for these files in the `man` directory. As you will notice, these files follow syntax that is reminiscent to LaTeX. If your functions are documented via the standards of `roxygen`, these files will be created automatically, but remain editable.

Open the file `R/diversity.R` and insert right before the function the following comments.

```
#' Function to calculate a diversity index based on entropy.
#'
#' This function returns entropy of a system with discrete states
#' @param x numeric vector, observed probabilities
#' @param prior numeric vector, the prior probabilities
#' @keywords diversity, entropy
#' @return the entropy or diversity measure
#' @examples
#' x <- c(0.4, 0.6)
#' diversity(x)
```

`document()`

To process this documentation, and make it available, use the function `document()`:

```
setwd("./div")  # or your choice of package directory
document()
```

`document()`

At this point, the `document()` function is likely to generate the following errors:

```
Warning: The existing 'NAMESPACE' file was not generated by roxygen2,
and will not be overwritten.
Warning: The existing 'diversity.Rd' file was not generated by
roxygen2, and will not be overwritten.
```

If you prefer to use `roxygen`, it is safe to delete the `./div/man/diversity.Rd` as well as the `./div/NAMESPACE` files, and then execute again the `document()` command. This command will now recreate both files and from now on over-write whenever we run the `document()` command again.

Alternatively, you can use the parameter `roclets` of the function `document` to list what R exactly should do when building your package.

A.4 Loading the Package

The package can be loaded directly from the source code that resides on our hard-disk.

```
setwd(".") # the directory in which ./div is a subdirectory
install("div")
```

The package is now loaded[3] and the functions are available as is its documentation. The code below illustrates the use of the package.

```
x  <- c(0.3, 0.2, 0.5)
pr <- c(0.33, 0.33, 0.34)
diversity(x, prior = pr)
## [1] 0.9411522
```

Test the documentation by executing the following.

```
?diversity
```

This will invoke the help file of that particular function in the usual way for your system setup (a `man`-like environment in the CLI, a window in RStudio or even a page in a web-browser).

[3] Note that the output of the `install()` function is quite verbose and is therefore, suppressed in the book.

A.5 Further Steps

github

If your package does not contain confidential information or propriety knowledge, and you have all rights to it, then you might want to upload it to "github" so it can be shared with the Internet community. `devtools` has even a function that does this for you:

```
install_github('div','your_github_username')
```

And of course – in agreement with the philosophy of agile programming – you will continue to add functions, improve them, improve documentation, etc. From time to time update the version number, document, and upload again.

Also consider to read the `Read-and-delete-me` file and grab Hadley Wickham's book about R-packages. Thanks to his thought leadership packages have indeed become the easiest way to share R-code.

Further information – Further reading

The guide by MIT is very helpful to take a few more steps and this guide also includes more information about other platforms such as Windows. It is here: `http://web.mit.edu/insong/www/pdf/rpackage_instructions.pdf`.
Another great reference is Hadley Wickham's book 2015 book "R packages: organize, test, document, and share your code." It is also freely available on the Internet: `http://r-pkgs.had.co.nz`.

♣ B ♣

Levels of Measurement

level of measurement

B.1 Nominal Scale

Some things in this world can be ordered in a meaningful way. Numbers, such as sales, turnover, etc. are such examples. Other things such as colours, people, countries, etc. might be ordered in some way or another, but do not have an inherent dominant order.

For example, when making a model for credit applications, we have usually information about the category of job. People that apply for a loan, choose one of the following that applies best: law enforcement, military, other government, blue collar worker, white collar worker, teacher, employee, self-employed, student, retired. This makes sense, because some professions are more risky and hence this will impact the quality of the credit. However, is there an order? This is an example of a nominal scale: we have only names, but no order imposes itself on these labels.

scale – nominal

The nominal scale is the simplest form of classification. It contains labels that do not assume an order. Examples include asset classes, first names, countries, days of the month, weekdays, etc. It is not possible to use statistics such as average or median, and the only thing that can be measured is which label occurs the most ("modus" or "mode").

This scale can be characterised as represented in Table B.1

Aspect	Nominal scale
Characterization	Labels (e.g. asset classes, stock exchanges)
Permissible Statistics	Mode (not median or average), chi-square
Permissible Scale Transformation	Equality
Structure	Unordered set

Table B.1: *Characterization of the nominal scale of Measurement.*

Note that it is possible to use numbers as labels, but that this is very misleading. When using an nominal scale, none of the traditional metrics (such as averages) can be used.

The Big R-Book: From Data Science to Learning Machines and Big Data, First Edition. Philippe J.S. De Brouwer.

Companion Website: www.wiley.com/go/De Brouwer/The Big R-Book

scale – ordinal

B.2 Ordinal Scale

A typical feedback is these days done by stars, usually a user can rate a service, driver, product or seller between 1 to 5 stars. This is, of course, equivalent to "Rate our product with a number between 1 and 5." The numbers are a little misleading here. Numbers as such can be added, subtracted, multiplied, etc. However, is this still meaningful in this example?

It does not really work. Is 2 stars twice as good as one star? Are 2 customers that provide 2 stars together as happy as one that provides 4 stars? It seems that for most people there is little difference between 3 and 4 stars, but there is a wider gap between 4 and 5 stars. So, while there is a clear order, this scale is not equivalent to using numbers. An average, for example has little meaning. However, calculating it and monitoring how it evolves can provide some insight in how customers appreciate recent changes.

This scale type assumes a certain order. An example is a set of labels such as very safe, moderate, risky, and very risky. Bond rating such as AAA, BB+, etc. also are ordinal scales: they indicate a certain order, but there is no way to determine if the distance between them is the same or different. For example, it is not really clear if the probability of default difference between AAA and AA is similar to the distance between BBB and BB.

For such variables, it may make sense to talk about a median, but it does not make any sense to calculate an average (as is sometimes done in the industry and even in regulations).

These characteristics are summarised in Table B.2

Aspect	Ordinal scale
Characterization	Ranked labels (e.g. ratings for bonds from rating agencies)
Permissible statistics	Median, percentile
Permissible scale Transformation	Order
Structure	(Strictly) ordered set

Table B.2: *Characterization of the ordinal scale of measurement.*

Ordinal labels can be replaced by others if the strict order is conserved (by a strict increasing or decreasing function). For example, AAA, AA-, and BBB+ can be replaced by 1, 2 and, 3 or even by -501, -500, and 500 000. The information content is the same, the average will have no meaningful interpretation.

B.3 Interval Scale

scale – interval

In some cases. we are able to be sure about the order as well as the distance between the different units. This type of scale will not only have meaningful order but also differences.

This scale can be used for many quantifiable variables: temperature (in degrees Celsius). In this case, the difference between 1 and 2 degrees is the same as the difference between 100 and 101 degrees, and the average has a meaningful interpretation. Note that the zero point has only an arbitrary meaning, just like using a number for an ordinal scale: it can be used as a name, but it is only a name.

Aspect	Interval Scale
Characterization	difference between labels is meaningful (e.g. the Celsius scale for temperature)
Permissible Statistics	mean, standard deviation, correlation, regression, analysis of variance
Permissible Scale Transformation	affine
Structure	affine line

Table B.3: *Characterization of the Interval Scale of Measurement.*

Rescaling is possible and remains meaningful. For example, a conversion from Celsius to Fahrenheit is possible via the following formula, $T_f = \frac{9}{5}T_c + 32$, with T_c the temperature in Celsius and T_f the temperature in Fahrenheit.

An affine transformation is a linear transformation of the form $\boldsymbol{y} = \boldsymbol{A}.\boldsymbol{x} + \boldsymbol{b}$. In Euclidean space an affine transformation will preserve collinearity (so that lines that lie on a line remain on a line) and ratios of distances along a line (for distinct collinear points p_1, p_2, p_3, the ratio $||p_2 - p_1||/||p_3 - p_2||$ is preserved).

In general, an affine transformation is composed of linear transformations (rotation, scaling and/or shear) and a translation (or "shift"). An affine transformation is an internal operation and several linear transformations can be combined into one transformation.

scale – ratio

B.4 Ratio Scale

Using the Kelvin scale for temperature allows us to use a ratio scale: here not only the distances between the degrees but also the zero point is meaningful. Among the many examples are profit, loss, value, price, etc. Also a coherent risk measure is a ratio scale, because of the property translational invariance implies the existence of a true zero point.

Aspect	Ratio Scale
Characterization	a true zero point exists (e.g. VAR, $V@R$, ES)
Permissible Statistics	geometric mean, harmonic mean, coefficient of variation, logarithms, etc.
Permissible Scale Transformation	multiplication
Structure	field

Table B.4: *Characterization of the Ratio Scale of Measurement.*

♣ C ♣

Trademark Notices

We do not claim any ownership to trademarks or registered names, we want to thank all those people and companies that spend money and time to produce great software and make it available for free to allow others to stand on the shoulders of giants. We used also names of well known companies and some of their commercial products. Also these words are only used as a citation and their rightful owner is the source.

The Big R-Book: From Data Science to Learning Machines and Big Data, First Edition. Philippe J.S. De Brouwer.

Companion Website: www.wiley.com/go/De Brouwer/The Big R-Book

C.1 General Trademark Notices

Some often used words are here:

- RStudio and Shiny are registered trademarks of RStudio, Inc.
- Excel and R are trademarks of Microsoft, Inc.
- Linux is a trademark owned by Linus Torvalds.
- Oracle and Java are registered trademarks of Oracle and/or its affiliates.
- SAS is owned by the SAS Institute for advanced analytics
- Tableau is ownde by Tableau Software, Inc.
- Spark, MySQL, Hadoop are owned by the Apache Software Foundation.
- Excel is owned by Microsoft.

All other references to names, people, software and companies are made without any claim of ownership. For example: IBM, FORTRAN, BASIC, CODASYL, COBOL, CALC, SQL, INGRES, DS, DB, CPUs, noSQL, CAP, NewSQL, Nvidia, AMD, Intel, PHP, Python, Perl, etc. all are companies or registered trademarks by other companies. This holds also for all other names that we might have omited in this list.

C.2 R-Related Notices

C.2.1 Crediting Developers of R Packages

R has a built in mechanism to cite the use of packages in scientific work. So, let us use those and honour all the packages that allowed this book to bring you interesting content.

```
# When you use a R-package, you should cite it in your work.
# The information can be found as follows:
citation('base')
##
## To cite R in publications use:
##
##   R Core Team (2018). R: A language and environment
##   for statistical computing. R Foundation for
##   Statistical Computing, Vienna, Austria. URL
##   https://www.R-project.org/.
##
## A BibTeX entry for LaTeX users is
##
##   @Manual{,
##     title = {R: A Language and Environment for Statistical Computing},
##     author = {{R Core Team}},
##     organization = {R Foundation for Statistical Computing},
##     address = {Vienna, Austria},
##     year = {2018},
##     url = {https://www.R-project.org/},
##   }
##
## We have invested a lot of time and effort in
## creating R, please cite it when using it for data
## analysis. See also 'citation("pkgname")' for citing
## R packages.

# You can even extract the BibTex information:
toBibtex(citation('base'))
## @Manual{,
##   title = {R: A Language and Environment for Statistical Computing},
##   author = {{R Core Team}},
##   organization = {R Foundation for Statistical Computing},
##   address = {Vienna, Austria},
##   year = {2018},
##   url = {https://www.R-project.org/},
## }
```

C.2.2 The R-packages used in this Book

Here are the packages that we have used in this book (in alphabetical order). Details about the citation can be found in the Bibliography, as with all other references to bibliography.

1. `base`: R Core Team (2018)
2. `binr`: Izrailev (2015)
3. `carData`: Fox et al. (2018)
4. `class`: Venables and Ripley (2002a)

5. `cluster`: Maechler et al. (2019)
6. `compiler`: R Core Team (2018)
7. `datasets`: R Core Team (2018)
8. `DBI`: R Special Interest Group on Databases (R-SIG-DB) et al. (2018)
9. `devtools`: Wickham et al. (2019)
10. `diagram`: Soetaert (2017a)
11. `DiagrammeR`: Iannone (2019)
12. `e1071`: Meyer et al. (2019)
13. `FactoMineR`: Lê et al. (2008)
14. `flexdashboard`: Iannone et al. (2018)
15. `forecast`: Hyndman et al. (2019) and Hyndman and Khandakar (2008)
16. `ggfortify`: Tang et al. (2016)
17. `ggplot2`: Wickham (2016)
18. `ggrepel`: Slowikowski (2018)
19. `ggvis`: Chang and Wickham (2018)
20. `googleVis`: Gesmann and de Castillo (2011)
21. `grid`: Rupp et al. (0 27)
22. `grid`: R Core Team (2018)
23. `gridExtra`: Auguie (2017)
24. `Hmisc`: Harrell Jr et al. (2019)
25. `InformationValue`: Prabhakaran (2016)
26. `knitr`: Xie (2018, 2015, 2014)
27. `latex2exp`: Meschiari (2015)
28. `MASS`: Venables and Ripley (2002b)
29. `mi`: Gelman and Hill (2011)
30. `mice`: van Buuren and Groothuis-Oudshoorn (2011)
31. `missForest`: Stekhoven (2013) and Stekhoven and Buehlmann (2012)
32. `msm`: Jackson (2011)
33. `neuralnet`: Fritsch et al. (2019)
34. `nFactors`: Raiche (2010)

35. `parallel`: R Core Team (2018)
36. `plot3D`: Soetaert (2017b)
37. `plotly`: Sievert (2018)
38. `plyr`: Wickham (2011)
39. `pROC`: Robin et al. (2011)
40. `profr`: Wickham (2018a)
41. `proftools`: Tierney and Jarjour (2016)
42. `pryr`: Wickham (2018b)
43. `psych`: Revelle (2018)
44. `quantmod`: Ryan and Ulrich (2019)
45. `randomForest`: Liaw and Wiener (2002)
46. `RColorBrewer`: Neuwirth (2014)
47. `RCurl`: Lang and the CRAN team (2019)
48. `Rcpp`: Eddelbuettel and Balamuta (2017)
49. `reshape`: Wickham (2007)
50. `rex`: Ushey et al. (2017)
51. `RMySQL`: Ooms et al. (2018)
52. `ROCR`: Sing et al. (2005)
53. `roxygen2`: Wickham et al. (2018)
54. `rpart`: Therneau and Atkinson (2019)
55. `rpart.plot`: Milborrow (2019)
56. `sandwich`: Zeileis (2004, 2006) and Berger et al. (2017)
57. `shiny`: Chang et al. (2019)
58. `shinydashboard`: Chang and Borges Ribeiro (2018)
59. `SnowballC`: Bouchet-Valat (2019)
60. `sodium`: Ooms (2017)
61. `sparklyr`: Luraschi et al. (2019)
62. `SparkR`: Venkataraman et al. (2019)
63. `stats`: R Core Team (2018)
64. `stringr`: Wickham (2019)

65. `tidyverse`: Wickham (2017)
66. `titanic`: Hendricks (2015)
67. `tm`: Feinerer and Hornik (2018) and Feinerer et al. (2008)
68. `VIM`: Kowarik and Templ (2016)
69. `vioplot`: Adler and Kelly (2018)
70. `viridisLite`: Garnier (2018)
71. `wordcloud`: Fellows (2018)
72. `XML`: Lang and the CRAN Team (2019)
73. `xts`: Ryan and Ulrich (2018)
74. `zoo`: Zeileis and Grothendieck (2005)

♣ D ♣

Code Not Shown in the Body of the Book

In rare occasions, we did not show all code. This was done out of concern that the code would get in the way of understanding the subject discussed. However, we do not want to deprive you from this code and will reproduce it here for all the occurrences where certain code was not omitted.

Code from Chapter 19 on page 343

In this section we introduce a simple method to visualize some aspects of the data that are helpful for deciding on how to bin the data and have used base-R. It might be useful to know that in `ggplot2` these results are equally easy to obtain and might suit better your needs. The output of this code is in Figure D.1 on page 840.

```
# -- remind the data fabrication ---
set.seed(1865)
age <- rlnorm(1000, meanlog = log(40), sdlog = log(1.3))
y <- rep(NA, length(age))
for(n in 1:length(age)) {
  y[n] <- max(0,
              dnorm(age[n], mean= 40, sd=10)
                 + rnorm(1, mean = 0, sd = 10 * dnorm(age[n],
                   mean= 40, sd=15)) * 0.075)
}
y <- y / max(y)
dt <- tibble (age = age, spending_ratio = y)

# -- plotting loesss and histogram with ggplot2 ---
library(ggplot2)
library(gridExtra)
p1 <- ggplot(dt, mapping = aes(x = age, y = spending_ratio)) +
     geom_point(alpha=0.2, size = 2) +
     geom_smooth(method = "loess")
p2 <- qplot(dt$age, geom="histogram", binwidth=5)
grid.arrange(p1, p2, ncol=2)
```

The Big R-Book: From Data Science to Learning Machines and Big Data, First Edition. Philippe J.S. De Brouwer.

Companion Website: www.wiley.com/go/De Brouwer/The Big R-Book

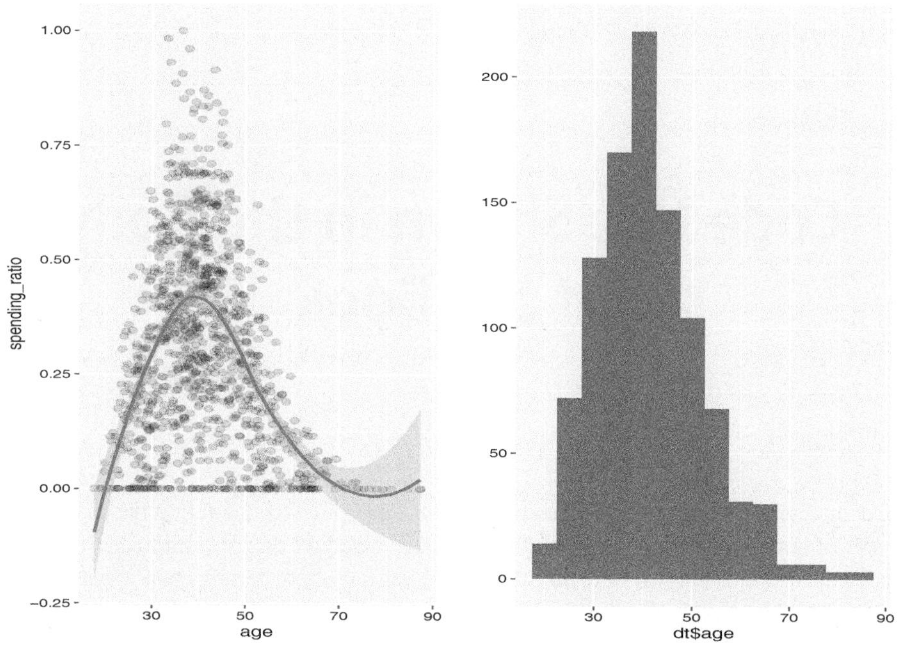

Figure D.1: *A visual aid to select binning borders is plotting a non-parametric fit and the histogram.*

Code from Section 9.1 on page 161

This code fragment produces a plot that visualizes all possible `pch` arguments for the plot function as shown in Figure 9.3 on page 162.

```
# This sets up an empty plotting field:
plot(x = c(0, 4.5),
     y = c(0, 5),
     main = "Some pch arguments",
     xaxt = "n",
     yaxt = "n",
     xlab = "",
     ylab = "",
     cex.main = 2.6,
     col = "white"
)

# This will plot all of the standard pch arguments:
y = rep(5:0, each=5)
for (i in 0:25) {
  points(x = i %% 5, y = y[i+1], pch = i,cex = 2, col="blue", bg="khaki3")
  text(0.3 + i %% 5, y = y[i+1], i, cex = 2)
}
for (i in 1:2) {
  ch <- LETTERS[i]
  points(x = i, y = 0, pch = ch,cex = 2, col="red")
  text(0.3 + i, y = 0, ch, cex = 2)
}
for (i in 1:2) {
  ch <- letters[i]
  points(x = i + 2, y = 0, pch = ch,cex = 2, col="red")
  text(0.3 + i + 2, y = 0, ch, cex = 2)
}
```

Code from Section 22.2.5 on page 398

The following code produces the plot that is presented in Figure 22.6 on page 398. It uses `ggplot2` and requires a fair amount of fiddling to get the line in the correct place and the words next to the line.

```
# Using model m and data frame t2:
predicScore <- predict(object=m,type="response")
d0 <- data.frame(
       score = as.vector(predicScore)[t2$Survived == 0],
       true_result = 'not survived')
d1 <- data.frame(
       score = as.vector(predicScore)[t2$Survived == 1],
       true_result = 'survived')
d  <- rbind(d0, d1)
d  <- d[complete.cases(d),]

cumDf0 <- ecdf(d0$score)
cumDf1 <- ecdf(d1$score)
x      <- sort(d$score)
cumD0  <- cumDf0(x)
cumD1  <- cumDf1(x)
diff   <- cumD0 - cumD1
y1     <- gdata::first(cumD0[diff == max(diff)])
y2     <- gdata::first(cumD1[diff == max(diff)])
x1     <- x2 <- quantile(d0$score, probs=y1, na.rm=TRUE)

# Plot this with ggplot2:
p <- ggplot(d, aes(x = score)) +
     stat_ecdf(geom = "step", aes(col = true_result), lwd=2) +
     ggtitle('Cummulative distributions and KS') +
     geom_segment(aes(x = x1, y = y1, xend = x2, yend = y2),
                  color='navy', lwd=3) +
     ggplot2::annotate("text",
              label = paste0("KS=",round((y1-y2)*100,2),"%"),
              x = x1 + 0.15, y = y2+(y1-y2)/2, color = "navy")
p
```

The aforementioned code includes a way to calculate KS. Since it is a little long, there are functions in available to help us with that, as explained in Section 22.2.5 *"Kolmogorov-Smirnov (KS) for Logistic Regression"* on page 398.

Code from Section 27.7.5 on page 540

The following code produces the plot Figure 27.15 on page 551. Note especially the following:

- the definition of the error function and related functions, they are not readily available in R, but are closely related to `qnorm()` – see for example Section 29 in Burington et al. (1973);
- the library `latex2exp` that provides the function `TeX()`, which allows to convert text to be type-setted by a LaTeX-like engine (note that it is not LaTeX it converts LaTeX syntax to R's plotmat functions);

`latex2exp`
`TeX()`
plotmat

theme()

- the function theme() acting on ggplot2 and more in particular:
  - legend.key.size = unit(1.5, "cm") which increases the spacing between the items in the legend,
  - legend.box.margin=ggplot2::margin(rep(20, times=4)) which regulates how far the legend is placed from the border of the plot — also note that the function margin() was overwritten by a function from randomForest and hence we need to make clear that we want to call the function of ggplot2 — also note that the function margin uses a vector of four numbers and since we wanted them to be all the same we use the function rep() to produce a repetition of the same number, this is just a little easier to change to find the perfect setting,
  - the rest of the theme function positions the legend on the plot area.

margin()

```
library(ggplot2)
library(latex2exp)
d <- seq(from = -3, to = +3, length.out = 100)

## error function family:
erf      <- function(x) 2 * pnorm(x * sqrt(2)) - 1
# (see Abramowitz and Stegun 29.2.29)
erfc     <- function(x) 2 * pnorm(x * sqrt(2), lower = FALSE)
erfinv  <- function (x) qnorm((1 + x)/2)/sqrt(2)
erfcinv <- function (x) qnorm(x/2, lower = FALSE)/sqrt(2)

## Gudermannian function
gd <- function(x) asin(tanh(x))

f1 <- function(x) erf( sqrt(pi) / 2 * x)
f2 <- function(x) tanh(x)
f3 <- function(x) 2 / pi * gd(pi / 2 * x)
f4 <- function(x) x / sqrt(1 + x^2)
f5 <- function(x) 2 / pi * atan(pi / 2 * x)
f6 <- function(x) x / (1 + abs(x))

df <- data.frame(d = d, y = f1(d),
                 Function = "erf( sqrt(pi) / 2 * d)")
df <- rbind(df, data.frame(d = d, y = f2(d), Function = "tanh(d)"))
df <- rbind(df, data.frame(d = d, y = f3(d),
                           Function = "2 / pi * gd(pi / 2 * d)"))
df <- rbind(df, data.frame(d = d, y = f4(d),
                           Function = "d / (1 + d^2)"))
df <- rbind(df, data.frame(d = d, y = f5(d),
                           Function = "2 / pi * atan(pi / 2 * d)"))
df <- rbind(df, data.frame(d = d, y = f6(d),
                           Function = "x / (1 + abs(d))"))

fn <- ""
fn[1] <- "erf \\left(\\frac{\\sqrt{\\pi} d}{2}\\right)"
fn[2] <- "tanh(x)"
fn[3] <- "\\frac{2}{\\pi} gd\\left( \\frac{\\pi d}{2} \\right)"
fn[4] <- "\\frac{d}{1 + d^2}"
fn[5] <- "\\frac{2}{\\pi} atan\\left(\\frac{\\pi d}{2}\\right)"
fn[6] <- "\\frac{x}{1+ |x|}"


ggplot(data = df, aes(x = d, y = y, color = Function)) +
  geom_line(aes(col = Function), lwd=2) +
  guides(color=guide_legend(title=NULL)) +
```

```
scale_color_discrete(labels=lapply(sprintf('$\\pi(d) = %s$', fn),
                                   TeX)) +
theme(legend.justification = c(1, 0),
     legend.position = c(1, 0),    # south east
     legend.box.margin=ggplot2::margin(rep(20, times=4)),
     # increase vertical space between legend items:
     legend.key.size = unit(1.5, "cm")
     ) +
ylab(TeX('Preference --- $\\pi$'))
```

The following code, produces the plot Figure 27.9 on page 541. Pay attention to:

- the function `deparse()` that converts objects to vectors of strings – in our case the functions have only one line, so we can capture the whole function in its second line – alternatively, one could use `collapse(deparse(f))` to capture the text of functions that span more than one line.
- for most functions we would not need to vectorize, however, for example `min()` and `max()` would fail, so we vectorize all of them. This creates a version of the function that can work on vectors;
- note also that the functions such as the `min`, `max`, and `gaus` versions require additional parameters – here some choices have been made.

```
f <- function(x) x^2 + 1
f
## function(x) x^2 + 1

Vectorize(f)
## function (x)
## {
##     args <- lapply(as.list(match.call())[-1L], eval, parent.frame())
##     names <- if (is.null(names(args)))
##         character(length(args))
##     else names(args)
##     dovec <- names %in% vectorize.args
##     do.call("mapply", c(FUN = FUN, args[dovec], MoreArgs = list(args[!dovec]),
##         SIMPLIFY = SIMPLIFY, USE.NAMES = USE.NAMES))
## }
## <environment: 0x56342f0a9c80>


f_curve <- function(f) {
  g <- Vectorize(f)
  s <- deparse(f)[2]
   curve(g, xlab = '', ylab = '', col = 'red', lwd = 3,
         from = -1, to = +1,
#-1-         main = bquote(bold(.(s))))
          main = s
         )
   }

gaus <- function(x) exp (-(x-0)^2 / 0.5)
f1 <- function(x) - 3/2 * x^5 + 5/2 * x^3
f2 <- function(x) sin(pi * x / 2)
f3 <- function(x) min(1, max(2*x, -1))
f4 <- function(x) x
f5 <- function(x) ifelse(x < 0 , gaus(x) - 1, 1 - gaus(x))
f6 <- function(x) tanh(3 * x)
```

```
par(mfrow=c(3,2))
f_curve(f1)
f_curve(f2)
f_curve(f3)
f_curve(f4)
f_curve(f5)
f_curve(f6)
par(mfrow=c(1,1))
```

Note that the part that is commented out with `#-1-` produces nicer expressions in all titles except for the one using the functions `min()` and `max()`. These functions work unfortunately different in `plotmath` and R itself, causing the typesetting to be messed up. Using the function TeX such as in previous plot would solve the issue. However, we wanted to point out the possibilities of the functions `expression()` and `bquote()` in this plot.

♣ E ♣

Answers to Selected Questions

Question 1 on page 14

A suggestion could be the distribution of IQ (as in intelligence quotient). This seems to follow more or less a normal distribution with a standard deviation of 15 points.

Question 2 on page 31

All operations in R are by default done on objects element by element. Hence, the following code is sufficient to create an object `nottemC` that contains all temperatures in degrees Celsius.

```
nottemC <- 5/9 * nottem - 32
```

While this concept makes such calculations very easy, it must be noted that this will – by default – not produce a matrix or vector product as you might expect.

Question 3 on page 34

There are many ways to tackle this problem. Here is one.

```
dotproduct <- function (m1, m2) {
  if(!is.matrix(m1) || !is.matrix(m2)) {
    print("ERROR 1: m1 and m2 must be matrices.");
    return(-1);
  }
  if (ncol(m1) != nrow(m2)) {
    print("ERROR 2: ncol(m1) must match nrow(m2).")
    return(-2);
  }
  # Note that we do not check for values NA or Inf. In R they work
  # as one would expect.
  m <- matrix(nrow=nrow(m1), ncol=ncol(m2))
  for (k in 1:nrow(m1)) {
    for (l in 1:ncol(m1)){
```

The Big R-Book: From Data Science to Learning Machines and Big Data, First Edition. Philippe J.S. De Brouwer.

Companion Website: www.wiley.com/go/De Brouwer/The Big R-Book

```
      m[k,l] <- 0
      for (n in 1:ncol(m1)) {
        m[k,l] <- m[k,l] + m1[k,n] * m2[n,l]
        }
      }
  }
  return(m);
}

M1 <- matrix(1:6, ncol=2)
M2 <- matrix(5:8, nrow=2)

# Try our function
dotproduct(M1, M2)
##      [,1] [,2]
## [1,]   29   39
## [2,]   40   54
## [3,]   51   69

# Compare with
M1 %*% M2
##      [,1] [,2]
## [1,]   29   39
## [2,]   40   54
## [3,]   51   69

# The following must fail:
dotproduct("apple", M2)
## [1] "ERROR 1: m1 and m2 must be matrices."
## [1] -1

# This must fail too:
dotproduct(as.matrix(pi),M1)
## [1] "ERROR 2: ncol(m1) must match nrow(m2)."
## [1] -2
```

Question 4 on page 48

This question allows a lot of freedom, and there are many ways to tackle it. Here is one of the possible solutions, by using histograms, correlations and scatter plots.

```
# Loading mtcars
library(MASS)     # However, this is probably already loaded

# Exploring the number of gears
summary(mtcars$gear)
##    Min. 1st Qu.  Median    Mean 3rd Qu.    Max.
##   3.000   3.000   4.000   3.688   4.000   5.000

## show the histogram
# since the only valuse are 3,4 and 5 we format the
# breaks around those values.
hist(mtcars$gear, col="khaki3",
     breaks=c(2.5, 2.9,3.1,3.9,4.1,4.9,5.1, 5.5))
```

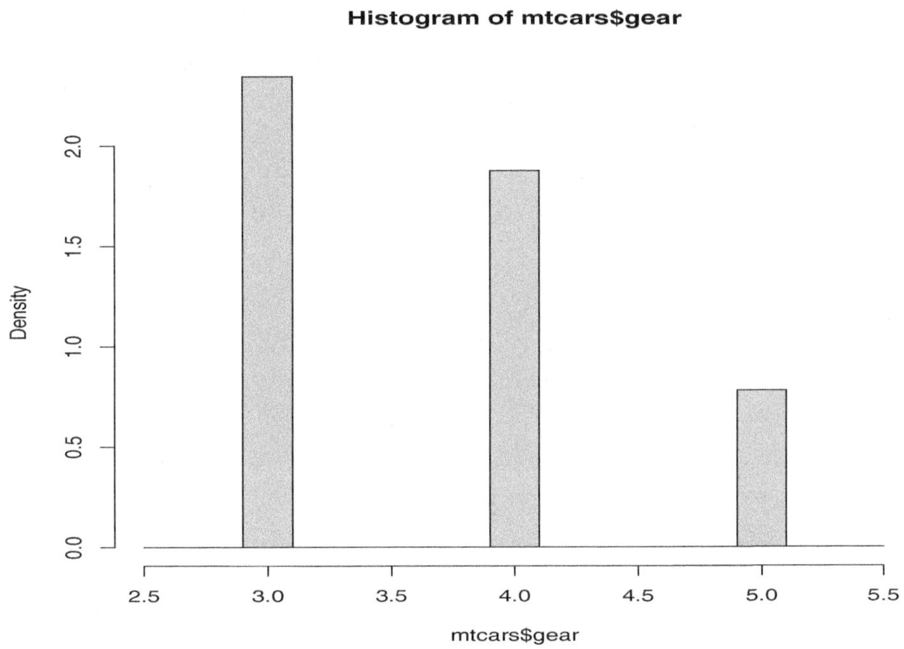


```
# Study the correlation between gears and transmission
# (see section on correlations)
cor(mtcars$gear, mtcars$am)
## [1] 0.7940588

cor(rank(mtcars$gear), rank(mtcars$am))
## [1] 0.807688

# We can also show the correlation between gears and transmission
# Since the data is mainly overlapping we use the function
# jitter() to add some noise so we can see individual points.
plot(jitter(mtcars$am,0.125), jitter(mtcars$gear,0.3),
            pch=21, col="blue", bg="red",
    xlab = "Transmission (0 = automatic, 1 = manual)",
    ylab = "Number of forward gears",
    main = "Jittered view of gears and transmission"
    )
# Add a blue line (linear fit -- see linear regression)
abline(lm(mtcars$gear ~ mtcars$am),
       col='blue',lwd=3)
```

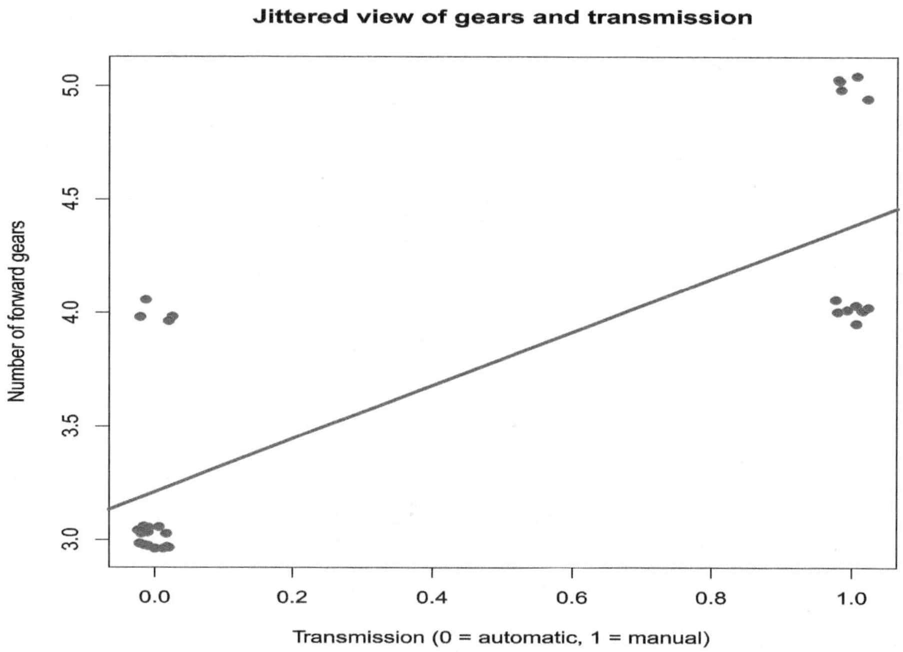

jitter()

Note the function `jitter()`, which adds random noise to each observation so that overlapping points get just a little separated. This make clear that in some places the density of dots is higher.

Question 5 on page 48

```
# Create the factors, taking care that we know what factor will
# be manual and what will be automatic.
f <- factor(mtcars$am,
            levels = c(0,1),
            labels = c("automatic", "manual")
            )

# To manually inspect the result:
head(cbind(f,mtcars$am))
##      f
## [1,] 2 1
## [2,] 2 1
## [3,] 2 1
## [4,] 1 0
## [5,] 1 0
## [6,] 1 0

# Show off the result
plot(f, col="khaki3")
```

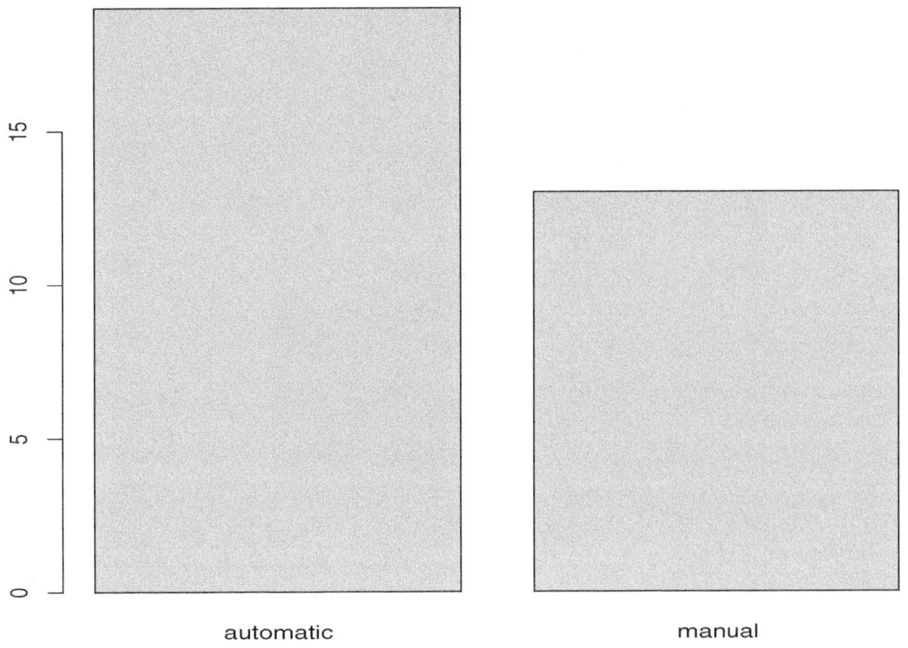

Question 6 on page 48

The first part of this question is overlapping with previous, so we will just show the histogram. This step is always important to get a good understanding of the data. Is there a car with 24 forward gears or 1000 horse power? Then – unless it is a serious truck – this is probably wrong. Not only we want to understand outliers but also get a feel with the data.

In this case, the reader will notice that the distribution of horsepower is skewed to the right. If we now are asked to provide labels low, medium. and high, we have to understand why we are doing this. Do we want an equal number of cars in each group? Or maybe it makes sense to have a very small group “high”?

```
# show the distribution
hist(mtcars$hp, col="khaki3")
```

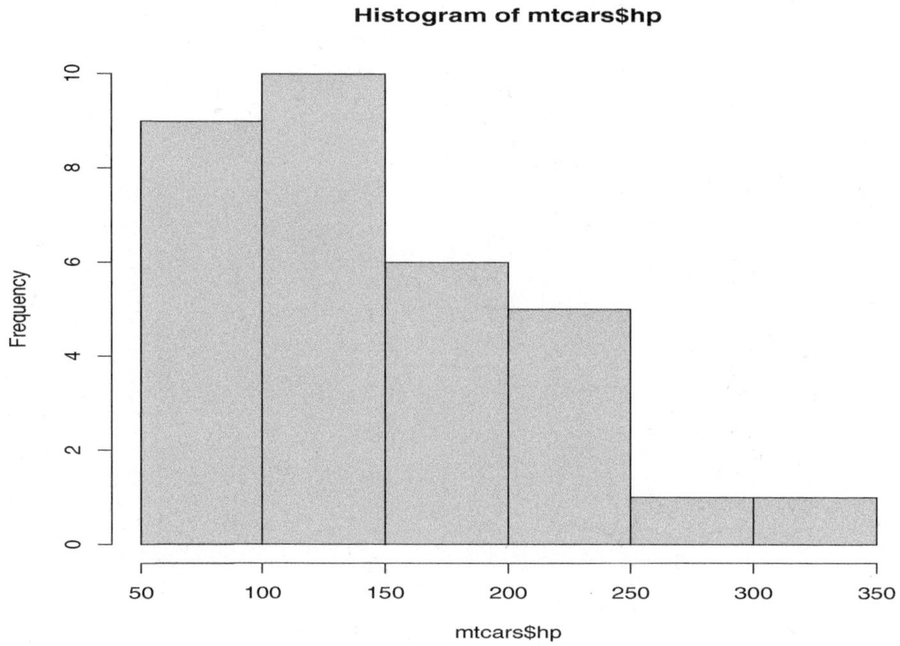


```
# Try a possible cut:
table(cut(mtcars$hp, breaks=c(50,100,175,350)))
##
##  (50,100] (100,175] (175,350]
##         9        13        10

# Assume we like the cut and use this one:
c <- cut(mtcars$hp, breaks=c(50,100,175,350)) # add the cut
l <- unique(c)                                # find levels
l                                             # check levels order
## [1] (100,175] (50,100]  (175,350]
## Levels: (50,100] (100,175] (175,350]

# Note that we provide the labels in the order of levels:
f <- factor(c, levels=l, labels=c("M", "L", "H"))
plot(f, col="khaki3",
     main="Horsepower as a factor",
     xlab="bins: L=low, M=Medium, H=high",
     ylab="number of cars in that bin")
```

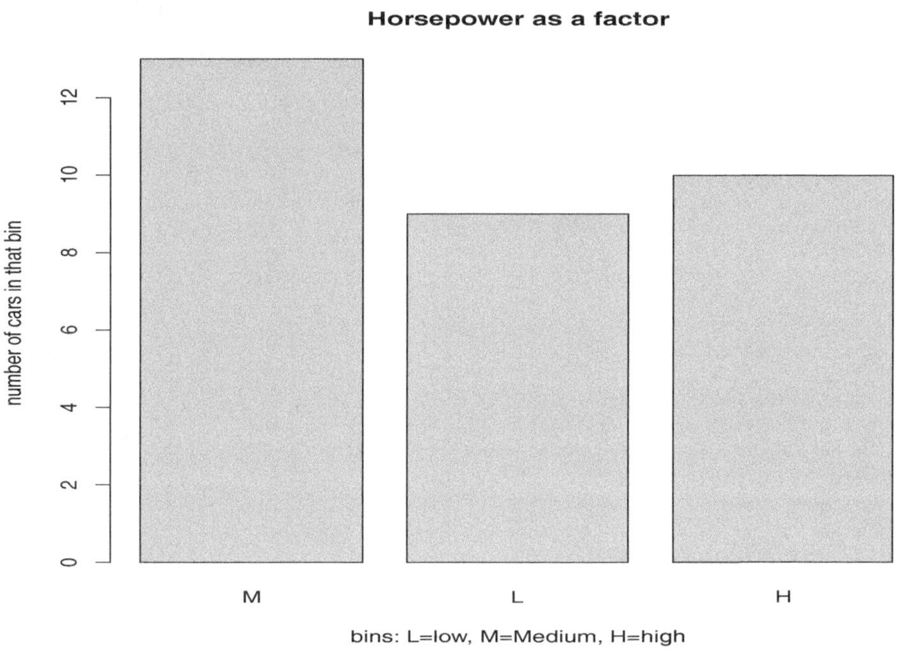

Question 7 on page 54

```
M <- matrix(c(1:9),nrow=3);
D <- data.frame(M);
rownames(D) <- c("2016","2017","2018");
colnames(D) <- c("Belgium", "France", "Poland");
cbind(D,rowSums(D));
##      Belgium France Poland rowSums(D)
## 2016       1      4      7         12
## 2017       2      5      8         15
## 2018       3      6      9         18

D <- D[,-2]
```

Question 12 on page 271

```
# Start a standard input
t <- read_csv(readr_example("challenge.csv"))

## Parsed with column specification:
## cols(
##   x = col_double(),
##   y = col_logical()
## )

# Then, to see the issues, do:
problems(t)
```

```
## # A tibble: 1,000 x 5
##      row col   expected     actual  file
##    <int> <chr> <chr>        <chr>   <chr>
##  1  1001 y     1/0/T/F/TRU~ 2015-0~ '/home/philippe/R/x86~
##  2  1002 y     1/0/T/F/TRU~ 2018-0~ '/home/philippe/R/x86~
##  3  1003 y     1/0/T/F/TRU~ 2015-0~ '/home/philippe/R/x86~
##  4  1004 y     1/0/T/F/TRU~ 2012-1~ '/home/philippe/R/x86~
##  5  1005 y     1/0/T/F/TRU~ 2020-0~ '/home/philippe/R/x86~
##  6  1006 y     1/0/T/F/TRU~ 2016-0~ '/home/philippe/R/x86~
##  7  1007 y     1/0/T/F/TRU~ 2011-0~ '/home/philippe/R/x86~
##  8  1008 y     1/0/T/F/TRU~ 2020-0~ '/home/philippe/R/x86~
##  9  1009 y     1/0/T/F/TRU~ 2011-0~ '/home/philippe/R/x86~
## 10  1010 y     1/0/T/F/TRU~ 2010-0~ '/home/philippe/R/x86~
## # ... with 990 more rows

# Notice that the problems start in row 1001, so
# the first 1000 rows are special cases. The first improvement
# can be obtained by increasing guess_max
## compare
spec_csv(readr_example("challenge.csv"))

## Parsed with column specification:
## cols(
##  x = col_double(),
##  y = col_logical()
## )
## cols(
##   x = col_double(),
##   y = col_logical()
## )

## with
spec_csv(readr_example("challenge.csv"), guess_max = 1001)

## Parsed with column specification:
## cols(
##  x = col_double(),
##  y = col_date(format = "")
## )
## cols(
##   x = col_double(),
##   y = col_date(format = "")
## )

# second step:
t <- read_csv(readr_example("challenge.csv"), guess_max = 1001)

## Parsed with column specification:
## cols(
##  x = col_double(),
##  y = col_date(format = "")
## )

# Let us see:
head(t)
## # A tibble: 6 x 2
##       x y
##   <dbl> <date>
## 1   404 NA
## 2  4172 NA
## 3  3004 NA
## 4   787 NA
## 5    37 NA
## 6  2332 NA
```

```
tail(t)
## # A tibble: 6 x 2
##       x y
##   <dbl> <date>
## 1 0.805 2019-11-21
## 2 0.164 2018-03-29
## 3 0.472 2014-08-04
## 4 0.718 2015-08-16
## 5 0.270 2020-02-04
## 6 0.608 2019-01-06
```

Question 14 on page 358

The case were the dependent variable is binary, is very important because it models any yes/no decision. In a corporate setting many decision will have that form: for example: go/no-go, yes/no, send/not_send, hire/not_hire, give_loan/not_give_loan, etc.

```
# We use the same seed so our results will be comparable.
set.seed(1865)

# We use the function mutate() from dplyr:
library(dplyr)

# For completeness sake, we generate the same data again.
N <- 500
age_f   <- rlnorm(N, meanlog = log(40), sdlog = log(1.3))
x_f <- abs(age_f + rnorm(N, 0, 20))     # Add noise & keep positive
x_f <- 1 - (x_f - min(x_f)) / max(x_f) # Scale between 0 and 1
x_f <- 0.5 * x_f / mean(x_f)           # Coerce mean to 0.5
# This last step will produce some outliers above 1
x_f[x_f > 1] <- 1                      # Coerce those > 1 to 1

age_m   <- rlnorm(N, meanlog = log(40), sdlog = log(1.3))
x_m <- abs(age_m + rnorm(N, 0, 20))     # Add noise & keep positive
x_m <- 1 - (x_m - min(x_m)) / max(x_m) # Scale between 0 and 1
x_m <- 0.5 * x_m / mean(x_m)           # Coerce mean to 0.5
# This last step will produce some outliers above 1
x_m[x_m > 1] <- 1                      # Coerce those > 1 to 1
x_m <- 1 - x_m                         # Make the relation increasing

p_f <- x_f
p_m <- x_m

tf <- tibble("age" = age_f, "sex" = "F", "is_good" = p_f)
tm <- tibble("age" = age_m, "sex" = "M", "is_good" = p_m)
t  <- full_join(tf, tm, by = c("age", "sex", "is_good")) %>%
      mutate("is_good" = if_else(is_good >= 0.5, 1L, 0L))%>%
      mutate("sexM"    = if_else(sex == "F", 0, 1))

###########
# Model 1 #
###########
regr1 <- glm(formula = is_good ~ age + sexM,
             family = binomial,
             data = t)

# assess the model:
summary(regr1)
```

```
##
## Call:
## glm(formula = is_good ~ age + sexM, family = binomial, data = t)
##
## Deviance Residuals:
##     Min       1Q   Median       3Q      Max
## -1.233   -1.159   -1.113    1.158    1.271
##
## Coefficients:
##              Estimate Std. Error z value Pr(>|z|)
## (Intercept)  0.185179   0.262594   0.705    0.481
## age         -0.002542   0.005968  -0.426    0.670
## sexM        -0.169578   0.126664  -1.339    0.181
##
## (Dispersion parameter for binomial family taken to be 1)
##
##     Null deviance: 1386.3  on 999  degrees of freedom
## Residual deviance: 1384.3  on 997  degrees of freedom
## AIC: 1390.3
##
## Number of Fisher Scoring iterations: 3


pred1 <- 1 / (1+ exp(-(coef(regr1)[1] + t$age * coef(regr1)[2]
                       + t$sexM * coef(regr1)[3])))
SE1 <-  (pred1 - t$is_good)^2
MSE1 <- sum(SE1) / length(SE1)

###########
# Model 2 #
###########
# make the same cut
t <- as_tibble(t)                                                  %>%
    mutate(is_LF = if_else((age <= 35) & (sex == "F"), 1L, 0L)) %>%
    mutate(is_HF = if_else((age >  50) & (sex == "F"), 1L, 0L)) %>%
    mutate(is_LM = if_else((age <= 35) & (sex == "M"), 1L, 0L)) %>%
    mutate(is_HM = if_else((age >  50) & (sex == "M"), 1L, 0L))

regr2 <- glm(formula = is_good ~ is_LF + is_HF + is_LM + is_HM,
             family = binomial,
             data = t)

# assess the model:
summary(regr2)
##
## Call:
## glm(formula = is_good ~ is_LF + is_HF + is_LM + is_HM, family = binomial,
##     data = t)
##
## Deviance Residuals:
##     Min       1Q    Median       3Q      Max
## -1.6942  -1.1363  -0.8424   1.2191   1.5546
##
## Coefficients:
##             Estimate Std. Error z value Pr(>|z|)
## (Intercept) -0.09754    0.08841  -1.103 0.269880
## is_LF        1.26069    0.21288   5.922 3.18e-09 ***
## is_HF       -0.62546    0.22984  -2.721 0.006502 **
## is_LM       -0.75595    0.19701  -3.837 0.000125 ***
## is_HM        1.21190    0.26657   4.546 5.46e-06 ***
## ---
## Signif. codes:
## 0 '***' 0.001 '**' 0.01 '*' 0.05 '.' 0.1 ' ' 1
##
```

```
## (Dispersion parameter for binomial family taken to be 1)
##
##     Null deviance: 1386.3  on 999  degrees of freedom
## Residual deviance: 1281.8  on 995  degrees of freedom
## AIC: 1291.8
##
## Number of Fisher Scoring iterations: 4

pred2 <- 1 / (1 + exp(-(coef(regr2)[1] +
                      + t$is_LF * coef(regr2)[2]
                      + t$is_HF * coef(regr2)[3]
                      + t$is_LM * coef(regr2)[4]
                      + t$is_HM * coef(regr2)[5]
                      )))
SE2 <-  (pred2 - t$is_good)^2
MSE2 <- sum(SE2) / length(SE2)

# Finally, we also note that the MSE has improved too.
MSE1
## [1] 0.2495116

MSE2
## [1] 0.2249014
```

Unsurprisingly the results are as follows.

1. For both models the MSE and SE are higher. That is normal, because with a logistic regression we will forecast a probability that is continuous between 0 and 1. If the explained variable is binary the distances will be necessarily be higher.
2. The model with matrix-bins performs better regardless binary or continuous parameter to be foretasted. Note in particular that
   - model 1 has no stars any more (general deterioration as explained previously),
   - model 2 has still three stars for each variable with only one exception where we still have two stars.

Note also that because we have now a value of our dependent variable that is strict equal to zero and others strictly equal to one that we can use `family = binomial` rather than `quasibinomial`.

Question 15 on page 358

One way to approach such situation could be to split the population in two parts and make two models: one for each part of the population.

Question 16 on page 358

The reason why we try to predict a variable that is between 0 and 1 (and linked to this why we use a logistic regression) is related to the nature of the bins that we create. Since they combine 2 (or more) units[1] we cannot reasonably only use a binary variable as independent variable.

[1] With units we mean things like meter, hours, minutes, dollars, number of people, produced units, etc. In our example we mix sex and age. In such case, the only thing that can be meaningful is a binary value: belongs or does not belong to this bin.

This in its turn makes logistic regressions theoretically more compelling and practically easier to understand.

Although, note that instead of using a logistic regression that it is also possible to use a linear regression and then cap the results. This is not elegant, and does not have the link with the underlying odds any more. While the model might still make good predictions, performance measures like AUC will be more appropriate than MSE for example.

Question 17 on page 361

```
library(InformationValue)
IV(X = factor(mtcars$vs), Y = factor(mtcars$am))
## [1] 0.1178631
## attr(,"howgood")
## [1] "Highly Predictive"

WOETable(X = factor(mtcars$vs), Y = factor(mtcars$am))
##   CAT GOODS BADS TOTAL     PCT_G     PCT_B        WOE
## 1   0     6   12    18 0.4615385 0.6315789 -0.3136576
## 2   1     7    7    14 0.5384615 0.3684211  0.3794896
##          IV
## 1 0.05333448
## 2 0.06452860
```

We can conclude that the gearbox type (automatic or manual) is a good predictor for the cylinder layout in the database mtcars.

Question 22 on page 545

One possible way of obtaining this is creating one object that holds the three matrices. For example, create an S3 object and return this.

```
# mcda_promethee_list
# delivers the preference flow matrices for the Promethee method
# Arguments:
#    M       -- decision matrix
#    w       -- weights
#    piFUNs -- a list of preference functions,
#              if not provided min(1,max(0,d)) is assumed.
# Returns (as side effect)
# phi_plus <<- rowSums(PI.plus)
# phi_min  <<- rowSums(PI.min)
# phi_     <<- phi_plus - phi_min
#
mcda_promethee_list <- function(M, w, piFUNs='x')
{
  if (piFUNs == 'x') {
       # create a factory function:
       makeFUN <- function(x) {x; function(x) max(0,x) }
       P <- list()
       for (k in 1:ncol(M)) P[[k]] <- makeFUN(k)
       } # else, we assume a vector of functions is provided
# initializations
```

```
PI.plus  <<- matrix(data=0, nrow=nrow(M), ncol=nrow(M))
PI.min   <<- matrix(data=0, nrow=nrow(M), ncol=nrow(M))
# calculate the preference matrix
for (i in 1:nrow(M)){
  for (j in 1:nrow(M)) {
    for (k in 1:ncol(M)) {
      if (M[i,k] > M[j,k]) {
        PI.plus[i,j] = PI.plus[i,j] + w[k] * P[[k]](M[i,k] - M[j,k])
      }
      if (M[j,k] > M[i,k]) {
        PI.min[i,j] = PI.min[i,j] + w[k] * P[[k]](M[j,k] - M[i,k])
      }
    }
  }
}
# Till this line the code was exactly the same.
# In the following line we create the list to return.
L <- list("phi_plus" <- rowSums(PI.plus),
          "phi_min"  <- rowSums(PI.min),
          "phi_"     <- phi_plus - phi_min
          )
return(L)
}
```

This code can now be used as follows.

```
set.seed(1492)
M <- matrix (runif(9), nrow = 3)
w <- c(runif(3))
L <- mcda_promethee_list(M, w)
```

Question 23 on page 595

A response on a request to rate a service or product on a scale from 1 to 5 is an ordinal scale – see Chapter B.2 *"Ordinal Scale"* on page 830 – for an average to make sense, one needs an interval scale – see Chapter B.3 *"Interval Scale"* on page 831.

Question 24 on page 598

After one year one the following interest accumulated:

$$(1 + r_y) = (1 + r_m)^{12}$$

This can be rewritten as

$$r_m = (1 + r_y)^{\frac{1}{12}} - 1$$

Bibliography

(1793, May). Collection générale des décrets rendus par la convention nationale, date: May 8, 1793 (du 8 mai 1793). available in Google Books Full View.

Adler, D. and S. T. Kelly (2018). *vioplot: violin plot.* R package version 0.3.0.

Andersen, E., N. Jensen, and N. Kousgaard (1987). *Statistics for economics, business administration, and the social sciences.* Springer-Verlag.

Artzner, P., F. Delbaen, J.-M. Eber, and D. Heath (1997). Thinking coherently. *Risk 10*(11), 68–71.

Artzner, P., F. Delbaen, J.-M. Eber, and D. Heath (1999). Coherent measures of risk. *Mathematical finance 9*(3), 203–228.

Auguie, B. (2017). *gridExtra: Miscellaneous Functions for "Grid" Graphics.* R package version 2.3.

Badal, S. and J. K. Harter (2014). Gender diversity, business-unit engagement, and performance. *Journal of Leadership & Organizational Studies 21*(4), 354–365.

Baesens, B. (2014). *Analytics in a big data world: The essential guide to data science and its applications.* John Wiley & Sons.

Barta, T., M. Kleiner, and T. Neumann (2012). Is there a payoff from top-team diversity. *McKinsey Quarterly 12*, 65–66.

Bengio, Y. and Y. Grandvalet (2004). No unbiased estimator of the variance of k-fold cross-validation. *Journal of machine learning research 5*(Sep), 1089–1105.

Berger, S., N. Graham, and A. Zeileis (2017, July). Various versatile variances: An object-oriented implementation of clustered covariances in R. Working Paper 2017-12, Working Papers in Economics and Statistics, Research Platform Empirical and Experimental Economics, Universität Innsbruck.

Bernoulli, D. (1738). Specimen theoriae novae de mensura sortis. *Comentarii Academiae Scientiarum Imperialis Petropolitanae Tomus V*, 175–192.

Bertsimas, D., G. Lauprete, and A. Samarov (2004). Shortfall as a risk measure: properties, optimization and applications. *Journal of Economic Dynamics and Control 28*(7), 1353–1381.

The Big R-Book: From Data Science to Learning Machines and Big Data, First Edition. Philippe J.S. De Brouwer.

Companion Website: www.wiley.com/go/De Brouwer/The Big R-Book

Black, F. and M. Scholes (1973). The pricing of options and corporate liabilities. *The journal of political economy*, 637–654.

Botev, Z. I., J. F. Grotowski, D. P. Kroese, et al. (2010). Kernel density estimation via diffusion. *The Annals of Statistics 38*(5), 2916–2957.

Botta-Dukát, Z. (2005). Rao's quadratic entropy as a measure of functional diversity based on multiple traits. *Journal of vegetation science 16*(5), 533–540.

Bouchet-Valat, M. (2019). *SnowballC: Snowball Stemmers Based on the C 'libstemmer' UTF-8 Library*. R package version 0.6.0.

Bowman, A. W. (1984). An alternative method of cross-validation for the smoothing of density estimates. *Biometrika 71*(2), 353–360.

Burington, R. S. et al. (1973). *Handbook of mathematical tables and formulas*. McGraw-Hill New York.

Chang, W. and B. Borges Ribeiro (2018). *shinydashboard: Create Dashboards with 'Shiny'*. R package version 0.7.1.

Chang, W., J. Cheng, J. Allaire, Y. Xie, and J. McPherson (2019). *shiny: Web Application Framework for R*. R package version 1.3.2.

Chang, W. and H. Wickham (2018). *ggvis: Interactive Grammar of Graphics*. R package version 0.4.4.

Chen, S. (2008). Nonparametric estimation of expected shortfall. *Journal of financial econometrics 6*(1), 87.

Cohen, M. B., S. Elder, C. Musco, C. Musco, and M. Persu (2015). Dimensionality reduction for k-means clustering and low rank approximation. In *Proceedings of the forty-seventh annual ACM symposium on Theory of computing*, pp. 163–172. ACM.

Cyganowski, S., P. Kloeden, and J. Ombach (2001). *From elementary probability to stochastic differential equations with MAPLE®*. Springer Science & Business Media.

De Brouwer, P. (2016). Zarzadzania ryzykiem w podejsciu zintegrowanym. In M. Postuła and R. Cieślik (Eds.), *Projekty inwestycyjne finansowanie, budżetowanie, ocena efektywności*, Chapter 8, pp. 231–298. Warsaw: Difin SA.

De Brouwer, P. J. S. (2012). *Maslowian Portfolio Theory, a Coherent Approach to Strategic Asset Allocation*. Brussels: VUBPress.

Dietterich, T. G. (1998). Approximate statistical tests for comparing supervised classification learning algorithms. *Neural computation 10*(7), 1895–1923.

Ding, C. and X. He (2004). K-means clustering via principal component analysis. In *Proceedings of the twenty-first international conference on Machine learning*, pp. 29. ACM.

Drury, C. M. (2013). *Management and cost accounting*. Springer.

Eddelbuettel, D. and J. J. Balamuta (2017, aug). Extending extitR with extitC++: A Brief Introduction to extitRcpp. *PeerJ Preprints 5*, e3188v1.

Eddelbuettel, D. and J. J. Balamuta (2018). Extending R with c++: A brief introduction to rcpp. *The American Statistician 72*(1), 28–36.

Epanechnikov, V. A. (1969). Non-parametric estimation of a multivariate probability density. *Theory of Probability & Its Applications 14*(1), 153–158.

Fang, J., A. L. Varbanescu, and H. Sips (2011). A comprehensive performance comparison of cuda and opencl. In *2011 International Conference on Parallel Processing*, pp. 216–225. IEEE.

Feinerer, I. and K. Hornik (2018). *tm: Text Mining Package.* R package version 0.7-6.

Feinerer, I., K. Hornik, and D. Meyer (2008, March). Text mining infrastructure in r. *Journal of Statistical Software 25*(5), 1–54.

Fellows, I. (2018). *wordcloud: Word Clouds.* R package version 2.6.

Fermanian, J. and O. Scaillet (2005). Sensitivity analysis of var and expected shortfall for portfolios under netting agreements. *Journal of Banking & Finance 29*(4), 927–958.

Fox, J., S. Weisberg, and B. Price (2018). *carData: Companion to Applied Regression Data Sets.* R package version 3.0-2.

Fritsch, S., F. Guenther, and M. N. Wright (2019). *neuralnet: Training of Neural Networks.* R package version 1.44.2.

Garnier, S. (2018). *viridisLite: Default Color Maps from 'matplotlib' (Lite Version).* R package version 0.3.0.

Gelman, A. and J. Hill (2011). Opening windows to the black box. *Journal of Statistical Software 40.*

Gesmann, M. and D. de Castillo (2011, December). googlevis: Interface between r and the google visualisation api. *The R Journal 3*(2), 40–44.

Gini, C. (1912). *Variabilità e mutabilità.*

Goldratt, E. M., J. Cox, and D. Whitford (1992). *The goal: a process of ongoing improvement*, Volume 2. North River Press Great Barrington, MA.

Good, I. J. (1966). Speculations concerning the first ultraintelligent machine. In *Advances in computers*, Volume 6, pp. 31–88. Elsevier.

Greene, W. H. (1997). *Econometric analysis (International edition).* Prentice Hall International Inc.

Grolemund, G. (2014). *Hands-On Programming with R: Write Your Own Functions and Simulations.* " O'Reilly Media, Inc.".

Grolemund, G. and H. Wickham (2011). Dates and times made easy with lubridate. *Journal of Statistical Software 40*(3), 1–25.

Hall, P., J. Marron, and B. U. Park (1992). Smoothed cross-validation. *Probability Theory and Related Fields 92*(1), 1–20.

Hara, A. and Y. Hayashi (2012). Ensemble neural network rule extraction using re-rx algorithm. In *Neural Networks (IJCNN), The 2012 International Joint Conference on*, pp. 1–6. IEEE.

Harrell Jr, F. E., with contributions from Charles Dupont, and many others. (2019). *Hmisc: Harrell Miscellaneous*. R package version 4.2-0.

Hastie, T., R. Tibshirani, and J. Friedman (2009). *The elements of statistical learning*. Springer.

Hendricks, P. (2015). *titanic: Titanic Passenger Survival Data Set*. R package version 0.1.0.

Hintze, J. L. and R. D. Nelson (1998). Violin plots: a box plot-density trace synergism. *The American Statistician 52*(2), 181–184.

Hyndman, R., G. Athanasopoulos, C. Bergmeir, G. Caceres, L. Chhay, M. O'Hara-Wild, F. Petropoulos, S. Razbash, E. Wang, and F. Yasmeen (2019). *forecast: Forecasting functions for time series and linear models*. R package version 8.6.

Hyndman, R. J. and Y. Khandakar (2008). Automatic time series forecasting: the forecast package for R. *Journal of Statistical Software 26*(3), 1–22.

Iannone, R. (2019). *DiagrammeR: Graph/Network Visualization*. R package version 1.0.1.

Iannone, R., J. Allaire, and B. Borges (2018). *flexdashboard: R Markdown Format for Flexible Dashboards*. R package version 0.5.1.1.

Izrailev, S. (2015). *binr: Cut Numeric Values into Evenly Distributed Groups*. R package version 1.1.

Jackson, C. H. (2011). Multi-state models for panel data: The msm package for R. *Journal of Statistical Software 38*(8), 1–29.

Jacobsson, H. (2005). Rule extraction from recurrent neural networks: Ataxonomy and review. *Neural Computation 17*(6), 1223–1263.

James, G., D. Witten, T. Hastie, and R. Tibshirani (2013). *An introduction to statistical learning*, Volume 112. Springer.

Jones, C., J. Marron, and S. Sheather (1996a). Progress in data-based bandwidth selection for kernel density estimation. *Computational Statistics* (11), 337–381.

Jones, M. C., J. S. Marron, and S. J. Sheather (1996b). A brief survey of bandwidth selection for density estimation. *Journal of the American Statistical Association 91*(433), 401–407.

Jost, L. (2006). Entropy and diversity. *Oikos 113*(2), 363–375.

Kahneman, D. (2011). *Thinking, fast and slow*. Macmillan.

Kaplan, R. S. and D. P. Norton (2001a). Transforming the balanced scorecard from performance measurement to strategic management: Part i. *Accounting horizons 15*(1), 87–104.

Kaplan, R. S. and D. P. Norton (2001b). Transforming the balanced scorecard from performance measurement to strategic management: Part ii. *Accounting Horizons 15*(2), 147–160.

Kardashev, N. S. (1964). Transmission of information by extraterrestrial civilizations. *Soviet Astronomy 8*, 217.

Karimi, K., N. G. Dickson, and F. Hamze (2010). A performance comparison of cuda and opencl. *arXiv preprint arXiv:1005.2581*.

Keylock, C. (2005). Simpson diversity and the shannon–wiener index as special cases of a generalized entropy. *Oikos 109*(1), 203–207.

Killough, L. N. and W. E. Leininger (1977). *Cost Accounting for Managerial Decision Making.* Dickenson Publishing Company.

Kondratieff, N. and W. Stolper (1935). The long waves in economic life. *The Review of Economics and Statistics 17*(6), 105–115.

Kondratieff, N. D. (1979). The long waves in economic life. *Review (Fernand Braudel Center),* 519–562.

Kowarik, A. and M. Templ (2016). Imputation with the R package VIM. *Journal of Statistical Software 74*(7), 1–16.

Kraut, R. (2002). Aristotle: political philosophy.

Kumar Nayak, I. (1985). On diversity measures based on entropy functions. *Communications in Statistics-Theory and Methods 14*(1), 203–215.

Kurzweil, R. (2010). *The singularity is near.* Gerald Duckworth & Co.

Lamport, L. (1994). *LATEX: a document preparation system: user's guide and reference manual.* Addison-wesley.

Lang, D. T. and the CRAN team (2019). *RCurl: General Network (HTTP/FTP/...) Client Interface for R.* R package version 1.95-4.12.

Lang, D. T. and the CRAN Team (2019). *XML: Tools for Parsing and Generating XML Within R and S-Plus.* R package version 3.98-1.19.

Lawrie, G. and I. Cobbold (2004). Third-generation balanced scorecard: evolution of an effective strategic control tool. *International Journal of Productivity and Performance Management 53*(7), 611–623.

Lê, S., J. Josse, and F. Husson (2008). FactoMineR: A package for multivariate analysis. *Journal of Statistical Software 25*(1), 1–18.

Lencioni, P. (2002). *The Five Dysfunctions of a Team.* Jossey-Bass.

Liaw, A. and M. Wiener (2002). Classification and regression by randomforest. *R News 2*(3), 18–22.

Liker, J. and G. L. Convis (2011). *The Toyota way to lean leadership: Achieving and sustaining excellence through leadership development.* McGraw Hill Professional.

Little, R. J. (1988). Missing-data adjustments in large surveys. *Journal of Business & Economic Statistics 6*(3), 287–296.

Luraschi, J., K. Kuo, K. Ushey, J. Allaire, and The Apache Software Foundation (2019). *sparklyr: R Interface to Apache Spark.* R package version 1.0.2.

Mackay, C. (1841). *Memoirs of extraordinary Popular Delusions and the Madness of Crowds* (First ed.). New Burlington Street, London, UK: Richard Bentley.

Maechler, M., P. Rousseeuw, A. Struyf, M. Hubert, and K. Hornik (2019). *cluster: Cluster Analysis Basics and Extensions*. R package version 2.0.8 — For new features, see the 'Changelog' file (in the package source).

Malliaris, A. and W. Brock (1982). Stochastic methods in economics and finance.

Markowitz, H. M. (1952). Portfolio selection. *Journal of Finance 6*, 77–91.

Marquet, D. L. (2015). *Turn the Ship Around!* Penguin Books Ltd.

Marr, B. (2016). *Key Business Analytics: The 60+ Business Analysis Tools Every Manager Needs To Know*. Pearson UK.

Meschiari, S. (2015). *latex2exp: Use LaTeX Expressions in Plots*. R package version 0.4.0.

Meyer, D., E. Dimitriadou, K. Hornik, A. Weingessel, and F. Leisch (2019). *e1071: Misc Functions of the Department of Statistics, Probability Theory Group (Formerly: E1071), TU Wien*. R package version 1.7-0.1.

Mikosch, T. (1998). *Elementary stochastic calculus with finance in view*. World Scientific Pub Co Inc.

Milborrow, S. (2019). *rpart.plot: Plot 'rpart' Models: An Enhanced Version of 'plot.rpart'*. R package version 3.0.7.

Monsen, R. J. and A. Downs (1965). A theory of large managerial firms. *The Journal of Political Economy*, 221–236.

Mossin, J. (1968). Optimal multiperiod portfolio policies. *Journal of Business 41*, 205–225.

Murtagh, F. and P. Legendre (2011). Ward's hierarchical clustering method: Clustering criterion and agglomerative algorithm. *arXiv preprint arXiv:1111.6285*.

Neter, J., W. Wasserman, and G. Whitmore (1988). *Applied statistics*. New York: Allyn and Bacon.

Neuwirth, E. (2014). *RColorBrewer: ColorBrewer Palettes*. R package version 1.1-2.

Nilsson, N. J. (1965). Learning machines.

Norreklit, H. (2000). The balance on the balanced scorecard a critical analysis of some of its assumptions. *Management accounting research 11*(1), 65–88.

Ooms, J. (2017). *sodium: A Modern and Easy-to-Use Crypto Library*. R package version 1.1.

Ooms, J., D. James, S. DebRoy, H. Wickham, and J. Horner (2018). *RMySQL: Database Interface and 'MySQL' Driver for R*. R package version 0.10.15.

Parzen, E. (1962). On estimation of a probability density function and mode. *The annals of mathematical statistics*, 1065–1076.

Peracchi, F. (2001). *Econometrics*. Wiley.

Prabhakaran, S. (2016). *InformationValue: Performance Analysis and Companion Functions for Binary Classification Models*. R package version 1.2.3.

Provost, F. and T. Fawcett (2013). *Data Science for Business: What you need to know about data mining and data-analytic thinking.* " O'Reilly Media, Inc.".

R Core Team (2018). *R: A Language and Environment for Statistical Computing*. Vienna, Austria: R Foundation for Statistical Computing.

R Special Interest Group on Databases (R-SIG-DB), H. Wickham, and K. MÃ¼ller (2018). *DBI: R Database Interface.* R package version 1.0.0.

Raiche, G. (2010). *an R package for parallel analysis and non graphical solutions to the Cattell scree test.* R package version 2.3.3.

Reichheld, F. F. (2003). The one number you need to grow. *Harvard business review 81*(12), 46–55.

Revelle, W. (2018). *psych: Procedures for Psychological, Psychometric, and Personality Research.* Evanston, Illinois: Northwestern University. R package version 1.8.12.

Robin, X., N. Turck, A. Hainard, N. Tiberti, F. Lisacek, J.-C. Sanchez, and M. MÃ¼ller (2011). proc: an open-source package for r and s+ to analyze and compare roc curves. *BMC Bioinformatics 12*, 77.

Rosenblatt, M. et al. (1956). Remarks on some nonparametric estimates of a density function. *The Annals of Mathematical Statistics 27*(3), 832–837.

Rudemo, M. (1982). Empirical choice of histograms and kernel density estimators. *Scandinavian Journal of Statistics*, 65–78.

Rupp, K., P. Tillet, F. Rudolf, J. Weinbub, T. Grasser, and A. JÃijngel (2016-10-27). Viennacl-linear algebra library for multi- and many-core architectures. *SIAM Journal on Scientific Computing.*

Russell, S. J. and P. Norvig (2016). *Artificial intelligence: a modern approach.* Malaysia; Pearson Education Limited.

Ryan, J. A. and J. M. Ulrich (2018). *xts: eXtensible Time Series.* R package version 0.11-2.

Ryan, J. A. and J. M. Ulrich (2019). *quantmod: Quantitative Financial Modelling Framework.* R package version 0.4-14.

Scaillet, O. (2004). Nonparametric estimation and sensitivity analysis of expected shortfall. *Mathematical Finance 14*(1), 115–129.

Scaillet, O. (2005). Nonparametric estimation of conditional expected shortfall. *Insurance and Risk Management Journal 74*(1), 639–660.

Scott, D. W. (1979). On optimal and data-based histograms. *Biometrika 66*(3), 605–610.

Scott, D. W. (2015). *Multivariate density estimation: theory, practice, and visualization.* John Wiley & Sons.

Setiono, R. (1997). Extracting rules from neural networks by pruning and hidden-unit splitting. *Neural Computation 9*(1), 205–225.

Setiono, R., B. Baesens, and C. Mues (2008). Recursive neural network rule extraction for data with mixed attributes. *IEEE Transactions on Neural Networks 19*(2), 299–307.

Sharpe, W. F. (1964). Capital asset prices: A theory of market equilibrium under conditions of risk. *Journal of Finance 19*(3), 425–442.

Sheather, S. J. and M. C. Jones (1991). A reliable data-based bandwidth selection method for kernel density estimation. *Journal of the Royal Statistical Society. Series B (Methodological)*, 683–690.

Sievert, C. (2018). *plotly for R.*

Simonoff, J. S. (2012). *Smoothing methods in statistics.* Springer Science & Business Media.

Sing, T., O. Sander, N. Beerenwinkel, and T. Lengauer (2005). Rocr: visualizing classifier performance in r. *Bioinformatics 21*(20), 7881.

Slowikowski, K. (2018). *ggrepel: Automatically Position Non-Overlapping Text Labels with 'ggplot2'.* R package version 0.8.0.

Smith, B. M. (2004). *A history of the global stock market: from ancient Rome to Silicon Valley.* University of Chicago press.

Soetaert, K. (2017a). *diagram: Functions for Visualising Simple Graphs (Networks), Plotting Flow Diagrams.* R package version 1.6.4.

Soetaert, K. (2017b). *plot3D: Plotting Multi-Dimensional Data.* R package version 1.1.1.

Stekhoven, D. J. (2013). *missForest: Nonparametric Missing Value Imputation using Random Forest.* R package version 1.4.

Stekhoven, D. J. and P. Buehlmann (2012). Missforest - non-parametric missing value imputation for mixed-type data. *Bioinformatics 28*(1), 112–118.

Stevens, S. S. (1946). On the theory of scales of measurement. *Science 103*(2684), 677–680.

Szekely, G. J. and M. L. Rizzo (2005). Hierarchical clustering via joint between-within distances: Extending ward's minimum variance method. *Journal of classification 22*(2), 151–183.

Tang, Y., M. Horikoshi, and W. Li (2016). ggfortify: Unified interface to visualize statistical result of popular r packages. *The R Journal 8.*

Thaler, R. H. (2016). Behavioral economics: Past, present and future. *Present and Future (May 27, 2016).*

Therneau, T. and B. Atkinson (2019). *rpart: Recursive Partitioning and Regression Trees.* R package version 4.1-15.

Tickle, A. B., R. Andrews, M. Golea, and J. Diederich (1998). The truth will come to light: Directions and challenges in extracting the knowledge embedded within trained artificial neural networks. *IEEE Transactions on Neural Networks 9*(6), 1057–1068.

Tierney, L. and R. Jarjour (2016).*proftools: Profile Output Processing Tools for R.* R package version 0.99-2.

Treynor, J. L. (1961). Market value, time, and risk. unpublished paper.

Treynor, J. L. (1962). Toward a theory of market value of risky assets. A final version was published in 1999, in Asset Pricing and Portfolio Performance: Models, Strategy and Performance Metrics. Robert A. Korajczyk (editor) London: Risk Books, pp. 15–22.

Triantaphyllou, E. (2000). *Multi-criteria decision making methods a comparative study*. Springer.

Ushey, K., J. Hester, and R. Krzyzanowski (2017). *rex: Friendly Regular Expressions*. R package version 1.1.2.

van Buuren, S. and K. Groothuis-Oudshoorn (2011). mice: Multivariate imputation by chained equations in r. *Journal of Statistical Software 45*(3), 1–67.

van der Merwe, A. and B. D. Clinton (2006). Management accounting-approaches, techniques, and management processes. *Journal of cost management 20*(3), 14–22.

Van Tilborg, H. C. and S. Jajodia (2014). *Encyclopedia of cryptography and security*. Springer Science & Business Media.

Venables, W. N. and B. D. Ripley (2002a). *Modern Applied Statistics with S* (Fourth ed.). New York: Springer. ISBN 0-387-95457-0.

Venables, W. N. and B. D. Ripley (2002b). *Modern Applied Statistics with S* (Fourth ed.). New York: Springer. ISBN 0-387-95457-0.

Venkataraman, S., X. Meng, F. Cheung, and The Apache Software Foundation (2019). *SparkR: R Front End for 'Apache Spark'*. R package version 2.4.3.

Wand, M. P. and M. C. Jones (1994). *Kernel smoothing*. Crc Press.

Wickham, H. (2007). Reshaping data with the reshape package. *Journal of Statistical Software 21*(12).

Wickham, H. (2011). The split-apply-combine strategy for data analysis. *Journal of Statistical Software 40*(1), 1–29.

Wickham, H. (2014). *Advanced r*. Chapman and Hall/CRC.

Wickham, H. (2015). *R packages: organize, test, document, and share your code*. " O'Reilly Media, Inc.".

Wickham, H. (2016). *ggplot2: Elegant Graphics for Data Analysis*. Springer-Verlag New York.

Wickham, H. (2017). *tidyverse: Easily Install and Load the 'Tidyverse'*. R package version 1.2.1.

Wickham, H. (2018a). *profr: An Alternative Display for Profiling Information*. R package version 0.3.3.

Wickham, H. (2018b). *pryr: Tools for Computing on the Language*. R package version 0.1.4.

Wickham, H. (2019).*stringr: Simple, Consistent Wrappers for Common String Operations*. R package version 1.4.0.

Wickham, H. et al. (2014). Tidy data.*Journal of Statistical Software 59*(10), 1–23.

Wickham, H., P. Danenberg, and M. Eugster (2018). *roxygen2: In-Line Documentation for R*. R package version 6.1.1.

Wickham, H. and G. Grolemund (2016). *R for data science: import, tidy, transform, visualize, and model data.* " O'Reilly Media, Inc.".

Wickham, H., J. Hester, and W. Chang (2019). *devtools: Tools to Make Developing R Packages Easier.* R package version 2.1.0.

Wolfgang, P. and J. Baschnagel (1999). *Stochastic Processes: From Physics to Finance.* Springer.

Xie, Y. (2014). knitr: A comprehensive tool for reproducible research in R. In V. Stodden, F. Leisch, and R. D. Peng (Eds.), *Implementing Reproducible Computational Research.* Chapman and Hall/CRC. ISBN 978-1466561595.

Xie, Y. (2015). *Dynamic Documents with R and knitr* (2nd ed.). Boca Raton, Florida: Chapman and Hall/CRC. ISBN 978-1498716963.

Xie, Y. (2018). *knitr: A General-Purpose Package for Dynamic Report Generation in R.* R package version 1.20.

Zeileis, A. (2004). Econometric computing with HC and HAC covariance matrix estimators. *Journal of Statistical Software 11*(10), 1–17.

Zeileis, A. (2006). Object-oriented computation of sandwich estimators. *Journal of Statistical Software 16*(9), 1–16.

Zeileis, A. and G. Grothendieck (2005). zoo: S3 infrastructure for regular and irregular time series.*Journal of Statistical Software 14*(6), 1–27.

Nomenclature

$\bar{x}$	mean, page 140.
BV	book value, page 620.
CapEx	capitial expenses, page 621.
CoverageR	coverage ratio, page 580.
DPR	$:= \frac{D}{E}$, dividend payout ratio, page 619.
DSCR	debt service coverage ratio, page 581.
EVA	economic value added, page 630.
GR	gearing ratio, page 581.
ι	continuously compounded income from an asset, page 638.
Λ_1	cut-off level for C_1 (for ELECTRE I), page 532.
Λ_2	cut-off level for C_2 (for ELECTRE I), page 532.
log()	ln or the natural logarithm, page 411.
F	the scaled decision matrix after application of the $f_j()$, so $\mathbf{F}_{ij} = f_j(x_{ij})$, page 527.
w	the vector of all w_j, page 527.
MVA	market value added, page 630.
NOPAT	net operational profit after tax, page 621.
PBR	$:= 1 - \text{DPR}$, plowback ratio, page 619.
$\Phi(a)$	the preference flow between an alternative a and all others over all criteria, equals: $\Phi(a) = \sum_{x \in \mathcal{A}} \sum_{j=1}^{k} \pi_j(f_j(a), f_j(x))$, page 550.

The Big R-Book: From Data Science to Learning Machines and Big Data, First Edition. Philippe J.S. De Brouwer.

Companion Website: www.wiley.com/go/De Brouwer/The Big R-Book

$\Phi(a, b)$	the preference flow between an alternative a and b as used in PROMethEE II, page 550.
$\Phi^+(a)$	the preference flow between an alternative a and all others, equals: $\Phi^+(a) = \frac{1}{k-1} \sum_{x \in \mathcal{A}} \pi(a, x)$ in PROMethEE, page 544.
$\Phi^-(a)$	the preference flow that indicates how much other alternatives are preferred over alternative a, it equals $\Phi^-(a) = \frac{1}{k-1} \sum_{x \in \mathcal{A}} \pi(x, a)$ in PROMethEE, page 544.
$\pi(a, b)$	the "preference" of solution a over solution b, page 541.
R_{RF}	risk free interest rate, page 616.
R_{RF}	the risk free return (in calculations one generally uses an average of past risk free returns and not the actual risk free return), page 610.
ROE	$:= \frac{E}{P}$, return on equity, page 619.
σ	the volatility of the returns of the underlying asset, page 650.
$\mathrm{SS}_{\mathrm{reg}}$	sum of squares of the regression, page 387.
$\mathrm{SS}_{\mathrm{res}}$	sum of squares of the residuals, page 387.
$\mathrm{SS}_{\mathrm{tot}}$	total sum of squares, page 387.
τ	the time to maturity, page 650.
VAR	variance, page 610.
VAR(X)	the variance of the stochastic variable X, page 610.
WACC	weighted average cost of capital, page 579.
WC	working capital, page 621.
$\wedge$	logical "and operator", page 544.
$a \succ b$	alternative a is preferred over alternative b, page 544.
$C_1(a, b)$	index of comparability of Type 1 (for ELECTRE I), page 532.
$C_2(a, b)$	index of comparability of Type 2 (for ELECTRE I), page 532.
$C_\alpha(T)$	cost of complexity function for the tree T and pruning parameter α, page 409.
C_M	the marginal cost, page 587.
cf	Cash Flow, page 600.
D	total debt expressed in currency, page 580.
d_1	$:= \frac{\log\left(\frac{S}{X}\right) + \left(r + \frac{\sigma^2}{2}\right)(\tau)}{\sigma\sqrt{\tau}}$, page 650.

d_2	$:= \frac{\log(\frac{S}{X})+\left(r-\frac{\sigma^2}{2}\right)(\tau)}{\sigma\sqrt{\tau}} = d_1 - \sigma\sqrt{\tau}$, page 650.
$d_k(a,b)$	the difference in performance of alternative a over b for criterion k, equals $m_k(a) - m_k(b)$, page 540.
D_t	the dividend paid in year t, page 617.
E	earnings, page 619.
E	total equity expressed in currency, page 580.
$f()$	function, page 406.
$f_{est}(x)$	the estimator for the probability density function, $f(x)$, page 635.
$f_{est}(x;h)$	the estimator for the probability density function for a kernel density estimation with bandwidth h, page 635.
g	the (dividend) growth rate, page 617.
h	the bandwidth or smoothing parameter in a kernel density estimation, page 635.
$h()$	hypothesis, page 406.
I	Interest payment (in monetary units), page 598.
i_n	nominal interest rate, page 599.
i_r	real interest rate, page 599.
IV	intrinsic value, page 641.
K	Capital paid by investors, page 630.
K	delivery price of a forward contract, page 638.
K_d	cost of debt, page 580.
K_e	cost of equity, page 580.
K_h	the kernel (of a kernel density estimation) with bandwidth h, page 635.
$log(x)$	$:= log_e(x)$, page 650.
M_j	the ideal value for criterion j in the Goal Programming Method, page 560.
N	a given natural number, page 598.
N	number of negative observations, page 391.
$N(\cdot)$	$:= \frac{1}{\sqrt{2\pi}} \int_{-\infty}^{x} e^{-\frac{z^2}{2}}\, dz$ the cumulative distribution function of the standard normal distribution, page 650.

$N(x)$	the result vector of the WSM for alternative x, page 527.
P	number of positive observations, page 391.
P	the price of the product, page 587.
p	inflation rate, page 599.
$P(a)$	the "total score" of alternative a as used in the WPM, page 530.
$P(a, b)$	the "total score" of alternative a as used in the dimensionless WPM, page 530.
P_t	the (market) price of a stock at moment t, page 618.
PV	present value, page 598.
Q	the quantity produced, page 587.
r	interest rate, page 600.
r	the capitalization rate, $E[R_k]$, estimated via the CAPM, page 617.
r	the risk free rate (annual rate, expressed in terms of continuous compounding), page 650.
r_1	the risk free interest rate over period 1, page 655.
R_i	the return of asset i, page 580.
r_j	a conversion factor that removes the unit, and scales it, page 560.
R_k	the return of an arbitrary asset K, page 610.
R_M	the return of the market, page 610.
RP	risk premium, page 616.
S	the spot price (of the underlying asset), page 641.
S_T	spot price at maturity of a derivative contract, page 638.
t	counter, page 600.
V_0	the intrinsic value of the stock now, page 617.
V_0	the value of an asset at time 0 (now) = the present value = PV, page 598.
V_i	the market value of asset i, page 580.
V_t	the value of an asset at time t, page 598.
V_{market}	Market Value (of a company), page 630.
w_j	the scaling factor in the WSM, it removes the unit, and scales (weights) the jth criterion, page 527.

X	the strike price, page 650.
x_k	an observation of the stochastic variable X, page 140.
y	yield to maturity, page 607.
y_j	factors to be minimized in the Goal Programming Method, page 560.
*nix	Unix and/or Linux, page 132.
ABC	activity based costing, page 585.
ACID	atomicity, consistency, isolation and durability, page 217.
AER	effective annual rate or annual equivalent rate, page 598.
AI	artificial intelligence, page 405.
Altiplano:	option combined with a coupon, which only paid if the underlying security never reaches its strike price during a given period, page 679.
American	an American option can be executed from the moment it is bought till its maturity date, page 640.
ANN	artificial neural network, page 434.
Annapurna	option where payoff is paid only if all securities increase a given amount, page 678.
API	application programming interface, page 748.
APR	annual percentage rate, page 598.
ARIMA	autoregressive integrated moving average model, page 204.
ARMA	autoregressive moving average model, page 204.
Asian	the strike or spot is determined by the average price of the underlying taken at different moments, page 678.
Atlas	option in which the best and worst performing securities are removed before maturity, page 678.
ATM	an option is in the money if its Intrinsic value is zero, page 639.
AU	asset utilisation, page 576.
AUC	area under curve, page 396.
AUC	area under the curve, page 396.
Barrier Option	generic term for knock-in and knock-out options, page 678.
Bermuda Option	an option where the buyer has the right to exercise at a set (always discretely spaced) number of times, page 679.

BS	Black and Scholes (model), page 647.
BSC	balanced scorecard, page 589.
BV	book value, page 627.
Call Option	the right to buy an underlying asset at a pre-agreed price, page 638.
Canary Option	can be exercised at quarterly dates, but not before a set time period has elapsed, page 679.
CapEx	Capital Expenditure, page 573.
CAPM	Capital Asset Pricing Model, page 610.
CART	Classification and Regression Tree, page 411.
cash settlement	pay out the profit to the option buyer in stead of deliver the underlying, page 639.
cdf	cumulative density function, page 150.
CGDDM	constant growth dividend discount model, page 616.
CLI	command line interface, page 132.
CLTV	customer lifetime value, page 710.
CLV	customer lifetime value, page 592.
CoGS	Cost of Goods Sold, page 568.
CR	current ratio, page 576.
CRR	Cox–Ross–Rubinstein (option pricing method), page 654.
CVM	customer value metric, page 592.
DAOE	discounted abnormal operating earnings valuation model, page 622.
DBMS	database management systems, page 215.
DCF	discounted cash flow, page 621.
DCFM	discounted cash flow model, page 621.
DDM	dividend discount model, page 615.
DE	debt to equity ratio, page 581.
deliver	provide or accept the underlying from the option buyer, page 639.
DST	daylight savings time, page 314.

EA	environmental accounting, page 587.
EAT	earnings after taxes, page 568.
EBIT	earnings before interest and taxes, page 568.
EBITDA	earniCompany Value and ngs before interest, taxes and depreciation, amortization, page 568.
EBT	earnings before taxes, page 568.
ER	entity relationship, page 223.
European	a European option can only be executed at its maturity date, page 640.
Everest	option based on the worst-performing securities in the basket., page 678.
exercising	sell or buy via the option contracts, page 638.
expiry date	the maturity date, page 638.
f()	probability density function, page 140.
FCF	free cash flow, page 619.
FCFF	free cash flow to firm, page 620.
FK	foreign key, page 221.
FN	false negative, page 391.
FOSS	Free and Open Source Software, page xxxiii.
FP	false positive, page 391.
FRA	forward rate agreement, page 639.
FV	future value, page 598.
FX	foreign exchange, page 496.
GAAP	generally accepted accounting principles, page 565.
gaia	geometrical analysis for interactive aid, page 554.
GM	gross margin, page 575.
GNUPG	GNU Privacy Guard – also GPG, page 262.
GPG	GnuPG – GNU Privacy Guard, page 262.
GPK	Grenzplankostenrechnung, page 586.
GPU	graphical processing unit, page 752.

HAC	hierarchical agglomerative clustering, page 468.
Himalayan	option based on the performance of the best asset in the portfolio., page 674.
IDE	integrated development environment, page 23.
IERS	International Earth Rotation and Reference Systems Service, page 314.
Intrinsic value	the value of the option at maturity, page 641.
IoT	Internet of things, page 217.
Israeli	option callable by the option writer, page 679.
IT	information technology, page 4.
ITM	an option is in the money if its intrinsic value is positive, page 641.
IV	information value, page 359.
JIT	just in time, page 805.
KDE	kernel density estimation, page 634.
kNN	k nearest neighbours, page 468.
Knock-In	this option only becomes active when a certain level (up or down) is reached, page 679.
Knock-Out	a knock-out option becomes inactive after a certain level of the underlying is reached, page 679.
KPI	key performance indicator, page 591.
KS	Kolmogorov Smirnov, page 398.
LCC	life cycle costing, page 586.
LCV	lifetime customer value, page 593.
long position	the buyer of an option is said to have a long position on his books, page 640.
look-back	the spot is determined as the best price of some moments., page 679.
LR	liquid ratio, page 576.
LSE	London Stock Exchange, page 643.
MacD	Macaulay duration, page 606.
MAD	mean average deviation, page 387.
mad	median absolute deviation, page 145.

MAR	missing at random, page 339.
maturity	or "maturity date" is the expiry date of an option, that is the last moment in time that it can change value because of the movement of the underlying, page 640.
MCDA	multi criteria decision analysis, page 511.
MI	Management Information, page 725.
MIS	Management Information System, page 584.
MISE	mean integrated squared error, page 635.
ML	machine learning, page 405.
MSE	mean square error, page 385.
MTM	Marked-to-Market, page 641.
Nasdaq	National Association of Securities Dealers Automated Quotations, page 496.
NAV	net asset value, page 624.
NN	neural network, page 434.
NOA	net operating assets, page 578.
NOP	net operating profit, page 629.
NOPAT	Net Operating Income After Taxes, page 569.
NPS	Net Promoter Score, page 594.
NPV	Net Present Value, page 600.
NSS	net satisfaction score, page 595.
NYSE	New York Stock Exchange, page 496.
OA	operating assets, page 577.
ODBC	open database connectivity, page 764.
OL	operating liabilities, page 578.
OLAP	online analytical processing, page 713.
OLS	ordinary least squares, page 406.
OO	object oriented, page 87.
OOP	object oriented programming, page 113.

OpEx	Operational Expense, page 574.
OpEx	Operational Expenditure, page 574.
ORM	object-relational mappings, page 216.
OS	operating system, page 121.
OTC	Over the counter, page 642.
OTM	an option is in the money if its Intrinsic value is negative, page 641.
P()	probability, page 140.
P&L	Profit and Loss, page 631.
PC	personal computer, page 216.
PCA	principal component analysis, page 363.
PCs	principal components, page 363.
pdf	probability density function, page 150.
PE	price earnings ratio, page 626.
PGP	pretty good privacy, page 262.
PK	primary key, page 219.
PK	primary key, page 223.
PM	profit margin, page 575.
PMM	predictive mean matching, page 338.
PTB	$= \frac{P}{\mathrm{BV}} =$ price-to-book ratio with BV = book value, page 627.
PTCF	$= \frac{P}{\mathrm{CF}} =$ price-to-cash-flow ratio, with CF = (free) cash flow, page 628.
PTS	$= \frac{P}{S} =$ price-to-sales ratio, with $S =$ sales, page 628.
PU	processing unit, page 752.
Put	the right to sell an underlying asset at a pre-agreed price, page 640.
PVGO	present value of growth opportunities, page 618.
QR	quick ratio, page 577.
R5	reference class (OO system), page 113.
R6	reference class (OO system)., page 119.
R&D	research and development (costs), page 568.

RC	reference class (OO system), page 113.
RCA	Resource Consumption Accounting, page 586.
RDBMS	relational database management system, page 219.
RDD	Resilient Distributed Dataset, page 771.
ROC	receiver operator characteristic, page 393.
ROCE	return on capital employed, page 629.
ROE	Return on Equity, page 629.
ROI	return on invested capital, page 628.
ROIC	return on invested capital, page 628.
Russian	lookback over the whole life time of the option, page 679.
S	sales, page 586.
SCA	Standard Cost Accounting, page 585.
short position	the seller of an option is said to have a short position on his books, page 640.
SLA	service level agreement, page 714.
SML	security market line, page 610.
spot price	the actual value of the underlying asset, the price to be paid for the asset to buy it today and have it today, page 641.
STL	seasonal trend decomposition using Loess, page 206.
strike	the "execution price," the price at which an option can be executed (e.g. for a call the price at which the underlying can be sold), page 640.
SVM	support vector machine, page 447.
SWOT	strengths, weaknesses, opportunities and threats analysis, page 516.
T	throughput, page 586.
TA	throughput accounting, page 586.
TargetC	target costing, page 587.
TC	total capital, page 579.
TCE	total capital employed, page 579.
TCstE	total cost of equity, page 576.

TN	true negative, page 391.
TP	true positive, page 391.
TVC	total variable costs, page 586.
UCITS	Undertaking for Collective Investments in Transferable Securities, page 624.
UDF	user defined function, page 780.
url	universal resource locator, page 307.
VB	Visual Basic, page 632.
Verde Option	option that can be exercised at set dates, but not before a set time period has elapsed, page 680.
WC	working capital, page 578.
WOE	weight of evidence, page 359.
WPM	weighted product method, page 530.
WSM	weighted sum method, page 527.
YAML	yet another markup language, page 701.

Index

The Big R-Book: From Data Science to Learning Machines and Big Data, First Edition. Philippe J.S. De Brouwer.

Companion Website: www.wiley.com/go/De Brouwer/The Big R-Book